THE SUSTAINABLE URBAN DEVELOPMENT READER

Third edition

Building on the success of its second edition, the third edition of *The Sustainable Urban Development Reader* offers an unrivalled selection of classic and contemporary readings and case studies providing a broad introduction to this key topic. It begins by tracing the roots of the sustainable development concept in the nineteenth and twentieth centuries, before presenting readings on a number of dimensions of the sustainability concept.

Topics covered include land use and urban design, transportation, ecological planning and restoration, energy and materials use, economic development, social and environmental justice, and green architecture and building. All sections have a concise editorial introduction that places the selection in context and suggests further reading. Additional sections cover tools for sustainable development, sustainable development internationally, visions of sustainable community, and case studies from around the world. The book also includes educational exercises for individuals, university classes, or community groups, and an extensive list of recommended readings.

The anthology remains unique in presenting a broad array of classic and contemporary readings in this field, each with a concise introduction placing it within the context of this evolving discourse. *The Sustainable Urban Development Reader* presents an authoritative overview of the field using original sources in a highly readable format for university classes in urban studies, environmental studies, the social sciences, and related fields. It also makes a wide range of sustainable urban planning-related material available to the public in a clear and accessible way, forming an indispensable resource for anyone interested in the future of urban environments.

Stephen M. Wheeler is Associate Professor in the Department of Human Ecology at the University of California at Davis. Previously he taught at the University of New Mexico and the University of California at Berkeley. A member of the American Institute of Certified Planners (AICP), he has been a consultant in the areas of smart growth, urban design, and sustainable development. He received Ph.D. and MCP degrees from U.C. Berkeley and a B.A. from Dartmouth College. Previously he edited the quarterly journal *The Urban Ecologist* and served as a lobbyist for Friends of the Earth in Washington, D.C. His articles have appeared in the *Journal of the American Planning Association*, *Local Environment*, and the *Journal of Planning Education and Research*. Besides *The Sustainable Urban Development Reader*, he is the author of *Climate Change and Social Ecology* (Routledge 2012) and *Planning for Sustainability* (Routledge 2004, 2013).

Timothy Beatley is Teresa Heinz Professor of Sustainable Communities in the School of Architecture at the University of Virginia, where he has taught for the last 28 years. His primary teaching and research interests are in environmental planning and policy, with special emphasis on community sustainability. He has published extensively on these subjects; his books include *The Ecology of Place* (1997, with Kristy Manning), *Green Urbanism: Learning from European Cities* (2000), *Native to Nowhere* (2005), and *Biophilic Cities* (2011). His research has been funded by the National Science Foundation, the U.S. Environmental Protection Agency, and the Lincoln Institute for Land Policy, among others. Beatley holds a Ph.D. in City and Regional Planning from the University of North Carolina at Chapel Hill.

THE ROUTLEDGE URBAN READER SERIES

Series editors

Richard T. LeGates

Professor Emeritus of Urban Studies and Planning, San Francisco State University

Frederic Stout

Lecturer in Urban Studies, Stanford University

The Routledge Urban Reader Series responds to the need for comprehensive coverage of the classic and essential texts that form the basis of intellectual work in the various academic disciplines and professional fields concerned with cities and city planning.

The readers focus on the key topics encountered by undergraduates, graduate students, and scholars in urban studies, geography, sociology, political science, anthropology, economics, culture studies, and professional fields such as city and regional planning, urban design, architecture, environmental studies, international relations, and landscape architecture. They discuss the contributions of major theoreticians and practitioners and other individuals, groups, and organizations that study the city or practice in a field that directly affects the city.

As well as drawing together the best of classic and contemporary writings on the city, each reader features extensive introductions to the book, sections, and individual selections prepared by the volume editors to place the selections in context, illustrate relations among topics, provide information on the author, and point readers towards additional related bibliographic material.

Each reader contains:

Between thirty-five and sixty *selections* divided into six to eight sections. Almost all of the selections are previously published works that have appeared as journal articles or portions of books.

- ■ A *general introduction* describing the nature and purpose of the reader.
- ■ *Section introductions* for each section of the reader to place the readings in context.
- ■ *Selection introductions* for each selection describing the author, the intellectual background, and context of the selection, competing views of the subject matter of the selection and bibliographic references to other readings by the same author and other readings related to the topic.
- ■ One or more plate sections and illustrations at the beginning of each section.
- ■ An index.

The series consists of the following titles:

THE CITY READER

The City Reader, fifth edition – an interdisciplinary urban reader aimed at urban studies, urban planning, urban geography, and urban sociology courses – is the *anchor urban reader*. Routledge published a first edition of *The City Reader* in 1996, a second edition in 2000, a third edition in 2003, and a fourth edition in 2007. *The City Reader* has become one of the most widely used anthologies in urban studies, urban geography, urban sociology, and urban planning courses in the world.

URBAN DISCIPLINARY READERS

The series contains *urban disciplinary readers* organized around social science disciplines and professorial fields: urban sociology, urban geography, urban politics, urban and regional planning, and urban design. The urban disciplinary readers include both classic writings and recent, cutting-edge contributions to the respective disciplines. They are lively, high-quality, competitively priced readers which faculty can adopt as course texts and which also appeal to a wider audience.

TOPICAL URBAN ANTHOLOGIES

The urban series includes *topical urban readers* intended both as primary and supplemental course texts and for the trade and professional market. The topical titles include readers related to sustainable urban development, global cities, cybercities, and city cultures.

INTERDISCIPLINARY ANCHOR TITLE

The City Reader, fifth edition
Richard T. LeGates and Frederic Stout (eds)

URBAN DISCIPLINARY READERS

The Urban Geography Reader
Nick Fyfe and Judith Kenny (eds)

The Urban Politics Reader
Elizabeth Strom and John Mollenkopf (eds)

The Urban and Regional Planning Reader
Eugenie Birch (ed.)

The Urban Sociology Reader, second edition
Jan Lin and Christopher Mele (eds)

The Urban Design Reader, second edition
Michael Larice and Elizabeth Macdonald (eds)

TOPICAL URBAN READERS

The City Cultures Reader, second edition
Malcolm Miles, Tim Hall with Iain Borden (eds)

The Cybercities Reader
Stephen Graham (ed.)

The Sustainable Urban Development Reader, second edition
Stephen M. Wheeler and Timothy Beatley (eds)

The Global Cities Reader
Neil Brenner and Roger Keil (eds)

Cities of the Global South Reader
Faranak Miraftab and Neema Kudva (eds)

FORTHCOMING

The City Reader, sixth edition
Richard T. LeGates and Frederic Stout (eds)

▪ ▪ ▪ ▪ ▪ ▪

For further information on The Routledge Urban Reader Series please visit our website:
http://www.routledge.com/articles/featured_series_routledge_urban_reader_series/
or contact

Andrew Mould
Routledge
2 Park Square, Milton Park,
Abingdon, Oxon, OX14 4RN
England
andrew.mould@routledge.co.uk

Richard T. LeGates
Department of Urban Studies
and Planning
San Francisco State University
1600 Holloway Avenue
San Francisco, CA 94132
(510) 642-3256
dlegates@sfsu.edu

Frederic Stout
Urban Studies Program
Stanford University
Stanford, California 94305-2048
fstout@stanford.edu

The Sustainable Urban Development Reader

Third edition

Edited by

Stephen M. Wheeler

and

Timothy Beatley

Routledge
Taylor & Francis Group

LONDON AND NEW YORK

First published 2014
by Routledge
2 Park Square, Milton Park, Abingdon, Oxon OX14 4RN

and by Routledge
711 Third Avenue, New York, NY 10017

Routledge is an imprint of the Taylor & Francis Group, an Informa business

British Library Cataloguing in Publication Data
A catalogue record for this book is available from the British Library

Library of Congress Cataloging in Publication Data
Sustainable urban development reader / edited by Stephen M. Wheeler, Timothy Beatley. – Third edition.
    pages cm. – (routledge urban reader series)
  1. City planning. 2. Community development, Urban. 3. Sustainable urban development.   I. Wheeler, Stephen (Stephen Maxwell) editor of compilation.
II. Beatley, Timothy, editor of compilation.
  HT166.S9135 2014
  307.1'416–dc23

                                                    2014012605

ISBN: 978-0-415-70775-6 (hbk)
ISBN: 978-0-415-70776-3 (pbk)
ISBN: 978-1-315-77036-9 (ebk)

Typeset in Amasis and Berthold Akzidenz Grotesk
by Graphicraft Limited, Hong Kong

Printed in Great Britain by Ashford Colour Press Ltd

Dedication

To Mimi and to the late Dave Brower, whose example
and encouragement have been invaluable (SMW)

To Anneke and Carolena (TB)

Contents

Urban sustainability at the neighborhood or district scale 511

Urban sustainability at the city and regional scale 531

PART 7 SUSTAINABILITY PEDAGOGY AND CLASS EXERCISES 563

Acknowledgments

We would like to thank the many people who have made this volume possible, above all the contributors, whose work continues to inspire us.

Series editor Richard T. LeGates first approached us with the idea of doing such a reader and has been a wonderful source of support and guidance during the process. Routledge's *City Reader*, edited by LeGates and Frederick Stout, has been an excellent model and high standard for us to follow.

David Orr, Marcia McNally, Wicak Sarosa, Keiro Hattori, Kang-Li Wu, Herbert Girardet, Mark Roseland, Richard LeGates, and four anonymous reviewers supplied very helpful comments on the contents and structure of the original edition of this book. Jana Carp, Maria Manta Conroy, Michael Larice, Elizabeth Macdonald, Rafael Pizarro, and four anonymous reviewers supplied very useful comments regarding the second edition, and seven more anonymous reviewers suggested additional material and revisions for the third edition. Thanks to all. Andrew Mould at Routledge has played a central role in making this book possible, while editorial assistants Faye Leerink and Sarah Gilkes, senior production editor Emma Hart, and copy editor Judith Oppenheimer skillfully guided the third edition into print. For assistance with previous editions, we would like to thank Melanie Attridge, Nicola Cooper, Ann King, Jennifer Page, Michael Jones, Vicky Claringbull, Ray Offord, and Lisa Salonen.

Over the years our students have been a great source of feedback on particular readings, and have challenged us to find material that does a good job of communicating sustainability concepts to those not yet familiar with the field. In addition, Stephen Wheeler would like to thank Mimi especially for her constant love and support as well as astute editorial comments during the process of preparing this *Reader*. Tim Beatley would like to thank, as always, his wife Anneke for her patience and love, and his daughter Carolena for her great energy and spirit (that keep him going).

INTRODUCTION TO THE THIRD EDITION

Since the first edition of this book appeared 10 years ago millions more people worldwide have embraced the goal of sustainable development, and new programs and courses on sustainability have appeared on hundreds of university campuses. Younger individuals especially recognize the urgency of the topic. Rising consciousness of global warming is one motivating factor. But other needs as well have come to the fore. Many activists and organizations are now rethinking food systems and developing new opportunities for urban agriculture. Another movement is concerned about public health, in particular the rising tide of obesity and physical inactivity in many countries, which is linked in part to the design of communities and economies. Renewable energy, water systems, and green jobs merit increased attention. New concepts such as "biophilic cities," "slow cities," and "landscape ecological urbanism" have come on the scene, and, as political systems seem more dysfunctional, we must consider how societies can develop more effective governance for sustainability.

This new edition contains additional readings on these and other topics. At the same time, we further develop our emphasis on thinking critically about urban sustainability in light of five years' additional experience and information. There is a great deal of literature that enthusiastically promotes one aspect or another of sustainable cities. However, it is important to inquire deeply into how different concepts and practices may work, into how different parties view these subjects, and into how truly meaningful changes may be brought about. Sustainable urbanism is still a new and emerging field of study. Only through critical thinking and joint exploration can we deepen our collective understandings of it.

This edition benefits from the experience of many previous users of the *Reader*, both within and outside the classroom. In response to feedback we have added 16 new readings while cutting 4, have updated the introductions and suggested further readings throughout, and have revised and expanded the case study section. As with previous versions we have tried to frame each section and reading in a way that will help those new to the subject to understand it, while providing useful intellectual context and references for those readers who already have some background, and stimulating additional debate on questions to which there is no easy answer.

This edition still asks the same basic question: How can we plan and develop communities that will meet long-term human and environmental needs? This concept of sustainable urban development provides a way for citizens, planners, and policymakers worldwide to explore such questions.

The third edition of this book, like the previous ones, aims to provide readers with a wide, thought-provoking selection of writings on this timely subject. We now present an expanded selection of 63 readings related to sustainable community development, drawn from books, academic journals, and general interest magazines. Many of these are "classic" pieces which helped change their fields in various ways, and are important today in order to understand the sustainability discourse. Others are more contemporary readings reflecting current thought and activity. Extensive introductions put each reading in context, and more than two dozen case studies of sustainable urban development initiatives help illustrate the range of projects now underway. Since many of us learn most "by doing," a final section of exercises related to sustainability planning helps individuals, students, or community groups work out their own detailed understandings of sustainable community planning.

Though many of these writings are from North America, we have included pieces from Europe, Asia, Latin America, Africa, and Australia, and consider many of the urban sustainability challenges addressed here to be universal. Cities and towns worldwide are facing similar problems of climate change, excessive motor vehicle use, suburban sprawl, pollution, profligate use of natural resources, rising inequities, and loss of indigenous landscapes and ecosystems. Communities in most parts of the world are also now confronted by a global economic system that frequently undercuts local traditions, businesses, community, environment, and sense of place. Though the context of urban development varies considerably from country to country, many sustainability strategies will be the same, for example, seeking to coordinate transportation and land use, restore urban ecosystems, or design the public realm so as to be friendly for women, children, and the elderly. Moreover, every society these days can learn from innovations in other places. So we have tried to keep our perspective as global as possible.

Our first section highlights classic historic writings that have paved the way for more recent discussions of urban sustainability. Writers such as Ebenezer Howard, Lewis Mumford, and Aldo Leopold raised questions in the early twentieth century about the nature of the industrial city and the fundamental relationship between human development and the natural world. Later writers such as Jane Jacobs, Ian McHarg, Herman Daly, Andres Gunder Frank, and the *Limits to Growth* team spurred reevaluation of unsustainable development practices during the critical period of the 1960s and 1970s when many of the ecological and social implications of global development were first widely understood. Authors such as Anne Whiston Spirn pointed out that cities are not divorced from nature, but rather are an integral part of natural systems.

Subsequent United Nations conferences and commissions – especially the mid-1980s Brundtland Commission and the 1992 Earth Summit – helped call attention to the need for a new development paradigm, issuing declarations such as *Agenda 21* that were influential in stimulating local planning initiatives in many parts of the world. Meanwhile the *Millennium Development Goals* have been the international community's most explicit statement to date of sustainable development goals. These and other influences helped lay the groundwork for current sustainability planning. Understanding particular historical themes – which continue to be echoed today – is important in order to understand how the world's communities can become more sustainable in the future.

After this look at the origins of the sustainability concept, we next survey classic writings in issue areas important to urban development. Climate change planning now merits its own section, and we start this off with a selection from Bill McKibben's bestselling book *The End of Nature*, which brilliantly inserted global warming into public consciousness in 1989. Then comes a classic 2004 piece by Steven Pacala and Robert Socolow suggesting specific strategies for reducing greenhouse gas emissions in 15 different sectors. Lastly, a recent analysis by William Solecki, Cynthia Rosenzweig, Stephen Hammer, and Shagun Mehrotra looks at climate impacts on cities worldwide and possible adaptation responses.

Through selections from the American New Urbanist planner Peter Calthorpe, the Danish designer Jan Gehl, Reid Ewing and his colleagues, and one of us (Wheeler), the section on land use and urban design addresses questions such as, Is compact urban development necessary for sustainability? How can more walkable and community-oriented neighborhoods be designed? How can infill development help revitalize our communities? and How can streets and public spaces work better for people? Our introduction to this section points interested readers toward additional writings and resources, especially those connected with recent movements such as the New Urbanism and Smart Growth.

Transportation systems are fundamental in shaping the land use and physical form of urban areas, as well as determining much about the livability of our communities. Our section on this subject starts with a selection by Robert Cervero, who describes ways that urban regions might make progress toward his vision of the "transit metropolis." Cervero also outlines the huge variety of public transportation modes that might play a role in reducing automobile use. Australian planners Peter Newman and Jeffrey Kenworthy then describe the history of the international movement known as "traffic calming," and explore the techniques, goals, and results of this approach. John Pucher and Ralph Buehler conclude the section with lessons from European communities about how to make cities and towns more bicycle friendly.

Our discussion of environmental planning and restoration begins with a piece by one of us (Beatley) considering the concept of "biophilic cities" – communities that put nature first and provide people with daily contact with it. Next, stream-restoration pioneer Ann Riley investigates the concept of restoration as it applies to urban watershed features. Whereas "conservation" was the watchword of previous generations of environmentalists, "restoration" has become the mantra of many recent activists, and offers exciting possibilities for constructive, hands-on action in cities and towns everywhere. Frederick Steiner then takes a broader view of two recent conceptual approaches, landscape urbanism and urban ecology, and proposes a synthesis of the two, "landscape ecological urbanism," as a perspective that can show "how designing with nature can improve the quality of cities for people, plants, and animals."

One of the most unsustainable dimensions of current urban development has to do with energy and materials use, and the waste, pollution, and greenhouse gas emissions that usually result from this consumption. In our section on resource use, we include a selection from British sustainability pioneer Herbert Girardet analyzing the flow of raw materials through the urban system. To dramatize his points, Girardet calculates the metabolism of Greater London in terms of energy and resources consumed. A recent piece from Worldwatch Institute founder Lester Brown then looks at the potential to power communities from renewable energy, and Pacific Institute founder Peter Gleick investigates how changing paradigms of water use might lead toward sustainability in the twenty-first century. Finally, regenerative design pioneer John Tillman Lyle examines the unglamorous but fascinating topic of urban waste, and analyzes potential methods of ecological sewage treatment. Lyle profiles the ecological sewage-treatment marsh in the city of Arcata, California, which processes wastes for 15,000 people while also creating important wildlife habitat and an attractive and well-used recreational trail system.

The rising tide of inequity in many societies – in which some groups within society prosper while others suffer – is profoundly rooted in current patterns of urban development. Two selections serve to highlight some urban dimensions of equity issues. In an excerpt from his classic 1990 book *Dumping in Dixie*, African-American sociologist Robert Bullard describes the emergence and nature of the environmental justice movement. Bullard calls for a new environmentalism that takes into account equity impacts on particular urban communities, and that fights the institutional forces perpetuating environmental injustice. In a selection from her 1984 book *Redesigning the American Dream*, Dolores Hayden, one of the foremost feminist critics of urban design and planning, examines how women have been excluded from or made to feel uncomfortable within urban environments. Hayden calls for "small, commonsense improvements in urban design" as well as larger changes in the ways society views gender roles, nurturing, and the split between private and public life. Taking a global perspective, Janice E. Perlman and Molly O'Meara Sheehan consider poverty within the rapidly growing megacities of the developing world and propose a number of strategies for reducing it.

One of the most fundamental challenges to sustainable urban development is the need to redirect economies into paths that are restorative rather than exploitative, for example that are not reliant on ever-growing consumption of material products, long-distance trade, and replacement of local businesses by local branches of multinational corporations. In our section on economic development, British economists David Pearce and Edward B. Barbier first of all describe the basic failure of current market economics to take into account many aspects of the world around us, especially natural environments. David C. Korten then discusses the dynamics of economic globalization, and its effects on local communities. In the following selection California businessman Paul Hawken outlines his concept of "natural capitalism," in which the energies of capitalist markets are harnessed for constructive rather than destructive purposes. Michael Shuman describes a vision of community self-reliance and import replacement, a path that runs counter to the current emphasis on global free trade, but that can potentially offer many environmental and social benefits for local communities. Mark Roseland and Lena Soots profile a number of ways that local businesses can be strengthened. And Michael Renner and his colleagues analyze the growth of green jobs and likely prospects for creating more of these.

The buildings that we live and work in are one of the most basic features of urban environments, and so we continue our survey of sustainable city dimensions with a discussion of green architecture and building practices. In a sermon given at the Cathedral of St. John the Divine in New York City in 1993, ecological architect William McDonough eloquently describes his philosophy of placing building design within the context of all of nature. McDonough includes the needs of human users and surrounding communities within this context, and looks to vernacular local traditions for clues as to how to design for particular climates and cultures. Long-time British pioneers Brenda and Robert Vale then outline what they see as basic principles of a "green" approach to architecture, likewise stressing that the wisdom of historic cultures can be a powerful guide to improve the sustainability of modern building practices. In a different vein, David Eisenberg and Peter Yost analyze how modern building codes constrain green building practices and how these regulations might be revised so as to make ecological design more possible. These authors succinctly describe the nature, emergence, and limitations of building codes, and urge the environmental design community to become more involved in revising these basic frameworks within which urban construction takes place. Next we include a description of the LEED (Leadership in Energy and Environmental Design) rating system developed by the U.S. Green Building Council, which has become very extensively used in North America, along with the recently released list of credits for LEED v4. Finally, Leath Sharp, the former green development coordinator at Harvard, outlines ways to make green building cost-effective.

Concluding our "Dimensions" part of the *Reader* is a new set of pieces looking at Food Systems and Health. First, well-known journalist Michael Pollan chronicles the rise of the "Food Movement." Then, Indian activist and scientist Vandana Shiva analyzes changes in global food systems from a developing-world perspective. Howard Frumkin, Lawrence Frank, and Richard Jackson analyze the intersection between town design, physical activity, and public health. And finally Carl Honoré describes the emergence of the slow cities movement, and the general value of "slowness" within a sustainable cities framework.

Although many traditional urban planning techniques, such as the preparation of general plans and zoning codes, can be adapted to promote sustainability goals, certain new or revitalized planning tools can be useful for sustainable urban development. In Part 3 on Sustainability Planning Tools and Politics, we investigate the subject of sustainability indicators through Virginia Maclaren's analysis of these in the *Journal of the American Planning Association*. We next delve into the topic of ecological footprint analysis through a piece by two of the originators of this concept, William Rees and Mathis Wackernagel, from their 1996 book *Our Ecological Footprint*. In a selection from his classic 1985 volume, *Looking at Cities*, Allan Jacobs provides a helpful guide to that most basic and essential of urban analysis techniques, the process of simply observing the city. Too often ignored by planners holed up in rooms full of computers, skillful first-hand analysis is essential in order to analyze how urban places function and to see their current handicaps and future possibilities. Uno Svedin, Tim O'Riordan, and Andrew Jordan consider multilevel "governance" strategies that might help otherwise weak public sector institutions deal more effectively with sustainability needs. And finally, in a piece that is still highly relevant 20 years after it was written, Michael Lerner provides an inspiring description of a "politics of meaning" which might become a basis for changing the status quo. Lerner calls upon professionals and politicians to fundamentally reevaluate the spiritual foundation of their work, and to develop new commitment to meaningful collective challenges such as the task of creating more sustainable and livable communities.

Part 4 examines sustainable urban development efforts internationally, with the aim of giving readers a taste of the wide variety of opportunities and challenges facing communities in different parts of the world. One of the most celebrated examples of innovative urban planning is Curitiba, a city of 1.6 million in southern Brazil. Described here in an article by planners Jonas Rabinovich and Josef Leitman reprinted from *Scientific American*, Curitiba has reshaped its physical form and transportation network over more than four decades, and has also been on the cutting edge of creative social planning. Tim Beatley then updates us on European sustainable community initiatives, profiling projects in the Netherlands, Germany,

England, Denmark, and other nations. Looking at a different hemisphere, Hsin-Huang Michael Hsiao and Hwa-Jen Liu analyze dynamics of urban environmentalism in Taipei, highlighting the tensions between the environmental interests of relatively well-off urban classes and the basic survival needs of the truly poor who crowd many cities in the developing world. Kang-Li Wu contributes a new assessment of urban sustainability practices in China, while Alfonso Iracheta considers challenges of sustainable cities in Mexico, and Kempe Ronald Hope Sr. analyzes climate change and urban development in Africa. Finally, Martha Honey discusses strategies for evaluating ecotourism – the effort to develop sustainable patterns of international leisure travel.

Next we include a brief part with several utopian descriptions of more sustainable communities, in the belief that such visions are important in calling people's attention to alternative philosophies and the need for change. These visions start with Ernest Callenbach's *Ecotopia*, which inspired environmental activists in the 1970s and, like all ecological utopias, has yet to be realized. Part 5 continues with Ursula K. LeGuin's science fiction world of Annares in *The Dispossessed*, specifically her description of the city of Abbaney. LeGuin portrays an even more radical social transformation far in the future, when a secessionist movement seeks to build a society based on cooperation, equity, and modest consumption, in contrast to its corrupt capitalist home world. And, since recent ecological utopias are hard to find, we finish with a new piece by one of us (Wheeler) describing a world after global warming, in which New York and Bangladesh are under water but leaders and publics have finally learned to live sustainably on the planet.

Part 6 contains 32 case studies of innovative sustainable urban development practices. These take place on several different scales: that of the individual building or site, that of the neighborhood or district, and that of the city or region. The book's final portion, Part 7, considers sustainability pedagogy, opening with a fine piece by Debby Cotton and Jennie Winter on that topic, and including a selection of exercises that readers may find it interesting to complete either individually, in groups, or through classes. These exercises have been developed by one of us (Wheeler) in conjunction with courses at the University of California at Berkeley, the University of New Mexico, and the University of California at Davis, but can be adapted for many other types of groups or situations.

In this *Reader* we have sought selections that are classic, readable, diverse, and if possible relatively specific in their analysis and recommendations. We hope that readers of many types will find this book useful – students, academics, planning professionals, and architects certainly, but also environmental activists, community leaders, and urban and suburban residents of all sorts. Each of us, after all, is confronted on a daily basis with the problems resulting from current modes of community development. Whether the issue is traffic congestion, lack of parks and open space, unfriendly streets and public spaces, poor schools, lack of decent-paying, meaningful work, or the frequent absence of community, finding solutions often depends on an understanding of the urban systems around us – both the ways they have arisen in the past, and the ways that they can be improved in the future. This book aims to provide a foundation for that understanding.

PART ONE

Origins of the sustainability concept

INTRODUCTION TO PART ONE

The term "sustainable development" appears to have been used for the first time in the 1972 book *Limits to Growth*, in a passage excerpted later in this section, and has been applied widely to fields such as urban planning and architecture only since the early 1990s. However, concerns about the unsustainability of modern urban development patterns have a much longer history. To put current efforts to create more sustainable cities in perspective, it is important to be aware of this history and the various themes that have converged upon present debates about sustainability.

At least as far back as the early nineteenth century many commentators in Britain, continental Europe, and the United States were worried about the rapid growth of industrial cities. Urban expansion was a subtext underlying Henry David Thoreau's writings at Walden Pond in the 1840s. Thoreau's retreat was in large part a flight from the pace and pressures of urban life, and new rail lines emanating from Boston plus a growing suburban population were encroaching on the pond itself. In 1840s England, the deplorable social conditions of the working class in Manchester and the increasing spatial segregation between the suburban estates of wealthy mill owners and the urban tenements of their workers helped motivate the writings of Frederick Engels. At the same time Romantic poets such as Keats, Wordsworth, and Shelley extolled the virtues of nature in a reaction against industrial society, while novelists ranging from Charles Dickens beginning in the 1830s to D.H. Lawrence in the 1920s described the horrors of industrial cities and towns and the efforts of individuals to surmount or come to terms with these circumstances.

Cities of a million or more residents were virtually unknown before 1800, when London achieved this level of population (ancient Rome may have been home to a similar number). But coal-fired industrial factories drew workers from the countryside, while factors such as overpopulation, the privatization of formerly commonly held rural land, and increasingly centralized rural land ownership pushed country dwellers away from their traditional communities. For the first time in the middle and late nineteenth century large numbers of people lived in crowded urban environments far from the countryside, and new technological advances such as the streetcar, the railroad, macadam road paving, modern plumbing, and electric lights helped distance people from the natural world. These forerunners of late twentieth-century megacities had enormous problems related to public health, sanitation, residential overcrowding, and nonexistent infrastructure. Deforestation of countrysides and pollution of air and water also reached new, and in some cases still unmatched, heights. Not surprisingly, many observers felt that the balance between humans and the natural world had been tipped too far in one direction.

Late nineteenth- and early twentieth-century social reformers sought to call attention to the deterioration of urban conditions and the need for alternative living environments. One of the most influential of these writers was Ebenezer Howard, a court stenographer whose slim book on "garden cities," first published in 1898, inspired generations of urban planners and designers. Howard, Scotch visionary Patrick Geddes, and their American follower Lewis Mumford saw the extreme overcrowding of early industrial cities – with its accompanying problems of sanitation, services, pollution, and public health – as the main problem to be addressed. In response, they called for a new balance between city and country in which population was decentralized into carefully planned new communities in the countryside. While this idea had many merits, these authors wrote before automobile use became widespread or its implications understood, and before the huge wave of twentieth-century suburbanization turned Howard's "garden city" idea into much-simplified "garden suburbs" and created a whole new set of development problems in the process. Yet these early writers did much to focus public attention on the unsustainability of urban development trends at that time, the inability of private sector forces to deal with these problems, and the need for thoughtful planning of better alternatives.

The professions of landscape architecture and city planning emerged largely in reaction to the rapid nineteenth-century expansion of industrial cities. The former focused in large part on providing picturesque parks and living environments to urban residents, and in the process helped lay the aesthetic groundwork for twentieth-century suburbia. The latter sought to ensure the forms of infrastructure, housing, land use, and transportation that were viewed as necessary for orderly urban growth. However, in their response to nineteenth-century problems both professions inadvertently established the conditions for another set of sustainability problems in the twentieth century, those related to low-density suburban sprawl.

So one main planning theme that emerged in the nineteenth century – and that sustainability-oriented writers have returned to ever since – was the balance between city and nature. Another theme, much less acted upon, had to do with the challenge of promoting equity. From oppressive working conditions within factories to the mile after mile of dreary tenements and working-class suburbs that were con-structed in the early industrial era, the environment in which working individuals lived was often grim, unhealthy, and unjust. Politicians eventually enacted some reforms, for example in the form of housing codes ensuring adequate light and air, but many other inequities continued. In the post-World War II period, and especially in the 1960s and 1970s, a growing number of writers criticized the widening rifts between rich and poor, between different ethnic or racial groups, and between how men and women are affected by urban environments. Many critics began to realize that twentieth-century development practices, both within cities and worldwide, were worsening inequities rather than improving them. The need for a more equitable society became another cornerstone of sustainable development, one that is more difficult to address than environmental clean-up because it so directly challenges the structure of wealth and power within nations.

A third recurrent theme has to do with the notion of economic growth and the inadequacy of eco-nomics to regulate human and natural systems. British economist and social critic John Stuart Mill in the mid-nineteenth century first raised the notion that a steady-state economy might be desirable, as opposed to one based on endlessly growing production and consumption. This theme was taken up a century later by Kenneth Boulding, E.F. Schumacher, and Herman Daly, all economists who considered whether existing concepts of economic development were compatible with the notion of a limited planet. Daly in particular further developed the concept of steady-state economics, and this part includes one of his classic essays on the topic. The chapter on economic development in this book's second part explores other economic implications of sustainable urban development.

Various debates around these three concerns – environment, equity, and economy, frequently referred to as the "three E's" of sustainable development – have thus been gestating for a century or more. Sustainability advocates have sought ways to maximize all three value sets at once, rather than playing them off against one another, as more traditional development strategies have often done. But the process is not easy, and is likely to require looking closely at each of these areas.

The concept of "sustained" development itself emerged most directly from the field of natural resource management. In the late nineteenth century Germany faced severe problems with over-cutting of forests, and developed sustained-yield forest management techniques to compensate. Americans such as Gifford Pinchot learned these approaches at continental forestry schools and imported them into the United States, which despite its vast natural resource holdings was beginning to confront the notion of limits as well. The concept of managing ecosystems for sustained resource yield was quickly applied to wild-life species and fisheries as well as forestry. This "conservationist" perspective pioneered in the late nineteenth century – based on a view of humans as apart from nature and managing natural resources for their own use – is frequently contrasted with the "preservationist" perspective advocated by Sierra

Club founder John Muir and others at the same time. In the latter view, nature has intrinsic value and should be protected for its own sake. Both perspectives have played a role in recent sustainable development discourses. Aldo Leopold, whose pivotal essay on "the land ethic" is included in this part, was central between these two camps. Although his career began within the conservationist tradition, in later life he came to see humans as part of a larger organic whole, and so helped lay the groundwork for more radical environmental movements such as "deep ecology." His assertion of a profound human responsibility to care for and heal natural systems is an important philosophy behind many sustainable community initiatives. The understanding that nature is inextricably woven into cities, developed by writers such as Anne Whiston Spirn in the early 1980s, likewise underlies sustainable city efforts.

In recent years concern about global warming has motivated the search for sustainable development. Although the concept that carbon dioxide from the burning of fossil fuels might trap heat in the earth's atmosphere has been known since the late nineteenth century, and the danger that it might accumulate rapidly rather than being largely absorbed by the earth's oceans has been realized since the late 1950s, it is only recently that humanity has taken this threat seriously. Bill McKibben's bestselling 1989 book *The End of Nature* eloquently spelled out the problem for millions of readers, and we have included an excerpt from that influential volume here.

Selections from the pivotal reports of the *Limits to Growth* research team, which in 1972 presaged current concerns about global warming and peak oil in more general terms, the 1987 Brundtland Commission, the 1992 Rio Earth Summit, and the 2000 United Nations Millennium Development Goals round out this part. These classic documents have all been extremely important internationally in stimulating sustainable development activity, though each is open to criticism on various grounds. Unfortunately there is no single, universally acknowledged manifesto that by itself sets out a sustainable *urban* development agenda. The 1996 United Nations Habitat II Conference, the "City Summit," sought to produce such a document, but the so-called Habitat Declaration has not attracted a wide following. However, other declarations of sustainable development principles have been put forth by architects, urban designers, and activists, and there does seem to be consensus emerging on many directions for sustainable urban development. The 2000 United Nations Millennium Development Goals encapsulate many of these themes.

"The Three Magnets" and "The Town–Country Magnet"

from *Garden Cities of To-morrow* (1898)

Ebenezer Howard

Editors' Introduction

Perhaps the single most influential and visionary book in the history of urban planning has been Ebenezer Howard's slim 1898 volume originally simply entitled *To-morrow*, and four years later reissued as *Garden Cities of To-morrow*. In a more detailed fashion than had ever been attempted before, Howard outlined a strategy for addressing the problems of the industrial city, one that attempts to balance city and country in what we might view today as a sustainable fashion.

To be sure, previous visionaries had suggested or even built new towns outside of cities. Scottish mill owner Robert Owen, for example, had constructed the town of New Lanark in 1800–10 near Glasgow for his workers, British soap manufacturer William Lever had built Port Sunlight in 1888 near Liverpool to house *his* workers, chocolate manufacturer George Cadbury had created Bournville near Birmingham in the 1880s, and American railroad magnate George Pullman had developed the town of Pullman outside Chicago for his employees at about the same time. Meanwhile, French philosophers Pierre-Joseph Proudhon and Charles Fourier as well as the Russian Peter Kropotkin had suggested principles for utopian new communities. Catalán engineer Ildefons Cerdá had laid out a large new extension to Barcelona in 1859, and had authored a pioneering book of urban planning philosophy in 1863, calling for a holistic, integrated approach to urbanization. But Howard's vision of systematically deconcentrating the population of an industrial city such as London into a ring of carefully organized garden cities surrounded by countryside and connected by railroads went far beyond anything presented before, and spoke powerfully to the needs of the time.

Later urbanists including Raymond Unwin, John Nolen, Lewis Mumford, Patrick Abercrombie, Ian McHarg, and Peter Calthorpe would seek different implementations of this basic idea. Two English garden cities were actually built in the early twentieth century, Letchworth and Welwyn, and the concept inspired the British New Town program that constructed 11 satellite cities around London between the 1940s and the 1960s. Swedish new towns such as Vällingby and Farsta, Dutch new towns such as Houten, and German new towns near Frankfurt have been built following many of the same principles. For its part, the United States government sponsored three garden cities in the 1930s – Greenbelt, Maryland, Greenhills, Ohio, and Greendale, Wisconsin – while private developers built a handful of new towns along the garden city model, including Radburn, New Jersey (of which only one neighborhood was completed), Baldwin Hills Village in Los Angeles, and (much later) Reston, Virginia.

A court stenographer by profession, Howard exemplifies how some of the most revolutionary ideas in city planning have come from concerned citizens rather than professional planners or architects. Jane Jacobs and Lewis Mumford – both writers rather than planners – also fall into this category. Howard's style was cautious, pragmatic, and designed to appear reasonable to the average citizen. He quoted extensively from leading authorities of the day, provided conceptual graphics, and included financial information attempting to show

how garden cities could be developed economically. His "three magnets" diagram, presented here, was a simple but effective metaphor to get readers to see that a new concept of urban development was needed, one that balanced city and country. Howard included not just physical planning principles, but detailed social and economic proposals as well, including social control over land. He also tries to strike a balance between prescriptive solutions and creative, incremental development of communities, taking local conditions into account. Thus he does not diagram out the location of every building and road, but presents schematic diagrams "for conceptual purposes only."

Although his scheme is highly visionary and quite unlike any previously existing community, Howard does his best to present his idea in a pragmatic, reasonable way that might actually be implemented. He quotes leading authorities of the time, and is very specific about the finances, physical extent, design, economic base, and social structure of his proposed community. Howard's efforts paid off, as a Garden Cities Association dedicated to creating such communities was established after the book's appearance, and he himself advised in the creation of Letchworth (1911) and Welwyn (1926) outside of London.

Howard's search for a balance between city and country life is still central to the task of creating more sustainable communities. But the balance has shifted. Instead of the extremely dense nineteenth-century city with a frequent lack of decent housing, clean water, and basic sanitation, we now have in many parts of the world low-density, automobile-dependent suburbs with a much higher quality of housing and infrastructure but many new problems. One type of balance between city and country has been created, though not the one Howard envisioned. Moreover, as we shall see in later parts, environmentalists now advocate integrating nature into cities in ways not thought of a century ago.

The question now is, to what extent is Howard's vision still useful? Does it provide a roadmap for more sustainable cities, and which particular aspects of urban sustainability is it best at addressing? How might we revise his concepts to create new forms of garden city, and how might those garden cities be brought about institutionally and economically?

◼◼◼◼◼◼◼

THE THREE MAGNETS (FROM THE AUTHOR'S INTRODUCTION)

There is, however, a question in regard to which no one can scarcely find any difference of opinion. It is wellnigh universally agreed by men of all parties, not only in England, but all over Europe and America and our colonies, that it is deeply to be deplored that people should continue to stream into the already over-crowded cities, and should thus further deplete the country districts.

Lord Rosebery, speaking some years ago as Chairman of the London County Council, dwelt with very special emphasis on this point:

> There is no thought of pride associated in my mind with the idea of London. I am always haunted by the awfulness of London: by the great appalling fact of these millions cast down, as it would appear by hazard, on the banks of this noble stream, working each in their own grove and their own cell, without regard or knowledge of each other, without heeding each other, without having the slightest idea how the other lives – the heedless casualty of unnumbered thousands of men. Sixty years ago a great Englishman, Cobbett, called it a wen. If it was a wen then, what is it now? A tumour, an elephantiasis sucking into its gorged system half the life and the blood and the bone of the rural districts. (March 1891)

Sir John Gorst points out the evil, and suggests the remedy:

> If they wanted a permanent remedy of the evil they must remove the cause; they must back the tide, and stop the migration of the people into the towns, and get the people back to the land. The interest and the safety of the towns themselves were involved in the solution of the problem. (*Daily Chronicle*, 6 November 1891)

Dean Farrar says:

> We are becoming a land of great cities. Villages are stationary or receding; cities are enormously increasing. And if it be true that great cities tend more and more to become the graves of the physique of our race, can we wonder at it when we see the houses so foul, so squalid, so ill-drained, so vitiated by neglect and dirt?

. . . All, then, are agreed on the pressing nature of this problem, all are bent on its solution, and though it would doubtless be quite Utopian to expect a similar agreement as to the value of any remedy that may be proposed, it is at least of immense importance that, on a subject thus universally regarded as of supreme importance, we have such a consensus of opinion at the outset. . . .

Whatever may have been the causes which have operated in the past, and are operating now, to draw the people into the cities, those causes may all be summed up as "attractions"; and it is obvious, therefore, that no remedy can possibly be effective which will not present to the people, or at least to considerable portions of them, greater "attractions" that our cities now possess, so that the force of the old "attractions" shall be overcome by the force of new "attractions" which are to be created. Each city may be regarded as a magnet, each person as a needle; and, so viewed, it is at once seen that nothing short of the discovery of a method for constructing magnets of yet greater power than our cities possess can be effective for redistributing the population in a spontaneous and healthy manner.

So presented, the problem may appear at first sight to be difficult, if not impossible, of solution. "What," some may be disposed to ask, "can possibly be done to make the country more attractive to a workaday people than the town – to make wages,

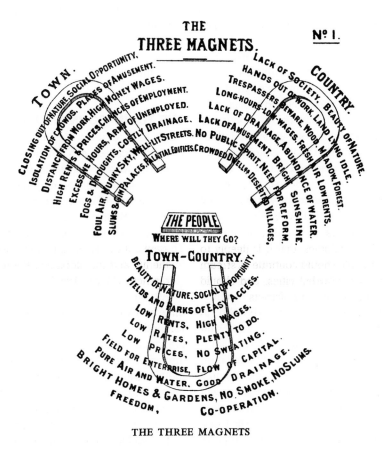

THE THREE MAGNETS

Figure 1 Howard's diagram of the "three magnets."

or at least the standard of physical comfort, higher in the country than in the town; to secure in the country equal possibilities of social intercourse, and to make the prospects of advancement for the average man or woman equal, not to say superior, to those enjoyed in our large cities?" . . .

There are in reality not only, as is so constantly assumed, two alternatives – town life and country life – but a third alternative, in which all the advantages of the most energetic and active town life, with all the beauty and delight of the country, may be secured in perfect combination; and the certainty of being able to live this life will be the magnet which will produce the effect for which we are all striving – the spontaneous movement of the people from our crowded cities to the bosom of our kindly mother earth, at once the source of life, of happiness, of wealth, and of power. The town and the country may, therefore, be regarded as two magnets, each striving to draw the people to itself – a rivalry which a new form of life, partaking of the nature of both, comes to take part in. . . .

Neither the Town magnet nor the Country magnet represents the full plan and purpose of nature. Human society and the beauty of nature are meant to be enjoyed together. The two magnets must be made one. As man and woman by their varied gifts and faculties supplement each other, so should town and country. The town is the symbol of society – of mutual help and friendly co-operation, of fatherhood, motherhood, brotherhood, sisterhood, of wide relations between man and man – of broad, expanding sympathies – of science, art, culture, religion. And the country! The country is the symbol of God's love and care for man. All that we are and all that we have comes from it. Our bodies are formed of it; to it they return. We are fed by it, clothed by it, and by it are we warmed and sheltered. On its bosom we rest. Its beauty is the inspiration of art, of music, of poetry. Its forces propel all the wheels of industry. It is the source of all health, all wealth, all knowledge. But its fullness of joy and wisdom has not revealed itself to man. Nor can it ever, so long as this unholy, unnatural separation of society and nature endures. Town and country *must be married*, and out of this joyous union will spring a new hope, a new life, a new civilization. It is the purpose of this work to show how a first step can be taken in this direction by the construction of a Town–Country magnet; and I hope to convince the reader that this is practicable, here and now, and that on principles which are the very soundest, whether viewed from the ethical or the economic standpoint.

I will undertake, then, to show how in "Town–country" equal, nay better, opportunities of social intercourse may be enjoyed than are enjoyed in any crowded city, while yet the beauties of nature may encompass and enfold each dweller therein; how higher wages are compatible with reduced rents and rates; how abundant opportunities for employment and bright prospects of advancement may be secured for all; how capital may be attracted and wealth created; how the most admirable sanitary conditions may be ensured; how beautiful homes and gardens may be seen on every hand; how the bounds of freedom may be widened, and yet all the best results of concert and co-operation gathered in by a happy people.

The construction of such a magnet, could it be effected, followed, as it would be, by the construction of many more, would certainly afford a solution of the burning question set before us by Sir John Gorst, "how to back the tide of migration of the people into the towns, and how to get them back upon the land."

THE TOWN–COUNTRY MAGNET

Howard included not just physical planning principles, but detailed social and economic proposals as well, including social control over land. He also tries to strike a balance between prescriptive solutions and creative, incremental development of communities taking local conditions into account. Thus he does not diagram out the location of every building and road, but presents schematic diagrams "for conceptual purposes only."

Although his scheme is highly visionary and quite unlike any previously existing community, Howard does his best to present his idea in a pragmatic, reasonable way that might actually be implemented. He quotes leading authorities of the time, and is very specific about the finances, physical extent, design, economic base, and social structure of his proposed community. Howard's efforts paid off, as a Garden Cities Association dedicated to creating such communities was established after the book's appearance, and he himself advised in the creation of Letchworth (1911) and Welwyn (1926) outside of London.

The reader is asked to imagine an estate embracing an area of 6,000 acres, which is at present purely agricultural, and has been obtained by purchase in the open market at a cost of £40 an acre, or £240,000. The purchase money is supposed to have been raised on mortgage debentures, bearing interest at an average rate not exceeding 4 percent. The estate is legally vested in the names of four gentlemen of responsible position and of undoubted probity and honour, who hold it in trust, first, as a security for the debenture-holders, and, secondly, in trust for the people of Garden City, the Town–Country magnet, which it is intended to build thereon. One essential feature of the plan is that all ground rents, which are to be based upon the annual value of the land, shall be paid to the trustees, who, after providing for interest and sinking fund, will hand the balance to the Central Council of the new municipality, to be employed by such Council in the creation and maintenance of all necessary public works roads, schools, parks, etc.

The objects of this land purchase may be stated in various ways, but it is sufficient here to say that some of the chief objects are these: To find for our industrial population work at wages of *higher purchasing power*, and to secure healthier surroundings and more regular employment. To enterprising manufacturers, co-operative societies, architects, engineers, builders, and mechanicians of all kinds, as well as to many engaged in various professions, it is intended to offer a means of securing new and better employment for their capital and talents, while to the agriculturists at present on the estate as well as to those who may migrate thither, it is designed to open a new market for their produce close to their doors. Its object is, in short, to raise the standard of health and comfort of all true workers of whatever grade – the means by which these objects are to be achieved being a healthy, natural, and economic combination of town and country life, and this on land owned by the municipality.

Garden City, which is to be built near the center of the 6,000 acres, covers an area of 1,000 acres, or a sixth part of the 6,000 acres, and might be of circular form, 1,240 yards (or nearly three-quarters of a mile) from center to circumference. (Figure 2 is a ground plan of the whole municipal area, showing the town in the center; and Figure 3 [The Figure 3 referred to is not included here – *Eds.*], which represents one section or ward of the town, will be useful in following the description of the town itself – *a description which is, however, merely suggestive, and will probably be much departed from.*)

Six magnificent boulevards – each 120 ft wide – traverse the city from center to circumference, dividing it into six equal parts or wards. In the center is a circular space containing about five and a half acres, laid out as a beautiful and well-watered garden; and, surrounding this garden, each standing in its own ample grounds, are the larger public buildings – town hall, principal concert and lecture hall, theatre, library, museum, picture-gallery, and hospital.

The rest of the large space encircled by the "Crystal Palace" is a public park, containing 145 acres, which includes ample recreation grounds within very easy access of all the people.

Running all round the Central Park (except where it is intersected by the boulevards) is a wide glass arcade called the "Crystal Palace," opening on to the park. This building is in wet weather one of the favorite resorts of the people, whilst the knowledge that its bright shelter is ever close at hand tempts people into Central Park, even in the most doubtful of weathers. Here manufactured goods are exposed for sale, and here most of that class of shopping which requires the joy of deliberation and selection is done. The space enclosed by the Crystal Palace is, however, a good deal larger than is required for these purposes, and a considerable part of it is used as a Winter Garden – the whole forming a permanent exhibition of a most attractive character, whilst its circular form brings it near to every dweller in the town – the furthest removed inhabitant being within 600 yards.

Passing out of the Crystal Palace on our way to the outer ring of the town, we cross Fifth Avenue – lined, as are all the roads of the town, with trees – fronting which, and looking on to the Crystal Palace, we find a ring of very excellently built houses, each standing in its own ample grounds; and, as we continue our walk, we observe that the houses are for the most part built either in concentric rings, facing the various avenues (as the circular roads are termed), or fronting the boulevards and roads which all converge to the center of the town. Asking the friend who accompanies us on our journey what the population of this little city

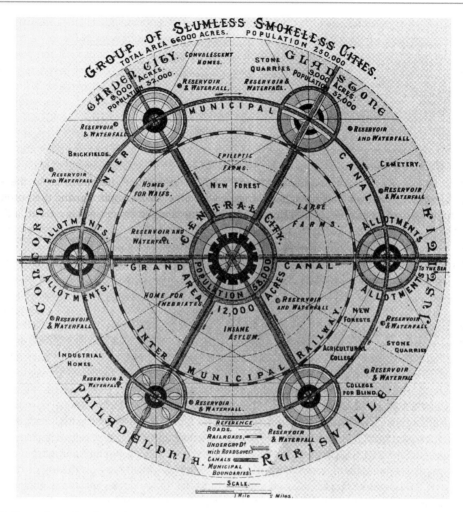

Figure 2 Howard's garden city project.

may be, we are told about 30,000 in the city itself, and about 2,000 in the agricultural estate, and that there are in the town 5,500 building lots of an average size of 20 ft × 130 ft – the minimum space allotted for the purpose being 20 ft × 100 ft. Noticing the very varied architecture and design which the houses and groups of houses display some having common gardens and co-operative kitchens – we learn that general observance of street line or harmonious departure from it are the chief points as to house building, over which the municipal authorities exercise control, for, though proper sanitary arrangements are strictly enforced, the fullest measure of individual taste and preference is encouraged.

Walking still toward the outskirts of the town, we come upon "Grand Avenue." This avenue is fully entitled to the name it bears, for it is 420 ft wide, and, forming a belt of green upwards of three miles long, divides that part of the town which lies outside Central Park into two belts. It really constitutes an additional park of 115 acres – a park which is within 240 yards of the furthest removed inhabitant. In this splendid avenue six sites, each of four acres, are occupied by public schools and their surrounding playgrounds and gardens, while other sites are reserved for churches, of such denominations as the religious beliefs of the people may determine, to be erected and maintained out of the funds of the worshippers and their friends.

We observe that the houses fronting on Grand Avenue have departed (at least in one of the wards – that of which Figure 3 is a representation) – from the general plan of concentric rings, and, in order to ensure a longer line of frontage on Grand Avenue, are arranged in crescents – thus also to the eye yet further enlarging the already splendid width of Grand Avenue.

On the outer ring of the town are factories, warehouses, dairies, markets, coal yards, timber yards, etc., all fronting on the circle railway, which encompasses the whole town, and which has sidings connecting it with a main line of railway which passes through the estate. This arrangement enables goods to be loaded direct into trucks from the warehouses and work shops, and so sent by railway to distant markets, or to be taken direct from the trucks into the warehouses or factories; thus not only effecting a very great saving in regard to packing and cartage, and reducing to a minimum loss from breakage, but also, by reducing the traffic on the roads of the town, lessening to a very marked extent the cost of their maintenance. The smoke fiend is kept well within bounds in Garden City; for all machinery is driven by electric energy, with the result that the cost of electricity for lighting and other purposes is greatly reduced.

The refuse of the town is utilized on the agricultural portions of the estate, which are held by various individuals in large farms, small holdings, allotments, cow pastures, etc.; the natural competition of these various methods of agriculture, tested by the willingness of occupiers to offer the highest rent to the municipality, tending to bring about the best system of husbandry, or, what is more probable, the best systems adapted for various purposes. Thus it is easily conceivable that it may prove advantageous to grow wheat in very large fields, involving united action under a capitalist farmer, or by a body of co-operators; while the cultivation of vegetables, fruits, and flowers, which requires closer and more personal care, and more of the artistic and inventive faculty, may possibly be best dealt with by individuals, or by small groups of individuals having a common belief in the efficacy and value of certain dressings, methods of culture, or artificial and natural surroundings.

This plan, or, if the reader be pleased to so term it, this absence of plan, avoids the dangers of stagnation or dead level, and, though encouraging individual initiative, permits of the fullest co-operation, while the increased rents which follow from this form of competition are common or municipal property, and by far the larger part of them are expended in permanent improvements.

While the town proper, with its population engaged in various trades, callings, and professions, and with a store or depot in each ward, offers the most natural market to the people engaged on the agricultural estate, inasmuch as to the extent to which the townspeople demand their produce they escape altogether any railway rates and charges; yet the farmers and others are not by any means limited to the town as their only market, but have the fullest right to dispose of their produce to whomsoever they please. Here, as in every feature of the experiment, it will be seen that it is not the area of rights which is contracted, but the area of choice which is enlarged.

The principle of freedom holds good with regard to manufacturers and others who have established themselves in the town. These manage their affairs in their own way, subject, of course, to the general law of the land, and subject to the provision of sufficient space for workmen and reasonable sanitary conditions. Even in regard to such matters as water, lighting, and telephonic communication – which a municipality, if efficient and honest, is certainly the best and most natural body to supply – no rigid or absolute monopoly is sought; and if any private corporation or any body of individuals proved itself capable of supplying on more advantageous terms, either the whole town or a section of it, with these or any commodities the supply of which was taken up by the corporation, this would be allowed. No really sound system of action is in more need of artificial support than is any sound system of thought. The area of municipal and corporate action is probably destined to become greatly enlarged; but if it is to be so, it will be because the people possess faith in such action, and that faith can be best shown by a wide extension of the area of freedom.

Dotted about the estate are seen various charitable and philanthropic institutions. These are not under the control of the municipality, but are supported and managed by various public-spirited people who have been invited by the municipality to establish these institutions in an open healthy district, and on land let to them at a pepper-corn rent, it occurring to the authorities that they can the better afford to be thus generous, as the spending power of these institutions greatly benefits the whole community. Besides, as those persons who migrate to the town are among its most energetic and resourceful members, it is but just and right that their more helpless brethren should be able to enjoy the benefits of an experiment which is designed for humanity at large.

"Cities and the Crisis of Civilization"

from *The Culture of Cities* (1938)

Lewis Mumford

Editors' Introduction

Howard's concern with the rapid nineteenth-century growth of industrial cities was taken up by many others in the United Kingdom, continental Europe, and the United States. British architect Raymond Unwin, for example, in 1909 lamented the dramatic expansion of urban areas in terms we might use today:

> The last century has been remarkable, not only in this country but in some others, for an exceedingly rapid and extensive growth of towns. In England this growth has produced most serious results. For many years social reformers have been protesting against the evils which have arisen owing to this rapid and disorderly increase in the size of towns and their populations. Miles and miles of ground, which people not yet elderly can remember as open green fields, are now covered with dense masses of buildings packed together in rows along streets which have been laid out in a perfectly haphazard manner, without any consideration for the common interests of the people . . .[1]

The solution of Unwin and his associate Barry Parker was a better-designed garden suburb, emphasizing the aesthetic, place-making themes pioneered a decade earlier by German architect Camillo Sitte. Many public officials took an even more pragmatic approach focused on regulation rather than physical design. Beginning in Germany in the 1890s and continuing in Britain and the United States during the early decades of the twentieth century, they enacted zoning codes designed to control densities and enforce segregation of housing, shops, and workplaces – presumably protecting neighborhood quality and property values. But visionaries such as Howard and Geddes sought a broader rethinking of development principles at a metropolitan scale.

Lewis Mumford was in the latter camp, a brilliant humanist critic of architecture and society. During his long life he played a central role as American popularizer of garden city ideas. Like Howard and Geddes, Mumford and his colleagues in the Regional Plan Association of America sought to respond to the problems of the overcrowded industrial city by advocating the decentralization of population so as to achieve a better balance of city and countryside. In his many books Mumford displayed an unparalleled ability to weave together an encyclopedic knowledge of history with an eloquent rhetorical style and a passionate concern for human culture and welfare. Although its prose style now seems from another era, books such as *The Culture of Cities* (New York: Harcourt, Brace, 1938), from which this passage is taken, inspired generations of later Urbanists. One of these was MIT urban design professor Kevin Lynch, who would continue Mumford's emphasis on developing a normative urban planning agenda in books such as *Good City Form* (Cambridge: MIT Press, 1981).

A writer rather than a professional architect or planner, Mumford nevertheless proved one of the single greatest influences on American planning in the twentieth century. An overview of his work is given by Mark

Luccarelli in his book *Lewis Mumford and the Ecological Region: The Politics of Planning* (New York: The Guilford Press, 1995); a leading biography is Donald L. Miller's *Lewis Mumford: A Life* (New York: Grove Press, 1989). Living through two world wars, a cold war, and the devastation of urban landscapes by the automobile and urban renewal, Mumford had a deep fear that mechanistic, war-like forces would subvert the humane values and small-scale relationships that he saw as characterizing pre-industrial cities. "Ours is an age in which the increasingly automatic processes of production and urban expansion have displaced the human goals they are supposed to serve," he wrote in 1961.[2] Taking a sweeping view of history, Mumford used terms such as "paleotechnic," "neotechnic," and "biotechnic" to denote different eras of human activity and mindsets. Optimistically, he saw the age of nineteenth-century coal-based industrialization giving way to a cleaner, less exploitative neotechnic era, based on electricity as a power source, and eventually to a more restorative biotechnic era, based on biological science and a more organic philosophy. Within this evolution, he saw cities as playing a central role in nurturing human culture. Although often not specific on details, Mumford was clear on the general change of direction needed, which was toward "the development of a more organic world picture, which shall do justice to all the dimensions of living organisms and human personalities."[3]

Now, more than 70 years later, we can still ask what a biotechnic era or a more organic view of urban development would look like. We can also debate to what extent large-scale, mechanistic forces run our lives, as Mumford feared, and to what extent a shift towards humane values and small-scale community is necessary for sustainable cities. And we can ask, How might cities truly become "man's greatest work of art" as Mumford saw them – a transformative environment for a sustainable future?

The city, as one finds it in history, is the point of maximum concentration for the power and culture of a community. It is the place where the diffused rays of many separate beams of life fall into focus, with gains in both social effectiveness and significance. The city is the form and symbol of an integrated social relationship: it is the seat of the temple, the market, the ball of justice, the academy of learning. Here in the city the goods of civilization are multiplied and manifolded; here is where human experience is transformed into viable signs, symbols, patterns of conduct, systems of order. Here is where the issues of civilization are focused: here, too, ritual passes on occasion into the active drama of a fully differentiated and self-conscious society.

Cities are a product of the earth. They reflect the peasant's cunning in dominating the earth; technically they but carry his skill in turning the soil to productive uses, in enfolding his cattle for safety, in regulating the waters that moisten his fields, in providing storage bins and barns for his crops. Cities are emblems of that settled life which began with permanent agriculture: a life conducted with the aid of permanent shelters, permanent utilities like orchards, vineyards, and irrigation works, and permanent buildings for protection and storage.

Every phase of life in the countryside contributes to the existence of cities. What the shepherd, the woodman, and the miner know, becomes transformed and "etherealized" through the city into durable elements in the human heritage: the textiles and butter of one, the moats and dams and wooden pipes and lathes of another, the metals and jewels of the third, are finally converted into the instruments of urban living: underpinning the city's economic existence, contributing art and wisdom to its daily routine. Within the city the essence of each type of soil and labor and economic goal is concentrated: thus arise greater possibilities for interchange and for new combinations not given in the isolation of their original habitats.

Cities are a product of time. They are the molds in which men's lifetimes have cooled and congealed, giving lasting shape, by way of art, to moments that would otherwise vanish with the living and leave no means of renewal or wider participation behind them. In the city, time becomes visible: buildings and monuments and public ways, more open than the written record, more subject to the gaze of many men than the scattered artifacts of the countryside, leave an imprint upon the minds even of the ignorant or the indifferent. Through the material fact of preservation, time challenges time, time clashes with time: habits and values carry over beyond the living group, streaking with different strata of time the character of any single generation. Layer upon layer, past times preserve themselves in the city until life itself is finally threatened with suffocation: then, in sheer defense, modern man invents the museum.

By the diversity of its time-structures, the city in part escapes the tyranny of a single present, and the monotony of a future that consists in repeating only a single beat heard in the past. Through its complex orchestration of time and space, no less than through the social division of labor, life in the city takes on the character of a symphony: specialized human aptitudes, specialized instruments, give rise to sonorous results which, neither in volume nor in quality, could be achieved by any single piece.

Cities arise out of man's social needs and multiply both their modes and their methods of expression. In the city remote forces and influences intermingle with the local: their conflicts are no less significant than their harmonies. And here, through the concentration of the means of intercourse in the market and the meeting place, alternative modes of living present themselves: the deeply rutted ways of the village cease to be coercive and the ancestral goals cease to be all-sufficient: strange men and women, strange interests, and stranger gods loosen the traditional ties of blood and neighborhood. . . .

The city is a fact in nature, like a cave, a run of mackerel or an ant-heap. But it is also a conscious work of art, and it holds within its communal framework many simpler and more personal forms of art. Mind *takes form* in the city; and in turn, urban forms condition mind. For space, no less than time, is artfully reorganized in cities: in boundary lines and silhouettes, in the fixing of horizontal planes and vertical peaks, in utilizing or denying the natural site, the city records the attitude of a culture and an epoch to the fundamental facts of its existence. The dome and the spire, the open avenue and the closed court, tell the story, not merely of different physical accommodations, but of essentially different conceptions of man's destiny. The city is both a physical utility for collective living and a symbol of those collective purposes and unanimities that arise under such favoring circumstance. With language itself, it remains man's greatest work of art. . . .

Today a great many things stand in the way of grasping the role of the city and of transforming this basic means of communal existence. During the last few centuries the strenuous mechanical organization of industry, and the setting up of tyrannous political states, have blinded most men to the importance of facts that do not easily fit into the general pattern of mechanical conquest, capitalistic forms of exploitation, and power politics. Habitually, people treat the realities of personality and association and city as abstractions, while they treat confused pragmatic abstractions such as money, credit, political sovereignty, as if they were concrete realities that had an existence independent of human conventions.

Looking back over the course of Western Civilization since the fifteenth century, it is fairly plain that mechanical integration and social disruption have gone on side by side. Our capacity for effective physical organization has enormously increased; but our ability to create a harmonious counterpoise to these external linkages by means of co-operative and civic associations on both a regional and a worldwide basis, like the Christian Church in the Middle Ages, has not kept pace with these mechanical triumphs. By one of those mischievous turns, from which history is rarely free, it was precisely during this period of flowing physical energies, social disintegration, and be-wildered political experiment that the populations of the world as a whole began mightily to increase, and the cities of the Western World began to grow at an inordinate rate. Forms of social life that the wisest no longer understood, the more ignorant were prepared to build. Or rather: the ignorant were completely unprepared, but that did not prevent the building.

The result was not a temporary confusion and an occasional lapse in efficiency. What followed was a crystallization of chaos: disorder hardened uncouthly in metropolitan slum and industrial factory districts; and the exodus into the dormitory suburbs and factory spores that surrounded the growing cities merely widened the area of social derangement. The mechanized physical shell took precedence in every growing town over the civic nucleus: men became dissociated as citizens in the very process of coming together in imposing economic organizations. Even industry, which was supposedly served by this planless building and random physical organization, lost seriously in efficiency: it failed to produce a new urban form that served directly its complicated processes. As for the growing urban populations, they lacked the most elementary facilities for urban living, even sunlight and fresh air, to say nothing of the means to a more vivid social life. The new cities grew up without the benefit of coherent social knowledge or orderly social effort: they lacked the useful urban folkways of the Middle Ages or the confident esthetic command of the Baroque period: indeed, a seventeenth-century Dutch peasant, in his little village, knew more about the art of living in communities than a nineteenth-century municipal councilor in London or Berlin. Statesmen who did not hesitate to weld together a diversity of regional interests into national states, or who wove together an empire that girdled the planet, failed to produce even a rough draft of a decent neighborhood.

In every department, form disintegrated: except in its heritage from the past, the city vanished as an embodiment of collective art and technics. And where, as in North America, the loss was not alleviated by the continued presence of great monuments from the past and persistent habits of social living, the result was a raw, dissolute environment, and a narrow, constricted, and baffled social life. Even in Germany and the Low Countries, where the traditions of urban life had lingered on from the Middle Ages, the most colossal blunders were committed in the most ordinary tasks of urban planning and building. As the pace of urbanization increased, the circle of devastation widened.

Today we face not only the original social disruption. We likewise face the accumulated physical and social results of that disruption: ravaged landscapes, disorderly urban districts, pockets of disease, patches of blight, mile upon mile of standardized slums, worming into the outlying areas of big cities, and fusing with their ineffectual suburbs. In short: a general miscarriage and defeat of civilized effort. So far have our achievements fallen short of our needs that even a hundred years of persistent reform in England, the first country to suffer heavily from disurbanization, have only in the last decade begun to leave an imprint. True: here and there patches of good building and coherent social form exist: new nodes of integration

can be detected, and since 1920 these patches have been spreading. But the main results of more than a century of misbuilding and malformation, dissociation and disorganization still hold. Whether the observer focuses his gaze on the physical structure of communal living or upon the social processes that must be embodied and expressed, the report remains the same.

Today we begin to see that the improvement of cities is no matter for small one-sided reforms: the task of city design involves the vaster task of rebuilding our civilization. We must alter the parasitic and predatory modes of life that now play so large a part, and we must create region by region, continent by continent, an effective symbiosis, or co-operative living together. The problem is to co-ordinate, on the basis of more essential human values than the will-to-power and the will-to-profits, a host of social functions and processes that we have hitherto misused in the building of cities and polities, or of which we have never rationally taken advantage.

Unfortunately, the fashionable political philosophies of the past century are of but small help in defining this new task: they dealt with legal abstractions, like Individual and State, with cultural abstractions, like Humanity, the Nation, the Folk, or with bare economic abstractions like the Capitalist Class or the Proletariat – while life as it was lived in the concrete, in regions and cities and villages, in wheatland and cornland and vineland, in the mine, the quarry, the fishery, was conceived as but a shadow of the prevailing myths and arrogant fantasies of the ruling classes – or the often no less shadowy fantasies of those who challenged them.

Here and there one notes, of course, valiant exceptions both in theory and in action. Le Play and Reclus in France, W.H. Riehl in Germany, Kropotkin in Russia, Howard in England, Grundtvig in Denmark, Geddes in Scotland, began half a century ago to lay the ideological basis for a new order. The insights of these men may prove as important for the new biotechnic regime, based on the deliberate culture of life, as the formulations of Leonardo, Galileo, Newton, and Descartes were for the more limited mechanical order upon which the past triumphs of our machine civilization were founded. In the piecemeal improvement of cities, the work of sanitarians like Chadwick and Richardson, community designers like Olmsted, far-seeing architects like Parker and Wright, laid the concrete basis for a collective environment in which the needs of reproduction and nurture and psychological development and the social processes themselves would be adequately served.

Now the dominant urban environment of the past century has been mainly a narrow by-product of the machine ideology. And the greater part of it has already been made obsolete by the rapid advance of the biological arts and sciences, and by the steady penetration of sociological thought into every department. We have now reached a point where these fresh accumulations of historical insight and scientific knowledge are ready to flow over into social life, to mold anew the forms of cities, to assist in the transformation of both the instruments and the goals of our civilization. Profound changes, which will affect the distribution and increase of population, the efficiency of industry, and the quality of Western Culture, have already become visible. To form an accurate estimate of these new potentialities and to suggest their direction into channels of human welfare, is one of the major offices of the contemporary student of cities. Ultimately, such studies, forecasts, and imaginative projects must bear directly upon the life of every human being in our civilization.

What is the city? How has it functioned in the Western World since the tenth century, when the renewal of cities began, and in particular, what changes have come about in its physical and social composition during the last century? What factors have conditioned the size of cities, the extent of their growth, the type of order manifested in street plan and in building, their manner of nucleation, the composition of their economic and social classes, their physical manner of existence and their cultural style? By what political processes of federation or amalgamation, co-operative union or centralization, have cities existed; and what new units of administration does the present age suggest? Have we yet found an adequate urban form to harness all the complex technical and social forces in our civilization; and if a new order is discernible, what are its main outlines? What are the relations between city and region? And what steps are necessary in order to redefine and reconstruct the region itself, as a collective human habitation? What, in short, are the possibilities for creating form and order and design in our present civilization? . . .

Today our world faces a crisis: a crisis which, if its consequences are as grave as now seems, may not fully be resolved for another century. If the destructive forces in civilization gain ascendancy, our new urban culture will be stricken in every part. Our cities, blasted and deserted, will be cemeteries for the dead: cold lairs given over to less destructive beasts than man. But we may avert that fate: perhaps only in facing such a desperate challenge can the necessary creative forces be effectually welded together. Instead of clinging to the sardonic funeral towers of metropolitan finance, ours to march out to newly plowed fields, to create fresh patterns of political action, to alter for human purposes the perverse mechanisms of our economic regime, to conceive and to germinate fresh forms of human culture.

Instead of accepting the stale cult of death that the fascists have erected, as the proper crown for the servility and the brutality that are the pillars of their states, we must erect a cult of life: life in action, as the farmer or the mechanic knows it: life in expression, as the artist knows it: life as the lover feels it and the parent practices it: life as it is known to men of good will who meditate in the cloister, experiment in the laboratory, or plan intelligently in the factory or the government office.

Nothing is permanent: certainly not the frozen images of barbarous power with which fascism now confronts us. Those images may easily be smashed by an external shock, cracked as ignominiously as fallen Dagon, the massive idol of the heathen: or they may be melted, eventually, by the internal warmth of normal men and women. Nothing endures except life: the capacity for birth, growth, and daily renewal. As life becomes insurgent once more in our civilization, conquering the reckless thrust of barbarism, the culture of cities will be both instrument and goal.

NOTES

1 Unwin, Raymond. 1909. *Town Planning in Practice: An Introduction to the Art of Designing Cities and Suburbs*. London: Ernest Benn, p. 2.

2 Mumford, Lewis. 1961. *The City in History: Its Origins, Its Transformations, and Its Prospects*. New York: Harcourt Brace & World, p. 570.

3 Ibid., p. 567.

"The Land Ethic"
from *A Sand County Almanac* (1949)

Aldo Leopold

Editors' Introduction

More than 60 years after his death, Aldo Leopold (1887–1948) is seen as one of the seminal figures in the development of modern environmentalism. His career included service as a conservation ecologist with the U.S. Forest Service in New Mexico and as a professor at the University of Wisconsin. Starting within the conservationist tradition, which emphasized managing natural resources for sustained yield, Leopold expanded his perspective towards an acknowledgement of the intrinsic value of ecosystems and a view of the world as an organic, evolving unity. This philosophy foreshadowed the worldview of deep ecologists and other more radical environmentalists of the 1970s and after.

Leopold was an ethicist as well as an environmentalist. His later work included a profound questioning of anthropocentric and economic values, and was based on a belief in the necessity of moral evolution in order for societies to live within "their sustained carrying capacity." In his essay "The Land Ethic," he equates the historic spread of ethical notions of human rights with the more recent growth of an understanding that entire ecosystems – not just certain elements of these – have value. He explicitly refutes the possibility of bringing about environmentally sound practices through economics alone, and believes that the only means through which this can be done is a process of social and moral growth. His land ethic – "A thing is right when it tends to preserve the integrity, stability, and beauty of the biotic community. It is wrong when it tends otherwise" – is a moral principle on a par with the Golden Rule, a mandate for ecological living.

In addition to *A Sand County Almanac* (New York: Oxford University Press, 1949), in which "The Land Ethic" is contained, an excellent collection of Leopold's writings is provided by *The River of the Mother Of God and other Essays by Aldo Leopold* (Madison: University of Wisconsin Press, 1991), edited by Susan L. Flader and J. Baird Callicott. Biographies of Leopold include Marybeth Lorbiecke's *Aldo Leopold: a Fierce Green Fire* (Helena, MT: Falcon Press, 1996); a biographical film entitled *A Fierce Green Fire* was released in 2013. More information on the development of environmental philosophies in the twentieth century and the range of these philosophies is contained in Carolyn Merchant's book *Radical Ecology: The Search for a Livable World* (New York: Routledge, 1993), Robert Gottlieb's history of the movement *Forcing the Spring: The Transformation of the American Environmental Movement* (Washington, D.C.: Island Press, 1993), Bill McKibben's edited collection *American Earth: Environmental Writing Since Thoreau* (Des Moines: Library of America, 2008), and Philip Shabecoff's books *A Fierce Green Fire: The American Environmental Movement* (New York: Hill and Wang, 2003) and *Earth Rising: American Environmentalism in the 21st Century* (Washington, D.C.: Island Press, 2000).

When God-like Odysseus returned from the wars in Troy, he hanged all on one rope a dozen slavegirls of his household whom he suspected of misbehavior during his absence.

This hanging involved no question of propriety. The girls were property. The disposal of property was then, as now, a matter of expediency, not of right and wrong.

Concepts of right and wrong were not lacking from Odysseus' Greece: witness the fidelity of his wife through the lonely years before at last his blackprowed galleys clove the wine-dark seas for home. The ethical structure of that day covered wives, but had not yet been extended to human chattels. During the three thousand years which have since elapsed, ethical criteria have been extended to many fields of conduct, with corresponding shrinkages in those judged by expediency only.

THE ETHICAL SEQUENCE

This extension of ethics, so far studied only by philosophers, is actually a process in ecological evolution. Its sequences may be described in ecological as well as in philosophical terms. An ethic, ecologically, is a limitation on freedom of action in the struggle for existence. An ethic, philosophically, is a differentiation of social from antisocial conduct. These are two definitions of one thing. The thing has its origin in the tendency of inter-dependent individuals or groups to evolve modes of co-operation. The ecologist calls these symbioses. Politics and economics are advanced symbioses in which the original free-for-all competition has been replaced, in part, by co-operative mechanisms with an ethical content.

The complexity of co-operative mechanisms has increased with population density, and with the efficiency of tools. It was simpler, for example, to define the antisocial uses of sticks and stones in the days of the mastodons than of bullets and billboards in the age of motors.

The first ethics dealt with the relation between individuals; the Mosaic Decalogue is an example. Later accretions dealt with the relation between the individual and society. The Golden Rule tries to integrate the individual to society; democracy to integrate social organization to the individual.

There is as yet no ethic dealing with man's relation to land and to the animals and plants which grow upon it. Land, like Odysseus' slave-girls, is still property. The land-relation is still strictly economic, entailing privileges but not obligations.

The extension of ethics to this third element in human environment is, if I read the evidence correctly, an evolutionary possibility and an ecological necessity. It is the third step in a sequence. The first two have already been taken. Individual thinkers since the days of Ezekiel and Isaiah have asserted that the despoliation of land is not only inexpedient but wrong. Society, however, has not yet affirmed their belief. I regard the present conservation movement as the embryo of such an affirmation.

An ethic may be regarded as a mode of guidance for meeting ecological situations so new or intricate, or involving such deferred reactions, that the path of social expediency is not discernible to the average individual. Animal instincts are modes of guidance for the individual in meeting such situations. Ethics are possibly a kind of community instinct in-the-making.

THE COMMUNITY CONCEPT

All ethics so far evolved rest upon a single premise: that the individual is a member of a community of interdependent parts. His instincts prompt him to compete for his place in the community, but his ethics prompt him also to co-operate (perhaps in order that there may be a place to compete for).

The land ethic simply enlarges the boundaries of the community to include soils, waters, plants, and animals, or collectively: the land.

This sounds simple: do we not already sing our love for and obligation to the land of the free and the home of the brave? Yes, but just what and whom do we love? Certainly not the soil, which we are sending helter-skelter downriver. Certainly not the waters, which we assume have no function except to turn turbines, float barges, and carry off sewage. Certainly not the plants, of which we exterminate whole communities without batting an eye. Certainly not the animals, of which we have already extirpated many of the largest and most beautiful species. A land ethic of course cannot prevent the alteration, management, and use of these "resources," but it does affirm their right to continued existence, and, at least in spots, their continued existence in a natural state.

In short, a land ethic changes the role of *Homo sapiens* from conqueror of the land-community to plain member and citizen of it. It implies respect for his fellow-members, and also respect for the community as such.

In human history, we have learned (I hope) that the conqueror role is eventually self-defeating. Why? Because it is implicit in such a role that the conqueror knows, *ex cathedra*, just what makes the community clock tick, and just what and who is valuable, and what and who is worthless, in community life. It always turns out that he knows neither, and this is why his conquests eventually defeat themselves.

In the biotic community, a parallel situation exists. Abraham knew exactly what the land was for: it was to drip milk and honey into Abraham's mouth. At the present moment, the assurance with which we regard this assumption is inverse to the degree of our education.

The ordinary citizen today assumes that science knows what makes the community clock tick; the scientist is equally sure that he does not. He knows that the biotic mechanism is so complex that its workings may never be fully understood.

That man is, in fact, only a member of a biotic team is shown by an ecological interpretation of history. Many historical events, hitherto explained solely in terms of human enterprise, were actually biotic interactions between people and land. The characteristics of the land determined the facts quite as potently as the characteristics of the men who lived on it.

Consider, for example, the settlement of the Mississippi valley. In the years following the Revolution, three groups were contending for its control: the native Indian, the French and English traders, and the American settlers. Historians wonder what would have happened if the English at Detroit had thrown a little more weight into the Indian side of those tipsy scales which decided the outcome of the colonial migration into the cane-lands of Kentucky. It is time now to ponder the fact that the cane-lands, when subjected to the particular mixture of forces represented by the cow, plow, fire, and axe of the pioneer, became bluegrass. What if the plant succession inherent in this dark and bloody ground had, under the impact of these forces, given us some worthless sedge, shrub, or weed? Would Boone and Kenton have held out? Would there have been any overflow into Ohio, Indiana, Illinois, and Missouri? Any Louisiana Purchase? Any transcontinental union of new states? Any Civil War?

Kentucky was one sentence in the drama of history. We are commonly told what the human actors in this drama tried to do, but we are seldom told that their success, or the lack of it, hung in large degree on the reaction of particular soils to the impact of the particular forces exerted by their occupancy. In the case of Kentucky, we do not even know where the bluegrass came from – whether it is a native species, or a stowaway from Europe.

Contrast the cane-lands with what hindsight tells us about the Southwest, where the pioneers were equally brave, resourceful, and persevering. The impact of occupancy here brought no bluegrass, or other plant fitted to withstand the bumps and buffetings of hard use. This region, when grazed by livestock, reverted through a series of more and more worthless grasses, shrubs, and weeds to a condition of unstable equilibrium. Each recession of plant types bred erosion; each increment to erosion bred a further recession of plants. The result today is a progressive and mutual deterioration, not only of plants and soils, but of the animal community subsisting thereon. The early settlers did not expect this: on the ciénegas of New Mexico some even cut ditches to hasten it. So subtle has been its progress that few residents of the region are aware of it. It is quite invisible to the tourist who finds this wrecked landscape colorful and charming (as indeed it is, but it bears scant resemblance to what it was in 1848).

This same landscape was "developed" once before, but with quite different results. The Pueblo Indians settled the Southwest in pre-Columbian times, but they happened *not* to be equipped with range live-stock. Their civilization expired, but not because their land expired.

In India, regions devoid of any sod-forming grass have been settled, apparently without wrecking the land, by the simple expedient of carrying the grass to the cow, rather than vice versa. (Was this the result of some deep wisdom, or was it just good luck? I do not know.)

In short, the plant succession steered the course of history; the pioneer simply demonstrated, for good or ill, what successions inhered in the land. Is history taught in this spirit? It will be, once the concept of land as a community really penetrates our intellectual life.

THE ECOLOGICAL CONSCIENCE

Conservation is a state of harmony between men and land. Despite nearly a century of propaganda, conservation still proceeds at a snail's pace; progress still consists largely of letterhead pieties and convention oratory. On the back forty we still slip two steps backward for each forward stride.

The usual answer to this dilemma is "more conservation education." No one will debate this, but is it certain that only the *volume* of education needs stepping up? Is something lacking in the *content* as well?

It is difficult to give a fair summary of its content in brief form, but, as I understand it, the content is substantially this: obey the law, vote right, join some organizations, and practice what conservation is profitable on your own land; the government will do the rest.

Is not this formula too easy to accomplish anything worthwhile? It defines no right or wrong, assigns no obligation, calls for no sacrifice, implies no change in the current philosophy of values. In respect of land use, it urges only enlightened self-interest. Just how far will such education take us? An example will perhaps yield a partial answer.

By 1930 it had become clear to all except the ecologically blind that southwestern Wisconsin's topsoil was slipping seaward. In 1933 the farmers were told that if they would adopt certain remedial practices for five years, the public would donate CCC labor to install them, plus the necessary machinery and materials. The offer was widely accepted, but the practices were widely forgotten when the five-year contract period was up. The farmers continued only those practices that yielded an immediate and visible economic gain for themselves.

This led to the idea that maybe farmers would learn more quickly if they themselves wrote the rules. Accordingly the Wisconsin Legislature in 1937 passed the Soil Conservation District Law. This said to farmers, in effect: *We, the public, will furnish you free technical service and loan you specialized machinery, if you will write your own rules for land-use. Each county may write its own rules, and these will have the force of law.* Nearly all the counties promptly organized to accept the proffered help, but after a decade of operation, *no county has yet written a single rule.* There has been visible progress in such practices as strip-cropping, pasture renovation, and soil liming, but none in fencing woodlots against grazing, and none in excluding plow and cow from steep slopes. The farmers, in short, have selected those remedial practices which were profitable anyhow, and ignored those which were profitable to the community, but not clearly profitable to themselves.

When one asks why no rules have been written, one is told that the community is not yet ready to support them; education must precede rules. But the education actually in progress makes no mention of obligations to land over and above those dictated by self-interest. The net result is that we have more education but less soil, fewer healthy woods, and as many floods as in 1937.

The puzzling aspect of such situations is that the existence of obligations over and above self-interest is taken for granted in such rural community enterprises as the betterment of roads, schools, churches, and baseball teams. Their existence is not taken for granted, nor as yet seriously discussed, in bettering the behavior of the water that falls on the land, or in the preserving of the beauty or diversity of the farm landscape. Land-use ethics are still governed wholly by economic self-interest, just as social ethics were a century ago.

To sum up: we asked the farmer to do what he conveniently could to save his soil, and he has done just that, and only that. The farmer who clears the woods off a 75 percent slope, turns his cows into the clearing, and dumps its rainfall, rocks, and soil into the community creek, is still (if otherwise decent) a respected member of society. If he puts lime on his fields and plants his crops on contour, he is still entitled to all the privileges and emoluments of his Soil Conservation District. The District is a beautiful piece of social machinery, but it is coughing along on two cylinders because we have been too timid, and too anxious for quick success, to tell the farmer the true magnitude of his obligations. Obligations have no meaning without conscience, and the problem we face is the extension of the social conscience from people to land.

No important change in ethics was ever accomplished without an internal change in our intellectual emphasis, loyalties, affections, and convictions. The proof that conservation has not yet touched these foundations of conduct lies in the fact that philosophy and religion have not yet heard of it. In our attempt to make conservation easy, we have made it trivial.

SUBSTITUTES FOR A LAND ETHIC

When the logic of history hungers for bread and we hand out a stone, we are at pains to explain how much the stone resembles bread. I now describe some of the stones which serve in lieu of a land ethic.

One basic weakness in a conservation system based wholly on economic motives is that most members of the land community have no economic value. Wildflowers and songbirds are examples. Of the 22,000 higher plants and animals native to Wisconsin, it is doubtful whether more than 5 percent can be sold, fed, eaten, or otherwise put to economic use. Yet these creatures are members of the biotic community, and if (as I believe) its stability depends on its integrity, they are entitled to continuance.

When one of these non-economic categories is threatened, and if we happen to love it, we invent subterfuges to give it economic importance. At the beginning of the century songbirds were supposed to be disappearing. Ornithologists jumped to the rescue with some distinctly shaky evidence to the effect that insects would eat us up if birds failed to control them. The evidence had to be economic in order to be valid.

It is painful to read these circumlocutions today. We have no land ethic yet, but we have at least drawn nearer the point of admitting that birds should continue as a matter of biotic right, regardless of the presence or absence of economic advantage to us.

A parallel situation exists in respect of predatory mammals, raptorial birds, and fish-eating birds. Time was when biologists somewhat overworked the evidence that these creatures preserve the health of game by killing weaklings, or that they

control rodents for the farmer, or that they prey only on "worthless" species. Here again, the evidence had to be economic in order to be valid. It is only in recent years that we hear the more honest argument that predators are members of the community, and that no special interest has the right to exterminate them for the sake of a benefit, real or fancied, to itself. Unfortunately this enlightened view is still in the talk stage. In the field the extermination of predators goes merrily on: witness the impending erasure of the timber wolf by fiat of Congress, the Conservation Bureaus, and many state legislatures.

Some species of trees have been "read out of the party" by economics-minded foresters because they grow too slowly, or have too low a sale value to pay as timber crops: white cedar, tamarack, cypress, beech, and hemlock are examples. In Europe, where forestry is ecologically more advanced, the non-commercial tree species are recognized as members of the native forest community, to be preserved as such, within reason. Moreover some (like beech) have been found to have a valuable function in building up soil fertility. The interdependence of the forest and its constituent tree species, ground flora, and fauna is taken for granted.

Lack of economic value is sometimes a character not only of species or groups, but of entire biotic communities: marshes, bogs, dunes, and "deserts" are examples. Our formula in such cases is to relegate their conservation to government as refuges, monuments, or parks. The difficulty is that these communities are usually interspersed with more valuable private lands; the government cannot possibly own or control such scattered parcels. The net effect is that we have relegated some of them to ultimate extinction over large areas. If the private owner were ecologically minded, he would be proud to be the custodian of a reasonable proportion of such areas, which add diversity and beauty to his farm and to his community.

In some instances, the assumed lack of profit in these "waste" areas has proved to be wrong, but only after most of them had been done away with. The present scramble to reflood muskrat marshes is a case in point.

There is a clear tendency in American conservation to relegate to government all necessary jobs that private landowners fail to perform. Government ownership, operation, subsidy, regulation is now widely prevalent in forestry, range management, soil and watershed management, park and wilderness conservation, fisheries management, and migratory bird management, with more to come. Most of this growth in governmental conservation is proper and logical, some of it is inevitable. That I imply no disapproval of it is implicit in the fact that I have spent most of my life working for it. Nevertheless the question arises: What is the ultimate magnitude of the enterprise? Will the tax base carry its eventual ramifications? At what point will governmental conservation, like the mastodon, become handicapped by its own dimensions? The answer, if there is any, seems to be in a land ethic, or some other force which assigns more obligation to the private landowner.

Industrial landowners and users, especially lumbermen and stockmen, are inclined to wail long and loudly about the extension of government ownership and regulation to land, but (with notable exceptions) they show little disposition to develop the only visible alternative: the voluntary practice of conservation on their own lands.

When the private landowner is asked to perform some unprofitable act for the good of the community, he today assents only with outstretched palm. If the act costs him cash this is fair and proper, but when it costs only forethought, open-mindedness, or time, the issue is at least debatable. The overwhelming growth of land-use subsidies in recent years must be ascribed, in large part, to the government's own agencies for conservation education: the land bureaus, the agricultural colleges, and the extension services. As far as I can detect, no ethical obligation toward land is taught in these institutions.

To sum up: a system of conservation based solely on economic self-interest is hopelessly lopsided. It tends to ignore, and thus eventually to eliminate, many elements in the land community that lack commercial value, but that are (as far as we know) essential to its healthy functioning. It assumes, falsely, I think, that the economic parts of the biotic clock will function without the uneconomic parts. It tends to relegate to government many functions eventually too large, too complex, or too widely dispersed to be performed by government.

An ethical obligation on the part of the private owner is the only visible remedy for these situations.

THE LAND PYRAMID

An ethic to supplement and guide the economic relation to land presupposes the existence of some mental image of land as a biotic mechanism. We can be ethical only in relation to something we can see, feel, understand, love, or otherwise have faith in.

The image commonly employed in conservation education is "the balance of nature." For reasons too lengthy to detail here, this figure of speech fails to describe accurately what little we know about the land mechanism. A much truer image is the one employed in ecology: the biotic pyramid. I shall first sketch the pyramid as a symbol of land, and later develop some of its implications in terms of land-use.

Plants absorb energy from the sun. This energy flows through a circuit called the biota, which may be represented by a pyramid consisting of layers. The bottom layer is the soil. A plant layer rests on the soil, an insect layer on the plants, a bird and rodent layer on the insects, and so on up through various animal groups to the apex layer, which consists of the larger carnivores.

The species of a layer are alike not in where they came from, or in what they look like, but rather in what they eat. Each successive layer depends on those below it for food and often for other services, and each in turn furnishes food and services to those above. Proceeding upward, each successive layer decreases in numerical abundance. Thus, for every carnivore there are hundreds of his prey, thousands of their prey, millions of insects, uncountable plants. The pyramidal form of the system reflects this numerical progression from apex to base. Man shares an intermediate layer with the bears, raccoons, and squirrels which eat both meat and vegetables.

The lines of dependency for food and other services are called food chains. Thus soil–oak–deer–Indian is a chain that has now been largely converted to soil–corn–cow–farmer. Each species, including ourselves, is a link in many chains. The deer eats a hundred plants other than oak, and the cow a hundred plants other than corn. Both, then, are links in a hundred chains. The pyramid is a tangle of chains so complex as to seem disorderly, yet the stability of the system proves it to be a highly organized structure. Its functioning depends on the co-operation and competition of its diverse parts.

In the beginning, the pyramid of life was low and squat; the food chains short and simple. Evolution has added layer after layer, link after link. Man is one of thousands of accretions to the height and complexity of the pyramid. Science has given us many doubts, but it has given us at least one certainty: the trend of evolution is to elaborate and diversify the biota.

Land, then, is not merely soil; it is a fountain of energy flowing through a circuit of soils, plants, and animals. Food chains are the living channels which conduct energy upward; death and decay return it to the soil. The circuit is not closed; some energy is dissipated in decay, some is added by absorption from the air, some is stored in soils, peats, and long-lived forests; but it is a sustained circuit, like a slowly augmented revolving fund of life. There is always a net loss by downhill wash, but this is normally small and offset by the decay of rocks. It is deposited in the ocean and, in the course of geological time, raised to form new lands and new pyramids.

The velocity and character of the upward flow of energy depend on the complex structure of the plant and animal community, much as the upward flow of sap in a tree depends on its complex cellular organization. Without this complexity, normal circulation would presumably not occur. Structure means the characteristic numbers, as well as the characteristic kinds and functions, of the component species. This interdependence between the complex structure of the land and its smooth functioning as an energy unit is one of its basic attributes.

When a change occurs in one part of the circuit, many other parts must adjust themselves to it. Change does not necessarily obstruct or divert the flow of energy; evolution is a long series of self-induced changes, the net result of which has been to elaborate the flow mechanism and to lengthen the circuit. Evolutionary changes, however, are usually slow and local. Man's invention of tools has enabled him to make changes of unprecedented violence, rapidity, and scope.

One change is in the composition of floras and faunas. The larger predators are lopped off the apex of the pyramid; food chains, for the first time in history, become shorter rather than longer. Domesticated species from other lands are substituted for wild ones, and wild ones are moved to new habitats. In this world-wide pooling of faunas and floras, some species get out of bounds as pests and diseases, others are extinguished. Such effects are seldom intended or foreseen; they represent unpredicted and often untraceable readjustments in the structure. Agricultural science is largely a race between the emergence of new pests and the emergence of new techniques for their control.

Another change touches the flow of energy through plants and animals and its return to the soil. Fertility is the ability of soil to receive, store, and release energy. Agriculture, by overdrafts on the soil, or by too radical a substitution of domestic for native species in the superstructure, may derange the channels of flow or deplete storage. Soils depleted of their storage, or of the organic matter which anchors it, wash away faster than they form. This is erosion.

Waters, like soil, are part of the energy circuit. Industry, by polluting waters or obstructing them with dams, may exclude the plants and animals necessary to keep energy in circulation.

Transportation brings about another basic change: the plants or animals grown in one region are now consumed and returned to the soil in another. Transportation taps the energy stored in rocks, and in the air, and uses it elsewhere; thus we fertilize the garden with nitrogen gleaned by the guano birds from the fishes of seas on the other side of the Equator. Thus the formerly localized and self-contained circuits are pooled on a world-wide scale.

The process of altering the pyramid for human occupation releases stored energy, and this often gives rise, during the pioneering period, to a deceptive exuberance of plant and animal life, both wild and tame. These releases of biotic capital tend to becloud or postpone the penalties of violence.

[...]

This thumbnail sketch of land as an energy circuit conveys three basic ideas:

1 That land is not merely soil.
2 That the native plants and animals kept the energy circuit open; others may or may not.

3 That man-made changes are of a different order than evolutionary changes, and have effects more comprehensive than is intended or foreseen.

These ideas, collectively, raise two basic issues: Can the land adjust itself to the new order? Can the desired alterations be accomplished with less violence?

Biotas seem to differ in their capacity to sustain violent conversion. Western Europe, for example, carries a far different pyramid than Caesar found there. Some large animals are lost; swampy forests have become meadows or plowland; many new plants and animals are introduced, some of which escape as pests; the remaining natives are greatly changed in distribution and abundance. Yet the soil is still there and, with the help of imported nutrients, still fertile; the waters flow normally; the new structure seems to function and to persist. There is no visible stoppage or derangement of the circuit.

Western Europe, then, has a resistant biota. Its inner processes are tough, elastic, resistant to strain. No matter how violent the alterations, the pyramid, so far, has developed some new *modus vivendi* which preserves its habitability for man, and for most of the other natives.

Japan seems to present another instance of radical conversion without disorganization.

Most other civilized regions, and some as yet barely touched by civilization, display various stages of disorganization, varying from initial symptoms to advanced wastage. In Asia Minor and North Africa diagnosis is confused by climatic changes, which may have been either the cause or the effect of advanced wastage. In the United States the degree of disorganization varies locally; it is worst in the Southwest, the Ozarks, and parts of the South, and least in New England and the Northwest. Better land uses may still arrest it in the less advanced regions. In parts of Mexico, South America, South Africa, and Australia a violent and accelerating wastage is in progress, but I cannot assess the prospects.

This almost worldwide display of disorganization in the land was to be similar to disease in an animal, except that it never culminates in complete disorganization or death. The land recovers, but at some reduced level of complexity, and with a reduced carrying capacity for people, plants, and animals. Many biotas currently regarded as "lands of opportunity" are in fact already subsisting on exploitative agriculture, i.e. they have already exceeded their sustained carrying capacity. Most of South America is overpopulated in this sense.

In arid regions we attempt to offset the process of wastage by reclamation, but it is only too evident that the prospective longevity of reclamation projects is often short. In our own West, the best of them may not last a century.

The combined evidence of history and ecology seems to support one general deduction: the less violent the man-made changes, the greater the probability of successful readjustment in the pyramid. Violence, in turn, varies with human population density; a dense population requires a more violent conversion. In this respect, North America has a better chance for permanence than Europe, if she can contrive to limit her density.

This deduction runs counter to our current philosophy, which assumes that because a small increase in density enriched human life, that an indefinite increase will enrich it indefinitely. Ecology knows of no density relationship that holds for indefinitely wide limits. All gains from density are subject to a law of diminishing returns.

Whatever may be the equation for men and land, it is improbable that we as yet know all its terms. Recent discoveries in mineral and vitamin nutrition reveal unsuspected dependencies in the up-circuit: incredibly minute quantities of certain substances determine the value of soils to plants, of plants to animals. What of the down-circuit? What of the vanishing species, the preservation of which we now regard as an esthetic luxury? They helped build the soil; in what unsuspected ways may they be essential to its maintenance? Professor Weaver proposes that we use prairie flowers to reflocculate the wasting soils of the dust bowl; who knows for what purpose cranes and condors, otters and grizzlies may some day be used? . . .

THE OUTLOOK

It is inconceivable to me that an ethical relation to land can exist without love, respect, and admiration for land, and a high regard for its value. By value, I of course mean something far broader than mere economic value; I mean value in the philosophical sense.

Perhaps the most serious obstacle impeding the evolution of a land ethic is the fact that our educational and economic system is headed away from, rather than toward, an intense consciousness of land. Your true modern is separated from the land by many middlemen, and by innumerable physical gadgets. He has no vital relation to it; to him it is the space between cities on which crops grow. Turn him loose for a day on the land, and if the spot does not happen to be a golf links or a "scenic" area, he is bored stiff. If crops could be raised by hydroponics instead of farming, it would suit him very well. Synthetic substitutes for wood, leather, wool, and other natural land products suit him better than the originals. In short, land is something he has "outgrown."

Almost equally serious as an obstacle to a land ethic is the attitude of the farmer for whom the land is still an adversary, or a taskmaster that keeps him in slavery. Theoretically, the mechanization of farming ought to cut the farmer's chains, but whether it really does is debatable.

One of the requisites for an ecological comprehension of land is an understanding of ecology, and this is by no means co-extensive with "education"; in fact, much higher education seems deliberately to avoid ecological concepts. An understanding of ecology does not necessarily originate in courses bearing ecological labels; it is quite as likely to be labeled geography, botany, agronomy, history, or economics. This is as it should be, but whatever the label, ecological training is scarce.

The case for a land ethic would appear hopeless but for the minority which is in obvious revolt against these "modern" trends.

The "key-log" which must be moved to release the evolutionary process for an ethic is simply this: quit thinking about decent land-use as solely an economic problem. Examine each question in terms of what is ethically and esthetically right, as well as what is economically expedient. A thing is right when it tends to preserve the integrity, stability, and beauty of the biotic community. It is wrong when it tends otherwise.

It of course goes without saying that economic feasibility limits the tether of what can or cannot be done for land. It always has and it always will. The fallacy the economic determinists have tied around our collective neck, and which we now need to cast off, is the belief that economics determines *all* land use. This is simply not true. An innumerable host of actions and attitudes, comprising perhaps the bulk of all land relations, is determined by the land-users' tastes and predilections, rather than by his purse. The bulk of all land relations hinges on investments of time, forethought, skill, and faith rather than on investments of cash. As a land-user thinketh, so is he.

I have purposely presented the land ethic as a product of social evolution because nothing so important as an ethic is ever "written." Only the most superficial student of history supposes that Moses "wrote" the Decalogue; it evolved in the minds of a thinking community, and Moses wrote a tentative summary of it for a "seminar." I say tentative because evolution never stops.

The evolution of a land ethic is an intellectual as well as emotional process. Conservation is paved with good intentions which prove to be futile, or even dangerous, because they are devoid of critical understanding either of the land, or of economic land use. I think it is a truism that as the ethical frontier advances from the individual to the community, its intellectual content increases.

The mechanism of operation is the same for any ethic: social approbation for right actions: social disapproval for wrong actions.

By and large, our present problem is one of attitudes and implements. We are remodeling the Alhambra with a steam-shovel, and we are proud of our yardage. We shall hardly relinquish the shovel has many good points, but we are in need of gentler and more objective criteria for its successful use.

WILDERNESS

Wilderness is the raw material out of which man has hammered the artifact called civilization.

Wilderness was never a homogeneous raw material. It was very diverse, and the resulting artifacts are very diverse. These differences in the end-product are known as cultures. The rich diversity of the world's cultures reflects a corresponding diversity in the wilds that gave them birth.

For the first time in the history of the human species, two changes are now impending. One is the exhaustion of wilderness in the more habitable portions of the globe. The other is the worldwide hybridization of cultures through modern transport and industrialization. Neither can be prevented, and perhaps should not be, but the question arises whether, by some slight amelioration of the impending changes, certain values can be preserved that would otherwise be lost.

To the laborer in the sweat of his labor, the raw stuff on his anvil is an adversary to be conquered. So was wilderness an adversary to the pioneer.

But to the laborer in repose, able for the moment to cast a philosophical eye on his world, that same raw stuff is something to be loved and cherished, because it gives definition and meaning to his life. This is a plea for the preservation of some tag-ends of wilderness, as museum pieces, for the edification of those who may one day wish to see, feel, or study the origins of their cultural inheritance.

ONE

4

"Orthodox Planning and The North End"

from *The Death and Life of Great American Cities* (1961)

Jane Jacobs

Editors' Introduction

Despite many good intentions, urban planning in the twentieth century often proceeded in directions that were profoundly unsustainable. Planners promoted freeways and other automobile infrastructure without considering their sprawl-inducing impacts, authorized the bulldozing of vibrant older urban neighborhoods for redevelopment into bland, modernist apartment blocks, allowed the destruction of natural landscape features such as streams and wetlands, and aided in the segregation of racial or socioeconomic groups through zoning and red-lining. Such actions were often camouflaged behind an image of the planner as detached, objective expert, with the authority of scientific method justifying decisions that were essentially subjective or political in nature.

Writers such as Mumford and William H. Whyte complained vigorously against modernist urban planning during the 1940s and 1950s. Paul and Percival Goodman's 1947 book *Communitas: Ways of Livelihood and Means of Life* (New York: Columbia University Press, 1947) was an especially profound critique of modernist city building, as well as a visionary exploration of alternatives, although this work did not receive wide circulation until reissued in the 1960s with a new introduction by Mumford. However, Jane Jacobs' 1961 book *The Death and Life of Great American Cities* was the bombshell that shocked many people around the world into questioning prevailing modes of urban planning. An editor at an architectural magazine, Jacobs had been living in New York City's Greenwich Village when plans were announced for an expressway through the community. At the time, the city's planning czar, Robert Moses, had bulldozed a network of expressways and bridges through dozens of other neighborhoods. However, Jacobs and her neighbors organized opposition and eventually defeated the city's plans. In the process she grew to appreciate even more the rich community life of this older urban neighborhood, and observed as well the effect of modernist redevelopment on other older neighborhoods such as Boston's West End, a tight-knit working-class community destroyed in order to create a cluster of bland, modernist apartment buildings for more affluent residents.

In *The Death and Life* Jacobs described in detail what makes dense urban neighborhoods work, and how modern city-building practices undermine many of the qualities that encourage pedestrian use of the street, neighborhood contacts, and a thriving local economy of small businesses. Her writing helped lay the groundwork for the field of environmental design research, in which later investigators carefully studied how people actually used buildings, streets, and neighborhoods, rather than simply following some abstract set of architectural criteria. Jacobs' specific emphasis on pedestrian-oriented urban form also served as an inspiration to many later urban activists, including the New Urbanists.

In her later writings Jacobs focused on questions related to local and regional economies. In her book *The Economy of Cities* (New York: Random House, 1969) she portrayed cities as the economic engines that have made possible broader development of societies. In *Cities and the Wealth of Nations: Principles of*

Economic Life (New York: Random House, 1984), she expanded this argument, arguing particularly that to grow and be successful cities and urban regions need to gradually replace the goods that they had previously imported by producing them internally. In her emphasis on the importance of networks of small firms, she foreshadowed much later work by academic economists, and comes close to the import-substitution method recommended by Shuman (see Part Two).

In this selection Jacobs describes her reaction to proposed redevelopment of Boston's North End, a dense, tight-knit Italian-American community that she held up repeatedly as an example of a vibrant urban neighborhood. She contrasts in particular the attitude of planners with the reality of life in the neighborhood as she observed it. Related critiques of mainstream urban planning have since been developed by many other writers. Beginning in the 1970s neo-Marxist authors such as David Harvey, Manuel Castells, and Robert Beauregard argued that urban planners have helped carry out the agendas of powerful class or economic interests, rather than working as a more independent force for social change. In her book *Dreaming the Rational City* (Cambridge MA: MIT Press, 1983), Christine Boyer provided a detailed historical analysis of how urban planning has been used to enforce the dominant discourses and power of capitalist elites. And in books such as *City of Quartz* (1992*)*, *Planet of Slums* (2006), and *Evil Paradises* (2007), Mike Davis has examined the often bizarre and inequitable urban landscapes created by capitalism and local governments working hand-in-hand.

The effect of all these analyses was to throw prevailing modes of urban planning into question during the second half of the twentieth century, to challenge the view of the planner as a detached, scientific expert, and to fuel calls for greater public participation and contextual, culturally informed understandings of urban problems. But Jacobs was the clearest, most down-to-earth, and most biting of these critics, and her critique laid the groundwork for much reevaluation of the field.

Jacobs' analysis raises many questions today about urban sustainability planning. If "an intricate and close-grained diversity of uses" is essential to a vibrant neighborhood, how do we ensure that? How do we preserve and enhance older neighborhood environments rather than rebuilding them? And most importantly, how do we avoid the trap of following "orthodox" planning ideas and pseudo-scientific expertise into sterile and counterproductive solutions, and instead base our action on an understanding of people's reality on the ground?

There is nothing economically or socially inevitable about either the decay of old cities or the fresh-minted decadence of the new unurban urbanization. On the contrary, no other aspect of our economy and society has been more purposefully manipulated for a full quarter of a century to achieve precisely what we are getting. Extraordinary governmental financial incentives have been required to achieve this degree of monotony, sterility and vulgarity. Decades of preaching, writing and exhorting by experts have gone into convincing us and our legislators that mush like this must be good for us, as long as it comes bedded with grass.

Automobiles are often conveniently tagged as the villains responsible for the ills of cities and the disappointments and futilities of city planning. But the destructive effects of automobiles are much less a cause than a symptom of our incompetence at city building. Of course planners, including the highwaymen with fabulous sums of money and enormous powers at their disposal, are at a loss to make automobiles and cities compatible with one another. They do not know what to do with automobiles in cities because they do not know how to plan for workable and vital cities anyhow – with or without automobiles.

The simple needs of automobiles are more easily understood and satisfied than the complex needs of cities, and a growing number of planners and designers have come to believe that if they can only solve the problems of traffic, they will thereby have solved the major problem of cities. Cities have much more intricate economic and social concerns than automobile traffic. How can you know what to try with traffic until you know how the city itself works, and what else it needs to do with its streets? You can't.

It may be that we have become so feckless as a people that we no longer care how things do work, but only what kind of quick, easy outer impression they give. If so, there is little hope for our cities or probably for much else in our society. But I do not think this is so.

Specifically, in the case of planning for cities, it is clear that a large number of good and earnest people do care deeply about building and renewing. Despite some corruption, and considerable greed for the other man's vineyard, the intentions going into the messes we make are, on the whole, exemplary. Planners, architects of city design, and those they have led along with them in their beliefs are not consciously disdainful of the importance of knowing how things work. On the contrary, they have gone to great pains to learn what the saints and sages of modern orthodox planning have said about how cities ought to work and what ought to be good for people and businesses in them. They take this with such devotion that when contradictory reality intrudes, threatening to shatter their dearly won learning, they must shrug reality aside.

Consider, for example, the orthodox planning reaction to a district called the North End in Boston. This is an old, low-rent area merging into the heavy industry of the waterfront, and it is officially considered Boston's worst slum and civic shame. It embodies attributes which all enlightened people know are evil because so many wise men have said they are evil. Not only is the North End bumped right up against industry, but worse still it has all kinds of working places and commerce mingled in the greatest complexity with its residences. It has the highest concentration of dwelling units, on the land that is used for dwelling units, of any part of Boston, and indeed one of the highest concentrations to be found in any American city. It has little parkland. Children play in the streets. Instead of super-blocks, or even decently large blocks, it has very small blocks; in planning parlance it is "badly cut up with wasteful streets." Its buildings are old. Everything conceivable is presumably wrong with the North End. In orthodox planning terms it is a three-dimensional textbook of "megalopolis" in the last stages of depravity. The North End is thus a recurring assignment for MIT and Harvard planning and architectural students, who now and again pursue, under the guidance of their teachers, the paper exercise of converting it into super-blocks and park promenades, wiping away its nonconforming uses, transforming it to an ideal of order and gentility so simple it could be engraved on the head of a pin.

Twenty years ago, when I first happened to see the North End, its buildings – town houses of different kinds and sizes converted to flats, and four- or five-story tenements built to house the flood of immigrants first from Ireland, then from Eastern Europe and finally from Sicily – were badly overcrowded, and the general effect was of a district taking a terrible physical beating and certainly desperately poor.

When I saw the North End again in 1959, I was amazed at the change. Dozens and dozens of buildings had been rehabilitated. Instead of mattresses against the windows there were Venetian blinds and glimpses of fresh paint. Many of the small, converted houses now had only one or two families in them instead of the old crowded three or four. Some of the families in the tenements (as I learned later, visiting inside) had uncrowded themevles by throwing two older apartments together, and had equipped these with bathrooms, new kitchens and the like. I looked down a narrow alley, thinking to find at least here the old, squalid North End, but no: more neatly repainted brickwork, new blinds, and a burst of music as a door opened. Indeed, this was the only city district I had ever seen – or have seen to this day – in which the sides of buildings around parking lots had not been left raw and amputated, but repaired and painted as neatly as if they were intended to be seen. Mingled all among the buildings for living were an incredible number of splendid food stores, as well as such enterprises as upholstery making, metal working, carpentry, food processing. The streets were alive with children playing, people shopping, people strolling, people talking. Had it not been a cold January day, there would surely have been people sitting.

The general street atmosphere of buoyancy, friendliness and good health was so infectious that I began asking directions of people just for the fun of getting in on some talk. I had seen a lot of Boston in the past couple of days, most of it sorely distressing, and this struck me, with relief, as the healthiest place in the city. But I could not imagine where the money had come from for the rehabilitation, because it is almost impossible today to get any appreciable mortgage money in districts of American cities that are not either high-rent, or else imitations of suburbs. To find out, I went into a bar and restaurant (where an animated conversation about fishing was in progress) and called a Boston planner I know.

"Why in the world are you down in the North End?" he said. "Money? Why, no money or work has gone into the North End. Nothing's going on down there. Eventually, yes, but not yet. That's a slum!"

"It doesn't seem like a slum to me," I said.

"Why, that's the worst slum in the city. It has two hundred and seventy-five dwelling units to the net acre! I hate to admit we have anything like that in Boston, but it's a fact."

"Do you have any other figures on it?" I asked.

"Yes, funny thing. It has among the lowest delinquency, disease and infant mortality rates in the city. It also has the lowest ratio of rent to income in the city. Boy, are those people getting bargains. Let's see ... the child population is just about average for the city, on the nose. The death rate is low, 8.8 per thousand, against the average city rate of 11.2. The TB death rate is very low, less than 1 per ten thousand, can't understand it, it's lower even than Brookline's. In the old days the North End used to be the city's worst spot for tuberculosis, but all that has changed. Well, they must be strong people. Of course it's a terrible slum."

"You should have more slums like this," I said. "Don't tell me there are plans to wipe this out. You ought to be down here learning as much as you can from it."

"I know how you feel," he said. "I often go down there myself just to walk around the streets and feel that wonderful, cheerful street life. Say, what you ought to do, you ought to come back and go down in the summer if you think it's fun now. You'd be crazy about it in summer. But of course we have to rebuild it eventually. We've got to get those people off the streets."

Here was a curious thing. My friend's instincts told him the North End was a good place, and his social statistics confirmed it. But everything he had learned as a physical planner about what is

good for people and good for city neighborhoods, everything that made him an expert told him the North End had to be a bad place.

The leading Boston savings banker, "a man 'way up there in the power structure,'" to whom my friend referred me for my inquiry about the money, confirmed what I learned, in the meantime, from people in the North End. The money had not come through the grace of the great American banking system, which now knows enough about planning to know a slum as well as the planners do. "No sense in lending money into the North End," the banker said. "It's a slum! It's still getting some immigrants! Furthermore, back in the Depression it had a very large number of foreclosures; bad record." (I had heard about this too, in the mean time, and how families had worked and pooled their resources to buy back some of those foreclosed buildings.)

The largest mortgage loans that had been fed into this district of some 15,000 people in the quarter-century since the Great Depression were for $3,000, the banker told me, "and very, very few of those." There had been some others for $1,000 and for $2,000. The rehabilitation work had been almost entirely financed by business and housing earnings within the district, plowed back in, and by skilled work bartered among residents and relatives of residents.

By this time I knew that this inability to borrow for improvement was a galling worry to North Enders, and that furthermore some North Enders were worried because it seemed impossible to get new building in the area except at the price of seeing themselves and their community wiped out in the fashion of the students' dreams of a city Eden, a fate which they knew was not academic because it had already smashed completely a socially similar – although physically more spacious – nearby district called the West End. They were worried because they were aware also that patch and fix with nothing else could not do forever. "Any chance of loans for new construction in the North End?" I asked the banker.

"Absolutely not!" he said, sounding impatient at my denseness. "That's a slum!"

Bankers, like planners, have theories about cities on which they act. They have gotten their theories from the same intellectual sources as the planners. Bankers and government administrative officials who guarantee mortgages do not invent planning theories nor, surprisingly, even economic doctrine about cities. They are enlightened nowadays, and they pick up their ideas from idealists, a generation late. Since theoretical city planning has embraced no major new ideas for considerably more than a generation, theoretical planners, financers and bureaucrats are all just about even today.

And to put it bluntly, they are all in the same stage of elaborately learned superstition as medical science was early in the last century, when physicians put their faith in bloodletting, to draw out the evil humors which were believed to cause disease. With bloodletting, it took years of learning to know precisely which veins, by what rituals, were to be opened for what symptoms. A superstructure of technical complication was erected in such dead-pan detail that the literature still sounds almost plausible. However, because people, even when they are thoroughly enmeshed in descriptions of reality which are at variance with reality, are still seldom devoid of the powers of observation and independent thought, the science of bloodletting, over most of its long sway, appears usually to have been tempered with a certain amount of common sense. Or it was tempered until it reached its highest peaks of technique in, of all places, the young United States. Bloodletting went wild here. It had an enormously influential proponent in Dr Benjamin Rush, still revered as the greatest statesman–physician of our revolutionary and federal periods, and a genius of medical administration. Dr Rush Got Things Done. Among the things he got done, some of them good and useful, were to develop, practice, teach and spread the custom of bloodletting in cases where prudence or mercy had heretofore restrained its use. He and his students drained the blood of very young children, of consumptives, of the greatly aged, of almost anyone unfortunate to be sick in his realms of influence. His extreme practices aroused the alarm and horror of European bloodletting physicians. And yet as late as 1851, a committee appointed by the State Legislature of New York solemnly defended the thoroughgoing use of bloodletting. It scathingly ridiculed and censured a physician, William Turner, who had the temerity to write a pamphlet criticizing Dr Rush's doctrines and calling "the practice of taking blood in diseases contrary to common sense, to general experience, to enlightened reason and to the manifest laws of the divine Providence." Sick people needed fortifying, not draining, said Dr Turner, and he was squelched.

Medical analogies, applied to social organisms, are apt to be farfetched, and there is no point in mistaking mammalian chemistry for what occurs in a city. But analogies as to what goes on in the brains of earnest and learned men, dealing with complex phenomena they do not understand at all and trying to make do with a pseudoscience, do have a point. As in the pseudoscience of bloodletting, just so in the pseudoscience of city rebuilding and planning, years of learning and a plethora of subtle and complicated dogma have arisen on a foundation of nonsense. The tools of technique have steadily been perfected. Naturally, in time, forceful and able men, admired administrators, having swallowed the initial fallacies and having been provisioned with tools and with public confidence, go on logically to the greatest destructive excesses, which prudence or mercy might previously have forbade. Bloodletting could heal only by accident or insofar as it broke the rules, until the time when it was abandoned in favor of the hard, complex business of assembling, using and testing, bit by bit, true descriptions of reality drawn not from how it ought to be, but from how it is. The pseudoscience of city planning and its companion, the art of city design, have not yet broken with the specious comfort of wishes, familiar superstitions, oversimplifications, and symbols and have not yet embarked upon the adventure of probing the real world. . . .

One principle emerges so ubiquitously, and in so many and such complex different forms, that [it] . . . becomes the heart of my argument. This ubiquitous principle is the need of cities for a most intricate and close-grained diversity of uses that give each other constant mutual support, both economically and socially. The components of this diversity can differ enormously, but they must supplement each other in certain concrete ways.

I think that unsuccessful city areas are areas which lack this kind of intricate mutual support, and that the science of city planning and the art of city design, in real life for real cities, must become the science and art of catalyzing and nourishing these close-grained working relationships.

5

"Plight and Prospect"

from *Design With Nature* (1969)

Ian L. McHarg

Editors' Introduction

Born in 1921 near Glasgow in Scotland, Ian McHarg grew up observing the spread of that industrial city, with all its poverty and grime, and found solace in long countryside rambles. The experience convinced him that nature was essential to humanize the city. After service in World War II and training in landscape architecture at Harvard he returned home to find many of his pastoral childhood haunts obliterated by urban development. Saddened, he embarked on a teaching and consulting career in which he continually sought ways to reintegrate nature with the city. He founded and for 30 years directed the University of Pennsylvania's Department of Landscape Architecture and Regional Planning, and also founded a well-known consulting firm, Wallace, McHarg, Roberts & Todd, which was involved in creating regional plans and large-scale design of new communities.

McHarg's 1969 book, *Design With Nature* (New York: Natural History Press, 1969), played a crucial role in bringing together environmental and urban planning concerns in the middle of the twentieth century. Part personal statement, part clarion call against environmental destruction, and part description of a method of environmental analysis using overlay maps, the book sold more than 250,000 copies. Along with other works such as Rachel Carson's *Silent Spring* (New York: Fawcett Crest, 1962), calling attention to the dangers of pesticides and other toxic chemicals, and Barry Commoner's *The Closing Circle* (New York: Knopf, 1971), likewise warning against the pollution and resource-consumption impacts of technological society, it helped catalyze the modern environmental movement. But it did so in a way that was more directly related to city planning, landscape architecture, and urban design, exhorting these professions to integrate elements of the natural world into their work. As such, McHarg can be seen as the inheritor of the ecological concerns of Mumford and the Regional Planning Association of America, trying on a more specific level to figure out how urban growth might coexist with the natural landscape. To do so he used the available technology of the time in the form of hand-drawn mapping overlays of information derived from the natural sciences to determine where development should go. This method was a precursor to today's Geographic Information Systems (GIS) computer-based methods.

In retrospect McHarg can be seen as taking an overly optimistic view of the power of ecological science to order urban development, a view better fitted with a technocratic approach toward sustainable development than Leopold's call for inner change. McHarg can also be criticized for ignoring the role of powerful social, cultural, and economic forces in promoting unsustainable development, and for a detached, Olympian rhetoric far different than Jacobs' emphasis on first-hand experience of life in a city's streets. But whatever his shortcomings, in *Design With Nature* – a book filled with lovely black-and-white photographs of life in all its forms – McHarg succeeded in issuing an eloquent alarm call about the unsustainability of twentieth-century development patterns, especially the rising tide of suburban growth that was the dominant and most unsustainable aspect of the century's urbanization. That alarm call and McHarg's invitation to coordinate design and nature resonated deeply with a generation.

Clearly the problem of man and nature is not one of providing a decorative background for the human play, or even ameliorating the grim city: it is the necessity of sustaining nature as source of life, milieu, teacher, sanctum, challenge and, most of all, of rediscovering nature's corollary of the unknown in the self, the source of meaning.

There are still great realms of empty ocean, deserts reaching to the curvature of the earth, silent, ancient forests and rocky coasts, glaciers and volcanoes, but what will we do with them? There are rich contented farms, and idyllic villages, strong barns and white-steepled churches, tree-lined streets and covered bridges, but these are residues of another time. There are, too, the silhouettes of all the Manhattans, great and small, the gleaming golden windows of corporate images – expressionless prisms suddenly menaced by another of our creations, the supersonic transport whose sonic boom may reduce this image to a sea of shattered glass.[1]

But what do we say now, with our acts in city and countryside? While I first addressed this question to Scotland in my youth, today the world directs the same question to the United States. What is our performance and example? What are the visible testaments to the American mercantile creed – the hamburger stand, gas station, diner, the ubiquitous billboards, sagging wires, the parking lot, car cemetery and that most complete conjunction of land rapacity and human disillusion, the subdivision. It is all but impossible to avoid the highway out of town, for here, arrayed in all its glory, is the quintessence of vulgarity, bedecked to give the maximum visibility to the least of our accomplishments.

And what of the cities? Think of the imprisoning gray areas that encircle the center. From here the sad suburb is an unrealizable dream. Call them no-place, although they have many names. Race and hate, disease, poverty, rancor and despair, urine and spit live here in the shadows. United in poverty and ugliness, their symbol is the abandoned carcasses of automobiles, broken glass, alleys of rubbish and garbage. Crime consorts with disease, group fights group, the only emancipation is the parked car.

What of the heart of the city, where the gleaming towers rise from the dirty skirts of poverty? Is it like midtown Manhattan where 20 percent of the population was found to be indistinguishable from the patients in mental hospitals? Both stimulus and stress live here with the bitch goddess Success. As you look at the faceless prisms do you recognize the home of *anomie*?

Can you find the river that first made the city? Look behind the unkempt industry, cross the grassy railroad tracks and you will find the rotting piers and there is the great river, scummy and brown, wastes and sewage bobbing easily up and down with the tide, endlessly renewed.

If you fly to the city by day you will see it first as a smudge of smoke on the horizon. As you approach, the outlines of its towers will be revealed as soft silhouettes in the hazardous haze. Nearer you will perceive conspicuous plumes which, you learn, belong to the proudest names in industry. Our products are household words but it is clear that our industries are not yet housebroken. Drive from the airport through the banks of gas storage tanks and the interminable refineries. Consider how dangerous they are, see their cynical spume, observe their ugliness. Refine they may, but refined they are not.

You will drive on an expressway, a clumsy concrete form, untouched by either humanity or art, testament to the sad illusion that there can be a solution for the unbridled automobile. It is ironic that this greatest public investment in cities has also financed their conquest. See the scars of the battle in the remorseless carving, the dismembered neighborhoods, the despoiled parks. Manufacturers are producing automobiles faster than babies are being born. Think of the depredations yet to be accomplished by myopic highway builders to accommodate these toxic vehicles. You have plenty of time to consider in the long peak hour pauses of spasmodic driving in the blue gas corridors.

You leave the city and turn towards the countryside. But can you find it? To do so you will follow the paths of those who tried before you. Many stayed to build. But those who did so first are now deeply embedded in the fabric of the city. So as you go you transect the rings of the thwarted and disillusioned who are encapsulated in the city as nature endlessly eludes pursuit.

You can tell when you have reached the edge of the countryside for there are many emblems – the cadavers of old trees piled in untidy heaps at the edge of the razed deserts, the magnificent machines for land despoliation, for felling forests, filling marshes, culverting streams, and sterilizing farmland, making thick brown sediments of the creeks.

Is this the countryside, the green belt – or rather the greed belt, where the farmer sells land rather than crops, where the developer takes the public resource of the city's hinterland and subdivides it to create a private profit and a public cost? Certainly here is the area where public powers are weakest – either absent or elastic – where the future costs of streets, sidewalks and sewers, schools, police and fire protection are unspoken. Here are the meek mulcted, the refugees thwarted.

[. . .]

Within the metropolitan region natural features will vary, but it is possible to select certain of these that exist throughout and determine the degree to which they allow or discourage contemplated land uses. While these terms are relative, optimally development should occur on valuable or perilous natural-process land only when superior values are created or compensation can be awarded.

A complete study would involve identifying natural processes that performed work for man, those which offered protection or were hostile, those which were unique or especially precious and those values which were vulnerable. In the first category fall natural water purification, atmospheric pollution dispersal, climatic amelioration, water storage, flood, drought, and erosion control, topsoil accumulation, forest and wildlife inventory increase. Areas that provided protection or were dangerous would include the estuarine marshes and the floodplains, among others. The important areas of geological,

ecological, and historic interest would represent the next category, while beach dunes, spawning and breeding grounds and water catchment areas would be included in the vulnerable areas.

No such elaborate examination has been attempted in this study [on open space preservation in the Philadelphia region]. However, eight natural processes have been identified and these have been mapped and measured. Each one has been described with an eye to permissiveness and prohibition to certain land uses. It is from this analysis that the place of nature in the metropolis will be derived.

Surface water. In principle, only land uses that are inseparable from waterfront locations should occupy them; and even these should be limited to those which do not diminish the present or pro-spective value of surface water for supply, recreation or amenity. Demands for industrial waterfront locations in the region are extravagantly predicted as fifty linear miles. Thus, even satisfying this demand, five thousand miles could remain in a natural condition.

Land uses consonant with this principle would include port and harbor facilities, marinas, water and sewage treatment plants, water-related and, in certain cases, water-using industries. In the category of land uses that would not damage these water resources fall agriculture, forestry, recreation, institutional and residential open space.

Marshes. In principle, land-use policy for marshes should reflect the roles of flood and water storage, wildlife habitat and fish spawning grounds. Land uses that do not diminish the operation of the primary roles include recreation, certain types of agriculture (notably cranberry bogs) and isolated urban development.

Floodplains. Increasingly, the fifty-year, or 2 percent, probability floodplain is being accepted as that area from which all development should be excluded save for functions which are unharmed by flooding or for uses that are inseparable from floodplains. In the former category fall agriculture, forestry, recreation, institutional open space and open space for housing. In the category of land uses inseparable from floodplains are ports and harbors, marinas, water-related industry and – under certain circumstances – water-using industry.

Aquifers. An aquifer is a water-bearing stratum of rock, gravel or sand, a definition so general as to encompass enormous areas of land. In the region

O
N
E

under study, the great deposits of porous material in the Coastal Plain are immediately distinguishable from all other aquifers in the region because of their extent and capacity. The aquifer parallel to Philadelphia in New Jersey has an estimated yield of one billion gallons per day. Clearly this valuable resource should not only be protected, but managed. Development that includes the disposal of toxic wastes, biological discharges or sewage should be prohibited. The use of injection wells, by which pollutants are disposed of into aquifers, should be discontinued.

Development using sewers is clearly more satisfactory than septic tanks, where aquifers can be contaminated, but it is well to recognize that even sewers leak significant quantities of material and are thus a hazard.

Land-use prescription is more difficult for aquifers than for any other category, as these vary with respect to yield and quality, yet it is clear that agriculture, forestry, recreation and low-density development pose no danger to this resource while industry and urbanization in general do.

All prospective land uses should simply be examined against the degree to which they imperil the aquifer; those which do should be prohibited. It is important to recognize that aquifers may be managed effectively by the impoundment of rivers and streams that transect them.

Like many other cities, Philadelphia derives its water supply from major rivers which are foul. This water is elaborately disinfected and is potable. In contrast to the prevailing view that one should select dirty water for human consumption and make it safe by superchlorination, it seems preferable to select pure water in the first place. Such water is abundant in the existing aquifers; it must be protected from the fate of the rivers.

Aquifer recharge areas. As the name implies, such areas are the points of interchange between surface water and aquifers. In any system there are likely to be critical interchanges. It is the movement of ground to surface water that contributes water to rivers and streams in periods of low flow. Obviously the point of interchange is also a location where the normally polluted rivers may contaminate the relatively clean – and in many cases, pure – water resources in aquifers. These points of interchange are then critical for the management and protection of groundwater resources. . . .

Steep lands. Steep lands, and the ridges which they constitute, are central to the problems of flood control and erosion. Slopes in excess of 12° are not recommended for cultivation by the Soil Conservation Service. The same source suggests that, for reasons of erosion, these lands are unsuitable for development. The recommendations of the Soil Conservation Service are that steep slopes should be in forest and that their cultivation be abandoned.

The role of erosion control and diminution of the velocity of runoff is the principle problem here. Land uses compatible with this role would be mainly forestry and recreation, with low-density housing permitted on occasion.

Prime agricultural land. Prime agricultural soils represent the highest level of agricultural productivity; they are uniquely suitable for intensive cultivation with no conservation hazards. It is extremely difficult to defend agricultural lands when their cash value can be multiplied tenfold by employment for relatively cheap housing. Yet the farm is the basic factory – the farmer is the country's best landscape gardener and maintenance work force, the custodian of much scenic beauty. Mere market values of farmland do not reflect the long-term value of the irreplaceable nature of these living soils. An omnibus protection of all farmland is difficult to defend; but protection of the best soils in a metropolitan area would appear not only defensible, but clearly desirable. . . .

Forests and woodlands. The natural vegetative cover for most of this region is forest. Where present, it improves microclimate and it exercises a major balancing effect upon the water regimen – diminishing erosion, sedimentation, flood and drought. The scenic role of woodlands is apparent, as is their provision of a habitat for game; their recreational potential is among the highest of all categories. In addition, the forest is a low-maintenance, self-perpetuating landscape.

Forests can be employed for timber production, water management, wildlife habitats, as airsheds, recreation or for any combination of these uses. In addition, they can absorb development in concentrations to be determined by the demands of the natural process they are required to satisfy.

[. . .]

The American dream envisioned only the single-family house, the smiling wife and healthy children, the two-car garage, eye-level oven, foundation planting and lawn, the school near by and the church of your choice. It did not see that a subdivision is not a community, that the sum of subdivisions that make a suburb is not a community, that the sum of suburbs that compose the metropolitan fringe of the city does not constitute community nor does a metropolitan region. It did not see that the nature that awaited the subdivider was vastly different from the pockmarked landscape of ranch and split-level houses.

And so the transformation from city to metropolitan area contains all the thwarted hopes of those who fled the old city in search of clean government, better schools, a more beneficent, healthy and safe environment, those who sought to escape slums, congestion, crime, violence and disease.

There are many problems caused by the form of metropolitan growth – the lack of institution which diminishes the power to effect even local decisions, the trauma that is the journey to work, the increasingly difficult problem of providing community facilities. Perhaps the most serious is the degree to which the subdivision, the suburb and the metropolitan area deny the dream and have failed to provide the smiling image of the advertisements. The hucksters made the dream into a cheap thing, subdivided we fell, and the instinct to find more natural environments became the impulse that destroyed nature, an important ingredient in the social objective of this greatest of all population migrations.

Let us address ourselves to this problem. In earlier studies we saw that certain types of land are of such intrinsic value, or perform work for man best in a natural condition or, finally, contain such hazards to development that they should not be urbanized. Similarly, there are other areas that, for perfectly specific reasons, are intrinsically suitable for urban uses. . . .

[I]t transpires, as we have seen before, that if one selects eight natural features, and ranks them in order of value to the operation of natural process, then that group reversed will constitute a gross order of suitability for urbanization. These are: surface water, floodplains, marshes, aquifer recharge areas, aquifers, steep slopes, forests and woodlands, unforested land.

As was discussed in the study of metropolitan open space, natural features can absorb degrees

of development – ports, harbors, marinas, water-related and water-using industries must be on riparian land and may occupy floodplains. Surface water, floodplains and marshes may be used for recreation, agriculture and forestry. The aquifer recharge areas may absorb development in a way that does not seriously diminish percolation or pollute groundwater resources. Steep slopes, when forested, may absorb housing of not more than one house per three acres, while forests on relatively flat land may support a density of development up to one-acre clusters. . . .

[. . .]

The application of this model requires elaborate ecological inventories. Happily, recent technological advances facilitate these. Earth satellites with remote scanning devices with high-level air photography and ground-level identification can provide rich data and time series information on the dynamism of many natural processes. When such inventories are completed they can be constituted into a value system. They can also be identified not only in degrees of value but of tolerance and intolerance. These data, together with the conception of fitness, constitute the greatest immediate utility of the ecological model. Ecosystems can be viewed as fit for certain prospective land uses in a hierarchy. It is then possible to identify environments as fit for ecosystems, organisms and land uses. The more intrinsically an environment is fit for any of these, the less work of adaptation is necessary. Such fitting is creative. It is then a maximum-benefit/minimum-cost solution.

These inventories would then constitute a description of the world, continent or ecosystem under study as phenomena, as interacting process, as a value system, as a range of environments exhibiting degrees of fitness for organisms, men and land use. It would exhibit intrinsic form. It could be seen to exhibit degrees of health and pathology. The inventories would include human artifacts as well as natural processes.

Certainly the most valuable application of such inventories is to determine locations for land uses and most particularly for urbanization. Urban growth in the United States today consists of emptying the continent toward its seaboard conurbations, which expand by accretion and coalesce. This offers the majority of future necropolitans the choice between the environments of Bedford Stuyvesant and Levittown. There must be other alternatives. Let us ask the land where are the best sites. Let us establish criteria for many different types of excellence responding to a wide range of choice. We seek not only the maximum range of differing excellences *between* city locations, but the maximum range of choice *within* each one. . . .

In the quest for survival, success and fulfillment, the ecological view offers an invaluable insight. It shows the way for the man who would be the enzyme of the biosphere – its steward, enhancing the creative fit of man–environment, realizing man's design with nature.

NOTE

1 When McHarg wrote in the late 1960s there were plans to develop and mass-produce supersonic passenger airplanes, which were bitterly opposed by environmentalists in the 1970s and ultimately succumbed to public opposition. Only one – the Concorde – was ever built, and that could be used supersonically only over water. – *Eds.*

6

"The Development of Underdevelopment"

from *Capitalism and Underdevelopment in Latin America* (1967)

Andre Gunder Frank

Editors' Introduction

In the 1960s a number of writers worldwide began to develop sophisticated analyses of why poverty persisted and even worsened despite the huge post-World War II international effort to promote global development. Development assistance by wealthy nations, they argued, was in many ways an extension of earlier colonization, and whatever its stated rationale, functioned to make poverty and inequality in what was then called the Third World worse rather than better. Many writers followed Marx in developing extensive "structuralist" critiques of the institutions and mechanisms of capitalist economics, and (we might say today) throwing the sustainability of global development patterns into question.

One of the most important strains of this growing chorus of criticism was "dependency theory," largely developed in the late 1960s by a group of economists living in Latin America – Andre Gunder Frank, Raul Prebisch, Fernando Cadoszo, and James Caporaso – known as the Economic Commission on Latin America. Their work emerged in part in reaction to the trend within mainstream economics to focus on indigenous factors as the reason for the failure of the least developed countries ("LDCs") to develop, thus "blaming the poor."

The German-born Frank had studied in the United States with noted Marxist economist Paul Baran, whose 1957 book *The Political Economy of Growth* (New York: Monthly Review Press, 1957) had begun analyzing global underdevelopment. But after moving to Latin America in 1962 Frank came to see conventional development economics in the United States as "widely used to defend social irresponsibility, pseudo-scientific scientism, and political reaction." "I had to learn from those who have been persecuted," he wrote.[1] He and others argued that the main effect of capitalist development efforts was to make Third World economies and societies increasingly dependent on wealthier nations in Europe and North America. Frank in particular argued that development created a hierarchy of "satellites," each dependent on others above it in the global system, which took physical form within the towns and metropolitan areas of developing nations.

At mid-century many other writers began to question prevailing theories of economic development around the world, laying the groundwork for later calls for more sustainable forms of development. The work of famed Swedish economist Gunnar Myrdahl during the 1950s and 1960s focused heavily on inequity within capitalist systems. American economist Kenneth Boulding sought as early as the late 1940s to develop an ethical economics that could promote peace, cooperation, and a one-world perspective. Greek historian L.S. Stavrianos described the gradual growth of global inequality in works such as *Global Rift: The Third World Comes of Age* (New York: Morrow, 1981). Egyptian-born writer Samir Amin promoted theories similar to Frank's in books such as *Accumulation on a World Scale* (New York: Monthly Review Press, 1974) and

Maldevelopment: Anatomy of a Global Failure (London: Zed Books, 1990). Famed linguist Noam Chomsky, in books such as *The Culture of Terrorism* (Boston: South End Press, 1988), *Deterring Democracy* (London: Verso, 1991), and *Profit Over People* (New York: Seven Stories Press, 2011, with Robert W. McChesney) has been a consistent critic of the role of powerful political and economic forces in creating violence and exploitation worldwide. Finally, Indian-born, Nobel Prize-winning economist Amartya Sen has studied the recurrence of famines and the role of women within development, and in books such as *Development as Freedom* (New York: Knopf, 1999) and *An Uncertain Glory* (Princeton University Press, 2013, with Jean Drèze) views development as a societal transformation guaranteeing liberty and humane living conditions for a greater number of people, rather than as an increase in economic production. These and other writers have thoroughly critiqued traditional economic development mechanisms as they have been applied internationally.

The style of critics such as Frank is far different than the prose of either Jacobs or McHarg. Schooled in academic Marxism, Frank adopted a more theoretical, formal language that found an audience among European and Third World intellectuals. Not surprisingly, given their Marxist orientation and theoretical style, he and his colleagues were not widely read in North America, although they drew extensively upon North American Marxist analysts such as Paul Baran and Paul Sweezy. They have also received criticism, such as for arguing that there is a single world capitalist system and that particular stages in development are inevitably to be found within it. Yet many anti-globalization activists today follow in the footsteps of such writers, and argue that recent trends for economic globalization, for example through the World Trade Organization and organizations such as the World Economic Forum which meets every January in Davos, Switzerland, are further increasing the power of global capitalism.

The question for urban sustainability activists worldwide is how the development of their communities is affected, in positive or negative ways, by global political and economic structures. How also might more local control be achieved, and "development" itself be redefined in better accord with sustainability?

Underdevelopment in Brazil, as elsewhere, is the result of capitalist development. The military coup of April 1964 and the political and economic events which followed are the logical consequences of this. My purpose here is to trace and to explain the capitalist development of underdevelopment in Brazil since its settlement by Portugal in the sixteenth century and to show how and why, within the metropolis-satellite colonialist and imperialist structure of capitalism, even the economic and industrial development that Brazil is capable of is necessarily limited to an under-developed development. My intent is not an exhaustive study of Brazil *per se*; it is rather an attempt to use the case of Brazil to study the nature of underdevelopment and the limitations of capitalist development.

To account for the underdevelopment and limited development of Brazil, and similar areas, it is common to resort to a dualist model of society. Thus the French geographer Jaques Lambert says in his book *Os dois Brasis* (The Two Brazils):

The Brazilians are divided into two systems of economic and social organization. . . . These two societies did not evolve at the same rate. . . . The two Brazils are equally Brazilian, but they are separated by several centuries. . . . In the course of the long period of colonial isolation, an archaic Brazilian culture was formed, a culture which keeps in isolation the same stability which still exists in the indigenous cultures of Asia and the Near East. . . . The dual economy and the dual social structure which accompanies it are neither new nor characteristically Brazilian – they exist in all unequally developed countries. (Lambert 1959, pp. 105–112)

The same view is shared by Arnold Toynbee (1962) and many others. Celso Furtado (1962), Brazil's Minister of Planning until the April 1964 coup, refers to the one Brazil, the modern capitalist and industrially more advanced Brazil, as an open society and to the archaic rural Brazil as a closed society.

The essential argument of all these students is that the modern Brazil is more developed because it is an open capitalist society; and the other archaic Brazil remains underdeveloped because it is not open, particularly to the industrial part and to the world as a while, and not sufficiently capitalist, but rather pre-capitalist, feudal or semi-feudal. Development is then often viewed as diffusion: "In Brazil, the motor of evolution is everywhere in the cities, from which it radiates change to the countryside" (Lambert, p. 108). The underdeveloped Brazil would develop if only it would open up, and the more developed Brazil would develop still more if the other Brazil would stop being a drag on it and would open its market to industrial goods. My analysis of Brazil's historical and contemporary experience contends that this dualist model is factually erroneous and theoretically inadequate and misleading.

An alternative model may be advanced instead. As a photograph of the world taken at a point in time, this model consists of a world metropolis (today the United States) and its governing class, and its national and international satellites and their leaders – national satellites like the Southern states of the United States, and international satellites like São Paulo. Since São Paulo is a national metropolis in its own right, the model consists further of its satellites: the provincial metropolises, like Recife or Belo Horizonte, and their regional and local satellites in turn. That is, taking a photograph of a slice of the world we get a whole chain of metropolises and satellites, which runs from the world metropolis down to the hacienda or rural merchant who are satellites of the local commercial metropolitan center but who in their turn have peasants as their satellites. If we take a photograph of the world as a whole, we get a whole series of such constellations of metropolises and satellites.

There are several important characteristics of this model: (1) Close economic, political, social and cultural ties between each metropolis and its satellites, which result in the total integration of the farthest outpost and peasant into the system as a whole. This contrasts with the supposed isolation and non-incorporation of large parts of society according to the dualist model. (2) Monopolistic structure of the whole system, in which each metropolis holds monopoly power over its satellites; the source of form of this monopoly varies from one case to another, but the existence of this monopoly is universal throughout the system. (3) As occurs in any monopolistic system, misuse and misdirection of available resources throughout the whole system and metropolis–satellite chain. (4) As part of this misuse, the expropriation and appropriation of a large part or even all of and more than the economic surplus or surplus value of the satellite by its local, regional, national or international metropolis.

Instead of a photograph at a point in time, the model may be viewed as a moving picture of the course of history. It then shows the following characteristics: (1) Expansion of the system from Europe until it incorporates the entire planet in one world system and structure. (If the socialist countries have managed to escape from this system, then there are now two worlds – but in no case are there three.) (2) Development of capitalism, at first commercial and later also industrial, on a world scale as a single system. (3) Polarizing tendencies generic to the structure of the system at world, national, provincial, local and sectoral levels, which generate the development of the metropolis and the underdevelopment of the satellite. (4) Fluctuations within the system, like booms and depressions, which are transmitted from metro-polis to satellite, and like the substitution of one metropolis by another, such as the passing of the metropolis from Venice to the Iberian peninsula to Holland to Britain to the United States. (5) Transformations within the system, such as the so-called Industrial Revolution. Among these transformations we give special emphasis below to important historical changes in the source or mechanism of monopoly which the capitalist world metropolis exercises over its satellites.

From this model in which metropolitan status generates development and satellite status generates underdevelopment, we may derive hypotheses about metropolis–satellite relations and their consequences which differ in important respects from some theses generally accepted, in particular those associated with the dualist model:

1 A metropolis (for example, a national metropolis) which is at the same time a satellite (of the world metropolis) will find that its

development is not autonomous; it does not itself generate or maintain its development; it is a limited or misdirected development; it experiences, in a word, underdeveloped development.

2 The relaxation, weakening or absence of ties between metropolis and satellite will lead to a turning in upon itself on the part of the satellite, an involution, which may take one of two forms:

(a) Passive capitalist involution toward or into a subsistence economy of apparent isolation and of extreme underdevelopment, such as that of the North and Northeast of Brazil. Here there may arise the apparently feudal or archaic features of the "other" sector of the dualist model. But these features are not original to the region, and they are not due to the region's or country's lack of incorporation into the system, as in the dualist model. On the contrary, they are due to and reflect precisely the region's ultra-incorporation, its strong (usually export) ties, which are followed by the region's temporary or permanent abandonment by the metropolis and by the relaxation of these ties.

(b) Weakening of ties together with active capitalist involution which may lead to more or less autonomous development or industrialization of the satellite, which is based on the metropolis–satellite relations of internal colonialism or imperialism. Examples of such active capitalist involution are the industrialization drives of Brazil, Mexico, Argentina, India and others during the Great Depression and the Second World War, while the metropolis was otherwise occupied. Development of the satellites thus appears not as the result of stronger ties with the metropolis, as the dualist model suggests, but occurs on the contrary because of the weakening of these ties. In the history of Brazil we find many cases of the first type of involution – in Amazonia, the Northeast, Minas Gerais, and Brazil as a whole – and one major instance of the second kind of involution in the case of São Paulo.

3 The renewal of stronger metropolis–satellite ties may correspondingly produce the following consequences in the satellite:

(a) The renewal of underdeveloped development consequent upon the reopening of the market for the retrenched region's export products, such as has occurred periodically in Brazil's Northeast. This apparent development is just as disadvantageous in the long run as was the satellite's initial metropolis-sponsored export economy: underdevelopment continues to develop.

(b) The strangulation and misdirection of the autonomous development undertaken by the satellite during the period of lesser ties through the renewal of stronger metropolis–satellite ties as the result of the recuperation of the metropolis after a depression, war, or other kinds of ups and downs. The inevitable result is the renewal of the generation of underdevelopment in the satellite, such as that which took place in the above-named countries after the Korean War.

4 There is a close interconnection of the economy and the socio-political structure of the satellite with those of the metropolis. The closer the satellite's links with and dependence on the metropolis, the closer is the satellite bourgeoisie, including the so-called "national bourgeoisie," linked and dependent on the metropolis. . . .

5 These ties, this growing interconnection, is accompanied by – no, produces – increasing polarization between the two ends of the metropolis–satellite chain in the world capitalist system. A symptom of this polarization is the growing international inequality of incomes and the absolute decline of the real income of the lower income recipients. Yet there is even more acute polarization at the lower end of the chain, between the national and/or local metropolises and their poorest rural and urban satellites whose absolute real income is steadily declining. This increasing polarization sharpens political tension, not so much between the international metropolis with its imperialist bourgeoisie and the national metropolises with national bourgeoisies, as between both of these and their rural

and city slum satellites. This tension between the poles becomes sharper until the initiative and generation of the transformation of the system passes from the metropolitan pole, where it has been for centuries, to the satellite pole.

NOTE

1 *Capitalism and Underdevelopment in Latin America*, p. xviii.

REFERENCES

Furtado, C. (1962) *A pre-revoluca Brasileiro*. Rio de Janeiro: Editora Fundo de Cultura.

Lambert, J. (1959) *Os dois Brasis*. Rio de Janeiro: Ministerio da Educacaoe Cultura.

Taynbee, A.J. (1962) America and the World Revolution. New York: Oxford University Press.

ONE

"Perspectives, Problems, and Models"

from *The Limits to Growth* (1972)

Donella Meadows, Dennis L. Meadows, Jörgen Randers, and William W. Behrens III

Editors' Introduction

One of the most influential books of the 1970s – and as far as we have been able to determine the first work ever to use the term "sustainable development" in its current sense – was *The Limits to Growth* (New York: Universe Books, 1972). This paperback bestseller by a team of MIT scientists catalyzed discussions worldwide about the future of human society. The book resulted from a study that the Club of Rome – an ad hoc group of global industrialists and humanitarians led by Italian economic consultant Aurelio Peccei – commissioned from Professor Jay Forester and his graduate students at MIT. Given the emergence of computers as a powerful analytic tool at this time, Peccei asked the researchers to use computer models for the first time to attempt to analyze the future of the world. The group developed a model known as World3, using a method of analysis they called systems dynamics.

The Forester team analyzed the basic factors most likely to limit growth: population, agricultural production, natural resources, industrial production, and pollution. They concluded that following then-current trends the limits to the growth of human society on the planet would be reached within 100 years, leading to a steep decline in global population and industrial capacity. Essentially, growing problems of resource depletion, pollution (including carbon dioxide concentration), loss of arable land, and declining food production would converge to halt progress. However, they also stated that it would be possible to alter these trends "to establish a condition of ecological and economic stability that is sustainable far into the future" (p. 24). The first of these conclusions shocked millions around the world, but fit with what many in the growing environmental movement had already concluded – that human development trends were headed in unsustainable directions. Other writers such as Rene Dubos, a French molecular biologist and originator of the phrase "think globally, act locally," Paul Ehrlich, a Stanford environmental scientist and author of the bestseller *The Population Bomb* (New York: Ballantine Books, 1968), Rachel Carson, Barry Commoner, and Ian McHarg had been saying much the same thing. The 1972 Stockholm Conference on Environment and Development and other United Nations events in the years that followed helped spread such ideas. But the *Limits to Growth* work was unique in that it for the first time used computer technology and scientific method to analyze the human future as well as questions such as whether growing human population and resource consumption were sustainable.

Not surprisingly, other writers vigorously opposed the *Limits to Growth* position, arguing that this approach was an alarmist recapitulation of arguments advanced by Thomas Malthus around 1800, comparing the linear increase in agricultural production with the geometric increase of population. These opponents argued that technology, economics, and human ingenuity would be able to help humanity through growth-related problems. In his 1981 book, *The Ultimate Resource* (Princeton, NJ: Princeton University Press, 1981), business and marketing professor Julian Simon for example advanced the view that the world would never run out of resources. Scarcity-fueled increases in resource prices would encourage conservation or resource substitution, in his view, thus avoiding any long-term problems. Simon even bet Ehrlich $5,000 that the prices of a certain set of metals would in fact fall over five years, and won the bet (Ehrlich later claimed that the time frame was too short).

The *Limits to Growth* team revisited their work two decades later in a second book, *Beyond the Limits* (Post Mills, VT: Chelsea Green, 1992), and then a third, *The Limits to Growth: The 30-year Update* (White River Junction, VT: Chelsea Green, 2004). After running an updated version of their model and considering additional evidence over the 30-year period, Meadows and her coauthors concluded that the fundamental themes of *Limits to Growth* had held up relatively well, and that the world had entered a period of "overshoot" in which it was well beyond sustainable levels of resource consumption, pollution, and population. Other conclusions remained the same – that it was still possible for humanity to change course, and that the sooner it did so, the better.

The language of this selection reflects the sudden sense of global environmental crisis that many felt in the late 1960s and 1970s, illustrated by the quote the authors start out with from United Nations Secretary-General U Thant. Previously, the end of World War II had ushered in a period of optimism based on faith that economic progress and the spread of new technologies would bring about continual human betterment. But by the end of the 1960s the world was in the midst of a cold war and a nuclear arms race, problems of global pollution and ecosystem degradation were being discovered, overpopulation was being recognized as a serious problem, and resources such as petroleum were suddenly seen to be limited. Events such as the 1973 energy crisis seemed to confirm this sense of crisis. Certainly the dangers of global catastrophe were overplayed by some. But certainly also the world's attention needed to be called to the challenge of living sustainably on a small planet in the long term. More than any other single work *The Limits to Growth* helped do this.

The style of this selection – especially that of principal author Donella Meadows – also reflects a rising awareness of the extent to which different worldviews or paradigms affect how problems are seen. The supposedly objective viewpoint of modern science, so strong at that time, was under increasing attack from many directions, especially from Thomas Kuhn, whose book *The Structure of Scientific Revolutions* (Chicago: University of Chicago Press, 1962) had argued that science, far from being completely objective and rational, in fact proceeds by moving through different paradigms dominant at different times. What is needed, Meadows and others argued, is an increased awareness of how our cognitive "lenses" affect our beliefs about global development. In particular, they believed that a new focus on the role of values in determining our beliefs and worldviews – and a rethinking of values themselves – was necessary in order to bring about more sustainable development practices.

More than 40 years after *Limits* originally appeared, it is still an open question in what ways "limits" will affect human development. The oil industry, for example, has proven remarkably adept at finding new sources through deep-water drilling, mining of tar sands, hydraulic fracturing of rock formations, and other technologies. So "peak oil" has not arrived as soon as some thought it would. However, other limits, for example to the amount of greenhouse gas pollution the Earth's climate can tolerate, may not be as forgiving. Debates over "limits" – on large scales and small – are woven throughout the sustainability discourse. How human attention might better focus on global problems is also a pressing question.

I do not wish to seem overdramatic, but I can only conclude from the information that is available to me as Secretary-General, that the Members of the United Nations have perhaps ten years left in which to subordinate their ancient quarrels and launch a global partnership to curb the arms race, to improve the human environment, to defuse the population explosion, and to supply the required momentum to development efforts. If such a global partnership is not forged within the next decade, then I very much fear that the problems I have mentioned will have reached such staggering proportions that they will be beyond our capacity to control. (U Thant, 1969)

The problems U Thant mentions – the arms race, environmental deterioration, the population explosion, and economic stagnation – are often cited as the central, long-term problems of modern man. Many people believe that the future course of human society, perhaps even the survival of human society, depends on the speed and effectiveness with which the world responds to these issues. And yet only a small fraction of the world's population is actively concerned with understanding these problems or seeking their solutions.

HUMAN PERSPECTIVES

Every person in the world faces a series of pressures and problems that require his attention and action. These problems affect him at many different levels. He may spend much of his time trying to find tomorrow's food for himself and his family. He may be concerned about personal power or the power of the nation in which he lives. He may worry about a world war during his lifetime, or a war next week with a rival clan in his neighborhood. These very different levels of human concern can be represented on a graph like that in Figure 1. The graph has two dimensions, space and time. Every human concern can be located at some point on the graph, depending on how much geographical space it includes and how far it extends in time. Most people's worries are concentrated in the lower left-hand corner of the graph. Life for these people is difficult, and they must devote nearly all of their efforts to providing for themselves and their families, day by day. Other people think about and act on problems farther out on the space or time axes. The pressures they perceive involve not only themselves, but the community with which they identify. The actions they take extend not only days, but weeks or years into the future.

A person's time and space perspectives depend on his culture, his past experience, and the immediacy of the problems confronting him on each level. Most people must have successfully solved the problems in a smaller area before they move their concerns to a larger one. In general the larger the

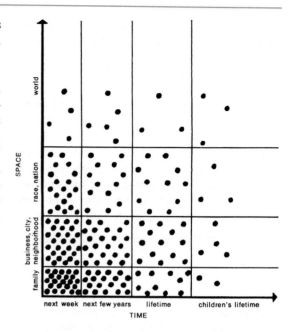

Figure 1 Human perspectives. Although the perspectives of the world's people vary in space and in time, every human concern falls somewhere on the space–time graph. The majority of the world's people are concerned with matters that affect only family or friends over a short period of time. Others look farther ahead in time or over a larger area – a city or a nation. Only a very few people have a global perspective that extends far into the future.

space and the longer the time associated with a problem, the smaller the number of people who are actually concerned with its solution.

There can be disappointments and dangers in limiting one's view to an area that is too small. There are many examples of a person striving with all his might to solve some immediate, local problem, only to find his efforts defeated by events occurring in a larger context. A farmer's carefully maintained fields can be destroyed by an international war. Local officials' plans can be overturned by a national policy. A country's economic development can be thwarted by a lack of world demand for its products. Indeed there is increasing concern today that most personal and national objectives may ultimately be frustrated by long-term, global trends such as those mentioned by U Thant.

Are the implications of these global trends actually so threatening that their resolution should take precedence over local, short-term concerns?

Is it true, as U Thant suggested, that there remains less than a decade to bring these trends under control?

If they are not brought under control, what will the consequences be?

What methods does mankind have for solving global problems, and what will be the results and the costs of employing each of them?

These are the questions that we have been investigating in the first phase of the Club of Rome's Project on the Predicament of Mankind. Our concerns thus fall in the upper right-hand corner of the space–time graph.

PROBLEMS AND MODELS

Every person approaches his problems, wherever they occur on the space–time graph, with the help of models. A model is simply an ordered set of assumptions about a complex system. It is an attempt to understand some aspect of the infinitely varied world by selecting from perceptions and past experience a set of general observations applicable to the problem at hand. A farmer uses a mental model of his land, his assets, market prospects, and past weather conditions to decide which crops to plant each year. A surveyor constructs a physical model – a map – to help in planning a road. An economist uses mathematical models to understand and predict the flow of international trade.

Decision-makers at every level unconsciously use mental models to choose among policies that will shape our future world. These mental models are, of necessity, very simple when compared with the reality from which they are abstracted. The human brain, remarkable as it is, can only keep track of a limited number of the complicated, simultaneous interactions that determine the nature of the real world.

We, too, have used a model. Ours is a formal, written model of the world.[1] It constitutes a preliminary attempt to improve our mental models of long-term, global problems by combining the large amount of information that is already in human minds and in written records with the new information-processing tools that mankind's increasing knowledge has produced – the scientific method, systems analysis, and the modern computer.

Our world model was built specifically to investigate five major trends of global concern – accelerating industrialization, rapid population growth, widespread malnutrition, depletion of nonrenewable resources, and a deteriorating environment. These trends are all interconnected in many ways, and their development is measured in decades or centuries, rather than in months or years. With the model we are seeking to understand the causes of these trends, their interrelationships, and their implications as much as one hundred years in the future.

The model we have constructed is, like every other model, imperfect, oversimplified, and unfinished. We are well aware of its shortcomings, but we believe that it is the most useful model now available for dealing with problems far out on the space–time graph. To our knowledge it is the only formal model in existence that is truly global in scope, that has a time horizon longer than thirty years, and that includes important variables such as population, food production, and pollution, not as independent entities, but as dynamically interacting elements, as they are in the real world.

Since ours is a formal, or mathematical, model it also has two important advantages over mental models. First, every assumption we make is written in a precise form so that it is open to inspection and criticism by all. Second, after the assumptions have been scrutinized, discussed, and revised to agree with our best current knowledge, their implications for the future behavior of the world system can be traced without error by a computer, no matter how complicated they become.

We feel that the advantages listed above make this model unique among all mathematical and mental world models available to us today. But there is no reason to be satisfied with it in its present form. We intend to alter, expand, and improve it as our own knowledge and the world data base gradually improve.

In spite of the preliminary state of our work, we believe it is important to publish the model and our findings now. Decisions are being made every day in every part of the world, that will affect the physical, economic, and social conditions of the world system for decades to come. These decisions cannot wait for perfect models and total understanding. They will be made on the basis of some model, mental or written, in any case. We feel that the model described here is already sufficiently developed to be of some use to decision-makers. Furthermore, the basic behavior modes we have already observed in this model appear to be so fundamental and general that we do not expect our broad conclusions to be substantially altered by further revisions.

It is not the purpose of this book to give a complete, scientific description of all the data and mathematical equations included in the world model. Such a description can be found in the final technical report of our project. Rather, in *The Limits to Growth* we summarize the main features of the model and our findings in a brief, nontechnical way. The emphasis is meant to be not on the equations or the intricacies of the model, but on what it tells us about the world. We have used a computer as a tool to aid our own understanding of the causes and consequences of the accelerating trends that characterize the modern world, but familiarity with computers is by no means necessary to comprehend or to discuss our conclusions. The implications of those accelerating trends raise issues that go far beyond the proper domain of a purely scientific document. They must be debated by a wider community than that of scientists alone. Our purpose here is to open that debate.

The following conclusions have emerged from our work so far. We are by no means the first group to have stated them. For the past several decades, people who have looked at the world with a global, long-term perspective have reached similar conclusions. Nevertheless, the vast majority of policy-makers seems to be actively pursuing goals that are inconsistent with these results.

Our conclusions are:

1   If the present growth trends in world population, industrialization, pollution, food production, and resource depletion continue unchanged, the limits to growth on this planet will be reached sometime within the next one hundred years. The most probable result will be a rather sudden and uncontrollable decline in both population and industrial capacity.

2   It is possible to alter these growth trends and to establish a condition of ecological and economic stability that is sustainable far into the future. The state of global equilibrium could be designed so that the basic material needs of each person on earth are satisfied and each person has an equal opportunity to realize his individual human potential.

3   If the world's people decide to strive for this second outcome rather than the first, the sooner they begin working to attain it, the greater will be their chances of success.

These conclusions are so far-reaching and raise so many questions for further study that we are quite frankly overwhelmed by the enormity of the job that must be done. We hope that this book will serve to interest other people, in many fields of study and in many countries of the world, to raise the space and time horizons of their concerns and to join us in understanding and preparing for a period of great transition – the transition from growth to global equilibrium.

NOTE

1   The prototype model on which we have based our work was designed by Professor Jay W. Forrester of the Massachusetts Institute of Technology. A description of that model has been published in his book *World Dynamics* (Cambridge, MA: Wright-Allen Press, 1971).

8

"The Steady-State Economy"

from *Toward a Steady-State Economy* (1973)

Herman Daly

Editors' Introduction

Although the field of economics provided much of the foundation for twentieth-century global development, a growing number of economists toward the middle and end of the century came to question its prevailing assumptions. Eloquent observers such as Boulding, Schumacher, and Hazel Henderson published works arguing that prevailing economic perspectives were deeply flawed, and that humanity needed to take environmental and social impacts of economic development into account as well as adopt a longer-term perspective than that supplied by economics. Particularly influential were Schumacher's book *Small Is Beautiful* (New York: Harper & Row, 1973), advocating "appropriate technology," and Boulding's call for an "economics for a spaceship earth" in books such as *Beyond Economics: Essays on Society, Religion, and Ethics* (Ann Arbor, MI: University of Michigan Press, 1968). In the 1980s and 1990s environmental economists and ecological economists worked in more specific ways to change the discipline. The former group has tried to integrate environmental and social concerns into existing economic tools, while the second has gone further to try to put economics into the context of a broader global picture.

However, it has been Herman Daly, a professor first at the University of Louisiana and later the University of Maryland, who has developed perhaps the most fundamental critique of traditional economics from a sustainability perspective. Picking up on an idea first mentioned by John Stuart Mill in the nineteenth century, Daly argued in the early 1970s that an economy based on endless growth in physical production was impossible, and called instead for a "steady state economy" based on qualitative but not quantitative growth. Daly's preferred method for achieving this steady state was through depletion quotas on resources, through which the government would essentially auction off the right to consume basic resources. The amount allowed to be consumed would decrease over time, setting in place market mechanisms (through higher prices, conservation, better technologies, substitution, etc.) to reduce consumption of that resource. Somewhat tongue-in-cheek, Daly even suggested that this mechanism be applied to the right to have children, so as to apply economics to the population problem.

Daly's ideas constituted a challenge to conventional economics that could not be ignored, and indeed questioning "growth" of all sorts – including population growth, metropolitan spatial growth, and growth in energy or resource consumption – has become a central feature of sustainability debates. For a time in the 1990s Daly was even hired by the World Bank as an in-house economic advisor, though his ideas appear to have had little impact on the Bank's overall policies. Although mainstream economists, politicians, and the global business community have yet to take the concept of a steady-state economy seriously, and assume continual growth in production and consumption to be essential for human welfare, Daly's ideas remain important as a leading philosophical alternative to endless growth, one that may eventually underpin a more sustainable society. Daly has also been a leading author in the field of ecological economics; leading works of his in this vein include *Ecological Economics: Principles and Applications* (Washington, D.C.: Island Press, 2004) and *Ecological Economics and Sustainable Development* (Northampton, MA: Edward Elgar, 2007).

Like Leopold and Meadows, Daly embraced notions of ethical and spiritual change that went far beyond his professional field. He has argued for changes in values and moral growth to serve as the underpinnings of a new economy, and developed these ideas further in collaboration with theologian John B. Cobb, Jr. in their book *For the Common Good* (Boston: Beacon Press, 1989). In that volume Daly and Cobb proposed an Index of Sustainable Economic Welfare that might serve as an alternative to the Gross Domestic Product (GDP) for measuring human well-being. Instead of a single measure of economic production, this indicator combines a wide range of measures of human welfare such as public expenditures on health and education, net capital growth, the value of household services, costs of pollution, depletion of non-renewable resources, and distributional inequality. The nongovernmental organization Redefining Progress, based in San Francisco, has produced a similar statistic called the Genuine Progress Indicator (for more information, see www.rprogress.org). Both measures show social welfare in the United States peaking in the early 1970s and declining ever since, due in part to rapidly growing inequities, environmental damage, and mounting personal costs for health and education.

Daly's writings raise the question of how we should best reconceptualize "growth" within a more sustainable society, and also how we should measure "progress." Should industrial production and household consumption be prime measures? Or new indicators of social welfare, environmental well-being, and collective happiness? And if we are to rethink conventional concepts of economic growth, what takes its place? How would sustainable urban and regional economies then be organized?

GROWTHMANIA

The fragmentation of knowledge and people by excessive specialization, the disequilibrium between the human economy and the natural ecosystem, the congestion and pollution of our spatial dimension of existence, the congestion and pollution of our temporal dimension of existence with the resulting state of harried drivenness and stress – all these evils and more are symptomatic of the basic malady of growthmania.

"Growthmania" is an insufficiently pejorative term for the paradigm or mind-set that always puts growth in first place – the attitude that there is no such thing as enough, that cannot conceive of too much of a good thing. It is the set of unarticulated preconceptions which allows the President's Council of Economic Advisers to say, "If it is agreed that economic output is a good thing it follows by definition that there is not enough of it." As a sop to environmentalists the Council does admit that "growth of GNP has its costs, and beyond some point they are not worth paying." But instead of raising the obvious question – "What determines this point of optimal GNP, and how do we know

when we have reached it?" – the Council merely pontificates that "the existing propensities of the population and the policies of the government constitute claims upon GNP itself that can only be satisfied by rapid economic growth." That of course is merely to restate the problem, not to give a solution. Apparently these "existing propensities and policies" are beyond discussion. That is growthmania. Brezhnev, Castro, and Franco receive much the same advice from their respective Councils of Economic Advisers. Growthmania is ecumenical.

The answer to the avoided question "When do the costs of growth in GNP outweigh the benefits?" is contained in the question itself. This occurs when the decreasing marginal benefit of extra GNP becomes less than the increasing marginal cost. The marginal benefit is measured by the market value of extra goods and services – i.e., the increment in GNP itself in value units. But what statistical series measures the cost? Answer: *none!* That is growthmania; literally not counting the costs of growth.

But the worst is yet to come. We take the real costs of increasing GNP as measured by the defensive expenditures incurred to protect ourselves from the *unwanted* side effects of production, and *add* these expenditures to GNP rather than subtract them. We count the real costs as benefits – this is hyper-growthmania. Since the net benefit of growth can never be negative with this Alice-in-Wonderland accounting system, the rule becomes "grow forever" or at least until it kills you – and then count your funeral expenses as further growth. This is terminal hyper-growthmania. Is the water table falling? Dig deeper wells, build bigger pumps, and up goes GNP! Mines depleted? Build more expensive refineries to process lower grade ores, and up goes GNP! Soil depleted? Produce more fertilizer, etc. As we press against the carrying capacity of our physical environment, these "extra-effort" and "defensive expenditures" (which are really costs masquerading as benefits) will loom larger and larger. As more and more of the finite physical world is converted into wealth, less and less is left over as nonwealth – i.e. the nonwealth physical world becomes scarce, and in becoming scarce it gets a price and thereby becomes wealth. This creates the illusion of becoming better off, when in actuality we are becoming worse off. We may already have passed the point where the marginal cost of growth exceeds the marginal benefit. This suspicion is increased by looking at who absorb the costs and who receive the benefits. We all get some of each, but not equal shares. Who buys a second car or a third TV? Who lives in the most congested, polluted areas? The benefits of growth go mainly to the rich, the costs go mainly to the poor. That statement is based on casual empiricism – we do not have social accounts which allow us to say precisely who receive the benefits and who absorb the costs of growth, a fact which is itself very revealing. Ignorance, if not blissful, is often politically expedient.

Growthmania is the paradigm upon which stand the models and policies of our current political economy. The answer to every problem is growth. For example:

1 Poverty? Grow more to provide more employment for the poor and more tax revenues for welfare programs.
2 Unemployment? Invest and grow to bolster aggregate demand and employment.
3 Inflation? Grow by raising productivity so that more goods will be chased by the same number of dollars and prices will fall.
4 Balance of payments? Grow more and increase productivity in order to increase exports. Cutting imports is seen only as a short-run stopgap, not a solution.
5 Pollution and depletion? Grow so we will be rich enough to afford the cost of cleaning up and of discovering new resources and technologies.
6 War? We must grow to be strong and have *both* guns and butter.

The list could be extended, but it can also be summarized in one sentence: The way to have your cake and eat it too is to make it grow.

Growthmania is the attitude in economic theory that begins with the theological assumption of infinite wants, and then with infinite hubris goes on to presume that the original sin of infinite wants has its redemption vouchsafed by the omnipotent savior of technology, and that the first commandment is to produce more and more goods for more and more people, world without end. And that this is not only possible, but desirable.

Environmental degradation is an iatrogenic disease induced by economic physicians who treat the basic malady of unlimited wants by prescribing unlimited economic growth. We experience environmental degradation in the form of increased scarcity of clean air, pure water, relaxed moments, etc. But the only way the growthmania paradigm knows to deal with scarcity is to recommend growth. Yet one certainly does not cure a treatment-induced disease by increasing the treatment dosage! Nevertheless the usual recommendation for combating pollution is to grow more because "a rising GNP will enable the nation more easily to bear the costs of eliminating pollution." Such a view is patently absurd.

The growth paradigm has outlived its usefulness. It is a senile ideology that should be unceremoniously retired into the history of economic doctrines. In the terminology of Thomas Kuhn's book, *The Structure of Scientific Revolutions*, the growth paradigm has been more than exhausted by the normal science puzzle-solving research done within its confines. Political economy must enter a period of revolutionary science to establish a new paradigm to guide a new period of normal science. Just as mercantilism gave way to physiocracy, physiocracy to classical *laissez-faire*, *laissez-faire* to Keynesianism, Keynesianism to the neoclassical growth synthesis –

so the current neoclassical growthmania must give way to a new paradigm. What will the new paradigm be? I submit that it must be very similar to an idea from classical economics that never attained the status of a paradigm, except for a brief chapter in John Stuart Mill's *Principles of Political Economy*. This idea is that of the steady-state economy.

THE STEADY STATE

What is meant by a "steady-state economy"?

Why is it necessary?

How can it be attained?

The first two questions are relatively easy and have been dealt with elsewhere. Hence they will be treated rapidly. The third question is extremely difficult, and will be the main focus of attention.

The steady state is defined as an economy in which the total population and the total stock of physical wealth are maintained constant at some desired levels by a "minimal" rate of maintenance throughput (i.e., by birth and death rates that are equal at the lowest feasible level, and by physical production and consumption rates that are equal at the lowest feasible level). The first part of the definition (constant stocks) goes back to John Stuart Mill, and the second part ("minimal" flow of throughput) goes back to 1949 vintage Kenneth Boulding. Minimizing throughput implies maximizing the average life expectancy of a member of the stock.

Why is the steady state necessary? Not for the reasons given by the classical economists who saw increasing rent and interest eliminating profit and thus the incentive for "progress." Rather, the necessity follows immediately from physical first principles. The world is finite, the ecosystem is a steady state. The human economy is a subset of the steady-state ecosystem. Therefore at some level and over some time period the subsystem must also become a steady state, at least in its physical dimensions of people and physical wealth. The steady-state economy is therefore a physical necessity. . . .

When we raise the third question, how to attain the steady state, things become more difficult. First, we must give operational definitions to the specific goals contained in the definition of steady state.

Second, we must specify the technologies, social institutions, and moral values which are in harmony with and supportive of the steady state.

To define more clearly the goal of the steady state we must face four questions.

1 *At what levels should the stocks of wealth and people be maintained constant?* Specifying the stock of wealth and of people also specifies the wealth per person or standard of living. In other words the question becomes the old one of what is the optimum population? So far no one has given a definite answer, and I certainly cannot. . . . [T]he optimum population is more likely to be discovered by experience than by a priori thought. We should attain a stationary population at some feasible nearby level. After experiencing it we could then decide whether the optimum level is above or below the current level. . . .

2 *What is the optimal level of maintenance throughput for a given level of stocks?* For the time being the answer is probably "as low as possible" or at least "less than at present". . . .

3 *What is the optimal time horizon or accounting period over which population and wealth are required to be constant?* Obviously we cannot mean day-to-day constancy and probably not even year-to-year constancy. Related to this is the question of the optimum amplitude of fluctuation around the steady-state mean during the accounting period.

Again, I cannot pretend to be able to answer this question. . . . But somehow [in standard economic theory] we manage to choose an accounting period and muddle through, and so we could also in a steady state. . . .

4 *What is the optimal rate of transition from the growing economy to the steady state?* We can never attain a steady state in the long run if our efforts to do so kill us in the short run. In the case of population there are interesting trade-offs between speed of attainment of a stationary population versus size of the stationary population and the amplitude of fluctuations in the birth rate induced by the current nonequilibrium age structure.

Once again I do not know the optimum rate of transition. But I think we are very unlikely to exceed it. In any case the sooner we begin deceleration to zero growth the longer we can afford to take and the less disruptive that adjustment will be. The important thing from all points of view is to begin the deceleration now. Later we can argue about the optimum rate. . . .

[. . .]

CONSTANT PHYSICAL WEALTH

. . . Let quotas be set on new depletion of each of the basic resources, both renewable and non-renewable, during a given time. The legal right to deplete to the amount of the quota for each resource would be auctioned off by the government at the beginning of each time period, in conveniently divisible units, to private firms, individuals, and public enterprises. After puchase from the government the quota rights would be freely transferable by sale or gift. As population growth and economic growth press against resources, the prices of the depletion quotas would be driven higher and higher. In the interests of conserving nonrenewable resources and optimal exploitation of renewable resources, quotas could then be reduced to lower levels, thereby driving the price of the quotas still higher. In this way, the increasing windfall rents resulting from increasing pressure of demand on a fixed supply would be collected by the government through the auctioning of the depletion rights. The government spends the revenues, let us say, by paying a social dividend. Even though the monetary flow is therefore undiminished, the real flow has been physically limited by the resource quotas. All prices of resources and of goods then increase, the prices of resource-intensive goods increase relatively more, and total resource consumption (depletion) is reduced. Moreover, in accordance with the law of conservation of matter-energy, reduction of initial inputs will result in reduction of ultimate outputs (pollution), reducing the aggregate through-put and with it the stress it puts on the ecosystem.

With depletion now made more expensive and with higher prices on final goods, recycling becomes more profitable. As recycling increases pollution is reduced even more. Higher prices make consumers more interested in durability and careful maintenance of wealth. Most importantly, prices now provide a strong incentive to develop new technologies and patterns of consumption that are resource saving. If there is any static inefficiency incurred in setting the rate of depletion outside the market (a doubtful

point), it is likely to be more than offset by the dynamic benefits of greater inducements to develop resource-saving technology.

Adjustment of the throughput of depletion and pollution flows to long-run ecologically sustainable levels can be effected gradually. At first depletion quotas could be set at the preceding year's levels, and if necessary gradually reduced by, say, two per cent per year until we reach the "optimal" throughput. Stocks will then adjust to equilibrium with the new throughput. Thereafter the constant stock would be maintained by the constant throughput. As we gradually exhaust nonrenewable resources, quotas for their depletion will approach zero and recycling will become the only source of inputs, at which time, presumably, the ever rising price of the resource will have led to the development of a recycling technology. . . .

[. . .]

CONTROL OF DISTRIBUTION

Distribution is the rock upon which most ships of state, including the steady state, are very likely to run aground. Currently we seek to improve distribution by establishing a minimum standard of living guaranteed by a negative income tax. In the growthmania paradigm there is no upper limit to the standard of living. In the steady-state paradigm there is an upper limit. Furthermore the higher the lower limit below which no one is allowed to fall, the lower must be the upper limit above which no one is allowed to rise. The lower limit has considerable political acceptance, the upper limit does not. But in the steady state the upper limit is a logical necessity. It implies confiscation and redistribution of wealth above a certain limit per person or per family. What does one say to the cries of "destruction of incentive"? Remember – we are no longer anxious to grow in the first place! Also one recalls Jonathan Swift's observation:

In all well-instituted commonwealths, care has been taken to limit men's possessions; which is done for many reasons, and, among the rest, for one which, perhaps, is not often considered; that when bounds are set to men's desires, after they have acquired as much as the laws will permit them, their private interest is at an end, and they have nothing to do but to take care of the public.[1]

The basic institution for controlling distribution is very simple: set maximum and minimum limits on wealth and income, the maximum limit on wealth being the most important. Such a proposal is in no way an attack on private property. Indeed, as John Stuart Mill argues, it is really a defense of private property:

> Private property, in every defense made of it, is supposed to mean the guarantee to individuals of the fruits of their own labor and abstinence. The guarantee to them of the fruits of the labor and abstinence of others, transmitted to them without any merit or exertion of their own, is not of the essence of the institution, but a mere incidental consequence which, when it reaches a certain height, does not promote, but conflicts with, the ends which render private property legitimate.[2]

According to Mill, private property is legitimated as a bastion against exploitation. But this is true only if everyone owns some minimum amount. Otherwise, when some own a great deal of it and others have very little, private property becomes an *instrument* of exploitation rather than a guarantee against it. It is implicit in this view that private property is legitimate only if there is some distributist institution (like, for example, the Jubilee year of the Old Testament) that keeps inequality of wealth within some tolerable limits. Such an institution is now lacking. The proposed institution of maximum and minimum wealth and income limits would remedy this severe defect and make private property legitimate again. Also it would go a long way toward legitimating the free market, since most of our blundering interference with the price system (e.g., the farm program, the minimum wage) has as its goal an equalizing alteration in the distribution of income and wealth. . . .

[. . .]

ON MORAL GROWTH

Is the above sketch of a steady state unrealistic and idealistic? On the contrary, it is in broad characteristics the only realistic possibility. The present economy is literally unrealistic because in its disregard for natural laws it is attempting the impossible. The steady-state paradigm, unlike growthmania, is realistic because it takes the physical laws of nature as its first premise.

Let us assume for a moment that the necessity of the steady state and the above outline of its appropriate technologies and social institutions are accepted. Logic and necessity are not sufficient to bring about social reform. The philosopher Leibnitz observed that,

> If geometry conflicted with our passions and interests as much as do ethics, we would contest it and violate it as much as we do ethics now, in spite of all the demonstrations of Euclid and Archimedes, which would be labeled paralogisms and dreams.[3]

Leibnitz is surely correct. However logical and necessary the above outline of the steady state, it is, on the assumption of static morality, nothing but a dream. The physically steady economy absolutely requires moral growth beyond the present level.

NOTES

1 Jonathan Swift, "Thoughts on Various Subjects," reprinted in G.B. Woods *et al.* (eds), *The Literature of England*, New York: Scott Foresman, 1958, p. 1003.

2 John Stuart Mill, "Of Property," in *Principles of Political Property*, Volume II, London: John Parker, 1857, Chapter 1.

3 Leibnitz quoted in A. Sauvy, *The General Theory of Population*, New York: Basic Books, 1970, p. 270.

9

"City and Nature"

from *The Granite Garden: Urban Nature and Human Design* (1984)

Anne Whiston Spirn

Editors' Introduction

Although landscape architects and park designers have long sought to bring nature into cities, this need was often ignored by developers and the nascent city planning profession in the nineteenth and twentieth centuries. Engineers and developers filled or paved over streams, wetlands, and shorelines to make way for urban expansion. Highways or railroad lines cut many cities off from their waterfronts. Hills were leveled and native vegetation removed. Landowners platted lots and roads without considering the implications for wildlife, native plant species, or human recreation. With the advent of central heating, electric lighting, air conditioning, long-distance food transport, and huge dams and pipelines bringing water from hundreds of miles away, urban residents became well insulated from nature in all its forms, and even from the limitations of climate and local geography. What remnants of "nature" still existed in the city were often highly manicured parks and yards.

Only with the environmental revolution of the 1960s did activists and policymakers come to think more systematically about integrating urban development with the natural world, as well as protecting humans from some of the worst abuses of urban environments. Efforts to restore damaged natural systems within cities gained speed in the 1980s and 1990s, and new fields such as landscape ecology provided conceptual tools for thinking about how reconstructed ecosystems might function. Communities experimented with watershed planning, citizens' groups worked to restore creeks and rivers, and use of native, climate-appropriate species soared within landscape architecture.

One of the classic pieces first calling attention to systematic relationships between nature and cities was Anne Whiston Spirn's book *The Granite Garden* (New York: Basic Books, 1984). While McHarg had focused on the interaction of new suburban or regional development with natural landscapes, Spirn looked at nature within densely built cities themselves. A professor of architecture at the University of Pennsylvania, she analyzed the role of different natural entities such as soil, water, wind, and light within urban landscapes, and argued that the city should be seen as part of nature, not something existing outside of it. If nature is welcomed into the city, in her view, a delightful urban environment can be created; if nature is ignored, disaster may result. As with McHarg's writings, Spirn's eloquent, passionate style inspired many in the environmental planning and landscape architecture professions who have since worked out specific ways to implement her philosophy.

Michael Hough, a landscape architect at the University of Toronto, took a very similar approach in his books *City Form and Natural Processes: Toward an Urban Vernacular* (New York: Van Nostrand Reinhold, 1984) and *Cities and Natural Process* (New York: Routledge, 1995). Other readings in a similar vein include *The Humane Metropolis: People and Nature in the 21st-Century City*, edited by Rutherford H. Platt (Amherst: University of Massachusetts Press, 2006); *Human Ecology: Following Nature's Lead* (Washington, D.C.: Island Press, 2002) and *The Living Landscape* (Washington, D.C.: Island Press, 2008) by Frederick Steiner; *Green City: People, Nature, and Urban Places*, by Mary Soderstrom (Montreal: Vehicule Books, 2006);

Richard Register's *Ecocities: Rebuilding Cities in Balance with Nature* (Gabriola Island, BC: New Society Publishers, 2006); and Tim Beatley's *The Ecology of Place: Planning for Environment, Economy, and Community* (Washington, D.C.: Island Press, 1997) and *Biophilic Cities: Integrating Nature into Urban Design and Planning* (Washington, D.C.: Island Press, 2011).

Nature pervades the city, forging bonds between the city and the air, earth, water, and living organisms within and around it. In themselves, the forces of nature are neither benign nor hostile to humankind. Acknowledged and harnessed, they represent a powerful resource for shaping a beneficial urban habitat; ignored or subverted, they magnify problems that have plagued cities for centuries, such as floods and landslides, poisoned air and water. Unfortunately, cities have mostly neglected and rarely exploited the natural forces within them.

More is known about urban nature today than ever before; over the past two decades, natural scientists have amassed an impressive body of knowledge about nature in the city. Yet little of this information has been applied directly to molding the form of the city – the shape of its buildings and parks, the course of its roads, and the pattern of the whole. A small fraction of that knowledge has been employed in establishing regulations to improve environmental quality, but these have commonly been perceived as restrictive and punitive, rather than as posing opportunities for new urban forms. Regulations have also proven vulnerable to shifts in public policy, at the mercy of the political concerns of the moment, whereas the physical form of the city endures through generation after generation of politicians.

[. . .]

THE GRANITE GARDEN

The city is a granite garden, composed of many smaller gardens, set in a garden world. Parts of the granite garden are cultivated intensively, but the greater part is unrecognized and neglected. To the idle eye, trees and parks are the sole remnants of nature in the city. But nature in the city is far more than trees and gardens, and weeds in sidewalk cracks and vacant lots. It is the air we breathe, the earth we stand on, the water we drink and excrete, and the organisms with which we share our habitat. Nature in the city is the powerful force that can shake the earth and cause it to slide, heave, or crumple. It is a broad flash of exposed rock strata on a hillside, the overgrown outcrops in an abandoned quarry, the millions of organisms cemented in fossiliferous limestone of a downtown building. It is rain and the rushing sound of underground rivers buried in storm sewers. It is water from a faucet, delivered by pipes from some outlying river or reservoir, then used and washed away into the sewer, returned to the waters of river and sea. Nature in the city is an evening breeze, a corkscrew eddy swirling down the face of a building, the sun and the sky. Nature in the city is dogs and cats, rats in the basement, pigeons on the sidewalks, raccoons in culverts, and falcons crouched on skyscrapers. It is the consequence of a complex interaction between the multiple purposes and activities of human beings and other living creatures and of the natural processes that govern the transfer of energy, the movement of air, the erosion of the earth, and the hydrologic cycle. The city is part of nature.

Nature is a continuum, with wilderness at one pole and the city at the other. The same natural processes operate in the wilderness and in the city. Air, however contaminated, is always a mixture of gasses and suspended particles. Paving and building stone are composed of rock, and they affect heat gain and water runoff just as exposed rock surfaces do anywhere. Plants, whether exotic or native, invariably seek a combination of light, water, and air to survive. The city is neither wholly natural nor wholly contrived. It is not "unnatural" but, rather, a transformation of "wild" nature by humankind to serve its own needs, just as agricultural fields are managed for food production and forests for timber. Scarcely a spot on the earth, however remote, is free from the impact of human activity. The human needs and the environmental issues that arise from them are thousands of years old, as old as the oldest city, repeated in every generation, in cities on every continent.

The realization that nature is ubiquitous, a whole that embraces the city, has powerful implications for how the city is built and maintained and for the health, safety, and welfare of every resident. Unfortunately, tradition has set the city against nature, and nature against the city. The belief that the city is an entity apart from nature and even antithetical to it has dominated the way in which the city is perceived and continues to affect how it is built. This attitude has aggravated and even created many of the city's environmental problems: poisoned air and water; depleted or irretrievable resources; more frequent and more destructive floods; in-creased energy demands and higher construction and maintenance costs than existed prior to urbanization; and, in many cities, a pervasive ugliness. Modern urban problems are no different, in essence, from those that plagued ancient cities, except in degree, in the toxicity and persistence of new contaminants, and in the extent of the earth that is now urbanized. As cities grow, these issues have become more pressing. Yet they continue to be treated as isolated phenomena, rather than as related phenomena arising from common human activities, exacerbated by a disregard for the processes of nature. Nature has been seen as a superficial embellishment, as a luxury, rather than as an essential force that permeates the city. Even those who have sought to introduce nature to the city in the form of parks and gardens have frequently viewed the city as something foreign to nature, have seen themselves as bringing a piece of nature to the city.

To seize the opportunities inherent in the city's natural environment, to see beyond short-term costs and benefits, to perceive the consequences of the myriad, seemingly unrelated actions that make up daily city life, and to coordinate thousands of incremental improvements, a fresh attitude to the city and the molding of its form is necessary. The city must be recognized as part of nature and designed accordingly. The city, the suburbs, and the countryside must be viewed as a single, evolving system within nature, as must every individual park and building within that larger whole. The social value of nature must be recognized and its power harnessed, rather than resisted. Nature in the city must be cultivated, like a garden, rather than ignored or subdued.

[. . .]

CITY AND NATURE

The rewards of designing the city in concert with nature apply equally to all cities, old and new, large and small. The investment required to up-grade the infrastructure of older cities will require billions of dollars in North America alone. The opportunities for a fresh approach to resources and waste are enormous, and the potential for costly blunders is equally vast. The challenge facing growing, smaller cities and New Towns is to learn from the mistakes of older cities and to design the city from the outset to exploit the opportunities of the natural environment. This challenge is particularly acute in fast-growing cities where entire new districts are springing up seemingly overnight.

Disregard of natural processes in the city is, always has been, and always will be both costly and dangerous. Many cities have suffered from a failure to take account of nature: Los Angeles and New York suffer poor air quality three days out of four, the result of both urban form and transportation modes; Mexico City has subsided 25 ft because it failed to recognize the relationship between water and ground stability; Los Angeles and Hong Kong are plagued by massive landslides, many of them triggered by urban development; Houston is devastated by floods caused by urbanization upstream, and Harrisburg by construction on the floodplains within the city; Boston and Detroit can no longer afford to maintain their parks and street trees; Niagara Falls is poisoned by its own accumulated wastes. The cost of disregarding nature extends also to quality of life. The newer parts of cities – across continents, climates, and cultures – are everywhere acquiring a boring sameness. The potential of the natural environment to contribute to a distinctive, memorable, and symbolic urban form is unrecognized and forfeited.

More fortunate are those few cities that have adapted ingeniously to nature: Stuttgart, West Germany, which has deployed its parkland to funnel clean, cool air into its congested downtown;

Woodlands, Texas, a New Town whose private and public open spaces function as an effective storm drainage system, soaking up floodwaters and preventing floods downstream; Boston, where wetlands upstream of the city were purchased for flood storage at a fraction of the cost of a new dam; Zurich and Frankfurt, which manage their urban forests for timber production as well as recreation; Philadelphia, which has transformed sewage sludge into a wide range of useful products. These cities have each dealt in a comprehensive way with at least one urban problem.

But comprehensive solutions are not the only means of improving the city. There are ingenious small projects as well: a tiny downtown park that provides a cool, calm retreat in the midst of Manhattan; plazas in Denver that detain storm-water to prevent floods; a project that has transformed the South Platte River into a resource for urban recreation and flood protection; parks in Delft, the Netherlands, that have exploited both the energy and beauty of wild landscapes. Incremental change through small projects is often more manageable, more feasible, less daunting, and more adaptable to local needs and values. When coordinated, incremental changes can have a far-reaching effect. Solutions need not be comprehensive, but the understanding of the problem *must* be.

Although many of the environmental challenges facing cities are more substantial than ever before, the understanding and the tools available to meet them are far more sophisticated. They need only be applied. Nature in the city must be cultivated and integrated with the varied pursuits and purposes of human beings; but first it must be recognized, and its power to shape human enterprises appreciated.

URBAN NATURE AND HUMAN DESIGN

In the natural environment of every city, there are elements of both the distinctive and the common. It is to the distinctive features of their natural envir-onment that many cities owe their location, their historic growth and population distribution, and even the character of their buildings, streets, and parks. Most cities occupy the sites of ancient villages, selected by the original inhabitants for east of defense, access to supplies of water, fuel, and building materials, and proximity to transportation

routes. The site of Washington, DC, for example, was not selected by chance. The falls of the Potomac at Georgetown mark the limit of navigation from the sea and the boundary between two physiographic regions, the piedmont and the coastal plain. These two physiographic regions, each with characteristic topography, building materials, and scenic qualities, bisect the city. The transition from steep hills to flat plains, from narrow rapids to broad rivers, and from rock quarries to clay pits delineates the boundary.

The flat coastal plain of northeastern and southern Washington, easily farmed and built upon, was settled a century before the hilly district of northwest Washington. L'Enfant laid out the capital's formal avenues across the level plain, siting monuments and major buildings on the higher elevations within it. Many of the earliest houses were built of brick from the abundant clay, but the poor drainage of the same clay soil eventually made that land undesirable and relatively inexpensive. Small row houses and large apartment complexes dominate this part of the city today, in contrast to the expensive, detached homes and mansions of northwest Washington. The erosion-resistant, metamorphic rocks of the piedmont give northwest Washington its distinctive character of steep slopes, incised stream valleys, and hilltops with views. Settled later than the coastal plain, it is now an area of large houses and embassies. Many of the houses are built from the rocks of the region – mica, schist, and gneiss.

Washington is not unique; many cities in the eastern United States, from New Jersey to Georgia, straddle the boundary between coastal plain and piedmont. Most of the major East Coast cities between Trenton and Macon and the railroads which connect them are on the "fall line" – Philadelphia, Wilmington, Baltimore, Washington, DC, and Richmond; and the same pattern of urban development recurs again and again. Like Washington, the oldest part of Philadelphia lies on the flat coastal plain where brick row houses are the dominant house type. The piedmont was settled later with larger homes built from the local schist.

Respect for the limitations imposed by nature and exploitation of its resources have led to memorable urban form. The ancient Greeks, for example, were masters at matching the buildings, squares, and streets of the city to its topography.

The urban form of Jerusalem enhances its spiritual significance. The entire city is composed of the local limestone; important monuments are sited atop the ridges and high points of the landscape, their silhouettes against the sky visible from afar. New York City owes the distinctive skyscraper skyline of Manhattan Island to the strength of the underlying bedrock and its proximity to the surface. The Manhattan schist that forms the spine of the island and provides the foundation for its tall buildings outcrops in Central Park. Further south, in midtown Manhattan, at Thirtieth Street, the bedrock plunges hundreds of feet below the ground, then rises up again to within 40 ft of the surface at the island's southern tip. Two clusters of skyscrapers, one in midtown between Thirty-fourth and Sixtieth streets and the other in the financial district near the tip, testify to the proximity of bedrock for foundations.

The resources afforded and the difficulties posed by each city's natural setting comprise a constant that successive generations in that city must address again and again, each in accordance with their own values and technology. Civilizations and governments rise and fall; traditions, values, and policies change; but the natural environment of each city remains an enduring framework within which the human community builds. A city's natural environment and its urban form, taken together, comprise a record of the interaction between natural processes and human purpose over time. Together they contribute to each city's unique identity.

Despite their differences, all cities have transformed their environments in similar fashion: certain urban natural features are as characteristic of ancient Babylon and Rome as they are of modern Boston and Chicago. The human activities that modify the natural environment are common to all cities: the need to provide security, shelter, food, water, and the energy to fuel human enterprises; the need to dispose of wastes, to permit movement within the city and into and out of it; and the ever-escalating demand for more space. The ancient cities of Asia and the Mediterranean and the old cities of Europe transformed nature into a characteristically urban environment many centuries ago. The younger cities of North America are equally urban, but the transition from wilderness to city took place more recently, over the past three centuries. The process continues today as new towns spring up in the countryside throughout the world and as existing cities expand on to adjacent farmland, forest, and desert.

The natural environments of London, Tokyo, and New York – all large cities with a temperate climate – have as much in common as each has with its own rural outskirts. All cities, by virtue of density of people and buildings and the combustion of fuel, alter the character of their original climate and pollute the air. The excavation and filling of the land necessary to secure abundant level ground for building, to find firm building foundations, and to exploit mineral resources transform the original landforms. The profusion of paved streets, sidewalks, and parking lots, and the storm sewers that drain them short-circuit the hydrologic cycle and change the character of streams and lakes. The disposal of wastes contaminates both surface water and groundwater, making the ever-increasing demand for clean water more difficult to satisfy. Fertilizers, herbicides, and pesticides applied to lawns and gardens, along with salt dumped on icy streets, further contaminate groundwater and diminish its value as a resource. Demand for water leads cities to seek resources many miles distant and has thus changed the water balance of entire regions and nations. Native vegetation is cleared and new plants are introduced (both intentionally and inadvertently), with the result that cities throughout the world with similar climates harbor virtually the same plant species.

All these interactions between human activities and the natural environment produce an ecosystem very different from the one that existed prior to the city. It is a system sustained by massive importation of energy and materials, a system in which human cultural processes create a place quite different from undisturbed nature, yet united to it through the common flow of natural processes. As cities grow in size and density, the changes they produce in the air, earth, water, and life within and around them trigger environmental problems that affect the well-being of every city resident.

10

"Towards Sustainable Development"

from *Our Common Future* (1987)

World Commission on Environment and Development

Editors' Introduction

No event did more to push sustainable development into the mainstream of worldwide policy debates than the 1987 release of the report of the World Commission on Environment and Development, commonly known as the Brundtland Commission. Widely distributed as a trade paperback entitled *Our Common Future* (New York: Norton, 1987), this volume formulated what has become the standard definition of sustainable development ("development that meets the needs of the present without jeopardizing the ability of future generations to meet their own needs"). A directive of the United Nations General Assembly established the commission in 1983, in the tradition of several previous influential U.N.-affiliated international commissions, the Palme Commission on Security and Disarmament, and the Brandt Commission on North–South Issues. Chaired by former Norwegian Prime Minister Gro Harlem Brundtland, the commission consisted of leading citizens from 21 nations. The group held public hearings on five continents, reviewed 10,000 pages of testimony, sought advice from numerous experts and advisory panels, and commissioned more than 75 studies and reports.

A rich compendium of analysis and strategies, the Brundtland Commission report succeeded remarkably well in calling global attention to the need for sustainable development and developing a common formulation of this concept. The Commission helped establish a strong foundation for the United Nations Summit on Environment and Development held in Rio de Janeiro in 1992 and many other subsequent events and programs. However, it has been criticized on many grounds as well, particularly for accepting conventional notions of continued economic growth as the path to improved human welfare, for insufficiently incorporating an analysis of global power relations, and for developing a definition of sustainable development that is highly anthropocentric and dependent on the difficult-to-define concept of "needs."

Other leading reports that called attention to the need for sustainable development in general, and sustainable urban development in particular, included the Worldwatch Reports from the Worldwatch Institute (www.worldwatch.org), an influential series of pamphlet-style analyses that began in 1975, the annual *State of the World* books published by the same organization beginning in 1984, the Global 2000 Report to U.S. President Jimmy Carter in 1980, and annual *World Conservation Strategy* reports from the World Conservation Union beginning in 1980. The establishment of national groups such as Canada's National Roundtable on the Environment and Economy emulated the work of the Brundtland Commission on a smaller scale, helping to place sustainability issues on public agendas.

A CALL FOR ACTION

Over the course of this century, the relationship between the human world and the planet that sustains it has undergone a profound change.

When the century began, neither human numbers nor technology had the power radically to alter planetary systems. As the century closes, not only do vastly increased human numbers and their activities have that power, but major, unintended changes are occurring in the atmosphere, in soils, in waters, among plants and animals, and in the relationships among all of these. The rate of change is outstripping the ability of scientific disciplines and our current capabilities to assess and advise. It is frustrating the attempts of political and economic institutions, which evolved in a different, more fragmented world, to adapt and cope. It deeply worries many people who are seeking ways to place those concerns on the political agendas.

The onus lies with no one group of nations. Developing countries face the obvious life-threatening challenges of desertification, deforestation, and pollution, and endure most of the poverty associated with environmental degradation. The entire human family of nations would suffer from the disappearance of rain forests in the tropics, the loss of plant and animal species, and changes in rainfall patterns. Industrial nations face the life-threatening challenges of toxic chemicals, toxic wastes, and acidification. All nations may suffer from the releases by industrialized countries of carbon dioxide and of gases that react with the ozone layer, and from any future war fought with the nuclear arsenals controlled by those nations. All nations will have a role to play in changing trends, and in righting an international economic system that increases rather than decreases inequality, that increases rather than decreases numbers of poor and hungry.

The next few decades are crucial. The time has come to break out of past patterns. Attempts to maintain social and ecological stability through old approaches to development and environmental protection will increase instability. Security must be sought through change. The Commission has noted a number of actions that must be taken to reduce risks to survival and to put future development on paths that are sustainable. Yet we are aware that such a reorientation on a continuing basis is simply beyond the reach of present decisionmaking structures and institutional arrangements, both national and international.

This Commission has been careful to base our recommendations on the realities of present institutions, on what can and must be accomplished today. But to keep options open for future generations, the present generation must begin now, and begin together.

To achieve the needed changes, we believe that an active follow-up of this report is imperative. It is with this in mind that we call for the UN General Assembly, upon due consideration, to transform this report into a UN Programme on Sustainable Development. Special follow-up conferences could be initiated at the regional level. Within an appropriate period after the presentation of this report to the General Assembly, an international conference could be convened to review progress made, and to promote follow-up arrangements that will be needed to set benchmarks and to maintain human progress.

First and foremost, this Commission has been concerned with people – of all countries and all walks of life. And it is to people that we address our report. The changes in human attitudes that we call for depend on a vast campaign of education, debate, and public participation. This campaign must start now if sustainable human progress is to be achieved.

The members of the World Commission on Environment and Development came from 21 very different nations. In our discussions, we disagreed often on details and priorities. But despite our widely differing backgrounds and varying national and international responsibilities, we were able to agree to the lines along which change must be drawn.

We are unanimous in our conviction that the security, well-being, and very survival of the planet depend on such changes, now.

A THREATENED FUTURE

The Earth is one but the world is not. We all depend on one biosphere for sustaining our lives. Yet each community, each country, strives for survival and prosperity with little regard for its impact on others. Some consume the Earth's resources at a rate that would leave little for future generations. Others, many more in number, consume far too little and live with the prospect of hunger, squalor, disease, and early death.

Yet progress has been made. Throughout much of the world, children born today can expect to live longer and be better educated than their parents. In many parts, the new-born can also expect to attain a higher standard of living in a wider sense. Such progress provides hope as we contemplate the improvements still needed, and also as we face our failures to make this Earth a safer and sounder home for us and for those who are to come.

The failures that we need to correct arise both from poverty and from the short-sighted way in which we have often pursued prosperity. Many parts of the world are caught in a vicious downwards spiral: Poor people are forced to overuse environmental resources to survive from day to day, and their impoverishment of their environment further impoverishes them, making their survival ever more difficult and uncertain. The prosperity attained in some parts of the world is often precarious, as it has been secured through farming, forestry, and industrial practices that bring profit and progress only over the short term.

Societies have faced such pressures in the past and, as many desolate ruins remind us, sometimes succumbed to them. But generally these pressures were local. Today the scale of our interventions in nature is increasing and the physical effects of our decisions spill across national frontiers. The growth in economic interaction between nations amplifies the wider consequences of national decisions. Economics and ecology bind us in ever-tightening networks. Today, many regions face risks of irreversible damage to the human environment that threaten the basis for human progress.

These deepening interconnections are the central justification for the establishment of this Commission. We traveled the world for nearly three years, listening. At special public hearings organized by the Commission, we heard from government leaders, scientists, and experts, from citizens' groups concerned about a wide range of environment and development issues, and from thousands of individuals – farmers, shanty-town residents, young people, industrialists, and indigenous and tribal peoples.

We found everywhere deep public concern for the environment, concern that has led not just to protests but often to changed behaviour. The challenge is to ensure that these new values are more adequately reflected in the principles and operations of political and economic structures.

We also found grounds for hope: that people can cooperate to build a future that is more prosperous, more just, and more secure; that a new era of economic growth can be attained, one based on policies that sustain and expand the Earth's resource base; and that the progress that some have known over the last century can be experienced by all in the years ahead. But for this to happen, we must understand better the symptoms of stress that confront us, we must identify the causes, and we must design new approaches to managing environmental resources and to sustaining human development.

SYMPTOMS AND CAUSES

Environmental stress has often been seen as the result of the growing demand on scarce resources and the pollution generated by the rising living standards of the relatively affluent. But poverty itself pollutes the environment, creating environmental stress in a different way. Those who are poor and hungry will often destroy their immediate environment in order to survive: They will cut down forests, their livestock will overgraze grasslands; they will overuse marginal land; and in growing numbers they will crowd into congested cities. The cumulative effect of these changes is so far-reaching as to make poverty itself a major global scourge.

On the other hand, where economic growth has led to improvements in living standards, it has sometimes been achieved in ways that are globally damaging in the longer term. Much of the improvement in the past has been based on the use of increasing amounts of raw materials, energy, chemicals, and synthetics and on the creation of pollution that is not adequately accounted for in figuring the costs of production processes. These trends have had unforeseen effects on the environment. Thus today's environmental challenges arise both from the lack of development and from the unintended consequences of some forms of economic growth. . . .

[. . .]

Sustainable development is development that meets the needs of the present without compromis-ing the ability of future generations to meet their own needs. It contains within it two key concepts:

■ the concept of 'needs', in particular the essential needs of the world's poor, to which overriding priority should be given; and
■ the idea of limitations imposed by the state of technology and social organization on the envir-onment's ability to meet present and future needs.

Thus the goals of economic and social development must be defined in terms of sustainability in all countries – developed or developing, market-oriented or centrally planned. Interpretations will vary, but must share certain general features and must flow from a consensus on the basic concept of sustainable development and on a broad strategic framework for achieving it.

Development involves a progressive transforma-tion of economy and society. A development path that is sustainable in a physical sense could the-oretically be pursued even in a rigid social and political setting. But physical sustainability cannot be secured unless development policies pay attention to such considerations as changes in access to resources and in the distribution of costs and benefits. Even the narrow notion of physical sustainability implies a concern for social equity between generations, a concern that must logically be extended to equity within each generation.

THE CONCEPT OF SUSTAINABLE DEVELOPMENT

The satisfaction of human needs and aspirations is the major objective of development. The essential needs of vast numbers of people in developing countries – for food, clothing, shelter, jobs – are not being met, and beyond their basic needs these people have legitimate aspirations for an improved quality of life. A world in which poverty and ineq-uity are endemic will always be prone to ecological and other crises. Sustainable development requires meeting the basic needs of all and extending to all the opportunity to satisfy their aspirations for a better life.

Living standards that go beyond the basic min-imum are sustainable only if consumption standards everywhere have regard for long-term sustainability. Yet many of us live beyond the world's ecological means, for instance in our patterns of energy use. Perceived needs are socially and culturally deter-mined, and sustainable development requires the promotion of values that encourage consumption standards that are within the bounds of the eco-logically possible and to which all can reasonably aspire.

Meeting essential needs depends in part on achieving full growth potential, and sustainable development clearly requires economic growth in places where such needs are not being met. Elsewhere, it can be consistent with economic growth, provided the content of growth reflects the broad principles of sustainability and non-exploitation of others. But growth by itself is not enough. High levels of productive activity and widespread poverty can coexist, and can endanger the environment. Hence sustainable development requires that societies meet human needs both by increasing productive potential and by ensuring equitable opportunities for all.

An expansion in numbers can increase the pressure on resources and slow the rise in living standards in areas where deprivation is wide-spread. Though the issue is not merely one of population size but of the distribution of resources, sustainable development can only be pursued if demographic developments are in harmony with the changing productive potential of the ecosystem.

A society may in many ways compromise its ability to meet the essential needs of its people in the future – by overexploiting resources, for example. The direction of technological developments may solve some immediate problems but lead to even greater ones. Large sections of the population may be marginalized by ill-considered development.

Settled agriculture, the diversion of water-courses, the extraction of minerals, the emission of heat and noxious gases into the atmosphere, commercial forests, and genetic manipulation are all examples of human intervention in natural systems during the course of development. Until recently, such interventions were small in scale and their impact limited. Today's interventions are more drastic in scale and impact, and more threatening to life-support systems both locally and globally.

This need not happen. At a minimum, sustainable development must not endanger the natural systems that support life on Earth: the atmosphere, the waters, the soils, and the living beings.

Growth has no set limits in terms of population or resource use beyond which lies ecological disaster. Different limits hold for the use of energy, materials, water, and land. Many of these will manifest themselves in the form of rising costs and diminishing returns, rather than in the form of any sudden loss of a resource base. The accumulation of knowledge and the development of technology can enhance the carrying capacity of the resource base. But ultimate limits there are, and sustainability requires that long before these are reached, the world must ensure equitable access to the constrained resource and reorient technological efforts to relieve the pressure.

Economic growth and development obviously involve changes in the physical ecosystem. Every ecosystem everywhere cannot be preserved intact. A forest may be depleted in one part of a watershed and extended elsewhere, which is not a bad thing if the exploitation has been planned and the effects on soil erosion rates, water regimes, and genetic losses have been taken into account. In general, renewable resources like forests and fish stocks need not be depleted provided the rate of use is within the limits of regeneration and natural growth. But most renewable resources are part of a complex and interlinked ecosystem, and maximum sustainable yield must be defined after taking into account system-wide effects of exploitation.

As for nonrenewable resources, like fossil fuels and minerals, their use reduces the stock available for future generations. But this does not mean that such resources should not be used. In general the rate of depletion should take into account the criticality of that resource, the availability of technologies for minimizing depletion, and the likelihood of substitutes being available. Thus land should not be degraded beyond reasonable recovery. With minerals and fossil fuels, the rate of depletion and the emphasis on recycling and economy of use should be calibrated to ensure that the resource does not run out before acceptable substitutes are available. Sustainable development requires that the rate of depletion of nonrenewable resources should foreclose as few future options as possible.

Development tends to simplify ecosystems and to reduce their diversity of species. And species, once extinct, are not renewable. The loss of plant and animal species can greatly limit the options of future generations; so sustainable development requires the conservation of plant and animal species.

So-called free goods like air and water are also resources. The raw materials and energy of production processes are only partly converted to useful products. The rest comes out as wastes. Sustainable development requires that the adverse impacts on the quality of air, water, and other natural elements are minimized so as to sustain the ecosystem's overall integrity.

In essence, sustainable development is a process of change in which the exploitation of resources, the direction of investments, the orientation of technological development, and institutional change are all in harmony and enhance both current and future potential to meet human needs and aspirations.

11

"The End of Nature"

from *The End of Nature* (1989)

Bill McKibben

Editors' Introduction

The threat that humanity would damage the entire global ecosystem in some profound way has been of concern since at least the late 1940s, when the people began to realize the pervasiveness of human impacts on the environment and the ways in which manmade chemicals and pollution spread throughout natural systems. Fairfield Osborn's 1948 book *Our Plundered Planet* was an early call for action; Rachel Carson's *Silent Spring* (1962) sounded a more specific alarm about chemicals such as DDT. In the 1950s social movements arose to protest above-ground nuclear testing, which was spreading radioactive materials into every living thing worldwide. As the United States and the Soviet Union amassed thousands of bombs far larger than those dropped on Hiroshima and Nagasaki, the prospect of nuclear war also became a clear danger to the survival of humanity on the planet. For the first time it became clear to many people that humanity had the power to end life as we know it.

Many of the most toxic industrial chemicals have by now been at least partially controlled by industrialized nations. Above-ground nuclear testing was ended by most nations following the 1963 Limited Test Ban Treaty. The danger of large-scale nuclear war waned after the end of Cold War rivalry between the U.S. and U.S.S.R. in the late 1980s. But another threat that began to be widely known in the late 1950s now seems far more likely to threaten the planet: global warming.

The fact that carbon dioxide can act as a greenhouse gas has been known at least since the late nineteenth century, and, as Bill McKibben explains in the following reading, the fact that oceans will not be able to absorb all the CO_2 produced by humans has been known since the late 1950s. But fossil fuel combustion, which produces CO_2, is so central to industrial societies that the problem has been largely ignored until very recently.

McKibben's *The End of Nature* (1989) was one of the first books to call attention to the global warming threat. It appeared first in serial form in *The New Yorker* magazine and then went on to become a *New York Times* bestseller. In it McKibben describes a range of ways, from burning fossil fuels to deforesting landscapes, that humanity is warming the planet, and paints a picture of a world profoundly and irreversibly altered by human activities. In his view global warming will literally be the "end of nature" as we know it.

Since this book appeared public knowledge of the problem has grown substantially. Release of the film *An Inconvenient Truth* in 2006, featuring former U.S. Vice President Al Gore, helped greatly to increase the sense of urgency. So did the Nobel Peace Prize subsequently awarded to Gore and the International Panel on Climate Change (IPCC). In 2007 the IPCC, a cautious, consensus-based organization of hundreds of scientists, concluded that the earth's temperature had risen 1.3 degrees from pre-industrial levels and that the earth's oceans had risen eight inches. Global warming, in other words, was no longer a threat but a reality.

Many if not most of the sustainable urban development strategies discussed later in this book help address global warming directly or indirectly, by reducing energy consumption in buildings and motor vehicles, limiting wasteful resource consumption and pollution, helping cities adapt to climatic changes, and bringing about landscape management policies that sequester carbon within the wood of forests and the humus of soils.

These days global warming is the most pressing threat driving the sustainability agenda. Many excellent books have been written on the topic. For readers wishing further background we particularly suggest Tim Flannery's *The Weather Makers* (New York: Grove Press, 2005), Elizabeth Kolbert's *Field Notes from a Catastrophe: Man, Nature, and Climate Change* (New York: Bloomsbury USA, 2006), Fred Pearce's *With Speed and Violence*: *Why Scientists Fear Tipping Points in Climate Change* (Boston: Beacon Press, 2007), George Monbiot's *Heat: How to Stop the Planet from Burning* (Cambridge, MA: South End Press, 2007), David Archer and Stefan Rahmstorf's *The Climate Crisis* (Cambridge, UK: Cambridge University Press, 2011), and Bill McKibben's *Earth: Making a Life on a Tough New Planet* (New York: St. Martin's Griffin, 2011). Original materials from the Intergovernmental Panel on Climate Change, including that body's 2001, 2007, and 2013 reports, are available through its website at http://www.ipcc.ch/.

We direct readers wishing further background on this topic to Tim Flannery's 2005 book *The Weather Makers* (New York: Grove Press), Fred Pearce's 2007 volume *With Speed and Violence: Why Scientists fear Tipping Points in Climate Change* (Boston, MA: Beacon Press), and George Monbiot's 2007 book *Heat: How to Stop the Planet from Burning* (Cambridge, MA: South End Press). The book version of *An Inconvenient Truth* (Emmaus, PA: Rodale Press, 2006.) introduces the subject of climate change to a popular audience with many illustrations. Original materials from the Intergovernmental Panel on Climate Change, including that body's 2001, 2007, and 2014 reports, are available through its web site at http://www.ipcc.ch/.

▪ ▪ ▪ ▪ ▪ ▪ ▪

Svante Arrhenius took his doctorate in physics at the University of Uppsala in 1884. His thesis earned him the lowest possible grade short of out-right refusal. Nineteen years later that thesis, which was on the conductivity of solutions, earned him the Nobel Prize. He subsequently accounted for the initial poor reception this way: "I came to my professor, Cleve, whom I admired very much, and I said, 'I have a new theory of electrical conductivity as a cause of chemical reactions.' He said, 'This is very interesting,' and then he said, 'Goodbye.' He explained to me later that he knew very well that there are so many different theories formed, and that they are almost certain to be wrong, for after a short time they disappeared; and therefore, by using the statistical manner of forming his ideas, he concluded that my theory also would not exist long."

Arrhenius's understanding of electrolytic conduction was not his only shrug-provoking new idea. As he surveyed the first few decades of the industrial revolution he realized that man was burning coal at an unprecedented rate, "evaporating our coal mines into the air." Scientists already knew that carbon dioxide, a by-product of fossil fuel combustion, trapped infrared radiation that would otherwise have reflected back out to space.

Jean-Baptiste Joseph Fourier, who developed the theory of heat conduction (and who was also one of the earliest students of Egyptian archeology), had speculated about the effect nearly a century before, and, indeed, had even used the hothouse as a metaphor. But it was Arrhenius who, employing measurements of infrared radiation from the full moon, did the first calculations of the possible effects of man's stepped-up production of carbon dioxide. The average global temperature, he concluded, would rise as much as 9°F if the amount of carbon dioxide in the air doubled from its preindustrial level. That is, heat waves in mid-American latitudes would run into the 110°s, the 120°s, and 130°s; the seas would rise many feet; crops would wither in the fields.

This idea floated in obscurity for a very long time. Now and then a few scientists took it up – the British physicist G.S. Callendar speculated in the 1930s, for instance, that increasing carbon dioxide levels could account for a warming of North America and northern Europe that meteorologists had begun to observe in the 1880s. But that warming seemed to be replaced by a temperature decline around 1940; anyway, most scientists were too busy creating better living through petroleum to be bothered with such long-term speculation.

And those few who did consider the problem concluded that the oceans, which hold much more carbon dioxide than the atmosphere, would soak up any excess that man churned out – that the oceans were an infinite sink down which to pour the problem.

Then, in 1957, two scientists at California's Scripps Institution of Oceanography, Roger Revelle and Hans Suess, published a paper in the journal *Tellus* on this matter of the oceans. What they found was dismaying. No, more than dismaying – what they found may turn out to be the single most important limit in an age of limits, the central awkward fact of a hot and constrained planet.

What they found was that the conventional wisdom was wrong: the upper layer of the oceans, where air and sea meet and transact their business, would absorb very little of the excess carbon dioxide produced by man.

To be precise, what they demonstrated was that "a rather small change in the amount of free carbon dioxide dissolved in seawater corresponds to a relatively large change in the pressure of carbon dioxide at which the oceans and the atmosphere are at equilibrium." To be dramatic, what they showed was that most of the carbon dioxide being pumped into the air by millions of smokestacks, furnaces, and car exhausts would stay in the air, where, presumably, it would gradually warm the planet. "Human beings are now carrying out a large-scale geophysical experiment of a kind that could not have happened in the past, nor be repeated in the future," they wrote. This experiment, they added with the morbid understatement of true scientists, "if adequately documented, may yield a far-reaching insight into the processes determining weather and climate."

While there are other parts to this story – the depletion of the ozone, acid rain, genetic engineering – the story of the end of nature really begins with that greenhouse experiment, with what will happen to the weather.

[. . .]

When we drill into an oilfield we tap into a vast reservoir of organic matter that has been in storage for millennia. We unbury it. When we burn that oil (or coal or natural gas) we release its carbon into the atmosphere in the form of carbon dioxide. This is not pollution in the normal sense of the word. Carbon *monoxide* is "pollution," an unnecessary

by-product. A clean-burning engine releases less of it. But when it comes to carbon dioxide, a clean-burning engine is no better than the motor on a Model T. It will emit about 5.6 lb of carbon in the form of carbon dioxide for every gallon of gasoline it consumes. In the course of about a hundred years, our various engines and fires have released a substantial amount of the carbon that has been buried over time. It is as if someone had scrimped and saved his entire life, and then spent every cent on one fantastic week's debauch. In this, if nothing else, wrote the great biologist A.J. Lotka in the 1960s, "the present is an eminently atypical epoch." We are living on our capital, as we began to realize during the gas crises of the 1970s. But it is more than waste, more than a binge. We are spending that capital in such a way as to alter the atmosphere. It is like taking that week's fling and, in the process, contracting a horrid disease.

There has always been, at least since the start of life, a certain amount of carbon dioxide in the atmosphere, and it has always trapped a certain amount of sunlight to warm the earth. If there were no carbon dioxide, our world might resemble Mars – it would probably be so cold as to be lifeless. A little bit of greenhouse is a good thing, then – the plant that is life thrives in its warmth. The question is: How much? On Venus the atmosphere is 97 percent carbon dioxide. As a result, it traps infrared radiation a hundred times more efficiently than the earth's atmosphere, and keeps the planet a toasty 700°F warmer than the earth. The earth's atmosphere is mostly nitrogen and oxygen; there's currently only about 0.035 percent carbon dioxide, hardly more than a trace. The worries about the greenhouse effect are actually worries about raising that figure from 0.035 percent to 0.055 or 0.06 percent, which is not very much. But plenty, it turns out, to make everything different.

In 1957, when Revelle and Suess wrote their paper, no one even knew for certain that carbon dioxide was increasing. The Scripps Institution hired a young scientist, Charles Keeling, and he set up monitoring stations at the South Pole and 11,150 ft above the Pacific on the side of Mauna Loa in Hawaii. His readings soon confirmed the Revelle–Suess hypothesis: the atmosphere was filling with carbon dioxide. Subsequent readings showed that each year the figure increased, and at a growing rate. Initially the annual increase was about 0.7 parts per million; now it is at least twice that, or 1.5 parts per million. Admittedly, 1.5 parts per million sounds absurdly small. But by drilling holes into glaciers and testing the air trapped in ancient ice, even by looking at the air sealed in old telescopes, scientists have calculated that the atmosphere prior to the Industrial Revolution contained about 280 parts per million carbon dioxide, and, in fact, that that was as high a level as had been recorded in the past 160,000 years. The current reading is near 360 parts per million. At a rate of 1.5 parts per million per year, the pre-Industrial Revolution concentration of carbon dioxide in the atmosphere would be doubled in the next 140 years. And since, as we have seen, carbon dioxide at a very low level helps determine the climate, carbon dioxide at double that very low level, even if it's still small in absolute terms, could have enormous effect. It's like misreading a recipe and baking bread for two hours instead of one: it matters.

But the 1.5 parts per million annual increase is not a given; it seems nearly certain to go higher. The essential facts here are demographic and economic, not chemical. The world's population has more than tripled in this century, and, according to UN statistics released in May of 1989, is expected to double and perhaps nearly triple again before reaching a plateau in the next century. (At the moment, after a decade or two of improving, the trends may be getting worse – China's fertility rate increased from 2.1 to 2.4 children per woman in 1986 and has remained there since.) And the tripled population has not contented itself with using only three times the resources. In the last century industrial production has grown fiftyfold. Four-fifths of that growth has come since 1950, and almost all of it, of course, has been based on fossil fuels. And in the next half-century, the United Nations predicts, this $13 trillion economy will grow another five to ten times larger.

These physical facts are almost as stubborn as the chemistry of infrared absorption. They mean that the world will use more energy – between 2 and 3 percent more each year by most estimates. And the largest increases may come in the use of coal. That is bad news, since coal spews more carbon dioxide into the atmosphere than any other sort of energy (twice as much as natural gas, for instance). China, which has the world's largest coal reserves and recently surpassed the Soviet Union

as the world's largest coal producer, has plans to almost double her coal consumption by the year 2000.

In other words, this is not something that has been happening for a long time. It is not a marathon or the twenty-four hours of Le Mans. It's a hundred-yard dash, a drag race, getting faster all the time. If energy use and other contributions to carbon dioxide levels continue to grow exponentially, a model devised by the World Resources Institute predicts that carbon dioxide levels will have doubled their level prior to the Industrial Revolution by about 2040; if they grow somewhat more slowly, as most estimates predict, the level would double some time around 2070. The leaders of the seven major industrial democracies agreed at their summit in mid-July of 1989 to "strongly advocate common efforts" to limit carbon dioxide, but nothing more, in part because the solutions are neither obvious nor easy. For example, installing some sort of scrubber on a power-plant smokestack to get rid of the carbon dioxide would seem an obvious fix. But a system that removed 90 percent of the carbon dioxide would also reduce the effective capacity of the plant 80 percent. One oft-heard suggestion is to use more nuclear power. But because so much of our energy use is for things like automobile fuel, even if we mustered the political will and economic resources to quickly replace every single electric generating station with a nuclear plant, our total carbon dioxide output would fall little more than a quarter. Ditto, at least initially, for cold fusion or hot fusion or any other clean method of producing electricity. So the sacrifices demanded may be on a scale we can't image and won't like.

[. . .]

Burning fossil fuels is not the only method human beings have devised to increase the level of atmospheric carbon dioxide. Burning down a forest also sends clouds of carbon into the air. Trees and scrubby forests still cover 40 percent of the land on earth, but this area has shrunk by about a third since preagricultural times, and that shrinkage, it almost goes without saying, is accelerating. In the Brazilian state of Pará, for instance, 180,000 km^2 was deforested between 1975 and 1986; in the hundred years preceding that decade, settlers had hacked away about 18,000 km^2. "At night, roaring and red, the forest looks to be at war," one correspondent wrote. The Brazilian government has tried to slow down the burning, but it employs only 900 forest wardens for an area larger than Europe.

This is not news – it's well known that the rain forests are disappearing, and are taking with them most of the world's plant and animal species. But forget for a moment that we are losing a unique resource, a cradle of life, irreplaceable grandeur, and so forth. The dense, layered rain forest contains three to five times more carbon per acre than an open, dry forest: an acre of Brazil up in flames equals three to five acres of Yellowstone. Deforestation currently adds between 1 billion and 2.5 billion tons of carbon to the atmosphere annually, 20 percent or more of the amount produced by fossil fuel burning. And that acre of rain forest, which has poor soil and can support crops for only a few years, soon turns to desert, or, at least, to pastureland.

And where there's pasture there are cows, and what cows support in their intestines are huge numbers of anaerobic bacteria, which break down the cellulose that cows chew. That is why cows, unlike people, can eat grass. Why does this matter? Because those bugs that digest the cellulose excrete methane. Methane, or natural gas, gives off carbon dioxide when burned, though only half as much as oil. When it escapes into the atmosphere without being burned, though, it is twenty times more efficient than carbon dioxide at trapping solar radiation and warming the planet. So, even though it makes up less than two parts per million of the atmosphere, it can have a significant effect. Even though much of the methane in the atmosphere comes from seemingly "natural" sources – the methanogenic bacteria – their present huge numbers are clearly man-made. Mankind owns 1.2 billion head of cattle, not to mention a large number of camels, horses, pigs, sheep, and goats, and together they belch about 73 million metric tons of methane into the air each year, a 435 percent increase in the last century. The buffalo and wildebeest they displaced belched as well, but their numbers were not as great.

We have raised the number of termites, too, and still more dramatically. Termites have the same bacteria in their intestines as cows; that is why they can digest wood. We tend to think of termites as house wreckers, but in most of the world they are house builders, erecting elaborate, rock-hard mounds twenty or thirty feet high. Inside these fortresses an elaborate hierarchy of termites guards the queens – some of the termites sport sharp pincers longer than their bodies; others have heads shaped like drain plugs, so they can block up the interior passages against intruders; still others explode when attacked, or squirt poison. If a bulldozer razes a mound, worker termites can rebuild it in hours. They are like most animals in that their numbers are limited only by the supply of food. And when we hack down a rain forest, all of a sudden there's dead wood everywhere – food galore. Termites have a "high digestion efficiency," much higher than earthworms, says Patrick Zimmerman of the National Center for Atmospheric Research. They can break down 65–95 percent of the carbon in the wood they ingest. (Wood is 50 percent carbon.) And they can excrete phenomenal amounts of methane – a single mound might give off five liters a minute. As the deforestation has proceeded, termite numbers have boomed. There is now, some scientists estimate, a half-ton of termites for every man, woman, and child on earth – that is, six or seven people's worth of termites for every actual person.

Researchers differ on the importance of termites as a methane source, but everyone agrees about the rice paddies. The oxygenless mud of marsh bottoms has always sheltered the methane-producing bacteria. (Methane is sometimes known as swamp gas.) But rice paddies may be even more efficient – the rice plants act almost like straws, venting as much as 115 million tons of the gas a year. Rice paddies, of course, increase in number and size every year, so that those 2.4 children per Chinese woman will have enough to eat.

And then there are landfills: 30 percent of a typical landfill is "putrescible," Zimmerman says – it rots, creating methane. At the main New York City landfill on Staten Island, gas is pumped from under the trash straight to thousands of homes, but in most places it just seeps out.

What's more, scientists have recently begun to think that these sources alone can't account for all the methane. "If you look more carefully," the Harvard physicist Michael McElroy says, "you do not come away with an awfully comfortable feeling." For one thing, scientists have begun to be able to measure isotopic concentrations: there is a "light" methane from cattle and termites and rice paddies, but also a "heavy" methane from someplace else. And here – forget the poison-squirting termites – is where the story begins to get a little scary. An enormous amount of methane is locked up as hydrates in the tundra and in the mud of the continental shelves. These are, in essence, methane ices; the ocean muds alone may hold 10 trillion tons of methane. If the greenhouse effect is beginning to warm the oceans, if it is starting to thaw the permafrost, then some scientists say that eventually those ices could start to melt. Some estimates of the potential methane release run as high as 0.6 billion tons a year, an amount that could more than double the present atmospheric concentration. This would be a nasty example of a feedback loop, with the altered atmosphere causing further alterations: Warm the atmosphere and release methane; release methane and warm the atmosphere; and so on.

When all sources are combined, we've done an even more dramatic job of increasing methane than of increasing carbon dioxide. Samples of ice from Antarctic glaciers show that the concentration of methane in the atmosphere has fluctuated between 0.3 and 0.7 parts per million for the last 160,000 years, reaching its highest levels during the earth's warmest period. In 1987 methane composed 1.7 parts per million of the atmosphere. That is, there is now two and a half times as much methane in the atmosphere as there was at any time through three glacial and interglacial periods. And concentrations are rising at a regular one percent a year.

Man is also pumping smaller quantities of many other greenhouse gases into the atmosphere. Nitrous oxide, chlorine compounds, and some others all trap warmth even more efficiently than carbon dioxide. Scientists now believe that methane and the rest of these gases, though their concentrations are small, will together account for 50 percent of the projected greenhouse warming – that is, taken together they are as much of a problem as carbon dioxide. And as all these compounds warm the atmosphere, it will be able to hold more water vapor, itself a potent greenhouse gas. The British Meteorological Office calculates that this extra water vapor will warm the earth two-thirds as much as the carbon dioxide alone.

[. . .]

So – we have increased the amount of carbon dioxide in the air by about 25 percent in the last century, and will almost certainly double it in the next; we have more than doubled the level of methane; we have added a soup of other gases. *We have substantially altered the earth's atmosphere.*

This is not like local pollution, not like smog over Los Angeles. This is the earth's entire atmosphere. If you'd climbed some remote mountain in 1960 and sealed up a bottle of air at its peak, and did the same thing this year, the two samples would be substantially different. Their basic chemistry would have changed. Most discussions of the greenhouse gases rush immediately to their future consequences – is the sea going to rise? – without pausing to let the simple fact of what has already happened sink in. The air around us, even where it is clean, and smells like spring, and is filled with birds, is *different*, significantly changed.

That said, the question of what this new atmosphere means must arise. If it means nothing, we'd soon forget about it, since the air would be as colorless and odorless as before and as easy to breathe. And, indeed, the direct effects *are* unnoticeable. Anyone who lives indoors breathes carbon dioxide at a level several times the atmospheric concentration without ill effects; the federal government limits industrial workers to a chronic exposure of 5,000 parts per million, or almost fifteen times the current atmospheric levels; a hundred years from now a child at recess will still breathe far less carbon dioxide than a child in a classroom.

This, however, is only mildly good news. The effects on us will be slightly less direct, but nevertheless drastic: changes in the atmosphere will change the weather, and *that* will change recess. The temperature, the rainfall, the speed of the wind will change. The chemistry of the upper atmosphere may seem an abstraction, a text written in a foreign language. But its translation into the weather of New York and Cincinnati and San Francisco will alter the lives of all of us.

The theories about the effects all begin with an estimate of expected warming. Arrhenius, you recall, said that doubling the preindustrial concentration of carbon dioxide would raise temperatures 9°F. The new wave of concern that began with Revelle and Suess's article and Keeling's Mauna Loa data has led to the development of vastly complex computer models of the entire globe. In those models the globe is divided into thousands of boxes, and each box is divided vertically into a large number of layers, usually ten or more, representing the various layers of the atmosphere and then of the land or ocean. The computer program, a sort of meteorological spreadsheet, first solves for each box the fundamental conservation laws of physics, and then goes on to calculate the transfer of mass, energy, and momentum from one box to the next – it "runs" the weather far into the future. You can change a variable – the amount of carbon dioxide in the air, for instance – and watch the result.

And the result, when increased carbon dioxide and other trace gases are taken as givens, does not differ all that much from what Arrhenius forecast. The models that have been constructed agree that when, as has been predicted, the level of carbon dioxide or its equivalent in other greenhouse gases doubles from pre-Industrial Revolution concentrations, the global average temperature will increase, and that the increase will be 1.5°C to 4.5°C, or 3°F to 8°F. The results of all the global climatic models are consistent within a factor of two.

Perhaps the most famous of the computer programs is in the hands of James Hansen and his colleagues at NASA's Goddard Institute for Space Studies, in, of all places, Manhattan. NASA used an early version of the model around 1970 to study the accuracy of predictions from satellite weather observations; when the Goddard weather group moved to Washington, Hansen, who was staying on in New York, decided he'd try the model on longer-range problems – on climate as opposed to weather. Over the years, he and his colleagues have fine-tuned the program, and even though it remains a rough simulation of the mightily complex real world, they have improved it to the point where they are willing to forecast not just the effects of a doubling of carbon dioxide but the incremental effects along the way – that is, the forecast not just for 2050, but for 2000.

In Dallas, for instance, a doubled level of carbon dioxide, or the equivalent combination of carbon dioxide and other gases like methane, would increase the number of days a year with temperatures above 100°F from nineteen to seventy-eight, according to Hansen's calculations. On sixty-eight days, as opposed to the current four, the

temperature wouldn't fall below 80°F at night. One hundred and sixty-two days a year – half the year, essentially – the temperature would top 90°F. New York City would have forty-eight days a year above the 90°F mark, up from fifteen at present. And so on. Such increases would quite clearly change the world as we know it. One of Hansen's colleagues observed to reporters, "It reaches one hundred twenty degrees in Phoenix now. Will people still live there if it's one hundred thirty degrees? One hundred forty?" (Heat waves like that are possible, even if the average global increase, figured over a year, is only a couple of degrees, for any average conceals huge swings.) But we need not wait decades for that doubling to occur. These changes, Hansen and his colleagues wrote in a paper published in the fall of 1988 in the *International Journal of Geophysics*, should begin to be obvious to the man in the street by the early 1990s at the latest – that is, the odds of a very hot summer will, thanks to the greenhouse effect, be better than even beginning now.

There are an infinite number of possible effects of such a temperature change. For example, the seas may well rise seven feet or more as polar ice melts and warmer water expands, while the interiors of the continents may dry up because of increased evaporation. Detailed studies have begun to emerge of what it will be like to live in this greenhouse world, and researchers have speculated on possible changes ranging from an increased spread of disease when insects spread north to the emergence of a warmer Canada as the globe's great power. Some of the figuring is quite insane – *Fortune* recently pointed out that if parts of the polar ice cap began to thaw, American and Soviet nuclear submarines would be deprived of cover. Not only that: "The effect would be more damaging to the USSR. Because American submarines are faster and can travel farther than their Soviet counterparts, they are less dependent on hiding places under the ice cap."

But a discussion of the effects is premature. First we should figure out if this is indeed going to happen, if the theory is valid. In recent years there have been, of course, any number of doom-laden prophecies that haven't come true – oil is selling, as I write these words, at $18 the barrel, half its price just a few years ago.

The obvious check is to measure the temperature to see if it's going up. But this is easier said than done, for, in the first place, the warming doesn't show up immediately. Although, as Revelle and Suess found, the oceans don't absorb much excess carbon dioxide, they can hold a lot of heat; the effects so far may be stored up in the seas, ready to re-radiate out to the atmosphere, the way a rock holds the sun's heat through the night. Such a "thermal lag" could be as little as ten years, as much as a hundred.

And when you go to check the thermometers it won't do to measure only a few places for only a few years because climate is "noisy" – full of random fluctuations and variability. To find what climatologists call the "warming signal" through this static of naturally cold and hot years requires a huge effort. Two such studies have been done, one at the University of East Anglia, the other by Hansen and his NASA colleagues. Each reached back almost a century, when scientists first began systematic weather observations. And, to find truly global averages, they included readings from thousands of land-based and shipboard monitoring stations. The two studies reached about the same conclusion: that the earth's temperature had increased about 1°F in the past hundred years, a number consistent with, though somewhat smaller than, the predictions of the greenhouse models. And both sets of readings show that the four warmest years on record occurred in the 1980s – that the rise is accelerating at the same time that more gases enter the air, just as the models forecast. Indeed, the British model now lists the six warmest years on record as, in order, 1988, 1987, 1983, 1981, 1980, and 1986.

[This classic piece was published in 1989. As of 2014, the five warmest years on record were 2010, 2005, 2007, 2002, and 1998, according to NASA. – *Eds.*]

12

"The Rio Declaration on Environment and Development" (1992)

Introduction to Chapter 7

from *Agenda 21* (United Nations Conference on Environment and Development) (1992)

"Millennium Development Goals" and "Millennium Declaration" (2002)

United Nations

Editors' Introduction

The 1992 United Nations Conference on Environment and Development held in Rio de Janeiro – the "Earth Summit" – was, like the Brundtland Commission, a watershed event that helped move the need for sustainable development in front of global media and the public. This conference, attended by leaders of 178 nations, produced a lengthy declaration known as *Agenda 21* that laid out sustainable development principles in many different areas. The Rio Declaration on Environment and Development below summarizes some of its major principles. The Earth Summit also led to the creation of a United Nations Commission on Sustainable Development which meets annually to review international implementation efforts, and a U.N. Division for Sustainable Development to coordinate the agency's work in this field. More information on the latter is available at www. un.org/esa/sustdev.

Chapter 7 of *Agenda 21* laid out directions for sustainable urban development and, at the urging of nongovernmental organizations (NGOs) such as the International Council on Local Environmental Initiatives (ICLEI), Chapter 28 on implementation stipulates that "by 1996, most local authorities in each country should have undertaken a consultative process with their population and achieved a consensus on a local Agenda 21 for the(ir) communities." This mandate for "Local Agenda 21" planning stimulated a large number of local planning initiatives, especially in the United Kingdom, northern Europe, and a number of developing nations.

The Earth Summit was followed in the early and mid 1990s by other United Nations conferences on global population, social development, women, and urban development. The 1996 Habitat II "City Summit," held in Istanbul, produced a lengthy consensus document on urban development principles, and perhaps just as importantly featured a huge exhibition of global best practices in sustainable city building. The 2002 World Summit on Sustainable Development, held in Johannesburg, South Africa, sought to review global progress in the 10 years following the Rio conference. Largely ignored by the U.S. administration of George W. Bush, which successfully persuaded delegates to avoid specific timetables for change, this event nonetheless produced some additional financial commitments to sustainable development programs from a number of European nations and the European Union. The 2012 Rio+20 United Nations Conference on Sustainable Development – a potential opportunity for renewed global commitment to sustainability – was not attended by US President Barack Obama, British Prime Minister David Cameron, or German Chancellor Angela Merkel, and was widely viewed as a disappointment.

Despite optimistic rhetoric from some agencies, international declarations such as *Agenda 21* and the Millennium Development Goals have often been downplayed or ignored by national governments. It is debatable to what extent the U.N. treaties, conferences, and declarations represent an important global process of education, consensus building, and action, and to what extent they are ineffectual gestures or window dressing for the status quo. Some might argue that the global trade institutions set up during the 1990s, such as the World Trade Organization, represent a second, more powerful global governance mechanism entirely bypassing the United Nations – one promoting the economic goals of powerful nations at the expense of sustainable development for all countries. However, it is clear that the Earth Summit and other U.N.-related events have indeed played at least some role in promoting global debates about sustainability and, even if ignored by many nations in the short term, have helped lay the groundwork for potential international action in the long run.

In the late 2000s and early 2010s, right-wing organizations such as the Tea Party movement in the United States sought to spread the belief that Agenda 21 is part of a global communist conspiracy to undermine American values. An example of this rhetoric is available at http://www.teaparty911.com/issues/what_is_agenda_21.htm, and a critique is at http://www.motherjones.com/politics/2010/11/tea-party-agenda-21-un-sustainable-development. We will leave it to our readers to decide whether or not the text below represents a radical threat to individual liberty.

In a further attempt to promote sustainable development globally – especially focusing on poverty, health, and education in developing nations – 189 United Nations member states and more than 23 international organizations voted to adopt a set of eight Millennium Development Goals on September 8, 2000. Countries promised to work together to meet these objectives by the year 2015. The goals and the related declaration are listed below. More detailed information on targets and actions is available at http://www.un.org/millenniumgoals/.

The good news is that as of 2014 the world has made substantial progress towards some of these goals. The proportion of people living in extreme poverty had been halved, more than 2 billion people had gained access to improved sources of drinking water, and hunger and mortality from diseases such as malaria and tuberculosis had been reduced. The bad news is that progress has been slow on other goals such as environmental sustainability, education, improved status for women, and AIDS prevention and treatment. Also, wealthy countries have fallen short of their promised financial contributions to less-developed countries, and many of the world's people even within relatively educated societies are unaware that the Millennium Development Goals even exist. Main questions for the future include the extent to which international action should take the place of initiatives at other levels of government, and if the former is desired, how greater political attention and funding can be leveraged for collective action.

THE RIO DECLARATION ON ENVIRONMENT AND DEVELOPMENT[1]

The United Nations Conference on Environment and Development, having met at Rio de Janeiro from 3 to 14 June 1992, reaffirming the Declaration of the United Nations Conference on the Human Environment, adopted at Stockholm on 16 June 1972,[2] and seeking to build upon it, with the goal of establishing a new and equitable global partnership through the creation of new levels of cooperation among States, key sectors of societies and people, working towards international agreements which respect the interests of all and protect the integrity of the global environmental and developmental system, recognizing the integral and interdependent nature of the Earth, our home, proclaims that:

Principle 1

Human beings are at the center of concerns for sustainable development. They are entitled to a healthy and productive life in harmony with nature.

Principle 2

States have, in accordance with the Charter of the United Nations and the principles of international law, the sovereign right to exploit their own resources pursuant to their own environmental and developmental policies, and the responsibility to ensure that activities within their jurisdiction or control do not cause damage to the environment of other States or of areas beyond the limits of national jurisdiction.

Principle 3

The right to development must be fulfilled so as to equitably meet developmental and environmental needs of present and future generations.

Principle 4

In order to achieve sustainable development, environmental protection shall constitute an integral part of the development process and cannot be considered in isolation from it.

Principle 5

All States and all people shall cooperate in the essential task of eradicating poverty as an indispensable requirement for sustainable development, in order to decrease the disparities in standards of living and better meet the needs of the majority of the people of the world.

Principle 6

The special situation and needs of developing countries, particularly the least developed and those most environmentally vulnerable, shall be given special priority. International actions in the field of environment and development should also address the interests and needs of all countries.

Principle 7

States shall cooperate in a spirit of global partnership to conserve, protect and restore the health and integrity of the Earth's ecosystem. In view of the different contributions to global environmental degradation, States have common but differentiated responsibilities. The developed countries acknowledge the responsibility that they bear in the international pursuit of sustainable development in view of the pressures their societies place on the global environment and of the technologies and financial resources they command.

Principle 8

To achieve sustainable development and a higher quality of life for all people, States should reduce and eliminate unsustainable patterns of production and consumption and promote appropriate demographic policies.

Principle 9

States should cooperate to strengthen endogenous capacity-building for sustainable development by improving scientific understanding through exchanges of scientific and technological knowledge, and by enhancing the development, adaptation, diffusion and transfer of technologies, including new and innovative technologies.

Principle 10

Environmental issues are best handled with the participation of all concerned citizens, at the relevant level. At the national level, each individual shall have appropriate access to information concerning the environment that is held by public authorities, including information on hazardous materials and activities in their communities, and the opportunity to participate in decision-making processes. States shall facilitate and encourage public awareness and participation by making information widely available. Effective access to judicial and administrative proceedings, including redress and remedy, shall be provided.

Principle 11

States shall enact effective environmental legislation. Environmental standards, management objectives and priorities should reflect the environmental and developmental context to which they apply. Standards applied by some countries may be inappropriate and of unwarranted economic and social cost to other countries, in particular developing countries.

Principle 12

States should cooperate to promote a supportive and open international economic system that would lead to economic growth and sustainable development in all countries, to better address the problems of environmental degradation. Trade policy measures for environmental purposes should not constitute a means of arbitrary or unjustifiable discrimination or a disguised restriction on international trade. Unilateral actions to deal with environmental challenges outside the jurisdiction of the importing country should be avoided. Environmental measures addressing transboundary or global environmental problems should, as far as possible, be based on an international consensus.

Principle 13

States shall develop national law regarding liability and compensation for the victims of pollution and other environmental damage. States shall also cooperate in an expeditious and more determined manner to develop further international law regarding liability and compensation for adverse effects of environmental damage caused by activities within their jurisdiction or control to areas beyond their jurisdiction.

Principle 14

States should effectively cooperate to discourage or prevent the relocation and transfer to other States of any activities and substances that cause severe environmental degradation or are found to be harmful to human health.

Principle 15

In order to protect the environment, the precautionary approach shall be widely applied by States according to their capabilities. Where there are threats of serious or irreversible damage, lack of full scientific certainty shall not be used as a reason for postponing cost-effective measures to prevent environmental degradation.

Principle 16

National authorities should endeavor to promote the internalization of environmental costs and the use of economic instruments, taking into account the approach that the polluter should, in principle, bear the cost of pollution, with due regard to the public interest and without distorting international trade and investment.

Principle 17

Environmental impact assessment, as a national instrument, shall be undertaken for proposed activities that are likely to have a significant adverse impact on the environment and are subject to a decision of a competent national authority.

Principle 18

States shall immediately notify other States of any natural disasters or other emergencies that are likely to produce sudden harmful effects on the environment of those States. Every effort shall be made by the international community to help States so afflicted.

Principle 19

States shall provide prior and timely notification and relevant information to potentially affected States on activities that may have a significant adverse transboundary environmental effect and shall consult with those States at an early stage and in good faith.

Principle 20

Women have a vital role in environmental management and development. Their full participation is therefore essential to achieve sustainable development.

Principle 21

The creativity, ideals and courage of the youth of the world should be mobilized to forge a global partnership in order to achieve sustainable development and ensure a better future for all.

Principle 22

Indigenous people and their communities and other local communities have a vital role in environmental management and development because of their knowledge and traditional practices. States should recognize and duly support their identity, culture and interests and enable their effective participation in the achievement of sustainable development.

Principle 23

The environment and natural resources of people under oppression, domination and occupation shall be protected.

Principle 24

Warfare is inherently destructive of sustainable development. States shall therefore respect international law providing protection for the environment in times of armed conflict and cooperate in its further development, as necessary.

Principle 25

Peace, development and environmental protection are interdependent and indivisible.

Principle 26

States shall resolve all their environmental disputes peacefully and by appropriate means in accordance with the Charter of the United Nations.

Principle 27

States and people shall cooperate in good faith and in a spirit of partnership in the fulfillment of the principles embodied in this Declaration and in the further development of international law in the field of sustainable development.

CHAPTER 7 OF *AGENDA 21*[2] (EXCERPT FROM INTRODUCTION)

Promoting sustainable human settlement development

7.1. In industrialized countries, the consumption patterns of cities are severely stressing the global ecosystem, while settlements in the developing world need more raw material, energy, and economic development simply to overcome basic economic and social problems. Human settlement conditions in many parts of the world, particularly the developing countries, are deteriorating mainly as a result of the low levels of investment in the sector attributable to the overall resource constraints in these countries. In the low-income countries for which recent data are available, an average of only 5.6 per cent of central government expenditure went to housing, amenities, social security and welfare. Expenditure by international support and finance organizations is equally low. For example, only 1 per cent of the United Nations system's total grant-financed expenditures in 1988 went to human settlements, while in 1991, loans from the World Bank and the International Development Association (IDA) for urban development and water supply and sewerage amounted to 5.5 and 5.4 per cent, respectively, of their total lending.

7.2. On the other hand, available information indicates that technical cooperation activities in the human settlement sector generate considerable public and private sector investment. For example, every dollar of UNDP technical cooperation expenditure on human settlements in 1988 generated a follow-up investment of $122, the highest of all UNDP sectors of assistance.

7.3. This is the foundation of the "enabling approach" advocated for the human settlement sector. External assistance will help to generate the internal resources needed to improve the living and working environments of all people by the year 2000 and beyond, including the growing number of unemployed – the no-income group. At the same time the environmental implications of urban development should be recognized and addressed in an integrated fashion by all countries, with high priority being given to the needs of the urban and rural poor, the unemployed and the growing number of people without any source of income.

Human settlement objective

7.4. The overall human settlement objective is to improve the social, economic and environmental quality of human settlements and the living and working environments of all people, in particular the urban and rural poor. Such improvement should be based on technical cooperation activities, partnerships among the public, private and community sectors and participation in the decision-making process by community groups and special interest groups such as women, indigenous people, the elderly and the disabled. These approaches should form the core principles of national settlement strategies. In developing these strategies, countries will need to set priorities among the eight programme areas in this chapter in accordance with their national plans and objectives, taking fully into account their social and cultural capabilities. Furthermore, countries should make appropriate

provision to monitor the impact of their strategies on marginalized and disenfranchised groups, with particular reference to the needs of women.

7.5. The programme areas included in this chapter are:

(a) Providing adequate shelter for all.
(b) Improving human settlement management.
(c) Promoting sustainable land-use planning and management.
(d) Promoting the integrated provision of environmental infrastructure: water, sanitation, drainage and solid-waste management.
(e) Promoting sustainable energy and transport systems in human settlements.
(f) Promoting human settlement planning and management in disaster-prone areas.
(g) Promoting sustainable construction industry activities.
(h) Promoting human resource development and capacity-building for human settlement development.

UNITED NATIONS MILLENNIUM DEVELOPMENT GOALS

Goal 1. Eradicate extreme poverty and hunger
Goal 2. Achieve universal primary education
Goal 3. Promote gender equality and empower women
Goal 4. Reduce child mortality
Goal 5. Improve maternal health
Goal 6. Combat HIV/AIDS, malaria, and other diseases
Goal 7. Ensure environmental sustainability
Goal 8. Develop a global partnership for development

UNITED NATIONS MILLENNIUM DECLARATION

The General Assembly
    Adopts the following Declaration:

I. Values and principles

1. We, heads of State and Government, have gathered at United Nations Headquarters in New

York from 6 to 8 September 2000, at the dawn of a new millennium, to reaffirm our faith in the Organization and its Charter as indispensable foundations of a more peaceful, prosperous and just world.

2. We recognize that, in addition to our separate responsibilities to our individual societies, we have a collective responsibility to uphold the principles of human dignity, equality and equity at the global level. As leaders we have a duty therefore to all the world's people, especially the most vulnerable and, in particular, the children of the world, to whom the future belongs.

3. We reaffirm our commitment to the purposes and principles of the Charter of the United Nations, which have proved timeless and universal. Indeed, their relevance and capacity to inspire have increased, as nations and peoples have become increasingly interconnected and interdependent.

4. We are determined to establish a just and lasting peace all over the world in accordance with the purposes and principles of the Charter. We rededicate ourselves to support all efforts to uphold the sovereign equality of all States, respect for their territorial integrity and political independence, resolution of disputes by peaceful means and in conformity with the principles of justice and international law, the right to self-determination of peoples which remain under colonial domination and foreign occupation, non-interference in the internal affairs of States, respect for human rights and fundamental freedoms, respect for the equal rights of all without distinction as to race, sex, language or religion and international cooperation in solving international problems of an economic, social, cultural or humanitarian character.

5. We believe that the central challenge we face today is to ensure that globalization becomes a positive force for all the world's people. For while globalization offers great opportunities, at present its benefits are very unevenly shared, while its costs are unevenly distributed. We recognize that developing countries and countries with economies in transition face special difficulties in responding to this central challenge. Thus, only through broad and sustained efforts to create a shared future, based upon our common humanity in all its diversity, can globalization be made fully inclusive and equitable. These efforts must include policies and measures, at the global level, which correspond to the needs of developing countries and economies in transition and are formulated and implemented with their effective participation.

6. We consider certain fundamental values to be essential to international relations in the twenty-first century. These include:

- *Freedom*. Men and women have the right to live their lives and raise their children in dignity, free from hunger and from the fear of violence, oppression or injustice. Democratic and participatory governance based on the will of the people best assures these rights.
- *Equality*. No individual and no nation must be denied the opportunity to benefit from development. The equal rights and opportunities of women and men must be assured.
- *Solidarity*. Global challenges must be managed in a way that distributes the costs and burdens fairly in accordance with basic principles of equity and social justice. Those who suffer or who benefit least deserve help from those who benefit most.
- *Tolerance*. Human beings must respect one other, in all their diversity of belief, culture and language. Differences within and between societies should be neither feared nor repressed, but cherished as a precious asset of humanity. A culture of peace and dialogue among all civilizations should be actively promoted.
- *Respect for nature*. Prudence must be shown in the management of all living species and natural resources, in accordance with the precepts of sustainable development. Only in this way can the immeasurable riches provided to us by nature be preserved and passed on to our descendants. The current unsustainable patterns of production and consumption must be changed in the interest of our future welfare and that of our descendants.
- *Shared responsibility*. Responsibility for managing worldwide economic and social development, as well as threats to international peace and security, must be shared among the nations of the world and should be exercised multilaterally. As the most universal and most representative

organization in the world, the United Nations must play the central role.

7. In order to translate these shared values into actions, we have identified key objectives to which we assign special significance.

[. . .]

19. We resolve further:

- To halve, by the year 2015, the proportion of the world's people whose income is less than one dollar a day and the proportion of people who suffer from hunger and, by the same date, to halve the proportion of people who are unable to reach or to afford safe drinking water.
- To ensure that, by the same date, children everywhere, boys and girls alike, will be able to complete a full course of primary schooling and that girls and boys will have equal access to all levels of education.
- By the same date, to have reduced maternal mortality by three quarters, and under-five child mortality by two thirds, of their current rates.
- To have, by then, halted, and begun to reverse, the spread of HIV/AIDS, the scourge of malaria and other major diseases that afflict humanity.
- To provide special assistance to children orphaned by HIV/AIDS.
- By 2020, to have achieved a significant improvement in the lives of at least 100 million slum dwellers as proposed in the "Cities without Slums" initiative.

20. We also resolve:

- To promote gender equality and the empowerment of women as effective ways to combat poverty, hunger and disease and to stimulate development that is truly sustainable.
- To develop and implement strategies that give young people everywhere a real chance to find decent and productive work.
- To encourage the pharmaceutical industry to make essential drugs more widely available and affordable by all who need them in developing countries.
- To develop strong partnerships with the private sector and with civil society organizations in pursuit of development and poverty eradication.

[. . .]

Eighth plenary meeting
8 September 2000

NOTES

1 New York: United Nations. 1992. (Report of the United Nations Conference on Environment and Development, Rio de Janeiro, 3–14 June 1992, Annex I; adopted by more than 178 governments.) Available at http://www.un.org/esa/sustdev/agenda21text.htm.

2 Report of the United Nations Conference on the Human Environment, Stockholm, 5–16 June 1972 (United Nations publication, Sales No. E.73.II.A.14 and corrigendum), chapter I.

PART TWO

Dimensions of sustainable urban development

INTRODUCTION TO PART TWO

After this historical overview, we now turn to specific topic areas related to sustainable urban development. Land use, urban design, transportation, environmental planning, resource use, environmental justice, local economic development, green building practices, food systems, and public health – these are among the dimensions of urban development in which thoughtful action can make a difference for sustainability. The following selections address these and other topics in turn. The editors' introduction to each selection suggests further directions for readers interested in exploring each topic at greater length.

Several points deserve emphasis as we enter Part Two.

1   **People and organizations conceptualize "sustainability" in different ways.** As is probably clear, based on Part One, different individuals and groups have very different perspectives on sustainable development. Some, following Aldo Leopold, might take a strongly "ecocentric" approach, viewing other species and nature as a whole as having intrinsic rights. These individuals might then want to sharply limit the growth of cities and towns and scale back the overall human footprint on the planet. Others affiliated with more traditional conservation organizations or bodies such as the Brundtland Commission see nature as a resource for humans, and in more anthropocentric fashion believe that humans should take stewardship over the natural world so as to sustain resources for future human generations. For this camp, well-managed urban expansion, logging, and mining would be acceptable.

Observers in the developing world such as Frank often prioritize dealing with human dimensions of sustainability such as global poverty and inequality, and see overconsumption in the developed world as a leading problem. Meanwhile, more science-oriented constituencies in wealthy nations may focus primarily on environmental dimensions of sustainability. Still others may see sustainability as a question of ethics and value change, or as primarily an economic problem. Some believe that the capitalist system is fundamentally at fault; others believe that market economies can be reformed in one way or another so as to operate more sustainably. The list of possible perspectives and theories related to sustainability could go on at some length.

At various times some of these perspectives may be more appropriate than others. Some may lead to stronger sustainability solutions than others ("deep green" vs. "light green" strategies); some may help reach certain important audiences; and some may include or omit consideration of important considerations such as social equity.

What is most crucial – following Donella Meadows' recommendation – is to understand the mental models and mindsets that each person and organization brings to sustainability debates, and to try to find common ground from which to address problems. It's important to think critically about the concept of urban sustainability, in other words, and about the underlying beliefs that individuals and organizations bring to it. Such analysis can help us forge consensus and set sustainable city priorities in any given setting.

2   **Actions in different areas fit together.** It is important to develop a sense of how sustainability actions in different topic areas relate to each other. How do transportation systems link to land use, housing, or environmental planning? How can particular economic development strategies reinforce

local land use planning visions or community livability? How can greener urban design or architecture promote environmental justice and equity? We encourage readers to think about the links between different readings and how action in each area can help create more balanced and sustainable urban environments overall. Indeed, a leading cause of unsustainable urban development in the past is that planners, elected leaders, and citizens often haven't made such linkages in their own minds. Engineers have planned road systems with very little attention to the ways they affect surrounding land use. Planners and developers have often mapped out new subdivisions without considering whether they promoted an equitable distribution of affordable housing in the region. Architects have at times designed buildings as though no ecological or social context existed. Economic development staffs have sought rapid expansion of jobs without asking whether housing was available for the new workers or whether this development would benefit existing business and residents. As Jacobs, McHarg, and Mumford have suggested, such compartmentalized thinking has helped create current urban problems.

3 **Actions can take place at different scales.** It is important to be aware of which scale is appropriate for action at any given time, and to integrate initiatives across different scales of action or levels of government. Urban land use decisions, for example, are primarily made by local governments in most countries, but are also profoundly influenced by policy and programs at national, state or provincial, and regional levels. Transportation planning and regulation of air and water pollution are often handled at a metropolitan regional level, but likewise take place within the context of state or national policy. Architecture and site design are quite local, but building codes – developed internationally and then tailored by national, state, and/or local governments – determine what can be done or not on the ground. So it is important to keep in mind how actions at different scales can reinforce or work against one another, and to help create an overall framework of governance in which sustainable urban development can occur.

4 **Urban planning and development actions are undertaken by a wide variety of actors.** In the world of twenty-first-century governance, many different individuals and groups participate. Nonprofit organizations, neighborhood associations, private developers, labor unions, businesses, the news media, and concerned individuals all influence the character and growth of cities. The public sector is also complex, including local governments, regional agencies, state departments of transportation, utility companies, national agencies, and many other actors. Even within local government many different players are involved, such as planning offices and staffs, public works departments, redevelopment agencies, appointed city commissions, and elected mayors or city council members. Each of these parties has opportunities to make a difference. For each of us there are many possible avenues for professional or activist involvement aimed at creating more sustainable communities. Learning the landscape of "who does what" is important in order to be able to best understand how change can be brought about.

It is important to bear in mind these themes – how different people and organizations conceptualize sustainability, how individual issues fit together, how actions at different scales relate, and who does what at each level – as we explore the various dimensions of sustainable urban development.

CLIMATE CHANGE PLANNING

13

"Stabilization Wedges: Solving the Climate Problem for the Next 50 Years with Current Technologies"

from *Science* magazine (2004)

Stephen Pacala and Robert Socolow

Editors' Introduction

The need to reduce greenhouse gas emissions and adapt to a changing climate is probably the main force driving the urban sustainability agenda as we move further into the twenty-first century. Many of the initiatives discussed later on in this book will help with climate planning – for example, promoting renewable energy, greening cities so as to make them cooler, and changing urban development patterns so as to reduce driving. But we start in this first chapter by considering comprehensive planning approaches directly addressing climate change.

Although societies have known that global warming is a growing problem at least since the 1950s (see the McKibben reading earlier), we still don't have a good sense of exactly how to accomplish the very steep greenhouse gas reductions that scientists have told us are necessary. One classic piece that meets this challenge head on is Stephen Pacala and Robert Socolow's proposed "wedges" strategy published in *Science* magazine in 2004. (They published a somewhat different version of this material in *Scientific American* in September 2006.) These two authors, professors of ecology and engineering, respectively, at Princeton University, consider how humanity might stabilize carbon dioxide emissions over the next 50 years, in theory holding atmospheric concentrations of this gas to approximately double the pre-industrial level (around 500 parts per million) and setting the stage for eventual reductions. Whether this would be sufficient to minimize global warming is debatable; the 350.org organization founded by McKibben calls for reducing atmospheric CO_2 to mid-twentieth-century levels, or around 350 parts per million. But the exercise of outlining specific strategies to manage emissions in a specific time period using existing technologies is still valuable, and illustrates the type of thinking that is needed.

Some of Pacala and Socolow's proposed strategies are controversial. They suggest nuclear power as an option, for example, despite nuclear weapons proliferation problems and the lack of any good solution to the disposal of spent radioactive fuel. They also propose use of biofuels, which if produced at a large scale could displace food crops and threaten food availability for the world's poorest citizens. However, their point is not to rank the desirability of each option so much as to outline a range of specific strategies that could set the world on a different course.

Other authors have also considered strategies for dramatically lowering greenhouse gas emissions. In his book *Heat: How to Stop the Planet from Burning* (Cambridge, MA: South End Press, 2007), journalist George Monbiot argues that emissions reductions of as much as 90 percent by 2030 should be sought in order to hold climate change to no more than 2 degrees Celsius, and proposes a variety of radical building, energy, transport, and lifestyle changes to do this. In *Reinventing Fire* (White River Junction, VT: Chelsea Green, 2011), energy guru Amory Lovins and colleagues from the Rocky Mountain Institute lay out a roadmap for weaning the US economy from fossil fuels by 2050. And in *World on the Edge* (New York: Norton, 2011), Lester Brown describes a broader approach to building an energy-efficient global economy, eradicating poverty, stabilizing population, and feeding eight billion people.

More general background about the problem of global warming can be found in Tim Flannery's book *The Weather Makers* (New York: Grove Press, 2005), Elizabeth Kolbert's *Field Notes from a Catastrophe: Man, Nature, and Climate Change* (New York: Bloomsbury USA, 2006), Fred Pearce's *With Speed and Violence: Why Scientists Fear Tipping Points in Climate Change* (Boston: Beacon Press, 2007), Mark Lynas' *Six Degrees: Our Future on a Hotter Planet* (Washington: National Geographic, 2008), David Archer's *Global Warming: Understanding the Forecast* (Hoboken, NJ: Wiley, 2011), and Bill McKibben's *Earth: Making a Life on a Tough New Planet* (New York: St. Martin's Griffin, 2011).

[...]

The debate in the current literature about stabilizing atmospheric CO_2 at less than a doubling of the preindustrial concentration has led to needless confusion about current options for mitigation. On one side, the Intergovernmental Panel on Climate Change (IPCC) has claimed that "technologies that exist in operation or pilot stage today" are sufficient to follow a less-than-doubling trajectory "over the next hundred years or more" [(1), p. 8]. On the other side, a recent review in *Science* asserts that the IPCC claim demonstrates "misperceptions of technological readiness" and calls for "revolutionary changes" in mitigation technology, such as fusion, space-based solar electricity, and artificial photosynthesis (2). We agree that fundamental research is vital to develop the revolutionary mitigation strategies needed in the second half of this century and beyond. But it is important not to become beguiled by the possibility of revolutionary technology. Humanity can solve the carbon and climate problem in the first half of this century simply by scaling up what we already know how to do.

WHAT DO WE MEAN BY "SOLVING THE CARBON AND CLIMATE PROBLEM FOR THE NEXT HALF-CENTURY"?

Proposals to limit atmospheric CO_2 to a concentration that would prevent most damaging climate change have focused on a goal of 500 ± 50 parts per million (ppm), or less than double the pre-industrial concentration of 280 ppm (3–7). The current concentration is ~375 ppm. The CO_2 emissions reductions necessary to achieve any such target depend on the emissions judged likely to occur in the absence of a focus on carbon [called a business-as-usual (BAU) trajectory], the quantitative details of the stabilization target, and the future behavior of natural sinks for atmospheric CO_2 (i.e., the oceans and terrestrial biosphere). We focus exclusively on CO_2, because it is the dominant anthropogenic greenhouse gas; industrial-scale mitigation options also exist for subordinate gases, such as methane and N_2O.

Very roughly, stabilization at 500 ppm requires that emissions be held near the present level of 7 billion tons of carbon per year (GtC/year) for the next fifty years, even though they are currently on course to more than double (Figure 1A). The next fifty years is a sensible horizon from several perspectives. It is the length of a career, the lifetime of a power plant, and an interval for which the technology is close enough to envision. The BAU and stabilization emissions in Figure 1A are near the center of the cloud of variation in the large published literature (8).

THE STABILIZATION TRIANGLE

We idealize the fifty-year emissions reductions as a perfect triangle in Figure 1B. Stabilization is

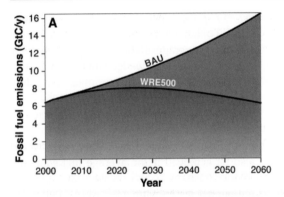

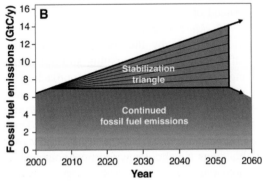

Figure 1 Wedges.

represented by a "flat" trajectory of fossil fuel emissions at 7 GtC/year, and BAU is represented by a straight-line "ramp" trajectory rising to 14 GtC/year in 2054. The "stabilization triangle," located between the flat trajectory and BAU, removes exactly one-third of BAU emissions.

To keep the focus on technologies that have the potential to produce a material difference by 2054, we divide the stabilization triangle into seven equal "wedges." A wedge represents an activity that reduces emissions to the atmosphere that starts at zero today and increases linearly until it accounts for 1 GtC/year of reduced carbon emissions in fifty years. It thus represents a cumulative total of 25 GtC of reduced emissions over fifty years. In this paper, to "solve the carbon and climate problem over the next half-century" means to deploy the technologies and/or lifestyle changes necessary to fill all seven wedges of the stabilization triangle.

Stabilization at any level requires that net emissions do not simply remain constant, but eventually drop to zero. For example, in one simple model (9) that begins with the stabilization triangle but looks beyond 2054, 500 ppm stabilization is

achieved by fifty years of flat emissions, followed by a linear decline of about two-thirds in the following fifty years, and a very slow decline thereafter that matches the declining ocean sink. To develop the revolutionary technologies in the second half of the century, enhanced research and development would have to begin immediately.

Policies designed to stabilize at 500 ppm would inevitably be renegotiated periodically to take into account the results of research and development, experience with specific wedges, and revised estimates of the size of the stabilization triangle. But not filling the stabilization triangle will put 500 ppm stabilization out of reach. In that same simple model (9), fifty years of BAU emissions followed by fifty years of a flat trajectory at 14 GtC/year leads to more than a tripling of the preindustrial concentration.

It is important to understand that each of the seven wedges represents an effort beyond what would occur under BAU. Our BAU simply continues the 1.5 percent annual carbon emissions growth of the past thirty years. This historic trend in emissions has been accompanied by 2 percent growth in primary energy consumption and 3 percent growth in gross world product (GWP). If carbon emissions were to grow 2 percent per year, then ~10 wedges would be needed instead of seven, and if carbon emissions were to grow at 3 percent per year, then ~18 wedges would be required. Thus, a continuation of the historical rate of decarbonization of the fuel mix prevents the need for three additional wedges, and ongoing improvements in energy efficiency prevent the need for eight additional wedges. Most readers will reject at least one of the wedges listed here, believing that the corresponding deployment is certain to occur in BAU, but readers will disagree about which to reject on such grounds. On the other hand, our list of mitigation options is not exhaustive.

WHAT CURRENT OPTIONS COULD BE SCALED UP TO PRODUCE AT LEAST ONE WEDGE?

Wedges can be achieved from energy efficiency, from the decarbonization of the supply of electricity and fuels (by means of fuel shifting, carbon capture and storage, nuclear energy, and renewable energy), and from biological storage in forests and agricultural soils. Below, we discuss fifteen different examples of options that are already deployed at an industrial scale and that would be scaled up further to produce at least one wedge (summarized in Table 1). Although several options could be scaled up to two or more wedges, we doubt that any could fill the stabilization triangle, or even half of it, alone.

Because the same BAU carbon emissions cannot be displaced twice, achieving one wedge often interacts with achieving another. The more the electricity system becomes decarbonized, for example, the less the available savings from greater efficiency of electricity use, and vice versa. Also, our focus is not on costs. In general, the achievement of a wedge will require some price trajectory for carbon, the details of which depend on many assumptions, including future fuels prices, public acceptance, and cost reductions by means of learning. Instead, our analysis is intended to complement the comprehensive but complex "integrated assessments" (1) of carbon mitigation by letting the full-scale examples that are already in the marketplace make a simple case for technological readiness.

Category I: efficiency and conservation

Improvements in efficiency and conservation probably offer the greatest potential to provide wedges. For example, in 2002, the United States announced the goal of decreasing its carbon intensity (carbon emissions per unit GDP) by 18 percent over the next decade, a decrease of 1.96 percent per year. An entire wedge would be created if the United States were to reset its carbon intensity goal to a decrease of 2.11 percent per year and extend it to fifty years, and if every country were to follow suit by adding the same 0.15 percent per year increment to its own carbon intensity goal. However, efficiency and conservation options are less tangible than those from other categories. Improvements in energy efficiency will come from literally hundreds of innovations that range from new catalysts and chemical processes, to more efficient lighting and insulation for buildings, to the growth of the service economy and telecommuting. Here, we provide four of many possible comparisons of greater and lesser efficiency in 2054.

Option 1: Improved fuel economy. Suppose that in 2054, 2 billion cars (roughly four times as many as today) average 10,000 miles per year (as they do today). One wedge would be achieved if, instead of averaging thirty miles per gallon (mpg) on conventional fuel, cars in 2054 averaged 60 mpg, with fuel type and distance traveled unchanged.

Option 2: Reduced reliance on cars. A wedge would also be achieved if the average fuel economy of the 2 billion 2054 cars were 30 mpg, but the annual distance traveled were 5,000 miles instead of 10,000 miles.

Option 3: More efficient buildings. According to a 1996 study by the IPCC, a wedge is the difference between pursuing and not pursuing "known and established approaches" to energy-efficient space heating and cooling, water heating, lighting, and refrigeration in residential and commercial buildings. These approaches reduce mid-century emissions from buildings by about one-fourth. About half of potential savings are in the buildings in developing countries (1).

Option 4: Improved power plant efficiency. In 2000, coal power plants, operating on average at 32 percent efficiency, produced about one-fourth of all carbon emissions: 1.7 GtC/year out of 6.2 GtC/year. A wedge would be created if twice today's quantity of coal-based electricity were produced at 60 percent instead of 40 percent efficiency.

Category II: Decarbonization of electricity and fuels

Option 5: Substituting natural gas for coal. Carbon emissions per unit of electricity are about half as large from natural gas power plants as from coal plants. Assume that the capacity factor of the average baseload coal plant in 2054 has increased to 90 percent and that its efficiency has improved to 50 percent. Because 700 GW of such plants emit carbon at a rate of 1 GtC/year, a wedge would be achieved by displacing 1,400 GW of baseload coal with baseload gas by 2054. The power shifted to gas for this wedge is four times as large as the total current gas-based power.

Table 1 Potential wedges: Strategies available to reduce the carbon emission rate in 2054 by 1 GtC/year or to reduce carbon emission from 2004 to 2054 by 25 GtC.

Option	Effort by 2054 for one wedge, relative to 14 GtC/year BAU	Comments, issues
Energy efficiency and conservation		
Economy-wide carbon-intensity reduction (emission/$GDP)	Increase reduction by additional 0.15% per year (e.g., increase US goal of 1.96% reduction per year to 2.11% per year)	Can be tuned by carbon policy
1 Efficient vehicles	Increase fuel economy for 2 billion cars from 30 mpg to 60 mpg	Car size, power
2 Reduced use of vehicles	Decrease car travel for 2 billion 30 mpg cars from 10,000 to 5,000 miles per year	Urban design, mass transit, telecommuting
3 Efficient buildings	Cut carbon emissions by one-fourth in buildings and appliances projected for 2054	Weak incentives
4 Efficient baseload coal plants	Produce twice today's coal power output at 60% instead of 40% efficiency (compared with 32% today)	Advanced high-temperature materials
Fuel shift		
5 Gas baseload power for coal baseload power	Replace 1,400 GW 50%-efficient coal plants with gas plants (four times the current production of gas-based power)	Competing demands for natural gas
CO_2 Capture and Storage (CCS)		
6 Capture CO_2 at baseload power plant	Introduce CCS at 800 GW coal or 1,600 GW natural gas (compared with 1,060 GW coal in 1999)	Technology already in use for H_2 production
7 Capture CO_2 at H_2 plant	Introduce CCS at plants producing 250 MtH_2/year from coal or 500 MtH_2/year from nature gas (compared with 40 MtH_2/year today from all sources)	H_2 safety, infrastructure
8 Capture CO_2 at coal-to-synfuels plant	Introduce CCS at synfuels plants producing 30 million barrels a day from coal (200 times Sasol), if half of feedstock carbon is available for capture	Increased CO_2 emissions, if synfuels are produced without CCS
Geological storage	Create 3,500 Sleipners	Durable storage, successful permitting
Nuclear fission		
9 Nuclear power for coal power	Add 700 GW (twice the current capacity)	Nuclear proliferation, terrorism, waste
Renewable electricity and fuels		
10 Wind power for coal power	Add 2 million 1 MW-peak windmills (50 times the current capacity) "occupying" 30×10^6 ha, on land or offshore	Multiple uses of land because windmills are widely spaced
11 PV power for coal power	Add 2,000 GW-peak PV (700 times the current capacity) on 2×10^6 ha	PV production cost
12 Wind H_2 in fuel-cell car for gasoline in hybrid car	Add 4 million 1 MW-peak windmills (100 times the current capacity)	H_2 safety, infrastructure
13 Biomass fuel for fossil fuel	Add 100 times the current Brazil or US ethanol production, with the use of 250×10^6 ha (one-sixth of world cropland)	Biodiversity, competing land use
Forests and agricultural soils		
14 Reduced deforestation, plus reforestation, afforestation, and new plantations	Decrease tropical deforestation to zero instead of 0.5 GtC/year, and establish 300 Mha of new tree plantations (twice the current rate)	Land demands of agriculture, benefits to biodiversity from reduced deforestation
15 Conservation tillage	Apply to all cropland (10 times the current usage)	Reversibility, verification

Option 6: Storage of carbon captured in power plants. Carbon capture and storage (CCS) technology prevents about 90 percent of the fossil carbon from reaching the atmosphere, so a wedge would be provided by the installation of CCS at 800 GW of baseload coal plants by 1054 or 1,600 GW of baseload natural gas plants. The most likely approach has two steps: (i) precombustion capture of CO_2, in which hydrogen and CO_2 are produced and the hydrogen is then burned to produce electricity, followed by (ii) geologic storage, in which the waste CO_2 is injected into subsurface geologic reservoirs. Hydrogen production from fossil fuels is already a very large business. Globally, hydrogen plants consume about 2 percent of primary energy and emit 0.1 GtC/year of CO_2. The capture part of a wedge of CCS electricity would thus require only a tenfold expansion of plants resembling today's large hydrogen plants over the next fifty years.

The scale of the storage part of this wedge can be expressed as a multiple of the scale of current enhanced oil recovery, or current seasonal storage of natural gas, or the first geological storage demonstration project. Today, about 0.01 GtC/year of carbon as CO_2 is injected into geologic reservoirs to spur enhanced oil recovery, so a wedge of geologic storage requires that CO_2 injection be scaled up by a factor of 100 over the next fifty years. To smooth out seasonal demand in the United States, the natural gas industry annually draws roughly 4,000 billion standard cubic feet (Bscf) into and out of geologic storage, and a carbon flow of 1 GtC/year (whether as methane or CO_2) is a flow of 69,000 Bscf/year (190 Bscf per day), so a wedge would be a flow to storage fifteen and twenty times as large as the current flow. Norway's Sleipner project in the North Sea strips CO_2 from natural gas offshore and reinjects 0.3 million tons of carbon a year (MtC/year) into a non-fossil-fuel-bearing formation, so a wedge would be 3,500 Sleipnersized projects (or fewer, larger projects) over the next fifty years.

A worldwide effort is under way to assess the capacity available for multicentury storage and to assess risks of leaks large enough to endanger human or environmental health.

Option 7: Storage of carbon captured in hydrogen plants. The hydrogen resulting from precombustion capture of CO_2 can be sent offsite to displace the consumption of conventional fuels rather than being consumed onsite to produce electricity. The capture part of a wedge would require the installation of CCS, by 2054, at coal plants producing 250 MtH_2/year. The former is six times the current rate of hydrogen production. The storage part of this option is the same as in Option 6.

Option 8: Storage of carbon captured in synfuels plants. Looming over carbon management in 2054 is the possibility of large-scale production of synthetic fuel (synfuel) from coal. Carbon emissions, however, need not exceed those associated with fuel refined from crude oil if synfuels production is accompanied by CCS. Assuming that half of the carbon entering a 2054 synfuels plant leaves as fuel but the other half can be captured as CO_2, the capture part of a wedge in 2054 would be the difference between capturing and venting the CO_2 from coal synfuels plants producing 30 million barrels of synfuels per day. (The flow of carbon in 24 million barrels per day of crude oil is 1 GtC/year; we assume the same value for the flow in synfuels and allow for imperfect capture.) Currently the Sasol plants in South Africa, the world's largest synfuels facility, produce 165,000 barrels per day from coal. Thus, a wedge requires 200 Sasol-scale coal-to-synfuels facilities with CCS in 2054. The storage part of this option is again the same as in Option 6.

Option 9: Nuclear fission. On the basis of the Option 5 estimates, a wedge of nuclear electricity would displace 700 GW of efficient baseload coal capacity in 2054. This would require 700 GW of nuclear power with the same 90 percent capacity factor assumed for the coal plants, or about twice the nuclear capacity currently deployed. The global pace of nuclear power plant construction from 1975 to 1990 would yield a wedge, if it continued for fifty years (10). Substantial expansion in nuclear power requires restoration of public confidence in safety and waste disposal, and international security agreements governing uranium enrichment and plutonium recycling.

Option 10: Wind energy. We account for the inter-
mittent output of windmills by equating 3 GW
of nominal peak capacity (3 GWp) with 1 GW
of baseload capacity. Thus, a wedge of wind
electricity would require the deployment of
2,000 GWp that displaces coal electricity in
2054 (or 2 million 1 MWp wind turbines).
Installed wind capacity has been growing at
about 30 percent per year for more than ten
years and is currently about 40 GW. A wedge
of wind electricity would thus require fifty times
today's deployment. The wind turbines would
"occupy" about 30 million ha (about 3 percent
of the area of the United States), some on land
and some offshore. Because windmills are
widely spaced, land with windmills can have
multiple uses.

Option 11: Photovoltaic electricity. Similar to a wedge
of wind electricity, a wedge from photovoltaic
(PV) electricity would require 2,000 GWp of
installed capacity that displaces coal electricity
in 2054. Although only 3 GWp of PV are currently
installed, PV elecdtricity has been growing at
a rate of 30 percent per year. A wedge of PV
electricity would require 700 times today's
deployment, and about 2 million ha of land in
2054, or 2 m^2 to 3 m^2 per person.

Option 12: Renewable hydrogen. Renewable electricity
can produce carbon-free hydrogen for vehicle
fuel by the electrolysis of water. The hydrogen
produced by 4 million 1 MWp windmills in
2054, if used in high-efficiency fuel-cell cars,
would achieve a wedge of displaced gasoline
or diesel fuel. Compared with Option 10, this
is twice as many 1 MWp windmills as would
be required to produce the electricity that
achieves a wedge by displacing high-efficiency
baseload coal. This interesting factor-of-two
carbon-saving advantage of wind-electricity
over wind–hydrogen is still larger if the coal
plant is less efficient or the fuel-cell vehicle is
less spectacular.

Option 13: Biofuels. Fossil-carbon fuels can also be
replaced by biofuels such as ethanol. A wedge
of biofuel would be achieved by the production
of about 34 million barrels per day of ethanol
in 2054 that could displace gasoline, provided
the ethanol itself were fossil-carbon-free. This
ethanol production rate would be about fifty
times larger than today's global production rate,

almost all of which can be attributed to Brazilian
sugarcane and United States corn. An ethanol
wedge would require 350 million ha committed
to high-yield (15 dry tons/ha) plantations by
2054, an area equal to about one-sixth of the
world's cropland. An even larger area would
be required to the extent that the biofuels
require fossil-carbon inputs. Because land suitable for annually harvested biofuels crops is
also often suitable for conventional agriculture,
biofuels production could compromise agricultural productivity.

Category III: Natural sinks

Although the literature on biological sequestration
includes a diverse array of options and some very
large estimates of the global potential, here we
restrict our attention to the pair of options that are
already implemented at a large scale and that could
be scaled up to a wedge or more without a lot of
new research.

Option 14: Forest management. Conservative assumptions lead to the conclusion that at least
one wedge would be available from reduced
tropical deforestation and the management of
temperate and tropical forests. At least one
half-wedge would be created in the current
rate of clear-cutting of primary tropical forest
were reduced to zero over fifty years instead
of being halved. A second half-wedge would
be created by reforesting or afforesting approximately 250 million ha in the temperate zone
(current areas of tropical and temperate forests
are 1,500 ha and 700 million ha respectively).
A third half-wedge would be created by establishing approximately 300 million ha of plantations on nonforested land.

Option 15: Agricultural soils management. When
forest or natural grassland is converted to
cropland, up to one-half of the soil carbon is
lost, primarily because annual tilling increases
the rate of decomposition by aerating unde-
composed organic matter. About 55 GtC, or
two wedges' worth, has been lost historically in
this way. Practices such as conservation tillage
(e.g., seeds are drilled into the soil without plow-
ing), the use of cover crops, and erosion control

can reverse the losses. By 1995, conservation tillage practices had been adopted on 110 million ha of the world's 1,600 million ha of cropland. If conservation tillage could be extended to all cropland, accompanied by a verification program that enforces the adoption of soil conservation practices that actually work as advertised, a good case could be made for the IPCC's estimate that an additional half to one wedge could be stored in this way.

CONCLUSIONS

In confronting the problem of greenhouse warming, the choice today is between action and delay. Here, we presented a part of the case for action by identifying a set of options that have the capacity to provide the seven stabilization wedges and solve the climate problem for the next half-century. None of the options is a pipe dream or an unproven idea. Today, one can buy electricity from a wind turbine, PV array, gas turbine, or nuclear power plant. One can buy hydrogen produced with the chemistry of carbon capture, biofuel to power one's car, and hundreds of devices that improve energy efficiency. One can visit tropical forests where clear-cutting has ceased, farms practicing conservation tillage, and facilities that inject carbon into geologic reservoirs. Every one of these options is already implemented at an industrial scale and could be scaled up further over fifty years to provide at least one wedge.

REFERENCES AND NOTES

1 IPCC, *Climate Change 2001: Mitigation*, B. Metz *et al.*, eds. (IPCC Secretariat, Geneva, Switzerland, 2001); available at www.grida.no/climate/ipcc_tar/wg3/index.htm.

2 M.I. Hoffert *et al.*, *Science* 298, 981 (2002).

3 R.T. Watson *et al.*, *Climate Change 2001: Synthesis Report*. Contribution to the Third Assessment Report of the Intergovernmental Panel on Climate Change (Cambridge University Press, Cambridge, UK, 2001).

4 B.C. O'Neill, M. Oppenheimer, *Science* 296, 1971 (2002).

5 Royal Commission on Environmental Pollution, *Energy: The Changing Climate* (2000); available at www.rcep.org.uk/energy.htm.

6 Environmental Defense, Adequacy of Commitments – Avoiding "Dangerous" Climate Change: A Narrow Time Window for Reductions and a Steep Price for Delay (2002); available at www.environmentaldefense.org/documents/2422_COP_time.pdf.

7 "Climate Options for the Long Term (COOL) synthesis report," NRP Rep. 954281 (2002); available at www.wau.nl/cool/reports/COOLVolumeAdef.pdf.

8 IPCC. *Special Report on Emissions Scenarios* (2001); available at www.grida.no/climate/ipcc/emission/index.htm.

9 R. Socolow, S. Pacala, J. Greenblatt, *Proceedings of the Seventh International Conference on Greenhouse Gas Control Technology*, Vancouver, Canada, 5–9 September, 2004, in press.

10 BP, *Statistical Review of World Energy* (2003), available at http://www.bp.com/productlanding.do?categoryId=6848&contentId=7033471.

The authors thank J. Greenblatt, R. Hotinski, and R. Williams at Princeton, K. Keller at Penn. State; and C. Mottershead at BP. This paper is a product of the Carbon Mitigation Initiative (CMI) of the Princeton Environmental Institute at Princeton University. CMI (www.princeton.edu/~cmi) is sponsored by BP and Ford.

Supporting online material at www.sciencemag.org/cgi/content/full/305/5686/968/.

14

"Towards Low Carbon Urbanism"

from *Local Environment* (2012)

Harriet Bulkeley, Vanesa Castan Broto, and Gareth Edwards

Editors' Introduction

Theoretical strategies such as those proposed by Pacala and Socolow are an important foundation for action, but a pressing question is who is going to implement them. Since national governments have often proven unwilling to take the initiative, cities and nonprofit organizations have moved to take action on their own. This piece by noted British climate policy researcher Harriet Bulkeley and her coauthors chronicles the rise of municipal action to fight global warming. Nongovernmental organizations such as ICLEI: Local Governments for Sustainability (originally the International Council on Local Environmental Initiatives) have been crucial in this effort. So have mayors, who have often taken the lead in establishing greenhouse gas reduction goals and policies. Some states and provinces also have initiated climate policy frameworks that support local action.

Bulkeley, Broto, and Edwards highlight the changing nature of urban climate policy, which over the past couple of decades has moved from a limited set of goals and strategies to reduce emissions toward a broader agenda including climate adaptation, concepts of "resilient" cities, attention to climate justice issues, and a broad ethic of "low carbon urbanism."

Many other books on local climate planning provide additional background. Another volume by Bulkeley, *Cities and Climate Change* (New York: Routledge, 2013), goes into somewhat more depth on local action from an international perspective. *Governing Climate Change* (New York: Routledge, 2010), by Bulkeley and Peter Newell, looks at institutional challenges of planning for global warming. *Local Climate Action Planning* (Washington, D.C.: Island Press, 2012), by Michael R. Boswell, Adrienne I. Greve, and Tammy L. Seale considers municipal action in the US context. Brian Stone's *The City and the Coming Climate: Climate Change in the Places We Live* (Cambridge University Press, 2012) emphasizes urban greening strategies for climate adaptation. *Planning for Climate Change: Strategies for Mitigation and Adaptation for Spatial Planners* (London: Earthscan, 2009), edited by Simin Davoudi, Jenny Crawford, and Abid Mehmood, considers physical planning implications of climate change, as does *Spatial Planning and Climate Change* (London: Routledge, 2010), by Elizabeth Wilson and Jake Piper. Peter F. Smith's *Building for a Changing Climate: The Challenge for Construction, Planning, and Energy* (London: Earthscan, 2010) looks more specifically at building design and energy systems issues. In the private sector, the consulting firm McKinsey & Company has produced important analyses of the economic feasibility of various greenhouse gas reduction strategies for societies, available through its report *Pathways to a Low-Carbon Economy* (2009) at https://solutions.mckinsey.com/climatedesk/default.aspx. This organization has produced a famous "cost curve" (reproduced at the end of this chapter) showing that a great many actions to reduce greenhouse gas emissions will actually save money.

Cities have been central to the evolving landscape of climate change responses (Betsill and Bulkeley 2007). In the late 1980s and early 1990s, mirroring growing international concern, individual cities and small groups of municipalities began to place climate change on their agendas. Focused on the issue of mitigation, that is of reducing GHG [greenhouse gas] emissions, these early pioneers argued that with their density of GHG emissions-producing activities and the potential role of municipal authorities in governing processes of regulation, planning, transportation, energy provision and waste collection through which such emissions were created, cities were an essential part of the response to climate change.

From the initial actions of just a few hundred cities, the key municipal associations that were formed at this time to address climate change – including ICLEI's Cities for Climate Protection (CCP) Program, the Climate Alliance, and EnergyCities – have grown to count several thousand cities among their membership. At the same time, new initiatives have been formed, including C40 Cities Climate Leadership Group, working with support of the Clinton Climate Initiative and targeting forty of the world's "global" cities, the US Mayors Climate Protection Agreement, and the European Covenant of Mayors. This growth in the scale and nature of municipal responses to climate change is one of the most significant features of the changing climate governance landscape over the past two decades.

FROM VOLUNTARISM TO STRATEGIC URBANISM

Urban responses that emerged in the immediate post-Rio period [after the 1992 United Nations sponsored Earth Summit conference in Rio de Janeiro] were characterized by a distinct focus on the mitigation agenda, concentrated among cities in North America and Europe and increasingly transnationally organized through the Cities for Climate Protection (CCP), Climate Alliance, and Energy-Cities networks.

During this phase, the membership of climate change networks in Europe grew steadily, though it had reached a plateau by the end of the 1990s, while elsewhere new regional campaigns by ICLEI CCP saw membership increase significantly in Australia and the USA, as well as in Asia and Latin America (Kern and Bulkeley 2009). Despite the growth in

membership, however, reported actions were primarily focused on the reduction of GHG emissions from within municipal operations – a "self-governing" approach (Bulkeley and Kern 2006). During this phase, municipal responses were driven by a sense of commitment to a global cause coupled with a realization of the additional benefits, in financial, health, and environmental terms, that acting to reduce GHG emissions might bring. In this sense, municipal responses were somewhat marginal to mainstream urban agendas, and usually voluntary in their nature. Action remained confined to a small number of municipalities and was primarily focused on issues of energy and efficiency.

By the beginning of the 2000s, with uncertainty about international commitment to the 1997 Kyoto Protocol, interest in urban responses to climate change seemed to be on the wane. Renewed momentum came from a combination of rather unlikely sources. In the USA and Australia, growing political recalcitrance at the national level spawned a growing movement of urban climate change responses, including the expansion of the CCP program and the emergence of the US Mayors Climate Protection Agreement, which now numbers more than 1000 members (Gore and Robinson 2009, US Mayors 2011). The actions of urban politicians were critical to this new wave of responses. For example, the development of the C40 Cities Climate Leadership Group owes much to the actions of the then-Mayor of London, Ken Livingstone, and his Deputy, Nicky Gavron, as well as to the engagement of mayors from other high profile cities including New York, Toronto, and Sao Paolo. Within the European Union, mayors have also become involved in the climate change agenda through the Covenant of Mayors, which requires signatories to pledge to go beyond the EU target of reducing CO_2 emissions by 20% by 2020 through the formation and implementation of a sustainable energy action plan (CoM 2011a), and in 2011 has more than two thousand members (CoM 2011b).

Alongside this, the private sector has been of growing importance in shaping urban responses to the issue. This includes both philanthropic organizations such as the Rockefeller Foundation and the Clinton Climate Initiative, as well as corporate organizations such as Cisco, HSBC and Arup, who are now actively developing urban responses to climate change (Hoffman 2011, Bulkeley and Schroeder 2012).

These developments have led to a growing engagement with issues of climate change in cities in the Global South. For example, some 50 local authorities in India have participated in ICLEI South Asia's Roadmap project to conduct emissions inventories and develop climate change action plans, while 20 of the 40 cities included in the C40 Cities Climate Leadership group are located in countries in the Global South.

While climate mitigation continues to attract the most attention, adaptation is increasingly on the urban agenda. Existing networks have begun to focus on climate adaptation and are seeking to engage cities through the concept of "resilience", epitomized in the annual conference on Resilient Cities first held by ICLEI in 2010. In addition, the Rockefeller Foundation has established the Asian Cities Climate Change Resilience Network, a network of ten cities explicitly focused on climate adaptation, and the UN-Habitat is working with cities in Asia, Africa and Latin America through its Cities and Climate Change Initiative to support urban adaptation responses.

At least for some cities, climate change is now a strategic concern and one that is more closely aligned to the concerns of urban growth and resource security that dominate urban agendas (Hodson and Marvin 2009, 2010, While *et al.* 2010). While the municipally based, voluntary response to the issue which dominated the 1990s remains pervasive, especially among smaller cities, this new phase of urban responses to climate change is creating an additional form of climate politics (Bulkeley and Betsill 2005, Bulkeley and Schroeder 2012).

BOX 1 The ICLEI 5-milestone process

An international non-governmental organization named ICLEI: Cities for Climate Protection (originally the International Council on Local Environmental Initiatives) has long assisted cities worldwide in tackling climate change. The group's Cities for Climate Protection (CCP) campaign helped local governments reduce greenhouse gas (GHG) emissions through a five-step process:

Milestone 1 *Conduct a baseline emissions inventory and forecast.* The local government calculates greenhouse gas emissions across many sectors for a base year (e.g. 1990 or 2000) and for a forecast year (e.g. 2020). This inventory and forecast represent benchmarks for evaluating progress.

Milestone 2 *Adopt an emissions reduction target for the forecast year.* Local leaders establish one or more GHG reduction goals, a step that helps build political commitment and guides policy development.

Milestone 3 *Develop a Local Action Plan.* The city creates a Local Action Plan setting out policies and programs the community will undertake to reduce emissions and achieve its goals. Stakeholders must be involved in this process. Plans must include timelines, financing mechanisms, public education, and responsibility for particular actions.

Milestone 4 *Implement policies and measures.* The community implements policies such as to construct green municipal buildings, improve public transit and human-powered transportation options, construct renewable energy facilities, and cap landfills to reduce methane emissions.

Milestone 5 *Monitor and verify results.* Last but certainly not least, municipalities must monitor implementation of climate policies, update emissions inventories, and revise policies to ensure that GHG reduction goals are met.

In the early 2010s ICLEI changed this program's name to the GreenClimateCities Network, and initiated other programs to help local governments develop renewable energy, become more climate resilient, and develop comprehensive sustainability policies. In particular, the organization has launched a global database of local climate actions that also serves as an online platform for cities to self-report GHG emissions and actions. More information is available at www.iclei.org.

LOW CARBON URBANISM: NEW AGENDAS?

If initial urban responses to climate change followed broadly similar lines, one of the marked characteristics of the situation some 20 years later is the sheer diversity of issues, initiatives and actors engaged within the climate-change fold. Climate change has become an issue that is affecting debates across the spectrum of urban sustainability issues. Of particular importance, however, has been the emergence of a specific focus on low carbon urbanism.

At the municipal level the idea of accounting for and reducing GHG emissions has been part of urban responses to climate change over the past two decades. But as efforts to mitigate climate change across urban communities have gathered pace, a variety of forms of low carbon urbanism have been articulated as the means through which cities may be able to foster both their climate change objectives and their long-term economic development.

On the one hand, this has given rise to what some have referred to as a politics of "secure urbanism and resilient infrastructure" (Hodson and Marvin 2010) and others describe as an era of "carbon control" (While *et al.* 2010). In such accounts, a range of urban actors, including those within municipal authorities but perhaps more significantly those in national governments, international agencies, and private sector corporations, have come to view cities as arenas within which new forms of low carbon economy can be developed.

Such projects may include the kinds of energy efficiency measures with which municipalities have sought to engage publics and private sector organizations over the past two decades, but they are also marked by a new interest in the development of low carbon and "self-sufficient" forms of energy infrastructure, including decentralized generation, smart grid projects, and "zero carbon" developments. Across a range of global cities, including, for example, London, New York, Los Angeles, Mexico City and Cape Town, new programs for reducing GHG emissions have been accompanied by overt references to enhancing the security and independence of energy supply for cities and reducing the costs of energy for residents (Hodson and Marvin 2010, Bulkeley and Schroeder 2012).

It is such principles that have also informed the development by the Chinese government of the "low carbon city" program, which was launched by China's Ministry of Construction and the World Wide Fund for Nature in 2008 (Liu and Deng 2011, p. 190), and which underpin the decision by the Taiwanese Environmental Protection Agency to build four pilot Low Carbon Cities (New Taipei City, Yilan County, Taichung City, Tainan City) by 2014 (EPA 2011).

Corporate actors, from banks to supermarkets, urban development companies, and utilities, are increasingly interested in the opportunities of low carbon urbanism. While this marriage of political and economic interest in the development of low carbon urbanism is becoming one dominant form, alternatives are also visible which often have a very different idea of how to secure and sustain urban communities. One such example is the Transition Towns movement. Initiated in the UK and now to be found in cities in North America, Asia, and Australia, the Transition Towns movement seeks to promote self-sufficiency as a means of achieving both community resilience and a response to the twin challenges of peak oil and rising GHG emissions (North 2010). Central to this vision is the notion of community engagement and involvement – rather than coming from the "top down" this is a vision of low carbon urbanism that is developed from the "bottom up" through local food growing projects, energy savings, and re-engagement with the local economy.

There is also evidence that alternative urban responses to climate change are emerging in cities in the Global South. One such example is the Kuyasa project in the Khayelitsha area of Cape Town. Financed through the Clean Development Mechanism and led by the NGO SouthSouthNorth, the project involved providing an energy upgrade to low-income housing which reduced energy use in households (hence yielding carbon savings) and energy poverty, providing direct financial benefits, as well as providing local employment opportunities. Another, example is the case of the ViDA (Vivienda de diseño ambiental) project in Monterrey, Mexico, led by Instituto de la Vivienda de Nuevo León, which has sought to develop a low carbon model for social housing that reduces energy consumption through passive design features and low-cost energy efficiency measures. While experiencing several challenges in terms of its maintenance and long-term energy savings on the ground, the ViDA project has become a "best practice" example for the national-wide "Hipoteca Verde" (Green Mortgage) program. In this manner, it has become part of a national response to demonstrate international commitment to addressing climate change, but one which is not solely concerned with economic security but where issues of social security are also regarded as important.

These different examples show that exactly what might constitute low carbon urbanism is still very much in the making. However, in most cases, rather than leading to the development of new forms of urban planning, or to systemic efforts to transform urban systems, what is emerging as a result of these multiple efforts is a patchwork mosaic of low carbon urbanisms – each different in its character, politics and possibilities. While those which are based on discourses of "secure urbanism" and "carbon control" suggest that low carbon urbanism is not only compatible with but essential to economic growth (Hodson and Marvin 2010, While *et al.* 2010), alternative approaches suggest other forms of development may be possible and that social change may be an equally important component to unlocking the potential of low carbon urbanism as are technological developments and new economic models.

REFERENCES

Betsill, M. and Bulkeley, H., 2007. Looking back and thinking ahead: a decade of cities and climate change research. *Local Environment: The International Journal of Justice and Sustainability*, 12 (5), 447–456.

Bulkeley, H. and Betsill, M.M., 2005. Rethinking sustainable cities: multilevel governance and the "urban" politics of climate change. *Environmental Politics*, 14 (1), 42–63.

Bulkeley, H. and Kern, K., 2006. Local government and the governing of climate change in Germany and the UK. *Urban Studies*, 43 (12), 2237–2259.

Bulkeley, H. and Schroeder, H., 2012. Global cities and the politics of climate change. In: P. Dauvernge, ed. *Handbook of global environmental politics*. Cheltenham, UK: Edward Elgar, 1–24.

Covenant of Mayors, 2011a. About the covenant [online]. Available from: http://www.eumayors.eu/about_the_covenant/index_en.htm [Accessed March 2011].

Covenant of Mayors, 2011b. Welcome [online]. Available from: http://www.eumayors.eu/home_en.htm [Accessed March 2011].

Environmental Protection Agency, 2011. Winners of low-carbon cities funding announced. Taiwan: Environmental Protection Agency Executive [online]. Available from: http://www.epa.gov.tw/en/NewsContent.aspx?path=426&NewsID=2735 [Accessed November 2011].

Gore, C. and Robinson, P., 2009. Local government response to climate change: our last, best hope? In: H. Selin and S.D. VanDeveer, eds. *Changing climates in North American politics: institutions, policymaking and multilevel governance*. Cambridge, MA: MIT Press, 138–158.

Hodson, M. and Marvin, S., 2009. "Urban ecological security": a new urban paradigm? *International Journal of Urban and Regional Research*, 33 (1), 193–215.

Hodson, M. and Marvin, S., 2010. *World cities and climate change: producing urban ecological security*. Milton Keynes: Open University Press.

Hoffman, M.J., 2011. *Climate governance at the crossroads: experimenting with a global response*. New York: Oxford University Press.

Kern, K. and Bulkeley, H., 2009. Cities, Europeanization and multi-level governance: governing climate change through transnational municipal networks. *Journal of Common Market Studies*, 47 (2), 309–332.

Liu, J. and Deng, X., 2011. Impacts and mitigation on climate change in Chinese cities. *Current Opinion in Environmental Sustainability*, 3 (3), 188–192.

North, P., 2010. Eco-localisation as a progressive response to peak oil and climate change – a sympathetic critique. *Geoforum*, 41 (4), 585–594.

US Mayors, 2011. Climate Protection Centre [online]. Available from: http://www.usmayors.org/climate protection/revised/ [Accessed May 2012].

While, A., *et al.*, 2010. From sustainable development to carbon control: eco-state restructuring and the politics of urban and regional development. *Transactions of the Institute of British Geographers*, 35, 76–93.

TWO

Version 2.1 Global GHG abatement cost curve beyond BAU – 2030

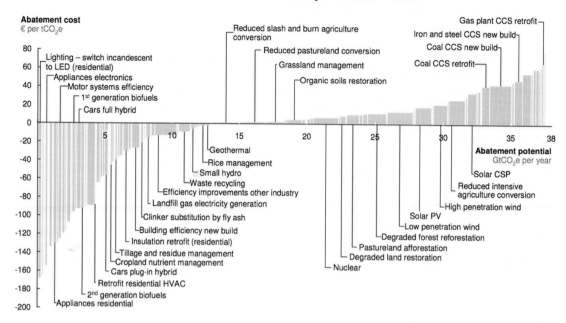

Note: The curve presents an estimate of the maximum potential of all technical GHG abatement measures below €80 per tCO₂e if each lever was pursued aggressively. It is not a forecast of what role different abatement measures and technologies will play.
Source: Global GHG Abatement Cost Curve v2.1

Figure 1 Global greenhouse gas abatement cost curve. Exhibit from "Impact of the financial crisis on carbon economies: Version 2.1 of the Global Greenhouse Gas Abatement Cost Curve", 2010, McKinsey & Company, www.mckinsey.com

15

"The Urbanization of Climate Change: Responding to a New Global Challenge"[1]

from *The Urban Transformation: Health, Shelter, and Climate Change* (2013)

William Solecki, Cynthia Rosenzweig, Stephen Hammer, and Shagun Mehrotra

Editors' Introduction

While many cities focused their climate planning efforts initially on efforts to reduce greenhouse gas emissions – the type of action known as "mitigation" – most have increasingly realized the importance of preparing communities for global warming impacts – strategies referred to as "adaptation." In this piece William Solecki, a geographer at Hunter College in New York, Cynthia Rosenzweig, a NASA climate scientist, and their colleagues emphasize the climate hazards faced by cities and policy responses. They portray climate adaptation as requiring cross-cutting initiatives in a variety of urban systems around land use, water, energy, transportation, and public health. Most of these steps help create sustainable cities in other ways as well.

Other resources on climate adaptation include Mark Pelling's *Adaptation to Climate Change: From Resilience to Transformation* (New York: Routledge, 2011), *Fairness in Adaptation to Climate Change*, edited by W. Neil Adger, Jouni Paavola, Saleemul Huq, and M.J. Mace (Cambridge: MIT Press, 2006), and *Adapting Cities to Climate Change: Understanding and Addressing the Development Challenges*, edited by Jane Bicknell, David Dodman, and David Satterthwaite (London: Earthscan, 2009), which focuses particularly on cities in the developing world.

Cities find themselves on the front lines of climate change.[2] The direct effects of a warming earth will exacerbate many longstanding urban ills, such as rapid population growth, sprawl, poverty and pollution, and, in general, climate change will stress the urban environment along multiple pathways (Rosenzweig *et al.*, 2011a; UN-Habitat, 2011). Its indirect impacts and feedbacks will also be most keenly felt in cities, simply because of their concentrated and integrated economic activities, their highly complex systems of infrastructure and social services, and their multilayered governance. Cities will need to find new ways to protect their citizens and assets, to determine how to set investment priorities for strengthening or replacing infrastructure, and to assess how climate change will affect their plans for long-term growth and development.

Cities must also be recognized as crucial elements in any global responses to climate change (Rosenzweig *et al.*, 2011a). Cities generate as much as 70 percent of global greenhouse-gas (GHG) emissions (International Energy Agency, 2008), and so they are obvious targets for mitigation efforts. But they also demand the special attention of policy-makers because of several factors, over-looked in early climate research, that make cities extremely vulnerable to climate change.

First, most people on the planet now live in cities, and urban growth is projected to continue well into the twenty-first century, nearly doubling to some 6.3 billion people by mid-century (UN, 2010). Second, cities are hubs of economic activity; they often support a larger metropolitan region or even the national economy. That, of course, also gives cities an advantage as well as a vulnerability – since, as centers of wealth and innovation, they often have the best tools and greatest resources for tackling the challenges of climate change.

Third, nearly all cities have grown up (and con-tinue to be built up) along coasts or riverbanks, exposing them to some of the most potentially damaging effects of climate change. For example, increases in sea level and large storm surges will threaten critical infrastructure. Also, more frequent and intense floods and droughts will put even greater demands on water supplies that are often scarce already. Fourth and finally, cities have outsize effects on their own environment: among their other environmental impacts, they create so-called urban heat islands (UHIs) and pollute their own air and water.

A generalized vulnerability and risk-management paradigm is emerging that provides a useful frame-work to city decision-makers for mitigating and adapting to climate change (Mehrotra et al., 2009; Rosenzweig et al., 2011b). City managers are increasingly recognizing that the environmental conditions of the past do not provide a particularly useful forecast of future environmental conditions. Instead, managers are depending increasingly on risk-based protocols for dealing with climate change (Yohe and Leichenko, 2010). Frameworks for assessing climate-change vulnerability and risk are typically developed out of three sets of indicators, on the basis of readily available data:

1   The climate hazards a city faces, such as more frequent and longer duration heat waves, more frequent heavy downpours, and more frequent and expanded coastal flooding.

2   Demographic and geographic features related to vulnerability, such as the size and density of a city's population, its topography, the portion of its population living in poverty, and the fraction of the national gross domestic product (GDP) that the city generates.

3   Indicators of adaptive capacity – in other words, data relevant to the ability of a city to act: What information about climate change is readily available? How many resources can be allocated for mitigation and adaptation? What institutions, governance bureaus and change agents are present and likely to be effective in helping the city to adapt?

CURRENT AND FUTURE CLIMATE HAZARDS IN CITIES

The scientific basis for action and policymaking by city decision-makers and other stakeholder groups comes from existing, city-specific climate data and down-scaled projections from global-climate models. Those models project that by the 2050s temperatures will increase around the world, on average, by between 1 and 4 Celsius degrees (Rosenzweig et al., 2011a). Cities, of course, will not be exempt from those increases.

The urban environment and the activities it supports will amplify those baseline effects. For one thing, cities are already hotter than their surrounding suburban and rural areas, both because concrete and other building materials absorb heat and because the landscape no longer benefits from the natural evaporative cooling of the trees and other vegetation that streets and buildings have displaced. The result is the now-familiar phenomenon of the UHI. Climate change will exacerbate UHI conditions in cities by intensifying summer heat waves, making them both hotter and longer than usual. Those heat waves in turn can lead to health crises and other negative effects on individuals, power shortages arising from increased use of air conditioning and other cooling technologies, and stress on machinery and other mechanical systems.

The concentration of residential, commercial, industrial and transportation systems (the last including roads, automobiles and railroads) in cities causes intense air pollution and its associated hazards to the health of urban residents. The warmer temperatures associated with climate change will aggravate those problems as well, producing greater amounts of secondary air pollution (e.g. oxides of nitrogen, NOx, and sulfur, SOx). Warmer temperatures also lead to greater energy use by heating, ventilation and air conditioning (HVAC) cooling systems and less-efficient machine functioning, further increasing the amount of primary and secondary air pollution.

Major contributors to weather and climate variability, such as tropical and extratropical storm systems as well as shifts in such global climate systems as the El Nino-Southern Oscillation and the North Atlantic Oscillation, affect climate extremes in cities. How the frequency of these events and systems will be altered, if at all, by anthropogenic climate change is still to be determined. What is clear, however, is that the greater the shift in a city's environmental baseline, the greater will be the impact of climate extremes.

Higher rates of sea-level rise, for instance, will lead to more severe coastal flooding. Changes in precipitation patterns are also projected to amplify local variability: the intensity of rain storms is expected to increase, causing more floods, but the time intervals between storms are expected to lengthen as well, leading to more droughts in many cities and regions.

CLIMATE CHANGE AND URBAN SECTORS AND SERVICES

As climate change emerged as an issue of global concern, cities focused on mitigation efforts to reduce energy consumption and carbon output. More recently, the emphasis has shifted to adaptation and resilience to climate change as well as to mitigation.[3] That shift makes it particularly important to understand how climate change will affect the ways urban systems operate within specific sectors and services: energy, water and wastewater, transportation and public health. Increased flooding, for instance, will degrade materials used in construction and infrastructure, which will particularly affect the energy and transportation sectors (Wilby, 2007).

Most impacts will be negative, as in the example of materials degradation, but not all: the demand for heat in the winter will likely decline. The take-home lesson is that analyzing how climate change is likely to affect specific sectors and services is critical if cities are to realistically consider their policy alternatives and develop effective strategies for adapting to (as well as mitigating) urban climate risks (Rosenzweig et al., 2011a). Here we briefly detail some of the most significant impacts across several sectors and services.

Energy

The effects of climate change on the energy sector will be felt on both supply and demand. Power plants are frequently located along bodies of water, hence are susceptible to both coastal and inland flooding – as the March, 2011, inundation of Japan's Fukushima Daiichi Nuclear Power Station made abundantly clear. As for hydropower, the projected changes in intensity and frequency of precipitation will increase the variability in both quantity and timing of the water available for conversion to energy. The likely increase in heat waves will lead to more peak load demands, which stress energy distribution systems and increase the chances of brownouts and blackouts. These interruptions of service will have negative effects both on local health and the local economy.

As we suggested earlier, climate change will generally reduce energy demand in cooler seasons, and it will increase demand in warmer seasons – but the overall impact will depend on the balance of seasonal effects. For any given city, analyses are needed to determine that balance; in general, the data recorded for such seasonal shifts show that the increased GHG emissions from increased cooling demands outweigh the GHG reductions from lower heating demands.[4]

For the energy sector, adaptation and mitigation strategies often overlap, and it is critical to emphasize both to help reduce the inevitable impacts of climate change. For example, programs for managing electricity demand to cut peak load blend elements of both adaptation and mitigation. So do projects for updating power plants and energy distribution networks, which aim to increase resilience to flooding, wind storms and extremes of temperature. A third example is diversifying the mix of fuels, including an increased share of renewable energy sources, that generate a city's power.

Seoul's Energy Declaration of 2007 well illustrates these parallel (and, at times, competing) sets of interest. The declaration focuses on improving the city's energy self-reliance, its use of alternative energy sources, and its commitment to demand-side management, as well as on building and enhancing strategies to cope with climate change (Kim and Choi, 2011). Yet in the cities affected by the declaration, climate-related concerns often take a back seat to the goals of reducing poverty, promoting economic development and improving social institutions by scaling up access to modern energy services. Adopting these mitigation measures could bring greater reliance on renewable sources of energy (including biomass-based fuels for cooking and heating), making cities even more vulnerable to climate change. After all, many sources of renewable energy are subject to changing climate regimes.

Water and wastewater

Cities are constantly trying to cope with water: maintaining supplies of fresh drinking water, managing excess water from flooding, controlling wastewater and sewerage flow.[5] Climate change will put all these systems under great stress. Both the quantity and quality of the water supply will be significantly affected by the projected increases in both floods and droughts (Aerts *et al.*, 2009; Case, 2008; Kirshen *et al.*, 2008).

Although precipitation is expected to increase in some areas, particularly in the mid and high latitudes, water availability is projected eventually to decrease in many regions, including cities whose water is supplied primarily by meltwater from mountain snow and glaciers.[6] The gap between water supply and demand will likely increase, as drought-affected areas expand (particularly for cities in the lower latitudes) and as floods intensify (Kamai-Chaoui, 2009). Within cities, impervious surfaces and increased precipitation intensity can overwhelm drainage systems.

In cities within developing countries, more than half the people rely on vendors, who make up an informal water supply system. As climate continues to change, both these informal services and the formal urban water supply will be highly vulnerable to drought, extreme precipitation and sea-level rise. Long-term planning for the impacts of climate

change on the formal and informal water-supply and wastewater-treatment sectors in cities is required, and plans should be monitored, reassessed and revised every five to ten years as climate science progresses and data improve.[7]

Several adaptation and mitigation strategies – often with co-benefits – are available for the water and wastewater sector, which make these systems more resilient in the face of increased supply and operational stress (Kirshen *et al.*, 2008; Nelson *et al.*, 2009). In the immediate future, programs for effective leak detection and repair, as well as for managing demand, should be undertaken in formal and, to the extent relevant, informal water-supply systems (Rosenzweig *et al.*, 2007). Demand can be managed, in part, through stronger water-conservation measures – beginning with low-flow toilets, showerheads and other fixtures.

Higher temperatures also bring higher evaporative losses, thereby both increasing demand for water and reducing its supply. Under those circumstances, among others, water reuse (e.g. the use of graywater) can play a key role in enhancing water-use efficiency, especially for landscape irrigation in urban open spaces.[8] Citywide water marketing through the informal private sector can also increase water-use efficiency, improve the robustness of the delivery system and help encourage reductions in water usage across various sectors of the economy. Where water is already becoming increasingly scarce, such as in Santiago, Chile, the capability of the water market to distribute water equitably – and thereby to resolve water conflicts – is expected to become increasingly important.[9] Water banking (whereby water collected during wet years is stored for use in dry years) is a way of hedging against uncertainties, and it, too, can improve system robustness. Capturing rainwater instead of pumping groundwater can substantially reduce the use of energy.[10]

Transportation

Because so many urban residents, particularly the urban poor, rely every day on transportation systems, any effects of climate change on urban transportation systems could have dramatic consequences for daily life (Revi, 2008).[11] Transport-related climate risks depend on the complex mix of transportation options unique to that city (Wilby, 2007), but certain broad distinctions among cities have outsize importance.

Cities tend to be built around either mass transit or the individually owned vehicle. And transportation systems may be built at ground level, underground or as elevated roads and railways. The potential impact of the various consequences of climate change, particularly flooding, evidently depends on those factors (Prasad *et al.*, 2009). Tunnels, vent shafts and ramps are clearly at risk. The possibility of flooding requires that large and numerous pumps be ready for use throughout these systems, and that a city maintain the capability to quickly remove debris and repair or replace such key infrastructure as motors, relays, resistors and transformers.

In addition to their vulnerability to rising sea levels and storm surges, transportation systems are particularly vulnerable to excessive heat. Overheating can cause buckling of steel rails, throwing them out of alignment, which can lead to train derailments.[12] Concrete roadways can also buckle or 'explode' and asphalt roads can melt. Excessive heat can also reduce the expected lifespans of train wheels and automobile tires. Whether a city's transportation system moves mainly people or goods also affects the risks – particularly the heat risks – associated with climate change.

Climate impacts on power and telecommunication systems can pose further risks in the transportation network. One of the most direct, and perhaps surprising, consequences of excess heat is that overhead wires can sag so low from expansion that they risk shorting out electrically.

Although urban transportation systems are exposed to many serious risks, they also have a key role to play in climate-change mitigation. We only scratch the surface of this role by highlighting such measures as adopting energy-efficient taxis and enhancing public transportation to reduce the use of individual vehicles. It is critical that climate considerations be incorporated into the plans, construction and management of transit systems, even as existing transportation assets are being retrofitted.

Public health

Cities are subject to serious health risks from climate change, since a large and high-density population amplifies the potential for negative outcomes.[13] The growth in the populations of urban poor and elderly compounds the threats of heat and vector-related diseases (that is, diseases transmitted by organisms, including humans, in the environment; Bartlett *et al.*, 2009). Cities with limited existing water services are also at greater risk of drought and vector-related diseases (Reid and Kovats, 2009). Other critical health-related issues can surface with rising sea levels and increased flooding of coastal zones (McGranahan *et al.*, 2007).

Because the infrastructure for health protection is already overburdened in many cities in developing countries, climate-change adaptation strategies need to focus on the most vulnerable urban residents. Adaptation and mitigation strategies associated with public-health issues in cities must be integrated with strategies for other sectors and services (Frumkin *et al.*, 2008; World Health Organization, 2009). Such strategies need to promote "cobenefits", such as reducing existing health hazards, particularly among those who bear the brunt of them, as well as helping to reduce people's health vulnerability to climate change (Bell *et al.*, 2007).[14] For example, efforts to ease the effects of UHIs via such passive approaches as tree planting and installing green roofs and permeable pavements not only will save energy (e.g. by reducing the need for air conditioning), but also promote better public health (Bell *et al.*, 2007).

Other strategies for adapting to climate change with minimal impact on public health include:

- improving and increasing water and energy service;
- regulating the growth of settlements in flood plains; and
- expanding health surveillance and early warning systems through technology and social networking.

Strategies for disaster risk-reduction also connect directly with strategies for maintaining public health (Solecki *et al.*, 2011). As we noted earlier, climate change is associated with extreme events, and such events have both direct and indirect implications for health. Cities in developing countries as well as in the developed world are actively working to link their climate change adaptation and disaster risk-reduction strategies, in hopes of enhancing resilience to climate change and adaptive capacity.

In Pune, India, for example, the Pune Municipal Corporation recently began a comprehensive disaster management-planning effort, focused on linking the resources of the city with those of regional management agencies. The resulting plan became the vehicle for a series of local adaptation-related activities, such as improved flood control and the development of early warning systems.[15]

CROSS-CUTTING ISSUES

It is entirely possible, of course, for inappropriate zoning, uninformed urban planning and general (mis)management of resources, combined with unchecked population growth, to exacerbate the effects of climate change. It is also possible (and clearly preferable) for cities to mitigate and adapt to it. City governments have many ways to strengthen their decision-making and take the latter course: through effective leadership, science-based policy-making, efficient and innovative financing, jurisdictional coordination – especially in cases of cross-boundary issues – and citizen participation and engagement (Bai, 2007).

We focus here on two points of discussion that deserve special attention: urban land-use planning and governance.

Urban land use and planning

The vulnerability of a city to climate change is not a fixed and unmanageable predicament. Rather, through the intelligent use of urban land, whatever vulnerability is present can be modified and, often, at least partly remedied. The remedies stem from an awareness of a city's natural setting, an understanding of how the design of urban form intentionally (or not) gives rise to the built environment, and an active effort to reduce the effects of UHIs (Blanco and Alberti, 2009).[16]

Adapting to climate change through urban land management involves many moving parts: the legal and political systems, planning departments, zoning regulations, infrastructure and urban services, land markets and fiscal arrangements. Planning and managing an effective response to climate change is highly dependent on coordinating these parts – the more so because many metropolitan areas

with seemingly common interests are politically fragmented. Cities in the Global South also have the additional disadvantages of lacking not only the extensive human and capital resources of their Northern counterparts, but also the traditional institutional mandate to implement and control local land use and development (Parnell *et al.*, 2009).

Several adaptation and mitigation strategies have been identified that reduce risk exposure, or vulnerability, promote reductions in energy use, or both (McEvoy *et al.*, 2006). Some of the strategies include relatively small-scale adjustments to existing building codes and land regulations that would bring big reductions in the hazards of climate change: elevating buildings in flood-prone areas, reducing energy use for heating and cooling, increasing the space planted in trees and vegetation to reduce the heat island effect (Condon *et al.*, 2009).

Other strategies would require more transformative shifts, many of which have been described in the hazard-mitigation literature: reducing sprawl by increasing the densities of people and of buildings, mixing land uses to reduce automobile traffic and increase the reliance on public transit, and restricting land use in areas subject to such predictable climate-change impacts as sea-level rise and riverine flooding.

Governance

Yet despite the abundance of good ideas in the literature, it is hard to overstate the many challenges local governments must face if their cities are to mitigate and adapt to climate change.[17] For any city, climate is just one of many issues on the local agenda. Governments must make trade-offs between current priorities and long-term risks. Often the uncertainties about the local impacts of climate change affect how a community sets priorities for investments and actions.

Local authorities can be constrained by policy and fiscal limitations. One of the most significant issues is how to finance climate-change action. Jurisdictional conflicts can also be a major challenge to local governments and the inclusion of multiple stakeholders adds a level of complexity that can prove hard to overcome. Yet despite all these difficulties, decision-makers and stakeholders in many cities are actively working on the challenges of implementing mitigation and adaptation strategies.

URBANIZATION AND THE ENVIRONMENTAL CRISIS OF CLIMATE CHANGE

Although cities may now find it daunting to respond to climate change, it is important to recognize that they have met environmental challenges in the past similar to the ones embodied in the risks of climate change. The responses of cities, furthermore, have been quite dynamic, and their residents have shown tremendous adaptive capacity. In many ways – as urban historian Thomas Bender notes – cities remain 'unfinished', and they are constantly being built and re-built, both literally and figuratively.

It is during this constant process of renewal that the opportunities for responding to past environmental challenges and promoting resilience can be seized. As cities grow, both resources and the capacity for continued expansion become limited. These limits occur frequently in a variety of contexts: environmental (e.g. limits on water and energy supply); societal (e.g. constraints imposed by ethnic and racial conflicts); political (e.g. limits of governance); and economic (e.g. limits on opportunity and employment). These limits can develop into crises that must be resolved if the city is to continue to grow.

The resolution of these crises often leads to fundamental transformations in the city, which are punctuated by transitions.

There are several root causes of environmental crises in cities. Many have to do with the natural setting and resources of the city (e.g. the lack of a natural supply of drinking water). Others derive from significant shifts in the character and structure of the connections between human and natural systems: a new and extensive reservoir system for supplying water, for example, or a major environmental disaster that alters the pattern of everyday life. Still other transformations are more explicitly societal in origin: a shift to a new mode of production and social reproduction, for example, as in the historical shift from a port-oriented, trading center to a production and manufacturing center.

Many drivers of environmental transitions and transformations have been documented, including the emergence of new technology, rapid population growth, the growth of poverty, the accumulation of wealth, the appearance of large-scale environmental hazards, the emergence of resource limitations involving local relative or absolute deficits of critical items such as water and energy, and dramatic shifts in institutional capacity and capability. But in all these cases, the role of crises is critical for understanding how the transitions occur.

What kinds of crises can drive transitions in cities? The list includes virtually any stress to which cities have been subjected: scarcity of ecological resources (water supply, food and energy); shifts in land use (e.g. the collapse of a peri-urban, agricultural community); demographic changes (e.g. ethnic and racial population shifts); socio-economic factors (e.g. war, civil disturbance, gentrification, capital disinvestment and reinvestment); and urban spatial development (e.g. congestion and lack of mobility).

The history of any city often includes many environmental crises and their associated transitions, each emerging over time and often overlapping with other crises and transitions. Crises often come to a head after several failed attempts to resolve an issue, during which time the issue deteriorates. When eventually the issue develops into a full-blown crisis, the economic stability, continued growth and future habitation of the city can come into question. In these situations, tipping points and thresholds are frequently passed and transitions occur. And that is not the end of the complications: the resolution of a crisis often leads to new conditions that contribute to future crises – at other spatial, temporal and social scales.

KEYS TO CLIMATE SUCCESS

Other municipalities and urban areas can draw from New York and other leading-innovator cities in devising their own decision-making support strategies and protocols for addressing climate change. Listing the elements that contribute to success is simple and straightforward: strong leadership; broad involvement among government, private sector and scientific communities; establishment of clear goals; creation of targeted tools to support the planning efforts; development of mechanisms for evaluating (and re-evaluating) the results; and creation of a knowledge network. As always, taking action that integrates all these elements is the hard part.

Ongoing assessments of climate change by cities themselves is critical if cities are to successfully adapt to climate change. By "owning" such assessments, cities can point to a process and cite a foundational knowledge base that confers the legitimacy, credibility and salience they need to assert their place at the table in global climate-change negotiations (Cash *et al.*, 2002). Legitimacy and credibility come about through interactions with scientific experts and city leaders (from both developed and developing countries); through membership in city associations such as ICLEI, C40, United Cities and Local Governments (UCLG) and Metropolis; through links with international development agencies such as Cities Alliance, the United Nations Environment Program (UNEP), UN-Habitat, and the World Bank; and through the support of nation-states themselves. We specifically include nation-states in this list, because it is vital for nations to recognize and support the role of cities as key partners in mitigation and adaptation.

The salience of urban climate assessments arises from the growing realization that cities are the 'first responders' to climate change, both in mitigation and in adaptation. The close connections between city leaders and their constituents – compared with the more distant relationships between national leaders and their citizens – make cities the appropriate level of government for a great many climate-change actions. That is not to say that other levels of government can stand aside: all government levels – local, provincial, national and international – are needed to respond to the immense challenges of climate change.

NOTES

1 Portions of this chapter were adapted from Rosenzweig *et al.* (2011a), Rosenzweig *et al.* (2010b) and Rosenzweig and Solecki (2010a).

2 We define "cities" here in the broad sense to be urban areas, including metropolitan and surrounding suburban regions. We conceptualize urbanization as ongoing and continuous system-level processes that take place in all cities, whether they be older, more established cities such as London and Tokyo or rapidly developing and growing cities such as Dkaha or Mumbai.

3–4 Hammer, S.A., Keirstead, J., Dhakal, S., Mitchell, J., Colley, M., Connell, R., Gonzalez, R., Herve-Mignucci, M., Parshall, L., Schulz, N., and Hyams, M. (2011) "Climate change and urban energy systems", in Rosenzweig *et al.* (2011a).

5–8 Major, D.C., Omojola, A., Dettinger, M., Hanson, R.T., and Sanchez-Rodriguez, R. (2011) "Water and wastewater", in Rosenzweig *et al.* (2011a).

9 Barton, J. and Heinrichs, D. (2011) "Santiago de Chile: adaptation, water management, and the challenges for spatial planning", in Rosenzweig *et al.* (2011a).

10 Major, D.C., Omojola, A., Dettinger, M., Hanson, R.T., and Sanchez-Rodriguez, R. (2011) "Water and wastewater", in Rosenzweig *et al.* (2011a).

11 See also Mehrotra, S., Lefevre, B., Zimmerman, R., Gercek, H., Jacob, K., Srinivasan, S., and Salon, D. (2011) "Climate change and urban transportation systems", in Rosenzweig *et al.* (2011a).

12 Mehrotra, S., Lefevre, B., Zimmerman, R., Gercek, H., Jacob, K., Srinivasan, S., and Salon, D. (2011) "Climate change and urban transportation systems", in Rosenzweig *et al.* (2011a).

13 Barata, M., Ligeti, E., De Simone, G., Dickinson, T., Jack, D., Penney, J., Rahman, M., and Zimmerman, R. (2011) "Climate change and human health in cities", in Rosenzweig *et al.* (2011a).

14 See also Barata, M., Ligeti, E., De Simone, G., Dickinson, T., Jack, D., Penney, J., Rahman, M., and Zimmerman, R. (2011) "Climate change and human health in cities", in Rosenzweig *et al.* (2011a).

15 Penney, J. (2011) "Disaster risk management in Pune, India", in Rosenzweig *et al.* (2011a).

16 See also Blanco, H., McCarney, P., Parnell, S., Schmidt, M., and Seto, K. (2011) "The role of urban land in climate change", in Rosenzweig *et al.* (2011a).

17 McCarney, P., Blanco, H., Carmin, J., and Colley, M. (2011) "Cities and climate change: the challenges for governance", in Rosenzweig *et al.* (2011a).

REFERENCES

Aerts, J., Major, D.C., Bowman, M., Dircke, P. and Marfai, M.A. (2009) *Connecting Delta Cities: Coastal Cities, Flood Risk Management, and Adaptation to Climate Change*, Free University of Amsterdam Press, Amsterdam.

Bai, X. (2007) "Integrating global concerns into urban management: the scale argument and the readiness argument", *Journal of Industrial Ecology*, 11, 51–92.

Bartlett, S., Dodman, D., Hardoy, J., Satterthwaite, D., Tacoli, C. (2009) Social Aspects of Climate Change in Urban Areas in Low- and Middle-Income Nations. World Bank Fifth Urban Research Symposium – Cities and Climate Change: Responding to an Urgent Agenda.

Bell, M.L., Goldberg, R., Hogrefe, C., Patrick, L.K., Knowlton, K., Lynn, B., Rosenthal, J., Rosenzweig, C. and Patz, J.A. (2007) "Climate change, ambient ozone, and health in 50 U.S. cities", *Climatic Change*, 82, 61–76.

Blanco, H. and Alberti, M. (2009) "Building capacity to adapt to climate change through planning", in H. Blanco and M. Alberti (eds) Hot, Congested, Crowded, and Diverse: Emerging Research Agendas in Planning, *Progress in Planning*, 71, 153–205.

Case, T. (2008) "Climate change and infrastructure issues. AWWA Research Foundation, drinking water research", *Climate Change* Special Issue, 18, 15–17.

Cash, D., Clark, W.C., Alcock, F., Dickson, N.M., Eckley, N. and Jager, J. (2002) "Salience, credibility, legitimacy and boundaries: linking research, assessment and decision making." KSG Working Papers Series RWP02-0 46, Harvard Kennedy School, Cambridge, MA.

Condon, P.M., Cavens, D. and Miller, N. (2009) "Urban planning tools for climate change mitigation", Policy Focus Report, Lincoln Institute of Land Policy, Cambridge, MA.

Frumkin, H., Hess, J., Luber, G., Malilay, J. and McGeehin, M. (2008) "Climate change: the public health response", *American Journal of Public Health*, 98, 435–445.

International Energy Agency (IEA) (2008) "Energy use in cities", in *World Energy Outlook*, IEA, Paris.

Kamai-Chaoui, L. (2009) "Competitive cities and climate change: an introductory paper", in OECD Regional Development Working Papers No.2, OECD, Paris.

Kim, K.-G. and Choi, Y.-S. (2011) "Seoul's efforts against climate change", in C. Rosenzweig, W.D. Solecki, S.A. Hammer and S. Mehrotra (eds) *Climate Change and Cities: First Assessment Report of the Urban Climate Change Research Network*, Cambridge University Press, New York.

Kirshen, P., Ruth, M. and Anderson, W. (2008) "Interdependencies of urban climate change impacts and adaptation strategies: a case study of Metropolitan Boston USA", *Climatic Change*, 86, 1–2, 105–122.

McEvoy, D., Lindley, S. and Handley, J. (2006) "Adaptation and mitigation in urban areas: synergies and conflicts", *Proceedings of the Institution of Civil Engineers*, 159, 4, 185–191.

McGranahan, G., Balk, D. and Anderson, B. (2007) "The rising tide: assessing the risks of climate change and human settlements in low elevation coastal zones", *Environment and Urbanization*, 19, 17–37.

Mehrotra, S., Natenzon, C., Omojola, A., Folorunsho, R., Gilbride, J. and Rosenzweig, C. (2009) "Framework for city climate risk assessment", World Bank Commissioned Research for Urban Research Symposium 5, World Bank, Washington, DC.

Parnell, S., Pieterse, E. and Watson, V. (2009) "Planning for cities in the global south: an Afrocam research agenda for sustainable urban settlements", in H. Blanco and M. Alberti (eds) "Shaken, shrinking, hot, impoverished and informal: emerging research agendas in planning", *Progress in Planning*, 72, 233–241.

Prasad, N., Ranghieri, F. and Shah, F. (eds) (2009) *Climate Resilient Cities*, WorldBank, Washington, DC.

Reid, H. and Kovats, S. (2009) "Special issue on health and climate change", *TIEMPO: A Bulletin on Climate and Development*, 71.

Revi, A. (2008) "Climate change risk: an adaptation and mitigation agenda for Indian cities", *Environment and Urbanization*, 20, 1, 207–229.

Rosenzweig, C. and Solecki, W. (eds) (2010a) Climate Change Adaptation in New York City: Building a Risk Management Response: New York City Panel on Climate Change 2010 Report, *Annals of the New York Academy of Sciences*, vol. 1196, New York Academy of Sciences, New York.

Rosenzweig, C., Solecki, W., Hammer, S.A. and Mehrotra, S. (2010b) "Cities lead the way in climate-change action", *Nature*, October 21, 467, 909–911.

Rosenzweig, C., Solecki, W., Hammer, S.A. and Mehrotra, S. (eds) (2011a) *Climate Change and Cities: First Assessment Report of the Urban Climate Change Research Network*, Cambridge University Press, New York.

Rosenzweig, C., Solecki, W.D., Blake, R., Bowman, M., Faris, C., Gornitz, V., Jacob, K., LeBlanc, A., Leichenko, R., Sussman, E., Yohe, G., Zimmerman. R. (2011b) "Developing coastal adaptation to climate change in the New York City infrastructure-shed: process, approach, tools, and strategies", *Climatic Change*, 106, 1, 93–127.

Rosenzweig, C., Major, D.C. and Demong, K. (2007) "Managing climate change risks in New York City's water system: assessment and adaptation planning", *Mitigation and Adaptation Strategies for Global Change*, 12, 8, 1391–1409.

Solecki, W., O'Brien, K. and Leichenko, R. (2011) "Disaster risk reduction and climate change adaptation strategies: convergence and synergies", *Current Opinion in Environmental Sustainability*, 3, 135–141.

UN-Habitat (United Nations Human Settlements Program) (2011) *Cities and Climate Change: Global Report on Human Settlements 2011*, Earthscan, London.

UN (United Nations) (2010) *World Population Prospects: 2009 Revision*, UN Department of Economic and Social Affairs, New York.

Wilby, R.L. (2007) "A review of climate change impacts on the built environment", *Built Environment*, 33, 1.

World Health Organization (WHO) (2009) *Protecting Health from Climate Change: Connecting Science, Policy and People*, WHO, Geneva.

Yohe, Gary and Leichenko, Robin (2010) "Chapter 2: Adopting a risk-based approach", in Cynthia Rosenzweig and William Solecki (eds) Climate Change Adaptation in New York City: Building a Risk Management Response: New York City Panel on Climate Change 2010 Report, *Annals of the New York Academy of Sciences*, 1196, 29–40, New York Academy of Sciences, New York.

LAND USE AND URBAN DESIGN

16

"The Next American Metropolis"

from *The Next American Metropolis: Ecology, Community, and the American Dream* (1993)

Peter Calthorpe

Editors' Introduction

In the early twentieth century, urban thinkers often focused on improving the physical form of the city, with specific proposals for new towns, improved neighborhoods, and dispersion of population from overcrowded industrial cities into regional constellations of communities. This tradition of visionary physical planning never entirely disappeared during the middle of the century – figures such as McHarg, American planning consultant Victor Gruen, and Greek visionary Constantine Doxiotis continued to explore new directions – but by and large urban planning became a more pragmatic field built on a foundation of scientific or economic analysis. Planning documents themselves no longer had as many maps, drawings, or graphic visions in them. Instead, many planners opted for the collection of quantitative data on economics, housing, or transportation, and relied on computer models and policy analysis. Some theorists such as University of California at Los Angeles urban geographer Edward Soja have argued that the dimension of "space" itself disappeared from planning discourses. Normative statements about what constitutes good city form also became scarce.

Toward the end of the century the pendulum began to swing back the other way, toward a renewed appreciation of the role of physical planning and urban design. Many observers came to see the need for new types of urban form that would make cities and towns more livable and ecologically oriented. Strong public movements to manage outward urban expansion ("growth management") and to create more coherent systems of parks, greenways, and open space also emerged. Jane Jacobs helped lay the groundwork for a renewed emphasis on "place-making" with her critique of the sterile, automobile-oriented urban landscapes created by much mid-twentieth-century modernist architecture and urban renewal. What was important, in her view, was the day-to-day life and vitality of urban places. MIT planning professor Kevin Lynch also helped catalyze a new interest in normative urban design values with books such as *Good City Form* (Cambridge, MA: MIT Press, 1981), which analyzed the physical form of human settlements throughout history and arrived at a set of design principles that Lynch argued were important for livable cities. University of California at Berkeley architecture professor Christopher Alexander and his colleagues likewise sought to determine features of what they called "the timeless way of building," and in their book *A Pattern Language* (New York: Oxford University Press, 1977) set forth a set of 50 characteristics of livable urban form throughout history that they argued could be combined to produce livable places.

These and other writers helped lay the groundwork for renewed attention to ways of creating livable, walkable places. But the leading movement in terms of actually changing community form came to be called the New Urbanism. This philosophy emerged in the late 1980s and 1990s as a number of architects and planners sought ways to create neighborhoods that emulated features of the traditional American small town. Early on, leaders of the movement used terms such as "traditional neighborhood design" to describe their work, and adopted many design concepts from towns laid out 100 years before such as grid-like street networks, mid-block alleys, village centers with small shops and workplaces, front porches, and garages in back of houses rather than in the front. (If these designers had used European small towns as a model instead, they might well have gravitated toward more winding, organic street patterns and more urban housing forms.)

Miami-based architects Andres Duany and Elizabeth Plater-Zyberk (designers of new communities such as Seaside and Kentlands), Bay Area-based designer Peter Calthorpe (designer of Laguna West and regional planning consultant for Portland, Salt Lake City, Minneapolis-St. Paul, and Chicago), and Los Angeles-based designers Stefanos Polyzoides and Elizabeth Moule were among the founders of the new movement. By taking the name Congress for the New Urbanism (CNU), they consciously positioned themselves as an alternative to the 1930s modernist architectural movement known as the Congrès Internationale des Artes Modernes (CIAM). The CNU held its first annual Congress in Alexandria, Virginia in 1993, and issued a Charter for the New Urbanism in 1996 (San Francisco: Congress for the New Urbanism, 2000). By the early twenty-first century several hundred New Urbanist-inspired neighborhoods had been built in North America, both on infill locations (within existing urban areas) and greenfield sites (unbuilt open land at the urban fringe). Equally importantly, New Urbanist design principles were diffusing into planning and design professions throughout the world. In Britain, Prince Charles' Prince of Wales Institute served as a vehicle for promoting similar sorts of urban design, and on the Continent architects such as Rob and Leon Krier designed relatively dense new urban additions to existing cities. Many New Urbanist projects can be seen as promoting sustainability, in that they help produce more compact, pedestrian-oriented, resource-efficient urban communities. Starting in 2007 the CNU has also developed an Urbanism +2030 project to promote a move towards low-carbon, resilient development (see http://www.cnu.org/urbanism2030). However, the New Urbanism can also be criticized on various grounds, such as for not focusing enough on affordable housing, infill development, and green building and site design practices.

Calthorpe, one of the leading New Urbanists, can be seen as an heir to Howard and Mumford in that through his regional and neighborhood planning work he has sought to develop a new version of the city–country balance. The co-editor (with ecological architect Sim Van der Ryn) of an earlier book entitled *Sustainable Communities* (San Francisco: Sierra Club Books, 1986), Calthorpe later sought a more pragmatic synthesis of pedestrian-oriented planning principles that could be adopted by the mainstream development industry. In works such as *The Next American Metropolis* (Princeton: Princeton Architectural Press, 1993), *The Regional City* (Washington, D.C.: Island Press, 2001; with William Fulton), and *Urbanism in the Age of Climate Change* (Washington, D.C.: Island Press, 2010), he has sought to promote coordinated physical planning changes on neighborhood, city, and regional scales. Calthorpe has also been a leading proponent of "transit-oriented development," clustering communities around a regional network of rail transit stations.

One of the greatest contributions of Calthorpe and other New Urbanists has been to develop consensus on specific design guidelines and place-making strategies. Calthorpe's graphics in this book represent some of these principles. More are provided by other New Urbanist designers such as Duany, Plater-Zyberk, and Jeff Speck in their book *Suburban Nation* (New York: North Point Press, 2000), and by organizations such as the Congress for the New Urbanism (www.cnu.org), the Sacramento-based Local Government Commission (www.lgc.org), and the Smart Growth Network (www.smartgrowth.org).

Although he speaks primarily to an American audience and talks of redefining the "American Dream," it is important to realize that Calthorpe is talking about a mode of development that has become common the world over – a suburban world of cul-de-sacs, detached single-family houses, single-use zoning, and dependence on automobiles. This "dream" is now sought with increasing frequency in Indonesia, South Africa, the Netherlands, Mexico, Eastern Europe, and countless other locations. Reasons for this include omnipresent American television, movies, and popular culture, the power of multinational corporations and their advertising to promote materialist lifestyles, and the employment of American planning consultants throughout the world.

The American Dream is an evolving image and the American Metropolis is its ever-changing reflection. The two feed one another in a complex, interactive cycle. At one point a dream moves us to a new vision of the city and community, at another the reflection of the city transforms that dream with harsh realities or alluring opportunities. We are at a point of transformation once again and the two, city and dream, are changing together. World War II created a distinct model for each: the nuclear family in the suburban landscape. That model and its physical expression is now stressed beyond retention. The family has grown more complex and diverse, while the suburban form has grown more demanding and less accessible. The need for change is blatant, with sprawl reaching its limits, communities fracturing into enclaves, and families seeking more inclusive identities. Clearly we need a new paradigm of development; a new vision of the American Metropolis and a new image for the American Dream.

The old suburban dream is increasingly out of sync with today's culture. Our household makeup has changed dramatically, the work place and work force have been transformed, average family wealth is shrinking, and serious environmental concerns have surfaced. But we continue to build post-World War II suburbs as if families were large and had only one breadwinner, as if the jobs were all downtown, as if land and energy were endless, and as if another lane on the freeway would end traffic congestion.

Over the last 20 years these patterns of growth have become more and more dysfunctional. Finally they have come to produce environments which often frustrate rather than enhance everyday life. Suburban sprawl increases pollution, saps inner-city development, and generates enormous costs – costs which ultimately must be paid by taxpayers, consumers, businesses, and the environment. These problems are not to be solved by limiting the scope, program, or location of development – they must be resolved by rethinking the nature and quality of growth itself, in every context.

This book attempts to map out a new direction for growth in the American Metropolis. It borrows from many traditions and theories: from the romantic environmentalism of Ruskin to the City Beautiful Movement, from the medieval urbanism of Sitte to the Garden Cities of Europe, from streetcar suburbs to the traditional towns of America, and from the theories of Jane Jacobs to those of Leon Krier. It is a work which has evolved from theory to practice in some of our fastest growing cities and regions. It is a search for a paradigm that combines the utopian ideal of an integrated and heterogeneous community with the realities of our time – the imperatives of ecology, affordability, equity, technology, and the relent-less force of inertia. The work asserts that our communities must be designed to reestablish and reinforce the public domain, that our districts must be human-scaled, and that our neighborhoods must be diverse in use and population. And finally, that the form and identity of the metropolis must integrate historic context, unique ecologies, and a comprehensive regional structure.

The net result is that we need to start creating neighborhoods rather than subdivisions; urban quarters rather than isolated projects; and diverse communities rather than segregated master plans. Quite simply, we need towns rather than sprawl.

Settlement patterns are the physical foundation of our society and, like our society, they are becoming more and more fractured. Our developments and local zoning laws segregate age groups, income groups, and ethnic groups, as well as family types. Increasingly they isolate people and activities in an inefficient network of congestion and pollution – rather than joining them in diverse and human scaled communities. Our faith in government and the fundamental sense of commonality at the center of any vital democracy is seeping away in suburbs designed more for cars than people, more for market segments than communities. Special interest groups have now replaced citizens in the political landscape, just as gated subdivisions have replaced neighborhoods.

REDEFINING THE AMERICAN DREAM

It is time to redefine the American Dream. We must make it more accessible to our diverse population: singles, the working poor, the elderly, and the pressed middle-class families who can no longer afford the "Ozzie and Harriet" version of the good life. Certain traditional values – diversity, community, frugality, and human scale – should be the foundation of a new direction for both the American

Dream and the American Metropolis. These values are not a retreat to nostalgia or imitation, but a recognition that certain qualities of culture and community are timeless. And that these timeless imperatives must be married to the modern condition in new ways.

The alternative to sprawl is simple and timely: neighborhoods of housing, parks, and schools placed within walking distance of shops, civic services, jobs, and transit – a modern version of the traditional town. The convenience of the car and the opportunity to walk or use transit can be blended in an environment with local access for all the daily needs of a diverse community. It is a strategy which could preserve open space, support transit, reduce auto traffic, and create affordable neighborhoods. Applied at a regional scale, a network of such mixed-use neighborhoods could create order in our balkanized metropolis. It could balance inner-city development with suburban investment by organizing growth around an expanding transit system and setting defensible urban limit lines and greenbelts. The increments of growth in each neighborhood would be small, but the aggregate could accommodate regional growth with minimal environmental impacts; less land consumed, less traffic generated, less pollution produced.

Such neighborhoods, called Pedestrian Pockets or Transit-Oriented Developments, ultimately could be more affordable for working families, environmentally responsible, and cost-effective for business and government. But such a growth strategy will mean fundamentally changing our preconceptions and local regulatory priorities, as well as redesigning the federal programs that shape our cities.

At the core of this alternative, philosophically and practically, is the pedestrian. Pedestrians are the catalyst which makes the essential qualities of communities meaningful. They create the place and the time for casual encounters and the practical integration of diverse places and people. Without the pedestrian, a community's common ground – its parks, sidewalks, squares, and plazas – become useless obstructions to the car. Pedestrians are the lost measure of a community, they set the scale for both center and edge of our neighborhoods. Without the pedestrian, an area's focus can be easily lost. Commerce and civic uses are easily decentralized into distant chain store destinations and government centers. Homes and jobs are isolated in subdivisions and office parks.

Although pedestrians will not displace the care anytime soon, their absence in our thinking and planning is a fundamental source of failure in our new developments. To plan as if there were pedestrians may be a self-fulfilling act; it will give kids some autonomy, the elderly basic access, and others the choice to walk again. To plan as if there were pedestrians will turn suburbs into towns, projects into neighborhoods, and networks into communities.

If we are now to reinvest in America, careful consideration should be given to what kind of America we want to create. Our investments in transit must be supported by land use patterns which put riders and jobs within an easy walk of stations. Our investments in affordable housing should place families in neighborhoods where they can save dollars by using their autos less. Our investments in open space should reinforce regional greenbelts and urban limit lines. Our investments in highways should not unwittingly support sprawl, inner-city disinvestments, or random job decentralization. Our investments in inner-cities and urban businesses ought to be linked by transit to the larger region, not isolated by gridlock. Our planning and zoning codes should help create communities, not sprawl.

Is such as transformation possible? Americans love their cars, they love privacy and independence, and they are evolving ever larger institutions. The goal of community planning for the pedestrian or transit is not to eliminate the car, but to balance it. In the 1970s the national love affair with the car was certainly hot, but we traveled on average 50 percent fewer miles per year than we do now. It *is* possible to accommodate the car and still free pedestrians. Practically, it means narrowing local roads and placing parking to the rear of buildings, not eliminating access for the car. Similarly, the suburban goals of privacy and independence do not have to be abandoned in the interests of developing communities with vital urban centers and neighborly streets. In fact, a walkable neighborhood may produce increased independence for growing segments of the population, the elderly and kids. The scale of our institutions may no longer fit the human scale proportions of an old village, but with careful design they could be integrated

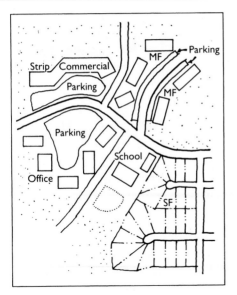

CONVENTIONAL SUBURBAN
DEVELOPMENT

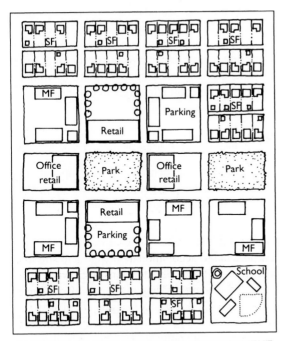

TRADITIONAL NEIGHBORHOOD DEVELOPMENT

Figure 1 Conventional suburban development (sprawl) versus traditional neighborhood development.

into mixed-use communities. Large businesses are quickly becoming aware of the benefits of being part of a neighborhood rather than an office park, with shared amenities and local services topping the list.

This new balance calls for the integration of seemingly opposing forces. Community and privacy, auto and pedestrian, large institution and small business, suburban and urban; these are the poles that must be fused in a new pattern of growth. The design imperatives of creating the post-suburban metropolis are complex and challenging. They are to develop a regional growth strategy which integrates social diversity, environmental protection, and transit; create an architecture that reinforces the public domain without sacrificing the variety and character of individual buildings; advance a planning approach that reestablishes the pedestrian in mixed-use, livable communities; and evolve a design philosophy that is capable of accommodating modern institutions without sacrificing human scale and memorable places.

DEFINITIONS

Transit-Oriented Development (TOD)

A Transit-Oriented Development (TOD) is a mixed-use community within an average 2,000 ft walking distance of a transit stop and core commercial area. TODs mix residential, retail, office, open space, and public uses in a walkable environment, making it convenient for residents and employees to travel by transit, bicycle, foot, or car.

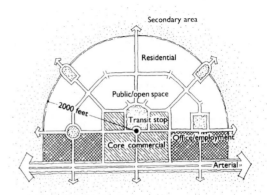

Figure 2 Transit-Oriented Development.

Figure 3 Housing types.

Residential areas

TOD residential areas include housing that is within a convenient walking distance from core commercial areas and transit stops. Residential density requirements should be met with a mix of housing types, including small lot single-family, townhomes, condominiums, and apartments.

Secondary areas

Each TOD may have a Secondary Area adjacent to it, including areas across and arterial, which are no further than one mile from the core commercial area. The Secondary Area street network must provide multiple direct street and bicycle connections to the transit stop and core commercial area, with a minimum of arterial crossings. Secondary Areas may have lower density single-family housing, public schools, large community parks, low intensity employment-generating uses, and park-and-ride lots.

Relationship to transit and circulation

The site must be located on an existing or planned trunk transit line or on a feeder bus route within 10 minutes transit travel time from a stop on the trunk line. Where transit may not occur for a period of time, the land use and street patterns within a TOD must function effectively in the interim.

Residential mix

A mix of housing densities, ownership patterns, price, and building types is desirable in a TOD. Average minimum densities should vary between

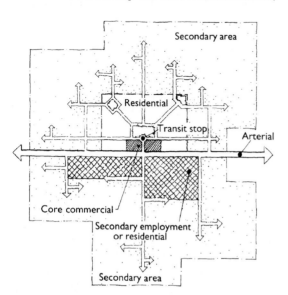

Figure 4 Secondary areas.

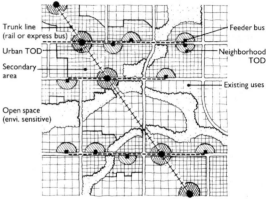

Figure 5 Relationship to transit.

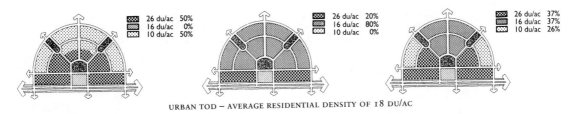

▨ 26 du/ac 50%	▨ 26 du/ac 20%
▨ 16 du/ac 0%	▨ 16 du/ac 80%
▨ 10 du/ac 50%	▨ 10 du/ac 0%

▨ 26 du/ac 37%
▨ 16 du/ac 37%
▨ 10 du/ac 26%

URBAN TOD – AVERAGE RESIDENTIAL DENSITY OF 18 DU/AC

Figure 6 Residential density mix.

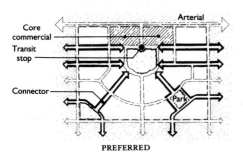

PREFERRED

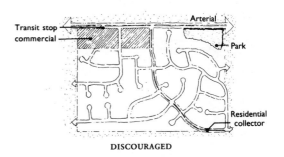

DISCOURAGED

Figure 7 Street and circulation system.

10 and 25 dwelling units/net residential acre (25 to 62 units/hectare), depending on the relationship to surrounding existing neighborhoods and location within the urban area.

Street and circulation system

The local street system should be recognizable, formalized, and inter-connected, converging to transit stops, core commercial areas, schools, and parks. Multiple and parallel routes must be provided between the core commercial area, residential, and employment uses so that local trips are not forced onto arterial streets. Streets must be pedestrian friendly; sidewalks, street trees, building entries, and parallel parking must shelter and enhance the walking environment.

Regional form

Regional form should be the product of transit accessibility and environmental constraints. Major natural resources, such as rivers, bays, ridgelands, agriculture, and sensitive habitat should be preserved and enhanced. An Urban Growth Boundary should be established that provides adequate area for growth while honoring these criteria.

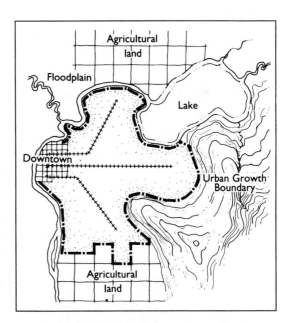

Figure 8 Regional form.

Criteria for New Towns

New Towns should only be planned if a region's growth is too large to be directed into Infill and adjacent New Growth Areas. They should be used to preserve the integrity of and separation between existing towns, as well as plan for a regional balance in jobs and housing. Appropriate sites should have

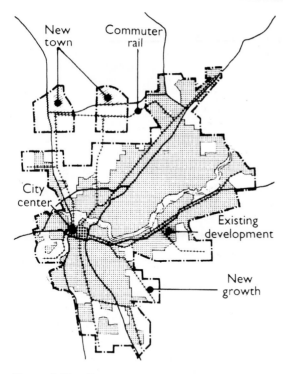

Figure 9 New Towns.

a viable commuter transit connection and are not on environmentally sensitive lands.

Open space resource protection

Major creeks, riparian habitat, slopes, and other sensitive environmental features should be conserved as open space amenities and incorporated into the design of new neighborhoods. Fencing and

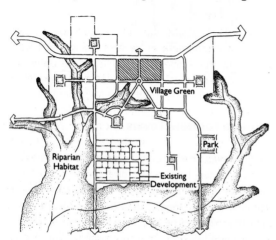

Figure 10 Open space.

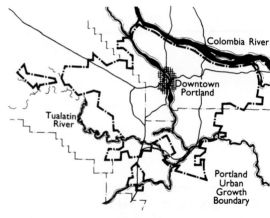

Figure 11 Urban growth boundaries.

piping of creeks should be avoided and channelization should be minimized.

Urban Growth Boundaries

Urban Growth Boundaries (UGBs) should be established at the edge of metropolitan regions to protect significant natural resources and provide separation between existing towns and cities. Lands within the UGB should be transit accessible, contiguous to existing development, and planned for long-term urbanization.

Wastewater treatment and water reclamation

On-site wastewater treatment facilities which use biological systems to reclaim water should be used

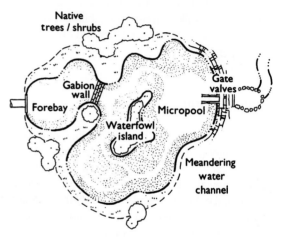

Figure 12 Waste water.

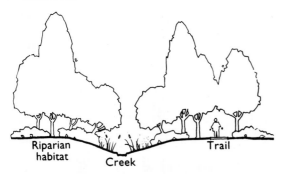

Figure 13 Drainage.

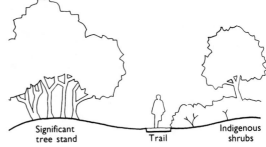

Figure 14 Landscaping.

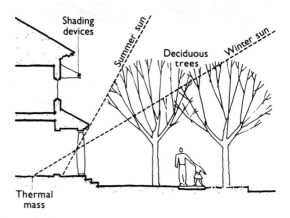

Figure 15 Energy conservation.

whenever possible. The reclaimed water should be used for on-site irrigation or for nearby farming.

Drainage and wetlands

Existing drainageways and wetlands should be maintained or enhanced in a natural state. In lower-density areas, drainage systems should recharge on-site groundwater by using swales and surface systems, rather than storm drains. All urban runoff must be treated on site with biological retention and filtration areas.

Indigenous and drought-tolerant landscaping

Landscape species used on public and private lands should be indigenous or proven adaptable to the local climate. In areas with water limitations, drought-tolerant species should be used in a majority of sites. Prominent stands of trees should be preserved.

Energy conservation

Energy conservation should be a goal of site as well as building design. Strategies such as passive solar,

natural ventilation, daylighting, and simple shading should be employed when cost-effective and appropriate to the climate. Micro-climate effects can be enhanced or mitigated through intelligent building configuration and landscape treatments.

Residential densities

Residential densities within Neighborhood TODs must be a minimum of seven units per net acre and a minimum average of at least ten units per net acre. Residential densities within Urban TODs must be a minimum of twelve units per net acre and have a minimum average of at least fifteen

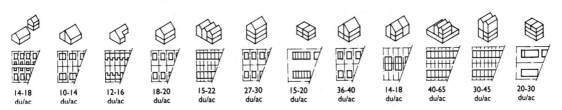

| 14-18 du/ac | 10-14 du/ac | 12-16 du/ac | 18-20 du/ac | 15-22 du/ac | 27-30 du/ac | 15-20 du/ac | 36-40 du/ac | 14-18 du/ac | 40-65 du/ac | 30-45 du/ac | 20-30 du/ac |

Figure 16 Residential densities.

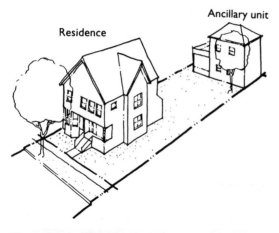

Figure 17 Ancillary units.

Figure 18 Building setbacks.

units per net acre. Maximum densities should be set by local plans.

Ancillary units

Ancillary units are encouraged to increase affordability and diversity. These units should be located in the single-family portion of residential areas. The additional unit will be counted toward meeting the minimum average density requirement.

Residential building setbacks

Residential building setbacks from public streets should be minimized, while maintaining privacy. Minimum and maximum front setbacks should be established that reflect the desired character of an

area and ensure that residences address streets and sidewalks.

Residential garages

Residential garages should be positioned to reduce their visual impact on the street. This will allow the active, visually interesting features of the house to dominate the streetscape. At a minimum, the garage should be set behind the front façade of the residential building. In single-family areas, garages may be sited in several ways: in the rear accessed from an alley, in the rear accessed by a side drive, or to the side recessed behind the front façade by at least 5 ft.

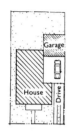

SIDE DRIVE (ATTACHED)

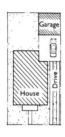

SIDE DRIVE (DETACHED)

RECESSED FRONT GARAGE

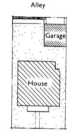

ALLEY
(ATTACHED OR DETACHED)

Figure 19 Garages.

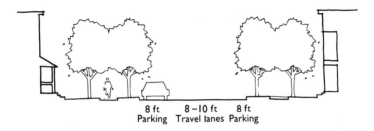

8 ft   8 – 10 ft   8 ft
Parking   Travel lanes   Parking

Figure 20 Streets.

Street dimensions and design speeds

Street widths, design speeds, and number of travel lanes should be minimized without compromising auto safety, on-street parking, or bike access. Streets should be designed for travel speeds of 15 mph. Travel lanes should be 8 ft to 10 ft wide.

Intersection design

Intersections should be designed to facilitate both pedestrian and vehicular movement. Intersection dimensions should be minimized while providing adequate levels of service.

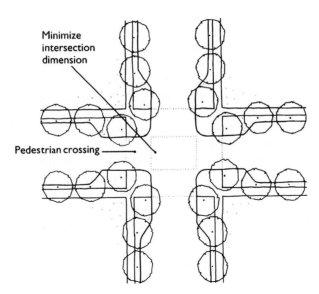

Minimize intersection dimension

Pedestrian crossing

Figure 21 Intersections.

17

"Compactness vs. Sprawl"
from *Companion to Urban Design* (2011)

Reid Ewing, Keith Bartholomew, and Arthur C. Nelson

Editors' Introduction

Compact development rather than suburban sprawl has often been thought essential for sustainable communities, but the debate is not necessarily simple. For one thing it has taken researchers many years to understand more exactly what "sprawl" is. The term doesn't just refer to low-density development, because there can be higher-density forms of sprawl as well. Rather, it refers to a package of qualities that can include a poor mix of land uses, poor street connectivity, discontiguous "leapfrog" development, and motor vehicle-oriented development. It is also far from clear exactly how much compactness or density to aim for in particular contexts. In parts of the world such as Asia, some existing development is extremely high density and arguably somewhat lower-density, greener, more walkable, more energy-efficient, and more mixed-use development is called for. (In recent years for example China has built enormous amounts of high-rise development with characteristics such as large block size that work against diverse, walkable urbanism.)

In North America, Europe, and elsewhere many planning initiatives have sought to encourage more compact urban development. Since the 1990s a nationwide "smart growth" movement has gathered steam in the US, essentially calling for denser, more mixed-use, transit-oriented, walkable communities. This movement borrows many principles from the New Urbanism and writers such as Calthorpe, but focuses also on reducing infrastructure costs for local governments and creating a fairer distribution of affordable housing. Many local governments, landowners, developers, and property rights advocates in turn have resisted this type of planning. Libertarians and free-market economists have argued that people freely choose to live in automobile-oriented, sprawling suburbs and that compact development is not necessarily a cure for traffic congestion.

A famous couple of articles in the mid-1990s in the *Journal of the American Planning Association* argued for and against compact urbanism. In opposing it, Peter Gordon and Harry Richardson asserted that North America still contains lots of undeveloped land, that the public prefers suburban living, and that various benefits of compact development were unproven. In supporting it, Reid Ewing argued that sprawl is not a natural consumer choice, but a product of market imperfections such as cheap gas and public subsidies for far-flung infrastructure. He also argued that many people do prefer compact centers, and that the environmental, social, and economic costs of sprawl aren't adequately taken into account.

In the following piece, Ewing and two co-authors from the University of Utah revisit the debate 15 years later. In light of much additional academic research they point out new costs of sprawl such as climate change and declining public health, and point out that the opponents' arguments about consumer choice hold less water now than they once did. They conclude by advocating "proactive planning" through which local governments can promote compact development. Other readings on this topic include *The Compact City: A Sustainable Urban Form?* (London: Spon Press, 1996), edited by Mike Jenks, Elizabeth Burton, and Katie Williams; *Compact Cities: Sustainable Urban Forms for Developing Countries* (London: Spon Press, 2000), edited by Mike Jenks and Rod Burgess; and *World Cities and Urban Form: Fragmented, Polycentric, Sustainable?* (London: Routledge, 2008), edited by Mike Jenks, Daniel Kozak, and Pattaranan Takkanon.

Some years ago the *Journal of the American Planning Association* published a pair of point-counterpoint articles now listed by the American Planning Association as "classics" in the urban planning literature. In the first article, "Are Compact Cities Desirable?" Peter Gordon and Harry Richardson argued in favor of urban sprawl as a benign response to consumer preferences (Gordon and Richardson, 1997). In the counterpoint article, "Is Los Angeles-Style Sprawl Desirable?" the lead author of this chapter argued for compact cities as an alternative to sprawl (Ewing, 1997). Gordon and Richardson and Ewing disagreed about nearly everything: the characteristics, causes, and costs of sprawl, and the cures for any costs associated with sprawl. This debate articulated the main disagreement in academic research about the nature and degree of interaction between land use and transportation.

In the intervening years, scholars have developed measures for sprawl and quantified its impacts for the first time. The smart growth, new urbanist, and transit-oriented development movements have come of age in reaction to sprawl. Life style changes and demographic shifts have begun to favor compact development, infill development, and walkable communities. Climate change, and its relationship to urban development patterns, has emerged as the main planning challenge of the twenty-first century. Concerns about rising obesity rates have led to a new partnership of public health professionals and planners under the rubric of active living.

Many states and cities in the US have undertaken smart growth initiatives. Light rail transit has been built in many American cities, while major highway expansion projects have been put on hold. Today, the case for sprawl seems dramatically weaker, and the case for compact communities dramatically stronger, than they did 12 years ago. It is time to reprise the debate based on new research and writing on these and related topics.

CHARACTERISTICS OF SPRAWL

Both articles used the term "compact" to describe one end of the development continuum, and "sprawl" to define the other end. Depending on the context, Gordon and Richardson equated compact development to high density or monocentric development,

arguing that a city like Los Angeles is in fact compact by virtue of its high average density. This is a most unfortunate characterization of compactness. High density is still not the preferred living arrangement for most Americans; and monocentric development is an anachronism, as central business districts have become just one of many centers in large metropolitan areas.

Density is only one dimension of sprawl, and that endless, uniform density is itself a hallmark of sprawl. Gordon and Richardson sometimes equated sprawl to "low density," and other times to "dispersed," "decentralized," "polycentric," or "suburban" development. In contrast, Ewing defined sprawl explicitly as one of three forms: first, leapfrog or scattered development, second, commercial strip development, or third, large expanses of low-density or single-use development. This definition comports with popular notions of sprawl. But even this definition has its limitations, and was expanded to include any development pattern characterized by poor accessibility and lack of functional open space. In Ewing's view, compact development was anything that didn't fit this definition, meaning a development pattern with contiguity, strong centers, mixed land uses, medium to high densities, good accessibility, and permanent open spaces.

CAUSES OF SPRAWL

Conceiving sprawl differently, the two earlier articles cited different reasons for its proliferation. To Gordon and Richardson, sprawl was a reflection of market forces. By their reasoning, consumers and businesses prefer outlying locations where land is inexpensive and congestion moderate. Modern telecommunications make clustering of businesses unnecessary. The low cost of automobile travel allows people to live far from their places of work and shopping. The resulting decentralized settlement patterns are economically efficient, and the only sources of market failure – that might render settlement patterns inefficient – are subsidies for the automobile (encouraging long-distance driving) and local land use regulations (discouraging higher densities and mixed uses).

In contrast, we view land markets as fraught with imperfections that induce sprawl. Perfectly functioning markets require many buyers and sellers, good information about prices and quality, homogeneous products in each market, no external costs or benefits, and so forth. Land markets meet none of these requirements. The rate of land appreciation is uncertain, causing land speculation and (where speculators guess wrong or land becomes legally encumbered) scattered development. Owner-occupied housing is subsidized through the tax code, a public policy that particularly benefits suburban residents who are primarily homeowners. Outlying development is subsidized through utility rate structures that are independent of distance from central facilities. The land market is rife with externalities, and government regulation may introduce additional market distortions by holding down densities and segregating land uses.

Given the choice between low-density suburban living and high-density urban living, most Americans will choose the former. However, compact alternatives to sprawl come in many forms, and these forms collectively have more than "boutique appeal" (Gordon and Richardson's dismissive term). Studies show that with a more complete set of housing choices compact development can hold its own in the marketplace (e.g. Belden *et al.* 2004; ULI and Pricewaterhouse Coopers 2005, 2006, 2007).

PUBLIC SUBSIDIES AND REGULATION

Consumer preferences help explain suburbanization and decentralization of activities within metropolitan areas, but they cannot explain the extent of dispersal, the absence of mixed land uses, and the loss of valuable natural areas. We must look to market failures to explain these phenomena.

The Office of Technology Assessment (1995) lists all manner of subsidies that result in urban sprawl. The biggest are subsidies for the highway system. If motorists had to cover the full social costs of automobile use – including vehicle emissions, free parking, uncompensated accident costs, military presence in the Middle East, and other external costs – they would likely opt for residential, work, shopping, and other locations that require a fraction of their current travel. This is what happens in Europe, where gasoline prices are about three times higher than in the US.

At least 14 studies have estimated the true social cost of automobile use. Arguably the most careful study was conducted by Mark Delucchi for the

Federal Highway Administration (Delucchi 1998). Nationwide totals from that study have been converted to a per vehicle mile basis by Todd Litman (2009) of the Victoria Transport Policy Institute. With a subsidy of just 30 cents per vehicle mile, the average auto user would have to pay an additional gas tax of $6 per gallon (20 mpg × 30 cents/gallon) to internalize the external costs of automobile use. We recently got a glimpse of what would happen if users were forced to pay the full cost of their automobile use out-of-pocket. With the 2007 spike in gasoline prices, the prices of suburban and exurban homes dropped considerably, but centrally located properties enjoyed price appreciation (Cortright 2008).

Does government regulation of land introduce market distortions? The literature since 1997 seems to suggest so, but there are also caveats. In his book *Zoned Out: Regulation, Markets, and Choices in Transportation and Metropolitan Land Use*, Jonathan Levine (2006) argues that extensive use of restrictive low-density zoning constrains the exercise of a free market in real estate development and that denser urban development would result from a more open market. In other words, one of the causes of sprawl is our current system of locally controlled zoning regulations that essentially mandate the construction of sprawl. This is a contention loudly voiced by new urbanist planners as well.

COSTS OF SPRAWL

Urban planners are ultimately less interested in development patterns, per se, than in the costs and benefits of one pattern versus another. That is to say, there are no inherently good or bad patterns, only good or bad outcomes. The loaded term "sprawl" has come to be applied to certain development patterns because of their documented negative outcomes. The following review focuses on costs of sprawl that have been extensively researched since the 1997 point-counterpoint articles.

Vehicle miles traveled

Gordon and Richardson claimed that ". . . the link between high-density development and reduced VMT (vehicle miles of travel), and hence reduced energy consumption, is by no means clear." This statement was simply incorrect. Even at that date a number of studies had linked density and other dimensions of compact development to lower VMT. The statement is even less defensible today. The potential to moderate travel demand through changes in the built environment is now the subject of more than 200 empirical studies. Indeed, it has become the most heavily researched subject in urban planning. There are at least 10 surveys of this literature (Handy 1996; Badoe and Miller 2000; Cao *et al.* 2009; Crane 2000; Ewing and Cervero 2001; Heath *et al.* 2006; McMillan 2005; Saelens *et al.* 2003; Stead and Marshall 2001; Saelens and Handy 2008).

In travel research, urban development patterns have come to be characterized by "D" variables. The original "three Ds," coined by Cervero and Kockelman (1997), are density, diversity, and design. The Ds have multiplied since Cervero and Kockelman's original article, with the addition of destination accessibility and distance to transit (Ewing and Cervero 2001). Areas that maximize these variables – such as center cities – may produce substantially lower VMT than dense mixed-use developments in the exurbs.

Oil dependence and climate change

Gordon and Richardson pointed to the "global energy glut," the weakness of the OPEC cartel, and the low real price of gasoline as evidence that energy impacts of sprawl were not worth worrying about. They argued that advances in vehicle emission control technology would solve our air quality problems. In contrast to this argument, Ewing (1997, 114) countered that:

> While the best case envisioned by [Gordon and Richardson] has the real price of gasoline holding steady, it is the worst case that worries others. . . . The fact that the most recent large-scale war fought was in the Persian Gulf is itself a testament to the risk of relying on the political stability of this region for a commodity [oil] so essential to economic activity. . . . Being unregulated, carbon dioxide emissions represent a bigger threat to national welfare than do regulated emissions. There is now a near-consensus

within the scientific community that carbon dioxide build-up in the atmosphere is causing global climate change, and that the long-term effects could be catastrophic.

Twelve years after the exchange, there seems to be little doubt that the "worst case" is upon us. Our dependence on foreign oil has never been greater. Gasoline recently peaked at an all-time high [in the US] of more than $4 per gallon, CO_2 concentration in the atmosphere is the highest it has been in the past 20 million years, and the fingerprints of climate change are everywhere (Emanuel 2005; Westerling *et al.* 2006; Madsen and Figdor 2007). Worldwide demand for oil continues to grow as Asia and the rest of the world follow the autocentric ways of the US.

The literature establishing connections between energy consumption and urban form, already well-established in 1997, has continued to expand (Alberti 1999; Andrews 2008; Bento *et al.* 2003; Burchell *et al.* 1998; Cooper *et al.* 2001; Ewing and Rong 2008; Kenworthy and Laube 1999; Saunders *et al.* 2008; US Environmental Protection Agency 2003). The results of these more recent studies further confirm that compact development patterns are substantially more energy-efficient than low-density sprawl. With the arrival of peak oil, compact development practices are likely to become more important for national energy policy. Among the benefits of compact development, perhaps the most important are greater energy security and reduced carbon footprint. Compact development can reduce fuel consumption and CO_2 emissions by 20 to 40 percent as compared to sprawl (Ewing *et al.* 2008).

Physical activity, obesity, and public health

The literature on the built environment and physical activity has expanded exponentially as funding has become available for active living research. The tremendous volume of research has generated a review of the many literature reviews (Gebel *et al.* 2007). Results of this research clearly show an association between the built environment and physical activity levels.

In 2003, sprawl was related for the first time to physical activity, obesity, and chronic diseases

(Ewing *et al.* 2003). After controlling for age, education, other sociodemographic and behavioral covariates, adults living in sprawling counties walked less, weighed more, were more likely to be obese and to suffer from high blood pressure than those living in compact counties. Seventeen of 20 subsequent studies have established statistically significant links between some aspect of the built environment and obesity (Papas *et al.* 2007), and correlations between sprawl and obesity have been affirmed (Committee on Physical Activity, Health, Transportation, and Land Use 2005; Frank *et al.* 2004; Kelly-Schwartz *et al.* 2004; Lopez 2004; Sturm and Cohen 2004; Cho *et al.* 2006; Doyle *et al.* 2006; Ewing *et al.* 2006; Rundle *et al.* 2007; Joshu *et al.* 2008).

Other costs

Recent research also shows that compact development outperforms sprawl in the following areas:

■ income growth (Nelson and Foster 1999; Nelson and Peterman 2000);
■ central city economic health (Nelson *et al.* 2004a, Dawkins and Nelson 2003);
■ protection of farmland (Nelson and Sanchez 2005);
■ racial integration (Nelson *et al.* 2004b; Nelson *et al.* 2005); and
■ residential neighborhood quality (Nelson *et al.* 2007)

CURES FOR SPRAWL

The only policy intervention endorsed by Gordon and Richardson was the imposition of congestion charges and emission fees as shadow prices for external costs of automobile use, specifically for delay and air pollution imposed on others. But while congestion pricing and emission fees have been touted by economists for decades, politicians have not exactly rushed to meter their constituents' travel.

A recent case in point is the failed effort of New York City to adopt an area-wide congestion pricing program. The program would have charged $8 for automobiles to enter a specific zone in a southern portion of Manhattan. Despite the power of Mayor Michael Bloomberg, as well as years of planning, the New York City Congestion Pricing Plan died a political death at the hands of the State Assembly. Area-wide congestion pricing is a good idea whose time, apparently, has not come [ed.: in the US; such pricing exists in cities such as London, Singapore, and Stockholm].

Our answer to sprawl is proactive planning of the type found almost everywhere except in the United States (but beginning to appear here out of necessity). What Gordon and Richardson refer to as "command-and-control" policies is really just planning. The common practice of local governments in the US, to wait for property owners to come forward with rezoning requests, is not planning but reacting, and hardly command-and-control.

Proactive plans should be supplemented by policies that reward good development and discourage bad. In the first wave of growth management nationally, the concern was how much growth would be allowed. In the second wave, the focus shifted to where and when growth would be permitted, and who would pay for it. The third wave is upon us, shifting the emphasis to what kind of growth is allowed or encouraged. Oregon's Transportation Planning Rule, New Jersey's State Plan, and California's Climate Change – Smart Growth Act (SB 375) are examples of initiatives to upgrade the quality of development, wherever and whenever it should occur.

Lest this answer to sprawl appear hopelessly European, Ewing (1997) cited an example from the United States. It wasn't Portland, OR, Arlington County, VA, or San Diego, CA, leaders in this type of planning. They could be dismissed as anomalies. It wasn't Charlotte, NC, Denver, CO, or Salt Lake City, UT. They hadn't embraced transit and transit-oriented development at that time. Rather it was Orlando, FL, home of Mickey Mouse and Shamu the killer whale. The point was that if Orlando could do it, anyone could.

Orlando government had entered into a partnership with the owners of multiple tracts southeast of the city. Through a cost-sharing arrangement, the partnership had prepared a master plan and development standards for the 12,000-acre site. The plan, development standards, and financial and administrative incentives are being used to encourage compact, mixed-use development where sprawl would otherwise almost surely occur.

An even better example of a public-private partnership is the new town-in-town of Baldwin Park, built on the site of the former Naval Training Center in Orlando (Ewing 2007). The city chose a master developer with a proven track record in reusing a military base. The two have been partnering ever since. In Baldwin Park, a public-private partnership has taken the place of Gordon and Richardson's unfettered market. Here, a handshake has replaced the invisible hand so revered by these two economists.

As the nation grows to 400 million by 2040, how it grows will affect the quality of life and economic well-being of Americans. The evidence is mounting that compact development, done right, can confer more benefits with fewer costs than urban sprawl. Concerns about energy, health, and climate change, along with shifting demographics, have resolved the debate of compactness vs. sprawl.

REFERENCES

Alberti, M. (1999). "Urban Patterns and Environmental Performance: What Do We Know?" *Journal of Planning Education and Research*, 19: 151–163.

Andrews, C.J. (2008). "Greenhouse Gas Emissions along the Rural–Urban Gradient." *Journal of Environmental Planning and Management*, 51(6): 847–870.

Badoe, D.A. and Miller, E.J. (2000). "Transportation–Land-Use Interaction: Empirical Findings in North America, and Their Implications for Modeling." *Transportation Research Part D*, 5: 235–263.

Belden, Russonello, and Stewart (2004). National Survey on Communities. Washington, D.C.: National Association of Realtors and Smart Growth America. www.brspoll.com/Reports/Smart%20Growth.pdf (accessed 18 August 2010).

Bento, A.M., Cropper, M.L, Mobarak, A.M., and Vinha, K. (2003). The Impact of Urban Spatial Structure on Travel Demand in the United States. Washington, DC: World Bank.

Burchell, R.W., Shad, N., Listokin, D., Phillips, H., Downs, A., Seskin, S., Davis, J.S., Moore, T., Helton, D., and Gall, M. (1998). The Costs of Sprawl – Revisited. Washington, DC: Transportation Research Board.

Cervero, R. and Kockelman, K. (1997). "Travel Demand and the 3Ds: Density, Diversity, and Design," *Transportation Research D*, 2, 199–219.

Cho, S., Chen, Z., Yen, S.T., and Eastwood, D.B. (2006). "The Effects of Urban Sprawl on Body Mass Index: Where People Live Does Matter," The 52nd Annual ACCI Conference, Baltimore, Maryland, March 15–18.

Committee on Physical Activity, Health, Transportation, and Land Use (2005). Does the Built Environment Influence Physical Activity? Examining the Evidence – Special Report 282. Washington, DC: National Academies of Science.

Cooper, J., Ryley, T., and Smyth, A. (2001). "Energy Trade-offs and Market Responses in Transport and Residential Land-use Patterns: Promoting Sustainable Development Policy." *Urban Studies*, 38(9): 1573–1588.

Cortright, J. (2008). Driven to the Brink. How the Gas Price Spike Popped the Housing Bubble and Devalued the Suburbs. CEOs for Cities, Portland, OR. http://brokerinsider.com/pdf/driventothebrinkfinal.pdf (accessed 18 August 2010).

Crane, R. (2000). "The Influence of Urban Form on Travel: An Interpretive Review." *Journal of Planning Literature*, 15(1): 3-23.

Dawkins, C.J. and Nelson, A.C. (2003). "Statewide Growth Management Policy and Central City Revitalization." *Journal of the American Planning Association*, 69(4): 381–396.

Delucchi, M. (1998). The Annualized Social Cost of Motor Vehicle Use in the United States, Based on 1990–1991 Data: Summary of Theory, Methods, Data, and Results, Institute of Transportation Studies, University of California, Davis.

Doyle, S., Kelly-Schwartz, A., Schlossberg, M., and Stockard, J. (2006). "Active Community Environments and Health: The Relationship of Walkable and Safe Communities to Individual Health." *Journal of the American Planning Association*, 72(1): 19–31.

Emanuel, K. (2005). "Increasing Destructiveness of Tropical Cyclones over the Past 30 Years." *Nature*, 436: 686–688.

Ewing, R. (1997). "Is Los Angeles-Style Sprawl Desirable?" *Journal of the American Planning Association*, 63: 107–126.

—— (2007). "Finding Happiness in Public-Private Partnerships: The Case for Case Studies." *Planning*, January, 53.

Ewing, R. and Cervero, R. (2001). "Travel and the Built Environment: A Synthesis." *Transportation Research Record*, 1780: 87–114.

Ewing, R. and Cervero, R. (2010). "Travel and the Built Environment: A Meta-Analysis." *Journal of the American Planning Association*, 76(3): 265–294.

Ewing, R. and Rong, F. (2008). "The Impact of Urban Form on US Residential Energy Use." *Housing Policy Debate*, 19(1): 1–30.

Ewing, R., Brownson, R.C., and Berrigan, D. (2006). "Relationship between Urban Sprawl and Weight of United States Youth." *American Journal of Preventive Medicine*, 31(6): 464–474.

Ewing, R., Schmid, T., Killingsworth, R., Zlot, A., and Raubenbush, S. (2003). "Relationship Between Urban Sprawl and Physical Activity, Obesity, and Morbidity." *American Journal of Health Promotion*, 18(1): 47–57.

Ewing, R., Bartholomew, K., Winkelman, S., Walters, J., and Chen, D. (2008). Growing Cooler: The Evidence on Urban Development and Climate Change, Washington, DC: Urban Land Institute.

Frank, L.D., Andersen, M.A., and Schmid, T.L. (2004). "Obesity Relationships with Community Design, Physical Activity and Time Spent in Cars." *American Journal of Preventive Medicine*, 27(2): 87–96.

Gebel, K., Bauman, A.E., and Petticrew, M. (2007). "The Physical Environment and Physical Activity: A Critical Appraisal of Review Articles." *American Journal of Preventive Medicine*, 32(5): 361–369.

Gordon, P. and Richardson, H. (1997). "Are Compact Cities a Desirable Planning Goal?" *Journal of the American Planning Association*, 63: 95–106.

Handy, S.L. (1996). "Understanding the Link between Urban Form and Non-work Travel Behavior." *Journal of Planning Education and Research*, 15(3): 183–198.

Handy, S., Sallis, J.F., Weber, D., Maibach, E. and Hollander, M. (2008). "Is Support for Traditionally Designed Communities Growing? Evidence from Two National Surveys." *Journal of the American Planning Association*, 74(2): 209–221.

Heath, G.W., Brownson, R.C., Kruger, J., Miles, R., Powell, K.E., Ramsey, L.T., and the Task Force on Community Preventive Services (2006). "The Effectiveness of Urban Design and Land Use and Transport Policies and Practices to Increase Physical Activity: A Systematic Review." *Journal of Physical Activity and Health*, 3: 55–76.

Joshu, C.E., Boehmer, T.K., Ewing, R., and Brownson, R.C. (2008). "An Examination of Personal, Neighborhood and Urbanization Correlates of Obesity in the United States." *Journal*

of Epidemiology and Community Health, 62: 202–208.

Kelly-Schwartz, A.C., Stockard, J., Doyle, S., and Schlossberg, M. (2004). "Is Sprawl Unhealthy: A Multilevel Analysis of the Relationship of Metropolitan Sprawl to the Health of Individuals." *Journal of Planning Education and Research*, 24: 184–196.

Kenworthy, J. and Laube, F. (1999). "A Global Review of Energy Use in Urban Transport Systems and Its Implications for Urban Transport and Land-Use Policy." *Transportation Quarterly*, 53(4): 23–48.

Leinberger, C.B. (2008). *The Option of Urbanism: Investing in a New American Dream*, Washington, DC: Island Press.

Levine, J. (2006). *Zoned Out: Regulation, Markets, and Choices in Transportation and Metropolitan Land Use*, Washington, DC: Resources for the Future.

Litman, T. (2009). Transportation Cost and Benefit Analysis II – Literature Review, Victoria Transport Policy Institute, Victoria, British Columbia. http://www.vtpi.org/tca/ (accessed 18 August 2010).

Lopez, R. (2004). "Urban Sprawl and Risk for Being Overweight or Obese." *American Journal of Public Health*, 94(9): 1574–1579.

Madsen, T. and Figdor, E. (2007). When It Rains, It Pours: Global Warming and the Rising Frequency of Extreme Precipitation in the United States. Boston, MA: Environment America Research & Policy Center. http://www.environmentamerica.org/uploads/oy/ws/oywshWAwZyEXPsabQKd4A/When-It-Rains-It-PoursUS-WEB.pdf (accessed 18 August 2010).

McMillan, T.E. (2005). "Urban Form and a Child's Trip to School: The Current Literature and a Framework for Future Research." *Journal of Planning Literature*, 19(4): 440–456.

Nelson, A.C. (2006). "Leadership in a New Era." *Journal of the American Planning Association*, 72(4): 393–407.

Nelson, A.C. and Foster, K. (1999). "Metropolitan Governance Structure and Economic Performance." *Journal of Urban Affairs*, 21(3): 309–324.

Nelson, A.C. and Peterman, D.R. (2000). "Does Growth Management Matter?" *Journal of Planning Education and Research*, 19(3): 277–286.

Nelson, A.C. and Sanchez, T.W. (2005). "The Effectiveness of Urban Containment Regimes in Reducing Exurban Sprawl." *DISP* 160: 42–47.

Nelson, A.C., Sanchez, T.W. and Dawkins, C.J. (2004a). "Urban Containment and Residential Segregation: A Preliminary Investigation." *Urban Studies*, 41(2): 423–440.

Nelson, A.C., Burby, R.J., Feser, E., Dawkins, C.J., Quercia, R., and Malizia, E. (2004b). "Urban Containment and Central City Revitalization." *Journal of the American Planning Association*, 70(4): 411–425.

Nelson, A.C., Dawkins, C.J. and Sanchez, T.W. (2005). "The Effect of Urban Containment and Mandatory Housing Elements on Racial Segregation in US Metropolitan Areas," *Journal of Urban Affairs*, 26(3): 339–350.

Nelson, A.C., Dawkins, C.J. and Sanchez, T.W. (2007). *Urban Containment and Society*. Hampshire: Ashgate.

Office of Technology Assessment (1995). *The Technological Reshaping of Metropolitan America*. Washington, DC: Congress of the United States, 193–218.

Papas, M.A., Alberg, A.J., Ewing, R., Helzlsouer, K.J., Gary, T.L., and Klassen, A.C. (2007). "The Built Environment and Obesity: A Review of the Evidence." *Epidemiologic Reviews*, 29(1): 129–143.

Rundle, A., Roux, A.V.D., Freeman, L.M., Miller, D., Neckerman, K.M., and Weiss, C.C. (2007). "The Urban Built Environment and Obesity in New York City: A Multilevel Analysis." *American Journal of Health Promotion*, 21(4): 326–334.

Saelens, B.E. and Handy, S. (2008). "Built Environment Correlates of Walking: A Review." *Medicine & Science in Sports & Exercise*, 40(S), S550–S567.

Saelens, B.E., Sallis, J.F., and Frank, L.D. (2003). "Environmental Correlates of Walking and Cycling: Findings from the Transportation, Urban Design, and Planning Literatures." *Annals of Behavioral Medicine*, 25(2): 80–91.

Saunders, M.J., Kuhnimhof, T., Chlond, B., da Silva, A.N. (2008). "Incorporating Transport Energy into Urban Planning." *Transportation Research Part A*, 42(6): 874–882.

Stead, D. and Marshall, S. (2001). "The Relationships between Urban Form and Travel Patterns. An International Review and Evaluation." *European Journal of Transport and Infrastructure Research*, 1(2): 113–141.

Sturm, R. and Cohen, D. (2004). "Suburban Sprawl and Physical and Mental Health." *Public Health*, 118(7): 488–496.

ULI – the Urban Land Institute and Pricewaterhouse Coopers LLP (2005). Emerging Trends in Real Estate 2005. Washington, DC.

—— (2006). Emerging Trends in Real Estate 2006. Washington, DC.

—— (2007). Emerging Trends in Real Estate 2007. Washington, DC.

US Environmental Protection Agency (2003). *Characteristics and Performance of Regional Transportation Systems*. Washington, DC.

Westerling, A.L., Hidalgo, H.G., Cayan, D.R., and Swetnam, T.W. (2006). "Warming and Earlier Spring Increase Western US Forest Wildfire Activity." *Science*, 313: 940–943. http://www.sciencemag.org/cgi/rapidpdf/1128834.pdf (accessed 18 August 2010).

18

"Infill Development"

from *Smart Infill* (2002)

Stephen M. Wheeler

Editors' Introduction

One of the main challenges to sustainable development of cities and towns is to build within the existing urbanized area rather than on "greenfield" land outside them. Such "infill" development often means reusing and restoring land that has already been built on in one way or another. It is a more complex and often more expensive form of development than suburban sprawl. But the rewards are great in terms of reducing motor vehicle use, providing needed homes, services, or jobs, taking advantage of existing infrastructure, reducing local government costs to service development, and restoring ecosystem elements such as creeks and wetlands that may have been inappropriately built on.

The following selection is part of an award-winning guidebook commissioned by the San Francisco Bay Area's Greenbelt Alliance. This open space advocacy group came to realize that one of the best ways to preserve open space at the edge of the region's cities is to encourage infill development. Other materials on this topic are available from the Urban Land Institute, a nonprofit association of development and real estate interests in the United States (www.uli.org); the Congress for the New Urbanism (www.cnu.org); and the Local Government Commission, a nonprofit based in California that promotes healthy, walkable, and resource-efficient communities (www.lgc.org).

"Infill" development refers to the construction of new housing, workplaces, shops, and other facilities within existing urban or suburban areas. This development can be of several types: building on vacant lots, reuse of underutilized sites (such as parking lots and old industrial sites), and rehabilitation or expansion of existing buildings. Through infill, communities can increase their housing, jobs, and community amenities without expanding their overall footprint into open space or otherwise undeveloped land.

Some infill development has always taken place within cities and towns. But the percentage of development that is infill instead of "greenfield" – on open space or agricultural land at the urban fringe – is relatively small in most US metro-politan regions. Instead of caring for and reusing our urban land, we have literally moved on to greener pastures. Meanwhile, as sprawl development draws jobs and people to the urban fringe, many older cities and suburbs have languished, with declining tax bases and little new investment.

Infill development can take many forms, described further below. In the past, not all infill has managed to create attractive places. But much has been learned in recent decades about how to design and build infill projects that add to quality of life for all members of the community. Successful infill development carefully integrates new projects into the urban context, adds needed housing and amenities, and attempts to meet needs of both existing neighbors and new workers or residents.

In many ways infill development represents the opposite of sprawl, in that it can help create compact and vibrant communities with a diverse mixture of land uses, well connected street patterns, and much-needed community resources such as parks, child care centers, shops, cafés, restaurants, schools, and walkable public spaces. Although infill development is sometimes thought of as a concern of older central cities, it is an important strategy for newer suburbs as well. Such development can help create active downtowns and a "sense of place" within suburban communities. It can also add a broader range of housing options, a better balance of workplaces, homes, and stores, and other community facilities.

Infill by itself won't solve regional growth problems. But combined with greenbelt protection, better public transit, more pedestrian-oriented street design, new congestion management efforts, measures to promote housing affordability, and protections for existing residents at risk of displacement, infill will be a central part of achieving regional smart growth and sustainability.

Infill development can take many forms tailored to the needs of particular communities. Residential infill can range from single-family detached homes to large multifamily developments. Mixed-use infill can vary from modest one- or two-story buildings on single lots to mid-sized complexes housing hundreds of residents to entire master-planned developments with housing, office, and commercial development. Compact infill can be achieved with many different housing forms, most of which still allow yards, patios, and/or a substantial amount of shared open space.

WHAT DENSITIES ARE APPROPRIATE?

Infill development often increases residential densities. Although "density" is often viewed as a negative, adding residents, jobs, and businesses to a community provides many advantages in terms of improving safety, increasing the viability of local businesses, adding cafes and restaurants, providing sufficient ridership for transit, and enhancing community interaction.

Recent US suburban densities have been relatively low, often six to eight dwelling units per net acre before local roads and public facilities are factored in (gross densities are even lower). By contrast, densities in many older American suburbs built a century ago are often ten to sixteen units per net acre. Densities for apartment buildings in downtown locations can range above 200 units per acre – yet such densities can fit well along existing streets. For example, a five-story, fifty-unit apartment building on a quarter-acre, 100 ft by 100 ft lot represents a net density of 200 units per acre, and can still have an attractive courtyard, entry plaza, and rooftop deck.

To make efficient use of well located sites, most infill buildings should be at least three to five stories tall with ground-floor shops or offices. These buildings represent net densities of 30–200 units per acre. In less central locations, townhouses, duplexes, small apartment buildings, and even small-lot detached houses can provide attractive housing choices at twelve to thirty units per acre, a level that can support public transit. In existing single-family neighborhoods, a very simple step that can double residential density with little or no change to neighborhood character is to allow homeowners to add second units behind or within existing houses.

MAKING INFILL AFFORDABLE

Cities and towns will need to take additional steps to ensure that infill housing is affordable to residents in all income categories. One main strategy to do this is to adopt "inclusionary zoning" requirements. These mandate that developers make a certain percentage of units in each project affordable to residents in specified income categories. Typically, inclusionary requirements only apply to developments of more than a certain size, often ten units, and require that 10 or 20 percent of units be affordable to households making 80 percent or less of the county median income. Some municipalities specify that some units also be affordable to very low-income households making 50 percent or less of the median.

Some cities allow developers to pay an in-lieu fee instead of actually creating affordable units. The city will presumably then use this money to construct its own affordable housing. However, such fees may be less effective in creating affordable housing, as the amount of the fee is often relatively low compared to the cost of constructing a unit. In-lieu fees also don't have the advantage of integrating affordable units into each new project that gets built.

Other financing strategies for affordable housing include charging fees to new retail or office development to support housing for service workers and other less affluent residents, increasing the level of funding set aside for affordable housing within municipal redevelopment programs, and bond financing measures to fund local government housing efforts.

Municipalities frequently provide low-interest loans or grants to non-profit housing providers to ensure that affordable units get created. A number of cities have established Affordable Housing Trust Funds for this purpose, and many use Community Development Block Grant (CDBG) monies from the federal government to support affordable housing. Such direct financial commitments are necessary to close the gap between what residents can afford and what housing costs to build in expensive areas. Other initiatives such as fast-tracking permits for affordable housing development, changing zoning standards to allow denser housing, and reducing parking requirements can also help make infill housing affordable.

AVOIDING DISPLACEMENT

One major risk of infill development is that it will gentrify neighborhoods and displace existing low-income communities in favor of affluent white residents. Often these long-time residents must then leave the region altogether to find affordable housing, or move to outlying locations. Infill development – especially of affordable units – can help fight such displacement by providing housing for existing lower-income residents. Creation of new jobs at infill locations can also give central city residents much-needed economic opportunities. Addition of stores and services can provide older urban neighborhoods with desperately needed amenities. (Often these neighborhoods lack basic services such as nearby supermarkets and banks.) In these ways, well planned infill can reduce displacement and not simply promote gentrification as investment returns to older neighborhoods and they gain popularity.

Municipalities can reduce displacement through a number of steps. They can adopt controls on conversion of rental properties to condominiums, a process that tends to decrease the supply of affordable rental housing. They can require that developers renovating or tearing down older housing replace any affordable units that would be lost on a one-for-one basis. They can support the construction of new affordable housing in infill neighborhoods. They can institute "inclusionary zoning" which requires that a certain percentage (usually between 10 and 20 percent) of units in all new developments be affordable to residents making 80 or even 50 percent of the local median income. They can institute "anti-flipping" taxes so that speculators holding properties for shorter periods pay larger percentages of their gains as taxes. And they can provide assistance directly to existing low-income residents to fix up their properties or to purchase new homes. Rent controls are another potential though controversial strategy as well.

A number of cities and towns have adopted or are considering other means to stabilize and preserve existing affordable housing, such as emergency rental assistance funds, mobile home park preservation programs, and landlord–tenant mediation requirements.

To help reduce displacement and meet the needs of existing residents, cities can make a strong commitment to public involvement in planning new infill development. Special outreach efforts may be needed to ensure that current residents are made aware of infill planning, and are invited to help decide the nature and design of new development.

STRATEGIES FOR PRODUCING INFILL

How do we increase the amount and quality of infill development? A number of strategies can help:

1. Ensuring land availability

One main challenge is ensuring that enough land exists to accommodate population growth in infill locations. Although there are still many vacant parcels in US cities, these are slowly disappearing as developers build new infill projects. Remaining vacant lots are often challenging in terms of size, location, or toxic contamination.

Although the supply of vacant infill parcels is diminishing, another large pool of infill land exists: underused parcels that can be redeveloped. These "refill" sites include surface parking lots, declining shopping centers, underused motels, decaying industrial districts, and low-intensity commercial strips. Old factories, sports stadiums, office parks, motor vehicle dealerships, military bases, and airports can also be redeveloped into compact, livable, mixed-use neighborhoods. In addition, many older buildings can be rehabilitated or expanded to create new infill housing, shops, or offices. Many cities have also zoned excessively for commercial development, on the theory that it will bring them greater tax revenue. Rezoning some of these areas for housing can help accommodate new residents within the existing urban footprint. Lastly, infill housing and stores could be added to many office parks and corporate campuses.

2. Reducing fiscal disincentives

One of the main reasons that cities and towns don't promote residential or mixed-use infill development currently is that there is very little economic reason for them to do so. Jurisdictions have little to gain from such development in terms of tax revenue, and much to lose in that they will need to provide services for residents. Much of this situation results from tax structures in the United States that emphasize sales taxes as a form of funding for local services. Action at the state level to address this situation is much needed, shifting the burden on to income or property taxes or promoting "tax base sharing" among local governments within urban regions. Increased federal support of local government, through expanded CDBG grants and increased funding for transit, could help as well.

3. Preparing Specific Area Plans

Often infill development may not occur in a particular area unless the municipality takes a lead in promoting it, working with neighbors, assembling parcels of land, and providing needed infrastructure

and amenities. Or infill may take place in a poorly designed and uncoordinated fashion, failing to help create an attractive, livable new neighborhood.

Specific Plans (also known as Area Plans or Precise Plans) help establish a framework for co-ordinating infill development and involving existing residents and businesses in developing a vision for a particular place. Essentially a city or other authority undertakes a planning process for a particular district, neighborhood, or large site. Workshops or design charrettes are held and public input solicited. Professional consultants may be hired to coordinate public involvement, prepare urban design guidelines, and produce an Environmental Impact Report if required by state law. The final plan is then approved by the local governing body as the framework for future development.

4. Revising zoning codes

Communities across the country have zoning codes in place that work against affordable, compact, mixed-use infill development. Particularly counterproductive are zoning rules limiting urban residential densities, prohibiting mixed-use development, setting one- or two-story height limits, and prohibiting secondary units in existing single-family home districts. Cities and towns can review their zoning codes for such limitations, and change them where appropriate. For example, minimum densities can be specified instead of maximum densities. Smaller lot sizes can be encouraged or required. And height limits can be raised to at least three stories in most locations.

5. Rethinking parking standards

Overly high parking standards for commercial or residential development waste valuable land, discourage transit use, reduce housing affordability, and diminish the comfort and safety of pedestrians. Excessive parking requirements add enormous costs to infill development. Surface parking frequently costs developers $5,000 or more per space; underground or structured parking can range between $20,000 and $50,000 per space. These costs require developers to charge higher prices for housing and may make entire developments financially infeasible.

Suburban jurisdictions – which paradoxically have the most on- and off-street space for parking – also usually have the toughest parking requirements. Many suburbs require two or more off-street spaces per unit, even if new homes are centrally located or near transit. Cities should review parking requirements, reduce them where possible, allow developers to adopt space-efficient methods such as tandem parking and stacked parking, and use market pricing to reduce parking demand.

6. Improving financing options

At certain times financing has been the biggest obstacle to infill development – in particular for affordable housing or mixed-use projects. Bankers are also often wary of mixed-use projects. Cities and towns can help address financing problems by making loans or grants available, creating or expanding Housing Trust Funds, lowering fees for infill as opposed to greenfield development, and pressuring banks to increase their lending to inner cities under the Community Investment Act or other regulation. State, federal, and regional action is also needed to improve financing options for infill, such as through expanded tax credit programs.

7. Establishing urban design guidelines

In the past infill development has frequently not been designed in ways that enhance overall neighborhood or community quality. New buildings have at times ignored the local architectural and historical context, blocked sun and views, or featured boring, monotonous facades. Landscaping has sometimes been poor or nonexistent. Inwardly facing complexes have turned their back on the street, reducing the quality of the pedestrian environment. Too often

Figure 1 Infill development comes in many forms. (1) Multifamily housing: a new five-story apartment building in Berkeley, CA.

Figure 2 Infill development. (2) Stapleton main street, part of an infill neighborhood, the center of the new Stapleton neighborhood in Denver, CO (on the site of the former airport).

little effort has been made to provide streetfront retail, community services, or pocket parks that could meet neighborhood needs.

The past decade or two have seen a revolution in knowledge about how to design successful infill development. Movements such as the New Urbanism have developed useful guidelines for designing livable, walkable communities. Municipalities can adopt urban design guidelines based on this knowledge. These guidelines can clarify for developers, planners, design review committees, and the public characteristics that the city or town would like to see in infill development.

8. Streamlining permitting processes

Lengthy and often difficult permitting processes can work against infill development. These procedures can require expensive studies or project redesign,

and related delays increase finance costs for developers, who often face additional risk and cost associated with unclear approval requirements. Timelines of a year and a half or more are common for infill development permitting.

It is important for local residents to have opportunity for input on infill projects. However, lengthy permitting processes help political opposition emerge and give neighbors numerous opportunities to derail projects. Meanwhile, the need to attend repeated hearings makes it difficult for advocates to support good infill development. To address this situation, cities can set time limits on permit processing, assign staff to shepherd each project through, carry out pre-application reviews to anticipate and address problems before they arise, adopt clear procedures for review, and establish "as of right" zoning under which projects meeting all zoning requirements and design guidelines are allowed to build without lengthy hearings.

Figure 3 Infill housing. (3) Stapleton residential street.

9. Working constructively with neighbors

One of the preeminent obstacles to infill development – and the single biggest concern of many developers – is community opposition, which in its worst forms is known as NIMBYism ("Not In My Back Yard"). Neighbor opposition can kill projects directly by turning zoning boards or city councils against them. Or neighbors can drag out the permitting process so long that developers lose money and projects no longer make financial sense.

Cities can adopt a number of strategies to reduce NIMBY opposition. They can require developers to meet with neighbors before submitting plans for a project. They can prepare Specific Plans in which residents have an opportunity to help set the overall framework for development in the neighborhood. They can organize meetings between developers and key neighborhood leaders to develop buy-in. They

can encourage community development corporations (CDCs), which have a strong neighborhood base, to undertake infill development, and they can encourage developers to add amenities that neighbors really need.

10. Cleaning up brownfields

"Brownfields" problems – having to do with toxic contamination of previously used sites – are common in most older cities and towns. In recent years a variety of programs has been put in place at state and federal levels to assist with brownfields cleanup. Cities can also set up offices to coordinate such cleanup, and map toxic problems within the city to provide information to future developers, can provide low-interest loans for cleanup, or can undertake such activities themselves, especially in cases where city groundwater is threatened.

11. Improving consistency and completeness

Often local plans, codes, and processes are not internally consistent regarding infill. Some support infill, while others work against it. For example, even though housing and land uses elements of a city's General Plan may call for infill, density and height limitations in the zoning code may be too low for infill to make economic sense. Decisions of planning commissions, zoning boards, design review committees, and even city councils may also not be consistent with such policies. To address this situation, cities and towns can direct planning staffs to review plans, codes, and processes to improve consistency, and can review actions of city boards and commissions.

12. Revitalizing communities and adding amenities

Infill development often doesn't take place because of systemic problems with community decline. Older cities may have lost residents, businesses, and tax base to newer suburbs, and have accumulated a host of problems including poor schools, crime, unemployment, pollution, lack of green spaces, and deteriorated infrastructure. Such systemic problems are not easy to fix. But local governments can do several things. They can focus attention on specific neighborhoods with high infill potential where the city can add a broad range of amenities. They can target infill projects at groups most likely to thrive in an upcoming central city environment – young professionals, singles, artists, and couples without children. They can locate public buildings within such areas to catalyze other development. They can use redevelopment powers to leverage up-front improvements with bonds that will be paid back through future increased tax revenues. And they can work with other local governments for regional tax sharing to address inequities in funding. Meanwhile, state and federal action may be required to improve schools and infrastructure in such areas.

T
W
O

19

"Outdoor Space and Outdoor Activities"

from *Life Between Buildings* (1980)

Jan Gehl

Editors' Introduction

Beginning in the 1960s writers such as Jacobs, Lynch, William H. Whyte, Clare Cooper Marcus, and Danish designer Jan Gehl emphasized the need to base urban design on the study of how people actually experience and use urban environments. A new discipline of environmental design emerged, devoted to researching how built environments work for people. Researchers developed methods using behavior observation, time-lapse photography, post-occupancy evaluation surveys, and cognitive mapping (in which people were asked to draw maps or images of how they perceived their urban environments) to provide factual information for improved urban design.

In his pioneering book *Life Between Buildings: Using Public Space* (New York: Van Nostrand Reinhold, 1980), Gehl took a remarkably perceptive look at different types of outdoor spaces and their social uses. What is most needed, he argued, is an increase in optional activities taking place in the public realm. The number and variety of human interactions, especially chance meetings in public spaces, was in his view the way to a healthier urban community. Analyzing public spaces within Copenhagen, he found places such as the Ströget (one of Europe's pioneering pedestrian streets) and the Tivoli Gardens particularly conducive to social life. Although many of Gehl's observations might seem common sense today, they represented a major departure from modernist urban design practices in which abstract architectural principles, rather than careful observation of how people actually use places, often dictated urban form. Other books in this vein include Lynch's *The Image of the City* (Cambridge, MA: MIT Press, 1960), Whyte's *The Social Life of Small Urban Spaces* (Washington, D.C.: The Conservation Foundation, 1980), Marcus and Wendy Sarkissian's *Housing as if People Mattered: Site Design Guidelines for Medium-Density Family Housing* (Berkeley: University of California Press, 1986), Marcus and Carolyn Francis' *People Places* (New York: Van Nostrand Reinhold, 1998), Gehl's *Cities for People* (Washington, D.C.: Island Press, 2010), and Elizabeth Dunham Jones' *Retrofitting Suburbia: Urban Design Solutions for Redesigning Suburbs* (Hoboken, NJ: Wiley, 2011).

THREE TYPES OF OUTDOOR ACTIVITIES

An ordinary day on an ordinary street. Pedestrians pass on the sidewalks, children play near front doors, people sit on benches and steps, the postman makes his rounds with the mail, two passersby greet on the sidewalk, two mechanics repair a car, groups engage in conversation. This mix of outdoor activities is influenced by a number of conditions. Physical environment is one of the factors: a factor that influences the activities to a varying degree and in many different ways. Outdoor activities, and a number of the physical conditions that influence them, are the subject of this book.

Greatly simplified, outdoor activities in public spaces can be divided into three categories, each of which places very different demands on the physical environment: *necessary activities*, *optional activities*, and *social activities*.

Necessary activities include those that are more or less compulsory – going to school or to work, shopping, waiting for a bus or a person, running errands, distributing mail – in other words, all activities in which those involved are to a greater or lesser degree required to participate.

In general, everyday tasks and pastimes belong to this group. Among other activities, this group includes the great majority of those related to walking.

Because the activities in this group are necessary, their incidence is influenced only slightly by the physical framework. These activities will take place throughout the year, under nearly all conditions, and are more or less independent of the exterior environment. The participants have no choice.

Optional activities – that is, those pursuits that are participated in if there is a wish to do so and if time and place make it possible – are quite another matter.

This category includes such activities as taking a walk to get a breath of fresh air, standing around enjoying life, or sitting and sunbathing.

These activities take place only when exterior conditions are optimal, when weather and place invite them. This relationship is particularly important in connection with physical planning because most of the recreational activities that are especially pleasant to pursue outdoors are found precisely in this category of activities. These activities are especially dependent on exterior physical conditions.

When outdoor areas are of poor quality, only strictly necessary activities occur.

When outdoor areas are of high quality, necessary activities take place with approximately the same frequency – though they clearly tend to take a longer time, because the physical conditions are better. In addition, however, a wide range of optional activities will also occur because place and situation now invite people to stop, sit, eat, plan, and so on.

| | Quality of the physical environment | |
	Poor	Good
Necessary activities	●	●
Optional activities	·	⬤
"Resultant" activities (Social activities)	•	●

Figure 1 Graphic representation of the relationship between the quality of outdoor spaces and the rate of occurrence of outdoor activities. When the quality of outdoor areas is good, optional activities occur with increasing frequency. Furthermore, as levels of optional activity rise, the number of social activities usually increases substantially.

In streets and city spaces of poor quality, only the bare minimum of activity takes place. People hurry home.

In a good environment, a completely different, broad spectrum of human activities is possible.

Social activities are all activities that depend on the presence of others in public spaces. Social activities include children at play, greetings and conversations, communal activities of various kinds, and finally – as the most widespread social activity – passive contacts, that is simply seeing and hearing other people.

Different kinds of social activities occur in many places: in dwellings; in private outdoor spaces, gardens, and balconies; in public buildings; at places of work; and so on; but in this context only those activities that occur in publicly accessible spaces are examined.

These activities could also be termed "resultant" activities, because in nearly all instances they evolve from activities linked to the other two activity categories. They develop in connection with the other activities because people are in the same space, meet, pass by one another, or are merely within view. . . .

[. . .]

LIFE BETWEEN BUILDINGS

It is difficult to pinpoint precisely what life between buildings means in relation to the *need for contact.*

Opportunities for meetings and daily activities in the public spaces of a city or residential area enable one to be among, to see, and to hear others, to experience other people functioning in various situations.

These modest "see and hear contacts" must be considered in relation to other forms of contact and as part of the whole range of social activities, from very simple and noncommittal contacts to complex and emotionally involved connections.

The concept of varying degrees of contact intensity is the basis of the following simplified outline of various contact forms:

High intensity   ↑   Close friendships
                     Friends
                     Acquaintances
                     Chance contacts
Low intensity   Passive contacts ("see and
                     hear" contacts)

In terms of this outline life between buildings represents primarily the low-intensity contacts located at the bottom of the scale. Compared with the other contact forms, these contacts appear insignificant, yet they are valuable both as independent contact forms and as prerequisites for other, more complex interactions.

Opportunities related to merely being able to meet, see, and hear others include:

- contact at a modest level
- a possible starting point for contact at other levels
- a possibility for maintaining already established contacts
- a source of information about the social world outside
- a source of inspiration, an offer of stimulating experience.

The possibilities related to the low-intensity contact forms offered in public spaces perhaps can best be described by the situation that exists if they are lacking.

If activity between buildings is missing, the lower end of the contact scale also disappears. The varied transitional forms between being alone and being together have disappeared. The boundaries between isolation and contact become sharper – people are either alone or else with others on a relatively demanding and exacting level.

Life between buildings offers an opportunity to be with others in a relaxed and undemanding way. One can take occasional walks, perhaps make a detour along a main street on the way more or pause at an inviting bench near a front door to be among people for a short while. One can take a long bus ride every day, as many retired people have been found to do in large cities. Or one can do daily shopping, even that it would be more practical to do it once a week. Even looking out of the window now and then, if one is fortunate to have something to look at, can be rewarding. Being among others, seeing and hearing others, receiving impulses from others, imply positive experiences, alternatives to being alone. One is not necessarily with a specific person, but one is, nevertheless, with others.

As opposed to being a passive observer of other people's experiences on television or video or film, in public spaces the individual himself is present, participating in a modest way, but most definitely participating.

Low-intensity contact is also a situation from which other forms of contact can grow. It is a medium for the unpredictable, the spontaneous, the unplanned. . . .

[. . .]

The trend from living in lifeless cities and residential areas that has accompanied industrialization, segregation of various city functions, and reliance on the automobile also has caused cities to become duller and more monotonous. This points up another important need, namely *the need for stimulation.*

Experiencing other people represents a particularly colorful and attractive opportunity for stimulation. Compared with experiencing buildings and other inanimate objects, experiencing people, who speak and move about, offers a wealth of sensual variation. No moment is like the previous or the following when people circulate among people. The number of new situations and new stimuli is limitless. Furthermore it concerns the most important subject in life: people.

Living cities, therefore, ones in which people can act with one another, are always stimulating because they are rich in experiences, in contrast to lifeless cities, which can scarcely avoid being poor in experiences and thus dull, no matter how many colors and variations of shape in buildings are introduced. . . .

[. . .]

OUTDOOR ACTIVITIES AND THE QUALITY OF OUTDOOR SPACE

Life between buildings is discussed here because the extent and character of outdoor activities are greatly influenced by physical planning. Just as it is possible through choice of materials and colors to create a certain palette in a city, it is equally possible through planning decisions to influence patterns of activities, to create better or worse conditions for outdoor events, and to create lively or lifeless cities.

The spectrum of possibilities can be described by two extremes. One extreme is the city with multistory buildings, underground parking facilities, extensive automobile traffic, and long distances between buildings and functions. This type of city can be found in a number of North American and "modernized" European cities and in many suburban areas.

In such cities one sees buildings and cars, but few people, if any, because pedestrian traffic is more or less impossible, and because conditions for outdoor stays in the public areas near buildings are very poor. Outdoor spaces are large and impersonal. With great distances in the urban plan, there is nothing much to experience outdoors, and the few activities that do take place are spread out in time and space. Under these conditions most residents prefer to remain indoors in front of the television or on their balcony or in other comparably private outdoor spaces.

Another extreme is the city with reasonably low, closely spaced buildings, accommodation for foot traffic, and good areas for outdoor stays along the streets and in direct relation to residences, public buildings, places of work, and so forth. Here it is possible to see buildings, people coming and going, and people stopping in outdoor areas near the buildings because the outdoor spaces are easy and inviting to use. This city is a living city, one in which spaces inside buildings are supplemented with usable outdoor areas, and where public spaces are allowed to function. . . .

In a survey recording all activities occurring in the center of Copenhagen during the spring and summer of 1986, it was found that the number of pedestrian streets and squares in the city center had tripled between 1968 and 1986. Parallel to this improvement of the physical conditions, a tripling in the number of people standing and sitting was recorded.

In cases where neighboring cities offer varying conditions for city activities, great differences can also be found.

In Italian cities with pedestrian streets and automobile-free squares, the outdoor city life is often much more pronounced than in the car-oriented neighboring cities, even though the climate is the same.

A 1978 survey of street activities in both trafficked and pedestrian streets in Sydney, Melbourne, and Adelaide, Australia, carried out by architectural students from the University of Melbourne and the Royal Melbourne Institute of Technology found a direct connection between street quality and street activity. In addition, an experimental improvement of increasing the number of seats by 100 percent on the pedestrian street in Melbourne resulted in an 88 percent increase in seated activities.

William H. Whyte, in his book *The Social Life of Small Urban Spaces*, describes the close connection between qualities of city space and city activities and documents how often quite simple physical alterations can improve the use of the city space noticeably.

Comparable results have been achieved in a number of improvement projects executed in New York and other US cities by the Project for Public Spaces.

In residential areas as well, both in Europe and the United States, traffic reduction schemes, courtyard clearing, laying out of parks, and comparable outdoor improvements have had a marked effect.

Figure 2 A sociable street.

Source: Photograph by Jan Gehl.

TRANSPORTATION

20

"Transit and the Metropolis: Finding Harmony"

from *The Transit Metropolis: A Global Inquiry* (1998)

Robert Cervero

Editors' Introduction

Rising traffic volume and congestion are leading citizen concerns in most cities and towns the world over, and of course produce other sustainability-related problems such as air pollution, greenhouse gas emissions, depletion of nonrenewable fossil fuels, destruction of open space by roads and suburban sprawl, and degradation of local neighborhood quality of life. How can this situation be changed? While there is no easy answer to this question, a number of strategies involving land use, alternative travel modes, pricing, and lifestyle change are likely to make a difference. This chapter explores some of these areas crucial to improving urban sustainability.

University of California at Berkeley professor Robert Cervero has studied relationships between transportation and land use the world over and is a leading authority on strategies to reduce motor vehicle use. In this selection from his book *The Transit Metropolis: A Global Inquiry* (Washington, D.C.: Island Press, 1998), he develops several potential solutions. Regions might adapt their land use to fit around major transit systems such as subways or light rail lines ("adaptive cities"). Or they might adapt their transit systems to fit their low-density land use by employing on-demand shuttles and vans and/or flexible bus systems ("adaptive transit"). Or various hybrid options are possible. Pricing of transportation and other "transportation demand management" policies will play a role as well. The long-term goal, in Cervero's view, is the "transit metropolis" where strong public transit alternatives exist to balance private vehicle use.

Other resources on the subject of reducing motor vehicle use include Peter Newman and Jeffrey Kenworthy's *Sustainability and Cities: Overcoming Automobile Dependence* (Washington, D.C.: Island Press, 1999; excerpted later in this section), Anthony Downs' *Still Stuck in Traffic: Coping With Peak-Hour Traffic Congestion* (Washington, D.C.: The Brookings Institution, 2004), Jeffrey Tumlin's *Sustainable Transportation Planning: Tools for Creating Vibrant, Healthy, and Resilient Communities* (Hoboken, NJ: Wiley, 2012), and *An Introduction to Sustainable Transportation Policy, Planning, and Implementation* (London: Earthscan, 2010), by Preston L. Schiller, Eric C. Bruun, and Jeffrey R. Kenworthy. An excellent study of land-use and transit integration in four developing world cities is provided by *Transforming Cities with Transit: Transit and Land-Use Integration for Sustainable Urban Development* (Washington, D.C.: The World Bank, 2013), by Hiroaki Suzuki, Robert Cervero, and Kanako Iuchi. Two excellent internet resources on transportation are the Surface Transportation Policy Project (www.transact.org) and the Victoria Transportation Policy Institute (www.vtpi.org), both of which offer an impressive array of materials on transportation policy and how it might be reformed.

Public transit systems are struggling to compete with the private automobile the world over. Throughout North America, in much of Europe, and even in most developing countries, the private automobile continues to gain market shares of motorized trips at the expense of public transit systems. In the United States, just 1.8 percent of all person trips were by transit in 1995, down from 2.4 percent in 1977 and 2.2 percent in 1983.[1] Despite the tens of billions of dollars invested in new rail systems and the underwriting of more than 75 percent of operating expenses, ridership figures for transit's bread-and-butter market – the work trip – remain flat. Nationwide, 4.5 percent of commutes were by transit in 1983; by 1995, this share had fallen to 3.5 percent.

The declining role of transit has been every bit as alarming in Europe, prompting some observers to warn that it is just a matter of time before cities like London and Madrid become as automobile-oriented as Los Angeles and Dallas. England and Wales saw the share of total journeys by transit fall from 33 percent in 1971 to 14 percent in 1991.[2] Since 1980, transit's market shares of trips have plummeted in Italy, Poland, Hungary, and former East Germany. Eroding market shares have likewise been reported in such megacities as Buenos Aires, Bangkok, and Manila.

Numerous factors have fueled these trends. Part of the explanation for the decline in Europe has been sharp increases in fares resulting from government deregulation of the transit sector. Public disinvestment has left the physical infrastructure of some transit systems in shambles in Italy and parts of Eastern Europe. However, transit's de-cline has been more an outcome of powerful spatial and economic trends that have been unfolding over the past several decades than of overt government actions (or inaction). Factors that have steadily chipped away at transit's market share worldwide include rising personal incomes and car ownership, declining real-dollar costs for motoring and parking, and the decentralization of cities and regions. Of course, these forces have partly fed off each other. Rising wealth and cheaper motoring, for instance, have prompted firms, retailers, and households to exit cities in favor of less dense environs. Spread-out development has proven to be especially troubling for mass transit. With trip origins and destinations today spread all over the map, mass transit is often no match for the private automobile and its flexible, door-to-door, no-transfer features.

Suburbanization has not crippled transit systems everywhere, however. Some cities and regions have managed to buck the trend, offering transit services that are holding their own against the automobile's ever-increasing presence, and in some cases even grabbing larger market shares of urban travel. These are places, I contend, that have been superbly adaptive, almost in a Darwinian sense. Notably, they have found a harmonious fit between mass transit services and their cityscapes.

Some, like Singapore and Copenhagen, have adapted their settlement patterns so that they are more conducive to transit riding, mainly by rail transit, whether for reasons of land scarcity, open space preservation, or encouraging what are viewed as more sustainable patterns of growth and travel. This has often involved concentrating offices, homes, and shops around rail nodes in attractive, well-designed, pedestrian-friendly communities. Other places have opted for an entirely different approach, accepting their low-density, often market-driven lay of the land, and in response adapting mass transit services and technologies to better serve these spread-out environs. These are places, such as Karlsruhe in Germany and Adelaide, Australia, that have introduced flexible forms of mass transit that begin to emulate the speedy, door-to-door service features of the car.

Still other places, like Ottawa, Canada, and Curitiba, Brazil, have struck a middle ground, adapting their urban landscapes so as to become more transit-supportive while at the same time adapting their transit services so as to deliver customers closer to their destinations, minimize waits, and expedite transfers. It is because these places have found a workable nexus between their mass transit services and urban settlement patterns that they either are or are on the road to becoming great transit metropolises.

What these areas have in common – adaptability – is first and fundamentally a calculated process of making change by investing, reinvesting, organizing, reorganizing, inventing, and reinventing. Adaptability is about self-survival in a world of limited resources, tightly stretched budgets, and ever-changing cultural norms, lifestyles, technologies, and personal values. In the private sector, any business that resists adapting to changing consumer wants and preferences is a short-lived business. More and more, the public sector is being held to similar standards. There is no longer the public largesse or patience to allow business as usual. Transit authorities must adapt to change, as must city and regional governments. Trends like suburbanization, advances in telecommunications, and chained trip-making require that transit agencies refashion how they configure and deliver services and that builders and planners adjust their designs of communities and places. In the best of worlds, these efforts are closely coordinated. This will most likely occur when and where there is the motivation and the means to break out of traditional, entrenched practices, which, of course, is no small feat in the public realm. Yet even transit's most ardent defenders now concede that steadily eroding shares of metropolitan travel are a telltale sign that fresh, new approaches are needed. Places that appropriately adapt to changing times, I contend, are places where transit stands the best chance of competing with the car well into the next millennium.

It bears noting that a functional and sustainable transit metropolis is not equated with a region whereby transit largely replaces the private automobile or even captures the majority of motorized trips. Rather, the transit metropolis represents a built form and a mobility environment where transit is a far more respectable alternative to traveling than currently is the case in much of the industrialized world. It is an environment where transit and the built environment harmoniously co-exist, reinforcing and enhancing each other in the process. Thus, while automobile travel might still predominate, a transit metropolis is one where enough travelers opt for transit riding, by virtue of the workable transit–land use nexus, to place a region on a sustainable course.

It is also important to emphasize . . . connections between transit and urbanization at the regional scale versus the local one. While considerable attention has been given to transit-oriented development (TOD) and the New Urbanism movement in recent years, both by scholars and the popular press, much of this focus has been at the neighborhood and community levels. Micro-scale designs that encourage walking and promote community cohesion have captivated the attention of many proponents of TODs and New Urbanism. While good quality designs are without question absolutely essential to creating places that are physically conducive to transit riding, they are clearly not sufficient in and of themselves. Islands of TOD in a sea of freeway-oriented suburbs will do little to change fundamental travel behavior or the sum quality of regional living. The key to making TOD work is to make sure that it is well coordinated across a metropolis. While land use planning and urban design are local prerogatives, their impacts on travel are felt regionally. . . .

[. . .]

TYPES OF TRANSIT METROPOLISES

[There are] four classes of transit metropolises:

- *Adaptive cities.* These are transit-oriented metropolises that have invested in rail systems to guide urban growth for purposes of achieving larger societal objectives, such as preserving open space and producing affordable housing in rail-served communities. All feature compact, mixed-use suburban communities and new towns concentrated around rail nodes. . . . examples are Stockholm, Copenhagen, Tokyo, and Singapore.
- *Adaptive transit.* These are places that have largely accepted spread-out, low-density patterns of growth and have sought to appropriately adapt transit services and new technologies to best serve these environs. [Models include] technology-based examples (e.g. dual-track systems in Karlsruhe, Germany), service innovations (e.g. track-guided buses in Adelaide, Australia), and small-vehicle, entrepreneurial services (e.g. colectivos in greater Mexico City).
- *Strong-core cities.* [Cities such as] Zurich and Melbourne have successfully integrated transit and urban development within a more confined, central city context. They have done so by providing integrated transit services centered around mixed-traffic tram and light rail systems. In these places, trams designed into streetscapes coexist nicely with pedestrians and bicyclists. These cities' primacies (high shares of regional jobs and retail sales in their cores) and healthy transit patronage are testaments to the success of melding together the renewal of both central city districts and traditional tramways.

■ *Hybrids: adaptive cities and adaptive transit.*
[Cities such as] Munich, Ottawa, and Curitiba
are best viewed as hybrids, in the sense that
they have struck a workable balance between
concentrating development along mainline transit corridors and adapting transit to efficiently
serve their spread-out suburbs and exurbs.
Greater Munich's hybrid of heavy rail trunkline
services and light rail and conventional bus
feeders – all coordinated through a regional
transit authority – has strengthened the central
city while also serving suburban growth axes.
Both Ottawa and Curitiba have introduced flexible transit centered around dedicated busways,
and at the same time have targeted considerable
shares of regional commercial growth around key
busway stations. The combination of flexible
bus-based services and mixed-use development
along busway corridors has given rise to unusually high per capita transit ridership rates in
both cities.

[. . .]

TRANSIT SERVICES AND TECHNOLOGIES

I have opted for the term *transit* to describe
generically the collective forms of passenger-
carrying transportation services – ranging from
vans and minibuses serving multiple origins and
destinations (many-to-many) over nonfixed routes
to modern, heavy rail trains operating point to point
(one-to-one) over fixed guideways. *Transit* is the
catchall used in the United States and Canada;
however, almost everywhere else, *public transport*
is the vernacular. And while in much of North
America, *public transport* or *public transit* is associated with mass transit services provided by the
public sector, almost everywhere else it means
services that are available to the public at large,
whether publicly or privately deployed. It is this
broader, more inclusive definition of public transport that is adopted [here].

Types or classes of transit services can be
defined along a continuum according to types of
vehicles, passenger-carrrying capacities, and operating environments. The following sections elaborate
on the forms of common-carrier transit services –
i.e., those available to the general public. . . .

Paratransit

The smallest carriers often go by the name of
paratransit, representing the spectrum of vans,
jitneys, shuttles, microbuses, and minibuses that fall
between the private automobile and conventional
bus in terms of capacities and service features.
Often owned and operated by private companies
and individuals, paratransit services tend to be
flexible and highly market-responsive, connecting
multiple passengers to multiple destinations within
a region, sometimes door-to-door and, because of
multiple occupants, at a price below a taxi (but
enough to more than cover full operating costs).
Driven by the profit motive, paratransit entrepreneurs aggressively seek out new and expanding
markets, innovating when and where necessary.
Much of their success lies in their flexibility and
adaptability. Unencumbered by strict operating
rules, jitney drivers will sometimes make a slight
detour to deliver someone hauling groceries to his
or her front door in return for an extra charge.
Besides being more human-scale, jitneys and minibuses can offer service advantages over bigger
buses – often, they take less time to load and un-
load, arrive more frequently, stop less often, and
are more maneuverable in busy traffic, and, studies
show, passengers tend to feel more secure since
each one is closer to the driver.[3]

In many parts of the developing world, jitneys
and minibuses are the mainstays of the transit
network. The archetypal service consists of a
constellation of loosely regulated owner-operated
collective-ride vehicles that follow more or less fixed
routes with some deviations as custom, traffic,
and hour of day permit. Jitney drivers respond to
curbside hails pretty much anywhere along a route.
Every paratransit system, however – whether the
2,000 *matatus* of Nairobi, the 15,000 *carros por
puesto* minibuses in Caracas, or the 40,000-plus
jeepneys of Manila – differs in some way. Some
load customers in the rear of vehicles and others
on the side; some are governed by federations of
jitney owners while others engage in daily head-to-
head competition; some have comfortable padded
seats and others have hard wooden benches.
Manila's jeepneys (converted US army jeeps that
serve up to twelve riders on semifixed routes) carry
about 60 percent of all peak-period trips in the
region. They cost 16 percent less per seat mile than

standard buses and generally provide a higher quality service (e.g., greater reliability, shorter waits) at a lower fare. Jeepney operations have historically been the last to petition for fare increases.[4]

Although banned in most wealthy countries, a handful of US cities today allow private minibus and jitney operators to ply their trade as long as they meet minimum safety and insurance requirements. New York City has the largest number of privately operated van services of any American city – an estimated 3,000 to 5,000 vehicles (seating 14 to 20 passengers) operate, both legally and illegally, on semifixed routes and variable schedules to subway stops and as connectors to Manhattan. Surveys show that more than three-quarters of New York's commuter van customers are former transit riders who value having a guaranteed seat and speedy, dependable services. Miami also has a thriving paratransit sector that caters mainly to recent immigrants from Cuba and the West Indies who find jitney-vans a more familiar and congenial form of travel than buses. Today, virtually all US cities allow private shuttle vans to serve airports.

Studies consistently show that jitneys and minibuses, whether in United States or Southeast Asia, confer substantial economic and financial benefits, both to the public sector and to private operators – namely, they are more effective at coaxing motorists out of cars than conventional transit in many settings, and do so without costly public subsidies.[5] However, as passenger volumes rise above a certain threshold (usually 4,000 or more per direction per hour), the economic advantages of paratransit begin to plummet, reflecting the limitations of smaller vehicles in carrying large line-haul loads. In both the developing and developed worlds, paratransit best operates in a supporting and supplement rather than substituting, role.

Bus transit

Urban *bus transit* services come in all shapes and sizes, but in most places they are characterized by 45- to 55-passenger pneumatic-tire coaches that ply fixed routes on fixed schedules. Buses are usually diesel propelled, though in some larger metropolises (e.g., Mexico City, Toronto), electric trolley buses powered by overhead wires also operate. Because they share road space, buses tend to be cheaper

and more adaptive than rail services. However, on a per passenger kilometer basis, bus transit is generally a less efficient user of energy and emits more pollution than urban rail services. It is partly because of environmental concerns, as well as image consciousness, that some cities have sought to trade in their bus routes for urban rail services.

Bus transit is particularly important in developing countries, such as India, where some 40 percent of all urban trips are by bus. In the Third World, the private sector serves more than 75 percent of bus trips. In Karachi, Pakistan, private enterprises operating medium-size buses handle 82 percent of transit journeys.[6] Because they are highly vulnerable to traffic congestion, buses are notoriously slow in megacities such as Shanghai, China, where it is generally faster to pedal a bike for trips under 14 kilometers in length.[7] One remedy is to reward high-occupancy travel through preferential treatment, such as reserved bus lanes and traffic signal preemptions. Bangkok, Thailand, has opened some 200 kilometers of reserved, contraflow bus lanes to expedite bus flows in a city where rush-hour speeds often fall below 10 kilometers/hour.

In most developed countries, bus transit falls largely under the domain of the public sector, though concerns over rising subsidies have prompted more and more public transit agencies to competitively tender services to private contractors. In much of the United Kingdom and Scandinavia, public bus services have been turned over to the private sector outright. For many small to medium-size metropolitan areas of the United States, Canada, and Europe, conventional coaches (operating over fixed routes on published schedules) are the predominant transit carriers; in larger areas, buses often function mainly as feeders into mainline rail corridors. Providing exclusive busways can allow buses to integrate feeder and line-haul functions in a single vehicle. In . . . Ottawa and Curitiba, dedicated passageways are provided for buses, enabling rubber-tire vehicles to emulate the speed advantages of conventional steel-wheel trains on line-haul segments, yet perform as regular buses on surface streets as well. Guided busways, or O-Bahns, introduced so far in Essen, Germany; Adelaide, Australia; and two British cities, Leeds and Ipswich, are particularly suited to corridors (such as freeway medians) with restricted right-of-ways. Because of faster operating speeds, the theoretical maximum passenger throughputs of busways are as high as 20,000 persons per direction per hour, more than twice that of conventional surface-street buses.[8]

Trams and light rail transit

Rail transit systems are mass transit's equivalents to motorized expressways, providing fast, trunkline connections between central business districts, secondary activity centers, and suburban corridors. The oldest and slowest rail services – *streetcars* in the United States and *tramways* in Europe – functioned as mainline carriers in an earlier era, but as metropolitan areas grew outward, those that remained intact were relegated to the role of central city circulators. In cities such as Zurich, Munich, and Melbourne, aging tramways have been refurbished in recent times to improve vehicle comfort, safety, and maneuverability. Trams are enjoying a renaissance in a number of European cities because their slower speeds, street-scale operations, and Old World character blend nicely with a pedestrian-oriented, car-free central city.

The modern-day version of the electric streetcar, *light rail transit* (LRT), has gained popularity as a more affordable alternative to expensive heavy rail systems, particularly in medium-size metropolitan areas of under 3 million population. Compared to tram services, LRT generally operates along exclusive or semi-exclusive right-of-ways using modern, automated train controls and technologies. The LRT vehicles tend to be roomier and more comfortable than tram cars, with more head clearance and lower floors. In the United States, where the most LRT trackage has been laid since the early 1980s, costs are often saved by building along disused railroad corridors. Medium-size US cities with fairly low densities, such as Sacramento, California, have managed to build LRT for as low as US$10 million per route mile; in Sacramento's case, costs were slashed by sharing a freight railroad right-of-way, building no-frills side-platform stations, and relying predominantly on single-track services. Light rail transit is generally considered safer than heavy rail because electricity comes from an overhead wire instead of a middle third rail. There is thus no need to fence in the track, not only saving costs but also allowing LRT cars to mix with traffic on city streets.

Today there are more than 100 tramways and LRT systems worldwide (mostly in Europe and North America), with the number continually rising. Among the factors behind the growing popularity of LRT and refurbished tramways are their lower costs relative to heavy rail investments and their ability to adapt to the streetscapes of built-up areas without much disruption. Other advantages include: they operate relatively quietly, thus are fairly environmentally benign and unobtrusive; they are electrically propelled, thus are less dependent than buses on the availability of petrochemical fuels; and they can be developed incrementally, a few miles at a time, eliminating the need for the long lead times associated with heavy rail construction.

. . . With four-car trains running as closely as three minutes apart, LRT can carry some 11,000 passengers per direction per hour; cutting the headways to ninety seconds (as found in some German cities, including Karlsruhe), maximum capacity can be doubled to more than 20,000. Advanced light rail transit (ALRT) systems – such as the skytrains in Vancouver, Toronto, and London's Docklands propelled by linear induction motors – can accommodate more than 25,000 passengers per direction per hour because of their higher engineering and design standards (though automated train control in lieu of on-board drivers constrains carrying capacities). It is for this reason they are also called intermediate capacity transit systems (ICTS).

Heavy rail and metros

In the world's largest cities, the big-volume transit carriers are the *heavy rail* systems, also called *rapid rail transit*, and known as *metros* in Europe, Asia, and Latin America. Metros . . . work best in large, dense cities. Indeed, the relationship is symbiotic. The densities found on Hong Kong's Victoria Island and New York's Manhattan Island could not be sustained without heavy rail services. And heavy rail service could not be sustained without very high densities. Presently, more than 90 percent of all peak-period trips to and from central London are by transit, mainly via the underground "tube"; for the remainder of greater London, transit serves fewer than a quarter of all peak-hour trips.[9]

Today, worldwide, there are some 80 metro systems, including 27 in Europe, 17 in Asia, 17 in the former Soviet Union, 12 in North America, seven in Latin America, and one in Africa. Some metros have been enormously successful, including Moscow's and Tokyo's, each of which carries 2.6 billion to 2.8 billion customers a year, more than twice as many as London's or Paris's metro systems, both of which are double the size of Moscow's and Tokyo's. On a riders per track kilometer basis, the world's most intensively used metros are, in order, São Paulo, Moscow, Tokyo, St. Petersburg, Osaka, Hong Kong, and Mexico City. Most Western European, Canadian, and US metros have one-third to one-quarter the passenger throughput per track kilometer of these cities, in large part because more of their residents own cars and the cost of driving is relatively low.

In contrast to light rail systems, few new metros are being built today, partly for fiscal reasons and partly because most areas that can economically justify the costly outlays already have them. Except for Southern California, no new heavy rail lines or extensions are being planned, designed, or constructed in North America. The World Bank lending for metro systems ceased completely in 1980 and has resumed again only recently. The Bank generally frowns on funding rail projects, even in megacities paralyzed by traffic congestion, viewing them as cost-ineffective means of achieving the Bank's principal missions of alleviating poverty and stimulating economic growth.[10]

The niche market of heavy rail services is high-volume, mainline corridors. Accommodating more than 50,000 passengers per hour in each direction, heavy rail services provide high-speed, high-performance connections within built-up cities as well as between outlying areas and central business districts. In city cores, heavy rail systems almost always operate below ground, thus the names undergrounds (in Great Britain and its former colonies) and subways. To justify the high costs for right-of-way acquisitions, relocations, and excavation, undergrounds require very high traffic volumes (toward the upper end of the capacity threshold). Outside the core, metro lines are normally either above ground (called elevated or aerial alignments) or at-grade within expressway medians. Most heavy rail stations are far more substantial and sited farther apart than LRT stops, usually two or more kilometers from each other, except in downtowns, where they might be three or four blocks away.

Because heavy rail systems are often the most expansive metropolitan rail services and operate at the highest speeds, their impacts on accessibility, and accordingly on urban development, tend to be the greatest.[11]

Heavy rail systems are almost universally electrically propelled, usually from a third rail, and each car has its own motor. Since contact with the high-voltage third rail can be fatal, rapid rail stations usually have high platforms and at-grade tracks are fenced.

Commuter and suburban railways

In terms of operating speed and geographic reach, *commuter rail* or *suburban rail*, stands at the top of the rail transit hierarchy. In Germany and central Europe, where suburb-to-city rail links are widespread, these services go by the name *S-Bahn*. Today, commuter rail services can be found on five continents in over 100 cities in more than 100 countries. Japan dominates the world's commuter rail market. In 1994, Tokyo carried almost six times the number of suburban rail commuters as Bombay, the largest commuter rail market outside Japan. Metropolitan New York's suburban rail is today only 2 percent of Tokyo's. Nevertheless, metropolitan New York, along with a dozen or so other North American metropolises, is in the midst of a commuter rail renaissance. More commuter rail tracks are currently being planned, designed, and constructed in the United States and Canada than any form of rail transit. In all, twenty-one US and Canadian cities either have commuter rail services or hope to have them within the next decade. This would raise the total US and Canadian commuter rail trackage to some 8,000 kilometers, more than five times as long as LRT and seven times as long as heavy rail.

Commuter rail services typically link outlying towns and suburban communities to the edge of a region's central business district. They are most common in big metropolitan areas or along highly urbanized corridors and conurbations, such as the Richmond–Boston axis in the northeastern United States. Commuter rail is characterized by heavy equipment (e.g., locomotives that pull passenger coaches), widely spaced stations (e.g., 5 to 10 kilometers apart), and high maximum speeds that compete with cars on suburban freeways (although trains are slow in acceleration and deceleration). Services tend to be of a high quality, with every passenger getting a comfortable seat and ample leg room. Routes are typically 40 to 80 kilometers long and lead to a stub-end downtown terminal. Outlying depots are normally surrounded by surface parking lots that enable suburbanites and exurbanites to access stations conveniently by car. With the exception of the greater New York area (along the MetroNorth corridor to Connecticut), relatively little land-use concentration or redevelopment can be found around US commuter rail stations – after all, the very premise of commuter rail is to serve the low-density lifestyle preferences of well-off suburban professionals who work downtown. Serving commuter trips almost exclusively also means that ridership is highly concentrated in peak hours, more so than any other form of mass transit service.

NOTES

1 Urban Mobility Corporation. 1997. The 1995 Nationwide Personal Transportation Survey, *Innovation Briefs*, 8 (7), p. 1; Pisarski, A. 1992. *Travel Behavior Issues in the 90's*. Washington, DC: Federal Highway Administration, US Department of Transportation.

2 Pucher J. and Lefèvre, C. 1996. *The Urban Transport Crisis in Europe and North America*. Basingstoke, UK: Macmillan Press.

3 Cervero, R. 1997. *Paratransit in America: Redefining Mass Transportation*. Westport, CT: Praeger.

4 Roth, G. and Wynne, G. 1982. *Learning from Abroad: Free Enterprise Urban Transportation*. New Brunswick, NJ: Transaction Books.

5 Roth and Wynne, op. cit.; Walters, A. 1979. The Benefits of Minibuses, *Journal of Transport Economics and Policy*, 13, pp. 320–334; Takyi, I.

1990. An Evaluation of Jitney Systems in Developing Countries, *Transportation Planning and Technology*, 44 (1), pp. 163–177.

6 Armstrong-Wright, A. 1993. *Public Transport in Third World Cities*. London: HMSO Publications.

7 Shen, Q. 1997. Urban Transportation in Shang-hai, China: Problems and Planning Implications, *International Journal of Urban and Regional Research*, 21 (4), pp. 589–606.

8 Under ideal conditions (e.g., very light traffic, flat terrain, straight lanes, no interruptions to flow such as traffic signals), buses operating on a conventional highway can move as many as 9,000 passengers per lane per direction. Sources: Trolley, R. and Turton, B. 1995. *Transport Systems and Policy Planning: A Geographical Approach*. Harlow, Essex: Longman Scientific & Technical; Vuchic, V. 1992. Urban Passenger Transportation Modes, in G. Gray and L. Hoel (eds), Public Transportation in the United States. Englewood Cliffs, NJ: Prentice-Hall, pp. 79–113.

9 Dasgupta, P. and Bly, P. 1995. *Managing Urban Travel Demand: Perspectives on Sustainability*. London: Department of Transportation.

10 The International Institute for Energy Conservation. 1996. *The World Bank & Transportation*. Washington, DC: The International Institute for Energy Conservation; Gutman, J. and Scurfield, R. 1990. Towards a More Realistic Assessment of Urban Mass Transit, in *Rail Mass Transit for Developing Countries*, Institute of Civil Engineers, London: Thomas Telford, pp. 327–338.

11 Knight, R. and Trygg, L. 1977. Evidence of Land Use Impacts of Rapid Transit Systems, *Transportation*, 6 (3), pp. 231–247; Cervero, R. 1984. Light Rail Transit and Urban Development, *Journal of the American Planning Association*, 50 (2), pp. 133–147; and Huang, H. 1996. The Land-Use Impacts of Urban Rail Transit Systems, Journal of Planning Literature, 11 (1), pp. 17–30.

21

"Traffic Calming"

from *Sustainability and Cities: Overcoming Automobile Dependence* (1999)

Peter Newman and Jeffrey Kenworthy

Editors' Introduction

Australian researchers Peter Newman and Jeffrey Kenworthy touched off an international debate with their analysis of the relation between urban density and petroleum consumption in their book *Cities and Automobile Dependence* (Brookfield, VT: Gower Technical, 1989). This work showed both the enormous range of urban densities worldwide and the very strong correlation between higher densities and decreased resource use. In their later book *Sustainability and Cities* (Washington, D.C.: Island Press, 1999), they place transportation squarely at the center of the urban sustainability challenge, and outline various strategies for moving away from automobile dependence. Newman and Kenworthy argue that five key policies are needed to overcome automobile dependence:

1   Traffic calming "to slow auto traffic and create more urban and humane environments better suited to other transportation modes,"
2   Improved transit, bicycling, and walking "to provide genuine options to the car,"
3   Improved land use, especially "urban villages" that can "create multinodal centers with mixed, dense land use that reduce the need to travel,"
4   Growth management "to prevent sprawl and redirect development into urban villages," and
5   Economic incentives, such as "taxing transportation better."

   In this selection Newman and Kenworthy discuss approaches to calming traffic, and provide historical background on the global traffic-calming movement that began in Europe in the 1970s. This effort to reclaim automobile-dominated streets for human use is now worldwide and goes far beyond simply improving public safety. It can be seen as part of an effort to humanize public space and reclaim cities for people instead of cars. Other authors have made this point as well, such as Bernard Rudofsky in his classic *Streets for People: A Primer for Americans* (New York: Doubleday, 1969), Donald Appleyard in the equally classic *Livable Streets* (Berkeley: University of California Press, 1981), David Engwicht in *Street Reclaiming: Creating Livable Streets and Vibrant Communities* (Gabriola Island, BC: New Society Press, 1999), Barbara McCann and Suzanne Rynne's *Complete Streets: Best Policy and Implementation Practices* (Chicago: American Planning Association, 2010), Jeff Speck's *Walkable City* (New York: Farrar, Straus and Giroux, 2012), and *Pedestrian & Transit-oriented Design* (Washington, D.C.: Urban Land Institute, 2013), by Reid Ewing and Keith Bartholomew. Also recommended are the Portland, Oregon Metro Council's *Creating Livable Streets: Street Design Guidelines* (Portland: Metropolitan Council, 2002) and Dan Burden's *Street Design Guidelines for Healthy Neighborhoods* (Sacramento: Local Government Commission, 2002). For a historic overview of how street design evolved, see Eran Ben-Joseph and Michael Southworth's *Streets and the Shaping of Cities and Towns* (Washington, D.C.: Island Press, 2003).

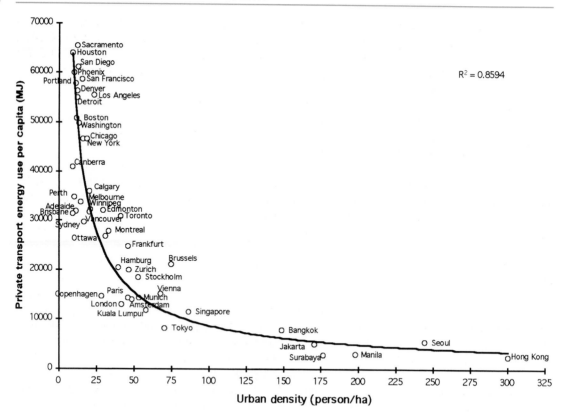

Figure 1 Energy use per capita in private passenger travel versus urban density in global cities. In a previous book Newman and Kenworthy developed this classic diagram showing the relation between urban density and energy consumption for transportation.

Traffic calming (from the German *Verkehrsberuhigung*) is the process of slowing down traffic so that the street environment is safer and more conducive to pedestrians, cyclists, shoppers, and residential life. Traffic calming is best done by physically altering the street environment through different road textures; changing the geometry of the road through chicanes (also known as S-shaped diverters), neckdowns (also known as chokers), speed plateaus and bumps, and other traffic engineering devices; introducing new street furniture designed to create a more human, safe environment; and planting attractive landscaping.

Together, these changes make drivers slow down by causing them to see less open black-top and to perceive the road as a space that is to be shared with pedestrians, cyclists, and transit vehicles. Through the avenues of trees and street gardens that accompany good traffic-calming schemes, urban wildlife habitats and corridors through cities can be created and soft surfaces can be increased so there

is less stormwater pollution. Traffic calming has the potential not only to lessen the direct negative impacts of road traffic but to foster urban environments that are more human and interactive, more beautiful, and more economically successful due to the greater social vitality possible in a city's public spaces.

It is not known exactly where or when the concept of traffic calming originated, but the German term is believed to have first been used in German federal government reports in the early 1970s. The late John Roberts of Transport and Environment Studies[1] in London was the first to translate the word into English and to bring the concept to the attention of transportation planners in other parts of the world. The idea of traffic calming, however, has its roots in earlier movements to protect city environments from the worst excesses of the automobile. This reached a watershed in the early 1960s with the publication of the major report entitled "Traffic in Towns," by Colin Buchanan.[2]

Although the British approach was to create more calmed city centers and protected residential precincts, the Buchanan report was used mostly to build large ring roads and bypasses that helped create automobile dependence. The report was used to justify major road proposals in Australian and North American cities as well. However, the European approach is based more on the organic integrity of the urban street and this approach is now gaining currency in the United Kingdom.[3]

Traffic calming emerged in Europe in the late 1960s from a number of sources: the Dutch *woonerf* or "living yard," created streets that had one shared surface with much planting to slow speeding traffic through inner-city streets and the original pedestrianization schemes in cities such as central Copenhagen.[4] Traffic calming gained rapid growth and acceptance in Europe in the 1980s through the successful action of many environmental groups trying to curtail the impacts of the automobile on European cities.[5]

Traffic calming's major objectives are to:

- Reduce the severity and number of accidents in urban areas;
- Reduce local air and noise pollution and vehicle fuel consumption;
- Improve the urban street environment for non-car-users;
- Reduce the car's dominance on roads by reclaiming road space for living space;
- Reduce the barrier effects of motor traffic on pedestrian and cycle movement; and
- Enhance local economic activity by creating a better environment for people.

With these broad objectives, traffic calming can also be of benefit to urban regeneration, housing renovation schemes, and city beautification programs (e.g., Freiburg, in southern Germany). These assist more deeply in reducing automobile dependence by bringing urban activity back to areas of the city that are inherently less dependent on the automobile (i.e., denser central and inner areas of cities built more around transit and nonmotorized modes). Traffic calming in Germany was in fact pioneered and promoted much more aggressively by the housing and urban development ministries than by the transportation ministry. This was primarily because of the positive impact traffic calming can have on the character and environmental quality of neighborhoods, making them much more desirable urban redevelopment and residential areas, while a significant number of transportation planners viewed traffic calming changes with suspicion.[6]

TECHNIQUES OF TRAFFIC CALMING AND THEIR IMPLEMENTATION

Traffic calming was originally restricted mainly to improving residential streets, and this is still a major focus. Traffic calming seeks to alter road layout and design without actually totally rebuilding a street system. It does this through a whole suite of possible techniques such as narrowed entries to streets, plantings of trees with strong vertical elements, variable street surfaces, speed restricting devices, and visual barriers that encourages cautious driving. . . . However, it has been recognized that to be really effective and to not just shift traffic problems from one area to another, traffic calming must be applied more on an area-wide basis,[7] which means involving arterial or main roads.

There are now many examples of traffic calming on through roads and in other busy areas throughout Europe (e.g., Frankfurt, Hamburg, Nürnberg, Berlin, and Copenhagen). Denmark has a nationwide program of traffic calming on main roads called Environmentally Adapted Through Roads.[8]

The approach to traffic calming has to be somewhat different on main roads because of the volumes of traffic involved, although there is overlap in the basic techniques used. In busier areas where there is a need to better balance the needs of motor vehicles with the needs of pedestrians and cyclists, the main goal is to be able to reclaim road space for other uses by reducing the speed of traffic and its impact. In most cases roads are simply reduced from six to four traffic lanes, or from four to two lanes, through critical areas of a city. . . .

In some cases the reductions in road space are accompanied by significant improvements to transit such as new rail links (e.g., Nürnberg), and in others no major changes are made but incremental improvements are implemented. Road capacity is not necessarily reduced because the loss of lanes is offset by slower speeds that reduce vehicle headways and enable more vehicles to pass. Similarly, parking supply is not necessarily reduced and in some cases may be increased nominally. Often, parallel parking on two sides of a road is converted to angle parking on alternate sides separated by landscaped strips.

The implementation of traffic calming, however, is not just a technical process but a wide-ranging community process whereby local residents can have a strong input into identifying the problems and helping to find the solutions. It has been repeatedly shown that consultation with and involvement of the community are essential to the widespread acceptance of traffic-calming schemes. In fact, an important aspect of traffic calming is the way it has been able to provide a focal point for mobilizing and galvanizing many communities around the world into developing and fighting for a vision of a more sustainable and socially acceptable solution to the problem of traffic in urban environments.[9]

EFFECTS OF TRAFFIC-CALMING SCHEMES

Many of the major traffic-calming schemes in Europe have been formally sponsored by national and local governments as demonstration projects, and one of the aims has been to test the effects of the traffic-calming schemes on key environmental indicators and safety factors. Much of the available evidence about the effects of traffic-calming schemes comes from before-and-after studies of these projects.

The following is a brief summary of the general effects of traffic-calming schemes, along with some specific examples:

Reduced accidents. Accidents, particularly the severity of accidents, are generally significantly reduced with traffic calming because speed is the most critical factor in road accidents – particularly regarding the risk of serious injury and the danger to pedestrians and cyclists. In Berlin, for example, an area-wide scheme resulted in the reductions shown in Table 1.

Most other schemes report similar kinds of data, such as in Heidelberg, which experienced average accident reductions of 31 percent and a 44 percent reduction for casualties after thirty-kilometer-per-hour [eighteen-mile-per-hour] residential speed limits were introduced along with selected physical traffic-calming measures.[10] Area-wide schemes in the Netherlands have reduced accidents involving injury by 50 percent in residential areas and 20 percent

Type of traffic	Accident measure	% reduction
All traffic	Fatal accidents	−57
	Serious accidents	−45
	Slight accidents	−40
	Accident costs	−16
Nonmotorized	Pedestrians	−43
	Cyclists	−16
	Children	−66

Table 1 Accident reductions in Berlin *moabit* (neighborhood) using comparable before and after periods

Source: Reported in Pharoah, T. and J. Russell. 1989. *Traffic Calming: Policy Evaluation in Three European Countries.* Occasional Paper 2/89. Department of Planning, Housing and Development. London: South Bank Polytechnic.

overall (measured per million vehicle kilometers) and no increase in accidents has occurred in surrounding areas.[11]

The Center for Livable Communities, in their Livable Places Update for March 1998, summarized some of the best US examples of traffic calming, and in relation to accidents, found the following:

■ The City of Seattle, where traffic-calming projects have been carried out for 20 years, surveyed the results of 119 completed projects and found an overwhelming 94 percent reduction in accidents.
■ In Portland, Oregon, 70 traffic circles and 300 speed bumps have been introduced and the number of reported accidents decreased by 50 percent.
■ A 1997 study of US street typology and accidents by Swift & Associates showed that as street width increases, accidents per mile per year increase exponentially. The safest residential street (curb to curb) turned out to be 24 feet (7.2 meters). Present US street regulations require 36 feet, primarily for access by fire vehicles, though the study found that fire vehicles can access 24-foot-wide roads when required. New Urbanism design guidelines are for 24-foot roads.

Noise reduced. Traffic calming generally results in a reduction in vehicle noise. Pharoah and Russell report that noise changes result from five factors: changes in traffic volume and composition, changes in carriageway layout, changes in carriageway surface, changes in vehicle speed, and changes in driving style.[12]

Air pollution benefits. Research in central Europe shows that in built-up areas, the higher the vehicle speed the more will be the proportion of acceleration, deceleration, and braking, and this increases air pollution. By contrast, traffic-calming schemes in some German residential areas have shown that idle times are reduced by 15 percent, gear changing by 12 percent, brake use by 14 percent, and fuel use by 12 percent.[13]

Evidence of the air pollution benefits of a slower, calmer style of driving comes from detailed work in Buxtehude, a German demonstration project (population 33,000). Table 2 shows the changes in the different types of emissions with a reduction of speed from 50 kilometers per hour (30 miles per hour) to 30 kilometers per hour (18 miles per hour) under two types of driving. In both aggressive and calm driving, emissions are reduced at the 30 kilometers per hour level, though the calm driving has a generally greater reduction and fuel use is lower.

It is also worth noting that even in instances when individual vehicles may experience an increase in fuel use and emissions (e.g., drivers do indulge in more acceleration, braking, and greater use of second gear), this may not result in an overall increase in local pollution and fuel use if the traffic-calming scheme has also resulted in lower traffic volumes.

Enhanced pedestrian and street activity. Traffic calming seeks to make the public environment safer and more attractive, so it is to be expected that traffic calming will result in a greater level of pedestrian and cycling activity in the area affected. In general, it can be expected that the results will be more noticeable in busier areas with a mix of land uses and the potential for people to make good use of reclaimed areas, such as for outdoor cafes and markets, children's facilities, etc.

Some formal measurements of the benefits are available from a summary of European experience by Pharoah and Russell (1989), such as in Berlin's federal demonstration project, where nonmotorized traffic on a wide range of streets in the scheme increased by between 27 percent and 114 percent; in Vinderup, a village in Denmark, where the main through route was traffic-calmed and out-door activities increased by up to 47 percent; and in Copenhagen, where traffic calming has led to immediate increases of pedestrian activity of between 20 percent to 40 percent, and in the long term, where central area activity is now 80 percent pedestrian and 14 percent by bike.[14] Where traffic calming reduces road capacity there is an overall decrease in traffic[15] and therefore better conditions are created for pedestrians.

Traffic calming also tends to increase the area used by pedestrians and cyclists and the extent to which streets are crossed by these users, since the severance effects of traffic are reduced. Pedestrians and cyclists tend not to confine themselves purely to walkways, but rather they extend their territory to the roadway in some instances.

Reduced crime rates. Appleyard (1981) showed that visiting among neighbors decreases when traffic increases,[16] and when neighboring ceases and people stop watching out for one another, then criminal activity can occur. The *Livable Places Update* (March 1998) overview on traffic calming quotes a Harvard University study that showed violent crimes in communities where residents willingly worked together were as much as 40 percent lower than in neighborhoods where such relationships were not as strong. Race and income were not factors in people's willingness to take part in such community activity. An example of a place where crime rates diminished after traffic calming is Weinland Park in Columbus, Ohio.

Positive economic implications. As pointed out in the objectives of traffic calming, economic revitalization of an area is an explicit aim in some schemes. A study by TEST (1989) attempted to

Emission	Driving style	
	Second gear, aggressive (%)	Third gear, calm (%)
Carbon monoxide	−17	−13
Hydrocarbons	−10	−22
Nitrogen oxides	−32	−48
Fuel consumption	+7	−7

Table 2 Changes in vehicle emissions and fuel use from 50 km/h to 30 km/h (%).

Source: Reported in Pharoah, T. and J. Russell, 1989, op cit.

confirm the hypothesis that "A good physical environment is a good economic environment" and examined ten European cities in detail. Roberts sums up the work by saying, "the message is simple: there is a strong likelihood that traffic restraint in all its forms, and environmental improvement, and a healthy economy, are causally related."[17]

The basis of this finding would appear to involve at least the following factors:

- People like to come to humanly attractive, green cities.
- Businesses like to locate in areas with a high quality urban environment.
- Car access is not banned, but it is not facilitated to the point of dominating everything else.
- Other modes are generally facilitated.

Hass-Klau (1993) shows conclusively that pedestrianization and traffic calming both have positive effects on the economic performance of an area; the more aggressive is the traffic calming, the more pronounced is the positive economic effect.[18]

In the United States, a West Palm Beach, Florida, neighborhood was economically depressed and bisected by fast-moving traffic. A traffic-calming scheme slowed the traffic through road narrowing and construction of speed bumps, traffic circles, and pedestrian islands. Then the city raised intersections, made sidewalks level with the street, and added a fountain, benches, and an amphitheater for "block parties." The development spurred new private investment and the cost of commercial space rapidly moved from five dollars per square foot to twenty-five dollars per square foot.[19] Similar case studies are given in the UK Friends of the Earth publication.[20]

TRAFFIC CALMING: A BROADER APPROACH

Traffic calming can be viewed as a broader transportation planning philosophy and not merely as a series of physical changes to roads.[21] Traffic calming in this broader sense is aimed at reducing total dependence on the automobile and promoting a more self-sufficient community with a transportation system more oriented to pedestrian, cycle, and transit use.

These broader objectives can be summarized as follows:

- A reduction of average motor vehicle speeds to discourage long-distance road travel in urban areas and promotion of a more compact urban form; traffic calming of main roads is included in this approach.
- Specific land use policies that better integrate transit and land development; the policies are directed at reducing the number, length, and need for motor vehicle trips.
- Strong promotion of walking, cycling, and transit.
- Restrictive measures against private traffic, including parking restrictions, limited major road building, and the direction of funds into transit and nonmotorized modes, as well as taxation policies on fuels and cars, including policies on company cars and road pricing.
- A shift in transportation planning philosophy from a traffic-generation approach of seeking to predict future traffic levels and the roads and parking needed to cope with them, to a traffic-dissolving approach of setting limits on motor vehicle growth and ensuring that transportation/ land use policies and practices are aimed at minimizing the need for more motor vehicle facilities.

A good example of a broader traffic-calming policy in action is the Dutch national policy from 1982 that openly promotes transit, walking, and cycling. It states that:

Henceforth other functions will be given priority over motor traffic [and] the car's dominance should be diminished by deliberately increasing travel times, by creating a less dense network of main roads, and by reducing speeds.[22]

NOTES

1 See Transport and Environment Studies (TEST). 1989. *Quality Streets – How Traditional Urban Centers Benefit from Traffic Calming*. London: TEST.
2 Ministry of Transport. 1963. *Traffic in Towns*. London: HMSO.

3 Standing Advisory Committee on Trunk Road Assessment (SACTRA). 1994. *Trunk Roads and the Generation of Traffic.* London: Department of Transport, United Kingdom; Department of Environment. 1994. *Planning Policy Guidance 13: Transport.* Whitehall, London: Department of Environment and Department of Transport.

4 Gehl, J. and Gemzøe, L. 1996. *Public Spaces, Public Life.* City of Copenhagen.

5 More detail on the evolution of traffic calming may be found in Hass-Klau, C. 1990. *The Theory and Practice of Traffic Calming: Can Britain Learn from the German Experience?* Discussion Paper 10, Rees Jeffreys Road Fund. Oxford: Transportation Studies Unit, Oxford University; Tolley, R. 1990. *Calming Traffic in Residential Areas.* Wales, UK: Brefi Press; Newman, P. and Kenworthy, J. 1991. *Towards a More Sustainable Canberra: An Assessment of Canberra's Transport, Energy and Land Use.* Institute for Science and Technology Policy, Murdoch University.

6 Hass-Klau. *The Theory and Practice of Traffic Calming.*

7 Hass-Klau, C. 1990. *The Pedestrian and City Traffic.* London: Belhaven Press.

8 Danish Road Data Laboratory. 1987. *Consequence Evaluation of Environmentally Adapted Through Road in Vinderup.* Report 52, Danish Road Data Laboratory, Danish Roads Directorate, Herlev, Copenhagen; *Consequence Evaluation of Environmentally Adapted Through Road in Skærbæk.* Report 63, Danish Road Data Laboratory, Danish Roads Directorate, Herlev, Copenhagen.

9 E.g. Tolley. 1990. *Calming Traffic in Residential Areas.*

10 Hass-Klau, C. 1990. *An Illustrated Guide to Traffic Calming: The Future Way of Managing Traffic.* London: Friends of the Earth.

11 Hass-Klau, C. (ed.) 1986. New Ways of Managing Traffic. *Built Environment,* 12 (1 and 2).

12 Pharoah, T. and Russell, J. 1989. *Traffic Calming: Policy Evaluation in Three European Countries.* Occasional Paper 2/89, Department of Planning, Housing and Development. London: South Bank Polytechnic.

13 Hass-Klau. 1990. *The Theory and Practice of Traffic Calming.*

14 Gehl and Gemzøe. 1996. *Public Spaces, Public Life.*

15 Goodwin, P.B. 1997. Solving Congestion. Inaugural Lecture for the Professorship of Transport Policy, University College, London, 23 October.

16 Appleyard, D. 1981. *Livable Streets.* Berkeley: University of California Press.

17 Roberts, J. 1988. Where's Downtown? "It Went Three Years Ago." *Town and Country Planning,* May, pp. 139–141.

18 Hass-Klau, C. 1993. Impact of Pedestrianization and Traffic Calming on Retailing: A Review of the Evidence from Germany and the UK. *Transportation Policy,* 1 (1), pp. 21–31.

19 Center for Livable Communities. 1998. Benefits of Traffic Calming Realized Across the Country. *Livable Places Update.* March.

20 Friends of the Earth. 1997. *Less Traffic, More Jobs: Direct Employment Impacts of Developing a Sustainable Transport System in the United Kingdom.* London: Friends of the Earth.

21 Hass-Klau. 1990. *The Theory and Practice of Traffic Calming.*

22 Ministry of Transport and Public Works. 1982. *From Local Traffic to Pleasurable Living.* The Hague: Ministry of Transport and Public Works, The Netherlands.

22

"Cycling for Everyone: Lessons from Europe"

from *Transportation Research Record* (2008)

John Pucher and Ralph Buehler

Editors' Introduction

Despite attempts to develop new devices such as the Segway scooter as an alternative to the automobile, the tried-and-true solution for short-distance personal mobility in many parts of the world has been the bicycle. Simple, cheap, pollution free, and easy to maintain, the bike has been used widely in nations ranging from China to Cuba. At rush hour waves of cyclists pass down the streets of European cities such as Copenhagen or Amsterdam, as well as those of countless cities in the developing world. Many nations have also sought to promote cycling as a convenient way for public transit patrons to reach transit stations.

Many European countries have been particularly successful at increasing bicycle use and safety in recent decades. In this selection transportation researchers John Pucher and Ralph Buehler explore why. Pucher is a professor in the Department of Urban Planning at Rutgers University in New Jersey, where he has written widely on transportation topics. Buehler is an associate professor at Virginia Tech. Other useful materials on bicycle use include *City Cycling* (Cambridge: MIT Press, 2012), edited by Pucher and Buehler; *Pedestrian and Bicycle Planning: A Guide to Best Practices*, by Todd Litman *et al.* (2002), available from the Victoria Transportation Policy Institute at www.vtpi.org; *Trails for the Twenty-First Century: Planning, Design, and Management Manual for Multi-Use Trails*, by Charles Flink, Kristine Olka, Robert Searns, and the Rails to Trails Conservancy (Washington, D.C.: Rails to Trails Conservancy and Island Press, 2001); and material from the Association of Pedestrian and Bicycle Professionals, at www.apbp.org.

The European Union has officially recognized the importance of cycling as a practical mode of urban transport that generates environmental, economic, and health benefits (1). The U.S. Department of Transportation emphasizes the same benefits and sets the specific goal of doubling the percentage of trips made by cycling and walking (2). Unfortunately, cycling levels in the United States lag far behind those in most European countries. Cycling in the US is primarily for recreation and not for daily, utilitarian travel. Equally important, from a social perspective, cycling is limited mainly to young men. In sharp contrast, cycling in northern Europe is as common for women as for men, and declines only slightly with age. In short, cycling in northern Europe is a normal way of getting around cities – for everyone and for all trip purposes.

If cycling is ever to become a viable way of getting around American cities, it must be made safe, convenient, and feasible for all ages, and for women as well as men. Cycling remains a marginal mode of transport in most American cities because it is widely viewed as requiring special equipment and training, physical fitness, and the courage and willingness to battle with motor vehicles on streets without separate bike lanes or paths. Cycling is a mainstream mode of urban travel in northern Europe precisely because it does not require any of those things. For example, most northern European cyclists ride on simple, inexpensive bikes; almost never wear special cycling outfits; and rarely use safety helmets.

The Netherlands, Denmark, and Germany have been especially successful in promoting safe and convenient cycling. Despite high rates of car ownership, these three countries have achieved high overall bike shares of urban travel, ranging from 9% in Germany to 19% in Denmark and 27% in the Netherlands – far higher than the 1% bike share of travel in the United States (3–6). Equally impressive, German, Danish, and Dutch women cycle as often as men, and all age groups make a considerable percentage of their daily trips by bike.

VARIATIONS BETWEEN COUNTRIES IN OVERALL CYCLING LEVELS

As shown in Figure 1, there are large differences between Australia, the United States, Canada, and European countries in the bike share of trips.

These differences roughly parallel differences in the average distance cycled per person per day, an alternative way of measuring and comparing cycling levels among countries. The European Conference of the Ministers of Transport estimated that per capita cycling per day ranges from a high of 2.5 km in the Netherlands to a low of only 0.1 km in Spain, Greece, and Portugal (1). The United States is also at the low end of the spectrum, averaging 0.1 km of cycling per person per day. Germany (0.9 km) and Denmark (1.6 km) are near the top, immediately following the Netherlands in distance cycled per inhabitant.

These statistics on cycling levels reflect data provided directly by federal ministries of transport and central statistical bureaus in each country. They are not entirely comparable because national travel surveys vary somewhat according to variable definitions, data collection method and frequency, target population, sample size, and response rates (11). At the very least, however, the national travel surveys facilitate approximate comparisons of different levels of cycling among countries, and whatever their limitations, they are the best available sources of information.

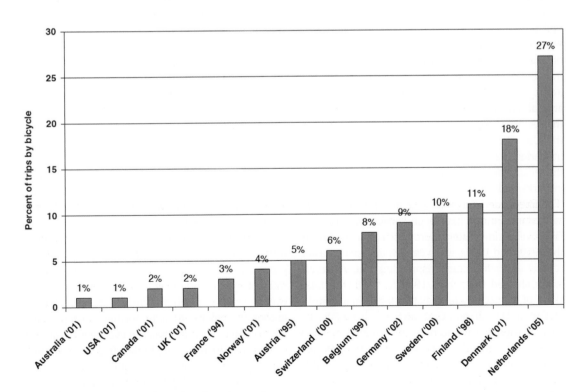

Figure 1 Bicycling share of trips in selected countries.

One might expect that Europeans cycle more than Americans due to shorter trip lengths in European cities. In fact, a considerably higher percentage of all trips in European cities are shorter than 2.5 km: 44% in the Netherlands, 37% in Denmark, and 41% in Germany, compared with 27% in the United States (1, 3–5, 12, 13). Even controlling for trip distance, however, northern Europeans make a much higher percentage of their local trips by bike. For example, Americans cycle for only 2% of trips shorter than 2.5 km compared with 37% in the Netherlands, 27% in Denmark, and 14% in Germany.

Not only do northern Europeans cycle more than Americans at every trip distance, but they are far more likely to cycle for practical, utilitarian purposes. Travel to work or school accounts for only 11% of all bike trips in the United States compared with 22% in Germany, 32% in the Netherlands, and 50% in Denmark. Similarly, shopping trips account for only 5% of all bike trips in the United States compared with 20% in Germany, 22% in the Netherlands, and 25% in Denmark (3–5, 12). Roughly three-fourths of all bike trips in the United States are for recreation compared with 35% in Germany, 27% in the Netherlands, and 24% in Denmark.

Those large differences in trip purpose suggest that the much higher levels of cycling in northern Europe are due to the use of bikes for a wide range of trip purposes, especially for daily utilitarian travel. Conversely, the very low levels of cycling in the United States might be due to mainly recreational cycling here, which is far less frequent. If that is true, then efforts to increase cycling in the United States should perhaps focus more on utilitarian, daily travel within cities as opposed to the recent emphasis on recreational cycling facilities in parks and rural areas.

Some readers might assume that bicycling levels in Western Europe have been consistently high. In fact, cycling in Western Europe fell sharply during the 1950s and 1960s, when car ownership increased rapidly and cities started spreading out. A Dutch study showed that, from 1950 to 1975, the bike share of trips fell by roughly two-thirds in a sample of Dutch, Danish, and German cities, from 50% to 85% of trips in 1950 to only 14% to 35% of trips in 1975 (6). Similarly, a study by the city of Berlin (14) found that the number of bike trips there fell by 78% from 1950 to 1975. During that 25-year period, cities throughout the Netherlands, Denmark, and Germany focused on accommodating and facilitating increased car use by vastly expanding roadway capacity and parking supply, while largely ignoring the needs of pedestrians and cyclists (15).

In the mid-1970s, transport and land use policies in all three countries dramatically shifted to favor walking, cycling, and public transport over the private car. To some extent, this was a reaction to the increasingly harmful environmental, energy, and safety impacts of rising car use (1, 6, 15–17). Most cities vastly improved their bicycling infrastructure while imposing ever more restrictions on car use, including making car use more expensive. That policy reversal led to turnarounds in the previous decline of bike use. From 1975 to 1995, the bicycling share of trips in the previously cited sample of Dutch, Danish, and German cities rose by roughly a fourth, resulting in 1995 bike shares of 20% to 43%. In Berlin, the total number of bike trips quadrupled from 1975 to 2001 (increasing by 275%), reaching 45% of the 1950 bicycling level (14). The rebound in cycling from 1975 is especially impressive given the continuing growth in per-capita income, car ownership, and suburban development in all three countries over the past three decades.

Analysis of nationwide aggregate data (as opposed to only a few sampled cities, as above) confirms a rebound in cycling since the 1970s. Average daily kilometers cycled per inhabitant rose in all three countries over the period 1978 to 2005: from 0.6 to 1.0 km in Germany, from 1.3 to 1.6 km in Denmark, and from 1.7 to 2.5 km in the Netherlands (1, 5, 13, 18). Not only do these three countries have high and growing levels of cycling, but cyclists comprise virtually all segments of society. For example, women are just about as likely to cycle as men. Women make 45% of all bike trips in Denmark, 49% in Germany, and 55% in the Netherlands (13, 18, 19). That compares with only 24% in the United States (3).

Another dimension of cycling's universality in northern Europe is the representation of all age groups. Children and adolescents have the highest rates of cycling in almost every country. However, cycling levels in northern Europe remain high even among the elderly. Also, in northern Europe there are no significant differences in cycling rates among income classes. For example, the 2002 national travel survey in Germany revealed that the lowest-income quartile was only slightly more likely to travel by bike (10% of trips) than the highest income quartile

(8.3%). Similarly, cyclists are distributed evenly among income classes in the Netherlands and Denmark (5, 18).

The remainder of this article examines possible reasons why cycling in Germany, the Netherlands, and Denmark is truly for everyone.

TRENDS IN CYCLING SAFETY

Perhaps the most important reason for the higher levels of cycling in northern Europe – especially among women, children, and the elderly – is that cycling is much safer there than in the United States. Both fatality and injury rates are much higher for cyclists in the United States compared with Germany, Denmark, and the Netherlands. Averaged over the years 2002 to 2005, the number of American bicyclist fatalities per 100 million km cycled was 5.8 compared with 1.7 in Germany, 1.5 in Denmark, and 1.1 in the Netherlands. Thus, cycling is over five times as safe in the Netherlands as in the United States, which probably explains why the Dutch do not perceive cycling as a dangerous way to get around. Cycling in Germany and Denmark is not quite as safe as in the Netherlands but still is three to four times safer than in the United States.

Germany, Denmark, and the Netherlands have greatly improved cycling safety since 1970. Although levels of cycling have increased in all three countries over the past 35 years, the total number of cycling fatalities has declined by over 70% (5, 12, 13, 18). By comparison, cycling fatalities have fallen by less than 30% in the United States over the same period (2). These trends in cycling safety correspond to overall traffic safety trends. Thus, total traffic fatalities for all modes have also declined much more in Europe than in the United States (19).

While safer cycling clearly encourages more cycling, there is also reason to believe that more cycling facilitates safer cycling. The phenomenon of safety in numbers has been consistently found to hold over time and across cities and countries. Fatality rates per trip and per kilometer are much lower for countries and cities with high bicycling shares of total travel, and fatality rates fall for any given country or city as cycling levels rise (20).

The perceived traffic danger of cycling is an important deterrent to more widespread cycling (1, 2, 5, 6, 18, 21–23). Women and the elderly appear to be especially sensitive to such traffic danger. Many American parents do not allow their children to cycle for the same reason. Thus, making cycling safer is surely one of the keys to increasing overall cycling levels in the United States, particularly among women, the elderly, and children.

It is important to emphasize that the much safer cycling in northern Europe is definitely not due to widespread use of safety helmets. On the contrary, in the Netherlands, with the safest cycling of any country, less than 1% of adult cyclists wear helmets, and even among children, only 3% to 5% wear helmets (5, 6). In 2002, 33% of German children aged 6–10 years wore helmets while cycling, compared with only 9% of adolescents aged 11–16, and only 2% of Germans aged 17 or older (18). In 2006, 66% of Danish school children aged 6–10 years wore helmets, compared with 12% among school children 11 years or older, and less than 5% among adults (22).

GOVERNMENT ROLES IN FUNDING AND PLANNING CYCLING FACILITIES AND PROGRAMS

Due to the mostly local, short-distance trips made by bike, policies and programs to promote safe and convenient cycling are usually carried out at the municipal level (1). Local governments in the Netherlands, Germany, and Denmark have been planning, constructing, and funding bicycling facilities for many decades, at least since the 1970s but much earlier in some cities. Cycling training, safety, and promotional programs are usually carried out at the local level as well, even if they are mandated and funded by higher levels. At the intermediate level, states, counties, and regional governments provide additional policy guidance, coordination, and funding, as well as some direct planning and construction of cycling facilities that serve rural areas or link different municipalities.

Central government involvement in cycling has been more recent, evolving gradually since about 1980 and providing overall goals, design guidelines, research support, model projects, coordination, and funding. The Netherlands, Denmark, and Germany all have official National Bicycling Master Plans (18, 21, 22). Each of these plans sets forth the overall goal of raising levels of cycling for daily travel while improving cycling safety. They also propose various strategies to achieve these dual goals: better design of lanes, paths, and intersections; more and better bike parking; coordination with public transport; and cycling safety and promotion campaigns.

Federal governments usually bear the cost of bicycling facilities built along national highways and contribute significantly to financing long-distance bicycling routes that cross state boundaries (1). In Germany, for example, the federal government contributed over €1.1 billion to doubling the extent of bikeways along federal highways from 1980 to 2000, and is now devoting €100 million per year for further bikeway extensions, cycling research, and demonstration projects (€1 = $1.37 in 2007 U.S. dollars). In addition, about €2 billion a year in revenues from the motor fuel tax are earmarked for a special urban transport investment fund that provides 70% to 85% federal matching funds for state and local governments wanting to build cycling facilities (e.g., paths, lanes, bridges, traffic signals, signs, and parking) (18). From 1990 to 2006, the Dutch central government contributed an average of €60 million per year to various cycling projects, including €25 million per year specifically for bike parking at train stations. In addition, the Dutch central government provides €1.8 billion a year for provinces to spend on transport projects, including cycling facilities (5). By comparison, the Danish central government has no regular funding for cycling projects but since 2000 contributed about €5 million year to various demonstration projects (1, 24).

The European Union has played a modest but increasing role in promoting cycling (1). It provides funding, for example, for transnational and cross-border bikeway projects through its EU Interreg Funding. The European Cycling Federation has already established a system of European-wide bicycling routes, and the European Union contributes toward the funding of missing bike route connections between countries and of cycling facilities in underdeveloped regions.

HOW TO MAKE CYCLING SAFE AND CONVENIENT FOR EVERYONE

Many policies and programs are necessary to make cycling safe and feasible for a broad spectrum of the population. Eight categories of measures that have been widely adopted in Dutch, Danish, and German cities are described below. Their success in making cycling possible for everyone is largely attributable to the coordinated implementation of all of these measures so that they reinforce the impact of each other in promoting cycling. Indeed, that is perhaps the key lesson to be learned: the necessity of a coordinated, multifaceted approach. Due to space limitations, only a few representative examples that illustrate the nature and extent of the eight types of measures can be provided.

Bike paths and lanes

Especially from the mid-1970s to the mid-1990s, separate facilities such as bike paths and lanes expanded greatly in all three countries. In Germany, for example, the bikeway network almost tripled in length, from 12,911 km in 1976 to 31,236 km in 1996. In the Netherlands, the bikeway network doubled in length, from 9,282 km in 1978 to 18,948 km in 1996 (25). Nationwide, aggregate statistics for the period since the mid-1990s are not available, but data for individual cities suggest continued expansion, albeit at a much slower rate than previously. The main focus now appears to be on improving the specific design of cycle paths and lanes to improve safety. The bicycling networks in cities also include numerous off-street shortcut connections for cyclists between streets and traversing blocks to enable them to take the most direct possible route from origin to destination. The result of all these facilities is a truly complete, integrated system of bicycling routes that permits cyclists to cover almost any trip either on completely separate paths and lanes or on lightly traveled, traffic-calmed residential streets.

Not only has the network of separate cycling facilities greatly expanded since the 1970s, but their design, quality, and maintenance have continually improved to ensure safer, more convenient, and more attractive cycling with each passing year. In addition, most cities have established a fully integrated system of directional signs for cyclists, color-coded to correspond to different types of bike routes. All large cities and most medium-sized cities provide detailed maps of their cycling facilities. Some cities have recently introduced internet bike route planning to assist cyclists in choosing the route that best serves their needs. In

Berlin and Odense, for example, cyclists can enter not only their origin and destination but also a range of personal preferences, such as speed, on-street or off-street facility, and avoiding intersections and heavy traffic. The internet program shows the optimal route on a map and provides all relevant information about time, average speed, bike parking, and public transport connections (26, 31).

The provision of separate cycling facilities is the cornerstone of Dutch, Danish, and German policies to make cycling safe and attractive to everyone. Policies are designed to provide safety, comfort, and convenience for both young and old, for women as well as men, and for all levels of cycling ability. Virtually all studies of the effects of separate facilities confirm that most cyclists prefer them (1, 7, 8, 32). Separate paths, in particular, are perceived as being much safer and more pleasant than cycling on the roadway, thus leading to significant growth in cycling volumes when such facilities are expanded.

Traffic calming

It is neither possible nor necessary to provide separate bike paths and lanes on lightly traveled residential streets, but they constitute an important part of the overall cycling route network. Thus, Dutch, Danish, and German cities have traffic calmed most streets in residential neighborhoods, reducing the legal speed limit to 30 km/h (19 mph) and often prohibiting any through traffic (6). In addition, many cities – especially in the Netherlands – introduced considerable alterations to the streets themselves, such as road narrowing, raised intersections and crosswalks, traffic circles, extra curves and zigzag routes, speed humps, and artificial deadends created by midblock street closures. Cycling is almost always allowed in both directions on all such traffic-calmed streets, even when they are restricted to one-way travel for cars. That further enhances the flexibility of bike travel. In the Netherlands, Denmark, and Germany, traffic calming is areawide and not for isolated streets.

Related to traffic calming, almost every city has created an extensive car-free zone in its center, mainly intended for pedestrian use, but generally permitting cycling during off-peak hours. In some Dutch cities, these car-free zones specifically include cycling facilities such as bike lanes and parking.

The combination of traffic calming of residential streets and prohibition of cars in city centers makes it virtually impossible in some cities for cars to traverse the city center to get to the other side.

Another kind of traffic calming is the so-called bicycle street, which has been increasingly adopted in Dutch and German cities. These are narrow streets where cyclists are given absolute traffic priority over the entire width of the street. Cars are usually permitted to use the streets as well, but they are limited to 30 km/h (or less) and cannot rush bicyclists or otherwise interfere with them (29).

Traffic-calmed residential neighborhoods, car-free city centers, and special bicycle streets all greatly enhance the overall bicycling network in all Dutch, Danish, and German cities. Most importantly, they offer much safer, less stressful cycling than streets filled with fast-moving motor vehicles. Since most bike trips start at home, traffic calming of neighborhood streets is crucial to enabling bike trips to start off in a safe, pleasant environment on the way to the separate bike paths and lanes that serve the rest of the trip.

All available evidence shows that traffic calming improves overall traffic safety. The benefits tend to be greatest for pedestrians, but serious cyclist injuries also fall sharply. Moreover, all studies report large increases in overall levels of walking and cycling. There are, of course, many different kinds of traffic calming. It is conceivable that one or another specific kind of traffic calming measure (perhaps traffic circles or speed humps) might detract from cycling safety in some circumstances. Overall, however, the evidence is overwhelming that traffic calming enhances both pedestrian and cyclist safety by reducing speeds on secondary roads (33–36).

Intersection modifications

While bike paths and lanes help protect cyclists from exposure to traffic dangers between intersections, they can pose safety problems when crossing intersections. Thus, Dutch, Danish, and German planners have worked continuously on perfecting the designs of intersections to facilitate safe cyclist crossings. The extent and specific design of intersection modifications vary, of course, from city to city, but they generally include most of the following:

- Special bike lanes leading up to the intersection, with advance stop lines for cyclists, far ahead of waiting cars;
- Advance green traffic signals for cyclists and extra green signal phases for cyclists at intersections with heavy cycling volumes;
- Turn restrictions for cars, while all turns allowed for cyclists;
- Highly visible, distinctively colored bike lane crossings at intersections;
- Special cyclist-activated traffic lights;
- Timing traffic lights to provide a green wave for cyclists instead of for cars, generally assuming 14- to 22-km/h bike speed; and
- Moving bike pathways a bit further away from their parallel streets when they approach intersections to help avoid collisions with right-turning cars.

Given the very nature of roadway intersections, it is virtually impossible to avoid all conflicts between motor vehicles and cyclists, but Dutch, Danish, and German planners have done a superb job of minimizing these dangers.

Bike parking

Extensive bike parking of various sorts is available throughout most Dutch, Danish, and German cities. Local governments and public transport systems themselves directly provide a large number of bike parking facilities (6, 29–31). Moreover, private developers and building owners are required by local ordinances to provide specified minimum levels of bike parking both within and adjacent to their buildings.

Aside from the large number of bike racks throughout these cities, the most visible and most innovative aspect of bike parking policy is the provision of state-of-the-art parking facilities at train stations. Immediately adjacent to Muenster's main train station, for example, there is a modern, attractive bike station (built in 1999) that offers secure parking for 3,300 bikes as well as bike repairs, bike rentals, and direct access to all train platforms (29). Amsterdam, Groningen, Odense, and Copenhagen offer similar, high-capacity bike parking facilities at their main train stations. Moreover, virtually every train station throughout

Dutch, Danish, and German metropolitan areas offers bike parking of some sort (6). In the Berlin region, there were 24,600 bike-and-ride parking spots at train stations in 2005 (including metro, suburban rail, and regional rail), with 7,000 additional bike parking spots planned by 2010 (26).

Many city centers also offer special bike parking facilities. The city of Odense, for example, recently added 400 sheltered bike racks near its main shopping area as well as a state-of-the-art automatic, secure parking station (31). Groningen offers 36 major bike parking facilities in its town center, including seven that are guarded to prevent bike theft (6). In 2007, Muenster added a secured, sheltered parking facility for 300 bikes adjacent to its main shopping district (29). The city of Copenhagen has 3,300 bike parking spaces in its center and added 400 between 2000 and 2002 (28). Amsterdam has 15 guarded bicycle parking facilities in its downtown area (27). The current policy focus in Dutch, Danish, and German cities is to increase the security of bike parking, since bike theft is a major problem.

Integration with public transport

Most Dutch, Danish, and German cities have integrated cycling with public transport (1, 6, 18, 21, 22). Public transport systems and city planners in northern Europe have increasingly recognized the key role that bicycling plays as a feeder and distributor service for public transport. Thus, copious bike parking is provided at train stations in the city center as well as at outlying stations along the rail network.

In cities such as Muenster, many suburban residents bike to reach the nearest suburban rail station, park the bike there, take the train to the city center, and then continue their trip with another bike they have parked at the main train station (29). Most rail systems permit bikes to be taken on vehicles only during off-peak hours and require an additional fee. Another form of bike–transit integration is the provision of bike rentals at virtually every major Dutch, Danish, and German train station and many suburban stations as well (6).

Most Dutch, Danish, and German cities surveyed do not permit bikes to be taken onboard regular city buses, and most buses do not come equipped with bike racks (6, 26, 27, 29). That contrasts starkly with the United States, where over 50,000 urban transit buses in 2007 had bike racks to facilitate bike-and-ride travel (37). It appears to be the one area where American transit systems do a better job of coordinating cycling with transit. The European approach is to provide bike parking facilities at major bus terminals, bus route interchanges, and even some suburban bus stops.

Training and education

Dutch, Danish, and German children receive extensive training in safe and effective cycling techniques as part of their regular school curriculum (6, 18, 21, 22). Most children complete such a course by the fourth grade. It includes both classroom instruction and road lessons, first on a cycling training track just for children, and then on regular cycling facilities throughout the city. Real police officers test the children, who receive official certificates, pennants, and stickers for their bikes if they pass the test. Since many children get to school by bike, training in safe cycling is considered essential to ensure their safety. But it also gets kids off to a lifetime of safe cycling skills. And since all school-children are included, it means that girls as well as boys start cycling at an early age.

Another crucial element in cyclist safety is training motorists to be aware of cyclists on the roadway and to avoid endangering them (6, 18, 21, 22). In general, motorist training in the Netherlands, Denmark, and Germany is far more extensive, more thorough, and more expensive than in the United States. By law motorists are required to anticipate and avoid situations that would endanger cyclists or pedestrians. Motorists are assumed to be responsible for any collisions with children, elderly, or disabled cyclists or pedestrians, even if they are jaywalking, cycling in the wrong direction, ignoring traffic signals, or otherwise behaving contrary to traffic regulations.

How do these training programs promote cycling for everyone? They ensure that by the age of 10, most children have the basic skills to cycle safely and that drivers of motor vehicles give special consideration to cyclists. Moreover, since motorists had to take the safe cycling lessons during their early school years, and since many of them still cycle, they are more likely to drive with respect and

consideration for cyclists. Cyclist and motorist training together enhance cycling safety, which has a major impact on cycling levels, especially among children, the elderly, and women.

Promotional events

Although the provision of safe and convenient cycling facilities is the key approach to promoting cycling, virtually all Dutch, Danish, and German cities have various programs to stimulate interest and enthusiasm for cycling by all groups. Below is a list of selected promotional measures used by six cities specifically surveyed [Amsterdam, Groningen, Copenhagen, Odense, Berlin, and Muenster (26–31)]:

- Well-signed and maintained bike routes both in the city and the surrounding countryside, with connections between different routes and color-coded, systematic numbering of paths for improved guidance;
- Comprehensive bike maps for every part of the city and the surrounding region;
- Bicycling websites with extensive information for cyclists on bicycling routes, activities, special programs, the health benefits of cycling, and bikes and bike accessories;
- Improved lighting and security of bike parking facilities, especially important for women concerned about their personal safety, often featuring priority bike parking exclusively for women;
- Cycling ambassador programs that send well-trained cyclists to residential neighborhoods to serve as role models of safe cycling and help with cycling promotion, distributing newsletters and information about cycling events;
- Annual bicycling festivals and car-free days that promote the environmental advantages of bicycling, display the latest bike models and accessories, disseminate various other relevant information for bike enthusiasts, and offer a range of bike races and mass bike rides;
- Wide range of cycling competitions for different ages and skill levels;
- Annual awards to firms that do the most to increase bicycling among their employees by providing showers, lockers, bike parking, bikes to borrow, and a flexible dress code;

■ A focus on the health benefits of cycling; and
■ Regular surveys of cyclists to assess their satisfaction with cycling facilities and programs and to gather specific suggestions for improvement.

These sorts of promotional activities tend to be more extensive in Denmark and Germany than in the Netherlands, where cycling levels are already so high that the focus is more on safer cycling than on more cycling, although the two are directly related, as noted earlier.

COMPLEMENTARY TAXATION, PARKING, AND LAND USE POLICIES

Most of the above policies refer to measures that make cycling safer and more convenient. But there are other important factors that indirectly encourage cycling (1, 6–8). For example, sales taxes on petrol and new car purchases, import tariffs, registration fees, license fees, driver training fees, and parking fees are generally much higher in Europe than in the United States (38–42). That results in overall costs of car ownership and use two to three times higher in Europe. That higher cost discourages car use to some extent and thus promotes alternative ways of getting around, including cycling, which is surely one of the cheapest of transport modes. In addition, there are many restrictions on car use and parking that reduce the relative speed, convenience, and flexibility of car travel compared with cycling.

Finally, land use and urban design policies in Dutch, Danish, and German cities are generally much stricter than in the United States and provide more government controls on low-density sprawl and the long trip distances that such sprawl usually generates (38–40, 43). Moreover, mixed-use zoning and transit-oriented developments have a long history in Europe. They facilitate the proximity of residential areas to commercial establishments, schools, churches, and a range of services. For the most part, these complementary taxation, parking, and land use policies are not specifically intended to promote cycling. Nevertheless, they provide dramatically more favorable preconditions for cycling than in the United States. Thus, one way that European cities promote cycling for everyone is by making car use more expensive, less convenient, and less necessary for everyone.

CONCLUSIONS: POLICIES TO MAKE CYCLING POSSIBLE FOR EVERYONE

It is impossible to provide rigorous, statistical proof that the eight categories of measures discussed above, either individually or in combination, lead to more and safer cycling. The available data do not permit multivariate analysis, and even if data were available, such analysis cannot determine direction of causation. Nevertheless, the aggregate national data and specific case study data presented in this paper strongly suggest an important link between the eight categories of measures and the desired outcomes of more and safer cycling.

The most important approach to making cycling safe, convenient, and attractive for everyone in Dutch, Danish, and German cities is the provision of separate cycling facilities along heavily traveled roads and at intersections, combined with extensive traffic calming of residential neighborhoods. Safe and relatively stress-free cycling routes are especially important for children, the elderly, women, and for anyone with special needs due to any sort of disability. Providing such separate facilities to connect practical, utilitarian origins and destinations also promotes cycling for work, school, and shopping trips as opposed to the mainly recreational cycling in the United States, where most separate cycling facilities are along urban parks, rivers, and lakes or in rural areas.

As noted in this chapter, separate facilities are only part of the solution. Dutch, Danish, and German cities reinforce the safety, convenience, and attractiveness of excellent cycling rights of way with extensive bike parking, integration with public transport, comprehensive traffic education and training of both cyclists and motorists, and a wide range of promotional events intended to generate enthusiasm and wide public support for cycling.

At the same time, car use is made expensive, less convenient, and less necessary through a host of taxes and restrictions on car ownership, use, and parking. And land use policies foster relatively compact, mixed-use developments that generate more bikeable, shorter trips.

The key to the success of cycling policies in northern Europe is the coordinated implementation of this multifaceted, self-reinforcing set of policies. Precisely because the policies are sensitive to the very different needs of different social groups, they also succeed in making cycling possible for virtually everyone. It is a lesson still to be learned in the United States.

REFERENCES

1 European Conference of the Ministers of Transport. National Policies to Promote Cycling. Organization for Economic Cooperation and Development, Paris, 2004.

2 National Bicycling and Walking Study: Ten-Year Status Report. FHWA, U.S. Department of Transportation, Washington, D.C., 2004.

3 National Household Travel Survey, 2001. FHWA, U.S. Department of Transportation, Washington, D.C., 2003.

4 German Federal Travel Survey 2002 (MiD). German Federal Ministry of Transport, Berlin, 2003.

5 Dutch Ministry of Transport. Cycling in the Netherlands. Ministry of Transport, Public Works, and Water Management, Rotterdam, Netherlands, 2006.

6 Dutch Bicycling Federation. Continuous and Integral: The Cycling Policies of Groningen and Other European Cities. Fietsberaad, Amsterdam, Netherlands, 2006. www.fietsberaad.nl. Accessed Oct. 20, 2007.

7 Tolley, R., Ed. Sustainable Transport: Planning for Walking and Cycling in Urban Environments. Woodhead Publishing Ltd., Cambridge, United Kingdom, 2003.

8 McClintock, H. Planning for Cycling: Principles, Practice, and Solutions for Urban Planners. Woodhead Publishing Ltd., Cambridge, United Kingdom, 2002.

9 European Cycling Federation. Bicycle Research Reports. www.ecf.com/37_1. Accessed Oct. 20, 2007.

10 Zegeer, C. FHWA Study Tour for Pedestrian and Bicyclist Safety in England, Germany, and the Netherlands. FHWA, U.S. Department of Transportation, Washington, D.C., 1994.

11 Kunert, U., J. Kloas, and H. Kuhfeld. Design Characteristics of National Travel Surveys: International Comparison for 10 Countries. In Transportation Research Record: Journal of the Transportation Research Board, No. 1804, Transportation Research Board of the National Academies, Washington, D.C., 2002, pp. 101–116.

12 National Travel Statistics. National Statistical Office, Copenhagen, Denmark, 2005.

13 Transportation Statistics. Statistics Netherlands, Amsterdam, Netherlands, 2005.

14 City of Berlin. Urban Transport in Berlin: Focus on Bicycling. Senatsverwaltung fuer Stadtentwicklung, Berlin, 2003.

15 Hass-Klau, C. The Pedestrian and City Traffic. Belhaven Press, New York, 1990.

16 Pucher, J. Bicycling Boom in Germany: A Revival Engineered by Public Policy. Transportation Quarterly, Vol. 51, No. 4, 1997, pp. 31–46.

17 Langenberg, P. Cycling in Amsterdam: Developments in the City. Velo Mondial, Amsterdam, Netherlands, 2000.

18 Ride Your Bike: National Bicycle Plan. German Federal Ministry of Transport, Berlin, 2002.

19 International Road Traffic Accident Database. Organisation for Economic Cooperation and Development, Paris, 2007. www.bast.de/htdocs/fachthemen/irtad. Accessed Oct. 20, 2007.

20 Jacobsen, P. Safety in Numbers: More Walkers and Bicyclists, Safer Walking and Bicycling. Injury Prevention, Vol. 9, 2003, pp. 205–209.

21 The Dutch Bicycle Master Plan. Ministry of Transport, Public Works, and Water Management, The Hague, Netherlands, 1999.

22 Promoting Safer Cycling: A Strategy. Danish Ministry of Transport, Copenhagen, Denmark, 2000.

23 Noland, R.B. Perceived Risk and Modal Choice: Risk Compensation in Transportation Systems. Accident Analysis and Prevention, Vol. 27, 1994, pp. 503–521.

24 Andersen, T. Odense: The National Cycle City of Denmark. Presented at Bicycling Federation of Australia Annual Conference, Brisbane, Australia, 2005.

25 Pucher, J., and L. Dijkstra. Promoting Safe Walking and Cycling to Improve Public Health: Lessons from the Netherlands and Germany. American Journal of Public Health, Vol. 93, 2003, pp. 1509–1516.

26 City of Berlin. Fahrradverkehr – Bicycling in Berlin. Senatsverwaltung fuer Stadtentwicklung, Berlin, 2007. www.stadtentwicklung.berlin.de/verkehr/radverkehr/index.shtml.

27 City of Amsterdam. The Amsterdam Bicycle Policy. Dienst Infrastructuur Verkeer en Vervoer, Gemeente Amsterdam, Netherlands, 2003.

28 Cycle Policy. City of Copenhagen, Denmark, 2002.

29 Verkehrsplanung in Muenster – Transportation Planning in Muenster. City of Muenster, Germany, 2007. www.muenster.de/stadt/stadtplanung/index_verkehr.html.

30 Key Figures. City of Groningen, Netherlands, 2007. www.groningen.nl/assets/pdf/kerncijfers_2006_engels.pdf.

31 National Cycle City (Cycleby). City of Odense, Denmark, 2007. www.cykelby.dk/eng/index.asp.

32 Jensen, S., C. Rosenkilde, and N. Jensen. Road Safety and Perceived Risk of Cycle Facilities in Copenhagen. Trafitec Research Center, Copenhagen, Denmark, 2007. www.Trafitec.dk.

33 Webster, D.C., and A.M. Mackie. Review of Traffic-Calming Schemes in 20-mph Zones. TRL Report 215. Transport Research Laboratory, Crowthorne, United Kingdom, 1996.

34 Morrison, D., H. Thomson, and M. Petticrew. An Evaluation of the Health Effects of a Neighbourhood Traffic-Calming Scheme. *Journal of Epidemiology and Community Health*, Vol. 58, 2004, pp. 837–840.

35 Herrstedt, L. Traffic-Calming Design – A Speed Management Method: Danish Experiences on Environmentally Adapted Through Roads. *Accident Analysis and Prevention*, Vol. 24, 1992, pp. 3–16.

36 Impacts of 20-mph Zones in London Boroughs. Transport for London, London, 2003.

37 Public Transportation Fact Book. American Public Transportation Association, Washington, D.C., 2007.

38 Nivola, P.S. Laws of the Landscape: How Policies Shape Cities in Europe and America. Brookings Institution Press, Washington, D.C., 1999.

39 Pucher, J. Urban Passenger Transport in the United States and Europe: A Comparative Analysis of Public Policies. Part 1. Travel Behavior, Urban Development and Automobile Use. *Transport Reviews*, Vol. 15, 1995, pp. 99–117.

40 Banister, D. *Unsustainable Transportation*. Routledge, New York, 2005.

41 Bratzel, S. Conditions of Success in Sustainable Urban Transport Policy: Policy Change in "Relatively Successful" European Cities. *Transport Reviews*, Vol. 19, 1999, 177–190.

42 Special Report 257: Making Transit Work: Insights from Western Europe, Canada, and the United States. Transportation Research Board, National Research Council, Washington, D.C., 2001.

43 Schmidt, S., and R. Buehler. The Planning Process in the U.S. and Germany: A Comparative Analysis. *International Planning Studies*, Vol. 12, 2007, pp. 55–75.

ENVIRONMENTAL PLANNING AND RESTORATION

23

"Biophilic Cities"

from *Biophilic Cities* (2011)

Timothy Beatley

Editors' Introduction

Many "green" buildings and communities don't seem particularly green to observers. They may be energy efficient or pedestrian oriented, but have little vegetation or wildlife present. In this selection Timothy Beatley argues that sustainable communities need to incorporate nature in ways that people can come into contact with daily. Building on E.O. Wilson's concept of "biophilia," which in his 1984 book *Biophilia* he defines as "the urge to affiliate with other forms of life," Beatley explores ways that green urban design can bring people into constant contact with other living things.

Beatley is the Teresa Heinz Professor of Sustainable Communities at the University of Virginia, and author of many books, including *The Ecology of Place: Planning for Environment, Economy, and Community* (Washington, D.C.: Island Press, 1997), *Native to Nowhere: Sustaining Home and Community in a Global Age* (Washington, D.C.: Island Press, 2004), and *Green Cities of Europe: Global Lessons on Green Urbanism* (Washington, D.C.: Island Press, 2012). This selection is taken from his book *Biophilic Cities: Integrating Nature into Urban Design and Planning* (Washington, D.C.: Island Press, 2011). Other works on this topic include *Biophilic Design: The Theory, Science, and Practice of Bringing Buildings to Life*, by Stephen R. Kellert, Judith H. Heerwagen, and Martin L. Mador (Hoboken, NJ: Wiley, 2008), and Sim Van der Ryn's *Design for an Empathic World: Reconnecting to People, Nature, and Self* (Washington, D.C.: Island Press, 2013).

While we are already designing biophilic buildings and the immediate spaces around them, we must increasingly imagine biophilic cities and should support a new kind of biophilic urbanism. Exactly what is a biophilic city, what are its key features and qualities? Perhaps the simplest answer is that it is a city that puts nature first in its design, planning, and management; it recognizes the essential need for daily human contact with nature as well as the many environmental and economic values provided by nature and natural systems.

A biophilic city is at its heart a biodiverse city, a city full of nature, a place where in the normal course of work and play and life residents feel, see, and experience rich nature – plants, trees, animals. The nature is both large and small – from treetop lichens, invertebrates, and even microorganisms to larger natural features and ecosystems that define a city and give it its character and feel. Biophilic cities cherish what already exists but also work hard to restore and repair what has been lost or degraded and to integrate new forms of nature into the design of every new structure or built project. We need contact with nature, and that nature can also take the form of shapes and images integrated into building designs, as we will see.

I have written much in the past about green cities and green urbanism, and I continue to argue for the importance of this broader agenda. Biophilic urban design and biophilic urban planning represent one particular, albeit critical, element of green urbanism – the connection with and designing-in of nature in cities. In recognizing the innate need for a connection to nature, biophilic cities tie the argument for green cities and green urbanism more directly to human wellbeing than to energy or environmental conservation.

For some the vision of green cities is not especially green – placing the emphasis on such things as investments in transit, renewable energy production, and energy-efficient building systems. Again, these are all important topics as we reimagine and redefine sustainable urban living in the twenty-first century. But biophilic cities place the focus squarely on the nature, on the presence and celebration of the actual green features, life-forms, and processes with which we as a species have so intimately coevolved.

Scale	Biophilic Design Elements
Building	Green rooftops
	Sky gardens and green atria
	Rooftop garden
	Green walls
	Daylit interior spaces
Block	Green courtyards
	Clustered housing around green areas
	Native species yards and spaces
Street	Green streets
	Sidewalk gardens
	Urban trees
	Low-impact development
	Vegetated swales and skinny streets
	Edible landscaping
	High degree of permeability
Neighborhood	Stream daylighting, stream restoration
	Urban forests
	Ecology parks
	Community gardens
	Neighborhood parks and pocket parks
	Greening grayfields and brownfields
Community	Urban creeks and riparian areas
	Urban ecological networks
	Green schools
	City tree canopy
	Community forest and community orchards
	Greening utility corridors
Region	River systems and floodplains
	Riparian systems
	Regional greenspace systems
	Greening major transport corridors

Table 1 Biophilic Urban Design Elements across Scales.

Source: Modified from Girling and Kellett, first appeared in Beatley, 2008.

While there is much overlap between biophilic cities and green urbanism, mostly complementary, there may also be ways in which these areas diverge or part. A biophilic city, as I will elucidate below, is even more than simply a biodiverse city: It is a place that learns from nature and emulates natural systems, incorporates natural forms and images into its buildings and cityscapes, and designs and plans with nature. Celebrating an urban building that assumes the shape of a form in nature, or encouraging ornamentation and textures that build connections to place and geology and natural history, are clearly biophilic but likely outside the usual rubric of green cities.

The love of and care for nature, the core value in biophilic cities, extends even beyond its borders to take steps and programs and actions that help to defend and steward over nature in other parts of the globe. And the green elements of cities serve many other important functions – they retain stormwater, sequester carbon, cool the urban environment, and moderate the impacts of air pollution, for example. For me, biophilic urbanism represents a creative mix of green urban design with a commitment to outdoor life and the protection and restoration of green infrastructure from the bioregional to the neighborhood level. The ability to reach on foot, by bicycle, or by transit a park or point of wild nature is essential. Parks are a part of the story, but we need to expand our notion of how a park is used.

Some cities, like New York, now encourage family camping in parks, and in many cities parks have become extended classrooms for schools. How much of a city's budget goes to actively restoring and repairing nature and to educating, celebrating, and actively working to bridge the nature disconnect? These are a few of the potential metrics.

In some cases we have good examples of cities that have established useful biophilic targets and are working toward them. New York has established the goal of providing a park or greenspace within a ten-minute walk of every resident. The city of Singapore has devoted approximately half of its ground area to nature and greenspace, an impressive achievement in what is a very dense city.

It is unlikely that a singular coherent vision of a biophilic city will emerge. Rather, perhaps there are many different kinds of biophilic cities, many different expressions of urban biophilia. And they might be expressed by different combinations and emphases of the qualities and conditions described here. At the simplest level, though, a biophilic city is a city that seeks to foster a closeness to nature – it protects and nurtures what it has (understands that abundant wild nature is important), actively restores and repairs the nature that exists, while finding new and creative ways to insert and inject nature into the streets, buildings, and urban living environments. And a biophilic city is an outdoor city, a city that makes walking and strolling and daily exposure to the outside elements and weather possible and a priority.

But as the above discussion also indicates, a biophilic city is not just about physical conditions or natural setting, and it is not just about green design and ecological interventions – it is just as much about a city's underlying biophilic spirit and sensibilities, about its funding priorities, and about the importance it places on support for programs that entice urbanites to learn more about the nature around them, for instance. A biophilic city might be measured and assessed more by how curious its citizens are about the nature around them and the extent to which they are engaged in daily activities to enjoy and care for nature than by the physical qualities or conditions or, for instance, the number or acres of parks and greenspaces per capita that exist in a city.

Natural and biophilic elements need to be central in everything and anything we design and build, from schools and hospitals to neighborhoods and urban blocks, to street systems and larger urban- and regional-scale design and planning. A "rooftop to region" or "room to region" approach is needed. The best biophilic cities are places where these different scales overlap and reinforce biophilic behaviors and lifestyles – children or adults should be able to leave their front door and move through a series of green features and biophilic elements, moving if they choose from garden and courtyard to green street and municipal forest and then to larger expanses of regional nature. Ideally, in a biophilic city these scales work together to deliver a nested nature that is more than the sum of its parts.

24

"What Is Restoration?"

from *Restoring Streams in Cities* (1998)

Ann L. Riley

Editors' Introduction

The early environmental movement in the late nineteenth and early twentieth centuries focused on "conservation" or "preservation" of natural lands, resources, and species. In contrast, many urban environmental groups after about 1980 came to focus on restoring previously damaged urban ecosystems. "Restoration" has thus become a catchword of the urban sustainability agenda. Restoration activities may take many forms, but often focus on cleaning up contaminated lands (often known as "brownfield" sites), replanting native vegetation, and restoring streams, wetlands, or other watershed elements.

In this selection, stream-restoration pioneer Ann L. Riley discusses some main issues surrounding urban environmental restoration, especially in the context of waterways. She describes what restoration is and is not, and gives examples in the context of creek restoration, a movement particularly active in the western United States. Examples include San Luis Creek through the center of San Luis Obispo, California, portions of Strawberry Creek in Berkeley, California, and the Guadalupe River in San Jose. Restoration of native species and habitats is a closely related movement, as is xeriscaping (use of drought-tolerant plants) in arid or semi-arid cities and towns. Other writings on the subject of restoration, use of native species, and permaculture (a philosophy of basing landscape design and "permanent agriculture" on sustainable natural systems) include *Ecological Restoration* by Susan M. Galatowitsch (Sunderland, MA: Sinauer Associates, 2012), *Ecological Restoration and Environmental Change: Renewing Damaged Ecosystems* by Stuart K. Allison (New York: Routledge, 2012), *Ecological Restoration: Principles, Values, and Structure of an Emerging Profession*, by Andre F. Clewell and James Aronson (Washington, D.C.: Island Press, 2013), and *Reclaiming American Cities: The Struggle for People, Place, and Nature Since 1900* by Rutherford H. Platt (Amherst, MA: University of Massachusetts Press, 2014).

The Society for Ecological Restoration defines ecological restoration as "the process of intentionally altering a site to establish a defined indigenous, historical ecosystem. The goal of this process is to emulate the structure, function, diversity, and dynamics of the specified ecosystem."

[Another] interesting definition that adds more of a human and social component is "the process of intentionally compensating for damage by humans to the biodiversity and dynamics of indigenous ecosystems by working with and sustaining natural regenerative processes in ways which lead to the reestablishment of sustainable and healthy relationships between nature and culture."

Using these definitions, the first problem the restorationist needs to address is what historical and indigenous (native to the location) conditions to restore to. In some circumstances it may be most practical to restore a waterway to its condition during a particular period of history, such as when it became formally integrated into the urban landscape as a 1930s Works Progress Administration (WPA) city park. The restoration project could include, for example, restoration of a creek's native vegetation and historical WPA rock work if the rock does not harm the waterway. Perhaps the history of a waterway from the late 1800s to the present has been as a degraded, polluted industrial channel. You may want to use records or maps from before this era to determine restoration goals.

It may be institutionally or ecologically impossible to restore a waterway to a landscape representing conditions before European settlers transformed the landscape to something else. For example, when we select objectives to restore the Chicago River, we cannot return it to a shallow, far-spreading prairie wetland as it was before its lowlands were dredged by humans for use as a shipping channel. Our options at this point are to use a riverine model to guide restoration attempts for the channel and to encourage, to the extent possible, the return of some of the pre-European-settlement prairie wetland species.

Restoration, particularly in urban settings, can require complicated compromises and trade-offs in establishing objectives based on the natural and human-built history that has shaped current land uses and ecological systems. A good practice is to refer to local experts who know the regional landscape well to see if any remnant natural rivers, streams, waterways, or wetlands can provide a restoration model for your degraded waterway.

Both ecological and human settlement needs will be met if you strive to create a landscape that is more self-sustaining than existing conditions. This means that the waterway is changed so that it is in greater balance. For a river or stream, this balanced condition usually means that it is not *excessively* eroding or depositing sediment. (Erosion and deposition are natural to streams; we intervene only when we establish that excessive conditions exist.) It also means that it has biologically diverse aquatic life and does not experience extremes in temperature, nutrients, algae growth, or other chemical parameters. If the natural physical features of the waterway are returned, it will not need as much intervention to correct for erosion, sedimentation, or pollution problems.

The physical features of rivers and streams include the streamside trees and shrubs, the channel and its width and depth, pools, riffles, and meanders. The river also includes its floodplain and may feature terraces, which are old, abandoned floodplains located above the current ones. These physical "structures" perform functions in the river ecosystem, including the transport of water and of sediment, the storage and conveyance of floodwaters, and the creation of terrestrial plant communities and wildlife habitat and aquatic habitat. Finally, stream dynamics include the transport of sediment; conveyance of water; formation of channels, floodplains, and terraces; and the interrelationships among these features and the land uses and vegetation in the watershed. Restoration attempts to return these structures, functions, and dynamics to the extent that it is possible given the constraints of our modern developed landscape.

Sometimes it helps to define what restoration is *not* as a way to clarify its objectives. Fisheries restoration is *not* a fish hatchery, where fish are raised at great expense in captivity and released or sometimes driven to rivers, streams, or lakes for release. Most rivers or lakes with stocked fish cannot support the life cycle of those fish. Consequently, the fish must continually be restocked. Fisheries restoration is reintroducing to a river, creek, or lake wild genetic stock that can maintain a self-sustaining population of fish that are genetically adapted to surviving in natural conditions. Restoration in that case means re-creating spawning and rearing habitats; removing barriers to migration; and restoring shelter, favorable temperatures, and water quality for the species that evolved in those conditions and therefore will survive in them on their own.

Restoration is *not* landscaping. Landscaping at its best has been a means to create new environments that provide sanctuary, adventure, symbolism, recreation, environment, and perhaps sustenance. Landscaping is also done to mitigate for a land-use change such as the building of a freeway; the construction of offices, parking lots, and housing developments; and the construction of water projects. Landscape professionals often use planting designs to screen structures, compensating for noise or lost shade or to cover up what we do not want to see. While those are all legitimate undertakings, they are not restoration. Stream restoration is also not the creation of a "native garden" with water running through it.

Planting trees and shrubs along a stream channelization project is *not* restoration – *even* if native species of plants are used. Planting that is done as an add-on to a flood-control channel, or to try to mitigate some of the lost values of the original river for wildlife or aesthetics, but does not function as a part of a natural riparian system, is landscaping. In such cases, we have not restored; we have only tried to mitigate or compensate for the project's environmental damages. However, if the vegetation functions as a component of a stream environment – if it helps slow the velocity of the water, strengthen stream banks, create vortexes to scour pools, shade the channel to prevent invasion of choking rushes and reeds, or re-create habitat for the species of birds, fish, and mammals that once used the site – then it *is* restoration.

[. . .]

Restoration can be knowing when not to act. Nature is resilient and often adjusts to changes in the watershed. A critical part of a restorationist's role is to know when to allow nature to make adjustments on its own. A variety of human changes might destabilize a stream, including the building of a dam, regulation of stream flows, diversion of water, urban development, fires or timber harvest, culverting, and channel relocation. Natural disasters, such as floods, tornadoes, earthquakes, and hurricanes, may also destabilize the stream's equilibrium. The stream will react to those changes, and its natural adjustments may or may not have unwanted consequences. A restorationist can give local residents insight into the merits and costs of intervention. In many situations, a stream will find a new equilibrium without intervention. In other situations, a stream will defy attempts to manipulate it by blowing out, eroding, or bypassing carefully designed bank protection projects. Sometimes native plant species will return naturally, coming back more quickly and vigorously as volunteers than we can replant them. The uncertainty of these natural changes underscores the importance of consulting with local geomorphologists, hydrologists, and other professionals knowledgable about local stream dynamics. There is a significant history

of mis-directed and make-work projects on streams that may do more harm than good to the correction of imbalances in channels and watersheds.

[. . .]

We are entering a new era of government engineering programs in which public works projects are going to be designed to accommodate a wider range of values and objectives. The concept of multi-objective floodplain management has gained wide acceptance in the past decade in the river engineering and management professions. This concept states that it is of greatest community benefit to manage river floodplains and flood-prone areas for a range of objectives including flood-damage reduction, protection of wildlife habitat, protection or improvement of water quality, ecological restoration, erosion control, provision of recreation, etc. This contrasts with the many older, single-objective public works projects for flood or erosion control.

[. . .]

Innovations are now being tried in the design of flood-control projects to avoid environmental impacts and performance and maintenance problems. River meanders are being kept, and floodplains are being restored to both better store and better convey large volumes of water. Revegetation systems . . . are replacing concrete, riprap, and sheet piling on stream banks, waterfronts, and lakesides. Restoration methods are providing an exciting alternative to old methods because they can often solve the important engineering problems of lowering property damages *and* provide environmental benefits. They attempt to return to the stream its structure (riparian forests, meanders, pools, riffles, and other physical features), its functions (instream habitat, flood storage, environmental balance, wildlife habitat), and its dynamics (which determine its shape, dimensions, and meander). By doing this, restoration can reduce excessive erosion, return fish habitat, help the stream recover from pollution, and even reduce flood damages. It becomes a win–win solution.

HOW URBANIZATION CAN CHANGE WATERSHEDS AND STREAMS

In urban areas the concept of watershed becomes more complex because the natural topography has been disturbed; the water may be drained through storm drains and in some cases may be diverted by drains into other basins. An area that contributes runoff either naturally or by storm drains to a particular point in a drainage is considered to be a part of that watershed. As an urban dweller in a flat part of town, I waited for it to rain and watched which way the water flowed to the storm sewers to identify what creek my storm drain flows to. That way I could figure out which watershed my house is in.

An urban stream at its worst may be muddy from sediments; have destabilized, sloughing banks and sparse vegetation; carry oils, gasoline, and urban waste from storm sewers; and be the dumping ground for residential garbage and yard debris including furniture, mattresses, and car batteries. The lack of canopy from large trees may expose the stream channel to more sunlight than normal, encouraging the growth of unwanted rushes, reeds, and algae. These can "choke" the channel, causing stagnant water and braiding. Shopping carts inhabit the urban stream with such predictability that the Urban Creeks Council of California has classified the shopping cart as the "indicator organism" of the urban creek. The council consequently has designated its highest award as the Golden Shopping Cart Award.

The worst physical modification of urban watersheds is the relegation of stream channels and tributaries to underground culverts. Riparian zones [areas of well watered vegetation along streams, with related animal life] are eliminated or separated from the stream channels. Removal of streamside vegetation results in the loss of nutrients to the aquatic organisms, loss of shade, increased bank erosion, lateral movement of the stream channel, increased sedimentation, and decreased pool depths. Floodplains become separated from the stream channels because the channels have become incised or deepened, or the previous land-use practices have added large layers of fill to floodplains, or both of these things have happened. Structural barriers such as levees, flood walls, and channelization can be added causes of this separation. Floodplains can be one of the most biologically productive parts of the watershed system as well as a storage and conveyance area for floodwater, but they are often impacted by urbanization.

Of all the land-use changes that can impact a watershed and its hydrology, urbanization is by far the most significant. Such development increases impervious surfaces, such as asphalt and concrete, producing greater volumes of runoff from storms, which "run off" the land quicker than if a natural watershed was absorbing rainfall. Urbanization tends to increase the volume and peak of stream flows. The delivery of runoff to streams after the beginning of rainfall becomes "flashier," reducing the lag time between the rainfall and the peak of a stream's flood stage.

Watershed management and restoration programs can address the impacts of urbanization and reduce increases in runoff, erosion, and sedimentation. On-site detention of storm water [near buildings and parking lots] using site design measures such as natural drainages, detention wetlands, reforestation, rainwater cisterns, and buffer zones can delay the timing and reduce the volume and peaks of runoff, while filtering the water before it enters stream and river channels.

Riparian restoration projects can help add stability to stream channels adjusting to greater urban flows. Buffer zones and greenways along waterways can prevent damages to structures from waterways that are adjusting and enlarging under the influence of urbanization. Native woodlands and vegetation can be returned to watershed slopes.

IDENTIFY WHAT YOUR STREAM NEEDS

There are five major areas of concern for those whose goal is to have a healthy stream that enhances their neighborhood or town. One of your first concerns should be to save existing healthy streams from the impacts of new urban development by putting land-use regulations in place. A common destroyer of streams is the placement of structures too close to stream banks, creating erosion and flood hazards – an expense that is usually borne by the community taxpayer. If you are in the common situation where you must reconcile conflicts between a stream and existing structures, your second concern will be to use the most environmentally and aesthetically sensitive technology available to protect both the stream and the structures. A third need you may have is negotiating for environmentally sensitive stream-channel maintenance practices by engineering officials. You may also find that in order to restore a stream, you need to remove culverts and concrete linings. Finally, you may need to address water pollution through conventional treatment facilities and restoration methods and address the need for an adequate water supply for life in the stream.

All five of these concerns will require that you interact with professionals from a number of disciplines, all of which have their own language, traditions, and even cultures and values.

First, however, you need to identify what problems your stream has or may face in the near future and your objectives in managing or avoiding these problems. Does your stream need protection from land-use changes, or is it already badly impacted by urbanization and you need to control bank erosion? You may need to consider one or more of the following as your management objectives:

PLAN AND REGULATE STREAM CORRIDORS

Land-use planning and site design can protect a natural waterway from the classic degradation caused by thoughtless urban development. Land-use planning locates developments away from hazard zones such as floodplains and river meander zones. Plans can designate hazard areas as neighborhood open space, parks, recreational areas, trails, hiking and bicycle paths, and transportation corridors.

Site-design standards and regulations provide guidelines for how to design a development once the development is properly located in the community by the land-use plan. Site-design measures can call for protection of stream-corridor buffer zones, minimal impervious surfaces and impacts to native vegetation, and sound storm-water management. Buffer zones of natural streamside vegetation and the use of natural swales and storm-water detention areas instead of sewer pipes greatly reduce the impacts of development on streams by reducing creek storm flows and runoff pollution. Setback requirements that site structures away from creeks lower the risk of future property damages from overbank flows and changing stream meanders. Adequate land-use planning and site-design measures create cost savings for a community by avoiding problems to begin with.

A second cousin to basic land-use planning is trail planning. It is easy to integrate recreational assets such as trails for walkers, joggers, and hikers along streams if an undeveloped zone follows along the stream. Planning to add a trail later without an existing undeveloped public right-of-way is much more difficult but is being done as part of the greenways movement to make urban and suburban centers more livable.

USE ENVIRONMENTALLY SENSITIVE FLOOD, EROSION, AND CHANNEL-INSTABILITY SOLUTIONS

Many areas do not have adequate setback requirements. If you inherit a situation where structures are in a flood-hazard area or are vulnerable to damage from meandering and eroding stream banks, you will want to seek out the most environmentally sensitive technology available to protect both the stream and the structures. Often, restoration alternatives can substitute for conventional engineering practices to reduce flood and erosion damage.

Streams in urban settings are typically out of balance so that there is excessive erosion or excessive deposition. Excessive erosion can be caused by increased runoff from paved surfaces, such as roads, driveways, and parking lots. Excessive deposition is frequently caused by upstream construction, grazing, or logging. Streams typically undergo two main cycles of adjustments to urbanization. They first tend to fill with sediment from construction sites. The channels later become enlarged in order to carry the increased runoff from the paved surfaces of the built-up city. River and lake waterfronts often have erosion problems because the natural vegetation once prolific along those waterways has been removed. The most appropriate remedy for most of these problems is to revegetate the waterway. Sometimes reshaping the waterway is also required.

USE ENVIRONMENTALLY SENSITIVE MAINTENANCE STRATEGIES

If you have a channelized stream that has been straightened and has had its natural tree and shrub species removed to increase space in the channel for flood flows, you will want to search for ways to return some of its environmental and aesthetic values. If your stream is now categorized by officials as a flood-control channel, you will have to work within the context of federal and local channel maintenance standards, which are generally designed to minimize the growth of native vegetation. As a flood-control channel, the primary purpose assigned to your stream is to be a conveyance for floodwaters. The old-style, conventional flood-control projects do not include aesthetics or ecological values among their objectives. This means you will have to propose a substitute stream-channel maintenance system, which allows you to integrate environmental objectives with the flood-control objectives. Instead of routinely removing most of the channel's vegetation every few years to increase channel capacity, it is more environmentally beneficial to allow 15–25 percent of the vegetation to remain and maintain it at that level over time. This can provide some environmental value while maintaining conveyance of flood flows. In addition, removal of garbage, debris, furniture, refrigerators, old boots, shopping carts, and other items not only improves the aesthetics of streams but also can remedy erosion problems and increase channel capacities for flood flows. Such cleanup projects are an obvious way to get restoration started.

REPLACE CULVERTS AND CONCRETE-LINED CHANNELS WITH MORE NATURAL ENVIRONMENTS

In many urban areas, only remnants of the former riparian environment remain. Streams are often relegated to culverts and buried underground to act as storm sewers. If not buried, they are contained in sterile concrete channels and locked behind chainlink fences. Many cities regret the loss of their streams and rivers as historical, aesthetic, and environmental assets and are trying to undo some of the damage. The city of Milwaukee, Wisconsin, is making plans to remove the concrete from Lincoln Creek in the center of town. Providence, Rhode Island, removed slabs of concrete bridging a river in order to restore the city's waterfront. The cities of Napa, Arcata, El Cerrito, and Berkeley, California; Salt Lake City, Utah; San Antonio, Texas; and Providence, Rhode Island, have dug up once buried streams and rivers and restored natural channels. Bellevue, Washington; St. Paul, Minnesota; Portland, Oregon; and Denver, Colorado, are among those cities that have plans to "daylight" creeks and jackhammer out concrete.

IMPROVE THE WATER QUALITY, WATER SUPPLY, AND HABITAT FOR STREAM LIFE

To date, most efforts to restore stream environments have focused on water quality and fish habitat improvement. In the past decade, stream restoration programs involving stream bank repair and revegetation have become part of state and local efforts to comply with nonpoint-source pollution control requirements of the 1987 amendments to the Clean Water Act. There is an extensive body of literature on monitoring and improving water quality and fish habitat restoration available through such groups as the Izaak Walton League in Washington, DC, the Adopt-a-Stream network in the Northwest, and the River Watch Network in the Northeast. The problems associated with securing adequate water supplies for instream life may involve a complicated system of water rights, with laws varying from state to state.

25

"Landscape Ecological Urbanism"

from *Landscape and Urban Planning* (2011)

Frederick Steiner

Editors' Introduction

In recent decades several important theoretical approaches have emerged that can help guide environmental planning and restoration activities. First came landscape ecology, a field pioneered by Richard Forman and others, which developed a language for describing landscapes in terms of "patches" of habitat, "edge" environments, "corridors" of wildlife movement, and "mosaics" of these features. Under this approach, the aim of environmental planning often becomes to preserve and restore these natural elements so as to improve habitat, enhance watershed function, and accommodate human activity.

Then in the late 1990s and 2000s Charles Waldheim and others developed an approach they called "landscape urbanism" which seeks to design layers of natural and human infrastructure for particular sites. Well-known examples include the High Line elevated-railway-turned-linear-park in New York City and Parc de la Villette in Paris. Finally, a number of ecological scientists have sought to study and model urban ecosystems, a field often referred to as "urban ecology." (There have been other uses of "urban ecology" historically for holistic approaches to analyzing cities and for specific organizations such as the Urban Ecology nonprofit in San Francisco, California.)

In this selection, noted environmental planner Frederick Steiner, dean of the School of Architecture at the University of Texas at Austin, surveys these different approaches and argues for a more integrated strategy which he calls "landscape ecological urbanism." This integration would combine aesthetics, ecological science, and human needs within the design of urban ecosystems. Related works include *Ecological Urbanism* (Zurich: Lars Müller Publishers, 2010) edited by Mohsen Mostafavi, Douglas Farr's *Sustainable Urbanism: Urban Design with Nature* (Hoboken, NJ: Wiley, 2008), and Patrick M. Condon's *Seven Rules for Sustainable Communities: Design Strategies for the Post-Carbon World* (Washington, D.C.: Island Press, 2010).

City design and planning are especially important in what has been called the "first urban century," with a majority of people on the planet living in city-regions for the first time in history. Since the mid-1990s, two ideas have emerged with implications for how we design and plan cities in the twenty-first century: landscape urbanism and urban ecology.

Landscape urbanism evolved from design theory within both architecture and landscape architecture. It melds high-style design and ecology. More traditional ecological design is perceived as messier (some detractors call ecological design practitioners "weedies") and, as a result, less appealing to international design elites. Thus far, landscape urbanism is largely theoretical, with a few, highly visible actual projects.

Urban ecology evolved from science-based research. Scholars apply ecological methods, largely developed in non-urban places, to metropolitan regions. To date, urban ecology exists primarily within the world of academic journals and books. Policy and design implications have been suggested but not yet implemented.

Landscape ecological urbanism offers a potential strategy to bring ideas from landscape urbanism and urban ecology together to create new territories that reflect cultural and natural processes. This synthesis also suggests some possible research directions.

LANDSCAPE URBANISM

The basic premise of landscape urbanism holds that landscape should be the fundamental building block for city design. In traditional urbanism, some structure – a wall, roads, or buildings – led development. Green spaces were relegated to left-over areas, unsuited for building, or were used for ornament. Through landscape urbanism, cultural and natural processes help the designer to organize urban form.

Landscape urbanism is largely the invention of Charles Waldheim, who coined the term (Waldheim, 2006, see also Almy, 2007). As a student of architecture at the University of Pennsylvania in the 1980s, Waldheim was influenced by both James Corner and Ian McHarg, who were at the time engaged in a vigorous debate about the future of landscape architecture. Waldheim identified common ground, integrating McHarg's ecological advocacy with Corner's urban design vision.

Landscape urbanism remains a relatively new concept with few realized works. The plan for New York City's Fresh Kills provides an example of a project moving toward realization. A key innovation is that James Corner and his Field Operations colleagues embraced long-term change in their design, eschewing a set end state for a more dynamic, flexible framework of possibilities grounded in an initial "seeding." Located in Staten Island, Fresh Kills covers some 2200 acres (890 ha) and was formerly the largest landfill in the world. Much of the debris resulting from the 11 September 2001 terrorist attacks on the World Trade Center was deposited there. The Field Operations plan suggests how the landfill can be converted into a park three times larger than Central Park. The 30-year plan involves the restoration of a large landscape and includes reclaiming much of the toxic wetlands that surround and penetrate the former landfill.

Another recent landscape urbanist example is the High Line Project in Manhattan. The Regional Plan Association and the Friends of the High Line advocated that an abandoned rail line weaving through 22 blocks in New York City be converted into a 6.7-acre (2.7-ha) park. They promote the 1.45-mile (2.33-km) long corridor as a recreational amenity, a tourist attraction, and a generator of economic development. In 2004, the Friends of the High Line and the City of New York selected Field Operations and Diller Scofidio + Renfro to design the project. The designers proposed a linear walkway that blurred the boundaries between paved and planted surfaces while suggesting evolutions in human use plus plant and bird life. The first phase of the High Line opened to much acclaim in June 2009. Its success suggests a model for how abandoned urban territories can be transformed into community assets.

As Field Operations advances landscape urbanism on the ground, others continue to refine the concept theoretically through competitions and proposals. For instance, Chris Reed and his StossLU colleagues presented many fresh ideas in their proposal for the 2007 Lower Don Lands invited design competition organized by the Toronto Waterfront Revitalization Corporation. The site covers 300 acres (121.4 ha) of mostly vacated, former port lands, just east of downtown Toronto. StossLU's approach considered flood protection, habitat restoration, and the naturalization of the Don River mouth. They also proposed new development areas and an integrated transportation system. The Canadian ecologist Nina-Marie Lister joined the StossLU team, and her contribution is evident in proposals for restoring the fish ecology. The approach suggested restoration and renewal strategies for both the Don River and Lake Ontario, with the river marsh envisioned as a breeding ground for fish.

The broader regional planning lessons of Ian McHarg (1969) are at the base of landscape urbanism. The approach involves understanding large-scale systems first and allowing them to inform and even structure proposals in order to develop schemes that engage and inaugurate ecological and social dynamics. However, landscape urbanism departs from McHarg in the ways its proponents allow multiple functions to be hybridized or to occupy the same territory simultaneously. McHarg's approaches brought people closer to nature. For example, McHarg's plan for The Woodlands in Texas successfully used storm drainage systems to structure the master plan, making water an organizing principle. Protected hydrologic corridors form green ribbons weaving through the urban fabric of The Woodlands.

In contrast, landscape urbanists are interested in having people and nature occupy the same space – and to construct new urban ecologies that tap into social, cultural, and environmental dynamics playing off one another. This is E.O. Wilson's concept of "consilience," insofar as urban natural systems and human systems interact and alter one another, producing an energetic synthesis in the process. Landscape urbanism adds to this the often unfathomable flows of cultural and economic data, updating, if not negating, McHarg's original vision.

URBAN ECOLOGY

Ecology is an evolving discipline with an increasing focus on landscapes and urban regions. Forman and Godron (1981, 1986) are responsible for defining the field of landscape ecology and illustrating its potential for planning. They explain: "Landscapes as ecological units with structure and function are composed primarily of patches in a matrix. Patches differ fundamentally in origin and dynamics, while size, shape, and spatial configuration are also important. Line corridors, strip corridors, stream corridors, networks, and habitations are major integrative structural characteristics of landscapes" (Forman and Godron, 1981, 733). Forman expanded the field to address regions and planning. His particular interest addresses the ecology of landscapes and regions "beyond the city." Meanwhile, ecologists have also begun to refocus their science inside the city.

The U.S. National Science Foundation (NSF) supports a network of 26 Long Term Ecological Research (LTER) projects. The NSF initiated the LTER program in 1980 to support research on long-term ecological phenomena. The LTER mission is to document, analyze, and understand ecological

processes and patterns that change over long temporal and large spatial scales. Until 1997, these LTERs were located outside urban regions. After an intense competition, the NSF selected the contrasting American cities of Phoenix (http://caplter.asu.edu) and Baltimore (http://www.beslter.org) for its first urban LTERs. Baltimore has a longer European settlement history and is located in a humid, coastal region. Although there were ancient native settlements, the Phoenix region has grown rapidly since World War II and is located in a desert.

The Baltimore LTER aims to understand the metropolitan region as an ecological system. The Baltimore Ecosystem Study team of cross-disciplinary researchers explores complex interactions between the built and the natural environments with ecological, social, economic, and hydrological processes (Pickett et al., 2007). The Baltimore LTER attempts to advance both ecological research and environmental policy. For example, "Our finding that urban riparian zones experiencing hydrologically induced drought are not sinks for nitrate, but in fact may be nitrate sources, helped lead policy makers concerned with the water quality of the Chesapeake Bay to reduce their reliance on stream corridor tree planting as a primary mitigation strategy" (Pickett et al., 2007, 51). In addition, the Baltimore LTER team has suggested how science might be used in urban landscape design.

The Central Arizona-Phoenix LTER also includes an interdisciplinary team of researchers at Arizona State University (ASU). They study the interactions of ecological and socio-economic systems in a rapidly growing urban environment. They have especially advanced our understanding of the impacts of land-use change on ecological patterns and processes (Grimm et al., 2000, 2008). Such understanding is important as cities in the Southwest United States continue to grow rapidly in an environmentally sensitive context.

In addition to the formal NSF-backed urban LTERs, other U.S. scholars are advancing urban ecology research across disciplines, most notably in the Puget Sound region of the Pacific Northwest (Alberti and Marzluff, 2004). The Puget Sound group from the University of Washington has contributed to our understanding of ecological resilience in urban ecosystems. Resilience, from the Latin resilire meaning to spring back or rebound, is a concept and a theory with growing appeal in the disciplines of ecology and planning. When rising from traditional concepts in ecology, resilience emphasizes equilibrium and stability. The United Nations defines resilience as the ability to absorb disturbances while retaining the same basic structure and ways of functioning, the capacity for self-organization, and the capacity to adapt to stress and change.

As a result of urban-based ecological studies, urban ecology is emerging as a field that emphasizes an interdisciplinary approach to understanding the drivers, patterns, processes, and outcomes associated with urban and urbanizing landscapes. Alberti (2008) conceives of urban ecosystems as complex coupled human–natural systems where people are the dominant modifiers of ecosystems, thus producing hybrid social–ecological landscape patterns and processes. Some urban ecology research focuses on the impact of habitat fragmentation on suburban and urban housing development patterns for avian species productivity; other research focuses on the integration of scientific analyses into growth-management strategies. Such diverse research agendas are united in their recognition that urban ecosystems are characterized by complexity, heterogeneity, and hybridity, and are therefore best analyzed within an interdisciplinary approach.

LANDSCAPE ECOLOGICAL URBANISM

Recently, Mohsen Mostafavi promoted the concept of "ecological urbanism" to imagine an approach "that has the capacity to incorporate the inherent conflictual conditions between ecology and urbanism" (Mostafavi and Doherty 2010, 17). Mostafavi and his colleagues draw strongly on landscape urbanism, but pay scant attention to the advances made in urban ecology. If those ecological advances were incorporated, then one might imagine a truly new synthesis: landscape ecological urbanism.

New ideas about city design and planning are necessary because urbanization poses significant social and environmental challenges. As the number of people in the world increases in this first urban century, the percentage of those dwelling in large city-regions is also expected to increase. The consequences of continuing to develop as we have in the past are clear: energy use and greenhouse gas production for buildings and transportation systems increase; water and air pollution spreads; valuable habitat and prime farmland are lost; social issues, such as crime and poverty, are exacerbated.

Urban ecology research indicates what should be obvious: people interact with other humans and with other species as well as their built and natural environments. The city is a human-dominated ecosystem. Landscape urbanism projects, such as the High Line and the Toronto waterfront, illustrate how designing with nature can improve the quality of cities for people, plants, and animals.

In doing so, ecosystem services can be enhanced. Ecosystem services can be defined as the benefits we receive from nature: resource services, such as food, water, and energy; regulatory services, such as purification of water, carbon sequestration and climate regulation, waste decomposition and detoxification, crop pollination, and pest and disease control; support services, such as nutrient dispersal and cycling, and seed dispersal; and cultural services, including cultural, intellectual, and spiritual inspiration, recreational experiences, ecotourism, and scientific discovery. The concept has evolved in the Unites States to provide a basis for measuring landscape design efficiency. For instance, the Sustainable Sites Initiative (SITES) has developed a measurement system for evaluating landscape performance. SITES is led by the American Society of Landscape Architects, the Lady Bird Johnson Wildflower Center of the University of Texas, and the U.S. Botanic Garden (www.sustainablesites.org). Its goal is to be the equivalent of the U.S. Green Building Council's LEED system for the outdoors. The SITES pilot projects currently underway suggest that ecosystem services can actually be enhanced and created through landscape design.

A goal of landscape ecological urbanism might be to design and plan cities to increase, rather than to decrease, ecosystem services. This suggests exciting new areas of research in landscape and urban planning, from ways to measure landscape performance to case studies of successful and not-so-successful projects.

CONCLUSIONS AND RESEARCH DIRECTIONS

Landscape ecological urbanism suggests three possible research directions: an evolution of aesthetic understanding, a deeper understanding of human agency in ecology, and reflective learning through practice. Humanities-based design theory can be a powerful force in how places are created. Traditional ecological design fell short in creating an alternative aesthetic to modernism (or its romantic offspring, postmodernism). Landscape urbanism, if nothing else, has succeeded in exciting architects, landscape architects, and urban designers about how city futures can be viewed.

Meanwhile, as ecological research has moved into cities, the role of people in urban ecosystems could not be ignored. Geographers and other social scientists have played a leadership role in urban ecology research, underscoring the dual cultural and natural foundations of human settlement. Concepts such as sustainability, regeneration, resilience, and ecosystem services hold the potential for advancing human ecology.

Projects such as Fresh Kills, the High Line, and the Lower Don Lands provide helpful lessons about what works and what does not through actual experience. Reflective practice and case studies have a strong heritage within city planning, landscape architecture, and urban design. Case studies can build on reflective practice by incorporating ecological research and design theory. In the process, new ways to design and plan city-regions with nature and culture can result.

REFERENCES

Alberti, M., Marzluff, J.M., 2004. Ecological resilience in urban ecosystems: linking urban patterns to human and ecological functions. *Urban Ecosystems* 7, 241–265.

Alberti, M., 2008. *Advances in Urban Ecology*. Springer Science, New York.

Almy, D.J. (Ed.), 2007. On Landscape Urbanism. *CENTER 14*. Center for American Architecture and Design, The University of Texas at Austin, Austin.

Forman, R.T.T., Godron, M., 1981. Patches and structural components for a landscape ecology. *Bioscience* 31, 733–740.

Forman, R.T.T., Godron, M., 1986. *Landscape Ecology*. John Wiley and Sons, New York.

Grimm, N.B., Grove, J.M., Redman, C.L., Pickett, S.T.A., 2000. Integrated approaches to long-term studies of urban ecological systems. *Bioscience* 50, 571–584.

Grimm, N.B., Faeth, S.H., Golubiewski, N.E., Redman, C.R., Wu, J., Bai, X., Briggs, J.M., 2008. Global change and the ecology of cities. *Science* 319, 756–760.

McHarg, I.L., 1969. *Design with Nature*. Natural History Press, Doubleday, Garden City, New York (2nd ed., John Wiley & Sons, New York, 1994).

Mostafavi, M., Doherty, G. (Eds.), 2010. *Ecological Urbanism*. Lars Müller Publishers, Basel, Switzerland.

Pickett, S.T.A., Belt, K.T., Galvin, M.F., Groffman, P.M., Grove, J.M., Outen, D.C., Pouyat, R.V., Stack, W.P., Cadenasso, M.L., 2007. Watersheds in Baltimore Maryland: understanding and application of integrated ecological and social processes. *J. Contemp. Water Res. Educ.* 136 (June), 44–55.

Waldheim, C. (Ed.), 2006. *The Landscape Urbanism Reader*. Princeton Architectural Press, New York.

ENERGY AND MATERIALS USE

26

"The Metabolism of Cities"

from *Creating Sustainable Cities* (1999)

Herbert Girardet

Editors' Introduction

The flow of natural resources into cities and wastes out of them represents one of the largest challenges to urban sustainability. Many argue that cities must "close the resource loop" by recycling, reusing, re-manufacturing, and otherwise diverting materials from their usual destination in landfills and incinerators. Reducing consumption in the first place is also seen as important. Likewise, more efficient urban uses of energy can be   sought to reduce dependence on nonrenewable fossil fuels, and renewable energy sources developed such as wind power (the world's fastest growing alternative energy source), solar power, geothermal energy, biomass conversion (the burning of organic materials for energy), and cogeneration (the use of waste energy or steam from one industrial process for heat or power).

Herbert Girardet has written eloquently on urban resource flows particularly as they affect the city of London. He has produced many television documentaries on urban sustainability, and has received a United Nations Global 500 Award "for outstanding environmental achievements." He is the author of *Surviving the Century: Facing Climate Change and Other Challenges* (London: Earthscan, 2007), *Cities People Planet: Urban Development and Climate Change* (Hoboken, NJ: Wiley-Academy, 2008), and *Creating Regenerative Cities* (London: Routledge, 2014). His analysis follows in the footsteps of other "appropriate technology" advocates including the great environmental philosopher E.F. Schumacher, author of *Small is Beautiful: Economics as if People Mattered* (New York: Harper & Row, 1973), and energy guru Amory Lovins, author of *Soft Energy Paths: Toward a Durable Peace* (San Francisco: Friends of the Earth, 1977), *Natural Capitalism: Creating the Next Industrial Revolution* (with Paul Hawken and L. Hunter Lovins, London: Earthscan, 1999), and *Reinventing Fire: Bold Business Solutions for the New Energy Era* (White River Junction, VT: Chelsea Green, 2011).

The growing understanding developed by the natural sciences of the way ecosystems function has a major contribution to make to solving the problems of urban sustainability. Cities, like other assemblies of organisms, have a definable metabolism, consisting of the flow of resources and products through the urban system for the benefit of urban populations. Given the vast scale of urbanization cities would be well advised to model themselves on the functioning of natural ecosystems, such as forests, to assure their long-term viability. Nature's own ecosystems have an essentially *circular* metabolism in which every output which is discharged by an organism also becomes an input which renews and sustains the continuity of the whole living environment of which it is a part. The whole web of life hangs together in a "chain of mutual benefit," through the flow of nutrients that pass from one organism to another.

The metabolism of most modern cities, in contrast, is essentially *linear*, with resources being 'pumped' through the urban system without much concern about their origin or about the destination of wastes, resulting in the discharge of vast amounts of waste products incompatible with natural systems. In urban management, inputs and outputs are considered as largely unconnected. Food is imported into cities, consumed, and discharged as sewage into rivers and coastal waters. Raw materials are extracted from nature, combined and processed into consumer goods that ultimately end up as rubbish which can't be beneficially reabsorbed into the natural world. More often than not, wastes end up in some landfill site where organic materials are mixed indiscriminately with metals, plastics, glass, and poisonous residues.

This linear model of urban production, consumption, and disposal is unsustainable and undermines the overall ecological viability of urban systems, for it has the tendency to disrupt natural cycles. In the future, cities need to function quite differently. On a predominantly urban planet, cities will need to adopt circular metabolic systems to assure their own long-term viability and that of the rural environments on whose sustained productivity they depend. To improve the urban metabolism, and to reduce the ecological footprint of cities, the application of ecological systems thinking needs to become prominent on the urban agenda. Outputs will also need to be inputs into the production system, with routine recycling of paper, metals, plastic and glass, and the conversion of organic materials, including sewage, into compost, returning plant nutrients back to the farmland that feeds the cities.

The *local* effects of the resource use of cities also need to be better understood. Urban systems accumulate vast quantities of materials. Vienna, for instance, with 1.6 million inhabitants, every day increases its actual weight by some 25,000 tons.[1] Much of this is relatively inert materials, such as concrete and tarmac which are part of the built fabric of the city. Other materials, such as lead, cadmium metals, nitrates, phosphates, or chlorinated hydrocarbons, build up and leach into the local environment in small, even minute quantities, with discernible environmental effects: they accumulate

1 *Inputs* (tons per year)	
Total tons of fuel, oil equivalent	20,000,000
Oxygen	40,000,000
Water	1,002,000,000
Food	2,400,000
Timber	1,200,000
Paper	2,200,000
Plastics	2,100,000
Glass	360,000
Cement	1,940,000
Bricks, blocks, sand, and tarmac	6,000,000
Metals (total)	1,200,000

2 *Wastes*	
CO_2	60,000,000
SO_2	400,000
NO_X	280,000
Wet, digested sewage sludge	7,500,000
Industrial and demolition wastes	11,400,000
Household, civic, and commercial wastes	3,900,000

Table 1 The metabolism of Greater London (population 7 million).

Source: Compiled by Herbert Girardet (1995, 1996); sources available from author.

in water and in the soil over time, with potential consequences for the health of present and future inhabitants. The water table under large parts of London, for instance, has become unusable for drinking water because of accumulations of toxins over the last 200 years. Much of its soil is polluted by the accumulation of heavy metals during the last 50 years.

The critical question today, as humanity moves to "full scale" urbanization, is whether living standards in our cities can be maintained whilst curbing their local and global environmental impacts. To answer this question, it helps to draw up balance sheets quantifying the environmental impacts of urbanization. We now need figures to compare the resource use by different cities. It is becoming apparent that similar-sized cities supply their needs with a greatly varying throughput of resources, and local pollution levels. The critical point is that cities and their people could massively reduce their throughput of resources, maintaining a good standard of living creating much needed local jobs in the process. I shall now discuss aspects of urban use of resources and energy in more detail.

WATER AND SEWAGE

Our cities consume vast amounts of water: in the UK, typically, some 400 liters per person per day. In the US the figure is as high as 600 litres. In older cities, such as London, water has to be pumped in from elsewhere because it is exceedingly costly to clean it to drinking water standards. Cities externalize the problem. The abstraction of river water, often many miles away from cities, has caused the destruction of river habitats and fisheries; today, many rivers' supply are a pale shadow of their former selves.

Water supplied to households, even if supplied from outside cities, goes through various treatment processes. River water, a source of supply in most countries, has to be cleaned of impurities, including pesticides, phosphates and nitrates from farming. The water is percolated through sand and charcoal filter beds before it is pumped into a city's network of water pipes. Chlorination, which is commonplace, disinfects drinking water, but its unpleasant taste causes many people to switch to bottled drinking water instead. This does not make financial sense, since bottled water, at up to 60p per litre, is often more expensive than the petrol we put in the tanks of our cars. Neither does it make sense environmentally, with vast quantities of bottled water being trucked in from hundreds or even thousands of miles away at great energy cost. It would be desirable to ensure that the quality of urban water could be high enough for it to be commonly used for drinking once again.

Unfortunately, a major function of urban water supply is as a carrier for household and commercial sewage. For this and other reasons, urban sewage systems are of an important issue in the quest for urban sustainability. Their main purpose is to collect human faeces and to separate it from people, to help prevent outbreaks of diseases such as cholera or typhoid. As a result, vast quantities of sewage are flushed away into rivers and coastal waters downstream from population centers. Coastal waters the world over are enriched both with human sewage and toxic effluents, as well as the run-off of mineral fertilizer and pesticides applied to the farmland feeding cities. The fertility taken from farms in the form of crops used to feed city people is not returned to the land. This open loop is not sustainable.

Whilst it is clear that cities need to have efficient sewerage systems, we need to redefine their purpose. Instead of building disposal systems we should construct recycling facilities in which sewage can be treated so that the main output is fertilizers suitable for farms, orchards, and market gardens. It has been too readily forgotten that sewage contains an abundance of valuable nutrients such as nitrates, potash, and phosphates. Returning these from cities to the land is an essential aspect of sustainable urban development.

A variety of new sewerage systems have been developed for this purpose using several new technologies: membrane systems that separate sewage from any contaminants; so-called "living machines" that purify sewage by biological methods; and drying technology which converts sewage into granules that can be used as fertilizer. These technologies can be used in combination with each other, making sewerage facilities into efficient fertilizer factories. These sorts of systems are now beginning to be used in cities all over the world.

In Bristol, the water and sewage company Wessex Water now dries and granulates all of the city's sewage. The annual sewage output of 600,000 people is turned into 10,000 tons of fertilizer granules. Most of it is currently used to revitalize the bleak slag heaps around former mining towns such as Merthyr Tydfil in South East Wales. In contrast, Thames Water in London is currently constructing incinerators for burning the sewage sludge produced by 4 million Londoners. This is a decision of historic short-sightedness given that phosphates – only available from North Africa and Russia – are likely to be in short supply within decades. Crops for feeding cities cannot be grown without phosphates.

There is an acknowledged problem with the contamination of sewage with heavy metals and chlorinated hydrocarbons. For this reason, there is growing concern about using sewage-derived fertilizer on farmland. However, the reduction of the use of lead in vehicle fuel and the de-industrialization of our cities is reducing this problem, lessening the load of contaminants that are flushed into sewage pipes. Also, more stringent environmental legislation is further reducing contamination of sewage. The quest for greater urban sustainability will certainly lead to a significant rethink on how we design sewerage systems. The aim should be to build systems to intercept the nutrients contained in sewage whilst assuring that it can be turned into safe fertilizers for the farmland feeding cities.

SOLID WASTE

Solid waste is the most visible output by cities. In recent decades there has been a substantial increase in solid waste produced per head, and the waste mix has become ever more complex. Today's "garbologists" see a vast difference between early and late twentieth century rubbish dumps. The former contain objects such as horse shoes, enamelled saucepans, pottery fragments and leather straps. The latter contain food wastes, plastic bags and containers, disposable nappies, mattresses, newspapers, magazines and transistor radios. But garbologists will also find plenty of discarded building materials, and crushed canisters containing various undefined, sometimes highly poisonous, liquids.

Urban wastes used to be dumped primarily in holes in the ground. Much of London's waste, for instance, is dumped in a few huge tips, such as at Mucking on the Thames in Essex. Household waste, as well as commercial and industrial waste, is taken here by central London, and "co-disposed" in pits lined with clay. The compacted rubbish is, eventually, sealed with a top layer of clay, which is then covered with soil and seeded with grass. Inside the dump, methane gas from the rotting waste is now intercepted in plastic pipes and used to run small power stations. However, their output is quite insignificant. Mucking receives the rubbish of some 2 million people, but its methane-powered generators supply electricity to just 30,000 people.[2]

More and more cities, including London, are seeing growing resistance from people in adjoining counties on receiving urban wastes; all the environmental implications of fleets of rubbish trucks, potential groundwater contamination, and stench in the vicinity of waste dumps are a growing concern. As the unwillingness to receive rubbish grows, other waste disposal options are urgently required. Dumping ever growing mounds of waste outside the cities where they originate is a waste of both space and resources that be used more beneficially.

We need to think again about the ways in which urban waste management systems work. Many cities all over the world have chosen incineration as the most convenient route for "modern" waste management. Incineration has the advantages of reducing waste materials to a small percentage of their original volume. Energy recovery can be an added bonus. But incineration is certainly not the main option for solving urban waste problems. The release of dioxins and other poisonous gases from the smokestacks of waste incinerators has given them a bad name. There have been great improvements in incineration and pollution control techniques, but only those wastes that cannot be recycled should be considered for incineration.

Recently, new objections to incinerators have been voiced in the United States because research has shown conclusively that incinerators compare badly with recycling in terms of energy conservation. Because of the high energy content of many manufactured products that end up in the rubbish bin, recycling paper, plastics, rubber and textiles is three to six [times] more energy-efficient than incineration. These are very significant figures, given that the energy and resource efficiency of urban systems is regarded as critical for future urban sustainability.[3] Many European cities are now deciding against investing in new incinerators, and in favor of a combination of recycling and composting facilities instead, with minimal incineration for waste products that cannot be further recycled.

It has been said that recycling is a red herring because it is so difficult to match the supply of materials to be recycled with regular demand for recycled products. But experiences in many European cities indicate that market incentives can make recycling economically advantageous and that the right policy signals and incentives at national and local level can transform prospects. Whilst not all waste materials can be recycled, much can be done to move in this direction. As concern grows about the continuing viability of the environments on which cities depend, the reuse and recycling of solid wastes is likely to become the rule rather than the exception. Deliberately constructing 'chains of use' that mimic natural ecosystems will be an important step forward for both industrial and urban ecology.

BOX 1 How Cairo recycles its waste

Cities in the developing world usually make highly efficient use of resources, particularly if people are supported in their recycling activities. Cairo and Manila actively encourage recycling and composting of wastes. There are a growing number of cities that are actually moving towards being zero-waste systems. Cairo, with 15 million people one of the world's largest cities, reuses and recycles most of its solid waste. Much of it is handled by a community of Coptic Christians called the Zaballeen. With the active support of the city authorities, the Zaballeen were able to acquire recycling and composting equipment. Metals and plastics are remanufactured into new products. Waste paper is reprocessed into new paper and cardboard. Rags are shredded and made into sacks and other products. Organic matter is composted and returned to the surrounding farmland as fertilizer.

The Zaballeen Environment and Development Programme has enabled the 10,000 strong community to substantially increase its income from its recycling and remanufacturing activities. In that way, social and environmental problems affecting Cairo are tackled simultaneously. Had the waste management of the city been given over to a conventional waste management company, thousands of waste collectors would have been out of work. By helping the Zaballeen with appropriate technology, they were able to improve on their traditional waste-handling methods, while Cairo could avoid putting vast waste dumps on the periphery of the city.

In the case of solid waste management, cities in the North have much to learn from the ingenuity of waste recycling in the South. Meanwhile, cities in the South could greatly benefit from the transfer of improved recycling technologies now available in the North.

Some modern cities have already made this a top priority. Cities across Europe are installing waste recycling and composting equipment. In German towns and cities, for instance, dozens of new composting plants are being constructed. In Sweden, Gothenburg has taken matters even further by setting up an ambitious program for developing "eco-cycles," minimizing the leakage of toxic substances into the local environment by helping companies develop advanced non-polluting production processes.[4] Vienna also has an impressive track record, currently recycling 43 percent of its domestic wastes.[5] This sort of figure is common to a growing number of European and American cities.

Most European cities exceed the household waste recycling performance of cities in the UK.[6] In some British cities, such as Bath and Leicester, where recycling has advanced a great deal, the benefits for people and the local environment are clearly apparent. The UK landfill tax, introduced in 1996, has increased recycling throughout the UK, helping to achieve the government target of 25 percent household waste recycling by 2000. This taxation should be extended to approximate a recycling rate of 50 percent, which is the target in other European countries.

In London, where currently only 7 percent of household waste is recycled, a proposal by the London Planning Advisory Council is intended to bring recycling up to unprecedented levels. By 2000, every London home would have a recycling box with separate compartments instead of conventional dustbins. Progressively more and more municipal waste would be recycled, establishing new reprocessing industries and creating 1,500 new jobs.[7] Early in the new century, this figure would increase further. Meanwhile, the composting of organic wastes is advancing well, with 'timber stations' that compost shredded branches of pruned trees and leaf litter being established in various locations.

Throughout the developing world, too, cities have made it their business to encourage recycling and composting of wastes.[8] Cairo, Manila, and Calcutta are interesting cases in point.

ENERGY

Looking down on the Earth from space at night, astronauts see an illuminated planet – vast city clusters lit up by millions of light bulbs as well as the flares of oil wells and refineries. Fossil fuels have made us what we are today – an urban–industrial species. Without the power stations they supply and the vehicles they power, our urban lifestyles and our astonishing physical mobility would not have developed.

Worldwide, fossil fuel use in the last 50 years has gone up nearly five times, from 1.715 billion tons of oil equivalent in 1950 to well over 8 billion tons today. Fossil fuels provide some 85 percent of the world's commercial energy, of which oil currently amounts to around 40 percent. The bulk of the world's energy consumption is *within* cities, and much of the rest is used for producing and transporting goods and people *to and from* cities. This realization is crucial for developing strategies for sustainable use of energy, particularly in the context of global warming.

Energy use is something most of us take for granted. As we switch on electric or gas appliances, we are hardly aware of the refinery, gas field, or the power station that supplies us. And despite publicity about acid rain and climate change, we rarely reflect on the impacts of our energy use on the environment because they are not experienced directly, except when we inhale exhaust fumes on a busy street.

Yet reducing urban energy consumption could make a major contribution to solving the world's air pollution problems. At the 1997 Kyoto conference on climate change, the industrialized nations agreed to cut CO_2 emissions by 5 percent by 2010, but a worldwide cut of some 60 percent is needed to actually *halt* global warming. As indicated in the Introduction, large cities and high levels of energy consumption are closely connected, particularly where routine use of motor cars, urban sprawl and air travel define urban lifestyles. Yet the potential exists for cities to be efficient users of energy.

London's 7 million people, for instance, use 20 million tons of oil equivalent per year (two supertankers a week), and discharge some 60 million tons of carbon dioxide. All in all, the per capita energy consumption of Londoners is amongst the highest in Europe. The city's electricity supply system, relying on remote power stations and long distance transmission lines, is no more than 30 to 35 percent efficient. The know-how exists to bring down London's energy use by between 30 and 50 percent without affecting living standards, and with the potential of creating tens of thousands of jobs in the coming decades. Significant energy conservation can be achieved by a combination of *energy efficiency* and by more efficient *energy supply systems*.

In the UK, national planning regulations have already substantially improved the energy efficiency of homes, but much more can be done. At the domestic level, two out of three low-income families lack even the most basic insulation in their homes. Eight million families cannot afford the warmth they need in the winter months. Treating cold-related illnesses costs the National Health Service over £1 billion per year.[9] Only one in twelve domestic properties in Britain have the level of energy efficiency currently required by law.[10] Yet energy efficiency's advantages are impressive:

- reduced fuel bills for everyone;
- benefits to the trade balance through curbing the need for imports;
- the creation of new jobs in the energy efficiency industry;
- the preservation of fossil fuel reserves;
- the alleviation of environmental problems, such as air pollution and global warming, contributed to by energy generation.

There are many examples, particularly from Scandinavia, of how energy efficiency combined with efficient supply systems can dramatically reduce the energy dependence of cities. There is no doubt that the energy supply systems in many cities of the world can be vastly improved. Take electricity: most cities are supplied by power stations located a long way away, fired mainly by coal, with electricity being transferred along high-voltage power lines. On average, these stations are only 34 percent efficient. Modern gas-fired stations are slightly better, at 40–50 percent efficiency. Combined heat and power (CHP) stations, in contrast, are about 80 percent efficient, because instead of wasting heat from combustion, they capture and distribute it through district heating systems.[11]

CHP systems are a very significant technology indeed. They can be fueled by a wide variety of sources – gas, geothermal energy, or even wood chips. CHP systems provide heat and chilled water, as well as electricity to urban buildings and factories. They are now commonplace in many

European cities. In Denmark 40 percent of electricity is produced by CHP; in Finland 34 percent, and in Holland 30 percent.

Helsinki has taken the development of CHP further than most cities. Waste heat from local coal-fired power stations is used to heat 90 percent of its buildings and homes. Its overall level of energy efficiency of 68 percent was achieved because its compact land use patterns made district heating a viable option. The compactness of the city also made the development of a highly effective public transport system economically viable.

In the development of CHP, the UK has been off to a slow start. Small-scale systems are being installed in some office blocks, schools, hospitals, and hotels, improving their energy efficiency considerably. All have the same high level of efficiency as large-scale systems.

The challenge for national governments and local authorities in the developed world is to put in place new energy policies, particularly to improve urban energy efficiency. The scenario includes the creation of municipally owned and operated energy systems. In some cities, such as Vienna and Stockholm, energy systems are operated by the "city works," which also supply water and run the transport and waste management systems. The synergies possible between these services are much harder to achieve in cities where privatization of services is the norm. It appears that the largest improvements in power distribution and consumption are realized by cities with a municipality-owned electricity company, such as Toronto and Amsterdam.[12]

The UK is just seeing the first schemes where greenhouse cultivation is being combined with CHP, utilizing their hot water and waste CO_2 to enhance crop growth for year-round cultivation.[13] Policies for encouraging CHP could thus also be used for enhancing urban agriculture, bringing producers closer to their markets instead of flying and trucking in vegetables from long distances. Once again, local job creation would result.

In addition to CHP, other significant new energy technologies are becoming available for use in cities. These include heat pumps, fuel cells, solar hot water systems and photovoltaic (PV) modules. In the near future, enormous reductions in fossil fuel use can be achieved by the use of PV systems, a technology particularly suited to cities. In the late 1990s there are only a few thousand buildings around the world using electricity from solar panels on their roofs or facades. Solar electricity could meet some of a building's requirements, with the rest of the power coming from the grid.

According to calculations by the oil company BP, London could supply most of its current summer electricity consumption from photovoltaic modules on the roofs and walls of its buildings. While this technology is still expensive, large scale automated production will dramatically reduce unit costs. And the only maintenance they require is cleaning once or twice a year.

Currently, solar energy is about eight times more expensive than conventional, but it is expected to be competitive as early as 2010 as the technology develops and the market grows. Major development programs have been announced in Japan, the USA, the Netherlands and the European Union to stimulate market growth. The technical potential for the generation of electricity from building integrated solar systems is very large indeed and could contribute significantly to the building energy requirements, even in a northerly climate like that in the UK. Of course, not all buildings will be suitable for the installation of a solar roof or facade, and adoption will be more rapid in countries with the highest sunshine level.

BOX 2 Solar energy in Saarbrücken, Germany

Saarbrücken, a city of 190,000 people, has a major investment program in solar energy. Since 1986 US$1.7 million has been spent on solar heating, PV systems, and other renewable energy sources. The state offers a 50 percent subsidy for technical assistance, and the local savings bank offers residential energy users favorable lending terms for the installations. The local energy utility owns the PV array, but the inhabitants of each house benefit from the solar electricity supply. In addition to domestic systems, there are also municipal PV installations, incorporated into highway noise barriers. The solar initiative has the support of the entire community because it is helping to lay the foundation for a sustainable future. A former coal-mining center, Saarbrücken has now become a center for the development of urban applications of solar energy systems.

Experimental solar buildings are springing up all over Europe. The new German government, elected in 1998, has a national program for installing 100,000 PV modules. PV programs in Japan and the USA are on a similar scale. In the UK, experimental systems have proved to be very promising. The Photovoltaics Centre at Newcastle University, a 1960s building recently clad with PV panels, has proved to be a great success. In Doxford, near Newcastle, Europe's largest solar-powered office building was completed in 1998.

In the new millennium, building designers will routinely incorporate this technology when designing a new building or refurbishing an existing one. In the meantime, to get experience with the technology, governments and urban authorities should vigorously encourage the installation of PV modules in our cities, enhancing the capacity to install PV systems. Every city should have buildings to test the potential of PV and to develop the local know-how.

Another energy technology of great promise, *fuel cells*, is fast coming of age. Fuel cells convert hydrogen, natural gas, or methanol into electricity by a chemical process without involving combustion. Fuel cells, like photovoltaic cells, have taken a long time to become commercially viable. Their development is now accelerating as the world searches for practical ways to produce cleaner electricity. Several companies have made great strides in making fuel cells competitive, in a variety of applications: from running generators and power stations, to buses, trucks and cars. Large-scale commercial production of fuel cells will be getting under way early in the new millennium. The combination of photovoltaic cells and fuel cells is a particularly compelling option. Electric energy from PV cells could split water into oxygen and hydrogen, and the latter could be stored and then used to run fuel cell power stations, or generators for individual buildings.

It is plausible that even large cities, whose genesis depended on the routine use of fossil fuels in the first place, may be able to make significant use of renewable energy in the future. To make their energy systems more sustainable, cities will require a combination of energy-efficient systems such as CHP with heat pumps, fuel cells and photovoltaic modules, and the efficient use of energy. Regulating the energy industry to improve generating efficiency, reduce discharge of waste gases and to adopt renewables will profoundly reduce the environmental impact of urban energy systems.

NOTES

1 Prof. Paul Brunner, TU, Vienna, personal communication.
2 Western Riverside Authority. 1991–2. *Annual Report*.
3 Worldwatch Institute. 1994. *Worldwatch Paper 121*. Washington, DC.
4 International Council on Local Government Initiatives. 1996. *The Local Agenda 21 Planning Guide*. Toronto.
5 Dr. Gerhard Gilnreiner, Vienna, personal communication.
6 Prof. Gerhard Vogel, Vienna, personal communication.
7 *Evening Standard*. 30 December 1996. London.
8 *Warmer Bulletin*. Summer 1995. London.
9 National Energy Action. 1997. Newcastle.
10 Energy Savings Trust. 1992. *Meeting the Challenge to Safeguard Our Future*. London.
11 Combined Heating and Power Association. 1998. London.
12 Nijkamp, Peter and Adriaan Perrels. 1994. *Sustainable Cities in Europe*. London: Earthscan.
13 Energy Savings Trust. 1992. *Meeting the Challenge to Safeguard Our Future*. London.

27

"Harnessing Wind, Solar, and Geothermal Energy"

from *World on the Edge* (2011)

Lester Brown

Editors' Introduction

Rethinking energy consumption and sources is a cornerstone of sustainable urbanism. The first (and most cost-effective) priority is usually to conserve energy and improve the efficiency with which we use it. But beyond that, a variety of renewable energy sources are coming into their own to help power the twenty-first-century metropolis.

In this selection, Lester Brown surveys renewable energy technologies and options worldwide. Brown is the founder of two influential nonprofit research organizations: the Worldwatch Institute and Earth Policy Institute. Since the 1970s he has taken a uniquely global approach to analyzing development trends through many publications. Recently he has focused on preparing a set of "Plan B" policy options for a sustainable global society. His books include *Full Planet, Empty Plates: The New Geopolitics of Food Scarcity* (New York: Norton, 2012), *Plan B 4.0: Mobilizing to Save Civilization* (New York: Norton, 2009), and the book from which this selection is taken, *World on the Edge: How to Prevent Environmental and Economic Collapse* (New York: Norton, 2011).

As fossil fuel prices rise, as oil insecurity deepens, and as concerns about climate change cast a shadow over the future of coal, a new world energy economy is emerging. The old energy economy, fueled by oil, coal, and natural gas, is being replaced with an economy powered by wind, solar, and geothermal energy. Despite the global economic crisis, this energy transition is moving at a pace and on a scale that we could not have imagined even two years ago.

The transition is well under way in the United States, where both oil and coal consumption have recently peaked. Oil consumption fell 8 percent between 2007 and 2010 and will likely continue falling over the longer term. During the same period, coal use also dropped 8 percent as a powerful grassroots anti-coal movement brought the licensing of new coal plants to a near standstill and began to work on closing existing ones.

While U.S. coal use was falling, some 300-wind farms with a generating capacity of 21,000 megawatts came online. Geothermal generating capacity, which had been stagnant for 20 years, came alive. In mid-2010, the U.S.-based Geothermal Energy Association announced that 152 new geothermal power plants were being developed, enough to triple U.S. geothermal generating capacity. On the solar front, solar cell installations are doubling every two years. The dozens of U.S. solar thermal power plants in the works could collectively add some 9,900 megawatts of generating capacity.

This chapter lays out the worldwide Plan B goals for developing renewable sources of energy by 2020. The goal of cutting carbon emissions 80 percent by 2020 is based on what we think is needed to avoid civilization-threatening climate change. This is not Plan A, business as usual. This is Plan B – a wartime mobilization, an all-out effort to restructure the world energy economy. To reach the Plan B goal, we replace all coal- and oil-fired electricity generation with that from renewable sources. Whereas the twentieth century was marked by the globalization of the world energy economy as countries everywhere turned to oil, much of it coming from the Middle East, this century will see the localization of energy production as the world turns to wind, solar, and geothermal energy.

The Plan B energy economy, which will be powered largely by electricity, does not rely on a buildup in nuclear power. If we used full-cost pricing – insisting that utilities pay for disposing of nuclear waste, decommissioning worn-out plants, and insuring reactors against possible accidents and terrorist attacks – no one would build a nuclear plant. They are simply not economical. Plan B also excludes the oft-discussed option of capturing and sequestering carbon dioxide (CO_2) from coal-fired power plants. Given the costs and the lack of investor interest within the coal community itself, this technology is not likely to be economically viable by 2020, if ever. Instead, wind is the centerpiece of the Plan B energy economy. It is abundant, low cost, and widely distributed; it scales up easily and can be developed quickly. A 2009 survey of world wind resources published by the U.S. National Academy of Sciences reports a wind-generating potential on land that is 40 times the current world consumption of electricity from all sources.

For many years, a small handful of countries dominated growth in wind power, but this is changing as the industry goes global, with more than 70 countries now developing wind resources. Between 2000 and 2010, world wind electric generating capacity increased at a frenetic pace from 17,000 megawatts to nearly 200,000 megawatts.

The United States, with 35,000 megawatts of wind generating capacity, leads the world in harnessing wind, followed by China and Germany with 26,000 megawatts each. Texas, long the leading U.S. oil-producing state, is now also the nation's leading generator of electricity from wind. It has 9,700 megawatts of wind generating capacity online, 370 megawatts more under construction, and a huge amount under development. If all of the wind farms projected for 2025 are completed, Texas will have 38,000 megawatts of wind generating capacity – the equivalent of 38 coal-fired power plants. This would satisfy roughly 90 percent of the current residential electricity needs of the state's 25 million people.

In July 2010, ground was broken for the Alta Wind Energy Center (AWEC) in the Tehachapi Pass, some 75 miles north of Los Angeles, California. At 1,550 megawatts, it will be the largest U.S. wind farm. The AWEC is part of what will eventually be 4,500 megawatts of renewable power generation, enough to supply electricity to some 3 million homes.

Since wind turbines occupy only 1 percent of the land covered by a wind farm, farmers and ranchers can continue to grow grain and graze cattle on land devoted to wind farms. In effect, they double-crop their land, simultaneously harvesting electricity and wheat, corn, or cattle. With no investment on their part, farmers and ranchers typically receive $3,000–10,000 a year in royalties for each wind turbine on their land. For thousands of ranchers in the U.S. Great Plains, wind royalties will dwarf their net earnings from cattle sales. In considering the energy productivity of land, wind turbines are in a class by themselves. For example, an acre of land in northern Iowa planted in corn can yield $1,000 worth of ethanol per year. That same acre used to site a wind turbine can produce $300,000 worth of electricity per year. This helps explain why investors find wind farms so attractive.

Impressive though U.S. wind energy growth is, the expansion now under way in China is even more so. China has enough onshore harnessable wind energy to raise its current electricity consumption 16-fold. Today, most of China's 26,000 megawatts of wind generating capacity come from 50- to 100-megawatt wind farms. Beyond the many other wind farms of that size that are on the way, China's new Wind Base program is creating seven wind mega-complexes of 10 to 38 gigawatts each in six provinces (1 gigawatt equals 1,000 megawatts). When completed, these complexes will have a generating capacity of more than 130 gigawatts. This is equivalent to building one new coal plant per week for two and a half years. Of these 130 gigawatts, 7 gigawatts will be in the coastal waters of Jiangsu Province, one of China's most highly industrialized provinces. China is planning a total of 23 gigawatts of offshore wind generating capacity. The country's first major offshore project, the 102-megawatt Donghai Bridge Wind Farm near Shanghai, is already in operation.

In Europe, which now has 2,400 megawatts of offshore wind online, wind developers are planning 140 gigawatts of offshore wind generating capacity, mostly in the North Sea. There is enough harnessable wind energy in offshore Europe to satisfy the continent's needs seven times over.

In September 2010, the Scottish government announced that it was replacing its goal of 50 percent renewable electricity by 2020 with a new goal of 80 percent. By 2025, Scotland expects renewables to meet all of its electricity needs. Much of the new capacity will be provided by offshore wind. Measured by share of electricity supplied by wind, Denmark is the national leader at 21 percent. Three north German states now get 40 percent or more of their electricity from wind. For Germany as a whole, the figure is 8 percent and climbing. And in the state of Iowa, enough wind turbines came online in the last few years to produce up to 20 percent of that state's electricity.

Denmark is looking to push the wind share of its electricity to 50 percent by 2025, with most of the additional power coming from offshore. In contemplating this prospect, Danish planners have turned conventional energy policy upside down. They plan to use wind as the mainstay of their electrical generating system and to use fossil-fuel-generated power to fill in when the wind dies down.

Spain, which has 19,000 megawatts of wind-generating capacity for its 45 million people, got 14 percent of its electricity from wind in 2009. On November 8th of that year, strong winds across Spain enabled wind turbines to supply 53 percent of the country's electricity over a five-hour stretch. London Times reporter Graham Keeley wrote from Barcelona that "the towering white wind turbines which loom over Castilla-La Mancha – home of Cervantes's hero, Don Quixote – and which dominate other parts of Spain, set a new record in wind energy production."

In 2007, when Turkey issued a request for proposals to build wind farms, it received bids to build a staggering 78,000 megawatts of wind generating capacity, far beyond its 41,000 megawatts of total electrical generating capacity. Having selected 7,000 megawatts of the most promising proposals, the government is issuing construction permits.

In wind-rich Canada, Ontario, Quebec, and Alberta are the leaders in installed capacity. Ontario, Canada's most populous province, has received applications for offshore wind development rights on its side of the Great Lakes that could result in some 21,000 megawatts of generating capacity. The provincial goal is to back out all coal-fired power by 2014. On the U.S. side of Lake Ontario, New York State is also requesting proposals. Several of the seven other states that border the Great Lakes are planning to harness lake winds.

At the heart of Plan B is a crash program to develop 4,000 gigawatts (4 million megawatts) of wind generating capacity by 2020, enough to cover over half of world electricity consumption in the Plan B economy. This will require a near doubling of capacity every two years, up from a doubling every three years over the last decade. This climate-stabilizing initiative would mean the installation of 2 million wind turbines of 2 megawatts each. Manufacturing 2 million wind turbines over the next 10 years sounds intimidating – until it is compared with the 70 million automobiles the world produces each year. At $3 million per installed turbine, the 2 million turbines would mean spending $600 billion per year worldwide between now and 2020. This compares with world oil and gas capital expenditures that are projected to double from $800 billion in 2010 to $1.6 trillion in 2015.

The second key component of the Plan B energy economy is solar energy, which is even more ubiquitous than wind energy. It can be harnessed with both solar photovoltaics (PV) and solar thermal collectors. Solar PV – both silicon-based and thin film – converts sunlight directly into electricity. A large-scale solar thermal technology, often referred to as concentrating solar power (CSP), uses reflectors to concentrate sunlight on a liquid, producing steam to drive a turbine and generate electricity. On a smaller scale, solar thermal collectors can capture the sun's radiant energy to warm water, as in rooftop solar water heaters.

The growth in solar cell production can only be described as explosive. It climbed from an annual expansion of 38 percent in 2006 to an off-the-chart 89 percent in 2008, before settling back to 51 percent in 2009. At the end of 2009, there were 23,000 megawatts of PV installations worldwide, which when operating at peak power could match the output of 23 nuclear power plants.

On the manufacturing front, the early leaders – the United States, Japan, and Germany – have been overtaken by China, which produces more than twice as many solar cells annually as Japan. Number three, Taiwan, is moving fast and may overtake Japan in 2010. World PV production has roughly doubled every two years since 2001 and will likely approach 20,000 megawatts in 2010.

Germany, with an installed PV power generating capacity of almost 10,000 megawatts, is far and away the world leader in installations. Spain is second with 3,400 megawatts, followed by Japan, the United States, and Italy. Ironically, China, the world's largest manufacturer of solar cells, has an installed capacity of only 305 megawatts, but this is likely to change quickly as PV costs fall.

Historically, photovoltaic installations were small-scale – mostly residential rooftop installations. Now that is changing as utility-scale PV projects are being launched in several countries. The United States, for example, has under construction and development some 77 utility-scale projects, adding up to 13,200 megawatts of generating capacity. Morocco is now planning five large solar generating projects, either photovoltaic or solar thermal or both, each ranging from 100 to 500 megawatts in size.

More and more countries, states, and provinces are setting solar installation goals. Italy's solar industry group is projecting 15,000 megawatts of installed capacity by 2020. Japan is planning 28,000 megawatts by 2020. The state of California has set a goal of 3,000 megawatts by 2017. Solar-rich Saudi Arabia recently announced that it plans to shift from oil to solar energy to power new desalination plants that supply the country's residential water. It currently uses 1.5 million barrels of oil per day to operate some 30 desalting plants.

With installations of solar PV climbing, with costs continuing to fall, and with concerns about climate change escalating, cumulative PV installations could reach 1.5 million megawatts (1,500 gigawatts) in 2020. Although this estimate may seem overly ambitious, it could in fact be conservative, because if most of the 1.5 billion people who lack electricity today get it by 2020, it will likely be because they have installed home solar systems. In many cases, it is cheaper to install solar cells for individual homes than it is to build a grid and a central power plant.

The second, very promising way to harness solar energy on a massive scale is CSP, which first came on the scene with the construction of a 350-megawatt solar thermal power plant complex in California. Completed in 1991, it was the world's only utility-scale solar thermal generating facility until the completion of a 64-megawatt power plant in Nevada in 2007.

Two years later, in July 2009, a group of 11 leading European firms and one Algerian firm, led by Munich Re and including Deutsche Bank, Siemens, and ABB, announced that they were going to craft a strategy and funding proposal to develop solar thermal generating capacity in North Africa and the Middle East. Their proposal would meet the needs of the producer countries and supply part of Europe's electricity via undersea cable. This initiative, known as the Desertec Industrial Initiative, could develop 300,000 megawatts of solar thermal generating capacity – huge by any standard. It is driven by concerns about disruptive climate change and by depletion of oil and gas reserves. Caio Koch-Weser, vice chair of Deutsche Bank, noted that "the Initiative shows in what dimensions and on what scale we must think if we are to master the challenges from climate change."

Even before this proposal, Algeria – for decades an oil exporter – was planning to build 6,000 megawatts of solar thermal generating capacity for export to Europe via undersea cable. The Algerians note that they have enough harnessable solar energy in their vast desert to power the entire world economy. This is not a mathematical error. A similar point often appears in the solar literature when it is noted that the sunlight striking the earth in one hour could power the world economy for one year. The German government was quick to respond to the Algerian initiative. The plan is to build a 1,900-mile high-voltage transmission line from Adrar deep in the Algerian desert to Aachen, a town on Germany's border with the Netherlands.

Although solar thermal power has been slow to get under way, utility-scale plants are being built rapidly now. The two leaders in this field are the United States and Spain. The United States has more than 40 solar thermal power plants operating, under construction, and under development that range from 10 to 1,200 megawatts each. Spain has 60 power plants in these same stages of development, most of which are 50 megawatts each.

One country ideally suited for CSP plants is India. The Great Indian Desert in its northwest offers a huge opportunity for building solar thermal power plants. Hundreds of large plants in the desert could meet most of India's electricity needs. And because it is such a compact country, the distance for building transmission lines to major population centers is relatively short.

One of the attractions of utility-scale CSP plants is that heat during the day can be stored in molten salt at temperatures above 1,000 degrees Fahrenheit. The heat can then be used to keep the turbines running for eight or more hours after sunset. The American Solar Energy Society notes that solar thermal resources in the U.S. Southwest can satisfy current U.S. electricity needs nearly four times over.

At the global level, Greenpeace, the European Solar Thermal Electricity Association, and the International Energy Agency's SolarPACES program have outlined a plan to develop 1.5 million megawatts of solar thermal power plant capacity by 2050. For Plan B we suggest a more immediate world goal of 200,000 megawatts by 2020, a goal that may well be exceeded as the economic potential becomes clearer.

The pace of solar energy development is accelerating as the installation of rooftop solar water heaters – the other use of solar collectors – takes off. China, for example, now has an estimated 1.9 billion square feet of rooftop solar thermal collectors installed, enough to supply 120 million Chinese households with hot water. With some 5,000 Chinese companies manufacturing these devices, this relatively simple low-cost technology has leapfrogged into villages that do not yet have electricity. For as little as $200, villagers can install a rooftop solar collector and take their first hot shower. This technology is sweeping China like wildfire, already approaching market saturation in some communities. Beijing's goal is to add another billion square feet to its rooftop solar water heating capacity by 2020, a goal it is likely to exceed.

Other developing countries such as India and Brazil may also soon see millions of households turning to this inexpensive water heating technology. Once the initial installment cost of rooftop solar water heaters is paid back, the hot water is essentially free.

In Europe, where energy costs are relatively high, rooftop solar water heaters are also spreading fast. In Austria, 15 percent of all households now rely on them for hot water. As in China, in some Austrian villages nearly all homes have rooftop collectors. Germany is also forging ahead. Some 2 million Germans are now living in homes where water and space are both heated by rooftop solar systems.

The U.S. rooftop solar water heating industry has historically concentrated on a niche market – selling and marketing 100 million square feet of solar water heaters for swimming pools between 1995 and 2005. Given this base, the industry was poised to mass-market residential solar water and space heating systems when federal tax credits were introduced in 2006. Led by Hawaii, California, and Florida, annual U.S. installation of these systems has more than tripled since 2005. The boldest initiative in the United States is California's goal of installing 200,000 solar water heaters by 2017. Not far behind is one launched in 2010 in New York State, which aims to have 170,000 residential solar water systems in operation by 2020.

Solar water and space heaters in Europe and China have a strong economic appeal, often paying for themselves from electricity savings in less than 10 years. With the cost of rooftop heating systems declining, many other countries will likely join Israel, Spain, and Portugal in mandating that all new buildings incorporate rooftop solar water heaters. The state of Hawaii requires that all new single-family homes have rooftop solar water heaters. Worldwide, Plan B calls for a total of 1,100 thermal gigawatts of rooftop solar water and space heating capacity by 2020.

The third principal component in the Plan B energy economy is geothermal energy. The heat in the upper six miles of the earth's crust contains 50,000 times as much energy as found in all of the world's oil and gas reserves combined – a startling statistic. Despite this abundance, as of mid-2010 only 10,700 megawatts of geothermal generating capacity have been harnessed worldwide, enough for some 10 million homes.

Roughly half the world's installed geothermal generating capacity is concentrated in the United States and the Philippines. Most of the remainder is generated in Mexico, Indonesia, Italy, and Japan. Altogether some 24 countries now convert geothermal energy into electricity. El Salvador, Iceland, and the Philippines respectively get 26, 25, and 18 percent of their electricity from geothermal power plants.

The geothermal potential to provide electricity, to heat homes, and to supply process heat for industry is vast. Among the geothermally rich countries are those bordering the Pacific in the so-called Ring of Fire, including Chile, Peru, Colombia, Mexico, the United States, Canada, Russia, China, Japan, the Philippines, Indonesia, and Australia. Other well-endowed countries include those along the Great Rift Valley of Africa, including Ethiopia, Kenya, Tanzania, and Uganda, and those around the Eastern Mediterranean. As of 2010, there are some 70 countries with projects under development or active consideration, up from 46 in 2007.

Beyond geothermal electrical generation, up to 100,000 thermal megawatts of geothermal energy are used directly – without conversion into electricity – to heat homes and greenhouses and to provide process heat to industry. For example, 90 percent of the homes in Iceland are heated with geothermal energy.

An interdisciplinary team of 13 scientists and engineers assembled by the Massachusetts Institute of Technology in 2006 assessed U.S. geothermal electrical generating potential. Drawing on the latest technologies, including those used by oil and gas companies in drilling and in enhanced oil recovery, the team estimated that enhanced geothermal systems could help the United States meet its energy needs 2,000 times over.

Even before this exciting new technology is widely deployed, investors are moving ahead with existing technologies. For many years, U.S. geothermal energy was confined largely to the Geysers project north of San Francisco, easily the world's largest geothermal generating complex, with 850 megawatts of generating capacity. Now the United States has more than 3,000 megawatts of existing geothermal electrical capacity and projects under development in 13 states. With California, Nevada, Oregon, Idaho, and Utah leading the way, and with many new companies in the field, the stage is set for a geothermal renaissance.

In mid-2008, Indonesia – a country with 128 active volcanoes and therefore rich in geothermal energy – announced that it would develop 6,900 megawatts of geothermal generating capacity; Pertamina, the state oil company, is responsible for developing the lion's share. Indonesia's oil production has been declining for the last decade, and in each of the last five years it has been an oil importer. As Pertamina shifts resources from oil to the development of geothermal energy, it could become the first oil company – state-owned or independent – to make the transition from oil to renewable energy.

Japan, which has 16 geothermal power plants with a total of 535 megawatts of generating capacity, was an early leader in this field. After nearly two decades of inactivity, this geothermally rich country – long known for its thousands of hot baths – is again building geothermal power plants.

Among the Great Rift countries in Africa, Kenya is the early geothermal leader. It now has 167 megawatts of generating capacity and is planning 1,200 more megawatts by 2015, enough to nearly double its current electrical generating capacity from all sources. It is aiming for 4,000 geothermal megawatts by 2030.

Beyond power plants, geothermal (ground source) heat pumps are now being widely used for both heating and cooling. These take advantage of the remarkable stability of the earth's temperature near the surface and then use that as a source of heat in the winter when the air temperature is low and a source of cooling in the summer when the air temperature is high. The great attraction of this technology is that it can provide both heating and cooling and do so with 25–50 percent less electricity than would be needed with conventional systems. In Germany, 178,000 ground-source heat pumps are now operating in residential or commercial buildings. At least 25,000 new pumps are installed each year.

Geothermal heat is ideal for greenhouses in northern countries. Russia, Hungary, Iceland, and the United States are among the many countries that use it to produce fresh vegetables in winter. With rising oil prices boosting fresh produce transport costs, this practice will likely become far more common. If the four most populous countries located on the Pacific Ring of Fire – the United States, Japan, China, and Indonesia – were to seriously invest in developing their geothermal resources, it is easy to envisage a world with thousands of geothermal power plants generating some

200,000 megawatts of electricity, the Plan B goal, by 2020.

As oil and natural gas reserves are being depleted, the world's attention is also turning to plant-based energy sources, including energy crops, forest industry byproducts, sugar industry byproducts, urban waste, livestock waste, plantations of fast-growing trees, crop residues, and urban tree and yard wastes – all of which can be used for electrical generation, heating, or the production of automotive fuels.

The potential use of energy crops is limited because even corn – the most efficient of the grain crops – can convert only 0.5 percent of solar energy into a usable form. In contrast, solar PV or solar thermal power plants convert roughly 15 percent of sunlight into electricity. And the value of electricity produced on an acre of land occupied by a wind turbine is over 300 times that of the corn-based ethanol produced on an acre. In this land-scarce world, energy crops cannot compete with solar-generated electricity, much less with wind power.

Yet another source of renewable energy is hydropower. The term has traditionally referred to dams that harnessed the energy in river flows, but today it also includes harnessing the energy in tides and waves as well as using smaller "in-stream" turbines to capture the energy in rivers and tides without building dams. Roughly 16 percent of the world's electricity comes from hydropower, most of it from large dams. Some countries, such as Brazil, Norway, and the Democratic Republic of the Congo, get the bulk of their electricity from river power.

Tidal power holds a certain fascination because of its sheer potential scale. The first large tidal generating facility – La Rance Tidal Barrage, with a maximum generating capacity of 240 megawatts – was built 40 years ago in France and is still operating today. Within the last few years interest in tidal power has spread rapidly. South Korea, for example, is building a 254-megawatt project on its west coast that would provide all the electricity for the half-million people living in the nearby city of Ansan. At another site to the north, engineers are planning a 1,320-megawatt tidal facility in Incheon Bay, near Seoul. And New Zealand is planning a 200-megawatt project in the Kaipara Harbour on that country's northwest coast.

Wave power, though a few years behind tidal power, is also now attracting the attention of both engineers and investors. Scottish firms Aquamarine Power and SSE Renewables are teaming up to build 1,000 megawatts of wave and tidal power off the coast of Ireland and the United Kingdom. Ireland is planning 500 megawatts of wave generating capacity by 2020, enough to supply 8 percent of its electricity. Worldwide, the harnessing of wave power could generate a staggering 10,000 gigawatts of electricity, more than double current world electricity capacity from all sources.

We project that the 980 gigawatts (980,000 megawatts) of hydroelectric power in operation worldwide in 2009 will expand to 1,350 gigawatts by 2020. According to China's official projections, 180 gigawatts should be added there, mostly from large dams in the southwest. The remaining 190 gigawatts in our projected growth of hydropower would come from a scattering of large dams still being built in countries like Brazil and Turkey, dams now in the planning stages in sub-Saharan Africa, a large number of small hydro facilities, a fast-growing number of tidal projects, and numerous smaller wave power projects.

The efficiency gains . . . more than offset projected growth in energy use to 2020. The next step in the Plan B 80-percent reduction of carbon emissions comes from replacing fossil fuels with renewable sources of energy. In looking at the broad shifts from the reference year of 2008 to the Plan B energy economy of 2020, fossil-fuel-generated electricity drops by 90 percent worldwide as the fivefold growth in renewably generated electricity replaces all the coal and oil and 70 percent of the natural gas now used to generate electricity. Wind, solar photovoltaic, solar thermal, and geothermal will dominate the Plan B energy economy, but as noted earlier wind will be the centerpiece – the principal source of the electricity to heat, cool, and light buildings and to run cars and trains.

The Plan B projected tripling of renewable thermal heating generation by 2020, roughly half of it to come from direct uses of geothermal energy, will sharply reduce the use of both oil and gas to heat buildings and water. And in the transportation sector, energy use from fossil fuels drops by some 70 percent. This comes from shifting to all-electric and highly efficient plug-in hybrid cars that will run almost entirely on electricity, nearly all of it from renewable sources. And it also comes from shifting to electric trains, which are much more efficient than diesel-powered ones.

Each country's energy profile will be shaped by its unique endowment of renewable sources of energy. Some countries, such as the United States, Turkey, and China, will likely rely on a broad base of renewables – wind, solar, and geothermal power. But wind, including both onshore and offshore, is likely to emerge as the leading energy source in all three cases.

Other countries, including Spain, Algeria, Egypt, India, and Mexico, will turn primarily to solar thermal power plants and solar PV arrays to power their economies. For Iceland, Indonesia, Japan, and the Philippines, geothermal energy will likely be the mother lode. Still others will likely rely heavily on hydro, including Norway, Brazil, and Nepal. And some technologies, such as rooftop solar water heaters, will be used virtually everywhere.

As the transition progresses, the system for transporting energy from source to consumers will change beyond recognition. In the old energy economy, pipelines and tankers carried oil long distances from oil fields to consumers, including a huge fleet of tankers that moved oil from the Persian Gulf to markets on every continent. In the new energy economy, pipelines will be replaced by transmission lines.

The proposed segments of what could eventually become a national U.S. grid are beginning to fall into place. Texas is planning up to 2,900 miles of new transmission lines to link the wind-rich regions of west Texas and the Texas panhandle to consumption centers such as Dallas–Fort Worth and San Antonio. Two high-voltage direct current (HVDC) lines will link the rich wind resources of Wyoming and Montana to California's huge market. Other proposed lines will link wind in the northern Great Plains with the industrial Midwest.

In late 2009 Tres Amigas, a transmission company, announced its plans to build a "SuperStation" in Clovis, New Mexico, that would link the country's three major grids – the Western grid, the Eastern grid, and the Texas grid – for the first time. This would effectively create the country's first national grid. Scheduled to start construction in 2012 and to be completed in 2014, the SuperStation will allow electricity, much of it from renewable sources, to flow through the country's power transmission infrastructure.

Google made headlines when it announced in mid-October 2010 that it was investing heavily in a $5-billion offshore transmission project stretching from New York to Virginia, called the Atlantic Wind Connection. This will facilitate the development of enough offshore wind farms to meet the electricity needs of 5 million East Coast residents.

A strong, efficient national grid will reduce generating capacity needs, lower consumer costs, and cut carbon emissions. Since no two wind farms have identical wind profiles, each one added to the grid makes wind a more stable source of electricity. With the prospect of thousands of wind farms spread from coast to coast and a national grid, wind becomes a stable source of energy, part of baseload power.

Europe, too, is beginning to think seriously of investing in a supergrid. In early 2010, a total of 10 European companies formed Friends of the Supergrid, which is proposing to use HVDC undersea cables to build the European supergrid offshore, an approach that would avoid the time-consuming acquisition of land to build a continental land-based system. This grid could then mesh with the proposed Desertec initiative to integrate the offshore wind resources of northern Europe and the solar resources of North Africa into a single system that would supply both regions. The Swedish ABB Group, which in 2008 completed a 400-mile HVDC undersea cable linking Norway and the Netherlands, is well positioned to help build the necessary transmission lines.

Governments are considering a variety of policy instruments to help drive the transition from fossil fuels to renewables. These include tax restructuring, lowering the tax on income and raising the tax on carbon emissions to include the indirect costs of burning fossil fuels. If we can create an honest energy market, the transition to renewables will accelerate dramatically.

Another measure that will speed the energy transition is eliminating fossil fuel subsidies. At present, governments are spending some $500 billion per year subsidizing the use of fossil fuels. This compares with renewable energy subsidies of only $46 billion per year. For restructuring the electricity sector, feed-in tariffs, in which utilities are required to pay set prices for electricity generated from renewable sources, have been remarkably successful. Germany's impressive early success with this measure has led to its adoption by some 50 other countries, including most of those in the European Union. In the United States, 29 states have adopted renewable portfolio standards requiring utilities to get up to 40 percent of their electricity from renewable sources. The United States has also used tax credits for wind, geothermal, solar photovoltaics, solar water and space heating, and ground-source heat pumps.

To achieve some goals, governments are simply using mandates, such as those requiring rooftop solar water heaters on all new buildings. Governments at all levels are adopting energy efficiency building codes. Each government has to select the policy instruments that work best in its particular economic and cultural setting.

In the new energy economy, our cities will be unlike any we have known during our lifetime. The air will be clean and the streets will be quiet, with only the scarcely audible hum of electric motors. Air pollution alerts will be a thing of the past as coal-fired power plants are dismantled and recycled and as gasoline- and diesel-burning engines largely disappear.

This transition is now building its own momentum, driven by an intense excitement from the realization that we are tapping energy sources that can last as long as the earth itself. Oil wells go dry and coal seams run out, but for the first time since the Industrial Revolution, we are investing in energy sources that can last forever.

Data, endnotes, and additional resources can be found on Earth Policy's website, at www.earth-policy.org.

28

"The Changing Water Paradigm: A Look at Twenty-First Century Water Resources Development"

from *Water International* (2000)

Peter H. Gleick

Editors' Introduction

Water is one of the most essential resources for human communities, but has been widely used in unsustainable ways. In this selection, Peter H. Gleick, an environmental scientist and founder of the Pacific Institute in Oakland, California, lays out a new and more sustainable water resources agenda for the current century. The new approach to water, in his view, will emphasize conservation, reclamation, and a better balancing of the needs of multiple stakeholders and the environment. Managing demand for water rather than increasing supplies will be the primary goal.

Other books by Gleick include *Bottled and Sold: The Story Behind Our Obsession with Bottled Water* (Washington, D.C.: Island Press, 2010) and *A Twenty-First Century U.S. Water Policy* (with Juliet Christian-Smith; Oxford: Oxford University Press, 2012). Other resources on water policy include *Rivers for Life: Managing Water for People and Nature* by Sandra Postel and Brian Richter (Washington, D.C.: Island Press, 2003); *Last Oasis: Facing Water Scarcity* by Sandra Postel (New York: Norton, 1997), and *Dry Run: Preventing the Next Urban Water Crisis* by Jerry Yudelson (Gabriola Island, BC: New Society Publishers, 2010).

Water resources development around the world has taken many different forms and directions since the dawn of civilization. Humans have long sought ways of capturing, storing, cleaning, and redirecting freshwater resources in efforts to reduce their vulnerability to irregular river flows and unpredictable rainfall. As the new millennium dawns, the dynamic process of managing freshwater resources and human demands for water is changing again.

I have previously described these shifts as "the changing water paradigm" (Gleick, 1998). There are many components to this change: a shift away from sole, or even primary, reliance on finding new sources of supply to address perceived new demands; a growing emphasis on incorporating ecological values into water policy; a re-emphasis on meeting basic human needs for water services; and a conscious breaking of the ties between economic growth and water use. The evidence of a true change in the way we think about water continues to accumulate.

A reliance on physical solutions continues to dominate traditional planning approaches, but these solutions are facing increasing opposition. At the same time, new methods are being developed to meet the demands of growing populations without requiring major new construction or new large-scale water transfers from one region to another. More and more water suppliers and planning agencies are beginning to shift their focus and explore efficiency improvements, implement options for managing demand, and reallocate water among users to reduce projected gaps and meet future needs. The connections between water and food are receiving increasing attention as the concerns of food experts begin to encompass the realities of water availability. These shifts have not come easily, they have not come without strong internal opposition, they are still not universally accepted, and they may not be permanent. Nevertheless, these changes represent a real shift in the way humans think about water use.

THE END OF TWENTIETH CENTURY PLANNING APPROACHES

The old paradigm of relying on ever-larger numbers of dams, reservoirs, and aqueducts to capture, store, and move ever-larger fractions of freshwater runoff is beginning to fail for environmental, economic, and social reasons. Basic human needs for water remain unmet. It is becoming harder and harder to find new, or even hold onto existing, water resources to supply croplands. And little attention has traditionally been paid to protecting natural ecosystems from which water supplies have been withdrawn.

Official water planning efforts usually make no attempt to analyze the details of what water is actually used for or how much water is required to meet different types of demands. Nor do they try to identify common goals for water development among conflicting stakeholders or to seek agreement on principles to resolve conflicts over water. The lack of consensus on a guiding ethic for water policy has led to fragmented policies and incremental changes that typically satisfy none of the many affected parties.

The search for new solutions is also being pushed along in some places by the changing nature of demand for water. Throughout the first three-quarters of the twentieth century, demand for water throughout the world increased. Water withdrawals were growing not only in an absolute sense, they were growing in a per-capita sense. In 1900 in the U.S., annual average per-capita freshwater use was less than 700 cubic meters per person. By the late 1970s and early 1980s, it had increased to nearly 2,300 m^3/p/yr (Solley et al., 1998; CEQ 1991; and Perlman, 1997). These increases in water demands in the U.S. and elsewhere, more than any other factor, drove the vast construction of water infrastructure.

Beginning in the mid-1980s and early 1990s, these trends ended in the United States. In a departure from the expectations and experience of water planners, water use started to drop, despite continued increases in population and economic wealth. Yet industrial output and productivity have continued to rise dramatically, clearly demonstrating that it is possible to break the link between water use and industrial production. The decline in water use is even more dramatic when per-capita withdrawals are analyzed. Per capita freshwater withdrawals peaked in 1980 and dropped more than 20 percent by 1995.

Though many regions continue to develop and need more water, long-term projections of future global water needs have been steadily declining. Without constantly increasing demand of three to four percent per year, the pressure to build new water infrastructure has diminished since existing supplies can be reallocated to other users. The changing philosophy away from new development has also been driven by two other important factors: the increasing concern about the environmental impacts of water projects, and their increasing economic costs.

There is growing grassroots opposition to big projects because of their serious local costs, including population displacement, land inundation, and ecological disruption. New water supply systems have also increasingly become expensive compared to non-structural alternatives. When the first major dam projects were being built, it was relatively unimportant that they be economically justifiable and economic analyses were done with incomplete information and questionable assumptions. For example, all non-market environmental and social costs were simply excluded because they were unquantified or unquantifiable. Economic games were also played with stretched-out repayment periods, high discount rates, low-interest loans, and a transfer of costs to non-dam parts of water developments.

Almost all past water infrastructure development was subsidized or fully paid for by governments and international financial organizations. Governmental budgets in Asia and many other regions are now under great pressure and there are serious constraints on new money for major water projects. While this pressure has been felt in every sector of society, it plays a particularly important and direct role in changing national water policies by limiting central governmental involvement in new capital-intensive projects and shifting more responsibility to regional and local governments.

A NEW PARADIGM FOR WATER PLANNING

Meaningful change towards a new approach and a new way of thinking has to begin with an open discussion of the ultimate ends of water resource policy. More people now place a high value on maintaining the integrity of water resources and the flora, fauna, and human societies that have developed around them. There are growing calls for the costs and benefits of water developments to be distributed in a more equitable manner and for unmet basic human needs to be addressed. And more and more, efforts are being made to understand and meet the diverse interests and needs of all affected stakeholders. If the next generation of water planners continues to try to integrate these principles, the present stalemate and paralysis on how to move forward will ease and a new era of innovative water management will ensue.

Traditional approaches to water planning, while still firmly entrenched in many water planning institutions, are beginning to change. Continued investments in huge systems that provide more water for some is being challenged by those who believe a higher priority should be assigned to projects that meet basic unmet human needs for water (Gleick, 1996). The question of whether we will be able to produce enough food to feed the world's burgeoning population, and get it to where it is needed, is now understood to be intricately connected to the question of where and when fresh water is available. Decisions made today about water policy will affect whether people continue to be undernourished in the coming decades. There is great potential for improving the "water efficiency" with which we produce food, by changing cropping patterns toward crops that require less water per calorie to produce, by reducing wasteful applications of water, by cutting losses between the field and the plate, and by altering diets and the functioning of international markets.

There is a new trend to take out or decommission dams that either no longer serve a useful purpose or have caused such egregious ecological impacts as to warrant removal.

Nearly 500 dams in the U.S. and elsewhere have already been removed and the movement toward river restoration is accelerating (Gleick, 2000). Within a few months of the Edwards Dam being removed in Maine in mid-1999, salmon, striped bass, alewives, and other affected fish returned to waters above the old dam site from which they had been absent for 162 years (New York Times, 1999).

As traditional water-supply approaches of building another dam or drilling another tubewell become less appropriate or more expensive, unconventional supply approaches are receiving more attention. More and more cities are discovering that wastewater can be a resource, not a liability, for purposes ranging from landscape irrigation to drinking water. Several other unusual approaches are receiving more attention, including large- and small-scale desalination technology, water reclamation and reuse, and techniques such as fog collection. Matching water demands with available waters of different quality can reduce water supply constraints, increase system reliability, and solve costly wastewater disposal problems. Even esoteric ideas are being explored, for example, the concept of transporting fresh water in large ocean-going plastic bags has moved from theory to reality, with small projects underway in the Mediterranean Sea (Gleick, 1998).

As an alternative to new infrastructure, efforts are now underway to rethink water planning and management. Many individual nations as well as international aid organizations are rethinking water policy and putting greater emphasis on development principles that reflect environmental, social, and cultural values. Among the major principles that appear to be common to all these new approaches are:

- Basic human needs for drinking water and sanitation services must be met.
- Basic ecosystem needs for water must be met.
- The use of non-structural alternatives to meet demands must receive higher priority.
- Economic principles must be applied more frequently and reliably to water use and management.

- New supply systems, if needed, must be flexible and maximally efficient.
- Non-governmental organizations, individuals, independent research organizations, and other affected stakeholders must all be involved in water management decisions.

A working definition of sustainable water use applying these principles is "the use of water that supports the ability of human society to endure and flourish into the indefinite future without undermining the integrity of the hydrological cycle or the ecological systems that depend on it."

Some new dams, aqueducts, and water infrastructure will certainly be built, particularly in developing countries where the basic water requirements for humans have still not been met. But even in these regions, new approaches are being developed, or old ones rediscovered, which permit water needs to be met with fewer resources, less ecological disruption, and less money. Successfully meeting human demands for water in the next century will increasingly depend upon non-structural solutions and a completely new approach to planning and management.

The most important single goal of this new paradigm is to re-integrate water use with maintaining ecological health and environmental well-being. On the water use side, we must refocus our efforts with the objective of increasing the productive use of water. Two approaches are needed: (1) increasing the efficiency with which current needs are met; and (2) increasing the efficiency with which water is allocated among different uses. Where new supplies are still needed, major new projects must now compete with innovative small-scale approaches, including micro-dams, run-of-river hydro, land management and protection methods, and other locally managed solutions. In addition to this, non-traditional sources of supply will play an increasing role, including reclaimed or recycled water and, in some limited circumstances, desalinated brackish water or seawater.

MEET BASIC HUMAN NEEDS FOR WATER

Universal access to basic water services is one of the most fundamental conditions of human development. Yet billions of people lack such access.

The numbers are stark, as more than a billion people in the developing world do not have safe drinking water and nearly three billion people live without access to adequate sanitation systems necessary for reducing exposure to water-related diseases. The failure of the international aid community, nations, and local organizations to satisfy these basic human needs has led to substantial, unnecessary, and preventable human suffering. An estimated 14 to 30 thousand people, mostly young children and the elderly, die every day from water-related diseases. At any given moment, approximately one-half of the people in the developing world suffer from diseases caused by drinking contaminated water or eating contaminated food (United Nations, 1997). A formal review of international law, declarations, and State practice supports the conclusion that access to a basic water requirement can be considered a fundamental human right (McCaffrey, 1992; Gleick, 1999).

RESTORE AND MAINTAIN ECOLOGICAL HEALTH

Perhaps the next greatest failing of twentieth century water policy was the failure to understand the connections between water and ecological health, and the links between the health of natural ecosystems and human wellbeing. As a result, among the most important objectives for twenty-first century water managers are to understand those links and to integrate ecological and human water needs in a comprehensive way. For some ecosystems, it is too late. The destruction of the Aral Sea and the extinction of the endemic species of fish there are largely irreversible. Invasions of aquatic systems by exotic species, once established, can be difficult or impossible to stop. But we have also learned that ecosystems can be resilient and restored, if proper resources and knowledge are applied.

Minimum water requirements must be determined, provided, and protected for natural ecosystems. Determining the nature and characteristics of these requirements can be very difficult; sometimes they are related to minimum flow requirements, or temperature limits, or a need for peak flows during certain periods, or water of a certain quality. But these requirements must be met as a fundamental condition of water resources development or we risk impoverishing ourselves of our natural resources and undermining the natural support structures on which we depend.

NON-STRUCTURAL WATER DEVELOPMENT: INCREASE THE EFFICIENT USE AND ALLOCATION OF WATER

A key component of non-structural approaches to water resources management is a focus on using water more efficiently and then reallocating saved water. In the mid-1970s, arguments against developing new supplies of energy began to gain favor, driven in part by concerns over nuclear power's high costs and the potential for catastrophic accidents, and over the accumulating environmental consequences of fossil fuel combustion. During this period, some analysts argued that more efficient use of energy could significantly reduce future demand and delay or avoid the need for the construction of economically and environmentally costly new construction (e.g., Lovins, 1977). These arguments have turned out to be largely true, and the proper incentives have led to tremendous drops in energy demand, while economic wellbeing has continued to improve.

The same arguments are now beginning to be heard over water. New sources of water supply can largely be avoided in many regions by implementing intelligent water conservation and demand-management programs, installing new efficient equipment, and applying appropriate economic and institutional incentives to shift water among users. Improvements in water use efficiency will come about through changes in technology, economics, and institutions. Vast improvements in water use efficiency are possible in almost all sectors. In both developed and developing countries, large losses occur in distribution systems, faulty or old equipment, and poorly designed or maintained irrigation systems. In California, "unaccounted" for water in urban systems is estimated at 10 percent, but many individual water districts have much higher losses (CDWR, 1998). Many irrigation canals in the western U.S. are unlined, leading to significant seepage losses. In Jordan, one estimate is that at least 30 percent

of all domestic water supply never reaches users because of flaws and inadequacies in the water-supply network, and the losses reach 50 percent in Jordan's capital, Amman (Salameh and Bannayan, 1993). It has been estimated that the amount of water lost in Mexico City's supply system is equal to the amount needed to supply a city the size of Rome (Falkenmark and Lindh, 1993). While such losses are difficult to estimate accurately, there is little disagreement that significant water savings improvements are possible.

Even where efforts to improve water use efficiency have begun, great potential still exists for reducing water use without sacrificing economic productivity or personal welfare. In a mundane but highly revealing example, the United States passed a law, effective in 1994, requiring that all new toilets use about one-third the amount of water traditionally used. Even today, however, this sector shows substantial untapped potential.

Technological change is a dynamic and ongoing process, even for a technology as mundane as toilets. Concerns over the reliability of water supply in Singapore, and [the city's] vulnerability to water supply disruptions, led [its leaders] to launch a campaign to improve water use efficiency, which included replacing what the U.S. now considers "ultra-low flow toilets" with even higher efficiency models. These kinds of changes have already led to further water and economic savings in Singapore (Zachary, 1997). And there are no technological reasons why toilets have to use any water at all. Some experts argue that in regions of the world where water is scarce, it makes little sense to use water for disposing of human wastes when other satisfactory alternatives exist (Kalbermatten et al., 1982; Rogers, 1997).

Water productivity can also be improved in outdoor gardens, municipal lawns, golf courses, and other urban landscapes. In some parts of the United States as much as half of all residential or institutional water demand goes to water gardens and lawns. Improvements in watering efficiency could reduce that demand substantially, as could changes in the composition of these gardens. Innovative garden designs, combined with new computer controllers, moisture sensors, and water technology all can reduce outdoor water use in homes by 25 to 50 percent or more depending on homeowners' preferences, the price of water, and the cost of alternatives (Gleick *et al.*, 1995). In some regions, outdoor municipal and institutional landscape irrigation is being done with reclaimed water, completely eliminating the use of potable water for this purpose.

The largest single use of water is in agriculture and a substantial amount of this water could be used more productively. Water is lost as it moves through leaky pipes and unlined aqueducts, as it is distributed to farmers, and as it is applied to grow crops. Some analysts estimate that the overall efficiency of agricultural water use worldwide is only 40 percent (Postel, 1997), meaning that more than half of all water diverted for agriculture never produces food. In most basins, overall irrigation efficiency may be higher than single farm efficiencies as downstream irrigators reuse water lost to seepage from inefficient irrigation. But even in these basins, efforts to reduce unproductive evaporative losses can produce new water for agricultural use.

In water-short areas, new techniques and new technologies are already changing the face of irrigation. New sprinkler designs, such as low-energy precision sprinklers and drip systems can increase irrigation efficiencies from 60 to 70 percent to as high as 95 percent (Postel, 1997). Not all crops are suitable for precision irrigation systems, although drip irrigation, formerly limited to orchards and vineyards, is increasingly being used for row crops. Even cotton is now being grown in parts of California with drip systems (Fidell *et al.*, 1999). Identifying technical and institutional ways of improving the efficiency of agricultural water systems will go a long way toward increasing agricultural production without having to develop new supplies of water. Examples of successful implementation of these kinds of strategies have been described in Owens-Viani *et al.* (1999).

ECONOMICS AND WATER PRICING

In the past, widespread subsidies have encouraged rapid development of supply systems and hindered water efficiency efforts. These subsidies have been very effective at accomplishing their goals in both urban and agricultural settings. Urban centers in the western U.S., Mexico City, Singapore, Beijing,

and many other cities now support very large urban populations where there would otherwise have been inadequate water supplies. Semi-arid deserts can now produce vast amounts of food where rainfall alone would be insufficient. But these subsidies have also been responsible for unplanned and undesired side effects. Subsidized cotton production in central Asia expanded so much that the inflows of water from the Amu and Syr Darya rivers to the Aral Sea were cut off, leading to a shrinking of the Sea, the extinction of endemic species, and adverse impacts on human health. Fossil groundwater in Saudi Arabia has been used unsustainably to grow subsidized wheat. Groundwater overdraft in India, encouraged by subsidized energy costs for pumping and a lack of groundwater regulation, now threatens that country's agricultural self-sufficiency.

The agricultural sector has particularly benefited from water subsidies. In much of the world, around 75 percent of all water consumed goes to agriculture. The extremely low cost of water encourages the production of crops that are both low valued and highly water intensive and it provides no incentive to use water efficiently. Even modest changes in agricultural practices would free up substantial amounts of water for other agricultural uses, urban needs, and environmental restoration.

Economic factors and pricing decisions also lead to inefficient water use in the urban sector. In many cities, water use is not measured or "metered," which leads to overuse of water and provides no incentives for efficient use. Even in regions with water metering, the inappropriate design of rate structures can lead to misuse of water. As a result, there is growing interest in the use of so-called "conservation" rates, such as increasing block rates, where increasing amounts of water are charged higher and higher rates. More and more water utilities are implementing such rate structures. In Beijing, China, a new pricing system links the cost of water to the amount of water used, thus encouraging conservation. A similar pricing system decreased average monthly residential water use by nearly 30 percent in Bogor, Indonesia (Postel, 1997). Regional water providers in South Africa have been able to delay the construction of new regional water-supply systems by imposing higher rates, distributing water-conservation equipment, and educating the public (Rand Water, 1996). In Hermanns, South Africa, a major conservation program included an 11-step rate structure that encouraged technological improvements and careful attention to wasteful practices. In just the first year of operation of the program, water use in the city dropped more than 30 percent (Gleick, 1998).

While the new emphasis on treating water as an economic good can eliminate wasteful practices and encourage increased efficiency and conservation, a purely market approach cannot adequately protect the natural ecosystems that also depend upon water. Nature provides services that help keep humans alive, but these services are not "purchased," rarely quantified, and routinely excluded from official economic accounts (Daily, 1997). In the drive toward economic rationality, care must be taken to preserve and protect those services that may fall outside of traditional economic measures.

ALTERNATIVE SUPPLIES

Rather than seeking new pristine sources from far away, a wide range of alternative water supplies will increasingly be used to meet certain demands. Meeting different needs with the appropriate quality of water may prove to be economically beneficial and at the same time reduce the need for new supplies at a higher and higher marginal cost. Reclaimed water, in particular, has some remarkable advantages, including a high reliability of supply, a known quality, and often, a centralized source near urban demand centers.

Reclaimed wastewater

Societies spend billions of dollars finding water, treating it to high standards, and then moving it to where it is needed. We then spend billions more to collect wastewater, treat it to reduce human and environmental health problems associated with sewage and industrial waste, and then dump it into the oceans or other sinks. The majority of urban water ends up being thrown away after being used once. More recently, attention has focused on treating this water and using it as a resource rather than considering it extraneous waste (Asano and Levine,

1998). Drought conditions limiting supply, environmental problems with sewage disposal, and growing demands have all made water reclamation more appealing.

Reclaimed water can be used to recharge groundwater aquifers, supply industrial processes, irrigate certain crops, or augment potable supplies. In the Middle East, parts of Africa, and the western U.S. there has been a significant increase in the availability and use of treated wastewater for a wide range of industrial, commercial, and institutional needs (Wong, 1999). Some agricultural water needs are now being met with treated wastewater. In Windhoek, Namibia, reclaimed water has been used to augment the potable water supply since 1968, and in drought years up to 30 percent of the city's drinking water supply is treated wastewater (Van der Merwe and Menge, 1996). The U.S. National Academy of Sciences has completed a study on appropriate uses of highly treated wastewater for indirect augmentation of drinking water supplies (U.S. National Research Council, 1998). Israel has extensive wastewater reclamation programs. Seventy percent of Israeli wastewater is treated and used for agricultural irrigation and Shuval (1996) estimates that 80 percent recycling is likely in the next few decades. Efforts to capture, treat, and reuse more wastewater are also being made in neighboring Jordan where overall water supplies are also highly constrained (Ahmad, 1989; Salameh and Bannayan, 1993).

Desalination

For decades, some water analysts and observers have held out desalination as the ultimate solution to the world's water woes. More than 97 percent of the water on the planet is too salty to drink or grow food. In theory, therefore, desalination offers a limitless supply of freshwater, freeing humans from the vagaries and inconsistencies of natural freshwater supplies. Yet like the unfulfilled promise of cheap nuclear power in the 1960s and 1970s, desalination remains a minor contributor to water supply, providing only a fraction of a percent of total human needs.

The desalination of seawater or brackish water is technologically well developed, but remains hindered by high economic costs in part because of the large amounts of energy required to strip salt ions from water. While technological optimists continue to predict declining costs with improving technology, desalination is only an option at present for extremely water-short countries with substantial energy or economic resources, such as in the Arabian Gulf and North Africa, where six of the top ten desalinating countries are located. In addition, the high cost of moving water from one place to another further constrains desalination developments to areas within a limited distance from the coasts.

SUMMARY: NEW THINKING, NEW ACTIONS

Basic concepts and philosophies of water development are undergoing fundamental changes. Whether water is plentiful or scarce, environmental, financial, and social constraints are slowing the construction of large projects and leading planners to extend limited supplies through reallocation and efficiency improvements. Even in regions where basic human needs for water have not been met, water conservation programs are becoming an integral part of practical solutions, since they permit overall water needs to be met with fewer resources, less disruption of ecosystems, and lower costs.

In many parts of the world there will continue to be strong pressures to provide major new water systems, especially if those systems are dedicated to meeting some of the basic human needs remaining in developing countries. Where providing drinking water supply or baseline hydroelectric power is still possible without enormous social dislocations or economic and environmental costs, large new projects may still be appropriate and necessary. But large-scale projects can no longer be expected to provide the answer to most water problems. In particular, in developed countries as well as in arid and semi-arid regions that cover another 30 percent of the Earth's land area, large-scale dams, reservoirs, and irrigation schemes are increasingly out of favor. Only four percent of sub-Saharan Africa's cropland is currently irrigated, yet few good dam sites remain and the economic, social, and environmental costs of large irrigation projects are high. Favorable conditions, such as high-yield groundwater basins, rivers with reliable flows, and large areas of uncultivated irrigable lands are increasingly rare.

Major new projects must now compete with innovative smaller-scale, locally managed technical, institutional, and economic solutions, including micro-dams, run-of-river hydro systems, shallow wells, low-cost pumps, water-conserving land management methods, and rainwater harvesting approaches. Such methods are often more cost effective and less disruptive to local communities, in part because of traditional experiences of these communities. There is already evidence that new applications of traditional methods can catalyze farmers into improving management techniques, stimulating local development, and meeting local water needs in many places where large-scale irrigation projects have failed (Clarke, 1991). One estimate is that 100 million people in Africa alone could benefit from the adoption of small-scale, low-cost traditional methods, but that lack of knowledge and technology continue to hinder their widespread adoption (Postel, 1997).

More rapid changes in water policy worldwide have not occurred because economic and institutional structures still encourage inefficient use of water. Part of the problem, however, also lies in the prevalence of old thinking among water planners and managers. An ethic of sustainability will require fundamental changes in how we think about water, and such changes come about slowly. Rather than endlessly trying to find the water to meet some projection of future desires, it is time to plan for meeting present and future human needs with the water that is available, to determine what desires can be satisfied within the limits of our resources, and to ensure that we preserve the natural ecological cycles that are so integral to human well-being. Water resource planning in a democratic society must involve more than simply deciding what big project to build next or evaluating which scheme is the most cost-effective from a narrow economic perspective. Planning must provide information that helps people to make judgments about which "needs" and "wants" can and should be satisfied. Water is a common good and community resource, but it is also used as a private good or economic commodity; it is not only a recreational resource but also a basic necessity of life; it is imbued with cultural values and plays a part in the social fabric of our communities. Applying new principles of sustainability and equity will help bridge the gap between such diverse and competing interests.

REFERENCES

Ahmad, A.A. 1989. Jordan Environmental Profile: Status and Abatement. USAID/Government of Jordan, Amman, Jordan.

Asano, T. and A.D. Levine. 1998. "Water Reclamation, Recycling, and Reuse: An Introduction." T. Asano, ed. *Wastewater Reclamation and Reuse 10*. Water Quality Management Library, Technomic Publishing Co. Inc. Lancaster, Pennsylvania: 1–56.

California Department of Water Resources (CDWR). 1998. The California Water Plan Update. Bulletin 160-98. Sacramento, California, USA.

Clarke, R. 1991. *Water: The International Crisis*. Cambridge, Massachusetts, USA. Earthscan Publications/MIT Press.

Council on Environmental Quality (CEQ). 1991. Environmental Quality: 21st Annual Report. Washington, D.C.

Daily, G.C., ed. 1997. *Nature's Services: Societal Dependence on Natural Ecosystems*. Washington, DC, USA. Island Press.

Falkenmark, M. and G. Lindh. 1993. "Water and Economic Development." In P.H. Gleick, ed. *Water in Crisis: A Guide to the World's Fresh Water Resources*. Oxford University Press, New York, New York, USA: 80–91.

Fidell, M., P.H. Gleick, and A.K. Wong. 1999. "Converting to Efficient Drip Irrigation: Underwood Ranches and High Rise Farms." In Owens-Viani, L., A.K. Wong, and P.H. Gleick, eds. *Sustainable Use of Water: California Success Stories*. Pacific Institute for Studies in Development, Environment, and Security, Oakland, California, USA: 165–177.

Gleick, P.H. 1996. "Basic Water Requirements for Human Activities: Meeting Basic Needs." *Water International* 21: 83–92.

Gleick, P.H. 1998. *The World's Water 1998–1999: The Biennial Report on Freshwater Resources*. Washington, DC, USA. Island Press.

Gleick, P.H. 1999. "A Human Right to Water." *Water Policy* l, No. 5: 487–503.

Gleick, P.H. 2000. *The World's Water 2000–2001: The Biennial Report on Freshwater Resources*. Washington, DC, USA. Island Press.

Gleick, P.H., P. Loh, S. Gomez, and J. Morrison. 1995. California Water 2020: A Sustainable Vision. Pacific Institute for Studies in Development, Environment, and Security, Oakland, California, USA.

Kalbermatten, J.M., D.S. Julius, C.G. Gunnerson, and D.D. Mara. 1982. "Appropriate Sanitation Alternatives: A Technical and Economic Appraisal," and "A Planning and Design Manual." World Bank Studies in Water Supply and Sanitation I and II. Baltimore, Maryland, USA. The Johns Hopkins University Press.

Lovins, A.B. 1977. *Soft Energy Paths: Toward a Durable Peace*. Friends of the Earth International, San Francisco, California and New York, New York, USA. Ballinger Publishing Company.

McCaffrey, S.C. 1992. "A Human Right to Water: Domestic and International Implications." *Georgetown International Environmental Law Review* 1: 1–24.

New York Times. 1999. "Dam Removal Project is Successful in Maine." *The New York Times*, National News Briefs, November 7: page 19.

Owens-Viani, L., A.K. Wong, and P.H. Gleick, eds. 1999. *Sustainable Use of Water: California Success Stories*. Pacific Institute for Studies in Development, Environment, and Security, Oakland, California, USA.

Perlman, H. 1997. Data for 1995 from the United States Geological Survey (USGS), via FTP from 144.47.32.102.

Postel, S. 1997. *Last Oasis: Facing Water Scarcity*. New York: Worldwatch Institute, W.W. Norton.

Rand Water. 1996. "Chief Executive's Review." Annual Report of Rand Water, Johannesburg, South Africa: 6–7.

Rogers, P. 1997. "Water for Big Cities: Big Problems Easy Solutions?" Draft paper. Harvard University. Cambridge Massachusetts, USA.

Salameh, E. and H. Bannayan. 1993. *Water Resources of Jordan: Present Status and Future Potentials*. Amman, Jordan. Friedrich Ebert Stiftung.

Shuval, H. 1996. "Sustainable Water Resources Versus Concepts of Food Security, Water Security, and Water Stress for Arid Countries." Background paper prepared for the Workshop on Chapter 4 of the Comprehensive Assessment of the Freshwater Resources of the World. May 18–19, 1996. New York, New York, USA.

Solley, W.B., R.R. Pierce, and H.A. Perlman. 1993. Estimated Use of Water in the United States in 1990. U.S. Geological Survey Circular 1200. Denver, Colorado, USA.

United Nations. 1997. Comprehensive Assessment of the Freshwater Resources of the World. Commission on Sustainable Development. United Nations, New York, New York, USA. Printed by the World Meteorological Organization for the Stockholm Environment Institute.

U.S. National Research Council. 1998. *Issues in Potable Reuse: The Viability of Augmenting Drinking Water Supplies with Reclaimed Water.* Washington, DC, USA. National Research Council, National Academy Press.

Van der Merwe, B. and J. Menge. 1996. "Water Reclamation for Potable Reuse in Windhoek, Namibia." Department of the City Engineer, City of Windhoek, Namibia.

Wong, A.K. 1999. "An overview to Water Recycling in California." In L. Owens-Viani, A.K. Wong, and P.H. Gleick, eds. *Sustainable Use of Water: California Success Stories.* Pacific Institute for Studies in Development, Environment, and Security, Oakland, California, USA.

Zachary, G.P. 1997. "Water Pressure: Nations Scramble to Defuse Fights Over Supplies." *The Wall Street Journal.* December 4: A17.

T
W
O

29

"Waste as a Resource"

from *Regenerative Design for Sustainable Development* (1994)

John Tillman Lyle

Editors' Introduction

One leading designer of sustainable resource systems was the late John Tillman Lyle, who taught at Cal Poly Pomona and founded the Center for Regenerative Studies there. Here, he discusses an aspect of industrial civilization that keenly illustrates its unsustainability – and that is widely ignored in daily life – the disposal of wastes. An architect and landscape architect, Lyle uses Mumford's terminology in labeling the current era "Paleotechic," that is, one based on primitive, exploitative technologies rather than regenerative ones. Much of Lyle's professional work involved creating "regenerative" systems such as sewage treatment marshes that use natural processes to improve the environment while processing human wastes, in contrast to "mechanical" or technology-based systems that try to intervene in or override natural systems.

Two other pioneers of ecological waste treatment are John and Nancy Todd, founders of the New Alchemy Institute on Cape Cod, who emphasize the creation of "living machines" that can process wastes. Such projects are described in their books *From Eco-Cities to Living Machines: Principles of Ecological Design* (Berkeley, CA: North Atlantic Books, 1993) and *A Safe and Sustainable World: The Promise of Ecological Design* (Washington, D.C.: Island Press, 2005). Bay Area architect Sim Van der Ryn has also created such devices, which frequently consist of a series of clear tanks with water hyacinths and other waste-processing plants within a greenhouse-type environment. Van der Ryn's books include *Sustainable Communities* (San Francisco: Sierra Club Books, 1986, with Peter Calthorpe), *Ecological Design* (Washington, D.C.: Island Press, 1996, with Stuart Cowan), and *Design for an Empathic World: Reconnecting People, Nature, and Self* (Washington, D.C.: Island Press, 2013).

Much of our difficulty with waste is embedded in the word itself. Waste is defined as material considered worthless and thrown away after use. In this sense it is a human invention, essential to the one-way flows of the throughput system; this definition depends on the assumption that energy and materials, having once served our immediate purposes, can simply cease to exist in any functional sense.

The laws of thermodynamics tell us otherwise. Energy continuously degrades and materials change form and state, but they are not destroyed and they do not disappear. In the functional order of natural ecosystems, materials are always reused. Natural processes have evolved a number of ways of accomplishing this on various time scales. In quantities matched to the evolved capacities of the landscape, materials are reintroduced after use into the processes of assimilation, filtration, storage, and production to continue their roles in nature's cycles. When the chemical composition of the waste materials is such that nature has not evolved a means of reprocessing them, or when their quantities are beyond the processing capacity of the landscape, then the sink side of the flow equation develops a serious problem of pollution or overload.

Throughout the Paleotechnic period and especially during the last half of the twentieth century, both overloading of sinks and introducing unassimilable materials into them have become regular, ongoing occurrences. There are several reasons for this. The populations of cities have multiplied, increasing the concentrations of people and thus of their wastes. Especially in the industrial nations, increasing levels of consumption have meant increasing amounts of waste produced per person. In the United States each person produces 50,000 pounds of waste each year and almost 20,000 gallons of sewage.

The preferred method in the United States for dealing with solid waste (or trash) in recent decades has been to bury it in landfills, which is the municipal equivalent of sweeping dirt under the rug; in Europe, incineration has been more common. Both have serious difficulties.

Historically, municipal landfills have been responsible for a great deal of soil and groundwater pollution. About 20 percent of the sites in the EPA's Superfund cleanup programs are municipal landfills. However, most of these are old sites. The technology of landfilling has improved considerably over the past few decades, but the improvements have been mostly palliative. Sealing the bottoms of landfills has reduced the chances of wastes getting into groundwater, at least temporarily. The heavy plastic liners now being used are expected to last at least 30 years. Decomposition processes, by contrast, are likely to go on for hundreds of years. Thus we might expect a plague of leaking landfills in the twenty-first century.

Effective drainage systems on the surfaces of finished landfills have reduced infiltration of water into the buried trash and thus the danger of chemicals being leached through the trash levels and through or around the sealed bottom. Problems with rodents, birds, odors, and blowing debris have been reduced to manageable levels by sanitary filling, that is, by covering each day's trash deposits with a layer of soil. Somewhat more regenerative in character are the methane collection systems that draw off the gases escaping from decomposition processes. These consist of networks of pipes buried in the layers of trash, which collect the methane as it is generated and convey it to boilers where it is burned to make steam. In most cases the steam is then used to generate electricity, which is fed into the electrical grid.

Such technological improvements have rendered landfilling a relatively harmless means for dealing with nontoxic, nonhazardous wastes. With methane collection, a small portion of the energy embedded in trash is returned for reuse. Even with these improvements, however, landfills remain a degenerative way of dealing with waste; they are a means for wasting waste. Material discarded and consigned to a landfill is effectively removed from the realm of human use. The materials, many of them nonrenewable, and the energy embedded within them, are no longer in the economy. Most of these materials are in fact reusable by some means.

Landfills do not facilitate nature's recycling processes either. Decomposition of buried trash is extremely slow. A number of researchers have dug up materials long ago deposited in landfills and found them hardly changed since the day they were covered over. Finding 30-year-old newspapers that are still readable and food items like hot dogs that are still recognizable is fairly common. Certain microorganisms are needed to decompose these materials and make their components available for reorganizing into new forms. The dry, anaerobic conditions inside a landfill provide a poor environment for them. Thus within a landfill, nature's continuous regeneration is slowed virtually to a standstill. In terms of ecological function, landfills provide not assimilation but storage.

Trash burning, which is commonly practiced in many cities in Europe and in a few in the United States, has appealing short-term advantages. It comes closer than any other technology to simply making waste disappear, and this makes it attractive within the Paleotechnic ethos. A second advantage is that, like methane collection, burning can recover some of the energy imbedded in waste materials by using the heat to generate electricity. The disadvantages of burning, on the other hand, are numerous. Even with the best pollution control equipment, incinerators release into the air considerable volumes of carbon monoxide, sulfur and nitrogen dioxides, dioxin (which is extremely toxic even in minute quantities), and numerous metals, including lead and mercury. The fluidized bed gasifier being tested by the Southern California Edison Company may be a pollution-free incineration technology, but we will not know that for sure for some time.

With burning, groundwater contamination also remains a problem. Incineration does not make the trash entirely disappear. At least 25 percent of the original weight and 10 percent of the original volume remain in the form of an ash residue that still must be disposed of, usually by landfilling. This ash still contains in concentrated form a considerable amount of the dangerous materials that were in the trash before burning, especially metals actually released in the burning process.

Moreover, incinerators are costly devices. For each ton of burning capacity per day, the cost has been estimated at $100,000 to $150,000, which is several times the capital cost of materials recovery systems. Furthermore, incinerators, like landfills, fail to make use of the potential utility that remains in a great deal of the material considered as waste.

For some public officials and even for some citizens' groups who opposed landfills and incinerators in their own environs, exporting trash also has the advantage of seeming disappearance. In Los Angeles there have been several proposals to ship it to the desert. For a great many people, the austere, sparsely vegetated, and almost unpopulated landscape of the desert is a wasteland and thus a suitable place for urban refuse. This has made it a likely location for facilities that would not be acceptable in cities, such as coal-fired generating stations. It sometimes happens that desert towns with economies as sparse as their landscape are quite willing to accept the exported urban pollution for a price. The effects on the fragile natural systems of the desert, however, can be considerably more damaging than they are in the more resilient urban environs.

In eastern US cities, where lack of landfill space has been an even more pressing problem than in western cities, exporting garbage has been an even more enticing solution. Philadelphia exports trash to eastern Ohio and northern Virginia. The famous garbage barge *Mobro* dramatized both the problem and some of the difficulties with the export solution:

It wandered about the world for 55 days in 1988 searching for a place to dump its cargo of urban refuse from New York and eventually ended the journey where it began. Eastern cities nevertheless routinely propose exporting trash to nonindustrial countries in South America and Africa. Whether intentionally or not, they provide a fitting metaphor for the relationship between the industrial and nonindustrial worlds.

Since Thomas Crapper invented the water closet in the nineteenth century, the prevailing means for dealing with human excrement in the United States and Europe has been to mix it with water in that device and then convey the mixture through underground pipes to the nearest sizable body of water, usually a river or bay. Along the way a highly mechanized sewage treatment plant separates the solids from the liquids. That is, after conventional treatment, the sewage is still there, though in different form.

As long as populations are small and dispersed in relation to the volume of water, rivers and bays can assimilate and dilute the nutrients and other materials in the sewage. However, at some point, as cities grow, the volume of nutrients and other materials becomes too great for the assimilative capacities of the water bodies, resulting in an excessive buildup of nutrients and other pollutants. This has happened in waters around most of the world's cities. Since the 1960s, Congress has tried repeatedly to solve the water pollution problem in the United States through a series of Water Quality Control acts. It is now illegal to dump treated sewage water into a body of water if the quality is lower than what is already there. Nevertheless, in 1991, 19 years after amendments to the Clean Water Act established this requirement, over 2000 beaches were reportedly closed along US coasts due to sewage pollution.[1]

In response to this requirement, many sewage plants have added a secondary level of treatment involving mechanical and biological devices to break down the organic solids still remaining in the water after primary screening. Some have also added various forms of advanced treatment to remove other specific materials. Nevertheless, despite hundreds of billions of dollars spent on treatment plants, pollution problems persist in the rivers and bays where cities dump their sewage, especially the larger ones. . . .

PRACTICES AND TECHNOLOGIES FOR MATERIALS REGENERATION

Among the most serious difficulties with waste management in the industrial nations is the immense quantities of materials to be dealt with. The industrial economies' high emphasis on productivity necessarily results in large volumes of waste. This is the essence of the throughput system. Given that the capacity of any environment – land or water – to assimilate waste is limited, large quantities create a basic conflict.

Regenerative design applies the [strategy of] letting nature do the work to increase the assimilative capacity of land and water. At the same time, thoughtful design can make more land available by multiple functions; that is, land used for processing wastes can often be used also for other purposes. However, these strategies can accomplish only so much. Regenerative goals make it clear that the volume of waste to be processed should be limited by the capacity of the environment to assimilate it.

Waste in other cultures

The success of industrial sewage systems in achieving their overriding goal, the control of disease vectors, has made it easy to assume that the western industrial way is the only way. In fact, there are numerous radically different ways of dealing with sewage in operation in other cultures. In China the excrement of a household, called night soil, is still left near the door in buckets to be collected during the night in some areas. In these places collectors take it to nearby farms where it is used as fertilizer. While this system does not provide the same protection from the spread of disease that western pipe-and-water systems do, it does have the advantage of returning nutrients to the soil. Whatever its shortcomings, it is regenerative. For obvious reasons, however, farmers find it objectionable and many have turned to other sources of fertilizer in recent years. In some parts of Japan, the night soil is stored in tanks in many cities and periodically collected by trucks that pump it from the holding tanks into their own tanks. This somewhat more sanitary process also returns the nutrients to the soil in nearby agricultural fields.[2]

Solid-waste practices are not so varied simply because few societies outside of contemporary western industrial societies have produced enough material goods to present serious problems in their disposal. Solid-waste issues were born of the one-way flow system and the extraordinary effectiveness of industrial technology in speeding the flow and thus the quantity of material collecting in the sinks at flow's end. In addition to reducing the volume of material by various means, regenerative practices can return these materials to the processes of natural and human ecosystems by two fundamentally different means: reuse and environmental reassimilation. There are two kinds of reuse: direct reuse and mechanical recycling.

Direct reuse

In industrial societies, the low cost of material goods often causes them to be discarded long before their usefulness is exhausted. In the reuse of these goods there is enormous potential for slowing the flow and thus for reducing both resource use and waste. Varied means for accomplishing this have appeared spontaneously, most of them operating outside the mainstream of the market economy. Among the examples are garage sales, flea markets, swap meets, junkyards, and thrift stores. In some third-world cities like Mexico City, small entrepreneurs have made a business of collecting usable items after they have been dumped in landfills and then reselling them. In cities like Los Angeles, complex underground economies have developed among ethnic minorities, dealing in secondhand goods, often on the basis of barter or credit. Thrift stores, usually operated for particular charities, also serve to keep reusable goods in circulation. Many of these do especially brisk business in clothing, furniture, and children's toys.

The sales volume of all of these combined is still minuscule in comparison with that of mainstream retailers. Nevertheless, they demonstrate that a reuse marketing network does function effectively even with no institutional incentives. Should social forces move away from one-way flows – should materials become suddenly less available or more expensive, for example, or the cost of disposal abruptly increase – then the importance of the reuse markets could become much greater. Each recession foreshadows such a trend, when thrift store and secondhand sales increase while those of conventional retailers decrease. A trend toward more durable, longer-lasting goods might also give impetus to this market.

Besides personal items, there are a great many materials commonly used on a large scale in industrial societies that might be reused to a far greater extent than they presently are. Prominent among these are building materials and containers, especially metal and plastic food and beverage containers.

Mechanical recycling

As compared to reuse, mechanical recycling requires the reshaping or remanufacturing of an old material into new form and thus involves energy use. Most of the items in the typical waste stream of an industrial society not suitable for reuse are suitable for recycling, through either mechanical or biological processes.

The composition of household trash varies considerably by region, by season, and even by district within a single city. The National Solid Wastes Management Association estimates the composition of household trash on a national basis as follows:

Paper	40%
Food	17%
Yard waste	13%
Glass	9%
Metals	9%
Wood	3%
Misc. organics	3%
Plastics	2%
Rubber and leather	2%
Textiles	2%

While all of these categories are recyclable by some means, the value of recycling is greater for some than others. Recycling metals is especially import-ant because of the nonrenewable materials and the energy used in manufacturing them. Making a can from recycled aluminum requires only about one-third the energy needed to make one from new aluminum. The ratio for steel is about two-thirds. The energy ratio for glass is roughly the same as for steel. However, the raw materials for glass are far more common than for aluminum or steel.

While the recycling of metals and glass became common practice in the 1980s and is increasing with growing government incentives, plastic recycling is still difficult and limited in its effectiveness. Most plastics are recycled into products far less valuable than those from which the material came. Low-grade packing and building materials are common uses.

The recycling potential of paper is limited by the fact that fibers are weakened with each remanufacture. Thus, with each recycling paper becomes weaker, lower in quality, and eventually loses its usefulness entirely. At that point, it can be biologically recycled, or composted.

As recycling becomes more common, the processes for it are increasingly superimposed on existing community structures which so far have evolved with no concern for such matters. However, to be truly effective, recycling will have to become an integral part of the community. Recycling cen-ters, composting sites, and separation facilities can become important activity nodes. Building design can also facilitate recycling. The renovated Audubon Society Headquarters ... features chutes running vertically through its nine stories for carrying used materials to a recycling center in the lower basement. There are separate chutes for aluminum, glass, organic materials, paper, and plastics. The Society's goal is to recycle 79 percent of the materials that enter the building.

As they become integral parts of our culture, recycling processes affect the built environment in myriad ways. With recycling, materials often become far more diverse in their uses than the specific functions of their first generation. Consider automobile tires for example. The best second generation function of tires is retreading and reuse. They may be retreaded a second and perhaps a third and fourth time as well, but eventually they become too worn for their original purpose. Then they can spread out into the environment. Without changing form, tires can become bumpers on boat docks or loading docks or crash barriers or swings on playgrounds. They are sometimes sunk in the ocean as artificial reefs or to support growth of crustaceans such as mussels. By cutting them up, it is possible to shape paving blocks, stair treads, or, commonly in nonindustrial countries, soles for sandals. Ground into crumbs, tire rubber makes an excellent roadbed or mulch for agricultural or sports fields. Tire crumbs can also be mixed with asphalt to make RUMAC, which is quieter, more resilient, holds more heat, and lasts twice as long as asphalt alone. Finally, tires can be decomposed by pyrolysis to yield fuel oil. While it is true that the economy has difficulty absorbing some of these products and tires continue piling up as waste, it seems almost certain that repeated and diverse recycling pathways will soon bring dramatic change to both the economy and the landscape.

Biological reassimilation

Biological reassimilation differs from reuse in that it follows the first strategy, letting nature do the work, by drawing on natural processes of decomposition to reintegrate materials into the landscape. Filtration and reassimilation of materials depend on the decomposing activity of countless bacteria and other microbes working unseen in our environment. Though we are hardly aware of them most of the time, these microbes account for most of the earth's biological activity. We can make use of their efforts in a number of ways, three of which are especially important: composting, natural sewage treatment, and bioremediation. These technologies are extremely significant for the future, likely to develop in effectiveness and sophistication.

Composting

The composting process biologically decomposes organic material under controlled conditions. The product is a loosely structured soil-like material that can be handled, stored, and applied to the land as a beneficial soil amendment without adversely affecting the environment. In some areas, compost can provide at least a partial antidote to prevalent conditions of soil degradation.

Important to the utility of compost is its very low, virtually nonexistent health risk. In the composting process, the activity of bacteria in decomposition causes the material to heat up. The high temperature kills pathogens and insect larvae in the mixture as well as most weed seeds.[3]

Most organic materials can be composted, though some decompose faster than others. A mass of organic material can be organized for composting in any number of ways. Piles and rows are common forms. For small operations on the scale of a backyard, it is common practice to load the organic material into a bin with openings in the sides for air movement. Since composting is an aerobic process involving a community of bacteria, fungi, and other microorganisms, providing for air movement is essential. There are several means for accomplishing this, ranging from periodically turning small piles and bins by hand, to force-air devices, to large machines that move between the rows of large composting operations, turning the material mechanically. Composting can also be accomplished in large containers which provide optimum conditions and also control odors.

Under optimum conditions the composting process generally takes from three weeks to two months, though the period can be as short as 12 days. . . .

Aquatic sewage treatment

Since the 1960s, researchers have developed an array of treatment systems that use the capacities of both plants and microorganisms to process sewage without using the elaborate, energy-intensive, often unreliable mechanical devices used in industrial sewage plants. That is, they let nature do the work of sewage treatment. Essentially, these systems replicate and intensify the processes of nature in organic recycling. They simultaneously filter the water and assimilate the solids into living organisms. They use water as a medium and treat sewage as it comes from the conventional collection systems.

The essential point concerning natural treatment systems is that they are landscapes in their basic character and operation, while the conventional treatment systems of the industrial period are basically machines. This is a fundamental difference with far-reaching implications that transcend the technological distinction to involve the design of environment, the shaping of cities, and even the character of societies. . . .

Before entering a natural treatment process, the solids are separated out of the water by settling or screening. The treatment systems fall into three general types: aquacultural ponds, wetlands, and rootzone beds, all of which have certain characteristics in common. The sewage water travels slowly among the roots and stems of aquatic plants, which take up some nutrients and other materials from the water in the process of supporting their own growth. However, the bulk of the work is done by bacteria and other microorganisms living on the roots and stems. Plants and microorganisms are capable of taking almost any materials out of the water, including nutrients, metals, and pathogens. The degree of treatment depends on the time; given enough time, natural systems can produce water suitable for human consumption from the densest raw sewage.

In natural treatment systems the limiting factor is usually biological oxygen demand, or the oxygen content of the water. The oxygenating activity of aquatic plants is important in maintaining oxygen levels as many of the plants take in oxygen through their leaves and release it through their roots. The oxygen-rich environment at the roots supports a rich microbial community that is very effective in the treatment process: bacteria, fungi, filter feeders, detritivores, and their predators. . . .

Aquatic treatment concepts are not limited to municipal or industrial facilities but can be applied in myriad ways at any scale in any situation where purer water or a richer aquatic environment is desirable. Water in a polluted stream, for example, might be diverted through a series of wetlands for treatment and then returned to the stream. A series of ponds and wetlands located at strategic points in a city's drainage system could treat urban runoff to a level suitable for return to natural waterways.

Aquatic treatment systems in all their forms are among the best and clearest examples of re-generative technologies. They are in themselves complex ecosystems which naturally do the work that human society needs done. They can treat water to any level of quality. They do not need the concrete and steel structures or the machinery, the pumps and pipes of conventional industrial treatment. They can work at any scale and are not subject to the breakdowns that plague mechanized treatment systems. They do not use fossil fuels or pollute the air. Finally, they cost far less than do mechanized systems.

The main disadvantage of aquatic systems is that they occupy more land than mechanized systems. In urban areas this can present serious problems. However, in this respect too, aquatic systems are quintessentially regenerative in that they incorporate all of the processing capabilities of the landscape: assimilation, filtering, storage, and production. They are also integrally related to other life-support processes. Besides supplying clean water for any number of uses, they can provide biomass for energy conversion, fertilizer for food production, and food for animals, both domestic and wild.

BOX 1 Arcata marsh treatment system

Arcata is a town of about 15,000 people located on the northern California coast. For decades its sewage was piped into an oxidation pond for a few days and then moved on to the Pacific Ocean. The federal Water Pollution Control Act of 1972 rendered this treatment system clearly inadequate. Some form of secondary treatment would have to be added, and the conventional technologies were very expensive. So in the late 1970s, Arcata developed a small pilot wetland system to function between the oxidation pond and ocean release. The results showed that the wetland could provide advanced treatment with capital and maintenance costs considerably lower than those of a mechanized second-ary treatment system. On the basis of those results, the city developed its present system.

This system begins with primary settling after which 2 to 3 million gallons of sewage move into three oxidation ponds each day and an equal amount moves out. After that, a 5.3 acre intermediate marsh, planted mostly with the hardstem bulrush (*Scirpus acutus*), reduces suspended solids. Mosquito fish control mosquito populations. Chlorination and dechlorination follow the intermediate marsh; then the water moves into the 154 acre Arcata Marsh and Wildlife Sanctuary and from there into Humboldt Bay. In this marsh, cattails grow along with bulrushes, and duckweed covers much of the surface. Analyses required by the state to maintain the facility's permit to discharge into the bay show that the water moving into the bay meets the standards for secondary treatment. Studies have shown that refinements in the design of the marsh could improve the treatment to the level of standards for advanced treatment.

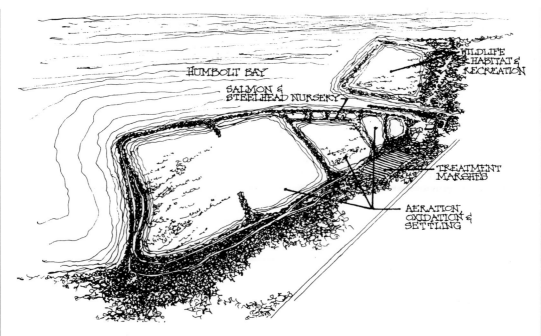

Arcata sewage treatment marsh.

As important as sewage treatment in the Arcata Marsh is the combination of other functions that it supports. Wildlife habitat is varied and extensive. This is due partly to the inherently diverse character of land–water interactions in a wetland and to the way in which this inherent quality was further augmented by design. In its final grading the levels of the marsh bottom were varied to provide conditions for a range of marsh plants that grow at different levels of submergence. These plants in turn provide conditions for a range of microbes, mollusks, fish, and birds. Over 220 bird species have been recorded here, many of them migratory species that stop over on their journey along the Pacific flyway.

Researchers have also established a fish hatchery in the lagoon. Salmon fry are imprinted to return here as adults to lay their own eggs.

The recreational value of the marsh has been proven as well. Over 100,000 people visit the area each year. Most of these are from the city of Arcata, since the marsh has become a local focal point, but many are tourists. A significant increase in the number of tourists visiting Arcata has been documented by the city.

NOTES

1 Stammer, Larry B. 1992. Sewage Forced Closure of 2000 Beaches in 1991. *Los Angeles Times*, 24 July.

2 Pradt, Louis A. 1971. Some Recent Developments in Night Soil Treatment. *Water Resources*, 5.

3 Hornick, S.B. *et al.* 1979. *Use of Sewage Sludge Compost for Soil Improvement and Plant Growth*. US Department of Agriculture ARM-NE-6.

SOCIAL EQUITY AND ENVIRONMENTAL JUSTICE

30

"People-of-Color Environmentalism"

from *Dumping in Dixie: Race, Class, and Environmental Quality* (1990)

Robert Bullard

Editors' Introduction

Sustainability goals are often presented in terms of the "three Es" – environment, economy, and equity – which in a sustainable society would all be enhanced over the long term. Of these, equity has been by far the least represented within public policy debates. There are relatively few well-organized groups advocating on behalf of low-income or otherwise disadvantaged communities. Even the environmental movement, with its relatively progressive middle-class constituency, developed initially with little consideration of the equity implications of its issues.

The link between social justice and environmental issues in the United States was forged in the 1980s in large part by working-class communities fighting against the location of garbage incinerators, landfills, and toxic chemical hazards near their neighborhoods. Black and Latino activists in particular criticized mainstream environmental groups for their lack of diversity, and demanded procedural changes to bring about more equitable public participation within environmental decision making. At the same time, activists in the developing world were calling attention to the inequitable impacts of development policies internationally – a separate but parallel set of equity debates.

Texas Southern University dean and sociology professor Robert D. Bullard has been at the forefront of chronicling and defining the environmental justice movement in the United States. Here he discusses the roots of the movement, links with gender issues, and prospects for future organizing. Other leading writings on the subject of environmental justice include *Growing Smarter: Achieving Livable Communities, Environmental Justice, and Regional Equity*, edited by Bullard (Cambridge, MA: MIT Press, 2007), *The Black Metropolis in the Twenty-first Century: Race, Power, and Politics of Place*, also edited by Bullard (Lanham, MD: Rowman and Littlefield, 2007), *Environmental Health and Racial Equity in the United States: Building Environmentally Just, Sustainable, and Livable Communities*, by Bullard, Glenn S. Johnson, and Angel O. Torres (Washington, D.C.: American Public Health Association, 2011), *Just Sustainabilities: Development in an Unequal World*, edited by Julian Agyeman, Bullard, and Bob Evans (Cambridge, MA: MIT Press, 2003), *Sustainable Communities and the Challenge of Environmental Justice*, by Agyeman (New York: New York University Press, 2005), and *Introducing Just Sustainability: Policy, Planning, and Practice* by Agyeman (London: Zed Books, 2013).

It is time for people to stop asking the question "Do minorities care about the environment?" The evidence is clear and irrefutable that white middleclass communities do not have a monopoly on environmental concern, nor are they the only groups moved to action when confronted with the threat of pollution. Although a "concern-and-action gap" may still exist between people of color and whites, communities of color are no longer being bullied into submission by industrial polluters and government regulators.[1]

Clearly, a "new" form of environmentalism has taken root in America and in communities of color. Since the late 1970s, a new grassroots social movement has emerged around the toxics threat. Citizens mobilized around the anti-waste theme. These social activists acquired new skills in areas where they had little or no prior experience. They soon became resident "experts" on toxics issues.... However, they did not limit their attacks to well-publicized toxic-contamination issues but sought remedial actions on problems like housing, transportation, air quality, and even economic development – issues the traditional environmental agenda had largely ignored.

Environmental justice embraces the principle that all people and communities are entitled to equal protection of environmental, health, employment, housing, transportation, and civil rights laws. Activists even convinced the EPA to develop a definition of environmental justice. The EPA defines environmental justice as:

> The fair treatment and meaningful involvement of all people regardless of race, color, national origin, or income with respect to the development, implementation, and enforcement of environmental laws, regulations and policies. Fair treatment means that no group of people, including racial, ethnic, or socio-economic group should bear a disproportionate share of the negative environmental consequences resulting from industrial, municipal, and commercial operations or the execution of federal, state, local, and tribal programs and policies.[2]

A major paradigm shift occurred in the 1990s. This shift created a new framework and a new leadership. Women led much of this grassroots leadership. The impetus behind this change included grassroots activism, redefinition of environmentalism as a "right," research documenting disparities, national conferences and symposia, emphasis on pollution and disease prevention, government initiatives, interpretation of existing laws and mandates, and grassroots alliances and coalitions.

Environmentalism has been too narrowly defined. Concern has been incorrectly equated with check writing, dues paying, and membership in environmental organizations. These biases have no doubt contributed to the misunderstanding of the grassroots environmental justice movement in people-of-color communities. People-of-color activists in this new movement focused their attention on the notion of deprivation. For example, when people of color compare their environmental quality with that of the larger society, a sense of deprivation and unequal treatment, unequal protection, and unequal enforcement emerges. Once again, institutional racism and discriminatory land-use policies and practices of government – at all levels – influence the creation and perpetuation of racially separate and unequal residential areas for people of color and whites. Too often the disparities result in groups fighting another form of institutional discrimination.[3]

All communities are not created equal. Institutional barriers have locked millions of people of color in polluted neighborhoods and hazardous, low-paying jobs, making it difficult for them to "vote with their feet" and escape these health-threatening environments. Whether in the ghetto or barrio, on the reservation, or in rural "poverty pockets," environmental injustice is making some people sick. Government has been slow to take these concerns as legitimate environmental and health problems. Mainstream environmentalists have also been slow in recognizing these grassroots activists as "real" environmentalists.[4]

The environmental justice movement is an extension of the social justice movement. Environmental justice advocates should not have to apologize for this historical fact. Environmentalists may be concerned about clean air but may have opposing views on public transportation, highway construction, industrial-facility siting, or the construction of low-income housing in white, middle-class suburban neighborhoods. On the other hand, environmental justice advocates also want clean air. People of color have come to understand that environmentalists are no more enlightened than nonenvironmentalists when it comes to issues of justice and social equity. But then, why should they be more enlightened? After all, we are all products of socialization and reflect the various biases and prejudices of this process. It is not surprising that mainstream environmental organizations have not been active on issues that disproportionately impact people of color, as in the case of toxics, workplace hazards, rural and urban housing needs, and the myriad of problems resulting from discriminatory zoning and strains in the urban, industrial complex. Yet people of color are the ones accused of being ill-informed, unconcerned, and inactive on environmental issues.

Environmental decision-making operates at the juncture of science, economics, politics, and ethics. It has been an uphill battle to try to convince some government and industry officials and some environmentalists that unequal protection, disparate impact, and environmental racism exist. Nevertheless, grassroots activists have continued to argue and in many instances have won their case. Working together, community stakeholders can assist government decision-makers in identifying "at-risk" populations, toxic "hot spots," research gaps, and action plans to correct existing imbalances and prevent future threats.[5] In order to accomplish their mission in an era of dwindling resources, environmental policymakers are increasingly turning to strategies that incorporate a *community-empowerment* approach. For example, community environmental protection (CEP) is being touted by the EPA as a "new" way of doing business.

Strengthening grassroots community groups can build a supportive social environment for decision-making. Residents and government authorities (local, state, and federal), often working together through creative partnerships with grassroots community groups, universities, nonprofit agencies, and other institutions, can begin solving environmental and health problems and design strategies to prevent future problems in low-income areas and communities of color. But the US Environmental Protection Agency and other governmental agencies cannot resolve all environmental problems alone. Communities also need to be in the position to assist in their own struggle for clean, safe, healthy, livable, and sustainable communities.

THE RIGHT TO BREATHE CLEAN AIR

Before the federal government stepped in, issues related to air pollution were handled primarily by states and local governments. Because states and local governments did such a poor job, the federal government established national clean-air standards. Congress enacted the Clean Air Act (CAA) in 1970 and mandated the EPA to carry out this law. Subsequent amendments (1977 and 1990) were made to the CAA that form the current federal program. The CAA was a response to states' unwillingness to protect air quality. Many states used their lax enforcement of environmental laws as lures for business and economic development.[6]

Transportation policies are also implicated in urban air-pollution problems. Automobile-choked highways create health-threatening air pollution.[7] Freeways are the lifeline for suburban commuters, and millions of central-city residents are dependent on public transportation as their primary mode of travel.[8] Are people of color concerned about air quality and transportation? The answer is yes. The air-quality impacts of transportation are especially significant to people of color, who are more likely than whites to live in urban areas with reduced air quality. . . .

Asthma is an emerging epidemic in the United States. The annual age-adjusted death rate from asthma increased by 40 percent between 1982 and 1991, from 1.34 to 1.88 per 100,000 population,[9] with the highest rates being consistently reported among blacks between the ages of 15 and 24 years during the period 1980–1993.[10] Poverty and minority status are important risk factors for asthma mortality. Children are at special risk from ozone.[11] Children also represent a considerable share of the asthma burden, that affliction being the most common chronic disease of childhood. Asthma affects almost 5 million children under 18 years of age. . . .

The public health community has insufficient information to explain the magnitude of some of the air pollution-related health problems. However, they do know that people suffering from asthma are particularly sensitive to the effects of carbon monoxide, sulfur dioxides, particulate matter, ozone, and nitrogen oxides.[12] Ground-level ozone may exacerbate health problems such as asthma, nasal congestion, throat irritation, respiratory-tract inflammation, reduced resistance to infection, changes in cell function, loss of lung elasticity, chest pains, lung scarring, formation of lesions within the lungs, and premature aging of lung tissues.[13]

African Americans, for example, have significantly higher prevalence of asthma than the general population.[14] A 1996 report from the federal Centers for Disease Control and Prevention shows hospitalization and death rates from asthma increasing for individuals 25 years old or younger.[15] The greatest increases occurred among African Americans. African Americans are two to six times more likely than whites to die from asthma.[16] Similarly, the hospitalization rate for African Americans is 3.4 times the rate for whites. . . . Air pollution, for many environmental justice advocates, translates into poor health, loss of wages, and diminished quality of life.

THE THREAT OF ECONOMIC EXTORTION

Why were people-of-color organizations late in challenging the environmental imbalance that exists in the United States? People-of-color organizations and their leaders have not been as sensitive to the environmental threats as they have been to problems in education, housing, jobs, drugs, and, more recently, the AIDS epidemic. In some cases, they have operated out of misguided fear and speculation that environmental justice will erode hard-fought civil rights gains or thwart economic development in urban core neighborhoods. There is no evidence that environmental justice or the application of Title VI of the Civil Rights Act of 1964 has hurt business or "brownfields" (abandoned properties that may or may not be contaminated) redevelopment opportunities in communities of color.[17] On the other hand, we do not have to speculate about the harm inflicted on the residents from racial red-lining by banks and insurance companies and the targeting of communities of color for polluting industries and locally unwanted land uses, or LULUS. The harm is real and measurable.

Grassroots groups in communities of color are beginning to take a stand against threatened plant closure and job loss as a trade-off for environmental risks. These threats are tantamount to economic extortion. This extortion has lost some of its appeal, especially in those areas where the economic incentives (jobs, taxes, monetary contributions, etc.) flow outside of the host community. People can hardly be extorted over economic benefits they never receive from the local polluting industry. There is a huge difference between the promise of a job and a real job. People will tell you, "You can't eat promises." Because of the potential to exacerbate existing environmental inequities, community leaders are now questioning the underlying assumptions behind so-called trade-offs as applied in poor areas.

In their push to become acceptable and credible, many mainstream environmental organizations adopted a corporate model in their structure, demeanor, and outlook. This metamorphosis has had a down side. These corporate-like environmental organizations have alienated many grassroots leaders and community organizers from the larger movements. The environmental justice movement – with its egalitarian worldview and social justice agenda – offers an alternative to the more staid traditional environmental groups.

Local community groups may be turned off by the idea of sitting around a table with a waste-disposal giant, a government regulator, and an environmentalist to negotiate the siting of a toxic-waste incinerator in their community. The lines become blurred in terms of the parties representing the interests of the community and those of business. Negotiations of this type fuel residents' perception of an "unholy trinity," where the battle lines are drawn along an "us-versus-them" power arrangement. Moreover, overdependence on and blind acceptance of risk-assessment analysis and "the best available technology" for policy setting serves to intimidate, confuse, and overwhelm individuals at the grassroots level.

Talk of risk compensation for a host community raises a series of moral dilemmas, especially where environmental imbalances already exist. Should risks be borne by a smaller group to spare the larger groups? Past discriminatory facility-siting practices should not guide future policy decisions. Having one polluting facility makes it easier to site another in the same general area. The "one more won't make a difference" logic often becomes the dominant framework for decision-making. Any saturation policy derived from past siting practices perpetuates equity impacts and environmental injustice. Facility siting becomes a ritual for selecting "victims for sacrifice."

MOBILIZING THE GRASS ROOTS

It is unlikely that the environmental justice movement will ever gain unanimous support in communities of color. Few social movements can count on total support and involvement of their constituent groups. All social movements have "free riders," individuals who benefit from the efforts of a few. Some people shake the trees, while others pick up the apples. People-of-color environmentalism has been and will probably remain wedded to a social-action and social-justice framework. The issues raised by environmental justice advocates challenge the very core of privilege in our society. Some people make money and profit off the misery of poisoning others. Some communities are spared environmental assaults because of industrial-siting practices of concentrating locally unwanted land uses in communities with little or no political power and limited resources. After all, American society has yet to achieve a race-neutral state where race- and ethnic-based organizations are no longer needed.

Although the color barrier has been breached in most professional groups around the country, blacks still find it useful to have their own organizations. The predominately black National Bar Association (NBA), National Medical Association (NMA), National Association of Black Social Workers (NABSW), Association of Black Psychologists (ABP), and Association of Black Sociologists (ABS) are examples of race-based professional organizations that will probably be around for some time in the new millennium.

Grassroots environmental organizations have the advantage of being closer to the people they serve and the problems they address. Future growth in the environmental movement is likely to come from the bottom up, with grassroots environmental groups linking up with social-justice groups for expanded spheres of influence and focus.

Communities of color do not have a long track record in challenging government decisions and private industries that threaten the environment and health of their residents. Many of the organizations and institutions were formed as a reaction to racism and dealt primarily with social-justice issues. Groups such as the NAACP, Urban League, Southern Christian Leadership Conference, and Commission for Racial Justice operate at the multistate level and have affiliates in cities across the nation. With the exception of Reverend Joseph Lowery of the Southern Christian Leadership Conference, Benjamin R. Chavis Jr. of the United Church of Christ's Commission for Racial Justice, and Reverend Jessie Jackson of the National Rainbow Coalition, few national black civil rights leaders and organizations embraced an ideology that linked environmental disparities with racism.[18] It was not until the 1980s that national civil rights organizations began to make such links. This linking of institutional racism with the structure of resource allocation (clean environments) has led people-of-color social-action groups to adopt environmental justice as a civil rights issue, an issue well worth "taking to the street."

NIMBYism [not-in-my-backyard politics] has operated to insulate many white communities from the localized environmental impacts of waste facilities while providing them the benefits of waste disposal. NIMBYism, like white racism, creates and perpetuates privileges for whites at the expense of people of color. Citizens see the siting and unequal protection question as an all-out war. Those communities that can mobilize political influence improve their chance of "winning" this war. Because people of color remain underrepresented in elected and appointed offices, they must, most often, rely on indirect representation, usually through white officials who may or may not understand the nature and severity of the community problem. Citizen redress often becomes a political issue. Often the only science involved in the government response and decision-making is political science.

Who are the frontline leaders in this quest for environmental justice? The war against environmental racism and environmental injustice has been waged largely by people of color who are indigenous to the communities. People-of-color grassroots community groups receive some moral support from outside groups, but few experts are down in the trenches fighting alongside the warriors. On the other hand, it was the mothers and grandmothers, ministers from the churches, and the activist leaders from community-based organizations, civic clubs, neighborhood associations, and parents' groups who mobilized against the toxics threat. Few of these leaders may identify themselves as environmentalists or see their struggle solely as an environmental problem. Their struggles embrace larger issues of equity, social justice, and resource distribution. Environmental justice activists question the fairness of the decision-making process and the outcome.

Many environmental justice disputes revolve around siting issues, involving government or private industry. Proposals for future sites are more likely to attract environmentalists' support than are existing sites. It is much easier to get outside assistance in fighting a noxious facility that is on paper than one that is in operation. Again, plant closure means economic dislocation. Because communities of color are burdened with a greater share of existing facilities – many of which have been in operation for decades – it is an uphill battle of convincing outside environmental groups to support efforts to close such facilities.

It makes a lot of sense for the organized environmental movement in the United States to broaden its base to include people-of-color, low-income, and working-class individuals and issues. Why diversify? People of color now form a potent voting bloc. Diversification makes good economic and political sense for the long-range survival of the environmental movement. However, it is not about selfishness or "quota filling." Diversification can go a long way in enhancing the national environmental movement's worldwide credibility and legitimacy in dealing with global environmental and development issues, especially in Third World nations.[19]

NOTES

1 See Bullard, R.D. 1996. *Unequal Protection: Environmental Justice and Communities of Color.* San Francisco: Sierra Club Books, pp. 3–22; Bryant, B. 1995. *Environmental Justice: Issues, Policies, and Solutions.* Washington, DC: Island Press, pp. 8–34.

2 US Environmental Protection Agency. 1998. *Guidance for Incorporating Environmental Justice Concerns in EPA's NEPA Compliance Analysis.* Washington, DC: EPA.

3 Bullard, R.D. 1999. Dismantling Environmental Racism in the USA. *Local Environment,* 4, pp. 5–19.

4 Westra, Laura and Wenz, Peter S. 1995. *Faces of Environmental Racism: Confronting Issues of Global Justice.* Lanham, MD: Rowman & Littlefield.

5 Bullard, R.D. 1999. Leveling the Playing Field Through Environmental Justice. *Vermont Law Review,* 23, pp. 453–478; Collin, Robert W. and Robin M. 1998. The Role of Communities in Environmental Decisions: Communities Speaking for Themselves. *Journal of Environmental Law and Litigation,* 13, pp. 37–89.

6 Reitze, Arnold W. Jr. 1991. A Century of Air Pollution Control Law: What Worked; What Failed; What Might Work. *Environmental Law,* 21, pp. 15–49.

7 Davis, Sid. 1997. Race and the Politics of Transportation in Atlanta. In R.D. Bullard and G.S. Johnson (eds). *Just Transportation: Dismantling Race and Class Barriers to Mobility.* Gabriola Island, DC: New Society Publishers, pp. 84–96; Environmental Justice Resource Center. 1999. *Sprawl Atlanta: Social Equity Dimensions of Uneven Growth and Development.* Atlanta, GA: Report prepared for the Turner Foundation, Clark Atlanta University.

8 For an in-depth discussion of transportation investments and social equity issues, see Bullard and Johnson (eds). *Just Transportation.*

9 Centers for Disease Control. 1995. Asthma – United States, 1982–1992. *Morbidity and Mortality Weekly Report,* 43, pp. 952–955.

10 Centers for Disease Control. 1996. Asthma Mortality and Hospitalization among Children and Young Adults – United States, 1980–1993. *Morbidity and Mortality Weekly Report,* 45, pp. 350–353.

11 Pribitkin, Anna E. 1994. The Need for Revision of Ozone Standards: Why Has the EPA Failed to Respond? *Temple Environmental Law and Technology Journal,* 13, p. 104.

12 See Mann, Eric. 1991. *L.A.'s Lethal Air: New Strategies for Policy, Organizing, and Action.* Los Angeles: Labor/Community Strategy Center.

13 US Environmental Protection Agency. 1996. *Review of National Ambient Air Quality Standards for Ozone, Assessment of Scientific and Technical Information.* Research Triangle Park, NC: OAQPS staff paper, EPA; American Lung Association. 1995. *Out of Breath: Populations-at-Risk to Alternative Ozone Levels.* Washington, DC: American Lung Association.

14 See Mak, H.P., Abbey, H., and Talamo, R.C. 1983. Prevalence of Asthma and Health Service Utilization of Asthmatic Children in an Inner City. *Journal of Allergy and Clinical Immunology*, 70, pp. 367–372; Goldstein, I.F. and Weinstein, A.L. 1986. Air Pollution and Asthma: Effects of Exposure to Short-Term Sulfur Dioxide Peaks. *Environmental Research*, 40, pp. 332–345; Schwartz, J., Gold, D., Dockey, D.W., Weiss, S.T., and Speizer, F.E. 1990. Predictors of Asthma and Persistent Wheeze in a National Sample of Children in the United States. *American Review of Respiratory Disease*, 142, pp. 555–562; Crain, F., Weiss, K., Bijur, J. *et al.*, 1994, An Estimate of the Prevalence of Asthma and Wheezing Among Innercity Children. *Pediatrics*, 94, pp. 356–362.

15 Centers for Disease Control and Prevention. 1996. Asthma Mortality and Hospitalization Among Children and Young Adults – United States, 1980–1993. *Morbidity and Mortality Weekly Report*, 45.

16 Centers for Disease Control. 1992. Asthma – United States, 1980–1990. *Morbidity and Mortality Weekly Report*, 39.

17 US Environmental Protection Agency. 1999. *Brownfields Title VI Case Studies: Summary Report*. Washington, DC: Office of Solid Waste and Emergency Response.

18 United Church of Christ Commission for Racial Justice. 1998. *From Plantation to Plant: Report of the Emergency National Commission on Environmental Justice in St. James Parish, Louisiana*. Cleveland: United Church of Christ.

19 See Bullard, R.D. 1993. *Confronting Environmental Racism: Voices from the Grassroots*. Boston: South End Press, chapter 1.

T
W
O

31

"Domesticating Urban Space"

from *Redesigning the American Dream: The Future of Housing, Work, and Family Life* (2002)

Dolores Hayden

Editors' Introduction

Just as current patterns of urban development have disproportionate impacts on communities of color, so they also profoundly affect women, children, the elderly, and other less powerful groups within society. Developing socially and environmentally healthy cities means recognizing and addressing these impacts. Although planners and designers have become more sensitive to the needs of women and other groups in recent years, much remains to be done to make cities friendly to all types of people.

In this selection from her classic book *Redesigning the American Dream: The Future of Housing, Work, and Family Life* (New York: Norton, original edition in 1984, revised in 2002), urban historian Dolores Hayden examines how women have been systematically excluded or made to feel unsafe or uncomfortable in urban environments. Similar writings include Daphne Spain's *Gendered Spaces* (Chapel Hill: University of North Carolina Press, 1992); *Women, the Environment and Sustainable Development: Towards a Theoretical Synthesis* (London: Zed Books, 1994), edited by Rosi Braidotti *et al.*; *Safe Cities: Guidelines for Planning, Design, and Management* (New York: Van Nostrand Reinhold, 1995) by Gerda R. Wekerle and Carolyn Whitzman; *Gendering the City* (Lanham, MD: Rowman & Littlefield, 2000), edited by Kristine B. Miranne and Alma H. Young; and *Space, Place, and Sex: Geographies of Sexualities* (Lanham, MD: Rowman and Littlefield, 2012), by Lynda Johnston and Robyn Longhurst. Vandana Shiva offers a broader analysis of the role of women in sustainable development within developing countries in her books such as *Close to Home: Women Reconnect Ecology, Health, and Development Worldwide* (Philadelphia: New Society Publishers, 1994); *Earth Democracy: Justice, Sustainability, and Peace* (Cambridge, MA: South End Press, 2005); and *Staying Alive: Women, Ecology, and Development* (Boston: South End Press, 2010).

The phrase "A woman's place is in the home" has defined much housing policy and urban design in American society. The query "What's a nice girl like you doing in a place like this?" has reflected the prevailing attitude toward women in public, urban space. Both phrases have their roots in a Victorian model of private and public life. The first involves the patriarchal home as haven; the second defines the Victorian male double standard of sexual morality.

Both are implicit rather than explicit principles of urban planning; neither will be found stated in large type in textbooks on land use. Both attitudes are linked to a set of nineteenth-century beliefs about female passivity and propriety in the domestic setting ("woman's sphere") versus male combativeness and aggression in the public setting ("man's world").

When nineteenth-century men (and women) argued that the good woman was at home in the

Figure 1 Running the gauntlet. Public space on the male double standard is shown in J.N. Hyde's *Running the Gauntlet, New York City* (1874). A respectable woman walks down the street with ten men ogling her as if she were a prostitute. Her body is tense and corseted, her eyes are averted; the men lounge and stare boldly, providing a graphic example of what Nancy Henley calls "body politics," the dominance of one race, class, or gender over another through positioning of bodies in space.

kitchen with her husband, they implied that no decent woman was out in city streets, going places where men went. Thus, it was "unladylike" for a woman to earn her own living. Because the working woman was no *one* urban man's property (her father or her husband had failed to keep her at home), she was *every* urban man's property. She was the potential victim of harassment in the factory, in the office, on the street, in restaurants, and in places of amusement such as theaters or parks.

While the numbers of employed women and women in active public life have increased, many of these spatial stereotypes and patterns of behavior remain. Just as the haven houses hobble employed women, so the double standard harasses them when they are alone or with their children.[1] Men do not escape the problem. As husbands and fathers they share the stresses of the isolated houses and the violent streets they and their wives and children must negotiate. But rarely do men attribute the problems of housing and the city to the Victorian

patriarchal views that reserved urban, public life for men only.[2]

THE FREEDOM OF THE CITY

At a theoretical level, women have never explicitly demanded or enjoyed the *droit de la ville*, the medieval right to the freedom of the city that distinguished urban citizens from feudal serfs.[3] The existing literature on urban history and theory conveys this in dozens of titles like *City of Man* or *The Fall of Public Man*. The experiences of women in urban space are absent from the content of many academic studies as well as from the titles.[4] The implicit assumption has been that either respectable women (and children) had the same urban experiences as men when they were with men, or else that women (and children) had no urban experiences, since their place was in the home or other segregated spaces. The closest that American women of all classes and races have come to challenging this view was in the Progressive Era, when Frances Willard attacked the double standard as expressed in saloons and the wardboss political culture of the late nineteenth century and attempted to domesticate the American city.

Some historians and critics have suggested that women failed to establish lasting power in the public sphere in the Progressive Era because they failed to develop an urban political theory suited to their needs.[5] Critics who complain of women's lack of "ideology" might first examine definitions of the nature of political theory and political activity. As long as the domestic world remains a romantic haven "outside" of public life and the political economy, politically active women can always be sent back to it, and men can justify the exclusion of women and children from their public debates and analyses. Yet the reverse is also true. If women can overcome what Lyn Lofland has called the "thereness" of women,[6] if they can transcend what Jessie Bernard has defined as "the female world of a segregated place,[7] new kinds of homes and neighborhoods might become the most powerful base in America for progressive political coalitions on urban issues.

A political program to overcome the "thereness" of women and win all female and male citizens, and their children, access to safe, public urban space requires that the presence of women (and their children) in public space be established as a political right; and that gender stereotypes be eliminated from architecture, urban design, and graphic design in public space. Such a political program would share many common features with Olmsted's attempt to create public space accessible to all in his parks; it would share many goals of the campaign Frances Willard launched in the 1890s for a "homelike world" in America's cities; it would require many new institutions like Jane Addams' Hull-House, a public center for community organizing on the model of a collective home open to all; it would link the campaigns of the anti-flirt club of the 1920s to the "Take Back the Night" marches of the 1970s.

Many professionals in the design fields are ready to support the political struggles necessary to bring domestic standards of amenity and safety into public space. In the recent past, disdainful speculators and politicians have claimed that the men engaged in these causes were not "real men," and the women were "little old ladies in tennis shoes," but on this subject, Lewis Mumford can be very reassuring. Calling for a serious study of resources and settlement design in 1956, Mumford challenged Americans to go beyond the old machismo of previous urban development patterns. "This new adventure," he said, "demands psychological maturity as the boyish heroism of the old adventures did not; for it is an exploration in depth, to fathom all the potential resources of a region . . . and to assess its possibilities for continued enjoyment."[8] The ultimate proof of maturity is the ability to nurture and protect human life, to develop public safety, public mobility, public amenities. Small, commonsense improvements in urban design can be linked to larger ideas about nurturing to help end the split between private life and public life.

RECAPTURING PUBLIC SPACE FOR THE NEEDS OF PARENTS

Anyone who has ever cared for babies or small children, even for a few hours away from a domestic environment, knows how little thought has been spent on making public space accessible to parents. When they begin a family, parents find that having a baby puts great spatial limits on their public, urban life. Although a first priority is to have adequate infant care and day care in residential neighborhoods for parents, it is also necessary to make it easier for adults to move in public space with their children. Other countries do this. In Denmark banks provide children's play areas with small furniture and toys while parents do their banking; in New Zealand, department stores offer day care to customers.[9]

To develop public facilities with the expectation that parents and grandparents with babies and young children will be using them is not a technically difficult task: it requires a commitment to better programming and a little imagination. Baby-changing spaces as a standard feature of both men's rooms and women's rooms would help; well-located seating and small children's furniture in banks, stores, and restaurants suggest that children are expected and welcomed with their fathers or mothers; windows at children's eye levels, as well as adults', are attractive in spaces where children represent a substantial number of users. Such changes in scale to accommodate children add live-liness and diversity to the urban scene. Concern for building materials and interior finishes neither too fragile nor too tough also helps define places that children can use. Play spaces can add a sense of joy for people of all ages, especially when they are organized to incorporate trees and flowers, public art or local landmarks such as ruins of old buildings, traces of past settlement patterns, or artifacts conveying economic history. Public space for children is, at its best, not only warm and educational but also fun. . . .

GREENLIGHTS AND SAFEHOUSES

Many adult Americans are afraid of public urban space. In 1967 the US President's Commission on Law Enforcement and the Administration of Justice concluded that "One-third of a representative sample of Americans say it is unsafe to walk alone at night in their neighborhoods."[10] Another study by a Congressional Committee in 1978 reported that fully one-half of all Americans were afraid to go out at night.[11] Yet another study by sociologists in Chicago states that one-half of all women and 20 percent of all men were afraid to walk alone in their neighborhoods at night.[12]

Two programs designed to bring a greater sense of security into the lives of citizens are the Greenlight Program developed by the Women's Safety Committee of a group called City Lights in Jamaica Plain, Massachusetts, and the Safehouse Program created by Tenderloin Tenants for Safer Streets in San Francisco. As one reporter put it, "Safehouses, the traditional refuges of intelligence agents, fugitive radicals – and more recently havens for battered wives – now are being established for senior citizens."[13]

Both programs attempt to extend a sense of domestic security into the public realm. Greenlight aims to make women feel more secure from mugging and rape. Any house in the neighborhood showing a green light in the window is a place where a frightened woman can find shelter, a telephone, and emergency counseling. The Safehouse Program, identified by signs showing a peaked roof and a dove, operates in stores, bars, and hotels in a very rough district of San Francisco. All people in distress are welcome to enter and ask for safety, help getting to the hospital, or assistance from the police. The group distributes maps of the area to show the locations of the refuges, and encourages residents to patronize them.

Obviously neither Greenlight nor Safehouse signs can prevent crime. However these two efforts do show how citizens and planners can begin to give public space a more homelike quality.

RECLAIMING ACCESS TO PUBLIC SPACE: BETTER PUBLIC TRANSPORTATION

[. . .]

Most of the United States' expenditures on transportation in the last thirty years have supported freeway construction and car culture rather than projects like the [Washington, DC] Metro. The United States urgently needs better public transit systems. In the meantime, the services of existing public transit – subways, commuter railroads, and buses – must be improved by better social programming and by recognition that minorities, the young, the elderly, and women are most often the citizens without private cars.

Access to the public domain is especially difficult for older women. After age sixty-five, many women reap the results of a lifetime of low earnings, limited mobility, and self-sacrifice. In a study of 82,000 widows in Chicago, Helena Lopata found that over half of them did not go to public places, and over a fifth did not even go visiting. While 82 percent were not in a position to offer transportation to others, 45 percent had no one, of any age, to rely on for transportation.[14]

There are many ways that bus schedules and fares can be tailored for older people. Security issues can be worked on for both female passengers and the elderly, who are apprehensive about long waiting periods, especially at night, in deserted bus stops, train stations, and subway stations. Child-oriented and elderly-oriented schedules offer a further bonus: they relieve adults, usually women, from the responsibility of driving the elderly and the young, and at the same time encourage independence and self-reliance.

Good transportation is also a key factor in rape prevention. Recent estimates suggest that one woman in three in the United States will experience an attempted or completed rape in her lifetime.[15] Of course, if most citizens, including politicians and police officers, believe that a woman's place is

Figure 2 Inhospitable environments. *Left* Older women wait for a bus while the car culture passes them by. *Right* A young woman waits for a bus while "Kim" advertises the double standard.

in the home to begin with, they will not necessarily be blamed for unsafe streets. Instead, they may blame the rape victim for being in urban public space. Several innovative transportation projects, documented by Rebecca Dreis and Gerda Wekerle, meet the demand for greater safety on the streets by adding needed, flexible transportation.[16] In 1973 the Madison, Wisconsin, Women's Transit Authority began a service operating two cars seven nights a week, serving 1,000 women per month, on a fixed route shuttle service plus a flexible service within a four mile radius of the University of Wisconsin campus. Volunteers drive; the university, city and county pay the costs of the vehicles. In White-horse, Alaska, the Yukon Women's Minibus Society reached an even broader constituency. Women concerned about access to public space and about security created a system of four minibuses with sixteen seats each that now serve 700 passengers per day. The women's project provided the first bus system for the whole community.[17]

Public transportation not only provides safe access to public space, it can also educate riders about their city and about political struggles to make public space hospitable for everyone. This can range from adult classes on commuter trains, such as those initiated by Michael Young in England, to exhibits in key places – bus shelters and subway stations, and on buses and subways – that can provide essential information about cities, infra-structure, and public safety. Riders are thus united as a constituency for better services.

ADVERTISEMENTS, PORNOGRAPHY, AND PUBLIC SPACE

Americans need to look more consciously at the ways in which the public domain is misused for spatial displays of gender stereotypes. These ap-pear in outdoor advertising, and to a lesser extent in commercial displays, architectural decoration, and public sculpture. While the commercial tone and violence of the American city is often criticized, there is little analysis of the routine way that crude stereotypes appear in public, urban spaces as the staple themes of commercial art. Most Americans are accustomed to seeing giant females in various states of undress smiling and caressing products such as whiskey, food, and records. Male models also sell goods, but they are usually active and clothed – recent ad campaigns aimed at gay men seem to be the first major exception. . . .

This double standard is the result of advertising practices, graphic design, and urban design. Sanctioned by the zoning laws, billboards are approved by the same urban planning boards who will not permit childcare centers or mother-in-law apartments in many residential districts. But the problem with billboards is not only aesthetic degradation. By presenting gender stereotypes in the form of non-verbal body language, fifty feet long and thirty feet high, billboards turn the public space of the city into a stage set for a drama starring enticing women and stern men.

NOTES

1 Saegert, S. 1981. Masculine Cities, Feminine Suburbs: Polarized Ideas, Contradictory Realities. In Stimpson, C. *et al.* (eds). 1980. *Women and the American City*. Chicago: University of Chicago Press; Chatfield-Taylor, A. 1973. Hitting Home. *Architectural Forum*, 138, pp. 58–61.

2 Katznelson, I. 1981. *City Trenches: Urban Politics and the Patterning of Class in the United States*. Chicago: University of Chicago Press. Discusses the split between "home" and "work" as an event of political importance but makes no gender analysis.

3 Lefebvre, H. 1974. *La Production de L'espace*. Paris: Anthropos. Suggests the political importance of this demand for men. On women's exclusion, see de Pizan, C. 1982. *The Book of the City of Ladies*, trans. E.J. Richards. New York: Persea Books.

4 See Gerde Werkerle's review essay in Stimpson *et al.*, 1980, *Women and the American City*, pp. 188–214. Also see Enjeu, C. and Savé, J. 1974. The City: Off-limits to Women. *Liberation*, 8, pp. 9–15; Kjaer, B. 1982. A Woman's Place. *Architect's Journal*, 176, p. 87. These last two essays from France and from Denmark (published in England) show that the double standard in public places is an international problem, not simply a result of American housing.

5 Degler, C. 1979. Revolution Without Ideology. *Daedalus*, 93, pp. 653–670, and also his 1979 *At Odds: Women and the Family*. New York: Oxford University Press.

6 Lofland, L. 1975. The "Thereness" of Women: A Selective Review of Urban Sociology. In Milliman, M. and Kanter, R.M. (eds). *Another Voice: Feminist Perspectives on Social Life and Social Science*. New York: Anchor.

7 Bernard, J. 1980. *The Female World*. New York: Simon & Schuster.

8 Mumford, L. 1956. *The Transformations of Man*. New York: Harper & Row.

9 For a description of one mother's problems in public, see Chesler, P. 1979. *With Child: A Diary of Motherhood*. New York: Thomas Crowell, pp. 149–150, 175.

10 Quoted in Edgerton, R.B. 1978. *Alone Together: Social Order on an Urban Beach*. Los Angeles: University of California Press, p. 4.

11 Ibid., p. 5.

12 Gordon, M., Riger, S., LeBailly, R.K., and Heath, L. "Crime, Women, and the Quality of Urban Life." In Stimpson *et al.*, 1980, *Women and the American City*, pp. 161–169.

13 Hager, P. 1982. Safehouses Ease Fears of Aged Residents of San Francisco. *Los Angeles Times*, 21 November, p. 1.

14 Lopata, H. The Chicago Woman: A Study of Patterns of Mobility and Transportation. In Stimpson *et al.*, 1980, *Women and the American City*, pp. 161–169.

15 Griswold Johnson, A. 1980. On the Prevalence of Rape in the United States. *Signs: Journal of Women in Culture and Society*, 6 (autumn), pp. 136–146. See also Brownmiller, S. 1975. *Against our Will: Men, Women, and Rape*. New York: Simon & Schuster, p. 15; Griffen, S. 1971. Rape: The All-American Crime. *Ramparts*, 10 (September), pp. 26–35; Herman, D. The Rape Culture. In Freeman (ed.) 1975, *Women: A Feminist Perspective*, Palo Alto, CA: Mayfield Pub. and Co., pp. 469–473.

16 Wekerle, G. 1979. Women's Self-help Projects in the City: Transportation and Housing. Paper presented at conference on Social Practice, UCLA Urban Planning Program, 2 March, p. 5.

17 Ibid., pp. 5–7. On women as providers of urban services, also see Cranz, G. Women and Urban Parks: Their Roles as Users and Suppliers of Park Services. In Keller (ed.). *Building for Women*. 1991, Lexington MA: Lexington Books, pp. 151–171; Women in Urban Parks. In Stimpson *et al.* (eds). *Women and the American City*, pp. 79–95.

32

"Fighting Poverty and Environmental Injustice in Cities"

from *State of the World 2007: Our Urban Future* (2007)

Janice E. Perlman with Molly O'Meara Sheehan

Editors' Introduction

Widening global gaps between rich and poor societies and the concentration of tens of millions of people within the rapidly growing "mega-cities" of developing nations are two of the most pressing equity issues related to current development patterns. Mexico City, Sao Paolo, Lima, Lagos, Johannesburg, Cairo, Karachi, Tehran, Mumbai, Dhaka, Shanghai, Seoul, Jakarta, and Manila are among these metropolises with more than 10 million residents, many living in great poverty in "informal" housing created illegally on public or private land out of whatever materials are at hand.

For more than two decades Janice E. Perlman has run the Mega-Cities Project, a nonprofit organization that since 1987 has created research and action teams in 21 of the world's largest cities, aiming to support innovative solutions to urban problems. In this selection she argues that ending environmental injustice in such mega-cities must become a core part of the sustainability agenda. Other related materials are available from the group's website www.megacitiesproject.org/.

Many other organizations provide information on conditions and sustainable development opportunities in large cities of the developing world, including the United Nations' Development Program (www.undp.org/) and the U.N.'s Human Settlements Program (www.unhabitat.org/). Excellent materials are also offered by The Worldwatch Institute (www.worldwatch.org/), from one of whose annual *State of the World* Reports this piece is excerpted, and the International Institute for Environment and Development (www.iied.org/), which publishes the journal *Environment and Urbanization*. Books on mega-cities include Aseem Inam's *Planning for the Unplanned: Recovering from Crises in Megacities* (New York: Routledge, 2005), David Westendorff's edited volume *From Unsustainable to Inclusive Cities* (Geneva: UN Research Institute for Social Development, 2004), and *Megacities: The Politics of Urban Exclusion and Violence in the Global South* (London: Zed Books, 2009), edited by Kees Koonings and Kirk Kruijt.

"Cities, like dreams, are made of desires and fears," wrote Italo Calvino in *Invisible Cities*. If so, our cities range from daydreams to nightmares, depending on the city, the moment in time, and a person's position within the social and physical landscape.

Cities, like regions and countries, have experienced uneven development exacerbated by government inability to offset the inequities produced by globalization. Within many cities, inequalities have deepened – between rich and poor, between included and excluded, and between the "formal" and the "informal" city.[1]

The informal city consists of squatter settlements, clandestine subdivisions, invaded residential and commercial buildings, provisional housing for refugees or migrant workers, and often degraded "social housing" complexes. These communities account for some 40 percent of the total urban population of the South, including 41 percent in Mumbai and 47 percent in Nairobi. Residents typically lack basic urban services (water and sanitation, electricity, paved roadways) and security of tenure, including official title to homes or land and freedom from eviction. Even where informal communities have urban infrastructure and de facto rights to use the land, they remain stigmatized spaces, while the low-cost labor of their residents helps sustain life's daydream for the privileged in the formal city.[2]

As housing prices in the formal city are prohibitive for the poor, they have no choice but to live in the most dangerous areas: on the streets, as is common in India; in alleyways outside wealthy homes, as in many Asian cities; on hillsides too steep for conventional construction, as in the favelas of Rio de Janeiro; on stilts in marshes, as in Bahia's alagados; on floodplains, as in many of Jakarta's kampungs; atop garbage dumps, as in Manila; or even in cemeteries, as in Cairo. Families often remain through several generations and upgrade their homes and communities over time. Even young people who manage to enter university often have no place to live outside these "marginalized" spaces.[3]

Poor urban neighborhoods face the worst of two worlds: the environmental health hazards of underdevelopment, such as lack of clean drinking water, and of industrialization, such as toxic wastes. Yet their residents tread lightly on the planet, using few resources and generating low levels of waste in comparison with their wealthier neighbors. The gap between rich and poor in cities from Nairobi to New York means that those with the fewest resources suffer most from pollution generated by the wealthiest.

Those who advocate "sustainable development" – meeting the needs of people today without despoiling the planet for future generations – too often overlook the striking environmental injustice in our cities. The logical sequence linking global sustainability to urban poverty is synthesized in what have become known as the Perlman Principles:

- There can be no global environmental sustainability without urban environmental sustainability: Economies of scale in cities create energy and resource efficiencies.
- Transforming the urban metabolism through circular rather than linear systems is the key to reversing global environmental deterioration.
- There can be no urban environmental solution without alleviating urban poverty: The urban poor tend to occupy the most ecologically fragile areas of cities and often lack adequate water, sewage, or solid waste management systems.
- There can be no solutions to poverty or environmental degradation without building on bottom-up, community-based innovations, which are small in scale relative to the magnitude of the problems.
- There can be no impact at the macro level without sharing what works among local leaders and scaling these programs up into public policy where circumstances permit.
- There can be no urban transformation without changing the old incentive systems, the "rules of the game," and the players at the table.
- There can be no sustainable city in the twenty-first century without social justice and political participation as well as economic vitality and ecological regeneration.

A casual look at book titles throughout the 1960s and 1970s shows that the call for attention to the urbanizing world has been sounded for decades: *The Urban Explosion in Latin America*, *The Exploding Cities*, *The Wretched of the Earth*, *Uncontrolled Urban Settlement*. There may have been slightly more interest in urban poverty during the cold war due to fear that migrants and squatters might lead to leftist regimes. But once it was recognized that squatters were more interested in better opportunities for their children than in social protest, this interest fell off. It has only recently resurfaced in light of urban violence and security issues.[4]

Gradually, international agencies have begun to acknowledge the significance of cities and urban poverty. In 1999 the World Bank and UN-HABITAT formed the Cities Alliance to coordinate slum upgrading, in 2001 UN-HABITAT became a full-fledged

United Nations program, in 2003 the international community agreed on a single definition of "slums," and in 2004 United Cities and Local Governments was formed, giving formerly competing local authority networks a unified voice.

These recent milestones are important, but the pace of change remains too slow.

One of the UN Millennium Development Goals explicitly focuses on urban poverty: improve the living conditions of 100 million slum dwellers by 2020 (Target 11, which is part of Goal 7 on environmental sustainability). Even if achieved, this would make but a small dent, as it aims at just 10 percent of the existing slum population, and an additional 1 billion people will be living in urban areas of developing countries by 2020. Why are we always playing catch-up?[5]

BARRIERS TO EQUITABLE CITIES

Among the obstacles to reducing urban poverty and promoting environmental justice are inept and corrupt governance, violence, anti-urban bias, skewed development assistance, counterproductive incentives and resistance to change, and a lack of reliable city-level data necessary to benchmark progress.

Weak governance. The urban poor, generally excluded from decision making, are the greatest untapped source of ideas about improving their cities and lives. Over the last several decades, mayors have been elected for the first time in many countries, arguably bringing the government closer to the people, but most poor people still do not have a voice in governance. Since decision makers tend to come from elite sections of society, they often have a vested interest in maintaining the status quo. Lawmakers in Kenya, for instance, who saw bicycles as toys for children rather than an important transportation mode, levied a luxury tax on bikes for years, keeping the cost too high for many low-income residents.[6]

Poor governance is reflected in uneven service delivery. World Bank Institute researchers have documented that people have less access to water, sewerage, and school-based Internet in cities that have a record of bribery in their utility companies, private firms that dictate local laws and regulations, and high levels of corruption nationwide. Ronald

MacLean-Abaroa discovered this reality after becoming the first elected mayor of Bolivia's capital, La Paz, in the mid-1980s: "Whenever I found problems in service delivery or the prompt completion of public works or the collection of revenues, they happened not just to be associated with inefficient organization but almost always with corruption."[7]

Corruption, clientelism, and cronyism – the "three Cs" that undermine democracy – worsen urban poverty. Corruption skews public spending toward sectors where bribing is easier, such as large public works construction, and generally away from education, health, and maintenance of existing infrastructure. When people need to bribe officials to get needed services, those who can least afford the bribes suffer the most. A survey in Indonesia found that bribes required by police, schools, electricity companies, and garbage collectors ate into the already tight budgets of the urban poor.[8]

Violence and stigma. The increase in urban violence that has accompanied the rise in drug and arms trafficking has created particularly high mortality rates among urban youth. When dealers purchase complicity from the police for their illegal activities, they can hold entire low-income neighborhoods hostage. The consequent association of the urban poor and violence only serves to deepen the stigma that already constrains life opportunities for these people.[9]

A multigenerational study of families in Rio confirmed this trend. Fear of violence keeps people homebound, while job interviews end when the applicant's address is identified as in a favela. Among people of equivalent educational levels, those who lived in favelas had less success finding jobs.[10]

Violent crime is a much greater threat in some places than others. Researchers have found the world's lowest levels of robbery and assault in Asia and the highest levels in Latin America and sub-Saharan Africa.[11]

Anti-urban bias. Environmentalists and development specialists have long portrayed cities as threats to nature. Many policymakers still adhere to this old mind-set, pitting environmental concerns against economic growth and preventing further growth of cities, especially megacities. In 1986, a study commissioned by the UN Fund for Population Activities found that almost every nation had made some attempt to limit urban growth by investing in rural development, creating "growth poles" or "new towns," forcing people to relocate to under-settled regions, moving their national capitals, or introducing closed-city policies.[12]

Countries spent a good deal of money and political capital in these efforts, but none succeeded in stemming the tide of migration. Investment in rural roads, electrification, education, health, and industrialization, while important for improving rural living standards, in many cases just increased the rate of migration to urban areas. As people became more aware, they used their new roads and skills to seek wider opportunities in the city.[13]

Even where freedom of movement was highly restricted – such as colonial governments, command-and-control economies, and police states – people nonetheless found ways to sneak into cities. The most "successful" efforts to prevent urban growth were in apartheid South Africa, which required passports for non-whites to enter cities; China, which used rice ration cards and a household reg-istration system; Russia, which used housing alloca-tions; and Cuba, where a national policy of keeping people in the countryside was backed by the use of force. Yet "floating populations" of migrants entered Chinese cities, Moscow apartments became crowded with friends and relatives trying to move to the city, and governments everywhere kept in-accurate records of city size to maintain the fiction of non-growth.[14]

As recently as 2005, scores of nations were still attempting to curb urbanization.

The United Nations Population Division recently analyzed migration policies of 164 countries and found that 70 percent aim to lower migration from rural to urban areas. This study confirmed that the impact on overall population distribution was "almost negligible."[15]

The myth that people will stop coming to cities if public housing is not built and squatter settle-ments are removed is unfounded and hurts the poor. Miloon Kothari, UN Special Rapporteur on Housing Rights, estimated in 2006 that the number of forced evictions had risen worldwide since 2000. "Without human rights safeguards," Kothari warns, "the commitment to the reduction of slums, including through the relevant MDGs, can easily become slum-eradication to the detriment of slum-dwellers."[16]

Skewed international assistance. Although virtually all of the world's population growth is expected in the cities of Africa, Asia, and Latin America, and most of this will be in low-income areas, develop-ment assistance has been reluctant to recognize the urbanization of poverty. From 1970 to 2000, all urban development assistance was estimated at $60 billion – just 4 percent of the total $1.5 trillion. Few bilateral aid agencies have any kind of urban housing program, or any serious urban program at all.[17]

The decisions of international development banks are important, even if aid is not the primary source of foreign investment in a given country. In recent years, aid has been roughly one-tenth the level of private capital flows in developing countries. Yet international aid agencies leverage additional funding and influence the research agendas and spending priorities of governments, universities, and nongovernmental organizations (NGOs).[18]

Aid that does flow to urban areas often misses the poor. The London-based International Institute for Environment and Development found that urban projects accounted for 20–30 percent of all lending at several agencies from 1981 to 1998. But housing, water, sanitation, and other services that improve conditions for the urban poor received just 11 per-cent of total lending at the World Bank, 8 percent at the Asian Development Bank, and 5 percent at Japan's Overseas Economic Cooperation Fund.[19]

The World Bank reports that its lending and staff devoted to urban areas continue to lag behind the resources invested in the rural sector. All country-wide World Bank investments and most bilateral aid in developing countries are guided by Poverty Reduction Strategy Papers prepared by govern-ments in consultation with the World Bank Group. These documents tend to neglect urban areas.[20]

The international assistance community cut its teeth on rural development and is professionally and structurally geared toward assisting rural peas-ants, not to continuing assistance for them when they move to cities. Development experts seem to prefer "missions" to attractive agricultural areas, fishing villages, and environmental preserves over those in the polluted, overcrowded, and often dangerous urban slums.

Pressure to bring more attention to urban poverty within the international development community has met with strong resistance. Every Executive Director of UN-HABITAT since its founding in 1978 has urged that we must act now, and yet none has succeeded in receiving funding parity with other UN agencies. When UN-HABITAT was headquartered in Nairobi along with the UN Environment Programme (UNEP) to assuage political pressures from African countries, it was no secret that these were considered the most expendable United Nations bodies. Although UNEP and UN-HABITAT have an impressive campus in Nairobi, both agencies remain politically and financially marginalized.

The voices for greater support for urban programs often face derision from those who believe that government spending is already pro-urban. But rigorous economic studies find that wealth generated in cities ends up subsidizing the countryside.[21]

On a more positive note, the World Bank is now considering subnational lending, which would allow loans to go directly to municipal governments, bypassing national Finance Ministries. This would help cities – particularly megacities – receive monies designated for them from international agencies that is now often held up by national governments that have imposed lending ceilings or that see political advantages in withholding funds. For instance, national governments in Brazil and Mexico do not want the mayors of their major cities to appear successful because they are potential competitors for the presidency and often from opposition parties.[22]

Counterproductive incentives and fear of change. The incentive systems of aid agencies are at odds with contextually specific, community-based, anti-poverty initiatives. Development professionals are promoted based on the size and rapidity of loans "pushed out the door," making one-size-fits-all approaches the best route to promotion, as opposed to smaller-scale projects where the priorities are set by local people. One analyst concluded that "the people whose needs justify the whole development industry are the people with the least power to influence development and to whom there is least accountability in terms of what is funded and who gets funded."[23]

The public sector is generally risk-averse. Elected and appointed officials often feel safer sticking with "the way things have always been done," even if results are suboptimal, rather than risk being fired or not re-elected for making a mistake. There is a high price to pay for an unsuccessful initiative – and little or no reward for innovation. As Alan Altshuler and Marc Zegans have noted, in the private sector the expectation that some ideas and initiatives will

not be successful is built into the process, and funds are set aside for R&D where the entire point is to experiment and innovate.[24]

Fragmentation and competition among public agencies and academic disciplines further limit cooperation in solving urban problems. Each area – water and sanitation, transportation, housing, land use planning, private-sector involvement, poverty alleviation – is in a separate department, often run by people who may compete for resources, attention, or staff. Yet these issues are intertwined, so that an apparent solution to one may lead to new problems in another. Even when universities or international agencies have created interdisciplinary structures, individuals still have the strongest allegiances to their home department, where appointments, promotions, and payments are set.

Inadequate data for benchmarking. Lack of city- and neighborhood-level data makes it difficult to measure progress and hold governments accountable. Most statistics are only available at the national level, at best broken down between rural and urban, but not by specific cities. Where cities have managed to mobilize resources to collect their own data, they often exclude informal settlements and are rarely comparable with earlier studies or with data from other cities. The Global Urban Observatory created by UN-HABITAT in 1998 to address this problem has had little success in finding indicators comparable across cities, despite its Web-accessible database of 237 cities that covers measures of poverty, environment, infrastructure, urban services, shelter, and land.[25]

Apparently neutral questions such as what to measure, what indicators to use, how to collect reliable data on these indicators, how to make sense of the results, and how to disseminate the results are in fact value-based issues that have political and social implications. Answers to these urban indicators questions will be a major topic of debate at the next World Urban Forum in Nanjing, China, in 2008. . . .

PROMISING NEW DIRECTIONS

Our global future will be urban, like it or not. People in developing countries will continue to vote with their feet, moving to cities or the urban fringe. It will be many decades before cities of the global

South reach a stable population. Although birth rates decline with urbanization, this happens more slowly with the urban poor.

What can be done to make our urban future a desirable and sustainable one? What kinds of cities foster conviviality and creativity? How can poverty and environmental degradation be alleviated and a voice for the disenfranchised be ensured?

There is no magic bullet for creating sustainable, equitable, and peaceful cities. But there are some necessary if not sufficient conditions for such transformations: transparent governance, decent work or a basic income, innovative infrastructure to conserve the environment, intelligent land use with integrated community development, and social cohesion along with cultural diversity.

Foster transparent governance. Effective governance is essential to scale up promising innovations into public policies, to provide basic services equitably, and to forge partnerships with the private and voluntary sectors. Addressing the World Bank Board in Singapore in September 2006, Bank President Paul Wolfowitz emphasized this point: "Without [good] governance, all other reforms will have limited impact. . . . It is the view I have heard on sidewalks and in taxis – in the marbled halls of Ministries and in the rundown shacks of shantytowns."[26]

Tackling the corruption that weakens governance requires fostering competition so that governments do not have monopoly power, reducing bureaucratic leeway, and increasing accountability. For instance, La Paz began to reduce bribery in construction permits by simplifying and publicizing the rules, contracting the permitting out to architects, and reducing the city's job to overseeing the contracts – a job that could be done by fewer municipal employees who could be paid more. Some promising efforts to foster transparent governance make government rules, purchases, and investments public knowledge by posting them on the Internet.[27]

Ensure decent work or a basic income. Jobs are a top priority for the urban poor. In 2001, the majority of interviewees in the multigenerational study of favela residents in Rio said that "the most important factor for a successful life" was a good job with a good salary. They want a chance to earn their livelihoods whether as employees or informal workers – that is the key to their dignity. What is needed is both job creation and preparation of people within their communities for jobs that exist in market sectors that are growing.[28]

Job and skills training, mentoring, and help in finding a first job can play a major role if done correctly. There is no use building people's capacity for jobs that no longer exist.

Savings and credit are key to job creation. Without access to these financial tools, would-be entrepreneurs cannot start small businesses. Many forms of microfinance, including community savings and credit funds as well as small loans to make improvements to housing, have been successful in urban settings.[29]

Larger companies can also play a role in strengthening the economies of poor communities. In Guadalajara, Mexico, where a good share of the population lives in unplanned settlements, a large, multinational cement company, Cemex, developed a savings-and-credit scheme to allow households that earn $5–$15 a day (low-income families, but not the poorest of the poor) to buy materials to build and improve their housing. The program has since expanded to twenty-three cities in Mexico.[30]

In some cases, governments can help create local jobs by hiring the urban poor to help solve pressing environmental problems, as reforestation in Rio demonstrated. In low-income communities in Dar-es-Salaam in Tanzania and Kampala in Uganda that were damaged in flooding, local governments used "community contracts" to get local labor for the necessary rebuilding.[31]

There is also a role for a "negative income tax" to help in periods between jobs or to supplement incomes too low to live on. Mexico and Brazil started innovative "conditional cash transfer programs" in which the government directly deposits funds into personalized debit cards for low-income people as an incentive for desirable action. For example, a family gets a certain amount of cash for children as long as they attend school regularly, gets other funds upon proof of immunization against contagious diseases, and receives further aid if there are elderly or infirm people in the family. Such programs have led to higher school enrollment and better preventive health care, especially in Latin American cities, and New York's mayor now plans to try the approach.[32]

Develop innovative infrastructure to conserve the environment. Cities that do not yet have full infrastructure have the chance to "leapfrog" over outmoded and wasteful systems created during the Industrial Revolution. They can take advantage of resource-conserving technologies, both low- and high-tech, to revolutionize the built environment. Examples include installing water-conserving toilets, separating drinking and gray water into different systems, using passive solar energy or biogas for heating, and adapting the recycling technologies developed by NASA for life in outer space. Architect William McDonough is working on low-cost housing projects in China that use biogas from wastewater for cooking, local compressed earth as building materials, and passive solar design for heating and cooling. These sorts of technologies are also being used in Johannesburg.[33]

One challenge in adopting low-impact or "alternative" technologies is that they do not carry the prestige of "modernity" seen on televisions and in houses of the rich around the world. While cutting-edge neighborhoods in places like Stockholm show that eco-friendly design is compatible with a high standard of living, most people in developing countries do not see these models and continue to aspire to the worst mistakes of the US and European systems.

Promote intelligent land use and integrated community development. Urban planning is making a comeback after decades of being dismissed as the province of useless colored maps. Creative urbanists have found new ways to involve communities in the trade-offs of physical planning decisions and to use old planning tools for progressive change. Zoning, building, and land use regulations have been adapted to foster mixed-use communities, with homes close to workplaces, commerce, and recreation. Incentives for development in areas with existing infrastructure have helped limit sprawl. "Areas of special interest" have been established to protect environmentally important areas, connect nature corridors, and allow the upgrading of informal settlements in flexible ways. Comprehensive transportation planning has now included investments in sidewalks, bicycle paths, and low-cost public transportation options with easy connections between local, regional, and longer-distance travel.

São Paulo has been among the leaders in advancing urban planning tools to create a more inclusive city. One of the biggest steps there was taxing building developers to create a fund for investments in the public interest, including public transportation, housing, and environmental improvements.

Low-income housing options are so limited in most cities that new migrants end up living in the most dangerous places. A logical response would be to lay out small plots of land with connection to basic urban services, available for small sums or loans. This "sites and services" approach, first tested in Dakar, Senegal, in 1972, has been little used, as it does not offer politicians the chance for ribbon cuttings and photo opportunities. Yet it makes monetary and environmental sense: the costs of infrastructure are small compared with retrofitting a squatter community, and the new settlements help reduce the invasion of environmentally sensitive areas.

Cultivate social cohesion and cultural diversity. Diversity makes natural ecosystems and human economies more resilient, yet prejudice and misunderstanding between different groups of people often squander the potential of cultural diversity to strengthen cities. Solving the complex problems that our cities face will require the greatest diversity of cultures and value sets possible.[34]

As violent crime rips apart the urban fabric and further isolates the poor, there is a need to focus on controlling the sale of arms and drugs, reducing the corruption that permits violence with impunity, and mobilizing the society at large to find solutions to the problem. Promising initiatives include efforts at community policing in low-income neighborhoods and all varieties of arts, culture, and sports programs for young people at risk. Some programs of arms amnesty and prevention of small arms sales have also been tried, with varying degrees of success in lowering urban violence.[35]

OUR URBAN FUTURE

The directions in politics, economics, environmental policy, and society just described require at least three fundamental changes. The first is to revise the architecture for support to cities and the urban poor, giving cities their due and reversing the reward systems to promote innovation. As the global population shifts toward cities, the agendas of aid agencies, national governments, foundations, research centers, and nonprofit groups need to reflect this reality. All too often in the discourse on the future of cities, the focus remains exclusively on global or world cities in their role as centers of capital and information flows and as corporate headquarters rather than on the more numerous and populous cities of the developing world.

To bridge the divide between official sources of assistance and the urban poor, governments and aid agencies might channel their support to a local fund or community foundation in each city. The fund would be earmarked for use by community organizations, have transparent decision making, and make it easy for groups that receive assistance to exchange ideas with each other. The Thai government's Community Organizations Development Institute offers one example of how national governments could contribute to local funds. The Swedish International Development Cooperation Agency has set up local organizations to run urban poverty programs in Costa Rica and Nicaragua. In Ecuador, the government reached agreements between 1988 and 1993 with foreign governments to restructure commercial debt instruments and then channeled the funds through local NGOs to finance development projects throughout the country.[36]

Rethinking development assistance includes finding ways for federations of the urban poor and NGOs working with them to bypass official aid channels. Two former World Bank staff members have started a web site, www.globalgiving.com, that allows individuals and institutions to support projects run by local people. Analyzing this initiative, former Bank economist William Easterly writes: "Think of the potential for creativity if thousands of potential donors, project proposers, technical advisers, and advocates for the poor were freed from the shackles of the large centralized bureaucracy and could find solutions that worked on the ground. This is not a panacea for redesigning all of foreign aid; it is just one promising experiment in how aid could reach the poor."[37]

A second change needed is to create systematic ways for benchmarking progress and measuring outcomes in cities. Without reliable, comparable indicators of poverty and environmental conditions, we will never know whether progress is being made or how to compare the impact of one set of practices and policies with another. As international and national efforts have not yielded city-level indicators, the need for local benchmarking is clear. One possibility is for local governments to hire residents

to collect health, housing, income, and environmental data. Federations of the urban poor, from Mumbai to Nairobi, have shown how this can be done, organizing communities to perform their own censuses. Cities can hold information collection fairs that would motivate people to collect data on their own areas of interest and responsibility.

Exchange of information is especially essential among those who are most directly involved in fighting urban poverty and among the urban poor themselves. The journal *Environment and Urbanization*, produced by the International Institute for Environment and Development, offers one important forum for researchers, NGO staff, and others to exchange ideas. There is also a need for face-to-face discussions among community leaders.

Some new projects aim to collect and share information on urban poverty and the environment. The International Development Research Centre in Canada has launched a program to study interventions in urban agriculture, water and sanitation, solid waste management, vulnerability to natural disasters, and land tenure as a cross-cutting theme in a handful of "focus cities" in developing countries. A new Urban Sustainability Initiative, supported by the Moore Foundation in the United States, is planning city partnership projects in China, Mexico, South Africa, and Tanzania, an exchange of ideas among cities, and a set of scientific and social indicators to gauge urban progress.[38]

In these and other initiatives it will be important to realize that rather than "best practices" and "competitive cities," what is needed is "better practices" and "collaborative cities." The "best practice" model is flawed by its implication of one ideal way for all communities. As the MegaCities Project discovered, each new innovation gives rise to new problems and contradictions, which require yet further innovations and revised solutions. Another lesson was that a "best practice" in one place may be useless or detrimental in another. Each location needs to adapt solutions to its own history, culture, and local circumstances. The current system of nominating, judging, and rewarding "best practices" allows for self-nomination and self-promotion but leaves little room for neutral external evaluation.

Despite apparent differences in politics, economics, and culture, cities in developing countries and the industrial world have many problems in common, often more than they share with small towns or villages in their own countries. Nearly every wealthy city contains within it neighborhoods with high infant mortality, malnutrition, homelessness, joblessness, and low life expectancy. And nearly every city in the developing world contains within it a world of high finance, high technology, and high fashion. If cities are to be used as laboratories for urban innovation, they can harvest ideas to exchange from the South to the North, since low-income cities have a lighter ecological footprint and have more experience in reuse. It is time to move from NIMBY (not in my backyard) and NOPE (not on planet earth) to the recognition that all by-products of human activities end up in someone's backyard and in the atmosphere that surrounds us.

The last basic shift needed is for people in positions of power to listen to the most vulnerable portions of the population, particularly young people and women. The cities of the future belong to the children of today. Unfortunately, a review of municipal efforts to incorporate children's concerns in decision making found that "there is generally more interest in showcase projects" than in broader changes.[39]

Cities could scale up programs that expose young people to arts and sports and develop areas in which they can excel and feel part of something worthwhile. In Rio, one such program – Afro-Reggae, started in the favela Vigário Geral – has used drumming, dance, and song lyrics that expressed the community's reality to attract youngsters, build solidarity, and develop a critical analysis of their situation. Its work, chronicled in the documentary *Favela Rising*, helped defuse a drug war with the adjacent favela and has spread to other communities.[40]

One of the most articulate explanations of the need to listen to the urban poor comes from Rose Molokoane of the South African Federation of the Urban Poor, or FED UP. She recently told an audience that included development professionals: "We are fed up of being the subject of the agenda. We are fed up with you not listening to us. . . . We are poor, but not hopeless. We have money, but no chance to come to the bank and open an account because we have no address. If you give me security of tenure, then I have an address, and I will open an account. We will show you we can do it. . . . The only thing we are concentrating on is

how to organize ourselves. If communities are organized, they are a tool to address issues that are giving you double stress."[41]

The gulf that Rose Molokoane identified between those who set "development goals" and those who are the target of that agenda is a subset of larger rifts between rich and poor, between the powerful and the powerless. Bridging these divides will require a new mind-set. Unless and until we are ready to expand our conception of "we" from "me and my family" to my community, city, country, and planet, the gap will continue to grow. We may have come this far through competition and survival of the fittest, but if we are to make the leap to a sustainable world for the centuries ahead, we will need to be intelligent enough to do it through collaboration and inclusion. In the words of Australian aboriginal elder Lilla Watson: "If you've come to help me, you're wasting your time. But if you've come because your liberation is bound up with mine, then let us work together."[42]

NOTES

1 Italo Calvino, *Invisible Cities*, trans. William Weaver (Orlando, FL: Harcourt, 1974); different types of inequality between countries in World Bank, *World Development Report 2006* (Washington, DC: Oxford University Press and World Bank, 2006).

2 Urban population in slums in developing countries from UN-HABITAT, *State of the World's Cities 2006/7* (London: Earthscan, 2006), pp. 16, 111; Mumbai and Nairobi from Gora Mboup, senior demographic and health expert, UN-HABITAT, Nairobi, e-mail to Molly Sheehan, 5 October 2006; Zuenir Ventura, Cidade Partida (São Paulo, Brazil: Companhia das Letras, 1994).

3 Janice E. Perlman, "Marginality: From Myth to Reality in the Favelas in Rio de Janeiro 1969–2002," in Ananya Roy and Nezar AlSayyad, eds., *Urban Informality: Transnational Perspectives from the Middle East, Latin America, and South Asia* (Lanham, MD: Lexington Books, 2004).

4 Glenn H. Beyer, ed., *The Urban Explosion in Latin America: A Continent in Process of Modernization* (Ithaca, NY: Cornell University Press, 1967); Peter Wilsher and Rosemary Righter, *The Exploding Cities* (London: A. Deutsch, 1975); Franz Fanon,

The Wretched of the Earth (London: MacGibbon & Kee, 1965); John F.C. Turner, *Uncontrolled Urban Settlement: Problems and Policies* (Pittsburgh, PA: University of Pittsburgh Press, 1966); squatters' interest in better opportunities for children from Janice Perlman, *The Myth of Marginality: Urban Poverty and Politics in Rio de Janeiro* (Berkeley, CA: University of California Press, 1976), and from Joan M. Nelson, *Access to Power: Politics and The Urban Poor in Developing Nations* (Princeton, NJ: Princeton University Press, 1979).

5 For Millennium Development Goals, see www.un.org/millenniumgoals; United Nations Population Division, *World Urbanization Prospects: The 2005 Revision* (New York, 2005).

6 Janice Perlman, "Re-democratization in Brazil: A View from Below, The Experience of Rio de Janeiro's Favelados 1968–2005," in Peter Kingstone and Timothy Power, eds., *Democratic Brazil Revisited* (Pittsburgh, PA: University of Pittsburgh Press, forthcoming); Kenya's tax on bicycles from VNG uitgeverij, *The Economic Significance of Cycling: A Study to Illustrate the Costs and Benefits of Cycling Policy* (The Hague, 2000), and from Jeffrey Maganya, Intermediate Technology Development Group, Nairobi, Kenya, discussion with Molly Sheehan, 8 May 2001.

7 Daniel Kaufmann, Frannie Léautier, and Massimo Mastruzzi, "Globalization and Urban Performance," in Frannie Léautier, ed., *Cities in a Globalizing World: Governance, Performance and Sustainability* (Washington, DC: World Bank Institute, 2006), pp. 38–49; Robert Klitgaard, Ronald MacLean-Abaroa, and H. Lindsey Parris, *Corrupt Cities: A Practical Guide to Cure and Prevention* (Oakland, CA: Institute for Contemporary Studies, 2000), p. 32.

8 Perlman, op. cit. note 6; T. Abed and Sanjeev Gupta, eds., *Governance, Corruption, and Economic Performance* (Washington, DC: International Monetary Fund (IMF), 2002); Vito Tanzi and Hamid Davoodi, *Corruption, Public Investment and Growth*, Working Paper 97/139 (Washington, DC: IMF, 1997); Sanjeev Gupta, Hamid Davoodi, and Rosa Alonso-Terme, *Does Corruption affect Income Inequality and Poverty?* Working Paper 98/76 (Washington, DC: IMF, 1998); Ratih Hardjono and Stefanie

Teggeman, eds., *The Poor Speak Up: Seventeen Stories of Corruption* (Jakarta: Partnership for Governance Reform, 2002).

9 UN-HABITAT, *The State of the World's Cities, 2004/2005* (London: Earthscan, 2004), pp. 134–57.

10 Ignacio Cano et al., *O impacto da violência no Rio de Janeiro*, Working Paper (Rio de Janeiro: Universidade do Estado do Rio de Janeiro, 2004); levels of violence from Luke Dowdney, *Children of the Drug Trade* (Rio de Janeiro: Viveiros de Castro Editoria, 2003); police provoking violence from "Law Enforcers on the Rampage: Brazil's Trigger-happy Police," *The Economist*, 9 April 2005.

11 UN-HABITAT, op. cit. note 9, pp. 134–57; François Bourguignon, "Crime, Violence, and Inequitable Development," paper prepared for the Annual World Bank Conference on Development Economics, Washington, DC, 28–30 April 1999.

12 For traditional views of environmentalists and development specialists, see Eugene P. Odum, *Fundamentals of Ecology*, 3rd ed. (Philadelphia: Saunders, 1971), and Michael Lipton, *Why Poor People Stay Poor: Urban Bias in World Development* (London: Temple Smith, 1977); Janice Perlman and Bruce Schearer, "Migration and Population Distribution Trends and Policies and the Urban Future," International Conference on Population and the Urban Future, UN Fund for Population Activities, Barcelona, Spain, May 1986.

13 Perlman and Schearer, op. cit. note 12.

14 Ibid.

15 United Nations Population Division, *World Population Policies 2005*, at www.un.org/esa/population/publications/WPP2005/Publication_index.htm.

16 United Nations High Commissioner for Human Rights, "Statement by Mr. Miloon Kothari, Special Rapporteur on adequate housing as a component of the right to an adequate standard of living, to the World Urban Forum III," Vancouver, 20 June 2006.

17 Martin Ravallion, *On the Urbanization of Poverty*, Development Research Group Working Paper (Washington, DC: World Bank, 2001); urban estimate comes from Michael Cohen, "Reframing Urban Assistance: Scale, Ambition,

and Possibility," *Urban Update*, Comparative Urban Studies Brief, Woodrow Wilson International Center for Scholars, No. 5, February 2004, p. 1; total assistance from OECD/DAC figures in Worldwatch Institute, *Worldwatch Global Trends*, CD-ROM, July 2005; lack of urban housing programs from Daniel S. Coleman and Michael F. Shea, "Assessment of Bilateral and Multilateral Development Assistance and Housing Assistance in Latin America, Asia, Africa and the Middle East," Interim Working Draft for the International Housing Coalition, 3 May 2006.

18 Aid from OECD/DAC, and private capital from UNCTAD, both in Worldwatch Institute, op. cit. note 17.

19 David Satterthwaite, "Reducing Urban Poverty: Constraints on the Effectiveness of Aid Agencies and Development Banks and some Suggestions for Change," *Environment and Urbanization*, April 2001, pp. 137–57.

20 Urban area expenditures from Frannie Léautier, World Bank Institute, e-mail to Molly Sheehan, July 2006; urban dimension missing from Diana Mitlin, *Understanding Urban Poverty: What the Poverty Reduction Strategy Papers Tell Us* (London: International Institute for Environment and Development (IIED), 2004).

21 William Alonso, "The Economics of Urban Size," *Papers in Regional Science*, December 1971, pp. 66–83; Rémy Prud'homme, "Antiurban Biases in the LDCs," Megacities International Conference, New York University, 1988; Rémy Prud'homme, "Managing Megacities," *Le Courrier du CNRS*, No. 82, 1996, pp. 174–76.

22 Alfredo Sirkis, Director of Urbanism, Rio de Janeiro, discussion with Janice Perlman, 25 August 2005.

23 Satterthwaite, op. cit. note 19, p. 140.

24 Alan Altshuler and Marc Zegans, "Innovation and Creativity: Comparisons between Public Management and Private Enterprise," *Cities*, February 1990, pp. 16–24.

25 UN-HABITAT, Global Urban Observatory, at hq/unhabitat.org/programmes/guo.

26 World Bank, *Assessing Aid* (Washington, DC: 1998); World Bank, *The Role and Effectiveness of Development Assistance* (Washington, DC: 2002); Paul Wolfowitz, President, World Bank Group, Address to Board of Governors of the World Bank Group, Singapore, 19 September 2006.

27 Klitgaard, MacLean-Abaroa, and Parris, op. cit. note 7; Winthrop Carty, Ash Institute for Democratic Governance and Innovation, Kennedy School of Government, Harvard University, "Citizen's Charters: A Comparative Global Survey," translation from Spanish of *Cartas compromiso: experiencias internacionales*, presented at the launch of the Mexican Citizen's Charter Initiative, June 2004.

28 Philip Amis, "Municipal Government, Urban Economic Growth, and Poverty Reduction – Identifying the Transmission Mechanisms between Growth and Poverty," in Carole Rakodi with Tony Lloyd-Jones, eds., *Urban Livelihoods: A People-centered Approach to Reducing Poverty* (London: Earthscan, 2002), pp. 97–111; Perlman, op. cit. note 10; Janice Perlman, "Violence as a Major Source of Vulnerability in Rio de Janeiro's Favelas," *Journal of Contingencies and Crisis Management*, winter 2005.

29 A. Zaidi, "Assessing the Impact of a Micro-finance Programme: Orangi Pilot Project, Karachi, Pakistan," in S. Coupe, L. Stevens, and D. Mitlin, eds., *Confronting the Crisis in Urban Poverty: Making Integrated Approaches Work* (Rugby, UK: Intermediate Technology Publications, 2006), pp. 171–88; Franck Daphnis and Bruce Fergus, eds., *Housing Microfinance: A Guide to Practice* (Bloomfield, CT: Kumarian Press, 2004).

30 C.K. Prahalad, *The Fortune at the Bottom of the Pyramid* (Upper Saddle River, NJ: Pearson, 2006); David L. Painter, TCG International, in collaboration with Regina Campa Sole and Lauren Moser, ShoreBank International, "Scaling up Slum Improvement: Engaging Slum Dwellers and the Private Sector to Finance a Better Future," paper presented at the World Urban Forum, Vancouver, June 2006.

31 UN Millennium Project Task Force on Improving the Lives of Slum Dwellers, *A Home in the City* (London: Earthscan, 2005), pp. 55–56; Jane Tournée and Wilma van Esch, *Community Contracts in Urban Infra-structure Works* (Geneva: International Labour Organization, 2001).

32 Bénédicte de la Brière and Laura B. Rawlings, *Examining Conditional Cash Transfer Programs: A Role for Increased Social Inclusion, Social*

Protection Discussion Paper No. 06083 (Washington, DC: World Bank, 2006); SYDGM and World Bank, the Third International Conference on Conditional Cash Transfer, Istanbul, June 2006, at info.worldbank.org/etools/icct06/welcome.asp; James Traub, "Pay for Good Behavior?" *New York Times Magazine*, 8 October 2006, pp. 15–16.

33 Janice Perlman, "Megacities and Innovative Technologies," *Cities*, May 1987, pp. 128–36; William McDonough, "China as a Green Lab," in Howard Gardner *et al.*, "The HBR List: Breakthrough Ideas for 2006," *Harvard Business Review*, February 2006, p. 35; Mara Hvistendahl, "Green Dawn: In China, Sustainable Cities Rise by Fiat," *Harper's*, February 2006, pp. 52–54.

34 Leonie Sandercock, *Cosmopolis II: Mongrel Cities in the Twenty-first Century* (London: Continuum, 2003).

35 Small Arms Survey, at www.smallarmssurvey.org; International Action Network on Small Arms, *Reviewing Action on Small Arms* (London: 2006), pp. 17–22.

36 Satterthwaite, op. cit. note 19, pp. 146, 148; Göran Tannerfeldt and Per Ljung, SIDA, *More Urban, Less Poor: An Introduction to Urban Development and Management* (London: Earthscan, 2006); Esquel Foundation, at www.synergos.org/latinamerica/ecuador.htm.

37 William Easterly, *The White Man's Burden: Why the West's Efforts to Aid the Rest Have Done So Much Ill and So Little Good* (New York: Penguin Press, 2006), p. 378.

38 Urban Sustainability Initiative, at bie.berkeley.edu/usi.

39 UN Millennium Project Task Force, op. cit. note 57, p. 101; Yves Cabannes, "Children and Young People build Participatory Democracy in Latin American Cities," *Environment and Urbanization*, April 2006, pp. 195–218; Sheridan Bartlett, "Integrating Children's Rights into Municipal Action: A Review of Progress and Lessons Learned," *Children, Youth, and Environments*, 15:2 (2005), pp. 18–40.

40 "YA! Youth Activism," *NACLA Report on the Americas*, May/June 2004, p. 48.

41 Rose Molokoane, presentation at Future of the Cities panel, World Urban Forum, Vancouver, 23 June 2006.

42 Watson, quoted in Ernie Stringer, *Action Research* (Thousand Oaks, CA: Sage Publications, 1999).

ECONOMIC DEVELOPMENT

33

"The Economic System and Natural Environments"

from the Introduction and Conclusion to *Blueprint for a Sustainable Economy* (2000)

David Pearce and Edward B. Barbier

Editors' Introduction

Economics and sustainable development coexist uneasily. The common denominator for many observers is a realization that current economic theory and practice – the driving forces behind much of the world's development – leave out many important social and environmental factors that represent the costs as well as the benefits of development. Traditional economic tools make it difficult to incorporate externalities such as pollution, resource depletion, and degradation of human living environments. They also have no good way to take long-term costs and benefits into account, assume endless growth in material consumption, inadequately take into account market distortions caused by subsidies and regulations, and in many other ways fail to promote sustainable development.

British authors David Pearce, Edward B. Barbier, Anil Markandya, and other colleagues have been at the forefront of developing the discipline known as environmental economics, which seeks ways to better incorporate ecological concerns into economic analysis. In 1989 they authored an influential book entitled *Blueprint for a Green Economy* (London: Earthscan, 1989), followed by *Blueprint 2: Greening the World Economy* (London: Earthscan, 1991). These books provided popular explanations of a wide range of eco-economic issues. In 2000 Pearce and Barbier updated these resources with *Blueprint for a Sustainable Economy* (London: Earthscan), and in 2013 Barbier and Anil Markandya produced a follow-on, *New Blueprint for a Green Economy* (London: Routledge). Here Pearce and Barbier review the challenges facing economists in incorporating environmental factors into their work. They focus particularly on the question of whether monetary values can be placed on clean air, clean water, and other environmental goods – important if the market is to be used as the main mechanism for sustainable development. And they argue that only by changing economics so as to properly incorporate environmental factors is sustainable development likely to come about.

These authors present features of the natural environment as "ecological services" that nature provides to humans – an anthropocentric approach to sustainability much in line with the Brundtland Commission but contrary to the more "deep ecology" perspectives first pioneered by writers such as Leopold. As such, this approach has sometimes been called "weak sustainability," as contrasted to "strong sustainability" in which nature is seen as having value in its own right. This reading also introduces the basic problem of "market externalities," those impacts on social and environmental welfare that are not internalized within the costs of doing business.

A second approach toward reconciling nature and economics has been that of ecological economics, which generally lends more intrinsic value to nature, has less faith in markets, casts a more skeptical eye on the ability of economics to incorporate environmental and social factors, and seeks new tools and techniques to balance social, environmental, and economic objectives. Useful readings in this field include Robert Costanza *et al.*, *An Introduction to Ecological Economics, Second Edition* (Boca Raton, FL: CRC Press, 2014), *The Local Politics of Global Sustainability*, by Tom Prugh, Robert Costanza, and Herman E. Daly (Washington, D.C.: Island Press, 2000), and Robert B. Richardson's edited volume *Building a Green Economy: Perspectives from Ecological Economics* (East Lansing, MI: Michigan State University Press, 2013).

The source of most environmental problems lies in the failure of the economic system to take account of the valuable services which natural environments provide for us. . . . Environmental assets are akin to other capital assets like machinery and roads and the stock of knowledge and skills. Increasing one form of asset while running another one down is likely to be a short-sighted prescription to increasing human well-being. Yet in many ways this has been the history of past economic development.

Environmental or ecological services and functions include:

■ the provision of non-renewable natural resources such as coal and oil, bauxite and iron;
■ the provision of renewable natural resources such as timber, fish and water;
■ the provision of waste sinks to receive and assimilate solid, liquid and gaseous wastes from economic systems;
■ the provision of amenity;
■ the provision of biogeochemical cycles which help stabilize climates, provide nutrients to living things, and purify water and air; and
■ the provision of information in the form of genetic blueprints and behavioural observation.

Economic systems are generally good at providing only the first of these. While some commentators still express concern about the exhaustion of fossil fuel energy and mineral resources, the functioning of market systems can, to a considerable extent, be relied upon to signal when these resources are approaching exhaustion. These signals will include rising prices, which constitute an early warning that we should be moving out of those resources and into others, such as into renewable energy sources. Moreover, the rising prices themselves stimulate new discoveries, substitution of now relatively cheaper resources, and encourage technological change. We can argue about the precise pace at which this transition should happen, but the evidence is that what we might call old-fashioned resource exhaustion is not a major environmental problem. But economic systems seem to go wrong when it comes to providing the rest of the environmental functions listed above. Why?

. . . One of the central messages of *Blueprint I* was that many environmental resources have no market. They are not bought and sold. Accordingly, there are no price signals to alert us to their scarcity or to induce discovery, substitution and technological change. The same feature of these missing markets also results in an uneven playing field between environmental conservation and the immediate factors which threaten conservation. To the slash-and-burn farmer there is little benefit in pointing to the many ecological functions served by the forest if he receives no income, in cash or kind, from those services. The fact that the trees act as a store of carbon is of no immediate consequence to the farmer, even though it is a matter of great concern if we believe in the science of global warming, for then the carbon is better stored in the biomass than released to the atmosphere as carbon dioxide through burning the trees.

So it is with a great many environmental assets. The upstream polluter has little incentive to take account of the downstream river user, unless forced to do so. Manufacturers of chlorofluoro-carbons had no incentive to be concerned about the effects of these chemicals on the stratospheric ozone layer and hence on excess skin cancers. European farmers have little incentive to take into account the loss of wildlife arising from hedgerow removal or pesticide and fertilizer applications. All of these cases are examples of externalities, uncompensated third-party effects. It is difficult enough to internalize the externality (that is to make the polluter regard the externality as a cost to himself or herself) when the externality results in cash losses to others. It is even harder when the losses show up as non-monetary losses, as for example in the loss of amenity or impaired health.

The missing-market phenomenon therefore biases our so-called economic development decisions against the environment and in favour of economic activity which harms the environment. Ultimately some of the development in question will be worthwhile – this is a matter of comparing costs and benefits. But it would be fairly obvious that much development will be far more doubtful once we recognize that it is at the cost of ecological functions which themselves have economic value. That economic value is simply not realized because there are no markets through which its value can be expressed.

In this central theme we have three of the main features of environmental economics and the way in which environmental economists seek to find solutions to environmental problems.

First, we see that without markets there is a clear bias towards economic activity which, at best, downgrades the environment and, at worst, ignores it altogether. Hence there is a need to establish or create markets where none exist, or to modify markets where they exist but fail adequately to reflect environmental impacts. This conclusion is mitigated by the extent to which those creating degradation may themselves be unaware of the impact they are having and be willing to change their behaviour because they do have wider social and environmental concerns.

Second, we see that these missing or imperfect markets are a cause of environmental degradation. This is a major advance on most casual discussions of the causes of degradation, which tend to focus on the agents of destruction – the farmer, the logging company, the multinational company – rather than on the reasons for their behaviour. To devise a policy we first have to know the underlying causes. In this case, the solution lies in creating markets.

Third, while we could simply catalogue environmental impacts and show these alongside the economic benefits of the economic activity in question, it is far more forceful to put a money value on the environmental damage done. Then, costs and benefits can be compared directly, using the same monetary language that is used to justify economic development. Moreover, we have a substantial and lengthy experience of the former non-monetary approach in the form of environmental impact assessments (EIA). An EIA is a critically important activity, but it is difficult to resist the view that much of it is cosmetic, more designed to say that the environment has been taken into account, than to overrule environmentally damaging developments. Going the one stage further and putting money values on environmental damage puts the environment on the same footing as the economic arguments in favour of development, provided, of course, that the exercise does not remain confined to the paper on which the analysis is written. There have to be incentives based on the monetary value analysis.

Posing the issue in terms of development versus conservation is a little dangerous. It implies that we can have one but not the other and, indeed, those who are opposed to economic growth see it in precisely those terms. We do not agree with that view. The comparison between development and environment has to be one of the types of economic development that can be secured at the same time as minimizing the risks to the environment. . . . The reality is that the world has to accommodate at least 50 percent more people than it has now and they will need space, food and water, shelter and infrastructure. That cannot be secured without environmental loss. While this may appear to lend support to those who argue against not only economic growth but also population growth, the population growth in question is unavoidable. A further 50 percent on top of that is avoidable and everything that can be done that is humane and respectful of individual liberties should be done. But the first 50 percent is demographically unavoidable. The debate has to be about the way in which we foster future economic development to meet the legitimate aspirations of people; that is, it has to be about sustainable development.

WHY IS MONETIZING THE ENVIRONMENT SO CONTROVERSIAL?

Blueprint 1 did place emphasis on the importance of placing money values on environmental assets and services. It proved to be probably the most controversial issue in terms of the popular discussion from the book in the media and in public forums. This perhaps diverted attention of the fact, also made clear in Blueprint 1, that the case for market-based approaches to solving environmental problems can be justified quite independently of whether valuation takes place. Nonetheless, we argued above that valuation is important because it places the environment in the same political dialogue as economic activity generally. Put another way, environment should be a core focus of a Ministry of Finance or Industry as much as a Ministry of Environment. It is appropriate therefore to address the question of the validity of the money valuation approach . . . .

Environmental economists have developed sophisticated techniques for measuring the total economic value of environmental assets. The rationale for applying these techniques lies in the need to ensure that environmental impacts are taken into account in decision making on the same basis as the conventional costs and benefits of economic activity. While there is an active debate about the reliability and validity of these techniques in eliciting money values, most of the controversy does not center on the techniques, but on the very idea of monetizing the environment. Some of the criticism is embedded in so much misunderstanding and ignorance of what environmental economists actually do that it is difficult to separate what may be intellectually interesting arguments and criticisms from misconception. We can start by asking what economic valuation is all about.

Economists begin with the notion that economic value arises from meeting the preferences of individuals. Thus, the economic value of a commodity A is greater than a commodity B if A is preferred to B. Elevating human preferences to this high-level status is usually summarized in the notion of consumer sovereignty, that what people want, matters. In this naïve form it is of course similar in concept to a popular interpretation of democracy.

The motives that people have for their preferences are various. Whereas economists might once have said the motives were immaterial, they are now the subject of fairly extensive analysis because of arguments about the moral standing of various motives, and about the extent to which we can add up preferences based on different kinds of motives.

But the essential point is that motives may vary. I may prefer A to B because it gives me personally more pleasure. I may prefer B because I think it is better for my children, for society as a whole, for all future generations, for other sentient beings, or for the good of the Earth. The reality is that motives vary enormously and more than one motive may be present for the choice of B over A. This is important, because some of the criticism of economic valuation rests on the idea that preferences have only one motive: self-interest. Self-interest quickly becomes equated with greed (even though they are not the same) and the misconceptions multiply rapidly. In fact it is extremely difficult to explain human behavior on the basis of self-interest alone, although self-interest is very powerful. It would be hard to explain our attitudes to charities, or even our savings behavior, if we were solely motivated by self-interest.

Preferences are revealed in various ways. When we vote for political parties and individuals we are expressing a preference. Economic decisions that affect the environment are sometimes the context for political votes, but the millions and millions of decisions that are made daily cannot be. We need another medium through which to identify preferences. The market is one such medium. In the marketplace preferences show up in our decisions to buy and not to buy goods and services. Since these goods and services have prices, the decision to buy can be equated with a willingness to pay at least the price that is asked. A decision not to buy is, conversely, equivalent to a willingness to pay less than the price. This is the link between preferences – the underlying tendency to vote "yes" by buying the goods – and willingness to pay (WTP).

Political votes tend to be yes/no expressions. We either vote for or against something. In reality political systems also have other pressures which tend to reflect intensity of preference: lobbies and campaigns, for example. Marketplace votes also reflect intensity of preference in that we could be willing to pay an amount only just above the asking price or very much more than the asking price. Recall that everyone with a WTP less than the price will not buy it. Those with a WTP greater than the price will buy. But some will have a substantial excess of WTP over price – they would have been willing to pay a lot more than the price. Others will have only a small excess of WTP over price. This excess is like getting something for nothing, a benefit or gain of individual well-being that exceeds the price paid. It is known as the consumer's surplus. This connect is the central measure of the individual's net benefit from a good or service.

A moment's reflection will show that WTP depends on ability to buy. In general, the higher an individual's income the higher the WTP. The indicator that economists use for measuring preferences – WTP – is therefore biased towards whatever the prevailing distribution of income happens to be. If income is redistributed it is very likely that a different configuration of goods and services will be provided. We can now see another concern that critics of economic valuation have, but this time there is some reason to it. Whether or not environmental assets are conserved on the economic approach will depend on the distribution of income. If rich people favor the environmentally destructive activity, their WTP could be instrumental in bringing about that destruction, even though a show of hands (political votes) might be against it. Of course, it could work the other way round, and often does. Richer people may favor the environment more and may therefore be instrumental in saving it by expressing a strong WTP. But let us grant the potential criticism. Can it be overcome?

The income bias is of course present in all markets. In some cases, markets differentiate goods so that people on different incomes can secure the same good but at different levels of quality or packaging, for example. To some extent this can be done with environmental assets: some wildlife safaris, for example, are geared to the higher-income market, some to the lower end. But it still leaves an uneasy feeling that the natural environment should, somehow, be freely available to all, or be available to all at the same price. Of course, making it freely available immediately falls foul of the analysis as to why environments degrade. They degrade precisely because they often are free to all comers. They are, effectively, open access resources with no prices for their use. The absence of a price means the same thing as a zero price and when things cost nothing to the user they tend to be abused. If energy were freely available, a great deal more of it would be used, and there would be a consequent increase in pollution. Surprisingly, there still are people who believe that the environment should be free to all comers. They

speak of rights to the countryside, for example, but with little awareness that free access spells certain disaster for the very thing they allegedly value.

In suggesting that environmental problems occur because of missing markets, and in noting that markets are biased in terms of the emphasis they give to those with higher incomes, one is effectively arguing that the environment should be allocated to users just like conventional goods. The next strand of criticism of the economic argument is that the environment is not like other economic goods. It is somehow different. Here the arguments vary but they might include reference to the fact that we all need the environment to survive: we need clean air and water, energy and even space and amenity. But we do not in fact allocate energy on a free-for-all basis: it is sold through markets. Some countries still adhere to the nominal idea that water should not be rationed by the market, but it is a fast disappearing idea precisely because it is widely recognized that water is often a scarce commodity. Why, then, should air or amenity be different? There is a source of debate here, but it is far from clear that the environment is so very different from other goods and services. . . .

We have to look to the misworkings of the economy to find the real causes of environmental degradation, and we have to attach price tags to environmental assets if we are to change the way we treat the environment. Everything we do has an environmental impact, and economic decisions pervade all that we do. Ultimately, the prospects for natural environments some 100 years hence are not good. They cannot be, because there is little or nothing to stop at least 50 percent more people inhabiting an earth where the impact of human activity on the environment is pervasive and profound. We can do the best we can, and that means addressing all those issues that are the subject of policy decisions by households, corporations and governments. That means getting rid of perverse subsidies, establishing property rights to the environment, creating markets, taxing polluters (and that is all of us), and changing our accounting systems.

WHAT HAS CHANGED IN TEN YEARS?

First, the language of environmental economics is now a common language, even down to "internalizing externalities," "demonstration and capture," and "creating markets." We doubt if we would have believed ten years ago that such a rapid change could have come about. We have both spent weeks, if not months, of our lives just trying to explain the basic ideas to an often intrigued but skeptical population of politicians, journalists and scientists. We claim little credit for the change. We are simply relieved that it has happened.

Second, the opponents of environmental economics are fewer. Environmentalists and politicians generally have seen at least some of the virtue in the economic approach. Skeptics remain and they are vocal, and that is good. Their vehemence in opposing the economic message is occasionally irritating, and that is good too. Perhaps the most telling comment, however, is that the critics still find it hard to come up with a better or more convincing story about causation and solutions. They speak of moral stances as saving the world and we would like them to be right in their faith in such solutions. But we think that so much of the moralizing is either whistling in the wind or counterproductive. One can take a moral view and also look to pragmatism for effecting those views. Above all, whatever route is taken has to be democratic and we share the wider concern that some modern environmentalism has lost its roots in what people, as opposed to elites, want.

Third, selling the message of environmental economics was one thing. Putting those messages into practice has been another, and we accept that it has all been more difficult than we might have anticipated. The problem is that, even when people are convinced, changing institutions, cultures and policies is extremely difficult in a world where those things have been built up over long periods of time – centuries in the case of the UK, for example. Nonetheless, as we have outlined, many of the changes we advocated have taken place or are taking place: environmental taxes are up and running, though perhaps not in quite the form we might have advocated; tradable permit systems exist; polluters are paying; and the rich are paying the poor to change their behavior towards the environment. Perhaps more important than anything is the change in corporate culture. We do not believe that corporations have gone green simply because it is the right thing to do, but, equally, we do not believe we can explain all the corporate initiatives on the environment on the basis of compliance, anticipation of regulation or cost cutting. There is a new commitment to the environment and we believe environmental economics has played its part in securing that change.

Fourth, it is part of human nature to search for new paradigms. Science would not progress but for this curiosity of human beings. But there is a risk that as fast as we discover solutions we reject them because they are no longer new. Huge energies are devoted to rethinking the problem rather than solving it. We do think we know how to solve environmental problems as far as anyone can against the backdrop of vast population change yet to come. The real challenge is perhaps the one most people find the least exciting. We know what to do. We need to get on and do it.

34

"Preparing for a New Economic Era"

from *Environment and Urbanization* (1996)

David C. Korten

Editors' Introduction

David C. Korten has presented a very different view of economics in works such as *When Corporations Rule the World* (San Francisco: Berrett-Koehler Publishers, 1996). In that international bestseller, translated into 15 languages, he indicts global capitalism as the main culprit behind growing injustice and unsustainable development world-wide. Through the process known as globalization, he argues, powerful corporations and elites have colonized the world for their own benefit, in the process establishing ideologies of market deregulation and individualism, corrupting political systems, and subverting the basic life economy that is in the end much more important.

Korten knows whereof he speaks. For some 25 years he worked as a development consultant for institutions such as the United States Agency for International Development and the Ford Foundation. He witnessed the destructive impacts of globalization in places such as Ethiopia, Nicaragua, the Philippines, and Southeast Asia. He also served as an Air Force captain during the Vietnam War and a high-ranking Pentagon aide.

The perspective that Korten articulates here is often known by labels such as "neo-Marxism," in that following Karl Marx's example in the nineteenth century it analyzes power and class relationships in society. In particular, such writers analyze the role of powerful economic institutions in structuring social and political systems to meet their own ends (an analytic perspective sometimes known as political economy). This approach is less hopeful about reforming capitalism to meet sustainable development goals than that of other authors in this section. It urges us to understand the nature of the system that has produced *un*sustainable development as a key step towards producing alternative institutions that can counter the power of global capital by asserting grassroots democracy and locally based economic development.

Korten's other books include *Agenda for a New Economy: From Phantom Wealth to Real Wealth* (San Francisco: Berrett-Koehler Publishers, 2010), which poses a hopeful alternative to global capitalism, and *The Great Turning: From Empire to Earth Community* (San Francisco: Berrett-Koehler Publishers, 2006), which traces a long contest between forces of empire (including capitalism) and more life-centered, sustainable, democratic alternatives. Another writer emphasizing political economy perspectives on urban issues is David Harvey, whose books include *Social Justice and the City* (Baltimore, MD: Johns Hopkins University Press, 1973), *Spaces of Hope* (Berkeley: University of California Press, 2000), and *The Enigma of Capital and the Crises of Capitalism* (Oxford: Oxford University Press, 2010). Jerry Mander's book *The Capitalism Papers: Fatal Flaws of an Obsolete System* (Berkeley: Counterpoint Press, 2012) and the essays in Martin O'Connor's classic volume *Is Capitalism Sustainable? Political Economy and the Politics of Ecology* (New York: The Guilford Press, 1994) likewise are generally skeptical about the fundamental ability of capitalism to change in directions that are not exploitative. Those interested in this perspective may also wish to explore authors associated with The International Forum on Globalization (www.ifg.org), who address economic globalization from a critical viewpoint.

UN summits have focused global attention on a wide range of critical human problems – including the environment, gender, human rights, food security, population, unemployment, poverty and social dis-integration. The fact that most of these problems continue to worsen at an alarming rate provides stark evidence that, for all the well-intentioned com-mitments the previous conferences have generated, they have failed to provide adequate solutions. The reason for the failure is both simple and profound. They have failed to identify and address the under-lying causes of the problems for which they sought solutions.

The search for these causes has been a con-tinuing concern for participants in the NGO [non-governmental organization] forums of previous UN conferences. Whilst the views of civil society are as diverse as its participants, a line of analysis has arisen from the unfolding dialogue that has a broad and growing base of support among people of widely varied backgrounds. This analysis represents a significant departure from the accepted world view that currently informs most policy discussions. For example, it suggests that:

■ The major barriers to providing healthy and sustainable living spaces for a growing world population are institutional rather than financial. Therefore, corrective action must be guided by a theory of why our present institutions are failing. Without such a theory, attention tends to focus on funding programs that treat the symptoms of failure whilst neglecting their un-derlying causes. The resulting actions are almost inevitably overly expensive, fragmented, incom-plete, often contradictory, and ineffectual.

■ The persistence of the problems addressed by the previous UN conferences is a warning signal telling us that twentieth-century approaches to solving our problems are not appropriate to our current historical circumstances. Adjustments at the margin of our present governing institutions are not adequate. To create a world of just and sustainable societies, more fundamental change is required.

■ Most human needs can be met only through appropriate local action. One reason appropriate action is not forthcoming is because of policy actions that concentrate control over productive resources in institutions that are unmindful of local needs and lack public accountability for the consequences of their actions. Enabling action is needed at global and national levels to restore to local people and communities control of the pro-ductive assets on which their livelihoods depend.

SUCCESS IN THE MONEY WORLD, CRISIS IN THE LIVING WORLD

During the latter part of the second millennium, and in particular the twentieth century, the lives of most of the world's people have become increasingly divided between two parallel and intertwined realities. One reality – the world of money – is governed by the rules set by governments and central banks and by the dynamics of financial markets. The other – the world of life – is governed by the laws of nature.

In the world of money, the health of society and of its institutions is measured by financial and economic indicators – by growth in such things as economic output, stock prices, trade, investment, and tax receipts. In the world of money, continuous, sustained growth seems to be the primary imperative of healthy function. Because they are structured to seek ever-increasing productivity and profits, modern economies either grow in terms of the monetary value of their output or they collapse. The growth imperative of the money world finds expression in the notion of development as an unending process of economic expansion – which has been the organizing principle of public policy for most of the last half of this century.

The living world is governed by quite different imperatives. Here, healthy function manifests itself in balance, diversity, sufficiency, synergy, and regenerative vitality. Growth is an integral part of the living world but only as a clearly defined segment of the life cycle of individual organisms. The sustained physical growth of any individual organism or the unlimited numerical expansion of any species is an indicator of system dysfunction and poses a threat to system integrity. Thus, growth in the living world tends to be self-limiting – as with a cancer that condemns its host or a species whose numbers upset the ecological balance and ultimately destroy their food supply.

Though, as living beings, we are creatures of the living world, we have yielded the power of decision in human affairs to the institutions of the money world and for these institutions the imperatives of the world of money take precedence over those of the living world. Money world institutions have been enormously successful in increasing total world economic output between five and seven times over the past fifty years. They have also brought unprecedented material wealth to approximately 20 percent of the world's people and vast riches to the most fortunate 1 percent. The world's power holders, who with few exceptions belong to the most fortunate 1 percent, understandably see this as a considerable accomplishment. Having experienced the possibilities of the underlying development model, they feel affirmed in their belief that the systems of governance that allocate the world's resources are fundamentally sound. Shielded by their wealth from sharing in the living world consequences of money world decisions, those consequences – for them – lack a compelling sense of reality.

From the perspective of the living world, however, the consequences of the economic development/growth agenda have been disastrous. Here, we see that each addition to economic output results in a comparable increase in the stress that humans place on the earth's ecosystem, deepens the poverty of those whose resources have been expropriated and labour exploited to fuel the engines of growth for the benefit of the few, and accelerates the destruction of non-human species. The terrible costs fall on those who are denied a political voice – the poor, the young and generations yet unborn.

In large measure, the crisis of global-scale social and environmental disintegration now under way can be explained in terms of a confrontation between the conflicting imperatives of the money world – which holds the power of decision – and the living world of people and nature – which bears the tragic consequences of those decisions.

A DANGEROUSLY FLAWED IDEOLOGY

Government officials, policy experts, corporate representatives, and the official declarations of past UN conferences continuously reaffirm their faith in the basic premise of neoliberal economic ideology that economic growth, market deregulation, privatization, and economic globalization are the irreducible

foundations of peace, equality, human rights, democracy, a healthy environment and social fabric, and universal prosperity. From the perspective of the money world, the logic of this assertion is impeccable. From the perspective of the living world the logic suffers from three serious flaws:

Continued economic growth on a finite planet with an already overtaxed ecosystem accelerates environmental breakdown, intensifies the competition for resources between rich and poor, and deprives future generations of the necessary means to meet their basic needs. This is confirmed by a growing body of evidence that many of the world's ocean fisheries, freshwater resources, and farm and forest lands are being exploited at rates substantially greater than their ability to regenerate.

Expansion of the market economy into even more of the activities once performed by households and communities is monetizing human relationships, weakening the social fabric and destroying livelihoods faster than new jobs offering more than poverty level compensation are being created.

The institutions of a globalized free market economy that control privatized assets respond only to the imperatives of the money world. They are vitally blind to the imperatives of the living world. Economic globalization is shifting control over resources, markets, and technology from people, communities, and governments to transnational financial markets and corporations – placing these institutions beyond the reach of public accountability, making responsible local action to meet local needs increasingly difficult, and creating dangerous financial instability. . . .

PREPARING FOR A NEW ERA

In dealing with the conflicting imperatives of money and life, we face an inescapable reality. Whilst the money world is our creation, we are creatures of the living world. Money is a useful means of facilitating certain kinds of transactions among people – nothing more. It is the living world that gave us birth and sustains our lives. Money world institutions are human creations and it is within our power to reconstruct them in ways that align their imperatives with the imperatives of living. Money must be the servant of life, not its master.

Acting in our citizen roles, we must engage in building citizen agendas for a life-centered reconstruction of our cities and societies. Ultimately, this agenda-building process must address three needs:

■ The need to organize human habitats in ways that support the right of all people to a place in society and on earth with access to the resources required to create a secure and fulfilling life for themselves, at peace with their neighbors, and in balance with the earth's natural systems.
■ The need to build – as complementary to the money economy – strong gender-balanced unmonetized household and community economies able to replenish the social capital that is essential to the healthy and efficient function of both economies and societies.
■ The need to create a global system of localized economies that root economic power and environmental responsibility in people and communities of place and encourage a substantial measure of local environmental self-reliance.

Some will note that this agenda is out of step with the prevailing conventional wisdom of economic policy – and they will be correct. That prevailing wisdom is taking us in the wrong direction. Our future depends on making a significant and conscious course change.

An essential first step towards giving precedence to the imperatives of the living world is to recognize that our challenge for the third millennium is not to accelerate and sustain economic growth – which is a money world imperative – but rather to create just, sustainable, and democratic societies that meet the basic material needs of all people while bringing the human species into balance with itself and the planet – a living world imperative. These three essential characteristics of healthy societies – just, sustainable, and democratic – address basic underlying causes of the larger crisis. Together they provide an essential framework for corrective action. Critical tasks include:

■ *Localizing economies and reclaiming citizen sovereignty.* The ideal of the market economy described by Adam Smith – markets served by small buyers and sellers and locally owned capital – bears little in common with the present system in which corporations plan and control internal economies larger than those of most nations, obtain their capital from globalized financial markets detached from the interests of any place or community, and dominate the political process to serve their special private ends. To reclaim control of our economic and political lives we must act to get corporations and big money out of politics, restore market competition by reducing the size of business units, and localize markets and capital ownership.

■ *Structuring human settlements to sustain balanced and fulfilling lives.* Equity and sustainability are impossible in societies that place greater value on the quantity of their consumption than on the quality of their living. Policies geared to increasing aggregate consumption have, in many instances, reduced the quality of our lives whilst making it ever more difficult to live in balance with one another and the earth. Our goal must be to create human settlements with physical and social structures suited to the needs of full and balanced living.

RECREATING SOCIAL CAPITAL

Healthy families are the foundation of healthy communities, which, in turn, are the foundation of healthy societies. Whilst they may take many forms, one mark of healthy families, communities, and societies is their capacity to nurture a dense fabric of caring relationships based on reciprocity, mutual trust, and cooperation. This must be an important consideration to how we organize the economies of human settlements in the third millennium.

Traditionally, most of the productive and reproductive activities that provided people with their basic needs for food, shelter, clothing, child care, health care, education, physical security, and entertainment were carried out within the framework of the caring economy of family and community, largely outside of the market. A substantial proportion of production/consumption activities took place within a single household or between people who related directly to one another. People met most of their needs through these non-market productive activities. In many societies, families and communities were the primary source of an individual's identity and livelihood and their importance provided a substantial incentive for people to invest in maintaining the social bonds of trust and obligation – the social capital – that is a necessary underpinning of any healthy society.

It is a fundamental, though often neglected, fact that social bonding is as essential to the healthy function of a modern society as it was to more traditional or tribal societies. Indeed, the market itself depends on the bonds of a well-developed social capital to maintain the ethical structure, social stability, and personal security essential to its efficient function.

The cultures of healthy societies maintain co-operation and competition in a state of dynamic but balanced tension. Social bonding without competition leads to stagnation and a lack of innovation. Competition without social bonding leads to an anarchy of violence. A dominance of money world values has contributed to creating an increasingly severe imbalance favoring competition in our modern world.

The more of the economic and social life of family and community that societies yield to the money world, the greater the power of the institutions that control access to money and to the things that money will buy. Yielding such power serves well the corporate interest because corporations are creatures of money. It serves poorly the human interest because people are creatures of community, nature, and spirit.

Forced to reexamine who we are by the limits of the planet's ability to sustain the demands of our greed, we find ourselves confronted with a remarkable truth. Whereas the pursuit of money world values has created material scarcity, the embrace of living world values brings a sense of social, spiritual, and even material abundance. People who experience an abundance of love in their lives rarely seek solace in compulsive, exclusionary personal acquisition. In contrast, for the emotionally deprived no extreme of materialistic indulgence can ever be "enough" and the material world becomes insufficient to sustain the demands placed upon it. A world starved of love becomes one of material scarcity. In contrast, a world of love is also one of material abundance.

When we are spiritually whole and experience the caring support of community, thrift is a natural part of a full and disciplined life. That which is sufficient to one's needs brings a fulfilling sense of nature's abundance.

Over-consumer societies are built on financial foundations. Just and sustainable societies are built on spiritual foundations.

ORGANIZING SETTLEMENT SPACES FOR SUSTAINABLE GOOD LIVING

Some of the differences between a human settlement organized for consumption and one organized for good living are suggested by the work of Alan Durning. [Table 1], which is based on Durning's analysis, makes a rough division of the world into three socio-ecological classes: over-consumers, sustainers, and the excluded. The over-consumers – the people whose lifestyles are unsustainable in a full world – are the some 20 percent of the world's people who consume roughly 80 to 85 percent of the world's available natural wealth – those whose lives are organized around automobiles, airplanes, meat-based diets, and the use of wastefully packaged disposable products. The excluded, a corresponding 20 percent of the world's people, live in absolute deprivation.

It is significant that roughly 60 percent of the world's people, although they face many hardships, are more or less meeting their basic needs in reasonably sustainable ways. We may refer to them as the world's sustainer class. Unfortunately, from a living world perspective, the goal of economic policy almost everywhere is to increase the consumption of the over-consumers and turn sustainers into over-consumers. This commitment is backed by World Bank and IMF lending and by billions of dollars in corporate advertising expenditure encouraging people to buy things they might not otherwise realize they want. Public policy backs this commitment with substantial investment in subsidies for over-consumer lifestyles – such as for airports, roads, and parking facilities – to the neglect of public transit, sidewalks, and bicycle paths. As a consequence, the lives of those who live sustainably become more difficult.

This suggests the need for a radical rethinking of public policy and market relationships. Rather than striving to increase the size and levels of consumption of the over-consumer class, the goal should be to improve the quality of living that a sustainer lifestyle potentially affords and to move both over-consumers and the excluded into the sustainer

Over-consumers 1.1 billion > US$7,500 per capita (cars, meat, disposables)	Sustainers 3.3 billion US$700–7,500 per capita (living lightly)	Excluded 1.1 billion < US$700 per capita (absolute deprivation)
Travel by car and air	Travel by bicycle and public surface transport	Travel by foot, maybe by donkey
Eat high-fat, high-calorie meat-based diets	Eat healthy diets of grains, vegetables and some meat	Eat nutritionally inadequate diets
Drink bottled water and soft drinks	Drink clean water plus some tea and coffee	Drink contaminated water
Use throw-away products and discard substantial wastes	Use unpackaged goods and durables and recycle wastes	Use local biomass and produce negligible wastes
Live in spacious, climate-controlled, single-family residences	Live in modest, naturally ventilated residences with extended/multiple families	Live in rudimentary shelters or in the open. Usually lack secure tenure
Maintain image-conscious wardrobes	Wear functional clothes	Wear second-hand clothing or scraps

Table 1 Earth's three socio-ecological classes.

Source: Based on Alan Durning, "Asking How Much is Enough," in Lester Brown *et al.*, *State of the World 1991* (New York: Norton, 1991), pp. 153–169.

class. Cities have an especially important role to play in this process because of the potential offered by high density human settlements to provide a high quality of living at a relatively low environmental cost.

The widespread belief that moving from an over-consumer to a sustainer lifestyle involves sacrifice and hardship is in part a consequence of misplaced public priorities embraced over the past fifty years in the name of development – such as when roads used by cyclists are converted to throughways reserved for cars.

While responsible individual choices play an important role in eliminating over-consumption, they are often limited by public choices beyond the individual's control. This point is readily demonstrated with regard to four major systems – urban space and transportation, food and agriculture, materials, and energy. Each system raises basic issues for how human settlements are organized and managed.

The goal of public policy should rather be to move both over-consumers and the excluded into the sustainer class. This is the only way we can create a world in which all the world's people have the opportunity to lead a decent life. Some may argue that this is a call for sacrifice and hardship on the part of the over-consumer class. That need not be the case. To the contrary, "sustainerism" can yield a higher quality of living than obsessive, shop-till-you-drop "over-consumerism" – if society is organized to support a high quality sustainer lifestyle.

Unfortunately, a great deal of what we do in the name of development eliminates the support systems that afford sustainers a high quality of living and replaces them with systems that make life in the sustainer class ever more difficult.

■ For example, anyone who uses a car as their primary means of transport is not living sustainably. Yet, we structure our cities and towns for the automobile, giving over ever more space and resources to what is arguably our most environmentally and socially destructive technology. In so doing we reduce the quality of urban living and make it ever more difficult to avoid devoting a major portion of our working lives to earning the money necessary to support one of these expensive and intrusive mechanical creatures. It need not be that way, as we know how to design settlement spaces for people that offer a far higher quality of life than do automobile-dominated settlements.

- We structure agriculture around the unsustainable intensive use of chemicals and long-haul transportation. This creates enormous environmental burdens and leaves us with few options other than to eat foods of dubious nutritional value and often contaminated with dangerous levels of poisons. We could be structuring agriculture around local bio-intensive production – producing more healthy food, creating more employment, and greatly reducing environmental burdens.

- We structure our production–consumption systems around the use of excessively packaged disposable products. A visitor from outer space might conclude, with considerable justification, that our species measures well-being by the amount of garbage we generate.

If we look carefully at why things are as they are, we may find that we have organized our habitats and production systems primarily to serve the growth imperative of the corporate world. Almost invariably this increases our quantity of consumption – and our burden on the environment – while all too often actually reducing the quality of our own lives – particularly if we take into account how much harder we have to work at often distasteful jobs to pay for it. If we were serious about improving both the sustainability and the quality of our living we would be seeking our world to virtually eliminate the automobile, the release of toxic and non-degradable substances, and the generation of garbage. We would also ban advertising that serves no other purpose than to entice us to buy and consume what we do not need.

It is entirely possible to progress towards each of these goals in ways that would improve the quality of our living while reducing the quantity of our consumption. The most important reason we do not is because doing so would conflict with money world imperatives for growth.

We face a basic choice in our vision of the cities of the future. Is our choice for cities, towns, and villages dedicated to serving as centers of economic growth in a competitive global economy? Places designed to cater to the money-making imperatives of the institutions of the money world? Places constructed of steel and concrete that shield us from contact with the wild forces of nature? Places that stimulate our competitive instincts and feed our desires for physical accumulation and consumption?

Or is our vision one of cities, towns, and villages that function as places for living? Places that serve our need for creative human interaction yet maintain our sense of connection to nature? Places that awaken and nurture our capacity for cooperation, caring, community building, and spiritual renewal?

These are real choices with important implications for how we choose to structure our human settlements and their relationship to rural spaces. One choice centers on the housing needs of global corporations and the rapid and efficient flow of automobiles. The other centers on creating spaces for people to live, create satisfying and meaningful livelihoods, and relate to nature and their neighbors. When the decisions are left to institutions driven by the imperatives of the money world, they invariably choose the former. When the decisions reside with ordinary people who retain a sense of connection to the living world, their inclination is commonly towards the latter. The question of power and with whom it resides is central. Civic engagement is about reclaiming this power from the institutions of the money world so that we can create spaces for living.

35

"Natural Capitalism"

from *Mother Jones*, March/April 1997

Paul Hawken

Editors' Introduction

Co-founder of the successful Smith & Hawken garden supply business, Paul Hawken has been a leading philosopher of ecological capitalism for several decades. His work includes *The Ecology of Commerce Revised Edition: A Declaration of Sustainability* (New York: HarperBusiness, 2010), *Natural Capitalism: Creating the Next Industrial Revolution* (London: Earthscan, 1999; with Amory and L. Hunter Lovins), and *Blessed Unrest: How the Largest Movement in the World Came into Being, and Why No One Saw It Coming* (New York: Viking, 2007). Hawken has also been chair of The Natural Step, an influential worldwide organization begun in Norway that works with businesses to develop more sustainable practices.

Adopting a more optimistic perspective than Korten, Hawken argues in this selection that powerful economic mechanisms can eventually be harnessed for restorative rather than exploitative ends. To bring this about, he believes that existing incentives and subsidies must be changed to eliminate those supporting unsustainable ways of doing business, and to encourage alternative economic activity. Doing this, he believes, can bring about a "natural capitalism" that is capable of transforming society. Other writings along this line include Ray Anderson's *Business Lessons From a Radical Industrialist* (New York: St. Martin's Press, 2010), Daniel C. Esty and Andrew S. Winston's *Green to Gold: How Smart Companies Use Environmental Strategy to Innovate, Create Value, and Build Competitive Advantage* (Hoboken, NJ: Wiley, 2006), and Lester Brown's *Eco-Economy: Building an Economy for the Earth* (New York: Norton, 2001).

Somewhere along the way to free-market capitalism, the United States became the most wasteful society on the planet. Most of us know it. There is the waste we can see: traffic jams, irreparable VCRs, Styrofoam coffee cups, landfills; the waste we can't see: Superfund sites, greenhouse gases, radioactive waste, vagrant chemicals; and the social waste we don't want to think about: homelessness, crime, drug addiction, our forgotten infirm and elderly.

Nationally and globally, we perceive social and environmental decay as distinct and unconnected. In fact, a humbling design flaw deeply embedded in industrial logic links the two problems. Toto, pull back the curtain: The efficient dynamo of industrialism isn't there. Even by its own standards, industrialism is extraordinarily inefficient.

Modern industrialism came into being in a world very different from the one we live in today: fewer people, less material well-being, plentiful natural resources. As a result of the successes of industry and capitalism, these conditions have now reversed. Today, more people are chasing fewer natural resources.

But industry still operates by the same rules, using more resources to make fewer people more productive. The consequence: massive waste – of both resources and people.

Decades from now, we may look back at the end of the twentieth century and ponder why business and society ignored these trends for so long – how one species thought it could flourish while nature ebbed. Historians will show, perhaps, how politics, the media, economics, and commerce created an industrial regime that wasted our social and natural environment and called it growth. As author Bill McKibben put it, "The laws of Congress and the laws of physics have grown increasingly divergent, and the laws of physics are not likely to yield."

The laws we're ignoring determine how life sustains itself. Commerce requires living systems for its welfare – it is emblematic of the times that this even needs to be said. Because of our industrial prowess, we emphasize what people can do but tend to ignore what nature does. Commercial institutions, proud of their achievements, do not see that healthy living systems – clean air and water, healthy soil, stable climates – are integral to a functioning economy. As our living systems deteriorate, traditional forecasting and business economics become the equivalent of house rules on a sinking cruise ship.

One is tempted to say that there is nothing wrong with capitalism except that it has never been tried. Our current industrial system is based on accounting principles that would bankrupt any company.

Conventional economic theories will not guide our future for a simple reason: They have never placed "natural capital" on the balance sheet. When it is included, not as a free amenity or as a putative infinite supply, but as an integral and valuable part of the production process, everything changes. Prices, costs, and what is and isn't economically sound change dramatically.

Industries destroy natural capital because they have historically benefited from doing so. As businesses successfully created more goods and jobs, consumer demand soared, compounding the destruction of natural capital. All that is about to change.

NATURAL CAPITAL

Everyone is familiar with the traditional definition of capital as accumulated wealth in the form of investments, factories, and equipment. "Natural capital," on the other hand, comprises the resources we use, both nonrenewable (oil, coal, metal ore) and renewable (forests, fisheries, grasslands). Although we usually think of renewable resources in terms of desired materials, such as wood, their most important value lies in the services they provide. These services are related to, but distinct from, the resources themselves. They are not pulpwood but forest cover, not food but topsoil. Living systems feed us, protect us, heal us, clean the nest, let us breathe. They are the "income" derived from a healthy environment: clean air and water, climate stabilization, rainfall, ocean productivity, fertile soil, watersheds, and the less-appreciated functions of the environment, such as processing waste – both natural and industrial. *Nature's Services*, a book . . . edited by Stanford University biologist Gretchen C. Daily, identifies trillions of dollars of critical eco-system services received annually by commerce.

For anyone who doubts the innate value of ecosystem services, the $200 million Biosphere II experiment stands as a reality check. In 1991, eight people entered a sealed, glass-enclosed, 3-acre living system, where they expected to remain alive and healthy for two years. Instead, air quality plummeted, carbon dioxide levels rose, and oxygen had to be pumped in from the outside to keep the inhabitants healthy. Nitrous oxide levels inhibited brain function. Cockroaches flourished while insect pollinators died, vines choked out crops and trees, and nutrients polluted the water so much that the residents had to filter it by hand before they could drink it. Of the original 25 small animal species in Biosphere II, 19 became extinct.

At the end of 17 months, the humans showed signs of oxygen starvation from living at the equivalent of an altitude of 17,500 ft. Of course, design flaws are inherent in any prototype, but the fact remains that $200 million could not maintain a functioning ecosystem for eight people for 17 months. We add eight people to the planet every three seconds.

The lesson of Biosphere II is that there are no man-made substitutes for essential natural services. We have not come up with an economical way to manufacture watersheds, gene pools, topsoil, wetlands, river systems, pollinators, or fisheries. Technological fixes can't solve problems with soil fertility or guarantee clean air, biological diversity, pure water, and climatic stability; nor can they increase the capacity of the environment to absorb 25 billion tons of waste created annually in America alone.

NATURAL CAPITAL AS A LIMITING FACTOR

Until the 1970s, the concept of natural capital was largely irrelevant to business planning, and it still is in most companies. Throughout the industrial era, economists considered manufactured capital – money, factories, etc. – the principal factor in industrial production, and perceived natural capital as a marginal contributor. The exclusion of natural capital from balance sheets was an understandable omission. There was so much of it, it didn't seem worth counting. Not any longer.

Historically, economic development has faced a number of limiting factors, including the availability of labor, energy resources, machinery, and financial capital. The absence or depletion of a limiting factor can prevent a system from growing. If marooned in a snowstorm, you need water, food, and warmth to survive. Having more of one factor cannot compensate for the absence of the other. Drinking more water will not make up for lack of clothing if you are freezing.

In the past, by increasing the limiting factor, industrial societies continued to develop economically. It wasn't always pretty: Slavery "satisfied" labor shortages, as did immigration and high birthrates. Mining companies exploited coal, oil, and gas to meet increased energy demands. The need for labor-saving devices provoked the invention of steam engines, spinning jennies, cotton gins, and telegraphs. Financial capital became universally accessible through central banks, credit, stock exchanges, and currency exchange mechanisms.

Because economies grow and change, new limiting factors occasionally emerge. When they do, massive restructuring occurs. Nothing works as before. Behavior that used to be economically sound becomes unsound, even destructive.

Economist Herman E. Daly cautions that we are facing a historic juncture in which, for the first time, the limits to increased prosperity are not the lack of man-made capital but the lack of natural capital. The limits to increased fish harvests are not boats, but productive fisheries; the limits to irrigation are not pumps or electricity, but viable aquifers; the limits to pulp and lumber production are not sawmills, but plentiful forests.

Like all previous limiting factors, the emergence of natural capital as an economic force will pose a problem for reactionary institutions. For those willing to embrace the challenges of a new era, however, it presents an enormous opportunity.

THE HIGH PRICE OF BAD INFORMATION

The value of natural capital is masked by a financial system that gives us improper information – a classic case of "garbage in, garbage out." Money and prices and markets don't give us exact information about how much our suburbs, freeways, and spandex cost. Instead, *everything else* is giving us accurate information: our beleaguered air and watersheds, our overworked soils, our decimated inner cities. All of these provide information our prices should be giving us but do not.

Let's begin with a startling possibility: The US economy may not be growing at all, and may have ceased growing nearly 25 years ago. Obviously, we are not talking about the gross domestic product (GDP), measured in dollars, which has grown at 2.5 percent per year since 1973. Despite this growth, there is little evidence of improved lives, better infrastructure, higher real wages, more leisure and family time, and greater economic security.

The logic here is simple, although unorthodox. We don't know if our economy is growing because the indices we rely upon, such as the GDP, don't measure growth. The GDP measures money transactions on the assumption that when a dollar changes hands, economic growth occurs. But there is a world of difference between financial exchanges and growth. Compare an addition to your home to a two-month stay in the hospital for injuries you suffered during a mugging. Say both cost the same. Which is growth? The GDP makes no distinction. Or suppose the president announces he will authorize $10 billion for new prisons to help combat crime. Is the $10 billion growth? Or what if a train overturns next to the Sacramento River and spills 10,000 gallons of atrazine, poisoning all the fish for 30 miles downstream? Money pours into cleanups, hatchery releases, announcements warning people about tainted fish, and lawsuits against the railroad and the chemical company. Growth? Or loss?

Currently, economists count most industrial, environmental, and social waste as GDP, right along with bananas, cars, and Barbie dolls. Growth includes *all* expenditures, regardless of whether society benefits or loses. This includes the cost of emergency room services, prisons, toxic cleanups, homeless shelters, lawsuits, cancer treatments, divorces, and every piece of litter along the side of every highway.

Instead of counting decay as economic growth, we need to subtract decline from revenue to see if we are getting ahead or falling behind. Unfortunately, where economic growth is concerned, the government uses a calculator with no minus sign.

WASTING RESOURCES MEANS WASTING PEOPLE

Industry has always sought to increase the productivity of workers, not resources. And for good reason. Most resource prices have fallen for 200 years – due in no small part to the extraordinary increases in our ability to extract, harvest, ship, mine, and exploit resources. If the competitive advantage goes to the low-cost provider, and resources are cheap, then business will naturally use more and more resources in order to maximize worker productivity.

Such a strategy was eminently sensible when the population was smaller and resources were plentiful. But with respect to meeting the needs of the future, contemporary business economics is pre-Copernican. We cannot heal the country's social wounds or "save" the environment as long as we cling to the outdated industrial assumptions that the summum bonum of commercial enterprise is to use more stuff and fewer people. Our thinking is backward: We shouldn't use more of what we have less of (natural capital) to use less of what we have more of (people). While the need to maintain high labor productivity is critical to income and economic well-being, labor productivity that corrodes society is like burning the furniture to heat the house.

Our pursuit of increased labor productivity at all costs not only depletes the environment, it also depletes labor. Just as overproduction can exhaust topsoil, overproductivity can exhaust a workforce.

The underlying assumption that greater productivity would lead to greater leisure and well-being, while true for many decades, has become a bad joke. In the United States, those who are employed, and presumably becoming more productive, find they are working 100 to 200 hours more per year than 20 years ago. Yet real wages haven't increased for more than 20 years.

In 1994, I asked a roomful of senior executives from Fortune 500 companies the following questions: Do you want to work harder in five years than you do today? Do you know anyone in your office who is a slacker? Do you know any parents in your company who are spending too much time with their kids? The only response was a few embarrassed laughs. Then it was quiet – perhaps numb is a better word.

Meanwhile, people whose jobs have been downsized, re-engineered, or restructured out of existence are being told – as are millions of youths around the world – that we have created an economic system so ingenious that it doesn't need them, except perhaps to do menial service jobs.

In parts of the industrialized world, unemployment and underemployment have risen faster than employment for more than 25 years. Nearly one-third of the world's workers sense that they have no value in the present economic scheme.

Clearly, when 1 billion willing workers can't find a decent job or any employment at all, we need to make fundamental changes. We can't – whether through monetary means, government programs, or charity – create a sense of value and dignity in people's lives when we're simultaneously developing a society that doesn't need them. If people don't feel valued, they will act out society's verdict in sometimes shocking ways. William Strickland, a pioneer in working with inner-city children, once said that "you can't teach algebra to someone who doesn't want to be here." He meant that urban kids don't want to be here at all, alive, anywhere on earth. They try to tell us, but we don't listen. So they engage in increasingly risky behavior – unprotected sex, drugs, violence – until we notice. By that time, their conduct has usually reached criminal proportions – and then we blame the victims, build more jails, and lump the costs into the GDP.

The theologian Matthew Fox has pointed out that we are the only species without full employment.

Yet we doggedly pursue technologies that will make that ever more so. Today we fire people, perfectly capable people, to wring out one more wave of profits. Some of the restructuring is necessary and overdue. But, as physicists Amory Lovins and Ernst von Weizsšcker have repeatedly advised, what we *should* do is fire the unproductive kilowatts, barrels of oil, tons of material, and pulp from old-growth forests – and hire more people to do so.

In fact, reducing resource use creates jobs and lessens the impact we have on the environment. We can grow, use fewer resources, lower taxes, increase per capita spending on the needy, end federal deficits, reduce the size of government, and begin to restore damaged environments, both natural and social.

At this point, you may well be skeptical. The last summary is too hopeful and promises too much. If economic alternatives are this attractive, why aren't we doing them now? A good question. I will try to answer it. But, lest you think these proposals are Pollyannaish, know that my optimism arises from the magnitude of the problem, not from the ease of the solutions. Waste is too expensive; it's cheaper to do the right thing.

RESOURCE PRODUCTIVITY

Economists argue that rational markets make this the most efficient of all possible economies. But that theory works only as long as you use financial efficiency as the sole metric and ignore physics, biology, and common sense. The physics of energy and mass conservation, along with the laws of entropy, are the arbiters of efficiency, not *Forbes* or the Dow Jones or the Federal Reserve. The economic issue is: How much work (value) does society get from its materials and energy? This is a very different question than asking how much return it can get out of its money.

If we already deployed materials or energy efficiently, it would support the contention that a radical increase in resource productivity is unrealistic. But the molecular trail leads to the opposite conclusion. For example, cars are barely 1 percent efficient in the sense that, for every 100 gallons of gasoline, only one gallon actually moves the passengers. Likewise, only 8 to 10 percent of the energy used in heating the filament of an incandescent lightbulb actually becomes visible light. (Some describe it as a space heater disguised as a lightbulb.) Modern carpeting remains on the floor for up to 12 years, after which it remains in landfills for as long as 20,000 years or more – less than 0.06 percent efficiency.

According to Robert Ayres, a leader in studying industrial metabolism, about 94 percent of the materials extracted for use in manufacturing durable products become waste before the product is even manufactured. More waste is generated in production, and most of that is lost unless the product is reused or recycled. Overall, America's material and energy efficiency is no more than 1 or 2 percent. In other words, American industry uses as much as 100 times more material and energy than theoretically required to deliver consumer services.

A watershed moment in the study of resource productivity occurred in 1976, when Amory Lovins published his now-famous essay "Energy Strategy: The Road Not Taken?" Lovins' argument was simple: Instead of pursuing a "hard path" demanding a constantly increasing energy supply, he proposed that the real issue was how best to provide the energy's "end use" at the least cost. In other words, consumers are not interested in gigajoules, watts, or Btu, he argued. They want well-illuminated workspaces, hot showers, comfortable homes, effective transport. People want the *service* that energy provides. Lovins pointed out that an intelligent energy system would furnish the service at the lowest cost. As an example, he compared the cost of insulation with that of nuclear power. The policy of building nuclear power plants represented the "supply at any cost" doctrine that still lingers today. He said it made no sense to use expensive power plants to heat homes, and then let that heat escape because the homes lack insulation. Lovins contended that we could make more money by saving energy than by wasting it, and that we'd find more energy in the attics of American homes than in all the oil buried in Alaska. His predictions proved correct, although his proposals remained largely unheeded by the government. Today, the nuclear power industry has become moribund, not because of anti-nuclear protests but because it is uncompetitive.

In 1976, energy experts used to argue about whether the United States could achieve energy savings of 30 percent. Twenty-one years later, having already obtained savings of more than 30 percent over 1976 levels – savings worth $180 billion a year – experts now wonder whether we can achieve an additional 50 to 90 percent. Lovins thinks we might possibly save as much as 99 percent. That may sound ridiculous, but certainly no more so than the claim that textile workers could use gears and motors to increase their efficiency a hundredfold would have sounded at the beginning of the Industrial Revolution.

The resource productivity revolution is at a similar threshold. State-of-the-shelf technologies – fans, lights, pumps, superefficient windows, motors, and other products with proven track records – combined with intelligent mechanical and building design, could reduce energy consumption in American buildings by 90 percent. State-of-the-art technologies that are just being introduced could reduce consumption still further. In some cases – wind power, for example – the technologies not only operate more efficiently and pollute less, they also are more labor-intensive. Wind energy requires more labor than coal-generated electricity, but has become competitive with it on a real-cost basis.

The resource revolution is starting to show up in all areas of business. In the forest products industry, clearinghouses now identify hundreds of techniques that can reduce the use of timber and pulpwood by nearly 75 percent without diminishing the quality of housing, the "services" provided by books and paper, or the convenience of a tissue. In the housing industry, builders can use dozens of local or composite materials, including those made from rice and wheat straw, waste-paper, and earth, instead of studs, plywood, and concrete. The Herman Miller company currently designs furniture that can be reused and remanufactured a number of times; DesignTex, a subsidiary of Steelcase, a leading manufacturer of office furniture, sells fabrics that can be easily composted.

Although a new "*hypercar*" is now in development, "new urbanist" architects, such as Peter Calthorpe, Andres Duany, Elizabeth Plater-Zyberk, and others, are designing communities that could eliminate 40 to 60 percent of driving needs. (A recent San Francisco study showed that communities can decrease car use by 30 percent when they double population density.) Internet-based transactions may render many shopping malls obsolete. Down

the road we'll have quantum semiconductors that store vast amounts of information on chips no bigger than a dot; diodes that emit light for 20 years without bulbs; ultrasound washing machines that use no water, heat, or soap; hyperlight materials stronger than steel; deprintable and reprintable paper; biological technologies that reduce or eliminate the need for insecticides and fertilizers; plastics that are both reusable and compostable; piezoelectric polymers that can generate electricity from the heel of your shoe or the force of a wave; and roofs and roads that do double duty as solar energy collectors. Some of these technologies, of course, may turn out to be impractical or have unwanted side effects. Nevertheless, these and thousands more are lining up like salmon to swim upstream toward greater resource productivity.

RESOURCE POLITICS

How can government help speed these entrepreneurial "salmon" along? The most fundamental policy implication is simple to envision, but difficult to execute: We have to revise the tax system to stop subsidizing behaviors we don't want (resource depletion and pollution) and to stop taxing behaviors we do want (income and work). We need to transform, incrementally but firmly, the sticks and carrots that guide business.

Taxes and subsidies are information. Everybody, whether rich or poor, acts on that information every day. Taxes make something more expensive to buy; subsidies artificially lower prices. In the United States, we generally like to subsidize environmental exploitation, cars, big corporations, and technological boondoggles. (We don't like to subsidize clean technologies that will lead to more jobs and innovation because that is supposed to be left to the "market.") Specifically, we subsidize carbon-based energy production, particularly oil and coal; we massively subsidize a transportation system that has led to suburban sprawl and urban decay; we subsidize risky technologies like nuclear fission and pie-in-the-sky weapons systems like Star Wars. (Between 1946 and 1961 the Atomic Energy Commission spent $1 billion to develop a nuclear-powered airplane. But it was such a lemon that the plane could not get off the ground. History's dustbin also includes a nuclear-powered ship, the

Savannah, that was retired after the Maritime Administration found she cost $2 million more per year than other ships.)

We subsidize the disposal of waste in all its myriad forms – from landfills, to Superfund clean-ups, to deep-well injection, to storage of nuclear waste. In the process, we encourage an economy where 80 percent of what we consume gets thrown away after one use.

As for farming, the US government covers all the bases: We subsidize agricultural production, agricultural nonproduction, agricultural destruction, and agricultural restoration. We provide price supports to sugarcane growers, and we subsidize the restoration of the Everglades (which sugarcane growers are destroying). We subsidize cattle grazing on public lands, and we pay for soil conservation. We subsidize energy costs so that farmers can de-plete aquifers to grow alfalfa to feed cows that make milk that we store in warehouses as surplus cheese that does not get to the hungry.

Then there is the money we donate to dying industries: federal insurance provided to floodplain developers, cheap land leases to ski resorts, deposit insurance given to people who looted US savings and loans, payments to build roads into wilderness areas so that privately held forest product com-panies can buy wood at a fraction of replacement cost, and monies to defense suppliers who have provided the Pentagon with billions of dollars in unnecessary inventory and parts.

Those are some of the activities we encourage. What we hinder, apparently, is work and social welfare, since we mainly tax labor and income, thereby discouraging both. In 1994, the federal government raised $1.27 trillion in taxes. Seventy-one percent of that revenue came from taxes on labor – income taxes and Social Security taxes. Another 10 percent came from corporate income tax. By taxing labor heavily, we encourage businesses not to employ people.

To create a policy that supports resource pro-ductivity will require a shift away from taxing the social "good" of labor, toward taxing the social "bads" of resource exploitation, pollution, fossil fuels, and waste. This tax shift should be "revenue neutral" – meaning that for every dollar of taxation added to resources or waste, one dollar would be removed from labor taxes. As the cost of waste and resources increases, business would save money by hiring less-expensive labor to save more-expensive resources. The eventual goal would be to achieve zero taxation on labor and income.

The purpose of this tax shift would be to change *what* is taxed, not *who* is taxed. But no tax shift is uniform, and without adjustments for lower incomes, a shift toward taxing resources would likely be regressive. Therefore, efforts should be made to keep the tax burden on various income groups more or less where it is now. (There are numerous means to accomplish this.) The important element to change is the *purpose* of the tax system because, other than generating revenue, the current tax system has no clear goal. The only incentive provided by the Internal Revenue Code, with its 9,000 sections, is to cheat or to hire tax lawyers.

A shift toward taxing resources would require steady implementation, in order to give business a clear horizon in which to make strategic investments. A time span of 15 to 20 years, for example, should be long enough to permit businesses to continue depreciating current capital investments over their useful lives.

Of course, a tax shift alone will not change the way business operates; a broad array of policy changes on issues of global trade, education, economic development, econometrics (including measures of growth and well-being), and scientific research must accompany it. For the tax shift to succeed, we must also reverse the wrenching break-down of our democracy, which means addressing campaign finance reform and media concentration.

It is easier, as the saying goes, to ride a horse in the direction it is going. Because the costs of natural capital will inevitably increase, we should start changing the tax system now and get ahead of the curve. Shifting taxes to resources won't – as some in industry will doubtless claim – mean diminishing standards of living. It will mean an explosion of innovation that will create products, techniques, and processes that are far more effective than what they replace.

Some economists will naturally counter that we should let the markets dictate costs and that using taxation to promote particular outcomes is inter-ventionist. But *all* tax systems are interventionist; the question is not whether to intervene but *how* to intervene.

A tax system should integrate cost with price. Currently, we dissociate the two. We know the price of everything but the cost of nothing. Price is what the buyer pays. Cost is what society pays. For example, Americans pay about $1.50 per gallon at the gas pump, but gasoline actually costs up to $7 a gallon when you factor in all the costs. Middle Eastern oil, for instance, costs nearly $100 a barrel: $25 to buy and $75 a barrel for the Pentagon to keep shipping lanes open to tanker traffic. Similarly, a pesticide may be priced at $35 per gallon, but what does it cost society as the pesticide makes its way into wells, rivers, and bloodstreams?

THE FUTURE

In 1750, few could imagine the outcome of industrialization. Today, the prospect of a resource productivity revolution in the next century is equally hard to fathom. But this is what it promises: an economy that uses progressively less material and energy each year and where the quality of consumer services continues to improve; an economy where environmental deterioration stops and gets reversed as we invest in increasing our natural capital; and, finally, a society where we have more useful and worthy work available than people to do it.

A utopian vision? No. The human condition will remain. We will still be improvident and wise, foolish and just. No economic system is a panacea, nor can any create a better person. But as the twentieth century has painfully taught us, a bad system can certainly destroy good people.

Natural capitalism is not about making sudden changes, uprooting institutions, or fomenting upheaval for a new social order. (In fact, these consequences are more likely if we don't address fundamental problems.) Natural capitalism is about making small, critical choices that can tip economic and social factors in positive ways.

Natural capitalism may not guarantee particular outcomes, but it *will* ensure that economic systems more closely mimic biological systems, which have successfully adapted to dynamic changes over millennia. After all, this analogy is at the heart of capitalism, the idea that markets have a power that mimics life and evolution. We should expand this logic, not retract it.

For business, the opportunities are clear and enormous. With the population doubling sometime

in the next century, and resource availability per capita dropping by one-half to three-fourths over that same period, which factor in production do you think will go up in value – and which do you think will go down? This basic shift in capital availability is inexorable.

Ironically, organizations like Earth First!, Rainforest Action Network, and Greenpeace have now become the *real* capitalists. By addressing such issues as greenhouse gases, chemical contamination, and the loss of fisheries, wildlife corridors, and primary forests, they are doing more to preserve a viable business future than are all the chambers of commerce put together. While business leaders hotly contest the idea of resource shortages, there are few credible scientists or corporations who argue that we are not losing the living systems that provide us with trillions of dollars of natural capital: our soil, forest cover, aquifers, oceans, grasslands, and rivers. Moreover, these systems are diminishing at a time when the world's population and the demand for services are growing exponentially.

Looking ahead, if living standards and population double over the next 50 years as some predict, and if we assume the developing world shared the same living standard we do, we would have to increase our resource use (and attendant waste) by a factor of 16 in five decades. Publicly, governments, the United Nations, and industries all work toward this end. Privately, no one believes that we can increase industrial throughput by a factor anywhere near 16, considering the earth's limited and now fraying life-support systems.

It is difficult for economists, whose important theories originated during a time of resource abundance, to understand how the decline in ecosystem services is laying the groundwork for the next stage in economic evolution. This next stage, whatever it may be called, is being brought about by powerful and much-delayed feedback from living systems. As we surrender our living systems, social stability, fiscal soundness, and personal health to outmoded economic assumptions, we are hoping that conventional economic growth will save us. But if economic "growth" does save us, it will be anything but conventional.

So why be hopeful? Because the solution is profitable, creative, and eminently possible. Societies may act stupidly for a period of time, but eventually they move to the path of least economic resistance.

The loss of natural capital services, lamentable as it is in environmental terms, also affects costs. So far, we have created convoluted economic theories and accounting systems to work around the problem.

You can win a Nobel Prize in economics and travel to the royal palace in Stockholm in a gilded, horse-drawn brougham believing that ancient forests are more valuable in liquidation – as fruit crates and Yellow Pages – than as a going and growing concern. But soon (I would estimate within a few decades), we will realize collectively what each of us already knows individually: It's cheaper to take care of something – a roof, a car, a planet – than to let it decay and try to fix it later.

While there may be no "right" way to value a forest or a river, there is a wrong way, which is to give it no value at all. How do we decide the value of a 700-year-old tree? We need only ask how much it would cost to make a new one. Or a new river, or even a new atmosphere.

Despite the shrill divisiveness of media and politics, Americans remain remarkably consistent in what kind of country they envision for their children and grandchildren. The benefits of resource productivity align almost perfectly with what American voters say they want: better schools, a better environment, safer communities, more economic security, stronger families and family support, freer markets, less regulation, fewer taxes, smaller government, and more local control.

The future belongs to those who understand that doing more with less is compassionate, prosperous, and enduring, and thus more intelligent, even competitive.

36

"Import Replacement"

from *Going Local: Creating Self-Reliant Communities in a Global Age* (1998)

Michael Shuman

Editors' Introduction

The engine that powers much unsustainable development around the world currently is rapid, export-oriented economic growth. Conventional wisdom holds that such growth is essential to raise standards of living and to enable societies the luxury of funding environmental protection and social welfare programs. But a growing chorus of critics – led by opponents of economic globalization – questions this export-driven model and calls instead for more locally oriented economic development rooted in the needs, resources, and skills of local communities.

Attorney Michael Shuman is the former director of the Institute for Policy Studies in Washington, D.C. and consults widely on community economics, international development, and citizen diplomacy. In this selection from his book *Going Local: Creating Self-Reliant Communities in a Global Age* (New York: The Free Press, 1998), Shuman argues for an import replacement approach towards economic development, in which communities seek to produce many basic goods and services themselves. The advantages of such a strategy in his view include vastly reduced transportation needs, lower pollution, more local jobs, local ownership of businesses, and retention of capital in the local community. Such import replacement strategies have in fact been a recurrent theme in alternative economic development philosophies since World War II. Jane Jacobs, for example, believed that development of local industries to replace imported goods is behind the rise of most successful cities. Some advocates of local self-sufficiency have even gone so far as to set up local exchange trading systems (LETS) in which community members barter goods and services from one another using local currencies. If long-distance trade is required for certain products, other activists believe that it should be "fair trade" in which producers receive decent wages and workers labor in healthy and safe conditions.

Shuman's other books include *The Small-Mart Revolution: How Local Businesses are Beating the Global Competition* (San Francisco: Berrett-Koehler, 2006) and *Local Dollars, Local Sense: How to Shift Your Money from Wall Street to Main Street and Achieve Real Prosperity* (White River Junction, VT: Chelsea Green, 2012). Other writings on this subject include David Morris' *Self-Reliant Cities* (San Francisco: Sierra Club Books, 1982), Jane Jacobs' *Cities and the Wealth of Nations* (New York: Random House, 1984), *The Case Against the Global Economy and for a Turn Toward the Local* (San Francisco: Sierra Club Books, 1996), edited by Jerry Mander and Edward Goldsmith, and David J. Hess' *Localist Movements in a Global Economy* (Cambridge, MA: MIT Press, 2009). More information is also available from the Institute for Local Self-Reliance (www.ilsr.org) and Green America (www.greenamerica.org). Information on the Fair Trade movement is available from the Fair Trade Federation (www.fairtradefederation.org) and the Fair Trade Foundation (www.fairtrade.org.uk).

Community self-reliance may be difficult to imagine, but it has been the norm for most of human history. Somewhere around 10,000 BC, *homo sapiens sapiens* pioneered agriculture and used crude tools to plow fields for millet and rice in Southeast Asia.[1] At roughly the same time, counterparts in France were building structures capable of sheltering 400 to 600 people. Three thousand people lived within the walls of the city of Jericho as early as 8,000 BC. Fragmentary records suggest that these early communities were capable of producing enough grains, vegetables, fruits, and livestock to feed their residents. Wood, grasses, stones, and mud provided the essentials for housing and furniture. Animal hides, furs, and fibers were fashioned into clothing. Locally harvested plants and minerals became sources of the first medications. The burning of wood and manure generated warmth and light, and even facilitated basic metallurgy. Mechanical energy from running water and from draft animals helped process raw ingredients like wheat into more usable products like flour. These "simple" communities, the first outposts of human civilization, left an impressive legacy of folklore, music, art, and science.

Few communities in the United States today are this self-reliant. They require oil brought in by truck, coal by rail, natural gas by pipeline, and electricity through expansive power grids. The typical food item travels 1,300 miles before it winds up on the dinner table, and is distributed through supermarkets and chain stores.[2] Wander through your house and you'll probably find more evidence of resources from places thousands of miles away than from your own bioregion: tuna caught in the Gulf of Mexico; beer brewed from water and hops in Germany and bottled in New Jersey; Evian water extracted from southern France. Your closets are filled with shirts and blouses that traveled from silk producers in China to production plants in Malaysia, on to packaging facilities in Guatemala, and finally to the racks at Macy's.

The variety of these well-traveled goods certainly enhances the quality of our lives, but our growing dependence on them carries profound risks. The more essential an item is for our survival, the more dangerous it is to depend on someone outside the community selling it to us. Basic necessities, of course, are difficult to define. But if you were to write down a list of what you need to survive, chances are good you would include what most nations assented to in Article 25 of the Universal Declaration of Human Rights: "Everyone has the right to a standard of living adequate for the health and well-being of himself and his family, including food, clothing, housing, and medical care and necessary social services...." To provide these basics, a community needs farmers, water suppliers, loggers, materials processors, and social-service providers. But is it really possible to structure a viable community around just these economic sectors? Can community-scale businesses meet local citizens' needs cost-effectively?

Nowhere are the dangers of depending on imports of basics more clear than with oil. Over the next decade the United States is expected to import 60 percent of its oil from foreign suppliers (up from 52 percent in 1994).[3] Just as OPEC's sudden boycott and price hikes in 1973 and 1979 triggered long queues at gas stations, shortages, inflation, and recessions, our continued dependence threatens economic instability in the years ahead. A 60 percent import rate means a trade deficit in oil of $100 billion per year.[4] Leaving aside the costs of spills from offshore oil-drilling and air pollution from oil-burning, the transportation of oil across oceans poses huge environmental risks to harbor and coastal ecosystems, as well as to the entire oceanic food chain. Another decade of imports will provide $1 trillion of revenue to Persian Gulf states, much of which will be spent on weapons, wars, and saber-rattling that are hardly in the interests of US foreign policy. America is once again in a position where a handful of governments, most of them undemocratic, can bring the nation's economy to its knees by turning off the spigots.

Dependency on necessities from outside the community means that a remote crisis can reverberate into a local one. A study for the Pentagon concluded that in a single night, without ever leaving Louisiana, a few saboteurs could cut off three-quarters of the natural gas supplies to the eastern United States for more than a year.[5] It also found that low-technology sabotage of any one of the nation's more than 100 power stations could trigger a catastrophic core meltdown, inflicting a Chernobyl-like accident, or worse, on the country. In November 1965, a faulty relay in Canada cascaded into a series of equipment failures that blacked out 30 million electricity customers in the Northeast for up to

13 hours.[6] On July 13, 1977, the chair of the Consolidated Edison Company in New York assured his customers that he could "guarantee" that a recurrence was extremely unlikely. Three days later, a combination of lightning and equipment failures blacked out nearly 9 million Con Ed customers for up to 25 hours, and set the stage for $121 million worth of looting. Two decades of reflection and planning did not prevent another major collapse of the electric grid on August 10, 1996. A tree falling into a transmission line near the Columbia River in Oregon toppled a series of mechanical dominoes and left millions of consumers in 14 states and two Canadian provinces without power for several hours. Dependence holds a community hostage to mistakes, misdeeds, and misfortunes totally outside its control.

A dependent community also loses the economic benefits of producing necessities for itself. A community that chooses not to generate its own electricity, not to grow its own food, not to process its own lumber, and so on, winds up losing the jobs and income that might have come from these commercial ventures. The more these activities are performed outside the community, the weaker the economic multiplier inside.

Economists are skeptical about the principles of self-reliance and import substitution, because they deprive a community of the benefits of trade. A community focusing inward narrows the range of goods and services available to its citizens, and deprives its businesses of the new machines, new production methods, and competitive forces necessary for progress. Moreover, if local goods and services cost more than those produced outside, going local means losing money, which condemns a community to less consumption and lower investment. Economies that have sought to seal themselves hermetically from the world – like the former Soviet Union, Albania, and Burma – soon crumble from obsolescence.

A better way to think of the goal of import substitution, however, is that it motivates a community to move the most important and valuable types of production back home, not to unplug completely from the national or international economy. And the means to accomplish this need not be tariffs or heavy-handed regulation, but simply smart choices by residents and local officials to buy, invest, and hire locally. Viewed this way, "the theo-retical case for emphasizing import substitution is strong," according to Joseph Persky, David Ranney, and Wim Wiewel, three economists at the University of Illinois at Chicago.[7]

Import-substituting growth facilitates the diversification of the local economy and the accumulation of its own capital, skills, and experience. Jane Jacobs, in her 1969 book *The Economy of Cities*, provides two examples of cities that were able to pump up their economies through import substitution:[8] The Los Angeles export industries that nourished the US military during World War II, like aircraft manufacturing, shipbuilding, and petroleum refining, all declined after the war ended in 1945. Yet the number of jobs in the city expanded because of the growth of new import-replacing businesses. Automobile companies in Detroit, for example, opened branch plants to serve customers in southern California. In Chicago, between 1845 and 1855, the city's population grew by nearly a factor of seven, as did the overall economy. The reason again, writes Jacobs, was import substitution:

> [A]t the beginning of the decade Chicago, like any other little Midwestern depot settlement, was importing most kinds of things that every town supplied. But by the end of that decade it was producing a very large range of the common city-made goods of the time and some of the luxuries too – clocks, watches, medicines, many kinds of furniture, stoves, kitchen utensils, many kinds of tools, most building components.[9]

Import substitution actually incorporates some of the export-oriented thinking of mainstream economists. As a community replacing imports grows in size, it naturally attracts more businesses that target national and global markets. Persky, Ranney, and Wiewel summarize the export-led theory of development in the following terms:

> [A]n area that is able to increase significantly its sales of a major export will experience population growth related to the new employment in the basic export sector. This population growth in turn implies that an area may pass thresholds for new products. Now the metropolis will provide some commodities for itself that formerly it imported. Export growth leads to import substitution.[10]

A community committed to import substitution, however, aims to *minimize* population growth. The goal is to expand the quantity and quality of jobs without drawing new people. University of Montana economist Thomas Michael Power has shown that the states in which jobs grew fastest such as Alaska, Arizona, Nevada, and Utah, also saw family income grow more slowly than the national average.[11] In states like New York and Rhode Island, where employment growth was slowest, income growth was above the national average. The reason for these seemingly paradoxical results is the inflow and outflow of people. Power concludes that "there is no reliable connection between mere quantitative expansion of the local economy and local economic well-being."[12] Expansion of economic activity should be targeted carefully at those people, institutions, and businesses that have a long-term commitment to the community. He winds up recommending a concerted local effort at import substitution:

> The import substitution can be direct, as in the case of energy or of a local bakery replacing imported bread, or it can be indirect. As the variety of goods and services produced locally expands, the richer commercial economy attracts and holds more of the residents' dollars. Local dollars that would have otherwise flowed out of the community, to purchase things that would add variety and quality to residents' lives, stay in the community to purchase local services. Live theater and music, instruction in skills, recreational facilities, and so on attract and hold dollars that otherwise would have flowed out to finance imports.[13]

If import replacement leads to greater exports, is the distinction from export-led development simply a matter of semantics? Hardly. A typical export-led development strategy targets just one or two industries. The World Bank pushed dozens of poor countries in the 1980s to specialize in the production of a few primary commodities like coffee, sugar, cocoa, copper, aluminum, and lumber. The fatal flaw of this approach was that each country's economy became so specialized that it was vulnerable to the collapse of the price of a targeted commodity. And this is exactly what happened. The nosedive of coffee prices in the late 1980s and early 1990s, for example, destabilized coffee-exporting countries worldwide, from Guatemala to Indonesia. In Rwanda, another country dependent on coffee exports, the consequent economic turmoil set off a chain of events that culminated in the Hutus massacring an estimated million Tutsis. . . .

Can a small community embark upon an import-replacement development strategy? Don't certain industries need a large enough local market to be set up in the first place? No, respond Persky, Ranney, and Wiewel: "[W]e do not observe a well-defined threshold population at which a given industry enters a community. For most industries, we only observe that the share of local demand supplied locally tends to rise with size."[14]

A small community like Eugene, Oregon, couldn't operate a factory to produce cars solely for its own needs, but it *could* build a plant that met the transport demand of the Pacific Northwest region. The key issue is what economists call the optimal *economy of scale*. Whenever the economy of scale of production is large and a plant needs a very high output to operate competitively, it must export to consumers outside the community. And any good or service for which this is true, by definition, can facilitate import substitution in only a limited number of communities. Because the economy of scale of automobile construction is large, it's neither possible nor desirable for every one of America's 36,000 municipalities to manufacture cars.

The prevailing wisdom among economists and businesspeople is that large economies of scale are the rules of thumb for industrial production. Huge factories with global distribution networks are assumed to deliver cheaper and better products than are small factories serving just local markets. This is because certain fixed costs, such as management, machinery, warehouses, marketing, and lawyers, can be spread over more and more units of production. Business consultants like the Boston Company and Wall Street investment houses like Drexel Burnham Lambert have accumulated vast fortunes acquiring, merging, and reorganizing firms to achieve larger economies of scale.

Yet, as the dinosaurs learned, bigger is not always better. Bigger businesses also tend to develop certain diseconomies of scale. The larger the distance between producers and consumers, the harder it is to fine-tune products to the particular tastes of local markets. Local businesses set up to serve the exacting demands of local consumers can be started more quickly, with smaller investments and smaller risks. Once a conscientious process of import replacement begins, a company may be surprised to discover other savings inherent in local production and distribution. Transportation costs go down. So do the costs of marketing.

So do the costs of excessive preservatives and packaging. A study in Europe found that a jar of strawberry yogurt traveled 2,166 miles before reaching the typical German consumer.[15] Most of the mileage wasn't logged by the food; the main ingredients – milk and sugar – came from the surrounding countryside, and the strawberries were grown in Poland. It was attributable to the packaging. Glass for the jar, paper for the label, paste for the paper, aluminum for the top all were produced from sites across Europe.

The bottom line depends on how these economies and diseconomies add up. For basic necessities, economies of scale appear to be shrinking to the point where hundreds or even thousands of US communities could move toward self-reliance. Recent breakthroughs in technology, workforce organization, and resource management are enabling local entrepreneurs to provide food, water, wood, energy, and materials in cost-effective ways. While few communities have deliberately tried to replace all their imports of basic necessities, a number of examples in each of these economic sectors testify to the possibility and promise of doing so.

FOOD INDUSTRIES

One of the saddest stories of the past century in the United States is the destruction of community-based family farming. The number of Americans on a farm today is less than one-fifth what it was in 1920.[16] Those involved in farming have substantially expanded their land holdings. In 1994, 2 percent of all the farms in the United States were so big that they were responsible for half of all food sales, while three-quarters of the farms were so small that they accounted for only 9 percent.[17] A closer look at recent developments in agriculture, however, suggests that small-scale systems to grow, process, and market food are becoming not only

cost-effective – in both rural and urban areas – but also essential to preserve the genetic integrity of the world's edible plants.

Leaving aside inflation, the value added to the economy by food-growing farms has remained constant throughout this century. What has changed is the explosive growth of the food-marketing business and of corporations providing inputs into farming (seeds, fertilizers, herbicides, pesticides, etc.). In 1910, for every dollar Americans spent for food, 41 cents went to farmers and 59 cents to marketers and input providers; now 9 cents go to farmers, 24 cents to input providers, and 67 cents to marketers.[18]

Economists celebrate these statistics, because fewer farmers can grow more food for American consumers at a lower price. But the costs have been the decimation of once-vibrant rural communities, and increased dependency of urban and suburban communities on far-off sources of nutrition. Another look at the numbers suggests that community-scale production might actually *lower* the cost of food for local consumers. Community-scale agriculture, even if it means higher-cost farming, might be able to bring down the cost of inputs through organic growing methods and the cost of marketing through local distribution.[19] When farmers are involved in distribution, they also have an opportunity to retain more of the "value added."

The economics of small-scale farming actually is improving, though it varies across the country. The valleys of the Mississippi, Missouri, and Ohio Rivers offer richer farming opportunities than do the Mohave desert or Fairbanks, Alaska, But the presence of *some* commercially viable agriculture in every one of America's fifty states (as well as in the less hospitable provinces of Canada) is a reminder that wherever there is land and water, growing one's own food is possible. One out of four fruits and vegetables distributed in today's commercial systems never makes it to the consumer's table, because it spoils during shipment or on the grocery-store shelf.[20] More locally grown produce, in contrast, needs to be thrown away less, and because it's closer to the customer (a typical item travels 200 miles instead of 1,300), transportation costs are lower. Finally, direct relationships between farmers and consumers can greatly reduce or eliminate the costs of packaging, marketing, middlemen, and supermarkets.

A recent study by the United Nations Development Program suggests that even dense urban areas hold enormous potential for cultivation.[21] Some 800 million people in the world who live in cities are engaged in urban agriculture, mainly for their own consumption.[22] In Hong Kong, which has an extraordinary population density, nearly half of all vegetables consumed are grown within the city limits, on 5 to 6 percent of the city's land. Squatters in Lusaka, Zambia, grow one-third of their food in the city. Residents of Kampala, Uganda, meet 70 percent of their poultry and egg consumption with local production. Data from the 1980s suggest that the 18 largest cities in China met over 90 percent of their vegetable needs, and half their meat needs, through urban farming. And Singapore raises 80 percent of its poultry, and a quarter of its vegetables, within city limits.

A nation's food system is rooted deeply in its history, culture, diet, and land-use policies, and therefore these experiences abroad may not be entirely transferable to the United States. But enough farming is occurring in or near US cities that the opportunities for expanded urban agriculture are certainly worth exploring. More than 30 percent of the dollar value of US agricultural production originates from farms within major metropolitan areas.[23] Over the past 20 years, New York City has opened a thousand community gardens on public land, and 18 public markets to sell produce grown in them.[24] Boston and Philadelphia have even more gardens per capita. Recognizing the potential of urban farming, several cities such as Chattanooga and Hartford, and states such as Massachusetts and Oregon, have developed comprehensive urban food policies.

Several trends in American cities make the economics of urban agriculture increasingly attractive. First, more land is becoming available for farming. Most US cities are spreading out over larger geographic areas, while their populations are remaining stable or declining. Some 40 percent of all the land in Detroit is now vacant. The exodus of industry and people from Pittsburgh has left more than 37,000 empty lots. These lots, once cleared of garbage, toxics, and abandoned buildings, have potential for cultivation, along with parkland, greenhouses, and even rooftops. Second, cities have a growing supply of idle labor. The highest unemployment rates in the country are in urban areas. In principle, much of the work required for municipal farming could be done by unskilled inner-city residents, though some training would be required and psychological barriers to manual labor would need to be overcome. Third, with recent cuts in Aid to Families with Dependent Children, cities will continue to have serious unmet demands for more food. The Food Research and Action Center estimates that nearly half the children under the age of 12 who live in urban households below the poverty line are "either hungry or at-risk of hunger." Developing new, cheap sources of nutrition for these young people is imperative – for their future and for the future of America's cities.

A handful of nonprofits have seized this opportunity in intriguing ways. In South Central Los Angeles, Reverend George Singleton's Hope LAS Horticulture Corps trains young and at-risk gang members to plant flowers, herbs, vegetables, and trees in vacant lots. Sales of the produce grown pay for the project, which combines agricultural coursework with hands on training. Singleton is now working with organizations to spread the program to five other sites in the Los Angeles area, and to the Bronx.

Farms need not be within a city's limits to contribute to its economic well-being. Linking farmers just outside a city with consumers inside it can boost the regional economy. One indication that this is happening is the revival of farmers' markets across the country, in which local growers set up stands (or use the backs of their station wagons) to sell fruits, vegetables, and nuts. This produce often is not as pretty as what's available at the local supermarket; it may well be discolored and irregular. But consumers are attracted to it because it tastes better, holds more nutrition, and often is free of pesticides.

Some clever farmers have organized groups of consumers (and vice versa) into subscribers' clubs. One model of what has become known as community-supported agriculture, or CSA, is that each member family agrees to pay a fee for the growing season, and in exchange the farmer promises to deliver a box of vegetables each week, enough to feed a family. The contents of the box vary throughout the season; one week it might be mostly asparagus and melon, another it might be potatoes and pumpkins. Some CSAs add farm products like eggs, milk, honey, spices, flowers, and firewood. Subscribers can pick up their box of produce at a nearby distribution point or have it delivered to their doorstep.

Some 600 such community-supported agricultural or horticultural operations now exist in 42 of the United States, with 100,000 members.[25] A typical participating farm has about 3 acres (plus grazing land) and serves the needs of 60 to 70 families who each pay about $400.[26] With the subscription fee paid up front, consumers share the risks of failure with the farmer. If consumers were to become co-owners of the land and equipment, effectively shareholders, this model could be seen as a community corporation.

If you can't imagine taking a box of produce chosen and packed by someone else, but still want to support local farmers, you might just try to shop only at community-friendly markets. Retailing food, of course, is another viable business for community corporations, which can set up outdoor farmers' markets or indoor stores, or take over and revitalize failing stores. The Village Retail Services Association (VIRSA) in England has a simple mission: to save small shops endangered by shopping malls and gigantic outlets. It organizes customers to invest in local shops, with the goal of strengthening their buying loyalty and the incentive of advertising by word by mouth. It also works with management to improve the efficiency and quality of operations. Sometimes VIRSA purchases the entire operation and hires new management; and occasionally, where there is very broad local commitment, it will organize the shop to be run entirely by volunteers. In VIRSA's first year it assisted 12 British communities.[27]

How far urban agriculture, CSAs, farmers' markets, and village stores can go in feeding a community is unclear. Even if these innovations can deliver fruits and vegetables cost-effectively, can they also provide grains, meat, and other kinds of food? The answer may only be positive if more and more value-added activity is undertaken by farmers themselves. One recent study found intriguing evidence that small-scale farmers do well supplementing their incomes with homestead chicken production, small dairy herds, and on-farm food processing.[28] Certainly if growing numbers of consumers continue to prefer food grown, raised, processed, packaged, and sold by local farmers, the economics of community production will steadily improve.

NOTES

1 Roberts, J.M. 1976. *History of the World*. London: Penguin, p. 45.

2 Imhoff, Daniel. 1996. Community Supported Agriculture. In Jerry Mander and Edward Goldsmith (eds). *The Case Against the Global Economy*. San Francisco: Sierra Club Books, p. 425.

3 Romm, Joseph J. and Curtis, Charles B. 1996. Mideast Oil Forever? *Atlantic Monthly*, April, p. 57.

4 Ibid., p. 60.

5 Lovins, Amory B. and Hunter, L. 1982. *Brittle Power: Energy Strategy for National Security*. Andover, MA: Brick House Publishers, p. 122; and Lovins, Amory B. and Hunter, L. 1983. The Fragility of Domestic Energy. *Atlantic Monthly*, November, pp. 118–126.

6 Lovins and Hunter, *Brittle Power*, pp. 51–67; Bolden, Tim. 1996. Blackout May Be Caution Sign on Road to Utility Deregulation. *New York Times*, 19 August, p. A14.

7 Persky, Joseph, Ranney, David, and Wievel, Wim. 1993. Import Substitution and Local Economic Development. *Economic Development Quarterly*, February, p. 18.

8 Jacobs, Jane. 1969. *The Economy of Cities*. New York: Vintage, pp. 145–179.

9 Ibid., p. 157.

10 Persky *et al.*, Import Substitution, p. 19.

11 Power, Thomas Michael. 1996. *Environmental Protection and Economic Well-Being*. Armonk, NY: M.E. Sharpe, pp. 155–180.

12 Ibid., p. 155.

13 Ibid., p. 194.

14 Persky *et al.* Import Substitution, p. 19.

15 Douthwaite, Richard. 1996. *Short Circuit*. Devon, UK: Resurgence, pp. 228–229.

16 Alperovitz, Gar and Faux, Jeff. 1984. *Rebuilding America*. New York: Pantheon, p. 196.

17 Lehman, Karen and Krebs, Al. 1996. In Mander and Goldsmith, *The Case Against the Global Economy*, p. 127.

18 Smith, Stewart. 1993. Sustainable Agriculture and Public Policy. *Maine Policy Review*, April, pp. 69–70.

19 Smith, Stewart. 1994. Farming Activities and Family Farms: Getting the Concepts Right. In *Symposium: Agricultural Industrialization and Family Farms: The Role of Federal Policy*. Hearing

before the Joint Economic Committee of the United States, 21 October 1992, pp. 57–929. Washington, DC: USGPO, pp. 117–133.

20 Imhoff, supra, p. 429.

21 United Nations Development Programme. 1996. *Urban Agriculture: Food, Jobs and Sustainable Cities.* New York: UNDP.

22 Ibid., pp. 26–27.

23 Ibid., p. 25.

24 Ibid., p. 26.

25 Keen, Elizabeth. 1997. Researcher for CSA of North America. Personal communication, 21 August.

26 Imhoff, supra, pp. 429–430.

27 Douthwaite, supra, pp. 315–318.

28 Integrity Systems Cooperative Co. 1997. *Adding Value to our Food System: An Economic Analysis of Sustainable Community Food Systems* (monograph). Everson, WA: Integrity Systems Cooperative Co., February.

37

"Strengthening Local Economies"

from *State of the World 2007: Our Urban Future* (2007)

Mark Roseland with Lena Soots

Editors' Introduction

Many specific tools, ranging from cooperatively owned or run businesses to fair trade networks to community-based financial institutions, can help bring about the sort of locally oriented economy that Shuman promoted in the previous reading. Here Mark Roseland provides examples from around the world of a number of these local economic development initiatives. Roseland is a professor of geography at Simon Fraser University in Burnaby, British Columbia and directs the University's Centre for Sustainable Community Development. His book *Toward Sustainable Communities: Resources for Citizens and Their Governments, Fourth Edition* (Gabriola Island, BC: New Society Publishers, 2012) has through many iterations been a valuable source of information about sustainable community topics. Other resources on green economic strategies include Joel Makower's *Strategies for the Green Economy: Opportunities and Challenges in the New World of Business* (New York: McGraw Hill, 2009) and *Sustainable Communities: Creating a Durable Local Economy* (London: Earthscan, 2013), by Rhonda Phillips, Bruce F. Seifer, and Ed Antczak.

The wealth of a nation depends in large measure on the economic health of its cities. Cities make countries rich. Highly urbanized countries have higher incomes than other nations, more stable economies, stronger institutions, and more ability to withstand the volatility of the global economy. Cities around the world are playing a growing role in creating wealth, enhancing social development, attracting investment, and harnessing both human and technical resources for achieving unprecedented gains in productivity and competitiveness. Cities are also engines of rural development. For example, improved infrastructure between rural areas and the cities that rely on them increases rural productivity and enhances rural residents' access to education, health care, markets, credit, information, and other services.

For the first time in history, more than half the world's people will soon live in urban areas. Cities are both surrounded by and made up of communities. Geographic or territorial communities have a shared destiny, represented by a municipal or local or, in indigenous communities, band or tribal form of government. They may include the built-up or densely populated area containing the city proper, suburbs, and continuously settled commuter areas. They may be larger or smaller than a metropolitan area.[1]

Strong local economies are the foundation of strong communities that can grow and withstand the pressures created by an increasingly urbanized world. And strong communities require a holistic approach that not only provides the traditional deliverables of economic development – jobs, income, wealth, security – but also protects the environment, improves community infrastructure, increases and develops local skills and capacity, strengthens the social fabric, and respects heritage and cultural identity. In this way, strong local economies also provide a foundation for strong national economies. Cities and towns provide enormous untapped opportunities to strengthen local economies by pioneering new approaches to sustainable development and community management.[2]

RECLAIMING LOCAL ECONOMIC CONTROL

There are few limits to what can be achieved when people work together for their mutual benefit. Since the 1800s, people have been forming cooperatives in order to meet a wide range of local needs. From agricultural producer cooperatives and consumer co-ops to worker co-ops and social cooperatives aimed at delivering health care and social services, cooperative enterprises can be found in nearly every country. In the context of increasing urbanization, cooperatives offer a community-based strategy for reducing poverty.

The International Labour Organization (ILO) defines a cooperative as "an autonomous association of persons united voluntarily to meet their common economic, social and cultural needs and aspirations through a jointly owned and democratically-controlled enterprise." Cooperatives thus practice a unique form of economic participation that is based on membership rather than amount of investment.[3]

Within cities, worker cooperatives – businesses owned and controlled by their employees – are the most common form of cooperative enterprise. Decisions for the operation of the business are made democratically on the basis of one member/one vote. Common forms of worker co-ops include manufacturing plants, retail stores, communications companies, technical firms, and various forms of service providers. Worker cooperatives typically form as a result of a group of people organizing to create employment for themselves and to overcome barriers to employment such as disabilities or racial, gender, or ethnic prejudices. Alternatively, worker co-ops form as a result of an existing company that has "mutualized" by selling shares to its employees, who then take over ownership and management of the business.[4]

Argentina provides a good example of the power of cooperation in saving the local economy. During the 2001 economic crisis, many Argentines decided to take collective action to save their jobs and their livelihoods. By occupying bankrupt factories and businesses, the workers turned their workplaces into cooperatives without upper management or unions. The National Movement of Recovered Companies spread across Argentina as a bottom-up approach to economic recovery with the motto: "Occupy, Resist, Produce." There are now roughly 200 worker-run factories and businesses in Argentina, employing over 15,000 people. Most of these cooperatives started during 2001.[5]

In 2002, the Ghelco ice cream factory in southern Buenos Aires went into bankruptcy. The workers were owed thousands of dollars in back pay. So they formed a co-op and stopped the owners from removing the machinery and dismantling the factory by protesting in front of it. Three months of protest led to an offer for the co-op to rent the factory. Five months after that, the factory was seized by the Buenos Aires legislature and given to the co-op. Now the factory is run by forty-three members of the co-op who earn more than they ever did, and they have better working conditions. The co-op does not have to pay high managerial salaries, unlike the previous owners; in addition, it gets to keep the large profits that used to end up in the owners' pockets.[6]

Historically, cooperatives have not only enabled people to lift themselves out of poverty, they have also become a way for low- and middle-income people to continue accumulating economic advantages they would not be able to achieve individually. Beyond this, co-ops can contribute to strengthening the social fabric of communities and can build social cohesion among community members as well as facilitate the equitable distribution of resources. In a press release on the occasion of the United Nations International Day of Cooperatives in 2001, Secretary General Kofi Annan underscored the role of cooperatives in development, noting "the values of cooperation – equity, solidarity, self-help and mutual responsibility – are cornerstones of our shared endeavour to build a fairer world. . . . The cooperative movement will be an increasingly adaptable and valuable partner of the United Nations in pursuing economic and social development for the benefit of all people."[7]

COMMUNITY-BASED FINANCE

Finance plays a fundamental role in the process of economic development – capital investments and growth in financial assets are important factors in the development of economies of low-income countries. Yet access to capital is not easy for those who live in poverty. People are often blocked from climbing out of poverty not by a lack of skills or motivation but by their lack of access to capital.

Poor people are often forced to rely on informal financial relationships, which are usually erratic, insecure, and costly to borrowers. Improving access to financial resources is thus an important component in the fight against poverty, and microfinance has proved to be an important tool in this regard. "Microfinance" refers to the provision of financial resources and services to people who are generally excluded from traditional financial systems based on their low socioeconomic status. Generally it includes the provision of loans, savings, and other basic financial services that poor people need to protect, diversify, and increase their sources of income.[8]

In 1997, delegates from 137 countries gathered in Washington, DC, for a Microcredit Summit and launched a campaign to reach 100 million of the world's poorest families, especially women, with credit for self-employment and other financial services by the end of 2005. According to the *State of the Microcredit Summit Report 2005*, by the end of 2004 a total of 3,164 microcredit institutions had reached 92,270,289 clients, 72 percent of whom were among the poorest of the poor when they took their first loan.[9] Nearly 84 percent of these poorest clients were women. Assuming five persons per family, the loans given to the 66.6 million poorest clients affected some 333 million people. Just over half of the 3,164 institutions reporting were in Asia, 31 percent were in Africa, 12 percent were in Latin America and the Caribbean, and just under 5 percent were in North America, Europe, the newly independent states, or the Middle East.[10]

Contrary to some opinions, poor people are very good at saving their money. In fact, their savings represent a higher proportion of their net assets than the savings of their higher-income counterparts do. With access to well-designed savings products and services, low-income people can accumulate wealth and begin to climb the ladder out of poverty.[11]

The Grameen Bank in Bangladesh has become the international model for microcredit programs. Its central feature is its credit program, which provides small loans to the poor for self-employment activities. The project was initiated in 1976 when Muhammad Yunus, head of the Rural Economics Program at the University of Chittagong, launched a research project on the possibility of designing a credit delivery system to provide banking services to the rural poor. The Grameen Bank Project began with the following objectives:

- to extend banking facilities and services to poor men and women;
- to eliminate the exploitation of the poor by moneylenders;
- to create opportunities for self-employment;
- to bring the disadvantaged, mostly women, within an organized format in which they can understand and manage themselves; and
- to reverse the vicious cycle of "low income, low savings and low investment."

After initial success in the village of Jobra, the project soon expanded to other villages. In 1983, government legislation transformed the project into an independent bank.[12]

The unique and innovative lending scheme of the Grameen Bank requires borrowers to voluntarily form small groups of five people to provide mutual, morally binding group guarantees in lieu of the collateral required by conventional banks. Initially, only two of the five group members are able to apply for a loan. Access to credit for the other members depends on the successful repayment of these initial loans. Although the Bank monitors borrowers fairly closely, organizing the borrowers into groups provides incentives for peer monitoring.

The success of the Grameen Bank is astounding, as the Nobel Committee noted when it awarded the 2006 Nobel Peace Prize to Muhammad Yunus for his pioneering work in this field. As of April 2006, it had a total of 6 million borrowers, 96 percent of whom are women. The bank has 2,014 branches working in 65,847 villages with a total staff of 17,816. Since it started, the total amount of loans dispersed is 271.94 billion taka ($5.46 billion). Of this, 241.63 billion taka ($4.83 billion) has been repaid. The loan recovery rate is 98.4 percent.[13] Because of the interconnection between financial power, poverty, and women, microfinance has an important role in empowering women and improving economic equality.

In many countries, particularly in low-income regions, women have few ownership rights. Cultural norms and expectations place additional constraints on women's access to assets and income-generating opportunities.[14] Studies have shown that income earned by female borrowers has more beneficial effects on the well-being of children and household members generally than income earned by male borrowers. And microcredit lending by the Badan Kredit Kecamatan in Indonesia has been found to increase women's participation in decision making, reduce fertility, and improve household nutrition.[15]

Particularly in low-income regions of the world where poverty rates are high, microfinance can be an important tool for improving the socio-economic conditions of communities. It helps to foster financially self-sufficient private sectors and create wealth for low-income people. As it does this, microfinance creates new consumers and markets for existing businesses as well, thus contributing to the overall integrity of local economies.

Community-based financial systems are not only found in low-income countries struggling with widespread poverty. In countries such as the United States and Canada, there are community-oriented financial institutions that have a vision of social and environmental as well as economic benefits.

In a financial climate dominated by big banks, community credit unions demonstrate that financial benefits can coincide with broader community values and objectives. A credit union is a non-profit financial institution that is cooperatively owned and controlled by its members and managed through the election of a volunteer Board of Directors. Credit unions offer the same financial services as banks (savings, investments, loans, and so on), but they generally market themselves as providing superior member services because of their community orientation. As cooperative institutions, their policies are set up to benefit the interests of their memberships as a whole. Credit unions typically pay higher interest rates on shares and charge lower interest rates on loans than traditional financial institutions.

Worldwide, there are over 157 million credit union members in ninety-two countries. Canada has the highest per capita use of credit unions, with over one-third of the population enrolled as members. As community-based financial institutions, credit unions ensure that financial investments are not only economically successful but also in line with the broader social objectives and values of the communities in which they operate. As such, credit unions strengthen local economies by providing positive social and economic returns to communities.[16]

The Downtown Eastside of Vancouver, Canada, is home to many people battling drug addictions, mental illness, and homelessness. Most local residents are low-income earners with little to no access to traditional financial services. Without identification to open a bank account or enough money to keep one open, people find themselves unable to attain any level of financial security, and many rely on costly private check-cashing facilities. With the help of VanCity Savings Credit Union, Pigeon Park Savings opened its doors in 2004 as a provider of low-cost, reliable financial services in a supportive environment. People can now cash or deposit checks and use a variety of financial services otherwise unavailable to Downtown Eastside residents. Pigeon Park's banking system runs entirely on VanCity's network, which also provides the operating infrastructure, technical support, administrative services, and security.[17]

Cooperatively owned credit unions are not the only form of community-based financial structures. Shorebank Pacific is a chartered, commercial bank in Washington State with a commitment to environmentally sustainable community development. It is the result of an innovative partnership between Ecotrust, a nonprofit environmental organization dedicated to fostering a conservation-based economy, and Shorebank Corporation of Chicago, a pioneer in developing inner-city community projects. To track financial progress, clients and loans are reviewed by a bank scientist who uses a scoring system of sustainability principles.[18]

BUY LOCAL, TRADE FAIR

Although a significant proportion of goods used in cities are also produced there, cities are by no means self-sustaining entities. In terms of local economic activity, urban centers rely heavily on rural areas for resources such as fuel and food. The trading systems that facilitate the production, distribution, and consumption of goods in urban areas thus affect more than just the economies within cities.

In the context of an increasingly globalized economy, strengthening local economies means having strong local networks and trading systems that support economic activity within and among communities and that contribute to the overall health and well-being of these areas. Promoting trading systems that contribute to strong local economies allows cities and communities to participate in and contribute to the larger, global economic system in ways that are sustainable, equitable, and just.

Fair trade is a rapidly growing movement that seeks to challenge unequal international trade relations and that makes trade beneficial to disadvantaged and vulnerable producers. It does this by establishing direct and positive links between producers of the South and consumers of the North. Fair trade attempts to level the global playing field by enabling poor producers to be part of a trading system that ensures a fair and stable price for their products. It also offers producers and their organizations support and services and promotes the use of ecologically sustainable production practices.[19]

Similar to certification in organics or labor standards, fair trade involves the implementation of voluntary global production standards. Fairtrade Labelling Organizations International (FLO) is the primary body that sets the labeling standards to be met by producer groups, traders, processors, wholesalers, and retailers. Although most commonly associated with rural production, FLO sets standards for hired labor situations as well, including standards for working conditions, employee returns, discrimination policies, and child labor. In the context of manufacturing and the production of goods within cities, the concept of fair trade can play a significant role in creating strong local economies.[20]

For many years, coffee was the main product sold with the fair trade label. Now at least twenty different products – from tea and chocolate to sporting goods, textiles, and handicrafts – are exported from developing countries to more than twenty countries in Europe and North America plus Australia, New Zealand, and Japan. There are now 531 producer organizations certified with FLO, representing some 1 million farmers and workers from fifty-eight countries in Africa, Asia, and Latin America. There are 667 registered traders, consisting of exporters, importers, processors, and manufacturers from fifty countries all over the world.[21]

Strongly linked to the cooperative movement, fair trade is helping communities build the financial resources to sustain livelihoods and alleviate poverty. It is also building community capacity, fostering strong relationships within communities, and promoting sustainable and equitable systems of production and trade.

As recent mass protests in Seattle, Quebec City, Prague, and Genoa demonstrated, concerns about corporate dominance in the global economy cover more than just the issue of fair trade. A localism movement has emerged promoting alternative approaches to global economic development, particularly with respect to energy, materials, and food. Campaigns to promote the localization (or "deglobalization") of business and trade are sprouting up all over North America and Europe, espousing the social, economic, and ecological benefits of more localized economies.

Although the definition of "local" varies – from bioregions to geopolitical boundaries – the idea is consistent: local is better. The benefits of localized economies are many: they support local businesses and keep money and profits within communities, they reestablish producer–consumer relationships and build social cohesion, and they reduce the negative ecological impacts of global trade, namely fossil fuel emissions from long-distance transport.

The Business Alliance for Local Living Economies (BALLE) is a growing alliance of business-people around the United States and Canada who join networks dedicated to building "local living economies" with a goal to green and strengthen their local economies. Businesspeople organize themselves into local business networks – each fully autonomous – and share a commitment to living economy principles. BALLE provides support and tools to catalyze, strengthen, and connect local business networks. Members of local networks join together to:

- support the development of community-based businesses;
- encourage local purchasing by consumers and businesses;
- create opportunities for business leaders to share best practices; and
- advocate public policies that strengthen independent local businesses and farms, promote economic equity, and protect the environment.[22]

Countless cities all over North America are working to promote the importance of strengthening local economies. The San Francisco Locally Owned Merchants Alliance, for example, has over fifty members and works to promote locally owned independent retailers in the area. The Buy Local Philly campaign is sponsored by the Sustainable Business Network of Greater Philadelphia and is a network of more than 200 local and independently owned businesses. The city of Portland, Oregon, has a Think Local First campaign to raise public awareness of the benefits of a more localized economy. Vancouver, Canada, has started a Buy Local, Support Yourself campaign to encourage local purchasing and support local fashion, food, art, and more.[23]

As the world faces the emergence of peak oil and related energy challenges, cities need to reduce their dependence on outside markets and fossil-fuel-dependent transportation systems. What is needed are more localized systems of production, distribution, and consumption. Local trading systems not only allow cities to increase self-sufficiency, they also contribute to strengthening local economies by giving communities the ability and resources to meet their own needs.

LOCAL ECONOMY ACTORS

Who are the local economic actors and what are their roles in strengthening local economies? Local authorities are clearly infrastructure that is essential for economic activity, and they set standards, regulations, taxes, and fees that determine the parameters for economic development. Local authorities procure large numbers of services and

products and can influence markets for goods and services.

Like private enterprises, local authorities serve as public enterprises to produce "products" that are sold on the market. These products include environmental services (such as water, waste management, and land-use control), economic services (transportation infrastructure, for example), and social services (such as health and education). Local authorities also operate under the considerable constraints of most public agencies: limited resources, jurisdiction, imagination, courage, time, and so on. For them to fulfill their potential in strengthening local economies, they need community organizations as partners.[24]

Local authorities are an influential employer and consumer in most communities. All community members have a legitimate interest in knowing what measures their local authority is, could, or should be taking to strengthen the local economy. These might include local studies of indicators, assets, imports, or subsidies; local training via entrepreneurship programs linked to incubators for locally owned businesses; help for those trying to purchase locally with a directory, a buy-local campaign, time dollars, or a local currency; local investing of municipal funds; and local public policy such as smart growth zoning or a living wage bylaw.[25]

In many low-income countries, local authorities have neither the resources nor the power to provide support of any kind to local business, let alone dedicate time and resources to overall localization initiatives. Cities with developed formal economies and stronger local authorities fortunately have more options.

Local tax shifting is one local authority strategy with enormous potential to strengthen local economies. Taxes generate revenue for governments, but they can also serve as an effective tool of governance, supporting community values and goals. The main principle of tax shifting is simple: tax "bads," not "goods," so that markets work to direct the economy where we want it to go. Tax pollution, waste, urban sprawl, and resource depletion, for example – not jobs, income, investment, good urban development, and resource conservation.

There are many successful examples of tax shifting in Scandinavian countries, which have been using this tool for at least a decade. Local authorities in

Canada and elsewhere are recognizing the benefits of tax shifting and are beginning to examine its potential.[26] For example, a 2005 study at Simon Fraser University in Vancouver examined six potential areas where the city could shift taxes in one municipal precinct: carbon emissions, drinking water, parking, solid waste, stormwater runoff, and sewage. The results were staggering: tax shifting on average could lead to a 23 percent reduction in environmentally harmful activities and could generate an extra $21 million in tax-based revenue. The surplus funds could in turn provide tax grants, subsidies, and rebates for environmentally friendly and sustainable development programs.[27]

The private sector also has a significant role to play in strengthening local economies. Business culture in general is starting to take on a different shape in light of public pressures and today's social and environmental realities. Over the last ten to fifteen years the socially responsible business movement has taken great strides in raising awareness that businesses need to serve the common good rather than simply maximize profits. There is growing recognition among companies and organizations that they have a social obligation to operate in ethically, socially, and environmentally responsible ways. Using "triple bottom line" accounting, some companies are measuring their performance in terms of social and environmental as well as economic results. Many business leaders are discovering the economic advantages to understanding and aligning business strategies with the values of stakeholders.[28]

Corporate social responsibility (CSR) is now a recognizable catchphrase in business circles, with standards that serve as a guide to conducting business in socially and environmentally responsible ways. General principles of CSR fall into the categories of ethics, accountability, governance, financial returns, employment practices, business relationships, products and services, community involvement, and environmental protection.[29] Most examples of CSR are found in high-income countries, although, as noted earlier, the fair trade movement is changing business practices in the developing world as companies realize the social and environmental costs of doing business. Fair trade standards embrace the principles of CSR, including environmental protection, community benefits, and fair wages.

Emerging business trends such as CSR represent a significant shift in thinking about business. Indeed, socially and environmentally responsible businesses play an important role in sustainable economic development, particularly at the local level. Given that most business transactions take place in cities, it is essential that sustainable economic development within urban centers include the commitment of businesses to adopt practices that consider people and the planet as well as profit.

In addition to local authorities and the private sector, people in their daily lives are of course important actors in local economic development in their multiple roles as workers, consumers, voters, volunteers, and advocates. People who are disengaged limit their constructive interaction and participation in their communities. Some may express their disengagement by looking elsewhere to do business or by sending their children away to learn and work; others may give up completely and relocate to another community, another city, or another country. It is through participating in local communities that people can take the necessary measures to create sustainable economies, so citizen engagement is an essential component of strengthening local economies.

COMMUNITY CAPITAL: USING ALL OUR RESOURCES

Alternative economic theories and ideas are not new: in 1973 E.F. Schumacher proposed the idea of "new economics" in his influential book *Small is Beautiful: Economics as if People Mattered*, promoting small-scale development based on meeting people's local needs. Since then, alternative local economic approaches have been put forward in both industrial and developing countries – approaches that are rooted in community and designed to meet local needs and objectives. These have emerged in response to the negative effects of globalization and as a policy approach to sustainable development at the community level. Community economic development (CED) and sustainable livelihoods are two examples of these alternative strategies.[30]

Community economic development provides a conceptual means of addressing sustainable economic development at the community level. Its core principles include a community-based approach to development; direct and meaningful community participation; integration of economic, ecological, and social aspects of local development; asset-based development based on community strengths and resources rather than deficiencies; and support for community self-reliance. Its distinguishing features are captured in this definition: CED "is a process by which communities can initiate and generate their own solutions to their common economic problems and thereby build long-term community capacity and foster the integration of economic, social, and environmental objectives."[31]

Just as sustainability has prompted a shift in transportation and energy planning away from traditional concerns with supply to a new focus on managing demand, the focus of economic development needs to shift from traditional concerns with increasing growth to one of reducing social dependence on economic growth – or what could be called EDM, economic demand management. This has distinct implications for sustainable community development, particularly regarding employment and community economic development.[32]

Community economic development not only promotes initiatives that contribute to the economic health and viability of communities, it also emphasizes environmental considerations and the importance of social considerations in broader economic thinking. Examples of sustainable CED initiatives include:

- car cooperatives to reduce the cost and necessity of car ownership (Bremen, Germany);
- sustainable employment plans to create jobs, spur private spending, and reduce pollution through public investment in energy conservation and audits (San José, California);
- new product development to encourage manufacturers to develop environmentally friendly products through municipal research and development assistance (Gothenberg, Sweden);
- increases in affordable housing supply through zoning codes that promote a variety of housing types, including smaller and multifamily homes (Portland, Oregon);
- experiments with local self-reliance through establishment of closed-loop, self-sustaining economic networks (St. Paul, Minnesota);
- community-supported agriculture projects to preserve farmland and help farmers while making

fresh fruits and vegetables available in city neighborhoods (Vancouver; London, Ontario; New York City);
- creation of local currencies such as LETS, Local Exchange Trading Systems, which seek to recirculate local resources and strengthen social ties (Toronto; Ithaca, New York; United Kingdom);
- a local ownership development project with a revolving loan fund to encourage employee-owned businesses, which are considered more stable over the long term and more likely to hire, train, and promote local residents (Burlington, Vermont); and
- a community beverage container recycling depot that employs street people – "dumpster divers" – and provides them with skills, training, and self-esteem (Vancouver).[33]

Closely related to CED is the sustainable livelihoods approach to poverty alleviation, which aims to address the immediate as well as the long-term needs of individuals and households and which takes into consideration the social and environmental as well as the economic sustainability of livelihood activities and strategies. The idea of sustainable livelihoods provides a framework to understand the practical realities and priorities of those struggling in poverty – that is, what they actually do to make a living, the assets they are able to draw on, and the everyday problems they face.[34]

Beyond income generation, successful strategies under a sustainable livelihoods approach should serve to improve access to and control over local assets and help to make individuals less vulnerable to shocks and stresses (such as illness, natural disasters, or job loss) that could otherwise exacerbate situations of debt and poverty.[35]

Historically, the sustainable livelihoods framework has been used primarily in the context of rural poverty alleviation. But the same framework can easily be applied to situations of urban poverty and livelihood generation. In fact, a sustainable livelihoods approach is necessary in order to tackle issues of urban poverty over the long term. According to the ILO, 184 million people in the world do not have jobs, although this figure reaches at least 1 billion if underemployment is also taken into account.[36]

The concepts of community economic development and sustainable livelihoods together provide a useful framework for an alternative approach to economic development that emphasizes the development of strong local economies.

Will the cumulative effects of the small initiatives described here be enough to create communities and economies with enough resilience and strength to withstand the pressures and problems of an increasingly urbanized world? As noted in the beginning of the chapter, conventional approaches to economic development leave little room for strengthening local economies. Strong local economies are those that not only generate revenue but also take into consideration the equitable distribution of wealth within communities and the environmental implications of economic activities. How does the overall approach to economic development need to change in order to facilitate the development of strong local economies?

Cities, communities, and local economies are multidimensional, with a complex interaction of social, economic, ecological, and cultural factors. Some analysts think of local economies in terms of assets or capital. The term community capital, conventionally used to refer just to economic or financial capital, has more recently been used to include natural, physical, economic, human, social, and cultural forms of capital. Strengthening local economies means focusing attention on these six forms of capital:

- Minimizing the consumption of essential natural capital means living within ecological limits, conserving and enhancing natural resources, using resources sustainably (soil, air, water, energy, and so on), using cleaner production methods, and minimizing waste (solid, liquid, air pollution, and so on).
- Improving physical capital includes focusing on community assets such as public facilities (hospitals and schools, for instance), water and sanitation provision, efficient transport, safe and high-quality housing, adequate infrastructure, and telecommunications.
- Strengthening economic capital means focusing on maximizing the use of existing resources (using waste as a resource, for example), circulating dollars within a community, making things locally to replace imports, creating a new product, trading fairly with others, and developing community financial institutions.

- Increasing human capital requires a focus on areas such as health, education, nutrition, literacy, and family and community cohesion, as well as on increased training and improved workplace dynamics to generate more productive and innovative workers; basic determinants of health such as peace and safety, food, shelter, education, income, and employment are necessary prerequisites.
- Multiplying social capital requires attention to effective and representative local governance, strong organizations, capacity-building, participatory planning, and access to information as well as collaboration and partnerships.
- Enhancing cultural capital implies attention to traditions and values, heritage and place, the arts, diversity, and social history.[37]

Strengthening local economies requires mobilizing people and their governments to shore up all these forms of community capital. Community mobilization is necessary to coordinate, balance, and catalyze community capital. This approach to stronger local economies requires some relatively new thinking about broad questions of community sustainability and self-reliance, as well as more-specific innovations concerning community ownership, management, finance, organization, capacity, and learning. This approach is increasingly referred to as sustainable community development and includes both community economic development and sustainable livelihoods strategies.

While individual actions and lifestyle choices, such as buying organic produce, are important personal contributions, strengthening local economies requires a collective shift in individual actions and political choices. Community mobilization has been effective in some contexts and some regions. The cooperative economy of Emilia Romagna in northern Italy, the Grameen Bank in Bangladesh, VanCity Credit Union in Vancouver, the Women's International Sewing Cooperative of Nueva Vida, and the campaigns for local trade across North America are all examples of the potential of community mobilization to help strengthen local economies.

Strong local economies are a fundamental part of sustainable communities. They give communities the capacity and resources to address specific and immediate problems such as the provision of health care, adequate housing, clean water and sanitation, and disaster prevention and response. Human settlements – large and small, rich and poor – need strong local economies to withstand the pressures created by an increasingly urbanized world.

NOTES

1 United Nations Population Division, World Urbanization Prospects: *The 2005 Revision*, online database at esa.un.org/unup, viewed September 2006.

2 UN-HABITAT, *State of the World's Cities, 2006/7* (London: Earthscan, 2006); International Council for Local Environmental Initiatives (ICLEI), *Accelerating Sustainable Development: Local Action Moves the World* (New York: United Nations Economic and Social Council, 2002).

3 Ian MacPherson, "Into the Twenty-first Century: Co-operatives Yesterday, Today and Tomorrow," in British Columbia Institute for Cooperative Studies, *Sorting Out: A Selection of Papers and Presentations, 1995–2005* (Victoria, BC: 2004); quote from Johnston Birchall, *Rediscovering the Co-operative Advantage: Poverty Reduction through Self-help* (Geneva: Cooperative Branch, International Labour Office, 2003), p. 3.

4 Julia Smith, *Worker Co-operatives: A Glance around the World* (Victoria, BC: British Columbia Institute for Cooperative Studies, 2003).

5 Benjamin Dangl, "Worker-run Cooperatives in Buenos Aires," *Z Magazine*, April 2005; Geoff Olson, "The Take – A Story of Hope," *Common Ground*, November 2004.

6 Smith, op. cit. note 18.

7 Birchall, op. cit. note 17; United Nations, "Co-operatives are Significant Actors in Development, says Secretary General," press release (New York, 7 July 2001).

8 For general information on microfinance, see Consultative Group to Assist the Poor, at www.cgap.org.

9 Microcredit Summit Campaign, at www.microcreditsummit.org, viewed 16 September 2006; Sam Daley-Harris, *State of the Microcredit Summit Campaign Report 2005* (Washington, DC: Microcredit Summit Campaign, 2005).

10 Daley-Harris, op. cit. note 23.

11 International Year of Microcredit 2005, *Microfinance and the Millennium Development Goals* (New York: UN Capital Development Fund, 2005).

12 *Grameen – Banking for the Poor*, at www.grameen-info.org/index.html, viewed 30 September 2006.

13 Celia W. Dugger, "Peace Prize to Pioneer of Loans for Those too Poor to Borrow," New York Times, 14 October 2006; *Grameen – Banking for the Poor*, op. cit. note 26; Alexandra Bernasek, "Banking on Social Change: Grameen Bank Lending to Women," *International Journal of Politics, Culture and Society*, spring 2003, pp. 369–85.

14 International Year of Microcredit 2005, op. cit. note 25; Box 8–3 from *Unitus – Innovative Solutions to Global Poverty* at www.unitus.com/sections/impact/impact_css_kenya.asp; Daley-Harris, op. cit. note 23.

15 Bernasek, op. cit. note 27; Rosintan D.M. Panjaitan-Drioadisuryo and Kathleen Cloud, "Gender, Self-employment and Microcredit Programs: An Indonesian Case Study," *Quarterly Review of Economics and Finance*, 39 (1999), pp. 769–79.

16 World Council of Credit Unions, at www.woccu.org, viewed 15 October 2006.

17 Information from Vancouver City Savings Credit Union, at www.vancity.com/MyCommunity, viewed 30 September 2006.

18 Shorebank Pacific, at www.eco-bank.com, viewed 30 September 2006.

19 Fair Trade Labelling Organizations International, at www.fairtrade.net/html, viewed 30 September 2006; William Young and Karla Utting, "Fair Trade, Business and Sustainable Development," *Sustainable Development*, 13 (2005), pp. 139–42.

20 Fair Trade Labelling Organizations International, op. cit. note 33.

21 TransFair Canada, at www.transfair.ca/en/fairtrade, viewed 30 September 2006; Fair Trade Labelling Organizations International, op. cit. note 33.

22 Business Alliance for Local Living Economies, at www.livingeconomies.org, viewed 14 October 2006.

23 Local campaigns from ibid.

24 ICLEI, International Development Research Centre (IDRC), and United Nations Environment Programme, *The Local Agenda 21 Planning Guide* (Toronto and Ottawa, ON: ICLEI and IDRC, 1996); Mark Roseland, *Toward Sustainable Communities:*

Resources for Citizens and their Governments (Gabriola Island, BC: New Society Publishers, 2005).

25 Shuman, op. cit. note 12.

26 Zane Parker, "Unravelling the Code: Aligning Taxes and Community Goals," Focus on Municipal Assessment and Taxation, June 2005, pp. 46–47.

27 Mark Roseland, ed., *Tax Reform as if Sustainability Mattered: Demonstrating Ecological Tax Shifting in Vancouver's Sustainability Precinct* (Vancouver, BC: Simon Fraser University, 2005).

28 Judy Wicks, *Local Living Economies: The New Movement for Responsible Business* (San Francisco: Business Alliance for Local Living Economies, 2006).

29 Social Venture Network, *Standards of Corporate Social Responsibility, 1999*, at www.svn.org/initiatives/standards.html, viewed 30 September 2006.

30 E.F. Schumacher, *Small is Beautiful* (New York: Harper & Row, 1973).

31 Markey *et al.*, op. cit. note 11, p. 2.

32 Roseland, op. cit. note 39.

33 All from ibid.

34 Lucy Stevens, Stuart Coupe, and Diana Mitlin, eds., *Confronting the Crisis in Urban Poverty: Making Integrated Approaches Work* (Bourton on Dunsmore: Intermediate Technology Publications, 2006); Robert Chambers and Gordon Conway, "Sustainable Rural Livelihoods: Practical Concepts for the Twenty-first Century," IDS Discussion Paper No. 296 (Brighton: Institute of Development Studies, December 1991).

35 Stevens, Coupe, and Mitlin, op. cit. note 49.

36 International Labour Office, Global Employment Trends Brief, February 2005.

37 Roseland, op. cit. note 39.

38

"Green Jobs"

from *Green Jobs: Working for People and the Environment* (2008)

Michael Renner, Sean Seeney, Jill Kubit, and Lisa Mastny

Editors' Introduction

On a practical level, moving towards more sustainable economies means creating large numbers of "green jobs" – types of work, as Michael Renner and his co-authors argue in this piece, that not only help protect and restore the environment but provide decent livelihoods for people. The proportion of green jobs has grown rapidly in recent years, but much more can be done to promote them. Sustainable cities and towns will be ones that emphasize this type of economic growth, through education, training, investment, regulation, and policies assisting businesses that provide them.

Renner is a long-time staff writer for the Worldwatch Institute, a Washington, D.C.-based thinktank that since the 1970s has taken a lead in researching and publicizing sustainable development strategies. This piece comes from the book *Green Jobs: Working for People and the Environment* (Washington, D.C.: Worldwatch Institute, 2008), based on a study done for the United Nations Environment Program and other organizations. Other readings on this topic include Van Jones' *The Green Collar Economy: How One Solution Can Fix Our Two Biggest Problems* (New York: HarperCollins, 2008) and, on a practical level, Scott M. Deitche's *Green Collar Jobs: Environmental Careers for the 21st Century* (Santa Barbara, CA: Praeger, 2010).

The pursuit of so-called "green jobs" – employment that contributes to protecting the environment and reducing humanity's carbon footprint – will be a key economic driver of the 21st century. "Climate-proofing" the global economy will involve large-scale investments in new technologies, equipment, buildings, and infrastructure, which will provide a major stimulus for much-needed new employment and an opportunity for retaining and transforming existing jobs.

The number of green jobs is on the rise. The renewable energy sector has seen rapid expansion in recent years, with current employment in renewables and supplier industries estimated at a conservative 2.3 million worldwide. The wind power industry employs some 300,000 people, the solar photovoltaics (PV) sector an estimated 170,000, and the solar thermal industry more than 600,000. More than 1 million jobs are found in the biofuels industry growing and processing a variety of feedstocks into ethanol and biodiesel.

Construction jobs can be greened by ensuring that new buildings meet high performance standards. And retrofitting existing buildings to make them more energy-efficient has huge job potential for construction workers, architects, energy auditors, engineers, and others. The weatherization of some 200,000 apartments in Germany created 25,000 new jobs and helped retain 116,000 existing jobs in 2002–04.

The transportation industry is a cornerstone of modern economies, but it also has the fastest-rising carbon emissions of any sector. Relatively green auto manufacturing jobs – those in manufacturing the most-efficient cars currently available – today number no more than about 250,000 out of roughly 8 million in the auto sector worldwide. Modern rail and urban transit systems offer a greener alternative, but they need fresh commitment and investments to reverse the job erosion of recent decades. In growing numbers of cities, good jobs are being generated by the emergence of bus rapid transit systems. There are also substantial green employment opportunities in retrofitting old diesel buses to reduce air pollutants and in replacing old equipment with cleaner compressed natural gas (CNG) or hybrid-electric buses. In New Delhi, the introduction of 6,100 CNG buses in the late 2000s created 18,000 new jobs.

The steel, aluminum, cement, and paper industries are highly energy-intensive and polluting. But increasing scrap use, greater energy efficiency, and reliance on alternative energy sources may at least render them a pale shade of green. Worldwide, more than 40 percent of steel output and one-quarter of aluminum production is based on recycled scrap, possibly employing more than a quarter million people.

Recycling and remanufacturing jobs worldwide number many millions, but incompatible definitions and a lack of data gathering make a global tally impossible. China alone is thought to have some 10 million jobs in this sector, and the United States has more than 1 million. In developing countries, recycling is often done by informal networks of scavengers. Brazil, which boasts a high rate of aluminum recycling, relies on some 500,000 scrap collectors. Cairo's 70,000 Zabaleen recycle as much as 85 percent of the materials they collect.

Agriculture and forestry often still account for the bulk of employment and livelihoods in many developing countries. Small farms are more labor- and knowledge-intensive than agroindustrial farms are, and they use fewer energy and chemical inputs. But relatively sustainable forms of smallholder agriculture are being squeezed hard by energy- and pesticide-intensive farms and by global supply chains. Organic farming is still limited. But because it is more labor-intensive than industrialized agriculture, it can be a source of growing green employment.

Afforestation and reforestation efforts, as well as better stewardship of critical ecosystems more generally, could support livelihoods among the more than 1 billion people who depend on forests, often through non-timber forest products. Planting trees creates large numbers of jobs, although these are often seasonal and low paid. Agroforestry, which combines tree planting with traditional farming, offers significant environmental benefits in degraded areas – including carbon sequestration. Some 1.2 billion people already depend on it to some extent.

There is additional job potential in efforts to adapt to, and cope with, climate change. Building flood barriers, terracing land, and rehabilitating wetlands is labor-intensive work. Efforts to protect croplands from environmental degradation and to adapt farming to climate change by raising water efficiency, preventing erosion, planting trees, using conservation tillage, and rehabilitating degraded crop and pastureland can also support rural livelihoods.

The potential for green jobs is immense. But much of it will not materialize without massive and sustained investments in the public and private sectors. Governments need to establish a firm framework for greening all aspects of the economy, with the help of targets and mandates, business incentives, and reformed tax and subsidy policies. It will also be critical to develop innovative forms of technology transfer to spread green methods around the world at the scale and speed required to avoid full-fledged climate change. Cooperative technology development and technology-sharing programs could help expedite the process of replicating best practices.

To provide as many workers as possible with the qualifications they will increasingly need, an expansion of green education, training, and skill-building programs in a broad range of occupations is crucial. Resource extraction and energy-intensive industries are likely to feel the greatest impact in transitioning to a low-carbon future, and regions and communities highly dependent on them will need assistance in diversifying their economic base, creating alternative jobs and livelihoods, and acquiring new skills. This is known as a "just transition."

Green jobs need to be decent jobs – offering good wages and income security, safe working conditions, dignity at work, and adequate workers' rights. Sadly, this is not always the case today. Recycling work is sometimes precarious, involving serious occupational health hazards and often generating less than living wages and incomes. Growing crops at biofuels plantations in countries like Brazil, Colombia, Malaysia, and Indonesia often involves excessive workloads, poor pay, exposure to pesticides, and oppression of workers.

These cautionary aspects highlight the need for sustainable employment to be good not only for the environment but also for the people holding the jobs. Still, an economy that reconciles human aspirations with the planet's limits is eminently possible.

DEFINING GREEN JOBS

All sectors – agriculture, manufacturing, and construction as well as scientific, technical, administrative, and service activities – need to undergo a greening process. Green jobs will eventually span a wide array of skills, educational backgrounds, and occupational profiles. In many existing industries and occupations, environmental awareness and applied green literacy will become increasingly important.

Some green jobs are easily identifiable – for example, people employed in installing solar panels or operating wind turbines. Other changes that help put the economy on a more sustainable footing – such as efforts to boost the efficiency of energy, water,

and materials use – also involve some degree of "green" employment. But because there is no clear threshold to define this efficiency, it can be difficult to decide which jobs are truly green. Moreover, a workplace may introduce green technologies and practices in ways that are hard to detect from the outside. Blue-collar workers may be transformed into green-collar workers fairly quietly as they respond to subtle changes in day-to-day practices and methods.

A narrow definition of green jobs might focus solely on the environmental credentials of a job. However, green jobs also need to be decent jobs – with regard to wages, career prospects, job security, occupational health and safety, and worker rights. (See Figure 1.) People's livelihoods, rights, and sense of dignity are bound up tightly with their employment; jobs need to provide equal hope for the environment and the jobholder. A job that is exploitative, harmful, or fails to pay a living wage (or worse, condemns workers to a life of poverty) is hardly reason for celebration. Decent work conditions must be as important to advocates for the environment as environmental concerns are to advocates for labor.

As the world moves toward a low-carbon, sustainable economy, those companies, countries, and regions that are leaders in green innovation, design, and technology will be more likely to retain and create new green jobs. The laggards, meanwhile, may incur substantial business and job penalties.

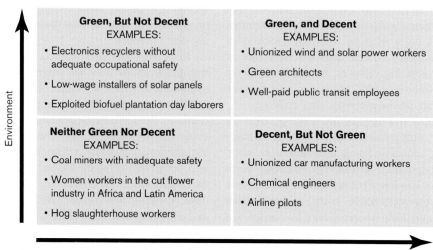

Figure 1 Green and decent jobs? A schematic overview.

Green jobs may arise in different locations than the old jobs in extractive and polluting industries, presenting a substantial challenge to places that depend heavily on these industries. But some cities and regions have begun to successfully reinvent themselves. Toledo, Ohio, a typical "rust-belt" city in the United States once dominated by automotive firms, has become a desirable location for solar companies. Glass manufacturers there have reoriented themselves from making car windshields to making solar panels.[1]

Resource-extractive and heavily polluting industries are likely to feel the greatest impact of the move toward sustainability. But blocking environmental action would not necessarily save jobs in these industries. Because of the rapid pace of automation and resource depletion, employment is already shrinking in many of these sectors, even as output grows. In fact, jobs are more likely to be at risk in industries where environmental standards are low and "clean-tech" innovation is lagging. As the urgency of sustainability rises, so does the cost of a do-nothing strategy that misses opportunities for early action.

PATHWAYS TO A SUSTAINABLE FUTURE

The potential for further green job growth is tremendous – from opportunities to address the accumulated environmental ills of the past, to improving our ability to cope with climate change, to creating more efficient and viable economies. The green employment that would result from these initiatives is many magnitudes larger than anything currently on the drawing board. Still, this optimistic assessment of the potential for future green job growth must be seen against the backdrop of some sobering realities that policymakers need to address. These include:

Green jobs are expanding, but not rapidly enough. This is especially true in light of the fact that the labor market is expanding by some tens of millions of people every year, but world unemployment is at record levels. Together, the unemployed and underemployed (those working hard without earning sufficient incomes) amount to 1 in 3 of the world's workers. Unemployment has hit young people, aged 15 to 24, the hardest, with 86.3 million youth representing 44 percent of the world's total unemployed in 2006.[2]

Green investment – and thus most of the green jobs in the foreseeable future – is found primarily in a relatively small number of countries. Countries like Japan and Germany that lead green technology development are likely to reap the bulk of the associated revenues and jobs. However, green jobs are still the exception in most developing countries, which account for some 80 percent of the world's workforce. More needs to be done to ensure that green employment becomes a truly global phenomenon.

The rising level of informality in the global economy constitutes a major challenge to green job growth. In addition, the chronic and worsening levels of inequality both within and between countries are a major impediment. The effort to advance decent work and pro-poor sustainable development is critical to building green jobs across the developing world in particular.

Unsustainable business practices remain prevalent and are often more profitable than green ways of doing business. Short-term pressures of shareholders and financial markets are not easily overcome. The early adopters of green business practices have to contend with companies – manufacturers and retailers – that command consumer loyalty through low prices achieved on the back of "externalized" costs. And surprisingly often, market failures, coupled with lack of green knowledge, impede action.

Striking the right balance between government and private sector action, financing a green jobs agenda, developing worker skills, and ensuring a "just transition" are critical to overcoming present obstacles. Private companies have an important role to play in green job creation. Green innovation helps businesses stay at the cutting edge and hold down costs by reducing wasteful practices. However, the risk and profit appraisals typical of modern business behavior, the ever-rising expectations of shareholders, as well as concerns about protecting intellectual property may impede the flow of capital into the green economy. Experience in various areas – from vehicle fuel economy to carbon trading – suggests that a purely market-driven process will not be able to deliver the changes needed at the scale and speed demanded by the climate crisis.

Today's business practices are too often driven by short-term considerations, whereas truly sustainable development requires a long-term approach.

Governments must therefore establish an ambitious and clear policy framework to reward, support, and drive sustainable economic and social activity, and be prepared to confront those whose business practices continue to pose a serious threat to a sustainable future. Timely action on the scale needed will occur only with a clear set of targets and mandates, business incentives, public investment, ecological tax reform, and genuine public-private partnerships.

Expediting the development and diffusion of green technologies is critical to a global green jobs future. Innovative public-private partnerships can be part of the solution. Cooperative R&D centers that anchor green technology development in the public realm are another. And an adequately endowed global fund to speed the spread of green technologies and climate adaptation measures also deserves urgent consideration.

INVESTMENT CREATES EMPLOYMENT

A sustainable economy cannot be built on "green for a few" – a few countries, a relatively limited number of workers, with regrettably few positive outcomes overall. It must mean "green for all" – creating decent work and stable communities and allowing for a fairer distribution of wealth.[3] To make the term "green jobs" meaningful, considerations such as wages, working conditions, and workers' rights will have to become an integral aspect of future policies and strategies. The shift to a low carbon and sustainable society must be as equitable as possible. It must, in a phrase, be a "just transition."

To achieve social solidarity and to mobilize political and workplace support for the needed changes, policies are needed to protect those who are likely to be negatively affected by the green jobs transition – such as through income support, retraining opportunities, and relocation assistance. Social dialogue is a critical component of a Just Transition, especially in the workplace where the worker/union voice is needed to help determine the design of sustainable production systems and work practices.

For example, joint labor–management committees and similar bodies could work to identify ways to improve energy and resource efficiency. In some instances, employers and unions are beginning to work together in greening the workplace, building on a long tradition of collaborating on occupational safety and health and other issues.

REFERENCES

1   Daniel McGinn, "Workers Find Jobs in Emerging Green Economy," *Newsweek*, 8 October 2007.
2   International Labour Organization, *Global Employment Trends: January 2008* (Geneva: 2008).
3   See www.greenforall.org.

GREEN ARCHITECTURE AND BUILDING

39

"Design, Ecology, Ethics and the Making of Things"

a sermon given at the Cathedral of St. John the Divine, New York City (1993)

William McDonough

Editors' Introduction

Solar energy technologies and passive solar heating of buildings (using the sun's energy to warm interior spaces) were hallmarks of the first wave of modern ecological architecture in the 1960s and 1970s, along with efforts to improve the energy efficiency of new construction. Solar strategies remained on the fringe of new construction, but energy conservation requirements were soon written into many building codes, in part to respond to the energy crises of the 1970s. Some pioneering architects also began experimenting with natural or recycled building materials, such as earthen or straw bale walls and remilled lumber, and employed new designs intended to emulate those found in nature.

By the 1990s and 2000s a much wider variety of sustainable design practices and building materials had begun to enter mainstream construction, encouraged by the emergence of green building standards such as the LEED (Leadership for Energy and Environmental Design) standards first codified by the U.S. Green Building Council in 1998. Even some relatively mainstream buildings began to contain green features. Along the way, many architects rediscovered the vernacular building practices that cultures have used historically to adapt their buildings to local climates, materials, and traditions. Designers also explored ancient systems such as the Chinese *feng shui*, an elaborate set of beliefs about the flow of energy within landscapes and buildings. While green building practices are still far from the mainstream of development, their use is gradually spreading, and there are an increasing number of examples of built projects.

Architect William McDonough has been a leader in the movement to rethink architecture and design in ways that can enhance urban sustainability. Former Dean of the School of Architecture at the University of Virginia, McDonough is the principal of William McDonough + Partners in Charlottesville, Virginia. Among his better-known projects are the headquarters of Environmental Defense in New York City, the Ford Motor Company's renovated River Rouge plant in Dearborn, Michigan, and a green Walmart store in Lawrence, Kansas. McDonough was also the author of the Hannover Principles, an influential set of ecological design principles that circulated widely in the early 1990s. His writings include the book *Cradle to Cradle: Remaking the Way We Make Things* (with Michael Braungart; New York: North Point Press, 2002) and *The Upcycle: Beyond Sustainability – Designing for Abundance* (with Michael Braungart; New York: North Point Press, 2013).

Other writings in the field of ecological design include Sim Van der Ryn and Cowan's *Ecological Design* (Washington, D.C.: Island Press, 2007), Van der Ryn and Jason McLennan's *Design for an Empathic World: Reconnecting People, Nature, and Self* (Washington, D.C.: Island Press, 2013), John Tillman Lyle's *Regenerative Design for Sustainable Development* (New York: Wiley, 1994), Janine M. Benyus' *Biomimicry: Innovation Inspired by Nature* (New York: HarperCollins, 1997), Nancy Todd's *A Safe and Sustainable World: the Promise of Ecological Design* (Washington, D.C: Island Press, 2005), and David Orr's *The Nature of Design* (New York: Oxford University Press, 2002), and *Design on the Edge: the Making of a High-performance Building* (Cambridge, MA: MIT Press, 2006).

It is humbling to be an architect in a cathedral because it is a magnificent representation of humankind's highest aspirations. Its dimension is illustrated by the small Christ figure in the western rose window, which is, in fact, human scale. A cathedral is a representation of both our longings and intentions. This morning, here at this important crossing in this great building, I am going to speak about the concept of design itself as the first signal of human intention and will focus on ecology, ethics, and the making of things. I would like to reconsider both our design and our intentions.

When Vincent Scully gave a eulogy for the great architect Louis Kahn, he described a day when both were crossing Red Square, whereupon Scully excitedly turned to Kahn and said, "Isn't it wonderful the way the domes of St. Basil's Cathedral reach up into the sky?" Kahn looked up and down thoughtfully for a moment and said, "Isn't it beautiful the way they come down to the ground?"

If we understand that design leads to the manifestation of human intention and if what we make with our hands is to be sacred and honor the earth that gives us life, then the things we make must not only rise from the ground but return to it, soil to soil, water to water, so everything that is received from the earth can be freely given back without causing harm to any living system. This is ecology. This is good design. It is of this we must now speak.

If we use the study of architecture to inform this discourse, and we go back in history, we will see that architects are always working with two elements, mass and membrane. We have the walls of Jericho, mass, and we have tents, membranes. Ancient peoples practiced the art and wisdom of building with mass, such as an adobe-walled hut, to anticipate the scope and direction of sunshine. They knew how thick a wall needed to be to transfer the heat of the day into the winter night, and how thick it had to be to transfer the coolness into the interior in the summer. They worked well with what we call "capacity" in the walls in terms of storage and thermal lags. They worked with resistance, straw, in the roof to protect from heat loss in the winter and to shield the heat gain in summer from the high sun. These were very sensible buildings within the climate in which they are located.

With respect to membrane, we only have to look at the Bedouin tent to find a design that accomplishes five things at once. In the desert, temperatures often exceed 120 degrees. There is no shade, no air movement. The black Bedouin tent, when pitched, creates a deep shade that brings one's sensible temperature down to 95 degrees. The tent has a very coarse weave, which creates a beautifully illuminated interior, having a million light fixtures. Because of the coarse weave and the black surface, the air inside rises and is drawn through the membrane. So now you have a breeze coming in from outside, and that drops the sensible temperature even lower, down to 90 degrees. You may wonder what happens when it rains, with those holes in the tent. The fibers swell up and the tent gets tight as a drum when wet. And of course, you can roll it up and take it with you. The modern tent pales by comparison to this astonishingly elegant construct.

Throughout history, you find constant experimentation between mass and membrane. This cathedral is a Gothic experiment intregrating great light into massive membrane. The challenge has always been, in a certain level, how to combine light with mass and air. This experiment displayed itself powerfully in modern architecture, which arrived with the advent of inexpensive glass. It was unfortunate that at the same time the large sheet of glass showed up, the era of cheap energy was ushered in, too. And because of that, architects no longer rely upon the sun for heat or illumination. I have spoken to thousands of architects, and when I ask the question, "How many of you know how to find true south?," I rarely get a raised hand.

Our culture has adopted a design stratagem that essentially says that if brute force or massive amounts of energy don't work, you're not using enough of it. We made glass buildings that are more about buildings than they are about people. We've used the glass ironically. The hope that glass would connect us to the outdoors was completely stultified by making the buildings sealed. We have created stress in people because we are meant to be connected with the outdoors, but instead we are trapped. Indoor air quality issues are now becoming very serious. People are sensing how horrifying it can be to be trapped indoors, especially with the thousands upon thousands of chemicals that are being used to make things today.

Le Corbusier said in the early part of this century that a house is a machine for living in. He glorified the steamship, the airplane, the grain elevator. Think about it: a house is a machine for living in. An office is a machine for working in. A cathedral is a machine for praying in. This has become a terrifying prospect, because what has happened is that designers are now designing for the machine and not for people. People talk about solar heating a building, even about solar heating a cathedral. But it isn't the cathedral that is asking to be heated, it is the people. To solar-heat a cathedral, one should heat people's feet, not the air 120 feet above them. We need to listen to biologist John Todd's idea that we need to work with living machines, not machines for living in. The focus should be on people's needs, and we need clean water, safe materials, and durability. And we need to work from current solar income.

There are certain fundamental laws that are inherent to the natural world that we can use as models and mentors for human designs. Ecology comes from the Greek roots Oikos and Logos, "household" and "logical discourse." Thus, it is appropriate, if not imperative, for architects to discourse about the logic of our earth household. To do so, we must first look at our planet and the very processes by which it manifests life, because therein lie the logical principles with which we must

work. And we must also consider economy in the true sense of the word. Using the Greek words Oikos and Nomos, we speak of natural law and how we measure and manage the relationships within this household, working with the principles our discourse has revealed to us.

And how do we measure our work under those laws? Does it make sense to measure it by the paper currency that you have in your wallet? Does it make sense to measure it by a grand summation called GNP? For if we do, we find that the foundering and rupture of the *Exxon Valdez* tanker was a prosperous event because so much money was spent in Prince William Sound during the clean-up. What then are we really measuring? If we have not put natural resources on the asset side of the ledger, then where are they? Does a forest really become more valuable when it is cut down? Do we really prosper when wild salmon are completely removed from a river?

There are three defining characteristics that we can learn from natural design. The first characteristic is that everything we have to work with is already here – the stones, the clay, the wood, the water, the air. All materials given to us by nature are constantly returned to the earth, without even the concept of waste as we understand it. Everything is cycled constantly with all waste equaling food for other living systems.

The second characteristic is that one thing allowing nature to continually cycle itself through life is energy, and this energy comes from outside the system in the form of perpetual solar income. Not only does nature operate on "current income," it does not mine or extract energy from the past, it does not use its capital reserves, and it does not borrow from the future. It is an extraordinarily complex and efficient system for creating and cycling nutrients, so economical that modern methods of manufacturing pale in comparison to the elegance of natural systems of production.

Finally, the characteristic that sustains this complex and efficient system of metabolism and creation is biodiversity. What prevents living systems from running down and veering into chaos is a miraculously intricate and symbiotic relationship between millions of organisms, no two of which are alike.

As a designer of buildings, things, and systems, I ask myself how to apply these three characteristics of living systems to my work. How do I employe the concept of waste equals food, of current solar income, of protecting biodiversity in design? Before I can even apply these principles, though, we must understand the role of the designer in human affairs.

In thinking about this, I reflect upon a commentary of Emerson's. In the 1830s, when his wife died, he went to Europe on a sailboat and returned in a steamship. He remarked on the return voyage that he missed the "Aeolian connection." If we abstract this, he went over on a solar-powered recyclable vehicle operated by craftspersons, working in the open air, practicing ancient arts. He returned in a steel rust bucket, spilling oil on the water and smoke into the sky, operated by people in a black dungeon shoveling coal into the mouth of a boiler. Both ships are objects of design. Both are manifestations of our human intention.

Peter Senge, a professor at MIT's Sloan School of Management, works with a program called the Learning Laboratory where he studies and discusses how organizations learn. Within that he has a leadership laboratory, and one of the first questions he asks CEOs of companies that attend is, "Who is the leader on a ship crossing the ocean?" He gets obvious answers, such as the captain, the navigator, or the helmsman. But the answer is none of the above. The leader is the designer of the ship because operations on a ship are a consequence of design, which is the result of human intention. Today, we are still designing steam-ships, machines powered by fossil fuels that have deleterious effects. We need a new design.

I grew up in the Far East, and when I came to this country, I was taken aback when I realized that we were not people with lives in America, but consumers with lifestyles. I wanted to ask someone: when did America stop having people with lives? On television, we are referred to as consumers, not people. But we are people, with lives, and we must make and design things for people. And if I am a consumer, what can I consume? Shoe polish, food, juice, some toothpaste. But actually, very little that is sold to me can actually be consumed. Sooner or later, almost all of it has to be thrown away. I cannot consume a television set. Or a VCR. Or a car. If I presented you with a television set and covered it up and said, "I have this amazing item. What it will do as a service will astonish you. But before I tell you what it does, let me tell you what it is made of and you can tell me if you want it in your house. It contains 4,060 chemicals, many of which are toxic, two hundred of which off-gas into the room when it is turned on. It also contains 18 grams of toxic methyl mercury, has an explosive glass tube, and I urge you to put it at eye-level with your children and encourage them to play with it." Would you want this in your home?

Michael Braungart, an ecological chemist from Hamburg, Germany, has pointed out that we should remove the word "waste" from our vocabulary and start using the word product instead, because if waste is going to equal food, it must also be a product. Braungart suggests we think about three distinct product types:

First, there are consumables, and actually we should be producing more of them. These are products that when eaten, used, or thrown away, literally turn back into dirt, and therefore are food for other living organisms. Consumables should not be placed in landfills, but put on the ground so that they restore the life, health, and fertility of the soil. This means that shampoos should be in bottles made of beets that are biodegradable in your compost pile. It means carpets that break down into carbon dioxide and water. It means furniture made of lignin, potato peels and technical enzymes that looks just like your manufactured furniture of today except it can be safely returned to the earth. It means that all "consumable" goods should be capable of returning to the soil from whence they came.

Second are products of service, also known as durables, such as cars and television sets. They are called products of service because what we want as customers is the service the product provides – food, entertainment, or transportation. To eliminate the concept of waste, products of service would not be sold, but licensed to the end-user. Customers may use them as long as they wish, even sell the license to someone else, but when the end-user is finished with, say, a television, it goes back to Sony, Zenith, or Philips. It is "food" for their system, but not for natural systems. Right now, you can go down the street, dump a TV into the garbage can, and walk away. In the process, we deposit persistent toxins throughout the planet. Why do we give people that responsibility and stress? Products of service must continue beyond their initial product life, be owned by their manufacturers, and be designed for disassembly, remanufacture, and continuous re-use.

The third type of product is called "unmarketables." The question is, why would anyone produce a product that no one would buy? Welcome to the world of nuclear waste, dioxins, and chromium-tanned leather. We are essentially making products or subcomponents of products that no one should buy, or, in many cases, do not realize they are buying. These products must not only cease to be sold, but those already sold should be stored in warehouses when they are finished until we can figure out a safe and non-toxic way to dispose of them.

I will describe a few projects and how these issues are implicit in design directions. I remember when we were hired to design the office for an environmental group. The director said at the end of contract negotiations, "By the way, if anybody in our office gets sick from indoor air quality, we're going to sue you." After wondering if we should even take the job, we decided to go ahead, that it was our job to find the materials that wouldn't make people sick when placed inside a building. And what we found is that those materials weren't there. We had to work with manufacturers to find out what was in their products, and we discovered that the entire system of building construction is essentially toxic. We are still working on the materials side.

For a New York men's clothing store, we arranged for the planting of 1,000 oak trees to replace the two English oaks used to panel the store. We were inspired by a famous story told by Gregory Bateson about New College in Oxford, England. It went something like this. They had a main hall built in the early 1600s with beams forty feet long and two feet thick. A committee was formed to try to find replacement trees because the beams were suffering from dry rot. If you keep in mind that a veneer from an English oak can be worth seven dollars a square foot, the total replacement cost for the oaks was prohibitively expensive. And they didn't have straight forty foot English oaks from mature forests with which to replace the beams. A young faculty member joined the committee and said, "Why don't we ask the College Forester if some of the lands that have been given to Oxford might have enough trees to call upon?" And when they brought in the forester he said,

"We've been wondering when you would ask this question. When the present building was constructed 350 years ago, the architects specified that a grove of trees be planted and maintained to replace the beams in the ceiling when they would suffer from dry rot." Bateson's remark was, "That's the way to run a culture." Our question and hope is, "Did they replant them?"

For Warsaw, Poland, we responded to a design competition for a high-rise building. When the client chose our design as the winner after seeing the model, we said, "We're not finished yet. We have to tell you about the building. The base is made from concrete and includes tiny bits of rubble from World War II. It looks like limestone, but the rubble's there for visceral reasons." And he said, "I understand, a phoenix rising." And we said the skin is recycled aluminum, and he said, "That's O.K., that's fine." And we said, "The floor heights are thirteen feet clear so that we can convert the building into housing in the future, when its utility as an office building is no longer. In this way, the building is given a chance to have a long, useful life." And he said, "That's O.K." And we told him that we would have opening windows and that no one would be further than twenty-five feet from a window, and he said that was O.K., too. And finally, we said, "By the way, you have to plant ten square miles of forest to offset the building's effect on climate change." We had calculated the energy costs to build the structure, and the energy cost to run and maintain it, and it worked out that 6,400 acres of new forest would be needed to offset the effects on climate change from the energy requirements. And he said he would get back to us. He called back two days later and said, "You still win. I checked out what it would cost to plant ten square miles of trees in Poland and it turns out it's equivalent to a small part of our advertising budget."

The architects representing a major retail chain called us a year ago and said, "Will you help us build a store in Lawrence, Kansas?" I said that I didn't know if we could work with them. I explained my thoughts on consumers with lifestyles, and we needed to be in the position to discuss their stores' impact on small towns. Click. Three days later we were called back and were told, "We have a question for you that is coming from the top. Are you willing to discuss the fact that people with lives have the right to buy the finest-quality products, even under your own terms, at the lowest possible price?" We said, "Yes." "Then we can talk about the impact on small towns."

We worked with them on the store in Kansas. We converted the building from steel construction, which uses 300,000 BTUs per square foot, to wood construction, which uses 40,000 BTUs, thereby saving thousands of gallons of oil just in the fabrication of the building. We used only wood that came from resources that were protecting biodiversity. In our research we found that the forests of James Madison and Zachary Taylor in Virginia had been put into sustainable forestry and the wood for the beams came from there and other forests managed this way. We also arranged for no CFCs to be used in the store's construction and systems, and initiated significant research and a major new industry in daylighting. We have yet to fulfill our concerns about the bigger questions of products, their distribution and the chain's impact on small towns, with the exception that this store is designed to be converted into housing when its utility as a retail outlet has expired.

For the City of Frankfurt, we are designing a daycare center that can be operated by the children. It contains a greenhouse roof that has multiple functions: it illuminates, heats both air and water, cools, ventilates, and shelters from the rain, just like a Bedouin tent. One problem we were having during the design process was the engineers wanted to completely automate the building, like a machine. The engineers asked, "What happens if the children forget to close the shade and they get too hot?" We told them the children would open a window. "What if they don't open a window?", the engineers wanted to know. And we told them that in that case the children would probably close the shade. And they wanted to know what would happen if the children didn't close the shade. And finally we told them the children would open windows and close shades when they were hot because children are not dead but alive. Recognizing the importance for children to look at the day in the morning and see what the sun is going to do that day and interact with it, we enlisted the help of teachers of Frankfurt to get this one across because the teachers had told us the most important thing was to find something for the children to do. Now the children have ten minutes of activity in the morning and have ten minutes of activity when they leave the building, opening and closing the system, and both the children and teachers love the idea. Because of the solar hot-water collectors, we asked that a public laundry be added to the program so that parents could wash clothes while awaiting their children in school. Because of advances in glazing, we are able to create a daycare center that requires no fossil fuels for operating the heating or cooling. Fifty years from now, when fossil fuels will be scarce, there will be hot water for the community, a social center, and the building will have paid back the energy "borrowed" for its construction.

As we become aware of the ethical implications of design, not only with respect to buildings, but in every aspect of human endeavor, they reflect changes in the historical concept of who or what has rights. When you study the history of rights, you begin with the Magna Carta, which was about the rights of white, English, noble males. With the Declaration of Independence, rights were expanded to all landowning white males. Nearly a century later, we moved to the emancipation of slaves, and during the beginnings of this century, to suffrage, giving the right to women to vote. Then the pace picks up with the Civil Rights Act in 1964, and then in 1973, the Endangered Species Act. For the first time, the right of other species and organisms to exist was recognized. We have essentially "declared" that Homo Sapiens are part of the web of life. Thus, if Thomas Jefferson were with us today, he would be calling for a Declaration of Interdependence which recognizes that our ability to pursue wealth, health, and happiness is dependent on other forms of life, that the rights of one species are linked to the rights of others and none should suffer remote tyranny.

This Declaration of Interdependence comes hard on the heels of realizing that the world has become vastly complex, both in its workings and in our ability to perceive and comprehend those complexities. In this complicated world, prior modes of domination have essentially lost their ability to maintain control. The sovereign, whether in the form of a king or nation, no longer seems to reign. Nations have lost control of money to global, computerized trading systems. The sovereign is also losing the ability to deceive and manipulate, as in the case of Chernobyl. While the erstwhile Soviet Republic told the world that

Chernobyl was nothing to be concerned about, satellites with ten-meter resolution showed the world that it was something to worry about. And what we saw at the Earth Summit was that the sovereign has lost the ability to lead even on the most elementary level. When Maurice Strong, the chair of the United Nations Conference on the Environment and Development, was asked how many leaders were at the Earth Summit, he said there were over 100 heads of state. Unfortunately, we didn't have any leaders.

When Emerson came back from Europe, he wrote essays for Harvard on Nature. He was trying to understand that if human beings make things and human beings are natural, then are all the things human beings make natural? He determined that Nature was all those things which were immutable. The oceans, the mountains, the sky. Well, we now know that they are mutable. We were operating as if Nature is the Great Mother who never has any problems, is always there for her children, and requires no love in return. When you think about Genesis and the concept of dominion over natural things, we realize that even if we want to get into a discussion of steward-ship versus dominion, in the end, the question is, if you have dominion, and perhaps we do have dominion, isn't it implicit that we have stewardship too, because how can you have dominion over something you've killed?

We must face the fact that what we are seeing across the world today is war, a war against life itself. Our present systems of design have created a world that grows far beyond the capacity of the environment to sustain life into the future. The industrial idiom of design, failing to honor the prin-ciples of nature, can only violate them, producing waste and harm, regardless of purported intention. If we destroy more forests, burn more garbage, drift-net more fish, burn more coal, bleach more paper, destroy more topsoil, poison more insects, build over more habitats, dam more rivers, produce more toxic and radioactive waste, we are creating a vast industrial machine, not for living in, but for dying in. It is a war, to be sure, a war that only a few more generations can surely survive.

When I was in Jordan, I worked for King Hussein on the master plan for the Jordan Valley. I was walking through a village that had been flattened by tanks and I saw a child's skeleton squashed into the adobe block and was horrified. My Arab host turned to me and said, "Don't you know what war is?" And I said, "I guess I don't." And he said, "War is when they kill your children." So I believe we're at war. But we must stop. To do this, we have to stop designing everyday things for killing, and we have to stop designing killing machines.

We have to recognize that every event and manifestation of nature is "design," that to live within the laws of nature means to express our human intention as an interdependent species, aware and grateful that we are at the mercy of sacred forces larger than ourselves, and that we obey these laws in order to honor the sacred in each other and in all things. We must come to peace with and accept our place in the natural world.

40

"Principles of Green Architecture"
from *Green Architecture* (1991)

Brenda Vale and Robert Vale

Editors' Introduction

Many sets of principles for an ecological or sustainable architecture have been proposed. Among these are McDonough's Hannover Principles, Van der Ryn and Cowan's Ecological Design Principles, and Lyle's Regenerative Design Principles. Most such manifestos set forth similar themes in slightly different ways – in particular encouraging designers to conserve energy and nonrenewable resources, to take account of local place characteristics, and to work with building users and surrounding communities.

British architects Brenda and Robert Vale here present one of the simplest and most straightforward frameworks for green architecture, contained in their book *Green Architecture: Design for an Energy-Conscious Future* (Boston: Little, Brown and Co., 1991). They illustrate these principles with extensive examples of building designs from Europe, the United Kingdom, and the United States. Like McDonough, the Vales emphasize learning from the vernacular architecture that already embodies generations of people's experience in living with a particular place and climate.

More detailed information on green building principles and implementation is available from many sources, including the American Institute of Architects' Committee on the Environment (www.aia.org/cote), the U.S. Green Building Council (www.usgbc.org), and, in the United Kingdom and Europe, Sustainable Homes (www.sustainablehomes.co.uk), the Ecological Design Association (www.edaweb.org), and the European Housing Ecology Network (www.ehen-europe.net/). Interested readers may also want to see David Pearson's writings such as *The New Natural House Book* (New York: Simon & Schuster, 1998), *New Organic Architecture* (Berkeley, CA: University of California Press, 2001), *In Search of Natural Architecture* (New York: Abbeville Press, 2005), and *Designing Your Natural Home: A Practical Guide* (New York: Collins Design, 2005). Other related works include James Steele's *Ecological Architecture: A Critical History* (London: Thames & Hudson, 2005) and Thomas Schröpfer's *Ecological Urban Architecture* (Basel: Birkhäuser, 2012).

The "green" approach to architecture is not a new approach. It has existed since people first selected a south-facing cave rather than one facing north to achieve comfort in a temperate climate. What is new is the realization that a green approach to the built environment involves a holistic approach to the design of buildings; that all the resources that go into a building, be they materials, fuels or the contribution of the users, need to be considered if a sustainable architecture is to be produced. Many buildings embody at least one of the various identifiable green characteristics. Few as yet embrace the holistic approach. . . .

Because the "green" argument is to suggest that issues are interdependent, and that the ramifications of any decision need to be considered, the notion of separate principles runs counter to a green approach. Inevitably principles will overlap. Nevertheless, the headings that follow are offered as a series of differing emphases among which a balance needs to he found for a green architecture to emerge.

PRINCIPLE 1: CONSERVING ENERGY

A building should be constructed so as to minimize the need for fossil fuels to run it.

Past societies accepted the necessity of this principle without question. It is only with the recent proliferation of materials and technologies that such a basis for ordinary building has been lost. Whether by the use of materials or the disposition of building elements, buildings modified climate to suit the needs of the users. Moreover, the very idea of community comes from the sheltering of people together, whether to provide maximum areas of shade and cooler air between buildings or to reduce the external surface area of the community as it faced the hostile weather. People constructed their buildings together because of the mutual benefit to be obtained. A policy of cheap energy removed this generator of traditional community as surely as did the automobile.

Recent buildings that have attempted to reduce dependence on fossil fuels have tended to stand alone as separate experiments rather than cluster in patterns that respond to local climate. Consequently, such experiments must be viewed as half-way attempts towards the creation of a green architecture. Many such experiments have also come from the committed individual rather than from the community as a whole, thereby further separating out the single achievement.

[. . .]

PRINCIPLE 2: WORKING WITH CLIMATE

Buildings should be designed to work with climate and natural energy sources. . . .

Building form and the disposition of building elements can alter internal comfort conditions, rather than demonstrating the reduction of fossil fuel demands through the use of insulation in the building fabric. Inevitably there will be some overlap in the two approaches.

In the days before the widespread exploitation of fossil fuels the main source of energy was wood. Firewood still provides about 15 percent of the world's energy today. As wood became scarce, it seemed natural to many civilizations to make use of the heat of the sun to help reduce the need for wood to provide heat. The ancient Greeks were well aware of the benefits of solar design, and commonly arranged their houses to collect the rays of the winter sun. Greek cities such as Priene, following its relocation to avoid flooding, were laid out on a grid plan with streets running east–west to allow a southerly orientation of the buildings.[1]

The Romans continued the solar design principles that they learned from their contacts with Greek examples, but they were able to make use of window glazing, developed in the first century AD, to increase the heat that could be gained. Growing shortage of wood for fuel made the use of south-facing glazing popular for the villas of the wealthy, and for the public baths.[2]

The tradition of designing with climate to achieve comfort in buildings is not confined to the provision of warmth. In many climates the problem that faces the architect is to cool spaces in order to achieve comfortable conditions. The conventional modern solution, the provision of air conditioning systems, is no more than a crude process of opposing climate with energy, which was foolish when energy was cheap and pollution not considered, but is now verging on the insane.

[. . .]

PRINCIPLE 3: MINIMIZING NEW RESOURCES

A building should be designed so as to minimize the use of new resources and, at the end of its useful life, to form the resources for other architecture.

Although this principle, like the others discussed, is directed towards new buildings, it acknowledges that immense resources are already a part of the existing built environment, and that the rehabilitation and upgrading of the existing building stock for minimal environmental impact is as important, if not more so, than the creation of a new green architecture. There are not sufficient resources in the world for the built environment to be reconstructed anew for each generation. Nor would it be right that it should be so, unless the layers of history that each generation applies to the built environment are to be ignored. So much of what is admired about buildings comes from associations, and an objective assessment of a building with no prior knowledge of its purpose or ownership is virtually impossible. Even a demolished building leaves traces, such as an entry point into the site, that need to be respected by anything new put in its place.

Re-use can take the form of recycling materials or recycling spaces. The recycling of both buildings and building components is part of the history of architecture. St Albans Abbey, for example, which was rebuilt between 1077 and 1115 on the site of a Benedictine abbey, used Roman bricks taken from the ruins of Verulamium at the bottom of the hill to reinforce the walls of flint, the only building stone available in Hertfordshire. The prefabrication methods of later medieval timber frame buildings whereby the pieces were cut and fitted together in the carpenters' yard, marked, disassembled and moved to site, meant that portions of medieval buildings could be moved if required, and even now often turn up in unexpected places. Sometimes whole structures have been moved to find a new purpose. For example, upon construction of the Victoria and Albert Museum in London, the Brompton Boilers building was no longer required, and in 1865 the iron building was offered to the authorities of North, South and East London with the intention that it should be re-erected as a local museum. East London accepted it, and the building opened as the Bethnal Green Museum in 1872. It is today the Museum of Childhood.

Often those whose access to resources is least have demonstrated the way in which structures designed for one purpose can be adapted to suit a different need. However, the alterations necessary can often more or less obliterate the original form of the structure or building. This produces a dilemma for those concerned with the conservation of buildings that has to be acknowledged. Should a building be preserved unchanged because it was once important, or should it, because it can still be made useful, be conserved in a changed state?

A green approach to the problem might suggest that the question be decided on resources alone. If the resources required to alter a building are less than for demolition and rebuilding then the former course is adopted. This, however, fails to acknowledge the historical importance of the structure, which might suggest that other values need to be considered. This dimension to the problem of changing existing buildings to make them meet present needs, especially in terms of upgrading buildings to improve thermal performance which may alter appearance, can be summarized as the paradox of Venice. If global warming produces even a small alteration in sea levels the future of Venice is again imperiled. . . . To preserve Venice, it may be necessary to conserve buildings in an altered state. It is important that these issues now form part of the debate of building conservation.

Some extraordinary schemes have been created through the re-use of large redundant buildings even without thought to thermal performance, such as the Gare d'Orsay in Paris, which is now the Museum of the Nineteenth Century. The difficulty of inserting a series of spaces to illustrate nineteenth-century paintings and sculpture in a vast volume that was intended to accommodate the then-new electric trains poses architectural problems which are, perhaps, not necessarily happily resolved in this instance. However, the benefits accruing from re-using a large piece of the urban fabric that has a presence in the city can override the internal considerations. The refurbishment of existing housing areas in cities and towns can also offer a considerable saving in resources over demolition and rebuilding, and avoid disruption of the community.

[. . .]

PRINCIPLE 4: RESPECT FOR USERS

A green architecture recognizes the importance of all the people involved with it.

This principle may appear to have little relevance to issues of pollution, global warming and the destruction of the ozone layer, but a green approach to architecture that includes respect for all the resources that contribute to making a building will not exclude human beings. All buildings are made by hand, but in some architectures this fact is acknowledged and appreciated, whereas others have attempted to deny the human dimension of the building process. Only in Japan have robots been developed to take over some of the human roles on building projects, but for a robot to work effectively the project-scale must be such that the same task, or a limited series of related tasks, can be repeated many times. At a very different scale, the small builder still has to rely on personal skill for a number of unrelated tasks, drawing on expert subcontractors where there is no alternative.

A greater respect for human needs and labor can be evidenced in two separate ways. For the professional builder, it is essential that the materials and processes that form the building are as little polluting and dangerous to the individual worker or user as they are to the planet. Architects have begun to realize the extent of the global or human poisons that may be found on building sites, and that it is no longer feasible to use insulating materials that contain CFCs, or to use methods of timber treatment that are carcinogenic. Alternative methods of detailing to protect timber physically become preferable to the chemical approach. Perhaps were designers to regard timber as living wood, as in the vernacular tradition, rather than as some squared and dimensionally stable material, the temptation to cover everything with chemicals might be lessened.

[. . .]

The other form of human involvement that needs to be considered is the positive involvement of the users in the design and construction process. If this energy remains uninvolved, a resource is being wasted which could inform the finished product and extend its usefulness. A range of buildings, some built by the owners, some involving large groups of people, have made use of this resource and the result has been a high level of satisfaction with the buildings created.

PRINCIPLE 5: RESPECT FOR SITE

A building will "touch-this-earth-lightly."

The Australian architect Glenn Murcutt quotes an Aboriginal saying, that "One must touch-this-

earth-lightly."[3] This saying embodies an attitude to the interaction of a building and its site that is essential to a green approach, but it also implies wider concerns. A building that guzzles energy, creates pollution, and alienates its users does not "touch-this-earth-lightly."

The most direct interpretation of the phrase to "touch-this-earth-lightly" would be the idea that a building could be removed from its site and leave it in the condition it was before the building was placed there. This relationship to site is seen in the traditional dwellings of nomads, but their lightness of touch is not just a matter of moving their homes, it also concerns the materials with which they build, and the possessions they carry with them.[4] The black tent of the Bedouin is woven from the hair of their goats, sheep, and camels. When erected, the tent cloth adopts a low, aerodynamically efficient profile to avoid damage by high winds; it is kept in place by long ropes, also woven from hair, and supported by a very few wooden poles, because wood is a scarce resource in the desert.

The Netsilik Inuit people of northern Canada carry their tents in the summer when they need to follow the game that are their food, but in the winter the skins that form the tent cover are dipped in water and wrapped round frozen fish. The long bundles freeze solid, and are joined in pairs with caribou bones. A mixture of moss and slush is rubbed in and allowed to freeze smooth to turn the tent skins and frozen fish into a sledge to carry the Inuit and their possessions over the snow. The parts of a nomadic structure must serve several functions, because only a minimum of possessions can be carried from place to place. Over generations the items necessary for survival, comfort, and the continuation of culture have been determined.

While societies have abandoned the nomadic life for one in fixed dwellings and architecture has come into being, there is still a continuing demand for temporary structures for exhibitions, performances, and other cultural manifestations. These structures frequently take the form of tents; however, an interesting example using very different technology is the sculpture pavilion designed by the Dutch architects Benthem Crouwel for the Sonsbeek '86 festival.[5] This building was designed to protect fragile works of sculpture placed outside, and so the structure was intended to be almost invisible. It used only four materials: precast concrete for the footings, laminated glass for the walls and roof, steel for trusses and connections, and silicone mastic to stick the panes of glass together. Fins of glass glued to the glass walls gave additional rigidity, and provided a place of attachment for the light steel trusses that carried the flat glass roof. The floor was the earth, merely covered with wood shavings to prevent it from becoming muddy. At the end of the event the building was unbolted and removed, the foundations were lifted and the soil replaced, the shavings were raked up, and the site was completely unaltered by the events that had taken place on it. The building could be taken away to be used elsewhere for another exhibition, or recycled into another structure.

PRINCIPLE 6: HOLISM

All the green principles need to be embodied in a holistic approach to the built environment.

It is not easy to find buildings that embody all the principles of green architecture, for a green architecture is yet to be realized. . . . A green architecture involves more than the individual building on its plot; it must encompass a sustainable form of urban environment. The city is far more than a collection of buildings; rather it can be seen as a set of interacting systems – systems for living, working, and playing – crystallized into built forms. It is by looking at systems that we can find the face of the city of tomorrow.

NOTES

1 K. Butti and J. Perlin, *A Golden Thread*, Cheshire Books, Palo Alto, CA, 1980.
2 Ibid.
3 G.L. Murcutt in Foreword to P. Drew, *Leaves of Iron: Glenn Murcutt: Pioneer of an Australian Architectural Form*, Law Book Company, North Ryde, New South Wales, 1985.
4 T. Faegre, *Tents: Architecture of the Nomads*, Anchor Press/Doubleday, Garden City, NY, 1979.
5 P. Buchanan, "Barely There," *Architectural Review*, No. 1087, 1987.

41

"Sustainability and Building Codes"

from *Environmental Building News* (2001)

David Eisenberg and Peter Yost

Editors' Introduction

Visionary ecological design principles are often confronted by the cold, hard reality of building codes and zoning regulations, which forbid many innovative design practices and materials. Most of these regulatory mechanisms didn't exist a century ago, but have become ever more complex and demanding in recent decades. They help ensure human health, safety, and welfare, to be sure, but also mandate particular modes of building that are now seen as unsustainable. For example, graywater systems (collecting sink or shower water for reuse in toilets or irrigation) and alternative building materials such as straw bale or rammed earth have been prohibited by codes in many locations until recently. Codes also often set minimum room sizes and require unnecessarily expensive construction materials and practices. Meanwhile, zoning regulations frequently require large amounts of parking, large lot sizes, substantial building setbacks from lot lines, and low building heights. All of these requirements constrain what ecological designers can do.

The following piece from the journal *Environmental Building News* (www.buildinggreen.com) addresses building codes specifically, tracing their development and some ways in which they might be changed to better promote sustainability. David Eisenberg is director of the Development Center for Appropriate Technology (DCAT) in Tucson, Arizona, and a member of the International Conference of Building Officials. His work has ranged from the steel and glass cover for Biosphere 2 to adobe, rammed-earth, and straw-bale structures. Coauthor of *The Straw Bale House* (White River Junction, VT: Chelsea Green, 1994; with Athena Swentzell Steen, Bill Steen, and David Bainbridge), he helped write the first load-bearing straw-bale construction code for the City of Tucson and the County of Pima, Arizona, and led DCAT in a collaborative effort called "Building Sustainability into the Codes." Peter Yost is a former editor of *Environmental Building News* and a research associate with the Building Science Corporation, a Boston-based architecture and building science consulting firm. Further information on this subject is available from the Developmental Center for Appropriate Technology in Tuscon, Arizona, at www.dcat.net.

Shallow frost-protected foundation, straw-bale walls, composting toilet, graywater system, rainwater harvesting. . . . An impressive array of green building features! From the foundation to the roof, these are exemplary systems and materials. But there is another commonality to these features: each represents a potential – if not likely – regulatory challenge. It can be frustrating to have the knowledge and skills required for building green, yet lack the approvals to do it.

This article takes an in-depth look at the inherent but largely unrecognized relationship between sustainability and building codes, and efforts under way to change this relationship. It also presents a

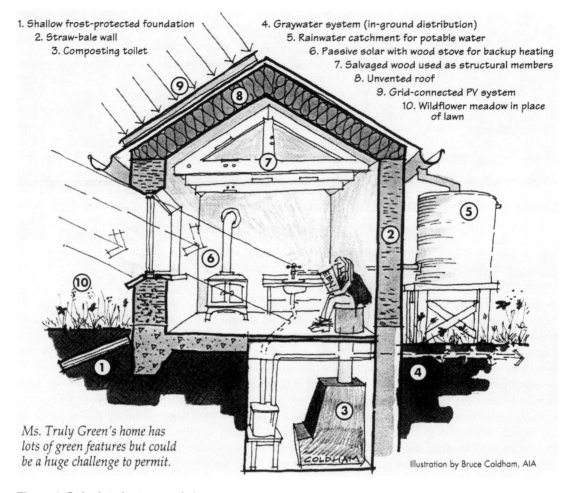

1. Shallow frost-protected foundation
2. Straw-bale wall
3. Composting toilet
4. Graywater system (in-ground distribution)
5. Rainwater catchment for potable water
6. Passive solar with wood stove for backup heating
7. Salvaged wood used as structural members
8. Unvented roof
9. Grid-connected PV system
10. Wildflower meadow in place of lawn

Ms. Truly Green's home has lots of green features but could be a huge challenge to permit.

Illustration by Bruce Coldham, AIA

Figure 1 Code obstacles to green design.

process for professionals to use in gaining approvals for alternative designs, systems, and materials within the existing regulatory framework. A sampling of code success stories demonstrates what is possible when this process is employed.

Though it is beyond the scope of this article, the issue of regulatory hurdles with green building is not restricted to buildings and building codes; a new approach is needed as well for the larger issues of land development, zoning, and planning.

A BRIEF HISTORY OF BUILDING CODES

Building codes have long been used by societies to protect individual and general welfare, and to hold practitioners accountable for their work. As long ago as 1750 BC, Hammurabi, the Babylonian king of Mesopotamia, created his famous Code of Laws covering a wide range of public and private matters. Number 229 of this Code states: "If a builder build a house for someone, and does not construct it properly, and the house which he built fall in and kill its owner, then that builder shall be put to death." This type of "performance" code must certainly have had an impact on quality of construction, but it very likely stifled innovation!

There were many intermediate steps on the way to our present codes. In 1189 AD, the city of London adopted regulations for the construction of common walls, rights to light access, drainage, and safe egress in case of fire. Historically, fire has been the most common concern driving interest in building regulations. Early in the Colonial period of the United States, concern about fire resulted in a ban on wood chimneys and thatch roofs. In 1860 the City of New York appointed a Superintendent of Building and provided staff for code enforcement.

In 1867, the Tenement House Act was enacted to regulate conditions in existing buildings, covering such things as fire escapes, ventilation, water supply, toilets, and stair railings. In 1905, the National Board of Fire Underwriters, an insurance industry group, wrote the first National Building Code.

This code led to the formation of organizations for building code officials and the next stage of code development in the United States. By 1940, three model code organizations were established: the Building Officials and Code Administrators International, Inc. (BOCA) in the northeastern US, which produced the National Building Code; the International Conference of Building Officials (ICBO) covering the western half of the United States, which produced the Uniform Building Code; and the Southern Building Code Congress International (SBCCI) in the southeastern United States, which published the Standard Building Code. Reflecting regional differences and different code philosophies, the three model codes also embodied variations that have made code compliance difficult for designers, builders, and manufacturers working across different code-enforcement areas.

Efforts to harmonize the three codes, initially through the Council of American Building Officials (CABO) and more recently by its successor, the International Code Council (ICC), have now resulted in the creation of a single national building code – or family of codes. The ICC codes (including the International Building Code, International Residential Code, and "International" versions of the Mechanical, Plumbing, Fire, and Energy Conservation codes) are replacing the BOCA, SBCCI, ICBO, and CABO codes, which are no longer being maintained. Instead, these groups now support and maintain the ICC codes, the first full edition of which was published in 2000. (Recently, the NFPA dealt a blow to this consolidation effort when it split from the ICC process and began developing its own building code to compete with the International family of codes.)

An important new development in the ICC process is creation of the International Performance Code (IPC). This code differs from the other International codes in that it is based on stating what must be accomplished, rather than describing in detail what must be done and how to do it. While the more typical *prescriptive* approach is straightforward and relatively easy to implement for both builder and code official (because everyone knows

what must be done), it can also be confining and thus limit innovation.

Though new to the US, the experience of other countries using performance codes has shown that they are viable. The greater flexibility provided by performance requirements is both liberating and problematic. The added freedom comes at a price because the performance approach requires that the proposed designs, materials, or methods be supported by calculation, test results, or other demonstrations of adequate performance. That often means more engineering services, testing, and time – both for designer and plan reviewer. It adds a burden for the building department because building officials must be able to analyze the project rather than just making sure it conforms to common practices with which they are familiar.

BUILDING CODES IN ACTION

One might assume that the creation of a single family of codes would bring about complete consolidation of building codes across the United States, but for several reasons this is not the case. First, unlike in many countries where code adoption takes place at the national level, in the United States it occurs at the local, county, or state level. Codes derive their legal authority from their enactment as laws, ordinances, or statutes. While it appears likely that most US jurisdictions will eventually employ the ICC system, in most cases each jurisdiction makes its own determination of which codes and which versions of those codes it will adopt. Some jurisdictions are still without any building codes.

Complicating the matter further, nearly all jurisdictions reserve – and often exercise – the right to add to or amend the codes they adopt. Local amendments may be in response to conditions such as high winds, wildfires, or earthquakes, and additions often include appendix chapters for traditional or regional building approaches – for example, adobe and rammed-earth in the southwestern United States.

At the other end of the spectrum, state or federal government can, as public policy, pass legislation or develop programs that either directly or indirectly supersede local codes. Two examples are the low-flow toilet requirements included in the 1992 Energy Policy Act[1] and the recent code requirement by the city of Frisco, Texas that all new homes be EPA ENERGY STAR-compliant.[2]

Just as important as the process by which codes are adopted is the process by which building codes are developed, changed, and enforced. Few people are aware that the building code development and code change processes are open to the public. Anyone – a business, interest group, or individual – can propose changes to the codes. On an annual basis, all filed proposals go through the same process – committee review, scheduled hearing, and voting. This process results in many changes to codes every year. Typically, supplements are published annually and then consolidated into a new edition of the code every three years.

At the other end of the process are local building officials who have the authority, granted by provisions in the codes, to approve alternative designs, materials, and methods of construction as long as they are deemed adequate to meet the intent of the building code. All codes have such provisions for dealing with building practices, materials, and systems not specifically addressed in the code. Understanding how to use this process can be of enormous benefit when proposing alternatives to standard practice.

THE CASE FOR INTEGRATING SUSTAINABILITY INTO THE CODES

A key to shifting the building regulatory system towards greater acceptance of more sustainable, alternative approaches is to create a context in which those alternatives can be seen both as positive and as representing a reduction of risk, rather than an increase in risk. That requires developing awareness of the inherent risk in the status quo: what is likely to happen or is already happening if we maintain our current practices. To see the risk requires shifting from the details of the codes to the larger context and intent of the codes – understanding how current practice jeopardizes the public welfare that the regulatory system was established to protect.

Historically, building codes were developed as a reaction to disasters and building failures. They derived their authority from a societal expectation that the public must be protected from these threats. This led to a focus on the protection of people in and around buildings and secondarily on protection of property. Over time, this focus has become ever more detailed and has expanded into nearly every aspect of buildings and their components and systems. It is no surprise that this focus, combined with our slow awakening to the scope and magnitude of the environmental impacts of the building industry, has resulted in a lack of concern for impacts that occur *away* from the actual building site, impacts that are cumulative or difficult to measure (such as climate impacts or the health effects of indoor air quality or toxicity of materials), or that extend into the future.

The idea of addressing such aggregated impacts through codes, though relatively new, has precedents in such areas as sewage systems, building energy codes, and water-efficiency requirements. Building energy codes provide a valuable, though still somewhat controversial, precedent for incorporating into building codes the larger, more distant, and cumulative consequences of buildings. It has been argued that energy-efficiency is not a safety issue and therefore has no place in the building codes. "I thought that [insulation requirements in building codes] was the dumbest idea I'd ever heard and that it had no place in the codes," admitted Bob Fowler in an interview in *Building Standards*. Good arguments were made for minimum insulation requirements for buildings exposed to extreme temperatures as part of the concern for health and safety of the occupants or users of buildings, and thus they were developed. But it took a combination of economic, environmental, health, and even national security issues to finally propel building energy codes into existence and widespread adoption. "Looking back," reflected Fowler in the same interview, "I see that the energy-efficiency requirements set a very important precedent for our learning to take responsibility for the full range of the consequences of our buildings. We now need to continue that learning process and open our eyes and our minds to the work of creating sustainable buildings."

The larger, ecologically based risks to public welfare must eventually be seen as risks that demand responsibility for protecting public welfare as much as structural integrity, fire safety, or means of egress. The current regulatory system requires a high degree of safety and certainty in each building project, while ignoring the unintended role it plays in encouraging the depletion of natural resources and the demise of the natural systems upon which everyone's health, safety, and survival ultimately depend.

It is not difficult to find evidence to support concerns about the environmental impacts of the built environment:[3]

■ Over 40 percent of the material resources entering the global economy today are related to the building industry.
■ Modern buildings use tremendous quantities of energy – in the United States (with less than 5 percent of the world's population) buildings alone account for a staggering 10 percent of *global* energy use.

Such statistics are all the more remarkable when one realizes that only about two billion of the world's more than six billion people live and work in resource-consumptive buildings – the sort of buildings described by modern building codes. The rest of the world's people today live in earthen buildings (adobe, rammed- or puddled-earth, cob, wattle-and-daub) or other types of indigenous buildings, shelters made of scavenged materials, or no buildings at all. Yet all over the world modern building methods, with their greater impacts and resource consumption, are replacing traditional – and often far more sustainable – ways of building. It is important not to romanticize indigenous buildings or dismiss the very real problems that are often associated with them (poor earthquake resistance, lack of insulation, etc.), but to recognize the value and viability of simple, low-tech materials and building methods when used wisely. At the same time, modern materials and building systems must be viewed with the same critical eye, acknowledging their real costs and impacts, not just their benefits. With projections of the world's population reaching at least eight or nine billion this century and with the needed development and construction that must accompany such growth, these issues cannot be ignored much longer. . . .

Relationships with leaders in the building codes community are important, but creating similar relationships locally and regionally is required in order to achieve the needed changes. That can only happen through the engagement of the environmental design and building community in

a proactive, constructive partnership with their building code officials, based on a very real, mutual interest in creating safe buildings. Then the definition of public health, safety, and welfare related to buildings can be expanded to include this larger set of responsibilities. . . .

The environmental design and construction community must become actively engaged in writing code change proposals and encouraging funding, research, and testing to support those changes. Additionally, standards-development activities, such as those in ASTM and ASHRAE, often result in requirements less than satisfactory in terms of the environment. The green building community needs to share their direct experience in contending with the realities of those standards by participating more fully in the standards-development process.

It is also time for the environmental design and construction community to seek representation on relevant building code development committees. The ICC code development process is now opening up representation on their committees to the public and industry. Organizations such as The American Institute of Architects Committee on the Environment (AIA-COTE), US Green Building Council (USGBC), Sustainable Building Industry Council (SBIC), Energy and Environmental Building Association (EEBA), and New Buildings Institute need to come together and focus on how to gain such representation. Other interest groups are well organized and funded to represent their interests; the green building community needs to take responsibility for bringing about changes, rather than simply lamenting the status quo.

Finally, local green building programs provide an ideal forum for education and exchange about alternative designs, materials, and methods and the building codes. Local code officials could be brought into these programs to share their existing skills and experience as well as for their education and enlightenment. Everyone would benefit from such an exchange.

NOTES

1 See *Environmental Building News*, 2 (1).
2 See *Environmental Building News*, 10 (6).
3 See *Environmental Building News* feature, 10 (5).

42

"Introduction to the LEED® Rating System"

United States Green Building Council

Editors' Introduction

Since the early 1990s a number of cities and nonprofit organizations have sought to develop rating systems that would recognize and reward environmentally friendly buildings. Among the best-known are the City of Austin, Texas' Green Building Program, BREEAM (Building Research Establishment's Environmental Assessment Method) in the United Kingdom, the Green Globes system in Canada, and the Leadership in Energy and Environmental Design (LEED) rating system developed by the U.S. Green Building Council. In the late 1990s and 2000s the LEED system has gained a great deal of attention in the United States, with many cities and states now requiring public buildings to meet LEED standards, and a few pioneering municipalities such as Boston and Los Angeles now requiring large public buildings to meet them as well.

Within the LEED system, projects are rated on many different aspects of green design, with a maximum available score of 46 points. Projects can achieve a basic level of certification, or higher Silver, Gold, and Platinum levels. Initially applied only to New Construction, the LEED system now features different versions for Existing Buildings, Commercial Interiors, Core & Shell, Schools, Retail, Healthcare, Homes, and Neighborhood Development. More information is available through www.usgbc.org.

WHY MAKE YOUR BUILDING GREEN?

The environmental impact of the building design, construction and operation industry is significant. Buildings annually consume more than 30 percent of the total energy and more than 60 percent of the electricity used in the US. Each day 5 billion gallons of potable water is used solely to flush toilets. A typical North American commercial construction project generates up to 2.5 lb of solid waste per square foot of completed floor space. Development shifts land usage away from natural, biologically diverse habitats to hardscape that is impervious and devoid of biodiversity. The far-reaching influence of the built environment necessitates action to reduce its impact.

Green building practices can substantially reduce or eliminate negative environmental impacts and improve existing unsustainable design, construction and operational practices. As an added benefit, green design measures reduce operating costs, enhance building marketability, increase worker productivity, and reduce potential liability resulting from indoor air quality problems. For example, energy efficiency measures have reduced operating expenses of the Denver Dry Goods building by approximately $75,000 per year. Students in day-lit schools in North Carolina consistently score higher on tests than students in schools using conventional lighting fixtures. Studies of workers in green buildings reported productivity gains of up to 16 percent, including reductions in absenteeism and improved work quality, based on "people-friendly" green design. At a grocery store in Spokane, Washington, waste management costs were reduced by 56 percent and forty-eight tons of waste was recycled during construction. In other words, green design has environmental, economic and social elements that benefit all building stakeholders, including owners, occupants and the general public.

LEED® GREEN BUILDING RATING SYSTEM

History of LEED®

Following the formation of the US Green Building Council (USGBC) in 1993, the membership quickly realized that a priority for the sustainable building industry was to have a system to define and measure "green buildings." The USGBC began to research existing green building metrics and rating systems. Less than a year after formation, the membership followed up on the initial findings with the establishment of a committee to focus solely on this topic. The diverse initial composition of the committee included architects, realtors, a building owner, a lawyer, an environmentalist and industry representatives. This cross-section of people and professions added a richness and depth both to the process and to the ultimate product.

The first LEED® Pilot Project Program, also referred to as LEED® Version 1.0, was launched at the USGBC Membership Summit in August 1998. After extensive modifications, the LEED® Green Building Rating System Version 2.0 was released in March 2000. Leed 2009 was launched in April 2009, building on the fundamental structure of the previous rating systems while ensuring that new technology and urgent priorities were addressed. LEED 4.0 was finalized in 2013, focusing on new market sectors, increased technical rigor, and streamlined services.

As LEED® has evolved and matured, the program has undertaken new initiatives. In addition to a rating system specifically devoted to building operational and maintenance issues, LEED® addresses the different project development/delivery processes that exist in the US building design and construction market.

Features of LEED®

The LEED® Green Building Rating System is a voluntary, consensus-based, market-driven building rating system based on existing proven technology. It evaluates environmental performance from a whole building perspective over a building's life cycle, providing a definitive standard for what constitutes a "green building." The development of the LEED® Green Building Rating System was initiated by the USGBC Membership, representing all segments of the building industry, and has been open to public scrutiny.

Version 4 of the rating system for New Construction and Major Renovation is organized into five categories: Location and Transportation, Sustainable Sites, Water Efficiency, Energy and Atmosphere, Materials and Resources, and Indoor Environmental Quality. There are additional categories for Innovation and Regional Priority credits.

LEED® is a measurement system designed for rating new and existing commercial, institutional and residential buildings. It is based on accepted energy and environmental principles and strikes a balance between known established practices and emerging concepts. It is a performance-oriented system where credits are earned for satisfying criteria designed to address specific environmental impacts inherent in the design, construction and operations and maintenance of buildings. Different levels of green building certification are awarded based on the total credits earned. The system is designed to be comprehensive in scope, yet simple in operation.

The future of LEED®

The green design field is growing and changing daily. New technologies and products are coming into the marketplace and innovative designs are proving their effectiveness. Therefore, the Rating System and the Reference Guide will evolve as well. Teams wishing to certify with LEED® should note that they will need to comply with the version of the rating system that is current at the time of their registration. USGBC will highlight new developments on its Web site on a continuous basis at www.usgbc.org.

LEED v4 for BD+C: New Construction and Major Renovation
Project Checklist

Project Name

Date

Y	?	N			
			Credi 1	Integrative Process	1

Y	?	N		**Location and Transportation**	**Possible Points:**	**16**
			Credit 1	LEED for Neighborhood Development Location		16
			Credit 2	Sensitive Land Protection		1
			Credit 3	High Priority Site		2
			Credit 4	Surrounding Density and Diverse Uses		5
			Credit 5	Access to Quality Transit		5
			Credit 6	Bicycle Facilities		1
			Credit 7	Reduced Parking Footprint		1
			Credit 8	Green Vehicles		1

Y	?	N		**Sustainable Sites**	**Possible Points:**	**10**
Y			Prereq 1	Construction Activity Pollution Prevention		Required
			Credit 1	Site Assessment		1
			Credit 2	Site Development--Protect or Restore Habitat		2
			Credit 3	Open Space		1
			Credit 4	Rainwater Management		3
			Credit 5	Heat Island Reduction		2
			Credit 6	Light Pollution Reduction		1

Y	?	N		**Water Efficiency**	**Possible Points:**	**11**
Y			Prereq 1	Outdoor Water Use Reduction		Required
Y			Prereq 2	Indoor Water Use Reduction		Required
Y			Prereq 3	Building-Level Water Metering		Required
			Credit 1	Outdoor Water Use Reduction		2
			Credit 2	Indoor Water Use Reduction		6
			Credit 3	Cooling Tower Water Use		2
			Credit 4	Water Metering		1

Y	?	N		**Energy and Atmosphere**	**Possible Points:**	**33**
Y			Prereq 1	Fundamental Commissioning and Verification		Required
Y			Prereq 2	Minimum Energy Performance		Required
Y			Prereq 3	Building-Level Energy Metering		Required
Y			Prereq 4	Fundamental Refrigerant Management		Required
			Credit 1	Enhanced Commissioning		6
			Credit 2	Optimize Energy Performance		18
			Credit 3	Advanced Energy Metering		1
			Credit 4	Demand Response		2
			Credit 5	Renewable Energy Production		3
			Credit 6	Enhanced Refrigerant Management		1
			Credit 7	Green Power and Carbon Offsets		2

Figure 1 LEED® portfolio.

			Materials and Resources	Possible Points:	13
Y			Prereq 1 Storage and Collection of Recyclables		Required
Y			Prereq 2 Construction and Demolition Waste Management Planning		Required
			Credit 1 Building Life-Cycle Impact Reduction		5
			Credit 2 Building Product Disclosure and Optimization - Environmental Product Declarations		2
			Credit 3 Building Product Disclosure and Optimization - Sourcing of Raw Materials		2
			Credit 4 Building Product Disclosure and Optimization - Material Ingredients		2
			Credit 5 Construction and Demolition Waste Management		2

			Indoor Environmental Quality	Possible Points:	16
Y			Prereq 1 Minimum Indoor Air Quality Performance		Required
Y			Prereq 2 Environmental Tobacco Smoke Control		Required
			Credit 1 Enhanced Indoor Air Quality Strategies		2
			Credit 2 Low-Emitting Materials		3
			Credit 3 Construction Indoor Air Quality Management Plan		1
			Credit 4 Indoor Air Quality Assessment		2
			Credit 5 Thermal Comfort		1
			Credit 6 Interior Lighting		2
			Credit 7 Daylight		3
			Credit 8 Quality Views		1
			Credit 9 Acoustic Performance		1

			Innovation	Possible Points:	6
			Credit 1 Innovation		5
			Credit 2 LEED Accredited Professional		1

			Regional Priority	Possible Points:	4
			Credit 1 Regional Priority: Specific Credit		1
			Credit 2 Regional Priority: Specific Credit		1
			Credit 3 Regional Priority: Specific Credit		1
			Credit 4 Regional Priority: Specific Credit		1

			Total	Possible Points:	110

Certified 40 to 49 points   Silver 50 to 59 points   Gold 60 to 79 points   Platinum 80 to 110

Figure 1 (*continued*)

BREEAM UK New Construction 2014 environmental assessment categories

Management

- Project brief and design
- Life cycle cost and service life planning
- Responsible construction practices
- Commissioning and handover
- Aftercare

Health and Wellbeing

- Visual comfort
- Indoor air quality
- Safe containment in laboratories
- Thermal comfort
- Acoustic performance
- Safety and security

Energy

- Reduction of energy use and carbon emissions
- Energy monitoring
- External lighting
- Low carbon design
- Energy efficient cold storage
- Energy efficient transportation systems
- Energy efficient laboratory systems
- Energy efficient equipment
- Drying space

Transport

- Public transport accessibility
- Proximity to amenities
- Cyclist facilities
- Maximum car parking capacity
- Travel plan

Water

- Water consumption
- Water monitoring
- Water leak detection
- Water efficient equipment

Materials

- Life cycle impacts
- Hard landscaping and boundary protection
- Responsible sourcing of materials
- Insulation
- Designing for durability and resilience
- Material efficiency

Waste

- Construction waste management
- Recycled aggregates
- Operational waste
- Speculative floor and ceiling finishes
- Adaptation to climate change
- Functional adaptability

Land Use and Ecology

- Site selection
- Ecological value of site and protection of ecological features
- Minimising impact on existing site ecology
- Enhancing site ecology
- Long term impact on biodiversity

Pollution

- Impact of refrigerants
- NO_x emissions
- Surface water run-off
- Reduction of night time light pollution
- Reduction of noise pollution

Innovation

- Innovation

Source: BRE Group, BREEAM UK New Construction 2014 Technical Manual. For more information, see http://www.breeam.org/.

Living Building Challenge "Imperatives" for Projects

IMPERATIVE		Preliminary Audit	Final Audit
01	Limits to Growth	x	
02	Urban Agriculture		x
03	Habitat Exchange	x	
04	Human Powered Living	x	
05	Net Positive Water		x
06	Net Positive Energy		x
07	Civilized Environment	x	
08	Healthy Interior Environment		x
09	Biophilic Environment	x	
10	Red List	x	
11	Embodied Carbon Footprint	x	
12	Responsible Industry	x	
13	Living Economy Sourcing	x	
14	Net Positive Waste		x
15	Human Scale + Humane Places		x
16	Universal Access to Nature and Place	x	
17	Equitable Investment		x
18	JUST Organizations	x	
19	Beauty + Spirit		x
20	Inspiration + Education	x	

Source: International Living Future Institute, *Living Building Challenge 3.0 Documentation Requirements 2014*. For more information, see http://living-future.org/lbc.

43

"The Ten Commandments of Cost-Effective Green Building Design"

from *Green Building Through Integrated Design* (2009)

Leith Sharp

Editors' Introduction

Many of the challenges of green building aren't technological or even economic in nature. Rather, they have to do with the attitudes, habits, and practices of professionals in the field. Simple, practical steps to change development processes can break through these barriers. Here, Leath Sharp, the former director of Harvard University's Green Campus Initiative, distills the wisdom she learned from dealing with design professionals in this role. Other resources on green development practices include *The Integrative Design Guide to Green Building: Redefining the Practice of Sustainability* (Hoboken, NJ: Wiley, 2009), by the 7group and Bill Reed. The Association for the Advancement of Sustainability in Higher Education (AASHE; www.aashe.org) provides information on other practical steps that can green university campuses and curricula.

When I first began working to introduce green building at Harvard, the most common perception I encountered was that green building was too expensive and that LEED was a costly point-chasing exercise with no value. Things reached their lowest point in 2001 when at one design team meeting, a faculty member acting as the project's client representative likened the belief that green building design could be cost effective to the belief that there were elephants in the hallway.

To help transcend these attitudinal barriers, I found three building project partners who agreed to pilot LEED at the university. By studying these projects, I was able to trace almost all of the criticisms leveled at LEED to *a range of failures in the design process itself rather than failures intrinsic to LEED.*

For example, the complaint that LEED certification was too expensive turned out to be the result of architects overcharging because they had little

experience and were trying to cover their own learning costs and perceived risks. The complaint that there were too many unexpected costs turned out to be the result of change orders that were in turn a result of poor integration of LEED requirements into the building construction documents. The accusation that LEED was a point-chasing exercise turned out to be the result of flawed sequencing of tasks such as the engineer doing the energy modeling after the design was already complete in order to satisfy the LEED documentation requirement, instead of doing it early enough to inform the design.

These pilot projects provided the necessary experiential evidence to prove that green building and the LEED framework in particular did have enormous value if utilized properly. Perhaps most importantly these projects proved to me that cost impact was largely subject to our own ability to properly manage the design process itself and that we needed to stop trying to answer the question "How much will green building and LEED cost us?" and start answering the question "How do we improve the design process to minimize or avoid additional costs for green building and LEED?"

By successfully working to answer this question, my team and I engaged the Harvard community in over 50 LEED projects, most striving for LEED Gold certification. Utilizing this momentum we were able to define and adopt a set of comprehensive green building guidelines that includes many key design process requirements, along with a minimum LEED Silver requirement. At the same time I worked to foster the capacities of both the community and the building profession that serves it by leading an effort to get everything that we had been learning about the process of green building into a publicly available web resource.[1]

Here are my Ten Commandments of Cost-Effective Green Building Design:

1 *Commitment.* The earlier the commitment is made, the better for everyone. This should be a formal, continuously improved, widely known, and detailed green building commitment for all building projects, integrated into capital project approval processes and related contracts.

2 *Leadership.* To minimize the risk of business as usual, the client and/or project manager must take an active and ongoing leadership role throughout the project, establishing project-specific environmental performance requirements in predesign (LEED is ideal for this), challenging, scrutinizing, and pushing the design team at every stage. The client and/or project manager should understand enough about LEED, integrated design, energy modeling, and life-cycle costing to ask the right questions at the right time.

3 *Accountability.* To avoid lost opportunities and unnecessary costs, establish all roles and responsibilities, sequencing and tracking requirements for every environmental performance goal. Use the LEED scorecard to empower the client to participate actively in holding the project team accountable. Utilize LEED's third-party verification process to keep the design team on track with documentation. Work to streamline LEED documentation procedures by paying attention to (and learning from) every project.

4 *Process Management.* The failure to properly manage tasks at each stage in the design process results in a wide range of missed opportunities and avoidable costs. Each green building performance goal requires a set of tasks to be identified, understood, allocated across the team, sequenced, and integrated properly into the design team process. At every stage in the design process, from predesign through to construction and occupancy, there are stage-specific activities that must be completed to maximize innovation and minimize added costs. For example, many design teams don't include the building operators, fail to get any real value from the energy modeling process (because it is done too late to inform the design), or fail to incorporate a life-cycle costing approach because cost estimations are either done too late and/or fail to include operating costs in the cost model.

5 *Integrated Design.* Effective integrated design can produce significant design innovations and cost savings. The client and project manager must commit to integrated design and apply constant pressure on the project team to comply. Commitment to the process must be included in all contracts, the selection process and any ongoing team performance evaluation and quality assurance processes. The right people must be included at the right time (e.g., future building operations staff, the cost estimator, commissioning agent, and controls vendor), and the team must be managed using a collaborative approach to optimize whole building systems rather than isolated components. Well-facilitated design charrettes during conceptual design and schematic design phases are essential.

6 *Energy Modeling.* Energy modeling should go hand-in-hand with the integrated design process and life-cycle costing. Energy modeling must be used at the right phases in the design process, such as schematic design and design development, to evaluate significant design alternatives, inform efforts to optimize building systems, and generate helpful life-cycle-costing data.

7 *Commissioning Plus!* [Ed. note: "Commissioning" is the process of ensuring that a building's systems perform as expected.] You should expect failures in both the installation and performance of new design strategies and technologies. Beyond making sure that the project team includes a

commissioning agent by the end of schematic design, you should undertake an additional effort to test the entire building to ensure that it is performing according to specifications. Projects should include metering, monitoring, and control strategies to support building performance verification and ongoing commissioning for the life of the building. For complex buildings such as laboratories, include the controls vendor by the end of schematic design to integrate the logic of the operating systems into the design. Be sure to train, support, and effectively hand the building over to the operations staff.

8   *Contracts and Specifications*. All green-building-associated process and LEED requirements must be effectively integrated into the owner's project requirements, requests for proposals, all contracts, and all design and construction documents.

9   *Life-Cycle Costing*. The commitment to utilize a life-cycle-costing approach should be made by the client before the project even begins. This commitment should be integrated into all related contracts and specifications. The cost estimator should be brought on board early in the projects, so that costs can be continuously evaluated, including operating cost projections. Energy modeling should be productively utilized to inform operating cost projections, and building operations staff should be engaged to assist in considering operating-cost alternatives. Ensure that a life-cycle-cost perspective is utilized during any value engineering activities.

10 *Continuous Improvement*. For organizations that own more than one building, lessons from every green building project experience should be intensively mined to inform continuous improvement in the building design process and the ready adoption of proven design strategies and technologies. Utilize LEED documentation to support continuous improvement. Where possible, have someone from your organization act as the clearinghouse for project lessons. Invest in deliberate mechanisms to transfer experience from one project to the next. Invest in measurement and verification strategies to evaluate the actual performance of building features.

It is still a challenge to successfully integrate all Ten Commandments into our projects, but with every experience we get closer. Harvard's Blackstone Office historic renovation has come the closest. As a direct result of utilizing many of these strategies, the renovation achieved its LEED Platinum certification in 2007 at no added cost to the project. The 40,000-square-feet project was completed on time in 2006 and on budget with a hard cost for construction of $250 per square foot. The client (owner) team did invest a significant amount of their time reviewing and guiding the project, a real cost that was absorbed by non-project budgets. Interestingly, even this investment of additional client time has resulted in the client group developing a range of spin-off campus service offerings such as an owner's acceptance program now offered by the facilities group, which provides building owners at Harvard with additional building systems testing and better training and support for building operations staff.

NOTE

1   See www.greencampus.harvard.edu.

FOOD SYSTEMS AND HEALTH

44

"The Food Movement, Rising"

from *The New York Review of Books* (2010)

Michael Pollan

Editors' Introduction

Since the first edition of this *Reader* was published major new areas of interest relate to urban food systems and public health. Many of us have come to realize that unsustainable patterns of development are embodied in the food we eat every day, the ways we are able to get physical activity into our lives, and our overall physical and mental health. Consequently communities are exploring ways to expand urban agriculture, improve availability of healthy food, rethink global food systems, and expand active living opportunities.

One of the leaders of the movement to rethink urban food systems is the writer Michael Pollan. Author of best-selling books such as *The Botany of Desire* (New York: Random House, 2001), *The Omnivore's Dilemma: A Natural History of Four Meals* (New York: Penguin, 2007), and *In Defense of Food: An Eater's Manifesto* (New York: Penguin, 2009), Pollan has critiqued the industrial food chain and called attention to simpler, more locally based ways of producing and consuming food. In this piece he provides an overview of the rise of the food movement in recent decades.

Other leading resources on this topic include Frances Moore Lappé's classic *Diet for a Small Planet* (New York: Random House, 1971), Eric Schlosser's *Fast Food Nation* (New York: Houghton Mifflin, 2001), John Robbins' *The Food Revolution: How Your Diet Can Help Save Your Life and Our World* (Berkeley: Conari Press, 2010), and Marion Nestle's *Food Politics: How the Food Industry Influences Nutrition and Health* (Berkeley: University of California Press, 2013). Films have been very instrumental in communicating the need for alternative food systems; these include Morgan Spurlock's *Super Size Me* (2004), *Food, Inc.* (2008), and *Forks Over Knives* (2011).

It might sound odd to say this about something people deal with at least three times a day, but food in America has been more or less invisible, politically speaking, until very recently. At least until the early 1970s, when a bout of food price inflation and the appearance of books critical of industrial agriculture (by Wendell Berry, Francis Moore Lappé, and Barry Commoner, among others) threatened to propel the subject to the top of the national agenda, Americans have not had to think very hard about where their food comes from, or what it is doing to the planet, their bodies, and their society.

Most people count this a blessing. Americans spend a smaller percentage of their income on food than any people in history – slightly less than 10 percent – and a smaller amount of their time preparing it: a mere thirty-one minutes a day on average, including clean-up. The supermarkets brim with produce summoned from every corner of the globe, a steady stream of novel food products (17,000 new ones each year) crowds the middle aisles, and in the freezer case you can find "home meal replacements" in every conceivable ethnic stripe, demanding nothing more of the eater than opening the package and waiting for the microwave to chirp. Considered in the long sweep of human history, in which getting food dominated not just daily life but economic and political life as well, having to worry about food as little as we do, or did, seems almost a kind of dream.

The dream that the age-old "food problem" had been largely solved for most Americans was sustained by the tremendous postwar increases in the productivity of American farmers, made possible by cheap fossil fuel (the key ingredient in both chemical fertilizers and pesticides) and changes in agricultural policies. Asked by President Nixon to try to drive down the cost of food after it had spiked in the early 1970s, Agriculture Secretary Earl Butz shifted the historical focus of federal farm policy from supporting prices for farmers to boosting yields of a small handful of commodity crops (corn and soy especially) at any cost.

The administration's cheap food policy worked almost too well: crop prices fell, forcing farmers to produce still more simply to break even. This led to a deep depression in the farm belt in the 1980s followed by a brutal wave of consolidation. Most importantly, the price of food came down, or at least the price of the kinds of foods that could be made from corn and soy: processed foods and sweetened beverages and feedlot meat. (Prices for fresh produce have increased since the 1980s.) Washington had succeeded in eliminating food as a political issue – an objective dear to most governments at least since the time of the French Revolution.

But although cheap food is good politics, it turns out there are significant costs – to the environment, to public health, to the public purse, even to the culture – and as these became impossible to ignore in recent years, food has come back into view. Beginning in the late 1980s, a series of food safety scandals opened people's eyes to the way their food was being produced, each one drawing the curtain back a little further on a food system that had changed beyond recognition. When BSE, or mad cow disease, surfaced in England in 1986, Americans learned that cattle, which are herbivores,

were routinely being fed the flesh of other cattle; the practice helped keep meat cheap but at the risk of a hideous brain-wasting disease.

The 1993 deaths of four children in Washington State who had eaten hamburgers from Jack in the Box were traced to meat contaminated with *E. coli* 0157:H7, a mutant strain of the common intestinal bacteria first identified in feedlot cattle in 1982. Since then, repeated outbreaks of food-borne illness linked to new antibiotic-resistant strains of bacteria (campylobacter, salmonella, MRSA) have turned a bright light on the shortsighted practice of routinely administering antibiotics to food animals, not to treat disease but simply to speed their growth and allow them to withstand the filthy and stressful conditions in which they live.

In the wake of these food safety scandals, the conversation about food politics that briefly flourished in the 1970s was picked up again in a series of books, articles, and movies about the consequences of industrial food production. Beginning in 2001 with the publication of Eric Schlosser's *Fast Food Nation*, a surprise best-seller, and, the following year, Marion Nestle's *Food Politics*, the food journalism of the last decade has succeeded in making clear and telling connections between the methods of industrial food production, agricultural policy, food-borne illness, childhood obesity, the decline of the family meal as an institution, and, notably, the decline of family income beginning in the 1970s.

Besides drawing women into the work force, falling wages made fast food both cheap to produce and a welcome, if not indispensable, option for pinched and harried families. The picture of the food economy Schlosser painted resembles an upside-down version of the social compact sometimes referred to as "Fordism": instead of paying workers well enough to allow them to buy things like cars, as Henry Ford proposed to do, companies like Wal-Mart and McDonald's pay their workers so poorly that they can afford *only* the cheap, low-quality food these companies sell, creating a kind of nonvirtuous circle driving down both wages and the quality of food. The advent of fast food (and cheap food in general) has, in effect, subsidized the decline of family incomes in America.

Cheap food has become an indispensable pillar of the modern economy. But it is no longer an invisible or uncontested one. One of the most interesting social movements to emerge in the last few years is the "food movement," or perhaps I should say "movements," since it is unified as yet by little more than the recognition that industrial food production is in need of reform because its social/environmental/public health/animal welfare/ gastronomic costs are too high.

As that list suggests, the critics are coming at the issue from a great many different directions. Where many social movements tend to splinter as time goes on, breaking into various factions representing divergent concerns or tactics, the food movement starts out splintered. Among the many threads of advocacy that can be lumped together under that rubric we can include school lunch reform; the campaign for animal rights and welfare; the campaign against genetically modified crops; the rise of organic and locally produced food; efforts to combat obesity and type 2 diabetes; "food sovereignty" (the principle that nations should be allowed to decide their agricultural policies rather than submit to free trade regimes); farm bill reform; food safety regulation; farmland preservation; student organizing around food issues on campus; efforts to promote urban agriculture and ensure that communities have access to healthy food; initiatives to create gardens and cooking classes in schools; farm worker rights; nutrition labeling; feedlot pollution; and the various efforts to regulate food ingredients and marketing, especially to kids.

It's a big, lumpy tent, and sometimes the various factions beneath it work at cross-purposes. For example, activists working to strengthen federal food safety regulations have recently run afoul of local food advocates, who fear that the burden of new regulation will cripple the current revival of small-farm agriculture.

But there are indications that these various voices may be coming together in something that looks more and more like a coherent movement. Many in the animal welfare movement, from PETA to Peter Singer, have come to see that a smaller-scale, more humane animal agriculture is a goal worth fighting for, and surely more attainable than the abolition of meat eating. Stung by charges of elitism, activists for sustainable farming are starting to take seriously the problem of hunger and poverty. They're promoting schemes and policies to make fresh local food more accessible to the poor, through programs that give vouchers redeemable at farmers'

markets to participants in the Special Supplemental Nutrition Program for Women, Infants, and Children (WIC) and food stamp recipients. Yet a few underlying tensions remain: the "hunger lobby" has traditionally supported farm subsidies in exchange for the farm lobby's support of nutrition programs, a marriage of convenience dating to the 1960s that vastly complicates reform of the farm bill – a top priority for the food movement.

The sociologist Troy Duster reminds us of an all-important axiom about social movements: "No movement is as coherent and integrated as it seems from afar," he says, "and no movement is as incoherent and fractured as it seems from up close." Viewed from a middle distance, then, the food movement coalesces around the recognition that today's food and farming economy is "unsustainable" – that it can't go on in its current form much longer without courting a breakdown of some kind, whether environmental, economic, or both.

For some in the movement, the more urgent problem is environmental: the food system consumes more fossil fuel energy than we can count on in the future (about a fifth of the total American use of such energy) and emits more greenhouse gas than we can afford to emit, particularly since agriculture is the one human system that *should* be able to substantially rely on solar energy. It will be difficult if not impossible to address the issue of climate change without reforming the food system. This is a conclusion that has only recently been embraced by the environmental movement, which historically has disdained all agriculture as a lapse from wilderness and a source of pollution. But in the last few years, several of the major environmental groups have come to appreciate that a diversified, sustainable agriculture – which can sequester large amounts of carbon in the soil – holds the potential not just to mitigate but actually to help solve environmental problems, including climate change. Today, environmental organizations like the Natural Resources Defense Council and the Environmental Working Group are taking up the cause of food system reform, lending their expertise and clout to the movement.

But perhaps the food movement's strongest claim on public attention today is the fact that the American diet of highly processed food laced with added fats and sugars is responsible for the epidemic of chronic diseases that threatens to bankrupt the health care system. The Centers for Disease Control estimates that fully three quarters of US health care spending goes to treat chronic diseases, most of which are preventable and linked to diet: heart disease, stroke, type 2 diabetes, and at least a third of all cancers. The health care crisis probably cannot be addressed without addressing the catastrophe of the American diet, and that diet is the direct (even if unintended) result of the way that our agriculture and food industries have been organized.

The political ground is shifting, and the passage of health care reform may accelerate that movement. The bill itself contains a few provisions long promoted by the food movement (like calorie labeling on fast food menus), but more important could be the new political tendencies it sets in motion. If health insurers can no longer keep people with chronic diseases out of their patient pools, it stands to reason that the companies will develop a keener interest in preventing those diseases. They will then discover that they have a large stake in things like soda taxes and in precisely which kinds of calories the farm bill is subsidizing. As the insurance industry and the government take on more responsibility for the cost of treating expensive and largely preventable problems like obesity and type 2 diabetes, pressure for reform of the food system, and the American diet, can be expected to increase.

It would be a mistake to conclude that the food movement's agenda can be reduced to a set of laws, policies, and regulations, important as these may be. What is attracting so many people to the movement today (and young people in particular) is a much less conventional kind of politics, one that is about something more than food. The food movement is also about community, identity, pleasure, and, most notably, about carving out a new social and economic space removed from the influence of big corporations on the one side and government on the other. As the Diggers used to say during their San Francisco be-ins during the 1960s, food can serve as "an edible dynamic" – a means to a political end that is only nominally about food itself.

One can get a taste of this social space simply by hanging around a farmers' market, an activity that a great many people enjoy today regardless of whether they're in the market for a bunch of carrots or a head of lettuce. Farmers' markets are thriving, more than five thousand strong, and there is a lot more going on in them than the exchange of money for food. Someone is collecting signatures on a petition. Someone else is playing music. Children are everywhere, sampling fresh produce, talking to farmers. Friends and acquaintances stop to chat. One sociologist calculated that people have ten times as many conversations at the farmers' market than they do in the supermarket. Socially as well as sensually, the farmers' market offers a remarkably rich and appealing environment. Someone buying food here may be acting not just as a consumer but also as a neighbor, a citizen, a parent, a cook. In many cities and towns, farmers' markets have taken on (and not for the first time) the function of a lively new public square.

Though seldom articulated as such, the attempt to redefine, or escape, the traditional role of consumer has become an important aspiration of the food movement. In various ways it seeks to put the relationship between consumers and producers on a new, more neighborly footing, enriching the kinds of information exchanged in the transaction, and encouraging us to regard our food dollars as "votes" for a different kind of agriculture and, by implication, economy. The modern marketplace would have us decide what to buy strictly on the basis of price and self-interest; the food movement implicitly proposes that we enlarge our understanding of both those terms, suggesting that not just "good value" but ethical and political values should inform our buying decisions, and that we'll get more satisfaction from our eating when they do.

That satisfaction helps to explain why many in the movement don't greet the spectacle of large corporations adopting its goals, as some of them have begun to do, with unalloyed enthusiasm. Already Wal-Mart sells organic and local food, but this doesn't greatly warm the hearts of food movement activists. One important impetus for the movement, or at least its locavore wing – those who are committed to eating as much locally produced food as possible – is the desire to get "beyond the barcode" – to create new economic and social structures outside of the mainstream consumer economy. Though not always articulated in these terms, the local food movement wants to decentralize the global economy, if not secede from it altogether, which is why in some communities, such as Great Barrington, Massachusetts, local currencies (the "BerkShare") have popped up.

It makes sense that food and farming should become a locus of attention for Americans disenchanted with consumer capitalism. Food is the place in daily life where corporatization can be most vividly felt: think about the homogenization of taste and experience represented by fast food. By the same token, food offers us one of the shortest, most appealing paths out of the corporate labyrinth, and into the sheer diversity of local flavors, varieties, and characters on offer at the farmers' market.

Put another way, the food movement has set out to foster new forms of civil society. But instead of proposing that space as a counterweight to an overbearing state, as is usually the case, the food movement poses it against the dominance of corporations and their tendency to insinuate themselves into any aspect of our lives from which they can profit. As Wendell Berry writes, the corporations will grow, deliver, and cook your food for you and (just like your mother) beg you to eat it. That they do not yet offer to insert it, pre-chewed, into your mouth is only because they have found no profitable way to do so. The corporatization of something as basic and intimate as eating is, for many of us today, a good place to draw the line.

The Italian-born organization Slow Food, founded in 1986 as a protest against the arrival of McDonald's in Rome, represents perhaps the purest expression of these politics. The organization, which now has 100,000 members in 132 countries, began by dedicating itself to "a firm defense of quiet material pleasure" but has lately waded into deeper political and economic waters. Slow Food's founder and president, Carlo Petrini, a former leftist journalist, has much to say about how people's daily food choices can rehabilitate the act of consumption, making it something more creative and progressive. Ever the Italian, Petrini puts pleasure at the center of his politics, which might explain why Slow Food is not always taken as seriously as it deserves to be. For why *shouldn't* pleasure figure in the politics of the food movement? Good food is potentially one of the most democratic pleasures a society can offer, and is one of those subjects, like sports, that people can talk about across lines of class, ethnicity, and race.

45

"The Hijacking of the Global Food Supply"

from *Stolen Harvest* (2000)

Vandana Shiva

Editors' Introduction

While writers such as Pollen look primarily at the nature of food systems within industrialized countries, developing world activists such as Vandana Shiva take a more global perspective. In this selection Shiva argues that agribusiness corporations, abetted by national policy and international agencies such as the World Bank, are literally stealing food from millions of hungry people in the developing world. At the same time these powerful forces are imposing unsustainable agricultural practices and seeking to patent the genetic diversity that Third World peoples have spent millennia creating.

Shiva's critique is part of a global movement against economic globalization, in large part on social justice grounds. Activists see economic globalization as concentrating power in the hands of a few while undermining the diverse economic, cultural, and environmental practices of local communities worldwide. Since the 1990s this movement has taken specific form in protests against the World Trade Organization, the World Bank, and other international development institutions. Although many "sustainability scientists" in developed countries want nothing to do with such grassroots confrontation, this movement is an important part of the world's sustainable development discourse. For further background, see *Alternatives to Economic Globalization: A Better World is Possible* (San Francisco: Berrett-Koehler, 2004) by John Cavanaugh and Jerry Mander, Walden Bello's *The Food Wars* (London: Verso, 2009), Raj Patel's *Stuffed and Starved: The Hidden Battle for the World Food System* (Brooklyn: Melville House, 2007), Maude Barlow's *Blue Future: Protecting Water for People and the Planet Forever* (New York: The New Press, 2013), and Vandana Shiva's other books such as *Making Peace with the Earth* (London: Pluto Press, 2013).

Food is our most basic need, the very stuff of life.

According to an ancient Indian Upanishad, "All that is born is born of anna [food]. Whatever exists on earth is born of anna, lives on anna, and in the end merges into anna. Anna indeed is the first born amongst all beings."[1]

More than 3.5 million people starved to death in the Bengal famine of 1943. Twenty million were directly affected. Food grains were appropriated forcefully from the peasants under a colonial system of rent collection. Export of food grains continued in spite of the fact that people were going hungry. As the Bengali writer Kali Charan Ghosh reports, 80,000 tons of food grain were exported from Bengal in 1943, just before the famine. At the time, India was being used as a supply base for the British military. "Huge exports were allowed to feed the people of other lands, while the shadow of famine was hourly lengthening on the Indian horizon."[2]

More than one-fifth of India's national output was appropriated for war supplies. The starving Bengal peasants gave up over two-thirds of the food they produced, leading their debt to double. This, coupled with speculation, hoarding, and profiteering by traders, led to skyrocketing prices. The poor of Bengal paid for the empire's war through hunger and starvation – and the "funeral march of the Bengal peasants, fishermen, and artisans."[3]

As the crisis began, thousands of women organized in Bengal in defense of their food rights. "Open more ration shops" and "Bring down the price of food" were the calls of women's groups throughout Bengal.[4] After the famine, the peasants also started to organize around the central demand of keeping a two-thirds, or *tebhaga*, share of the crops. At its peak the Tebhaga movement, as it was called, covered 19 districts and involved 6 million people. Peasants refused to let their harvest be stolen by the landlords and the revenue collectors of the British Empire. Everywhere peasants declared, "Jan debo tabu dhan debo ne" – "We will give up our lives, but we will not give up our rice." In the village of Thumniya, the police arrested some peasants who resisted the theft of their harvest. They were charged with "stealing paddy."[5]

A half-century after the Bengal famine, a new and clever system has been put in place, which is once again making the theft of the harvest a right and the keeping of the harvest a crime. Hidden behind complex free-trade treaties are innovative ways to steal nature's harvest, the harvest of the seed, and the harvest of nutrition.

THE CORPORATE HIJACKING OF FOOD AND AGRICULTURE

I focus on India to tell the story of how corporate control of food and globalization of agriculture are robbing millions of their livelihoods and their right to food both because I am an Indian and because Indian agriculture is being especially targeted by global corporations. Since 75 percent of the Indian population derives its livelihood from agriculture, and every fourth farmer in the world is an Indian, the impact of globalization on Indian agriculture is of global significance.

However, this phenomenon of the stolen harvest is not unique to India. It is being experienced in every society, as small farms and small farmers are pushed to extinction, as monocultures replace biodiverse crops, as farming is transformed from the production of nourishing and diverse foods into the creation of markets for genetically engineered seeds, herbicides, and pesticides.

As farmers are transformed from producers into consumers of corporate-patented agricultural products, as markets are destroyed locally and nationally but expanded globally, the myth of "free trade" and the global economy becomes a means for the rich to rob the poor of their right to food and even their right to life. For the vast majority of the world's people – 70 percent – earn their livelihoods by producing food. The majority of these farmers are women. In contrast, in the industrialized countries, only 2 percent of the population are farmers.

FOOD SECURITY IS IN THE SEED

For centuries Third World farmers have evolved crops and given us the diversity of plants that provide us nutrition. Indian farmers evolved 200,000 varieties of rice through their innovation and breeding. They bred rice varieties such as Basmati. They bred red rice and brown rice and black rice. They bred rice that grew 18 feet tall in the Gangetic floodwaters, and saline-resistant rice that could be grown in the coastal water. And this innovation by farmers has not stopped. Farmers involved in our movement, Navdanya, dedicated to conserving native seed diversity, are still breeding new varieties.

The seed, for the farmer, is not merely the source of future plants and food; it is the storage place of culture and history. Seed is the first link in the food chain. Seed is the ultimate symbol of food security. Free exchange of seed among farmers has been the basis of maintaining biodiversity as well as food security. This exchange is based on cooperation and reciprocity. A farmer who wants to exchange seed generally gives an equal quantity of seed from his field in return for the seed he gets.

Free exchange among farmers goes beyond mere exchange of seeds; it involves exchanges of ideas and knowledge, of culture and heritage. It is an accumulation of tradition, of knowledge of how to work the seed. Farmers learn about the plants they want to grow in the future by watching them grow in other farmers' fields.

New seeds are first worshipped, and only then are they planted. New crops are worshipped before being consumed. Festivals held before sowing seeds as well as harvest festivals, celebrated in the fields, symbolize people's intimacy with nature.[6] For the farmer, the field is the mother; worshipping the field is a sign of gratitude toward the earth, which, as mother, feeds the millions of life forms that are her children.

But new intellectual-property-rights regimes, which are being universalized through the Trade Related Intellectual Property Rights Agreement of the World Trade Organization (WTO), allow corporations to usurp the knowledge of the seed and monopolize it by claiming it as their private property. Over time, this results in corporate monopolies over the seed itself.

Corporations like RiceTec of the United States are claiming patents on Basmati rice. Soybean, which evolved in East Asia, has been patented by Calgene, which is now owned by Monsanto. Calgene also owns patents on mustard, a crop of Indian origin. Centuries of collective innovation by farmers and peasants are being hijacked as corporations claim intellectual-property rights on these and other seeds and plants.[7]

"FREE TRADE" OR "FORCED TRADE"

Today, ten corporations control 32 percent of the commercial-seed market, valued at $23 billion, and 100 percent of the market for genetically engineered, or transgenic, seeds.[8] These corporations also control the global agrochemical and pesticide market. Just five corporations control the global trade in grain. In late 1998, Cargill, the largest of these five companies, bought Continental, the second largest, making it the single biggest factor in the grain trade. Monoliths such as Cargill and Monsanto were both actively involved in shaping international trade agreements, in particular the Uruguay Round of the General Agreement on Trade and Tarriffs, which led to the establishment of the WTO.

This monopolistic control over agricultural production, along with structural adjustment policies that brutally favor exports, results in floods of exports of foods from the United States and Europe to the Third World. As a result of the North American Free Trade Agreement (NAFTA), the proportion of

Mexico's food supply that is imported increased from 20 percent in 1992 to 43 percent in 1996. After 18 months of NAFTA, 2.2 million Mexicans had lost their jobs, and 40 million had fallen into extreme poverty. One out of two peasants was not getting enough to eat. As Victor Suares has stated, "Eating more cheaply on imports is not eating at all for the poor in Mexico."[9]

In the Philippines, sugar imports have destroyed the economy. In Kerala, India, the prosperous rubber plantations were rendered unviable due to rubber imports. The local $350 million rubber economy was wiped out, with a multiplier effect of $3.5 billion on the economy of Kerala. In Kenya, maize imports brought prices crashing for local farmers who could not even recover their costs of production.

Trade liberalization of agriculture was introduced in India in 1991 as part of a World Bank/International Monetary Fund (IMF) structural adjustment package. While the hectares of land under cotton cultivation had been decreasing in the 1970s and 1980s, in the first six years of World Bank/IMP-mandated reforms, the land under cotton cultivation increased by 1.7 million hectares. Cotton started to displace food crops. Aggressive corporate advertising campaigns, including promotional films shown in villages on "video vans," were launched to sell new, hybrid seeds to farmers. Even gods, goddesses, and saints were not spared: in Punjab, Monsanto sells its products using the image of Guru Nanak, the founder of the Sikh religion. Corporate, hybrid seeds began to replace local farmers' varieties.

The new hybrid seeds, being vulnerable to pests, required more pesticides. Extremely poor farmers bought both seeds and chemicals on credit from the same company. When the crops failed due to heavy pest incidence or large-scale seed failure, many peasants committed suicide by consuming the same pesticides that had gotten them into debt in the first place. In the district of Warangal, nearly 400 cotton farmers committed suicide due to crop failure in 1997, and dozens more committed suicide in 1998.

Under this pressure to cultivate cash crops, many states in India have allowed private corporations to acquire hundreds of acres of land. The state of Maharashtra has exempted horticulture projects from its land-ceiling legislation. Madhya Pradesh is offering land to private industry on long-term leases, which, according to industry, should last for at least 40 years. In Andhra Pradesh and Tamil Nadu, private corporations are today allowed to acquire over 300 acres of land for raising shrimp for exports. A large percentage of agricultural production on these lands will go toward supplying the burgeoning food-processing industry, in which mainly transnational corporations are involved. Meanwhile, the United States has taken India to the WTO dispute panel to contest its restrictions on food imports.

In certain instances, markets are captured by other means. In August 1998, the mustard-oil supply in Delhi was mysteriously adulterated. The adulteration was restricted to Delhi but not to any specific brand, indicating that it was not the work of a particular trader or business house. More than 50 people died. The government banned all local processing of oil and announced free imports of soybean oil. Millions of people extracting oil on tiny, ecological, cold-press mills lost their livelihoods. Prices of indigenous oilseed collapsed to less than one-third their previous levels. In Sira, in the state of Karnataka, police officers shot farmers protesting the fall in prices of oilseeds.

Imported soybeans' takeover of the Indian market is a clear example of the imperialism on which globalization is built. One crop exported from a single country by one or two corporations replaced hundreds of foods and food producers, destroying biological and cultural diversity, and economic and political democracy. Small mills are now unable to serve small farmers and poor consumers with low-cost, healthy, and culturally appropriate edible oils. Farmers are robbed of their freedom to choose what they grow, and consumers are being robbed of their freedom to choose what they eat.

CREATING HUNGER WITH MONOCULTURES

Global chemical corporations, recently reshaped into "life sciences" corporations, declare that without them and their patented products, the world cannot be fed. As Monsanto advertised in its $1.6 million European advertising campaign:

Worrying about starving future generations won't feed them. Food biotechnology will. The world's population is growing rapidly, adding the equiv-

alent of a China to the globe every ten years. To feed these billion more mouths, we can try extending our farming land or squeezing greater harvests out of existing cultivation. With the planet set to double in numbers around 2030, this heavy dependency on land can only become heavier. Soil erosion and mineral depletion will exhaust the ground. Lands such as rainforests will be forced into cultivation. Fertilizer, insecticide, and herbicide use will increase globally. At Monsanto, we now believe food biotechnology is a better way forward.[10]

But food is necessary for all living species. That is why the Taittreya Upanishad calls on humans to feed all beings in their zone of influence.

Industrial agriculture has not produced more food. It has destroyed diverse sources of food, and it has stolen food from other species to bring larger quantities of specific commodities to the market, using huge quantities of fossil fuels and water and toxic chemicals in the process.

It is often said that the so-called miracle varieties of the Green Revolution in modern industrial agriculture prevented famine because they had higher yields. However, these higher yields disappear in the context of total yields of crops on farms. Green Revolution varieties produced more grain by diverting production away from straw. This "partitioning" was achieved through dwarfing the plants, which also enabled them to withstand high doses of chemical fertilizer.

However, less straw means less fodder for cattle and less organic matter for the soil to feed the millions of soil organisms that make and rejuvenate soil. The higher yields of wheat or maize were thus achieved by stealing food from farm animals and soil organisms. Since cattle and earthworms are our partners in food production, stealing food from them makes it impossible to maintain food production over time, and means that the partial yield increases were not sustainable.

The increase in yields of wheat and maize under industrial agriculture were also achieved at the cost of yields of other foods a small farm provides. Beans, legumes, fruits, and vegetables all disappeared both from farms and from the calculus of yields. More grain from two or three commodities arrived on national and international markets, but less food was eaten by farm families in the Third World.

The gain in "yields" of industrially produced crops is thus based on a theft of food from other species and the rural poor in the Third World. That is why, as more grain is produced and traded globally, more people go hungry in the Third World. Global markets have more commodities for trading because food has been robbed from nature and the poor.

Productivity in traditional farming practices has always been high if it is remembered that very few external inputs are required. While the Green Revolution has been promoted as having increased productivity in the absolute sense, when resource use is taken into account, it has been found to be counterproductive and inefficient.

Perhaps one of the most fallacious myths propagated by Green Revolution advocates is the assertion that high-yielding varieties have reduced the acreage under cultivation, therefore preserving millions of hectares of biodiversity. But in India, instead of more land being released for conservation, industrial breeding actually increases pressure on the land, since each acre of a monoculture provides a single output, and the displaced outputs have to be grown on additional acres, or "shadow" acres.[11]

A study comparing traditional polycultures with industrial monocultures shows that a polyculture system can produce 100 units of food from 5 units of inputs, whereas an industrial system requires 300 units of input to produce the same 100 units. The 295 units of wasted inputs could have provided 5,900 units of additional food. Thus the industrial system leads to a decline of 5,900 units of food. This is a recipe for starving people, not for feeding them.[12]

THE INSECURITY OF IMPORTS

It is repeatedly argued that food security does not depend on food "self-sufficiency" (food grown locally for local consumption), but on food "self-reliance" (buying your food from international markets). According to the received ideology of free trade, the earnings from exports of farmed shrimp, flowers, and meat will finance imports of food. Hence any shortfall created by the diversion of productive capacity from growing food for domestic consumption to growing luxury items for consumption by rich Northern consumers would be more than made up.

However, it is neither efficient nor sustainable to grow shrimp, flowers, and meat for export in countries such as India. In the case of flower exports, India spent Rs. 1.4 billion as foreign exchange for promoting floriculture exports and earned a mere Rs. 320 million.[13] In other words, India can buy only one-fourth of the food it could have grown with export earnings from floriculture.[14] Our food security has therefore declined by 75 percent, and our foreign exchange drain increased by more than Rs. 1 billion.

In the case of meat exports, for every dollar earned, India is destroying 15 dollars' worth of ecological functions performed by farm animals for sustainable agriculture. Before the Green Revolution, the byproducts of India's culturally sophisticated and ecologically sound livestock economy, such as the hides of cattle, were exported, rather than the ecological capital, that is, the cattle themselves. Today, the domination of the export logic in agriculture is leading to the export of our ecological capital, which we have conserved over centuries. Giant slaughterhouses and factory farming are replacing India's traditional livestock economy. When cows are slaughtered and their meat is exported, with it are exported the renewable energy and fertilizer that cattle provide to the small farms of small peasants. These multiple functions of cattle in farming systems have been protected in India through the metaphor of the sacred cow. Government agencies cleverly disguise the slaughter of cows, which would outrage many Indians, by calling it "buffalo meat."

In the case of shrimp exports, for every acre of an industrial shrimp farm, 200 acres of productive ecosystems are destroyed. For every dollar earned as foreign exchange from exports, six to ten dollars' worth of destruction takes place in the local economy. The harvest of shrimp from aquaculture farms is a harvest stolen from fishing and farming communities in the coastal regions of the Third World. The profits from exports of shrimp to U.S., Japanese, and European markets show up in national and global economic growth figures. However, the destruction of local food consumption, groundwater resources, fisheries, agriculture, and livelihoods associated with traditional occupations in each of these sectors does not alter the global economic value of shrimp exports; such destruction is only experienced locally.

In India, intensive shrimp cultivation has turned fertile coastal tracts into graveyards, destroying both fisheries and agriculture. In Tamil Nadu and Andhra Pradesh, women from fishing and farming communities are resisting shrimp cultivation through *satyagraha* [nonviolent civil disobedience]. Shrimp cultivation destroys 15 jobs for each job it creates. It destroys $5 of ecological and economic capital for every dollar earned through exports. Even these profits flow for only three to five years, after which the industry must move on to new sites. Intensive shrimp farming is a non-sustainable activity, described by United Nations agencies as a "rape and run" industry.

Since the World Bank is advising all countries to shift from "food first" to "export first" policies, these countries all compete with each other, and the prices of these luxury commodities collapse. Trade liberalization and economic reform also include devaluation of currencies. Thus exports earn less, and imports cost more. Since the Third World is being told to stop growing food and instead to buy food in international markets by exporting cash crops, the process of globalization leads to a situation in which agricultural societies of the South become increasingly dependent on food imports, but do not have the foreign exchange to pay for imported food. Indonesia and Russia provide examples of countries that have moved rapidly from food-sufficiency to hunger because of the creation of dependency on imports and the devaluation of their currencies.

STEALING NATURE'S HARVEST

Global corporations are not just stealing the harvest of farmers. They are stealing nature's harvest through genetic engineering and patents on life forms.

Genetically engineered crops manufactured by corporations pose serious ecological risks. Crops such as Monsanto's Roundup Ready soybeans, designed to be resistant to herbicides, lead to the destruction of biodiversity and increased use of agrochemicals. They can also create highly invasive "superweeds" by transferring the genes for herbicide resistance to weeds. Crops designed to be pesticide factories, genetically engineered to produce toxins and venom with genes from bacteria, scorpions, snakes, and wasps, can threaten non-pest species

and can contribute to the emergence of resistance in pests and hence the creation of "superpests." In every application of genetic engineering, food is being stolen from other species for the maximization of corporate profits.

To secure patents on life forms and living resources, corporations must claim seeds and plants to be their "inventions" and hence their property. Thus corporations like Cargill and Monsanto see nature's web of life and cycles of renewal as "theft" of their property. During the debate about the entry of Cargill into India in 1992, the Cargill chief executive stated, "We bring Indian farmers smart technologies, which prevent bees from usurping the pollen."[15] During the United Nations Biosafety Negotiations, Monsanto circulated literature that claimed that "weeds steal the sunshine."[16] A worldview that defines pollination as "theft by bees" and claims that diverse plants "steal" sunshine is one aimed at stealing nature's harvest, by replacing open, pollinated varieties with hybrids and sterile seeds, and destroying biodiverse flora with herbicides such as Monsanto's Roundup.

This is a worldview based on scarcity. A worldview of abundance is the worldview of women in India who leave food for ants on their doorstep, even as they create the most beautiful art in *kolams*, *mandalas*, and *rangoli* with rice flour. Abundance is the worldview of peasant women who weave beautiful designs of paddy to hang up for birds when the birds do not find grain in the fields. This view of abundance recognizes that, in giving food to other beings and species, we maintain conditions for our own food security. It is the recognition in the Isho Upanishad that the universe is the creation of the Supreme Power meant for the benefits of (all) creation. Each individual life form must learn to enjoy its benefits by farming a part of the system in close relation with other species. Let not any one species encroach upon others' rights.[17] The Isho Upanishad also says,

a selfish man over-utilizing the resources of nature to satisfy his own ever-increasing needs is nothing but a thief, because using resources beyond one's needs would result in the utilization of resources over which others have a right.[18]

In the ecological worldview, when we consume more than we need or exploit nature on principles of greed, we are engaging in theft. In the anti-life view of agribusiness corporations, nature renewing and maintaining herself is a thief. Such a world-view replaces abundance with scarcity, fertility with sterility. It makes theft from nature a market imperative, and hides it in the calculus of efficiency and productivity.

FOOD DEMOCRACY

Food totalitarianism can only be stopped through major citizen mobilization for democratization of the food system. This mobilization is starting to gain momentum in Europe, Japan, India, Brazil, and other parts of the world. We have to reclaim our right to save seed and to biodiversity. We have to reclaim our right to nutrition and food safety. We have to reclaim our right to protect the earth and her diverse species. We have to stop this corporate theft from the poor and from nature. Food democracy is the new agenda for democracy and human rights. It is the new agenda for ecological sustainability and social justice.

NOTES

1 *Taittreya Upanishad*, Gorakhpur: Gita Press, p. 124.
2 Kali Charan Ghosh, *Famines in Bengal, 1770–1943*, Calcutta: Indian Associated Publishing Company, 1944.
3 Bondhayan Chattopadhyay, "Notes Towards an Understanding of the Bengal Famine of 1943," *Transaction*, June 1981.
4 Peter Custers, *Women in the Tebhaga Uprising*, Calcutta: Naya prokash, 1987, p. 52.
5 Peter Custers, p. 78.
6 Festivals like Uganda, Ramanavami, Akshay Trateeya, Ekadashi Aluyana Amavase, Naga Panchami, Naolu Hunime, Ganesh Chaturthi, Rishi Panchami, Navartri, Deepavali, Rathasaptami, Tulsi Vivaha Campasrusti, and Bhoomi Puja all include religious ceremonies around the seed.
7 Vandana Shiva, Vanaja Ramprasad, Pandurang Hegde, Omkar Krishnan, and Radha Holla-Bhar, "The Seed Keepers," New Delhi: Navdanya, 1995.
8 These companies are DuPont/Pioneer (U.S.), Monsanto (U.S.), Novartis (Switzerland), Groupe Limagrain (France), Advanta (U.K. and Netherlands), Guipo Pulsar/Semins/ELM (Mexico), Sakata (Japan), KWS HG (Germany), and Taki (Japan).
9 Victor Suares, Paper presented at International Conference on Globalization, Food Security, and Sustainable Agriculture, July 30–31, 1996.
10 "Monsanto: Peddling 'Life Sciences' or 'Death Sciences'?" New Delhi: RFSTE, 1998.
11 ASSINSEL (International Association of Plant Breeders), "Feeding the 8 Billion and Preserving the Planet," Nyon, Switzerland: ASSINSEL.
12 Francesca Bray, "Agriculture for Developing Nations," *Scientific American*, July 1994, pp. 33–35.
13 *Business India*, March 1998.
14 T.N. Prakash and Tejaswini, "Floriculture and Food Security Issues: The Case of Rose Cultivation in Bangalore," in *Globalization and Food Security: Proceedings of Conference on Globalization and Agriculture*, ed. Vandana Shiva, New Delhi, August 1996.
15 Interview with John Hamilton, *Sunday Observer*, May 9, 1993.
16 Hendrik Verfaillie, speech delivered at the Forum on Nature and Human Society, National Academy of Sciences, Washington, DC, October 30, 1997.
17 Vandana Shiva, "Globalization, Gandhi, and Swadeshi: What is Economic Freedom? Whose Economic Freedom?" New Delhi: RFSTE, 1998.
18 Vandana Shiva, "Globalization, Gandhi, and Swadeshi."

46

"Physical Activity, Sprawl, and Health"

from *Urban Sprawl and Public Health: Designing, Planning, and Building for Healthy Communities* (2004)

Howard Frumkin, Lawrence Frank, and Richard Jackson

Editors' Introduction

Even as residents of communities in many developing countries struggle to get enough to eat, populations in industrialized nations are increasingly overweight, leading to a variety of public health problems including diabetes. Such problems are not just a question of diet. Rather, they stem also from the motor-vehicle-dependent nature of many cities and towns, the lack of safe and attractive places to walk or bike, the rise of new communications technologies such as television and the internet, and the patterns and pressures of daily life.

In this selection three leading public health and urban planning researchers consider the physical and mental costs of sedentary lifestyles, and identify some factors that might lead people to live more physically active lives. Their work is representative of that of many others during the 2000s who sought to zero in on the causes of physical inactivity and initiate programs to reverse this trend. Other writings on this topic include Richard Jackson's *Designing Healthy Communities* (San Francisco: Jossey-Bass, 2012), Jason Corburn's *Toward the Healthy City: People, Places, and the Politics of Urban Planning* (Cambridge: MIT Press, 2009), and *Making Healthy Places: Designing and Building for Health, Well-being, and Sustainability* (Washington, D.C.: Island Press, 2011), edited by Andrew L. Dannenberg, Howard Frumkin, and Richard J. Jackson.

A century ago, physical activity was woven into the fabric of life. Most jobs required physical exertion. Much of the population was agricultural, and farm life consisted of long days of hard work. Factory work, construction work, and even many service jobs required strenuous exertion. People walked to get from place to place, they used stairs instead of elevators and escalators, and household chores – cleaning, cooking, gardening, and repair work – were acts of manual labor.

In just a few generations, the built environment has changed profoundly, and with it, the levels of physical activity in daily life. With changes in technology and migration from the countryside to metropolitan areas, agricultural labor now accounts for less than 5 percent of the workforce. Machines have replaced muscle power, transforming manufacturing and construction work. In a "postindustrial" economy, the typical job now involves sitting at a desk or computer terminal.

Conveyor belts move us through airports, escalators move us up and down in stores, and elevators take us from lobbies to upper floors. At home, washing machines, dryers, dishwashers, blenders, vacuum cleaners, leaf blowers, gasoline-powered lawn mowers, and countless other appliances have eased the physical burden of household work. And changes in land use have radically changed the way we travel. Different land uses are separated from each other by large distances. The transportation infrastructure is increasingly planned and built for automobiles rather than for pedestrians. Travel by foot or bicycle has given way to driving.

The result is a nation of sedentary people. According to the Behavioral Risk Factor Surveillance System (BRFSS), an annual national survey, more than half of American adults are not physically active on a regular basis, and just over one in four reports no leisure-time physical activity at all.[1] In 2000, only 26.2 percent of adults were classified as meeting recommended levels of physical activity (defined as any physical activity for at least thirty minutes per day at least five days per week, or vigorous physical activity for at least twenty minutes at least three days per week). Among children aged nine to thirteen, the pattern is similar: 61.5 percent participate in no organized physical activity when not in school, and 22.6 percent engage in no free-time physical activity.[2]

Sedentary lifestyles have emerged as a pressing public health challenge, because some of the consequences – overweight, type 2 diabetes, and other conditions – have reached epidemic proportions. Public health advocates have worked hard to promote more physical activity, and researchers have worked hard to identify what factors will help in this effort. Clearly, there are many such factors. In their 1999 book, *Physical Activity and Behavioral Medicine*,[3] Sallis and Owen proposed an ecological model that included six categories: demographic and biological factors (such as age, gender, race, and socioeconomic status); psychological, cognitive, and emotional factors (such as knowledge, attitudes, beliefs about exercise, and stress levels); behavioral attributes and skills (such as past history of physical activity); social and cultural factors (such as family and social support); physical environment factors (such as the presence of sidewalks and attractive scenery); and physical activity characteristics (such as intensity). The

ecological model predicts that these categories of factors interact in complex ways.

THE VARIETIES OF PHYSICAL ACTIVITY

To consider the relationship between sprawl and physical activity, we need to understand the categories of physical activity. Three dimensions are especially relevant: recreational versus utilitarian activity, moderate versus intense activity, and activity as it varies among different groups of people.

Physical activity may be either recreational or utilitarian.[4] Recreational physical activity – a jog in the park, a game of tennis – is carried out with the intention of getting exercise. In contrast, utilitarian physical activity is done for a purpose, such as walking to the store, to the theater, or to work. The primary purpose of such a trip is to arrive at the destination, and the physical activity it involves is incidental. Physical activity done at work – lifting boxes, carrying tools, and so on – is also utilitarian, and for some people, it accounts for the majority of physical activity. (Some activities, such as gardening, have both recreational and utilitarian qualities.)

The distinction is important because the impetus for recreational physical activity is very different than the impetus for utilitarian physical activity. Recreational physical activity, or exercise, requires a high level of motivation, and even people who begin exercise programs often do not sustain them. Utilitarian physical activity, on the other hand, is secondary to other goals. For this reason, it may be easier to build into a daily routine and maintain over time. A person who walks three blocks from home to the subway each morning, rides to the station near his office, walks the final two blocks to his office, and reverses the commute at the end of the day, walks at least ten blocks a day (and even more if he walks to lunch or on errands at midday). Even if he gets no "exercise" at all, his daily routine includes a fair level of physical activity.

The built environment influences both recreational and utilitarian physical activity. Environments that provide facilities for active recreation, such as nearby parks, multiuse trails, and even appealing sidewalks or public spaces for evening strolls, may promote recreational physical activity. On the utilitarian side, environments that facilitate commuting by foot, bicycle, or transit (most transit riders are also walkers, since they have to travel to and from the transit stops) help incorporate walking or bicycling as a daily routine. Environments that locate stores, theaters, and other destinations within walking distance of home and work have the same potential, a strategic opportunity since nonwork trips account for the majority of trips people make.

Another important distinction is between moderate and vigorous physical activity. Moderate physical activity is defined as activity that raises the heart rate to 50 to 69 percent of its maximum capacity, whereas vigorous physical activity raises the heart rate to at least 70 percent of its maximum. A person's maximum heart rate is commonly estimated by subtracting his or her age from 220. For example, a fifty-year-old person's estimated maximum heart rate would be $220 - 50 = 170$ beats per minute. The 50 percent and 70 percent levels would be 85 and 119 beats per minute, respectively.

Brisk walking, bicycling, and even gardening qualify as moderate physical activities.[5] Current recommendations are for a half hour of moderate physical activity on at least five days per week,[6] although some experts have suggested higher levels.[7] Moderate physical activity is as beneficial as vigorous exercise in preventing cardiovascular disease, assuming that equivalent levels of energy are expended.[8]

Contrary to popular opinion, such activity does not need to be accumulated in one activity session, such as a gym workout. Multiple episodes during that day, as short as eight or ten minutes, offer the same benefit. This has implications for built environment design; places designed so that people walk on multiple occasions during the day may go a long way toward helping them reach recommended levels of physical activity.

A final distinction is not between different kinds of physical activity, but among different groups of people. Different people are active (or inactive) in different ways. Much research in recent years has characterized the physical activity patterns, and the reasons for inactivity, among various populations. Studies have examined physical activity patterns according to gender, age, and race and ethnicity.[9] Generally, these studies have suggested that inactivity is higher among members of minority groups, poor people, and women. Members of these groups cite a wide range of constraints on physical activity.

Some pertain to life circumstances, such as being too busy; juggling competing demands from job, family, or friends; being physically tired or ill; and major life changes or traumas. Poor people cite economic constraints to physical activity. Other constraints pertain more to the environment, including safety concerns, weather and environment, and a lack of facilities and opportunities to be physically active.

Those who want to promote physical activity through community design, then, face several questions. What design features promote utilitarian physical activity, which may be the most sustainable strategy? What design features promote recreational physical activity among those who might otherwise not exercise? How much physical activity results from various design interventions? And how should strategies be tailored to meet the needs of different groups of people?

THE HEALTH BENEFITS OF PHYSICAL ACTIVITY

These questions matter because physical activity is good for health, and being sedentary threatens health both directly and indirectly. A sedentary lifestyle increases the risk of cardiovascular disease, stroke, and all-cause mortality,[10] whereas physical activity prolongs life.[11] Men in the lowest quintile of physical fitness have a two- to threefold increased risk of dying overall, and a three- to fivefold increased risk of dying of cardiovascular disease, compared to men who are more fit.[12] Among women, walking ten blocks per day or more is associated with a 33 percent decrease in the risk of cardiovascular disease.[13] The risk of low physical fitness is comparable to, and in some studies greater than, the risk of hypertension, high cholesterol, diabetes, and even smoking.[14] Physical fitness prevents cardiovascular disease and prolongs life among people with diabetes, and the benefits are greatest among those with the highest blood sugar levels.[15] Physical activity also appears to be protective against cancer,[16] cognitive decline in the elderly,[17] depression,[18] osteoporosis,[19] and a range of other common diseases.

Magazines, newspapers, radio talk shows, television shows and ads, and billboards are full of weight loss supplements and miraculous new diets, suggesting that better eating will solve the problem of

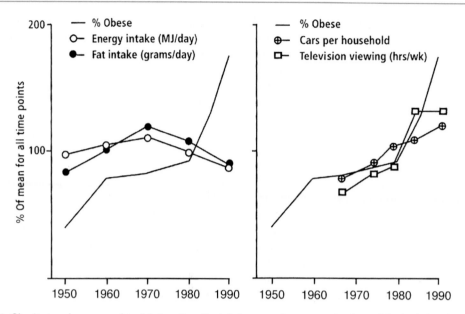

Figure 1 Obesity trends compared to "gluttony" on the left (measured as energy intake and fat intake) and to "sloth" on the right (measured by car ownership and television viewing). This comparison shows the powerful role played by lifestyle (in this case television viewing and driving) as opposed to diet.

Source: A.M. Prentice and S.A. Jebb, "Obesity in Britain: Gluttony or Sloth?" *British Medical Journal* 1995: 311: 437–9.

overweight. But physical activity is also a crucial part of the equation. This was graphically demonstrated in a *British Medical Journal* article on obesity in Britain that asked the provocative question, "gluttony or sloth?" (Figure 1). The article showed that from 1950 to 1990, obesity steadily increased, even as gluttony peaked and declined. Sloth, on the other hand, increased in tandem with obesity, suggesting an important causal role.[21]

PHYSICAL ACTIVITY AND THE BUILT ENVIRONMENT

In November 2003, the *Atlanta Journal-Constitution* profiled a sixty-seven-year-old retired couple, Carolyn and Norman Daniels, who had recently moved to a neighborhood on the Silver Comet Trail, a popular multiuse trail near Atlanta.[22] Both reported routinely using the trail, Norman on his bicycle and Carolyn on foot. For each of them, the regular activity was new, and both gave credit to the nearby trail. "I hadn't used that bike in how many years?" Norman wondered aloud to his wife and the reporter. "Six or seven or eight?" Carolyn, who had exercised

only occasionally on a stationary bike in her bedroom, was walking three times each week with other women in the neighborhood. "I like to walk with the girls," she said. "We just enjoy running our mouths. It's more sociable." As this story showed, a convenient, attractive trail could motivate previously sedentary people to become active. It also made clear that other factors – such as the company of friends – played an important role.

What features of community design encourage physical activity? In particular, what environmental factors get people out of cars for utilitarian trips, and motivate inactive people to start exercising? To what extent are these factors found in sprawling areas? In recent years, more and more evidence has become available to help answer these questions.[23]

Frank, Engelke, and Schmid[24] identify three dimensions of the built environment that help organize answers to these questions: land use patterns, design characteristics, and transportation systems. Each of these has a distinct role in shaping activity. Land use patterns operate at a large spatial scale, and determine the arrangement of physical activities across the metropolis. Design characteristics operate on a smaller spatial scale.

Examples include the architecture of buildings; the width, tree canopy, and placement of sidewalks; and the vistas in a park; which when taken collectively create a sense or feeling of place. Transportation systems connect different land uses, and define the relative ease and convenience of walking, bicycling, transit, and driving.

Pikora and colleagues in Australia[25] proposed a further framework, related primarily to design characteristics, for classifying the determinants of walking and cycling. They identified four categories: functional factors, safety factors, aesthetic factors, and destination factors. Functional factors relate to the physical attributes of the street or path, such as path continuity and design, street type and width, and traffic volume. Safety factors include crossing aids, lighting, and the level of passive surveillance of the path or sidewalk. Aesthetic factors include cleanliness, maintenance, the presence of trees, and architecture. And destinations are such places as parks, transit nodes, stores, and restaurants.

In interviews with experts, they asked which of these factors seemed most important in determining people's walking and bicycling behavior, for both recreational and utilitarian purposes. The experts identified several factors as being most important. These included safety factors, attractiveness of the streetscape, the presence of destinations (for walking), and the presence of a continuous route and traffic safety (for bicycling).

THE MENTAL HEALTH COSTS OF SPRAWL

Sprawl may also carry mental health costs. For example, who benefits by "getting away from it all"? Escaping to a suburban home may offer more to men than to women, since women still bear a disproportionate share of household responsibilities – amounting to between twenty-five and forty-five hours per week, according to various studies.[26] At the same time the nation's cities have sprawled, working hours have increased, both individually and on a household basis. For two-career households, if the woman has a full-time job, the travel time of a long commute, and the burden of household duties, including transporting children to school and after-school activities, the hours spent behind the wheel each week are likely to contribute significantly to stress.

And what of the nature contact available in suburban locations? That nature may be a highly constructed one – a carefully laid out grassy lawn with a limited number of trees, and perhaps a garden. While this is a restorative environment for many people, it comes at a cost. When thousands of acres are developed as suburban housing, with no preservation of forest, field, and farm, then large parks and natural areas become much less accessible. We gain some opportunities for nature contact even as we lose others.

In fact, the pleasant backyard is only one part of suburban sprawl. The highways and broad feeder roads, the vast parking lots, and the rows of big-box stores are a prominent part of the landscape as well. And for many people, these aspects of the environment are anything but a mental health asset. Country roads seem to be better for mental health than thoroughfares cluttered with road signs and billboards, strip malls and body shops, and large parking lots. In one study, volunteers looked at films of both country roads and commercial roads.[27] They showed less stress and quicker stress recovery when viewing the rural road scenes than when viewing the commercial roadway scenes.

Psychologists, geographers, architects, and planners have much to say about the form, scale, and speed of the environments we inhabit, and of how they make us feel.[28] The high speeds of suburban boulevards, on which everything rushes by quickly; the large scales of big-box stores and vast parking lots; the absence of tranquil and attractive "places of the heart" in daily travels: could these features undermine mental health, or at least forfeit important opportunities to promote it?

Finally, we need to consider the archetypal experience of living in a sprawling area: driving. Aside from the truncated access to large tracts of natural land, aside from the time pressure, aside from the alienating quality of some suburban landscapes, driving itself is a cardinal feature of sprawl, and one of the best understood in terms of its impact on mental health.

Researchers have known for years that driving may have effects on physiology and mood, and may even affect mental health. In the decades after World War II, physiologists and physicians increasingly came to view stress as a medical concern. At the same time, automobiles became more and more a central fixture in modern life, so it was no surprise that stress researchers turned their attention to driving. They studied drivers under various conditions, both on roads and in simulators. They found that driving caused "physiologic arousal" – a combination of elevated heart rate, electrocardiographic changes, increases in serum cortisol and catecholamine levels, and self-reports of anxiety, agitation, and similar feelings. In the language of stress researchers, the "stress" of driving resulted in "strain" among drivers.

REFERENCES

1 Macera CA, Jones DA, Yore MM, Ham SA, Kohl HW, Kimsey CD, Buchner D. Prevalence of physical activity, including lifestyle activities among adults – United States, 2000–2001. *Morbidity and Mortality Weekly Report* 2003; 52 (32): 764–69.

2 Duke J, Huhman M, Heitzler C. Physical activity levels among children aged 9–13 years – United States, 2002. *Morbidity and Mortality Weekly Report* 2003; 52 (33): 785–88.

3 Sallis JF, Owen N. *Physical Activity and Behavioral Medicine.* Thousand Oaks, CA: Sage Publications, 1999.

4 Frank LD, Engelke PE, Schmid TL. *Health and Community Design: The Impacts of the Built Environment on Physical Activity.* Washington, DC: Island Press, 2003.

5 Ainsworth BE, Haskell WL, Leon AS, *et al. Compendium of physical activities: Classification of energy costs of human physical activities.* Medicine and Science in Sports and Exercise 1993; 25: 71–80; Centers for Disease Control (CDC) and Prevention. *Physical Activity and Health: A Report of the Surgeon General.* Centers for Disease Control and Prevention, National Center for Chronic Disease Prevention and Health Promotion, 1996.

6 Pate RR, Pratt M, Blair SN, Haskell WL, Macera CA, Bouchard C, *et al.* Physical activity and public health: A recommendation from the Centers for Disease Control and Prevention and the American College of Sports Medicine. *Journal of the American Medical Association* 1995; 27 (3): 402–07; CDC, 1996, op. cit.

7 Institute of Medicine, Food and Nutrition Board. *Dietary Reference Intakes for Energy, Carbohydrate, Fiber, Fat, Fatty Acids, Cholesterol, Protein, and Amino Acids (Macronutrients).* Washington, DC: National Academy Press, 2002.

8 Manson JE, Greenland P, LaCroix AZ, Stefanick ML, Mouton CP, Oberman A, *et al.* Walking compared with vigorous exercise for the prevention of cardiovascular events in women. *New England Journal of Medicine* 2002; 347 (10): 716–25.

9 Eyler AA, Baker E, Cromer L, King AC, Brownson RC, Donatelle RJ. Physical activity and minority women: A qualitative study. *Health Education and Behavior* 1998; 25: 640–52; King AC, Rejeski WJ, Buchner DM. Physical activity interventions targeting older adults: A critical review and recommendations. *American Journal of Preventive Medicine* 1998; 15: 316–33; Ransdell LB, Wells CL. Physical activity in urban White, African-American, and Mexican-American women. *Medicine and Science in Sports and Exercise* 1998; 30: 1608–15; Taylor WC, Baranowski T, Young DR. Physical activity interventions in low-income, ethnic minority, and populations with disability. *American Journal of Preventive Medicine* 1998; 15 (3): 34–43; King AC, Castro C, Wilcox S, Eyler AA, Sallis JF, Brownson RC. Personal and environmental factors associated with physical inactivity among different racial-ethnic groups of US middle-aged and older aged adults. *Health Psychology* 2000; 19: 354–64; Booth MN, Owen A, Bauman A, Clavisi O, Leslie E. Social-cognitive and perceived environmental influences associated with physical activity in older Australians. *Preventive Medicine* 2000; 31: 15–22; Richter DL, Wilcox S, Greaney ML, Henderson KA, Ainsworth BE. Environmental, policy, and cultural factors related to physical activity in African American women. *Women & Health* 2002; 36 (2): 91–109; Henderson KA, Ainsworth BE. A synthesis of perceptions about physical activity among older African American and American Indian women. *American Journal of Public Health* 2003; 93 (2): 313–17; Crespo CJ, Smith E, Andersen RE, Carter-Pokras O, Ainsworth BE. Race, ethnicity, social class and their relation to physical inactivity during leisure time: Results from the Third National Health and Nutrition Examination Survey, 1988–1994. *American Journal of Preventive Medicine* 2000; 18 (1): 46–53; Eyler AA, Brownson RC, Donatelle RJ, King AC, Brown D, Sallis JF. Physical activity social support and middle- and older-aged minority women: Results from a US survey. *Social Science & Medicine* 1999; 49 (6): 781–89.

10 CDC, 1996, op. cit; National Institutes of Health Consensus Development Panel on Physical Activity and Cardiovascular Health. NIH Consensus Conference: Physical activity and cardiovascular health. *Journal of the American Medical Association* 1996; 27 (6): 241–46; Wannamethee SG, Shaper AG, Walker M, Ebrahim S. Lifestyle and 15-year survival free of heart attack, stroke, and diabetes in middle-aged British men. *Archives of Internal Medicine* 1998; 158 (22): 2433–40; Wannamethee SG, Shaper AG. Physical activity and the prevention of stroke. *Journal of Cardiovascular Risk* 1999; 6 (4): 213–16; Pate *et al.*, 1995, op. cit.

11 Lee IM, Paffenbarger RS Jr. Associations of light, moderate, and vigorous intensity physical activity with longevity. The Harvard Alumni Health Study. *American Journal of Epidemiology* 2000; 151: 293–99; Wannamethee SG, Shaper AG, Walker M. Changes in physical activity, mortality and incidence of coronary heart disease in older men. *Lancet* 1998; 351: 1603–08.

12 Wei M, Kampert JB, Barlow CE, Nichaman MZ, Gibbons LW, Paffenbarger RS, Blair SN. Relationship between low cardiorespiratory fitness and mortality in normal-weight, overweight, and obese men. *Journal of the American Medical Association* 1999; 282: 1547–53.

13 Sesso HD, Paffenbarger RS, Ha T, Lee IM. Physical activity and cardiovascular disease risk in middle-aged and older women. *American Journal of Epidemiololopy* 1999; 150 (4): 408–16.

14 Wei *et al.*, 1999, op. cit; Blair SN, Kampert JB, Kohl HW III, *et al.* Influences of cardiorespiratory fitness and other precursors on cardiovascular disease and all-cause mortality in men and women. *Journal of the American Medical Association* 1996; 276: 205–10.

15 Kohl HW III, Gordon NF, Villegas JA, Blair SN. Cardiorespiratory fitness, glycemic status, and mortality risk in men. *Diabetes Care* 1992; 15: 184–92.

16 Kampert JB, Blair SN, Barlow CE, Kohl HW III. Physical activity, physical fitness, and all-cause and cancer mortality: A prospective study of men and women. *Annals of Epidemiology* 1996; 6: 452–57; Lee IM, Sesso HD, Paffenbarger RS Jr. Physical activity and risk of lung cancer. *International Journal of Epidemiology* 1999; 28: 620–25; Slattery ML, Edwards SL, Boucher KM, Anderson K, Caan BJ. Lifestyle and colon cancer: An assessment of factors associated with risk. *American Journal of Epidemiology* 1999; 150: 869–77; Thune I, Brenn T, Lund E, Gaard M. Physical activity and the risk of breast cancer. *New England Journal of Medicine* 1997; 336 (18): 1269–75; Sesso HD, Paffenbarger RS Jr, Lee IM. Physical activity and breast cancer risk in the College Alumni Health Study (United States). *Cancer Causes & Control* 1998; 9: 433–39; Oliveria SA, Christos PJ. The epidemiology of physical activity and cancer. *Annals of the New York Academy of Sciences* 1997; 83 (3): 79–90.

17 Yaffe K, Barnes D, Nevitt M, Lui L-Y, Covinsky K. A prospective study of physical activity and cognitive decline in elderly women: Women who walk. *Archives of Internal Medicine* 2001; 161 (14): 1703–08.

18 Brosse AL, Sheets ES, Lett HS, Blumenthal A. Exercise and the treatment of clinical depression in adults: Recent findings and future directions. *Sports Medicine* 2002; 32 (12): 741–60; Dunn AL, Trivedi MH, O'Neal HA. Physical activity dose-response effects on outcomes of depression and anxiety. *Medicine & Science in Sports & Exercise.* 2001; 33 (6 Suppl): S587–97; Strawbridge WJ, Deleger S, Roberts RE, Kaplan GA. Physical activity reduces the risk of subsequent depression for older adults. *American Journal of Epidemiology* 2002; 156 (4): 328–34.

19 Bonaiuti D, Shea B, Iovine R, Negrini S, Robinson V, Kemper HC, Wells G, Tugwell P, Cranney A. Exercise for preventing and treating osteoporosis in postmenopausal women (Cochrane Methodology Review). In: *The Cochrane Library*, Issue 4, 2003. Chichester, UK: John Wiley & Sons, Ltd.

20 Nestle M. Food Politics: How the Food Industry Influences Nutrition and Health. Berkeley: University of California Press, 2002; Schell ER. *The Hungry Gene: The Science of Fat and the Future of Thin.* New York: Atlantic Monthly Press, 2002; Critser G. *Fat Land: How Americans Became the Fattest People in the World.* Boston: Houghton Mifflin, 2003; Brownell KD, Borgen KB. *Food Fight: The Inside Story of the Food Industry, the American Obesity Crisis, and What We Can Do About It.* New York: McGraw-Hill, 2003.

21 Prentice AM, Jebb SA. Obesity in Britain: Gluttony or Sloth? *British Medical Journal* 1995; 311: 437–39.

22 Frankston J. Health pros link sprawl with spread. Suburbs, obesity stir debate. *Atlanta Journal-Constitution*, 17 November 2003, p. F1.

TWO

23 Saelens B, Sallis J, Frank L. Environmental correlates of walking and cycling: Findings from the transportation, urban design, and planning literatures. *Annals of Behavioral Medicine* 2003; 25 (2): 80–91; Rumpel N, Owen N, Leslie E. Environmental factors associated with adults' participation in physical activity: A review. *American Journal of Preventive Medicine* 2002; 22 (3): 188–99; Trost SG, Owen N, Bauman AE, Sallis JF, Brown W Correlates of adults' participation in physical activity: Review and update. *Med Sci Sports Med* 2002; 34 (12): 1996–2001; Handy S, Boarnet M, Ewing R, Killingsworth R. How the built environment affects physical activity: Views from urban planning. *American Journal of Preventive Medicine* 2002; 23 (2S): 64–73; French SA, Story M, Jeffery RW. Environmental influences on eating and physical activity. *Ann Rev Public Health* 2001; 22: 309–35; Kahn EB, Ramsey LT, Brownson RC, Heath GW, Howze EH, Powell KE, *et al*. The effectiveness of interventions to increase physical activity: A systematic review. *American Journal of Preventive Medicine* 2002; 22 (4S): 73–107; Sallis JF, Bauman A, Pratt M. Physical activity interventions: Environmental and policy interventions to promote physical activity. *American Journal of Preventive Medicine* 1998; 15 (4): 3 79–97.

24 Frank *et al.*, 2003, op. cit.

25 Pikora T, Giles-Corti B, Bull F, Jamrozik K, Donovan R. Developing a framework for assessment of the environmental determinants of walking and cycling. *Social Science & Medicine* 2003; 56: 1693–1703.

26 Cowan RS. *More Work for Mother: The Ironies of Household Technology from the Open Hearth to the Microwave*. New York: Basic Books, 1983; Schor JB. *The Overworked American: The Unexpected Decline of Leisure*. New York: Basic Books, 1992.

27 Parsons R, Tassinary L, Ulrich R, Hebl M, Grossman-Alexander M. The view from the road: Implications for stress recovery and immunization. *Journal of Environmental Psychology* 1998; 18: 113–40.

28 Alexander C, Ishikawa S, Silverstein M, Jacobson M, Fiksdahl-King I, Angel S. *A Pattern Language: Towns, Buildings, Construction*. New York: Oxford University Press, 1977; Tuan Y-F. *Space and Place: The Perspective of Experience*. Minneapolis: University of Minnesota Press, 1977; Whyte WH. *The Social Life of Small Urban Spaces*. Washington, DC: The Conservation Foundation, 1980; Walter EV. *Placeways: A Theory of the Human Environment*. Chapel Hill: University of North Carolina Press, 1988; Gallagher W. *The Power of Place: How Our Surroundings Shape Our Thoughts, Emotions, and Actions*. New York: Poseidon Press, 1993; Kaplan *et al.*, 1998, op. cit.

47

"Slow is Beautiful"

From *In Praise of Slowness* (2004)

Carl Honoré

Editors' Introduction

One interesting question is the pace at which people in more sustainable communities might live their lives. The modern world is one of great speed – fast cars, fast food, overpacked schedules – but all this hurry may not be good for people individually or collectively. A growing movement promotes "slow food" and many other forms of more leisurely life. Of course, many non-industrial cultures still live according to different conceptions of time, and even within developed countries "colored people's time" is seen as different from the mainstream culture.

In this selection, Carl Honoré provides an historical overview of resistance to speed. Author of *In Praise of Slowness: Challenging the Cult of Speed* (New York: HarperCollins, 2004) and *The Slow Fix: Solve Problems, Work Smarter, and Live Better in a World Addicted to Speed* (New York: HarperCollins, 2013), Honoré argues that our sense of time is political, and that slower-paced lives may be healthier. Other resources include Jeremy Rifkin's book *Time Wars: The Primary Conflict in Human History* (New York: Touchstone, 1989), Robert Levine's *A Geography of Time: The Temporal Misadventures of a Social Psychologist* (New York: Basic Books, 1997), and *Slow Food: Collected Thoughts on Taste, Tradition, and the Honest Pleasures of Food* (White River Junction, VT: Chelsea Green, 2001), edited by Carlo Petrini.

Wagrain, a resort town nestled deep in the Austrian Alps, moves at a slow pace. People come here to escape the hurly-burly of Salzburg and Vienna. In the summer, they hike the wooded trails and picnic beside mountain streams. When the snow falls, they ski through the forests, or down the steep, powdery slopes. Whatever the season, the Alpine air fills the lungs with the promise of a good night's sleep back in the chalet.

Once a year, though, this small town does more than just live at a slow pace. It becomes a launch pad for the Slow philosophy. Every October, Wagrain hosts the annual conference of the Society for the Deceleration of Time.

Based in the Austrian city of Klagenfurt, and boasting a membership that stretches across central Europe, the Society is a leader in the Slow movement. Its more than one thousand members are foot soldiers in the war against the cult of doing everything faster. In daily life, that means slowing down when it makes sense to do so. If a Society member is a doctor, he might insist on taking more time to chat to his patients. A management consultant could refuse to answer work calls on the weekend. A designer might cycle to meetings instead of driving. The Decelerators use a German word – *eigenzeit* – to sum up their creed. *Eigen* means "own" and *zeit* means "time." In other words, every living being, event, process or object has its own inherent time or pace, its own *tempo giusto*.

As well as publishing earnest papers on man's relationship with time, the Society stirs up debate with tongue-in-cheek publicity stunts. Members patrol city centers wearing sandwich boards emblazoned with the slogan "Please hurry up!" Not long ago, the Society called on the International Olympic Committee to award gold medals to the athletes with the slowest times.

"Belonging to the Slow movement does not mean that you must always be slow – we take planes, too! – or that you must always be very serious and very philosophical, or that you want to spoil everybody else's fun," says Michaela Schmoczer, the Society's very efficient secretary. "Seriousness is okay, but you don't need to lose the humor."

With that in mind, the Decelerators regularly run "speed traps" in town centers. Using a stopwatch, they time pedestrians going about their daily business. People caught covering 50 meters in less than thirty-seven seconds are pulled over and asked to explain their haste. Their punishment is to walk the same 50 meters while steering a complicated turtle marionette along the pavement. "It is always a huge success," says Jurgen Adam, a schoolteacher who ran a speed trap in the German city of Ulm. "Most people have not even thought about why they are going so fast. But once we get them talking about speed and time, they are very interested. They like the idea of slowing down. Some even return later in the day asking to walk the turtle a second time. They find it so soothing."

The Society members are not alone. Around the world, people are banding together into pro-Slow groups. More than seven hundred Japanese people now belong to the Sloth Club, which advocates less hurried, more environmentally friendly living. The group runs a cafe in Tokyo that serves organic food, stages candlelight concerts and sells T-shirts and coffee mugs bearing the slogan "Slow is beautiful." Tables are deliberately spaced farther apart than is normal in Japan, to encourage people to relax and linger. Thanks in part to the Sloth Club, deceleration is now hip in Japan. The nation's advertisers use the English word "slow" to sell everything from cigarettes and holidays to apartment blocks. Admiration for the easygoing lifestyle of Mediterranean Europe is so widespread that one commentator talks of the "Latinization of the Japanese people."

In 2001, one of the Sloth Club's founders, an anthropologist and environmental activist named Keibo Oiwa, published a survey of the various campaigns for slowness around the world. The book was called *Slow Is Beautiful*, and is already into its twelfth print run. When I visit Oiwa at his office at the Meiji Gakuin University outside Tokyo, he is just back from a well-attended three-day workshop on slowness held by the Hyogo prefecture. "More and more people in Japan, especially young people, are realizing that it is okay to be slow," he says. "For us that represents a total sea change in attitudes."

On the other side of the Pacific, from its headquarters in San Francisco, the Long Now Foundation is adding to the groundswell. Its members warn that we are so busy sprinting to keep up with the daily grind that we seldom lift our gaze beyond the next deadline, the next set of quarterly figures. "Civilization is revving itself into a pathologically short attention span," they say. To make us slow down, to open our eyes to the long view and the big picture, the Foundation is building huge, intricate clocks that tick once a year and measure time over ten millennia. The first, a beautiful beast of bronze and steel, is already on display at the Science Museum in London, England. A second, much larger clock will eventually be carved into a limestone cliff near Great Basin National Park in eastern Nevada.

Many Long Now supporters work in the technology sector. Danny Hillis, who helped invent supercomputers, is on the board. Among the corporate donors are high-tech giants such as PeopleSoft, Autodesk and Sun Microsystems, Inc. Why are players from the fastest industry on earth backing an organization that promotes slowness? Because they, too, have realized that the cult of speed is out of hand.

Today's pro-Slow organizations belong to a tradition of resistance that started long before the industrial era. Even in the ancient world, our ancestors chafed against the tyranny of timekeeping. In 200 BC, the Roman playwright Plautus penned the following lament:

The Gods confound the man who first found out
How to distinguish the hours – confound him,
    too
Who in this place set up a sundial

To cut and hack my days so wretchedly
Into small pieces!
. . . I can't (even sit down to eat) unless the sun
    gives leave.
The town's so full of these confounded dials . . .

As mechanical clocks spread across Europe, protest was never far behind. In 1304, Daffyd ap Gwvilyn, a Welsh bard, fumed: "Confusion to the black-faced clock by the side of the bank that awoke me! May its head, its tongue, its pair of ropes, and its wheels moulder; likewise its weights and dullard balls, its orifices, its hammer, its ducks quacking as if anticipating day and its ever restless works."

As timekeeping wormed its way into every corner of life, satirists poked fun at the European devotion to the clock. In *Gulliver's Travels* (1726), the Lilliputians decide that Gulliver consults his watch so often that it must be his god.

As industrialization gathered pace, so, too, did the backlash against clock-worship and the cult of speed. Many denounced the imposition of universal time as a form of slavery. In 1884, Charles Dudley Warner, an American editor and essayist, gave vent to the popular unease, echoing Plautus in the process: "The chopping up of time into rigid periods is an invasion of individual freedom and makes no allowances for differences in temperament and feeling." Others complained that machines were making life too fast, too hectic, less humane. The Romantic movement of artists, writers and musicians that swept across Europe after 1770 was partly a reaction against the modern culture of hustle and bustle, a harking back to a lost idyllic era.

Right through the Industrial Revolution, people sought ways to challenge, restrain or escape the accelerating pace of life. In 1776, the bookbinders of Paris called a strike to limit their working day to fourteen hours. Later, in the new factories, unions campaigned for more time off. The standard refrain was: "Eight hours for work, eight hours for sleep, eight hours for what we will." In a gesture that underscored the link between time and power, radical unionists smashed the clocks above the factory gates.

In the United States, meanwhile, a group of intellectuals known as the Transcendentalists exalted the gentle simplicity of a life rooted in nature. One of their number, Henry David Thoreau, retired to a one-room cabin beside Walden Pond near Boston in 1845, from which he decried modern life as a treadmill of "infinite bustle . . . nothing but work, work, work."

In 1870, the British-based Arts and Crafts movement turned away from mass production to embrace the slow, meticulous handwork of the artisan. In cities across the industrial world, weary urbanites found solace in the cult of the rural idyll. Richard Jeffries made a career of writing novels and memoirs about England's green and pleasant land, while Romantic painters such as Caspar David Friedrich in Germany, Jean-Francois Millet in France and John Constable in England filled their canvases with soothing country scenes. The urban desire to spend a little time resting and recharging the batteries in Arcadia helped bring about the emergence of modern tourism. By 1845, there were more tourists than sheep in Britain's Lake District.

In the late nineteenth century, physicians and psychiatrists began calling attention to the deleterious effects of speed. George Beard got the ball rolling in 1881 with *American Nervousness*, which blamed fast living for everything from neuralgia to tooth decay and hair loss. Beard argued that the modern obsession with punctuality, with making every second count, made everyone feel that "a delay of a few minutes might destroy the hopes of a lifetime."

Three years later, Sir James Crichton-Browne blamed the high tempo of modern life for the sharp rise in the number of deaths in England from kidney failure, heart disease and cancer. In 1901, John Girdner coined the term "newyorkitis" to describe an illness whose symptoms included edginess, quick movements and impulsiveness. A year later, a Frenchman named Gabriel Hanotaux prefigured modern environmentalism by warning that the reckless pursuit of speed was hastening the depletion of the world's coal reserves: "We are burning our way during our stay in order to travel through more rapidly."

Some of the fears articulated by the early critics of speed were patently absurd. Doctors claimed that passengers travelling on steam trains would be crushed by the pressure, or that the mere sight of a speeding locomotive would drive onlookers insane. When bicycles first became popular in the 1890s, some feared that riding into the wind at high speed would cause permanent disfigurement, or "bicycle face." Moralists warned that bikes would corrupt the young by enabling them to enjoy romantic trysts far from the prying eyes of their guardians. However risible these misgivings turned out to be, it was nevertheless clear by the end of the nineteenth century that speed really did take a toll. Thousands were dying every year in accidents involving the new vessels of velocity – bicycles, cars, buses, trams, trains, and steamships.

Through the twentieth century, resistance to the cult of speed grew, and began to coalesce into broad social movements. The counterculture earthquake of the 1960s inspired millions to slow down and live more simply. A similar philosophy gave birth to the Voluntary Simplicity movement. In the late 1980s, the New York-based Trends Research Institute identified a phenomenon known as downshifting, which means swapping a high-pressure, high-earning, high-tempo lifestyle for a more relaxed, less consumerist existence. Unlike decelerators from the hippie generation, downshifters are driven less by political or environmental scruples than by the desire to lead more rewarding lives. They are willing to forgo money in return for time and slowness. Datamonitor, a London-based market research firm, expects the number of downshifters in Europe to rise from twelve million in 2002 to over sixteen million by 2007.

These days, many people are seeking refuge from speed in the safe harbor of spirituality. While mainstream Christian churches face dwindling congregations, their evangelical rivals are thriving. Buddhism is booming across the West, as are bookstores, chat rooms and healing centers dedicated to the eclectic, metaphysical doctrines of New Ageism. All of this makes sense at a time when people crave slowness. The spirit, by its very nature, is Slow. No matter how hard you try, you cannot accelerate enlightenment. Every religion teaches the need to slow down in order to connect with the self, with others and with a higher force. In Psalm 46, the Bible says: "Be still then, and know that I am God."

In the early twentieth century, Christian and Jewish clerics lent moral weight to the campaign for a shorter workweek, arguing that workers needed more time off in order to nourish their souls. Today, the same plea for slowness is once again emanating from pulpits around the world. A Google search turns up scores of sermons railing against the demon speed. In February 2002, at the First Unitarian Church in Rochester, New York, Reverend Gary James made an eloquent case for the Slow philosophy. In a sermon entitled "Slow Down!" he told his congregation that life "requires moments of intense exertion and quickened pace. . . . But it also requires a pause now and then – a Sabbath moment to assess where we are going, how quickly we wish to get there – and, more important, why. Slow can be beautiful." When Thich Nhat Hanh, a well-known Buddhist leader, visited Denver, Colorado, in 2002, more than five thousand people came to hear him speak. He urged them to slow down, "to take the time to live more deeply." New Age gurus preach a similar message.

If they are to make any headway at all, pro-Slow campaigners must root out the deep prejudice against the very idea of slowing down. In many quarters, "slow" remains a dirty word. Just look at how the Oxford English Dictionary defines it: "not understanding readily, dull, uninteresting, not learning easily, tedious, slack, sluggish." Hardly the sort of stuff you would put on your CV. In our hyped-up, faster-is-better culture, a turbocharged life is still the ultimate trophy on the mantelpiece. When people moan, "Oh, I'm so busy, I'm run off my feet, my life is a blur, I haven't got time for anything," what they often mean is, "Look at me: I am hugely important, exciting and energetic." Though men seem to like speed more than women, both sexes indulge in faster-than-thou one-upmanship. With a mixture of pride and pity, New Yorkers marvel at the slower pace of life elsewhere in the United States. "It's like they're on vacation all the time," sniffs one female Manhattanite. "If they tried to live like that in New York, they'd be toast."

Perhaps the greatest challenge of the Slow movement will be to fix our neurotic relationship with time itself. To teach us, in the words of Golda Meir, the former Israeli leader, how to ". . . govern the clock, not be governed by it." This may already be happening, below the radar. As the Curator of Time at the Science Museum in London, David Rooney oversees a splendid collection of five hundred timekeeping devices, ranging from ancient sundials and hourglasses to modern quartz watches and atomic clocks. Not surprisingly, the bespectacled twenty-eight-year-old has a claustrophobic relationship with time. On his wrist he wears a terrifyingly accurate radio-controlled watch. An antenna hidden in the wristband receives a daily update from Frankfurt. If the watch misses a signal, the number 1 appears in the lower left corner of the screen. If it misses the next day's signal, the number changes to 2, and so on. All of this accuracy makes Rooney very anxious indeed.

"I feel a real sense of loss when I miss my signal," he tells me as we wander round the museum's Measuring Time exhibit, raising our voices to be heard over the persistent tick-tock-tick-tock. "When the counter on the watch reaches 2, I get worried. Once it went to 3, and I had to leave it in a drawer at home. I get stressed knowing it's just a milli-second out."

Rooney knows this is not healthy behavior, but he sees hope for the rest of us. The historical trend towards embracing ever more accurate timepieces has finally come to an end with the radio-controlled watch, which failed to catch on as a consumer product. People would rather put style ahead of accuracy by wearing a Swatch or a Rolex. Rooney thinks this reflects a subtle shift in our feelings about time.

"In the Industrial Revolution, when life became ruled by work, we lost control over our use of time," he says. "What we're seeing now is maybe the beginnings of a reaction against that. People seem to have reached the point where they don't want to have their time diced up into smaller and smaller pieces, with greater and greater accuracy. They don't want to be obsessed with time, or a slave to the clock. There may be an element of 'the boss keeps time, so I don't want to.'"

A few months after our meeting, Rooney decided to tackle his own obsessive timekeeping. Instead of fretting over mislaid milliseconds, he now wears a 1960s windup watch that is usually around five minutes off. "It's my own reaction against too much accuracy," he tells me. Rooney deliberately chose a windup watch to symbolize regaining the upper hand over time. "If you don't wind it every day, it stops, so you're in control," he says. "I feel like time is working for me now, rather than the other way round, which makes me feel less pressured. I don't hurry so much."

Some people are going even further. On a recent trip to Germany, my interpreter raved about the benefits of not wearing a watch at all. He remains scrupulously punctual, thanks to the clock on his mobile phone, but his former obsession with minutes and seconds is waning. "Not having a watch on my wrist definitely makes me more relaxed about time," he told me. "It is easier for me to slow down, because time is not always there in my line of vision saying, 'No, you must not slow down, you must not waste me, you must hurry.'"

Time is certainly a hot topic these days. How should we use it? Who controls it? How can we be less neurotic about it? Jeremy Rifkin, the American economist, thinks it could be the defining issue of the twenty-first century. "A battle is brewing over the politics of time," he wrote in his book *Time Wars*. "Its outcome could determine the future course of politics around the world in the coming century." It will certainly help to determine the future of the Slow movement.

PART THREE

Sustainability planning tools and politics

INTRODUCTION TO PART THREE

Although the field of urban planning contains many existing mechanisms and tools, a somewhat different mix of skills and techniques will probably be needed to help bring about more sustainable cities and towns. An improved ability to analyze long-term trends and monitor progress toward healthier urban environments is essential, for one thing, as is development of processes for constructive public participation and achievement of a more constructive urban politics in general. This section highlights several tools that have been influential to date within the movement for sustainable urban development, and considers what a more satisfactory political and governance context for sustainability planning might look like.

We start with the topic of sustainability indicators, on which much work is being done around the world. Virginia Maclaren's classic article from the *Journal of the American Planning Association* defines some key characteristics of these measures and discusses several leading examples. Next, we learn about the concept of ecological footprint analysis from two of its originators, William Rees and Mathis Wackernagel. This method has been widely used to dramatize the resource impacts of cities and metropolitan regions. In our third selection, from Allan Jacobs' book *Looking at Cities*, we highlight careful, first-hand observation as a much-neglected tool through which planners and citizens can analyze urban environments and develop knowledge of their historical development and their future possibilities. Next we consider the role of various information technologies in assisting with sustainability planning, and through a short selection by Jessica Hsu focus on uses of social media. Moving towards a bigger-picture view of social change, a classic piece from Michael Lerner suggests that the ultimate mechanism for a more constructive politics may be a "politics of meaning" based on new values, understandings of interdependency, and a renewed spiritual commitment. Finally, a piece by noted video author Annie Leonard considers "The Story of Change," and in particular the role of organizing within social movements.

The large array of existing urban planning tools – General Plans, Specific Plans, Regional Plans, development-permitting processes, design charettes, Environmental Impact Reports, Geographic Information Systems (GIS), and so forth – will of course also be essential to sustainable urban development. Readers can gain more background on these through many standard urban planning texts, such as the International City/County Management Association's *Local Planning: Contemporary Principles and Practice* (Washington, D.C.: ICMA, 2009), and materials published by the American Planning Association (www.planning.org), the Canadian Institute of Planners (www.cip-icu.ca), the Royal Town Planning Institute (www.rtpi.org.uk), and other professional associations. However, bringing about more sustainable communities will require rethinking some of these traditional planning methods – at least to make them more effective at meeting goals such as the three E's of sustainability – and the aim of this section is to suggest some of these new techniques for the field.

48

"Urban Sustainability Reporting"

from the *Journal of the American Planning Association* (1996)

Virginia W. Maclaren

Editors' Introduction

One of the main questions facing those interested in bringing about more sustainable communities is: How do we recognize progress toward sustainability? Some method of measuring the direction of current trends and success or failure of particular initiatives is crucial. This is not a new topic within urban planning – performance measurement and evaluation techniques have been around for years – but the subject takes on new urgency when the aim is to substantially change current ways of developing cities and to justify substantial new initiatives. Indicators can also be extremely useful in educating the public about the direction of current trends, and in developing political support for change.

For such reasons sustainability indicators have become one of the central tools for sustainable urban development. Examples range from Sustainable Seattle's community-based indicator set, developed by a group of citizens who convened public meetings with local leaders in the mid-1990s, to the United Nations' Human Development Index, developed by a large international agency relying on national-level data from 125 countries. (For further information on both of these, see www.sustainableseattle.org and www.undp.org.)

The following review of sustainability indicator approaches by Canadian planner Virginia Maclaren is excerpted from the *Journal of the American Planning Association*. The author teaches at the University of Toronto and has also written on the topics of waste management and regional economics. Here she emphasizes the distinctive characteristics and types of sustainability indicators. For further examples of local, regional, or national indicator programs, interested readers might consult the Canadian International Development Agency's report *Indicators for Sustainability: How Cities are Monitoring and Evaluating Their Success* (Ottawa, 2012; available through www.sustainablecities.net). Additional materials are provided by Redefining Progress (www.rprogress.org), the Global Cities Institute at the University of Toronto (www.globalcitiesinstitute. org), and the Australian Conservation Foundation (www.acfonline.org.au), which created a Sustainable Cities Index for Australia's 20 largest cities in 2010.

The concept of sustainability is starting to have a significant influence on planning and policy at the local level. Previous research has identified numerous examples of urban sustainability initiatives in North America. A certain number of communities are starting to adopt sustainability as a goal in comprehensive plans and other planning activities (Maclaren 1993, Oullet 1993, Beatley 1995). Now, the important next step for sustainability initiatives at the local level is to determine whether or not these actions are leading a community to become more sustainable. A significant barrier to accomplishing this task is the absence of a clearly articulated method of reporting on urban sustainability.

Urban sustainability reports include a range of information about environmental, economic, and social conditions and policies in the local community and use that information to make judgments about whether the community is making progress towards sustainability. Evidence of positive progress is important for justifying past expenditures on sustainability initiatives and building support for new initiatives. Evidence of a lack of sustainability can provide ammunition for community groups in local government, other levels of government, or the private sector. Individuals in the community also can use sustainability reports to educate themselves about sustainability trends and evaluate how their own actions may improve sustainability.

The purpose of this paper is to present a structured process for urban sustainability reporting that improves upon the *ad hoc* reporting processes currently in use, and to explore some of the characteristics of urban sustainability indicators. In researching this paper, I examined some of the first efforts at urban sustainability reporting in North America and Europe and drew on local experiences with related types of reporting, namely state of the environment reporting, healthy city reporting and quality of life reporting. State of the environment (SOE) reports describe and analyze environmental conditions and trends of significance. Social or economic conditions are discussed only insofar as they relate to the biophysical environment (Campbell and Maclaren 1995). Thus SOE reporting is not broad enough to be called sustainability reporting. In contrast, healthy city reporting has just as broad a focus as sustainability reporting, but with a much stronger emphasis on human health. (See, for example, Healthy City Toronto 1993.) Quality of life reporting has evolved to the point where it, too, has become very similar to sustainability reporting in that it examines economic, environmental, and social conditions and the linkages among them (e.g., Murdie *et al.* 1992); but quality of life reporting does not have the same concern for issues of intergenerational equity.

The examples of urban sustainability reports that are referred to in this paper come from three different levels of government: (1) the city of Seattle, Washington; (2) the Regional Municipality of Hamilton-Wentworth, Ontario; and (3) the province of British Columbia. Each of these cases is described briefly below.

Sustainable Seattle is the name of a multi-stakeholder group that was established in 1990 as a volunteer network and civic forum for the promotion of community sustainability. It is administered by the local YMCA and governed by an independent board of trustees. In 1993, the group released an urban sustainability report for Seattle containing 20 sustainability indicators and an evaluation of Seattle's progress towards sustainability (Sustainable Seattle 1993). An additional 20 indicators were released two years later. The target audience for the report was primarily individual members of the community and the media, with businesses and local government being a secondary target.

The Sustainable Community Indicators project in the Regional Municipality of Hamilton-Wentworth, Ontario, is a continuation of the region's Sustainable Community Initiative, which began in 1990. At that time, the Regional Council appointed a citizen's Task Force on Sustainable Development with a mandate to examine the concept of sustainable development as a basis for reviewing all regional policies. In 1992, after consultation with over 400 individuals and 50 community groups, the Task Force released a document entitled "Vision 2020," describing the type of community that Hamilton-Wentworth could be in the year 2020 if it followed the principles of sustainable development (Regional Municipality of Hamilton-Wentworth 1992). As a follow-up to this document, the Council launched the Sustainable Community Indicators project in 1994, with the goal of developing sustainability indicators for measuring the region's progress towards Vision 2020. The output of the project will be an annual report card that identifies the status of the indicators as well as the way in which they can be influenced by individuals, organizations, business, local government, and the community as a whole.

The British Columbia Round Table's State of Sustainability Report examines urban sustainability at the provincial level. The Round Table is a multi-stakeholder group, funded by the provincial government, and was responsible for developing the province's first sustainability strategy. For its urban sustainability report, the Round Table chose a sample of five cities, accounting for over 60 per cent of the province's population, to represent the broad regions of the province as well as a variety of economic, environmental, and social conditions. The report, containing over 90 urban sustainability indicators, was released in 1994 (British Columbia Round Table 1994). Like the Hamilton-Wentworth initiative, the British Columbia report is meant to be a guide for both modifying personal behavior and informing planning and policy decisions. . . .

DEFINING URBAN SUSTAINABILITY

What is the meaning of the term "urban sustainability"? It may help to first compare it to "sustainable urban development." The meanings of these two terms are very close and are often used interchangeably in the literature (cf. Richardson 1994). One way of distinguishing them, however, is to think of sustainability as describing a desirable *state* or set of conditions that persists over time. In contrast, the word "development" in the term "sustainable urban development" implies a *process* by which sustainability can be attained.

Some of the key characteristics of urban sustainability that are often mentioned in the literature and in policy documents are: intergenerational equity, intragenerational equity (including social equity, geographical equity,[1] and equity in governance), protection of the natural environment (and living within its carrying capacity), minimal use of nonrenewable resources, economic vitality and diversity, community self-reliance, individual well-being, and satisfaction of basic human needs.[2]

There is considerable debate within the academic community, planning agencies, and other organizations over the relative importance of each of these urban sustainability characteristics, and there is even disagreement on whether all of them should be included when developing sustainability goals. Almost everyone who has tried to define urban sustainability agrees, however, that the concept points to the necessity of introducing environmental considerations to the policy debate over the future of our cities. Some maintain that environmental considerations should now be paramount in this debate, while others call for a more holistic approach that balances environmental, economic, and social concerns.

For the purposes of urban sustainability reporting, I contend that there is no single "best" definition of urban sustainability, since different communities are likely to develop slightly, or even significantly, different conceptualizations of urban sustainability, depending on their current economic, environmental, and social circumstances and on community value judgments. As a consequence, a set of indicators designed to measure progress towards achievement of one community's sustainability goals may not necessarily be appropriate for measuring progress in another community. Nevertheless, there are certain fundamental properties of sustainability indicators that all communities will wish to consider. These are described in the next section.

WHAT IS AN URBAN SUSTAINABILITY INDICATOR?

One definition of urban sustainability indicators is that they are "bellwether tests of sustainability and reflect something basic and fundamental to the long term economic, social or environmental health of a community over generations" (Sustainable Seattle 1993: 4). This definition provides a good starting point, but it requires considerable elaboration. Looking first at the "indicator" component of "urban sustainability indicators," it is important to remember that most indicators are simplifications of complex phenomena. The term "indicator" should therefore be taken literally in the sense that it provides only an indication of conditions or problems (Whorton and Morgan 1975, Clarke and Wilson 1994). Since a single indicator will seldom be able to give the full picture, it is often useful to employ a wide range of indicators to characterize the different dimensions or aspects of a situation. Unfortunately, this requirement can conflict with the need to identify a fairly limited set of indicators for purposes of decision-making, and to minimize double-counting.

Urban sustainability indicators can be distinguished from simple environmental, economic, and social indicators by the fact that they are:

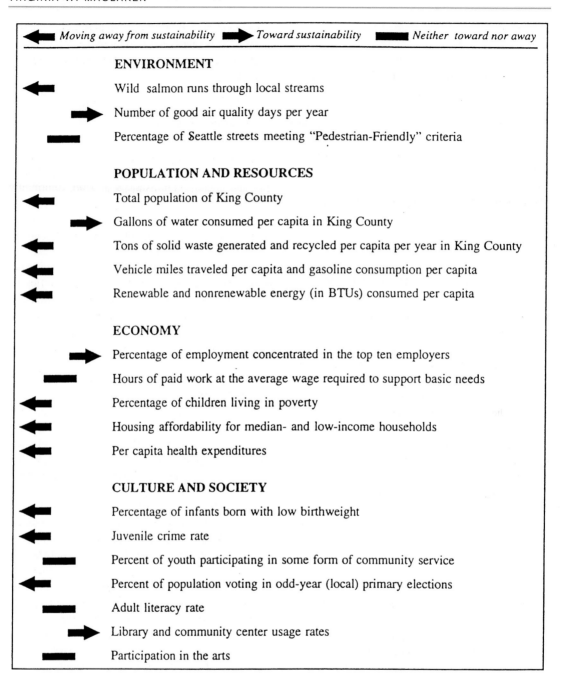

Figure 1 Sustainable Seattle indicators.

Source: Sustainable Seattle (1993).

- integrating
- forward-looking
- distributional
- developed with input from multiple stakeholders in the community.

All sustainability indicators should possess the last characteristic. It may not be possible to develop individual sustainability indicators that possess all of the first three characteristics, but they should possess at least one, and within a given set of sustainability indicators, all of these characteristics should be represented.

Integrating indicators

Sustainability indicators are integrating in the sense that they attempt to portray linkages among the economic, environmental, and social dimensions of sustainability. One example of an integrating indicator might be the amount of "brownfield" land found in an urban area. This could be considered both as an indicator of industrial activity loss and as an indicator of environmental constraints on redevelopment (if the lands are contaminated). Still another integrating indicator would be the unemployment rate, since it is a measure of both economic stress and social stress.

One of the integrating indicators used by Sustainable Seattle is the number of salmon returning to spawn in a representative sample of local salmon runs. This indicator is relevant for both an environmental condition (water quality) and an economic vitality condition (survival of one of the Seattle area's most important industries).

Composite indicators, which combine two or more individual indicators, can also be useful as integrative indicators. For example, the cost of recycling per ton of waste recycled is a simple composite indicator that integrates economic and environmental considerations. Unfortunately, the construction of more complex composite indicators faces a number of methodological problems, including such issues as deciding how to weight the individual indicators, how to standardize different measurement units, and whether to choose a multiplicative or additive aggregation technique (Ott 1978, Innes 1990). Despite these problems, some composite indicators, such as the Human

Development Index,[3] have gained considerable popularity because they reduce the information contained in several individual indicators down to a single number.

Forward-looking indicators

A second important characteristic of sustainability indicators is that they must be forward-looking if they are to be used in measuring progress towards achieving intergenerational equity. There are several different ways in which an indicator might be considered forward-looking. The simplest type of forward-looking indicator is a "trend indicator." A trend indicator describes historical trends and provides indirect information about future sustainability. For example, it is often obvious from examining historical trends that a development path followed in the past cannot possibly be sustainable into the future. However, because trend indicators provide only indirect information about the future, they are more useful for reactive than for proactive policy-making (Ruitenbeek 1991).

The forward-looking capabilities of trend indicators can be enhanced if they are linked to reference points that define intermediate or final steps in the move towards meeting sustainability goals. The two main types of reference points are targets and thresholds. Whereas targets are levels that must be met in the future if sustainability is to be achieved, thresholds are levels that should not be exceeded. Thresholds are scientifically determined and may possess regulatory status. Examples include air and water quality standards. Targets can be set in a fairly arbitrary manner either by using easily recognized numbers (e.g., reduce solid waste by 50 per cent by the year 2000), by comparison to higher order jurisdictions (e.g., national or state means), or by norms (e.g., the poverty level). A threshold, such as an air quality standard, also can be part of a target (e.g., zero exceedances of the standard by the year 2020).

The Oregon "benchmarks" are a well-known application of the use of targets for reviewing government accountability. In 1991, the Oregon Progress Board released its first benchmarks report, in which it identified 272 indicators of environmental, social, and economic well-being in that state (Oregon Progress Board 1991). The Board also specified a series of targets for each indicator, to be met at regular intervals up to the year 2010. They referred to these targets as benchmarks. The indicators in the report are primarily output indicators (e.g., number of households with drinking water that does not meet government standards) rather than input indicators (e.g., expenditures on water treatment facilities), and are being used to help set a broad range of program and budget priorities.

Both targets and thresholds are present in the Netherlands' national environmental policy indices. Each index has one or more policy targets set for specified future dates (e.g., the years 2000, 2010), and in some cases the index includes a longer-term "sustainability level" that is scientifically determined. For example, the Eutrophication Index, which measures releases of phosphates and nitrogen compounds to the environment, will reach a sustainable level when the excessive supply of phosphates and nutrients has been reduced enough that a balance has been achieved between supply and the removal from the environment of these two major contributors to eutrophication (Adriaanse 1993).

Another type of forward-looking indicator is the "predictive indicator." Predictive sustainability indicators rely on mathematical models for the future state and development of variables describing the environment, the economy, and society, or the linkages among them. Population levels and population growth are commonly used predictive indicators found in planning reports. Bratt (1991) notes that since all predictions are inherently disputable, the best that predictive indicators can do is to provide plausible information about future conditions. Only trend indicators provide scientifically reliable information, assuming that the data collection methods were reliable.

The uncertainty inherent in predictive indicators points to the need for a third type of forward-looking sustainability indicator known as the "conditional indicator." Conditional indicators depend on a form of scenario development; they answer the question: "If a given indicator achieves or is set at a certain level, what will the level of an associated indicator be in the future?" This type of indicator attempts to overcome the difficulty that predictive indicators have in forecasting, by developing a range of forecasts or predictions. Table 1, taken from the British Columbia Round Table's State of Sustainability Report (1994), provides an example of a conditional indicator of urban form. The "if" indicator is future residential density. The "then" indicator is the total amount of land that will be needed to accommodate the expected urban population of British Columbia in 2021 at each of these density levels. Two different measures of the land-area indicator are presented: the amount of land in hectares and the equivalent amount of land currently occupied by the City of Vancouver. The former measure may be most useful for planners, and the later measure is probably more meaningful to the general public.

Housing (units per ha)	Area needed for housing (ha)	Area needed for other urban functions	Total area needed (ha)	Density[a] City of Vancouver equivalents
1.4	479,000	240,000	719,000	64
2.3	290,000	145,000	435,000	38
6.5	103,000	52,000	155,000	14
9.5	70,000	35,000	105,000	9
18	37,000	19,000	56,000	5

Table 1 Land area needed for cities to serve additional British Columbia residents in the year 2021 at various residential densities.

Source: BC Round Table on Environment and Economy (1994).

Note: [a] From lowest to highest, these are the current densities, respectively, for the City of Kelowna, the City of Cranbrook, Greater Victoria, Greater Vancouver Regional District, and the City of Vancouver.

Distributional indicators

Sustainability indicators must be able to measure not only intergenerational equity but also intragenerational equity. They should be able to take into account the distribution of conditions (social, economic, environmental) within a population or across geographic regions. Typically, spatially aggregated indicators fail to account for distributive effects. An example is GNP, which may increase even though economic conditions for many groups or different regions in the country are declining (Liverman *et al.* 1988). Disaggregating certain indicators for a community by such factors as age, gender, and location can help to overcome this problem.

Sustainability indicators should also be able to distinguish between local and nonlocal sources of environmental degradation, and between local and nonlocal environmental effects. A downstream community may generate very little pollution and display all the characteristics of a sustainable community – except for the fact that it suffers from significant upstream water pollution or upwind air pollution. The development of indicators that can identify pollution sources outside the local community's control will facilitate the formulation of appropriate policy responses to geographical inequities. Similarly, sustainability indicators should also measure the extent to which a local community contributes to environmental degradation in other communities, regions, or the world at large.

Multi-stakeholder input

A final characteristic that distinguishes sustainability indicators from other types of indicators is the manner in which they are developed. The history of the social indicator movement suggests that the most influential, valid, and reliable indicators have been those that were developed with input from a broad range of participants in the policy process (Innes 1990). This lesson is especially applicable to the development of sustainability indicators, since sustainability is such a value-laden and context-sensitive concept. It therefore makes sense to seek input on sustainability concerns and priorities from a broad range of stakeholders. This can be accomplished by assigning significant responsibility for selecting sustainability indicators to a broadly-based, multi-stakeholder group or by consulting in some other way with multiple stakeholders from the earliest stages of indicator development.

NOTES

1 Geographical equity is a term coined by Haughton and Hunter (1994) to emphasize the undesirability of achieving economic growth of a higher quality of life in one community at the expense of environmental degradation in another. They assert that this form of development is inequitable unless some form of reparation or compensation takes place between the communities.

2 See, e.g., Alberta Round Table 1993, Jacobs 1991, Hardoy *et al.* 1992, Richardson 1992, British Columbia Round Table 1994, Haughton and Hunter 1994, Beatley 1995.

3 The Human Development Index was developed by the United Nations Development Program (UNDP) for comparing human welfare levels in different countries. The index is an aggregation of four indicators: life expectancy at birth, adult literacy, average years of schooling, and GDP per capita. The UNDP publishes the Index for all members of the United Nations in its annual "Human Development Report."

REFERENCES

Adriaanse, A. 1993. *Environmental Policy Performance Indicators: A Study of Indicators for Environmental Policy in the Netherlands.* The Hague: Sdu Uitgeverji Koninginnegracht.

Alberta Round Table on Environment and Economy. 1993. *Steps to Realizing Sustainable Development.* Edmonton: ARTEE.

Beatley, T. 1995. Planning and Sustainability: The Elements of a New (Improved?) Paradigm. *Journal of Planning Literature* 9, 4: 383–395.

Bratt, L. 1991. The Predictive Meaning of Sustainability Indicators. In *In Search of Indicators of Sustainable Development*, edited by O. Kuik and H. Vergruggen. Boston: Kluwer Academic Publishers, pp. 57–70.

British Columbia Round Table on the Environment and the Economy. 1994. *State of Sustainability: Urban Sustainability and Containment.* Victoria:

British Columbia Round Table on the Environment and the Economy.

Campbell, M. and V.W. Maclaren. 1995. *Municipal State of the Environment Reporting in Canada: Current Status and Future Needs*. Ottawa: Occasional Paper Series No. 6, State of the Environment Reporting, Environment Canada.

Clarke, G.P. and A.G. Wilson. 1994. Performance Indicators in Urban Planning: The Historical Context. In *Modelling the City: Performance, Policy and Planning*, edited by C.S. Bertuglia, G.P. Clarke and A.G. Wilson. London: Routledge, pp. 4–19.

Hardoy, J.E., D. Mitlin, and D. Satterthwaite. 1992. *Environmental Problems in Third World Cities*. London: Earthscan.

Haughton, G. and C. Hunter. 1994. *Sustainable Development and Geographical Equity*. Paper presented at the Annual Conference of the Association of American Geographers, Chicago.

Healthy City Toronto. 1993. *A Strategy for Developing Healthy City Indicators*. Toronto: Healthy City Toronto.

Innes, J.E. 1990. *Knowledge and Public Policy: The Search for Meaningful Indicators*, 2nd edn. New Brunswick, NJ: Transaction Publishers.

Jacobs, M. 1991. *The Green Economy*. London: Pluto Press.

Liverman, D., M.E. Hanson, B.J. Brown, and R.W. Meredith. 1988. Global Sustainability: Towards Measurement. *Environmental Management* 12, 2: 133–143.

Maclaren, V.W. 1993. *Sustainable Urban Development in Canada: From Concept to Practice*. Volumes I–III. Toronto: ICURR Press.

Murdie, R.A., D. Rhyne, and J. Bares. 1992. *Modelling Quality of Life Indicators in Canada: A Feasibility Analysis*. Ottawa: Canada Mortgage and Housing Corporation.

Ontario Round Table on Environment and Economy. 1995. *Sustainable Communities Resource Package*. Toronto: Ontario Round Table on Environment and Economy.

Oregon Progress Board. 1991. *Oregon Benchmarks: Setting Measurable Standards for Progress*. Report to the 1991 Legislature. Salem: Oregon Progress Board.

Ott, W. 1978. *Environmental Indices*. Ann Arbor, MI: Ann Arbor Science Publishing.

Oullet, P. 1993. *Environmental Policy Review of 15 Canadian Municipalities*. Toronto: ICURR Press.

Regional Municipality of Hamilton-Wentworth. Regional Chairman's Task Force on Sustainable Development. 1992. *Vision 2020: The Sustainable Region*. Hamilton: Regional Municipality of Hamilton-Wentworth.

Richardson, N. 1992. Canada. In *Sustainable Cities: Urbanization and the Environment in International Perspective*, edited by R. Stren, J.B.R. Whitney, and R. White. Boulder, CO: Westview Press, pp. 145–168.

Richardson, N. 1994. Making Our Communities Sustainable: The Central Issue is Will. In *Sustainable Communities Resource Package*, Ontario Round Table on Environment and Economy. Toronto: Ontario Round Table on Environment and Economy, pp. 15–44.

Roseland, M. 1992. *Toward Sustainable Communities*. Ottawa: National Round Table on the Environment and the Economy.

Ruitenbeek, J.H. 1991. *Indicators of Ecologically Sustainable Development: Towards New Fundamentals*. Ottawa: Canadian Environmental Advisory Council.

Sustainable Seattle. 1993. *Sustainable Seattle Indicators of Sustainable Community: A Report to Citizens on Long Term Trends in Their Community*. Seattle, WA: Sustainable Seattle.

Tomalty, R. and D. Pell. 1994. *Sustainable Development and Canadian Cities: Current Initiatives*. Ottawa: Canada Mortgage and Housing Corporation and the Royal Society of Canada.

Whorton, J.W., Jr. and D.R. Morgan. 1975. *Measuring Community Performance: A Handbook of Indicators*. Norman, OK: University of Oklahoma, Bureau of Government Research.

49

"What *Is* an Ecological Footprint?"

from *Our Ecological Footprint* (1996)

Mathis Wackernagel and William Rees

Editors' Introduction

One of the most intriguing tools with which to measure the sustainability of particular places or lifestyles is ecological footprint analysis, developed by University of British Columbia professor William Rees and his colleagues and graduate students. By converting resource needs and pollution into the equivalent land area that would be required to produce or offset these, the footprint model provides a dramatic indication of the impacts of modern life. Footprints can be calculated for individuals, entire cities, regions, and nations. Readers can create a simple, personal footprint online at the website of Redefining Progress, a San Francisco-based NGO that has worked extensively to develop new measurements of social and ecological health (www.rprogress.org).

Ecological footprint analysis may be most useful as an educational tool, to give a general indication of the sustainability of particular places or modes of living. Although researchers are attempting to calculate extremely detailed footprints for some cities and nations, these efforts run into the same problems that most ecological economics encounters, such as how to place a value on social and environmental costs in terms of either dollars or land area. Assumptions must be made about how to translate resource use and pollution into these other variables, and these are open to question. Still, the ecological footprint method has an intuitive appeal, similar to more general notions of "carrying capacity," that has made it attractive to many people.

This selection is drawn from Wackernagel and Rees' book *Our Ecological Footprint: Reducing Human Impact on the Earth* (Gabriola Island, BC: New Society Publishers, 1996). Swiss engineer and planner Mathis Wackernagel is director of the Global Footprint Network; Canadian urban planning professor William Rees has worked in the areas of ecological economics and biodiversity. For additional information on ecological or social accounting, see *Sharing Nature's Interest: Ecological Footprints as an Indicator for Sustainability* (London: Earthscan, 2000) by Nicky Chambers, Craig Simmons, and Mathis Wackernagel, the *Living Planet Report* (Gland, Switzerland: World Wildlife Federation, 2012; available at www.panda.org/livingplanet), and the Ecological Footprint Quiz from the Center for Sustainable Economy at www.myfootprint.org/.

Ecological footprint analysis is an accounting tool that enables us to estimate the resource consumption and waste assimilation requirements of a defined human population or economy in terms of a corresponding productive land area. Typical questions we can ask with this tool include: how dependent is our study population on resource imports from "elsewhere" and on the waste assimilation capacity of the global commons?, and will nature's productivity be adequate to satisfy the rising material expectations of a growing human population into the next century? William Rees has been teaching the basic concept to planning students for 20 years and it has been developed further since 1990 by Mathis Wackernagel and other

students working with Bill on UBC's Healthy and Sustainable Communities Task Force.

To introduce the thinking behind ecological footprint analysis, let's explore how our society perceives that pinnacle of human achievement, "the city." Ask for a definition, and most people will talk about a concentrated population or an area dominated by buildings, streets and other human-made artifacts (this is the architect's "built environment"); some will refer to the city as a political entity with a defined boundary containing the area over which the municipal government has jurisdiction; still others may see the city mainly as a concentration of cultural, social and educational facilities that would simply not be possible in a smaller

Figure 1 The ecological footprint is a measure of the "load" imposed by a given population on nature. It represents the land area necessary to sustain current levels of resource consumption and waste discharge by that population.

Source: Illustration by Phil Testemale.

settlement; and, finally, the economically-minded see the city as a node of intense exchange among individuals and firms and as the engine of production and economic growth.

No question, cities are among the most spectacular achievements of human civilization. In every country cities serve as the social, cultural, communications and commercial centers of national life. But something fundamental is missing from the popular perception of the city, something that has so long been taken for granted it has simply slipped from consciousness.

We can get at this missing element by performing a mental experiment based on two simple questions designed to force our thinking beyond conventional limits. First, imagine what would happen to any modern city or urban region – Vancouver, Philadelphia or London – as defined by its political boundaries, the area of built-up land, or the concentration of socioeconomic activities, if it were enclosed in a glass or plastic hemisphere that let in light but prevented material things of any kind from entering or leaving – like the "Biosphere II" project in Arizona (Figure 2). The health and integrity of the entire human system so contained would depend entirely on whatever was initially trapped within the hemisphere.

It is obvious to most people that such a city would cease to function and its inhabitants would perish within a few days. The population and the economy contained by the capsule would have been cut off from vital resources and essential waste sinks, leaving it both to starve and to suffocate at the same time! In other words, the eco-systems contained within our imaginary human terrarium would have insufficient "carrying capacity" to support the ecological load imposed by the contained human population. This mental model of a glass hemisphere reminds us rather abruptly of humankind's continuing ecological vulnerability.

The second question pushes us to contemplate this hidden reality in more concrete terms. Let's assume that our experimental city is surrounded by a diverse landscape in which cropland and pasture, forests and watersheds – all the different ecologically productive land-types – are represented in proportion to their actual abundance on the Earth, and that adequate fossil energy is available to support current levels of consumption using prevailing technology. Let's also assume our imaginary glass enclosure is elastically expandable. The question now becomes: how large would the hemisphere have to become before the city at its center could sustain itself indefinitely and exclusively on the land and

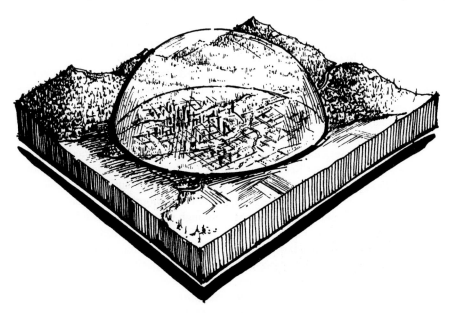

Figure 2 Living in a terrarium. How big would the glass hemisphere need to be so that every city under it could sustain itself exclusively on the ecosystems contained?

Source: Illustration by Phil Testemale.

Figure 3 What is an ecological footprint? Think of an economy as having an "industrial metabolism." In this respect it is similar to a cow in its pasture. The economy needs to "eat" resources, and eventually all this intake becomes waste and has to leave the organism – the economy – again. So the question becomes: How big a pasture is necessary to support that economy – to produce all its feed and absorb all its waste? Alternatively, how much land would be necessary to support a defined economy sustainably at its current material standard of living?

Source: Illustration by Phil Testemale.

water ecosystems and the energy resources contained within the capsule? In other words, what is the total area of terrestrial ecosystem types needed continuously to support all the social and economic activities carried out by the people of our city as they go about their daily activities? Keep in mind that land with its ecosystems is needed to produce resources, to assimilate wastes, and to perform various invisible life-support functions. Keep in mind too, that for simplicity's sake, the question as posed does not include the ecologically productive land area needed to support other species independent of any service they may provide to humans.

For any set of specified circumstances – the present example assumes current population, prevailing material standards, existing technologies, etc. – it should be possible to produce a reasonable estimate of the land/water area required by the city concerned to sustain itself. By definition, the total ecosystem area that is essential to the continued existence of the city is its *de facto* Ecological

Footprint on the Earth. It should be obvious that the Ecological Footprint of a city will be proportional to both population and per capita material consumption. Our estimates show for modern industrial cities the area involved is orders of magnitude larger than the area physically occupied by the city. Clearly, too, the Ecological Footprint includes all land required by the defined population wherever on Earth that land is located. Modern cities and whole countries survive on ecological goods and services appropriated from natural flows or acquired through commercial trade from all over the world. The Ecological Footprint therefore also represents the corresponding population's total "appropriated carrying capacity."

By revealing how much land is required to support any specified lifestyle indefinitely, the Ecological Footprint concept demonstrates the continuing material dependence of human beings on nature. For example, Table 1 shows the Ecological Footprint of an average Canadian, i.e., the amount

Cell entries = ecologically productive land in ha/capita	A Energy	B Degradation	C Garden	D Crop	E Pasture	F Forest	Total
1 Food	**0.33**		**0.02**	**0.60**	**0.33**	**0.02**	**1.30**
11 Fruit, vegetables, grain	0.14		0.02	0.18		0.01?	
12 Animal products	0.19			0.42	0.33	0.01?	
2 Housing	**0.41**	**0.08**	**0.002?**			**0.40**	**0.89**
21 Construction/maintenance	0.06					0.35	
22 Operation	0.35					0.05	
3 Transportation	**0.79**	**0.10**					**0.89**
31 Motorized private	0.60						
32 Motorized public	0.07						
33 Transport of goods	0.12						
4 Consumer goods	**0.52**	**0.01**		**0.06**	**0.13**	**0.17**	**0.89**
40 Packaging	0.10					0.04	
41 Clothing	0.11			0.02	0.13		
42 Furniture and appliance	0.06					0.03?	
43 Books/magazines	0.06					0.10	
44 Tobacco and alcohol	0.06			0.04			
45 Personal care	0.03						
46 Recreational equipment	0.10						
47 Other goods	0.00						
5 Services	**0.29**	0.01					**0.30**
51 Government/military	0.06						
52 Education	0.08						
53 Health care	0.08						
54 Social services	0.00						
55 Tourism	0.01						
56 Entertainment	0.01						
57 Bank/insurance	0.00						
58 Other services	0.05						
Total	2.34	0.20	0.20	0.66	0.46	0.59	4.27

Table 1 The consumption–land-use matrix for the average Canadian (1991 data).

Notes:
0.00 = less than 0.005 ha or 50 m²; blank = probably insignificant; ? = lacking data.
A Energy = fossil energy consumed expressed in the land area necessary to sequester the corresponding CO_2.
B Degradation = degraded land or built-up environment.
C Garden = gardens for vegetable and fruit production.
D Crop = crop land.
E Pasture = pastures for dairy, meat and wool production.
F Forest = prime forest area. An average roundwood harvest of 163 m³/ha every 70 years is assumed.

of land required from nature to support a typical individual's present consumption. This adds up to almost 4.3 ha, or a 207 m square. This is roughly comparable to the area of three city blocks. The column on the left shows various consumption categories and the headings across the top show corresponding land-use categories. "Energy" land as used in the table means the area of carbon sink land required to absorb the carbon dioxide released by per capita fossil fuel consumption (coal, oil and natural gas) assuming atmospheric stability as a goal. Alternatively, this entry could be calculated according to the area of cropland necessary to produce a contemporary biological fuel such as ethanol to substitute for fossil fuel. This alternative produces even higher energy land requirements. "Degraded Land" means land that is no longer available for nature's production because it has been paved over or used for buildings. Examples of the resources in "Services" are the fuel needed to heat hospitals, or the paper and electricity used to produce a bank statement.

To use Table 1 to find out how much agricultural land is required to produce food for the average Canadian, for example, you would read across the "Food" row to the "Crop" and "Pasture" columns. The table shows that, on average, 0.95 ha of garden, cropland and pasture is needed for the typical Canadian. Note that none of the entries in the table is a fixed, necessary or recommended land area. They are simply our estimates of the 1990s ecological demands of typical Canadians. The Ecological Footprints of individual and whole economies will vary depending on income, prices, personal and prevailing social values as they affect consumer behavior, and technologic sophistication – e.g., the energy and material content of goods and services.

SO WHAT? THE GLOBAL CONTEXT

Our economy caters to growing demands that compete for dwindling supplies of life's basics. The

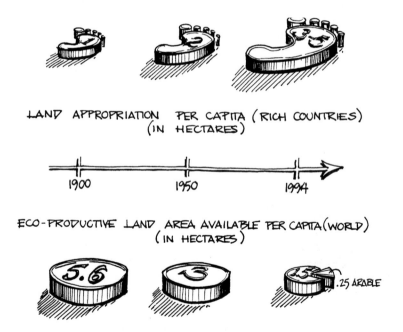

Figure 4 Our ecological footprints keep growing while our per capita "earth shares" continue to shrink. Since the beginning of the twentieth century the ecologically productive land available has decreased from over 5 ha to less than 1.5 ha in 1994. At the same time the average North American's footprint has grown to over 4 ha. These opposing trends are in fundamental conflict: the ecological demands of average citizens in rich countries exceed per capita supply by a factor of three. This means that the Earth could not support even today's population of 5.8 billion sustainably at North American material standards.

Source: Illustration by Phil Testemale.

Ecological Footprint of any population can be used to measure its current consumption and projected requirements against available ecological supply and point out likely shortfalls. In this way, it can assist society in assessing the choices we need to make about our demands on nature. To put this into perspective, the ecologically productive land "available" to each person on Earth has decreased steadily over the last century (Figure 4). Today, there are only 1.5 ha of such land for each person, including wilderness areas that probably shouldn't be used for any other purpose. In contrast, the land area "appropriated" by residents of richer countries has steadily increased. The present Ecological Footprint of a typical North American (4–5 ha) represents three times his/her fair share of the Earth's bounty. Indeed, if everyone on Earth lived like the average Canadian or American, we would need at least three such planets to live sustainably. Of course, if the world population continues to grow as anticipated, there will be 10 billion people by 2040, for each of whom there will be less than 0.9 ha of ecologically productive land, assuming there is no further soil degradation.

Figure 5 Wanted: two (phantom) planets. If everyone lived like today's North Americans it would take at least two additional planet Earths to produce the resources, absorb the wastes, and otherwise maintain life support. Unfortunately, good planets are hard to find ...

Source: Illustration by Phil Testemale.

Such numbers become particularly telling when used to compare selected geographic regions with the land they actually "consume." For example, we estimate the Ecological Footprint for the Lower Fraser Valley, east of Vancouver to Hope, BC. This valley bottom has 1.8 million inhabitants for a population density of 4.5 people per hectare. In short, the area is far smaller than needed to supply the ecological resources used by its population. If the average person in this basin needs the output of 4.3 ha (Table 2), then the Lower Fraser Valley depends on an area 19 times larger than that contained within its boundaries for food, forestry products, carbon dioxide assimilation and energy. Similarly, Holland has a population of 15 million people, or 4.4 people per hectare, and although Dutch people consume less than North Americans on average, they still require about 15 times the available land within their own country for food, forest products and energy use (Figure 6). In other words, the ecosystems that actually support typical industrial regions lie invisibly far beyond their political or geographic boundaries.

A world upon which everyone imposed an oversized Ecological Footprint would not be sustainable – the Ecological Footprint of humanity as a whole must be smaller than the ecologically productive portion of the planet's surface. This means that if every region or country were to emulate the economic example of the Lower Fraser Basin or the Netherlands, using existing technology, we would all be at risk from global ecological collapse.

The notion that the current lifestyle of industrialized countries cannot be extended safely to everyone on Earth will be disturbing to some. However, simply ignoring this possibility by blindly perpetuating conventional approaches to economic development invites both eco-catastrophe and subsequent geopolitical chaos. To recognize that not everybody can live like people do in industrialized countries today is not to argue that the poor should remain poor. It is to say that there must be adjustments all round and that, if our ecological analyses are correct, continuing on the current development path will actually hit the less fortunate hardest. Blind belief in the expansionists' cornucopian dream does not make it come true – rather it side-tracks us from learning to live within the means of nature and ultimately becomes ecologically and socially destructive.

To keep things simple, we consider only four important categories of domestic consumption: built-up land, food, forest products and fossil energy. This avoids any significant double counting, yet is sufficient to illustrate the strength of Ecological Footprint analysis.

BASIC DATA (Netherlands):

- 1991 population: 15,050,000; land area: 33,920 km^2.
- Built-up land: 538,000 ha.
- Commercial energy consumption in 1991: 3,197 PJ − 36 PJ from non-fossil fuel sources (mainly nuclear energy). Therefore, for this calculation, (3197 − 36)[PJ] / 15 million Dutch =) 210 GJ/cap./yr. is used to represent the fossil fuel consumption.

CALCULATIONS:

- **forest:** assuming a consumption of 1.1 m^3/cap./yr. and a forest productivity of 2.3 m^3/ha/yr., this consumption corresponds to (1.1 [m^3/cap./yr.] / 2.3 [m^3/ha/yr.]) = 0.47 [ha/cap.] of forest land.
- **fossil fuel:** 210 [GJ/cap./yr.] corresponds to (210 [GJ/cap./yr.] / 100[GJ/ha/yr.] =) 2.10 [ha/cap.].

RESULTS:

food:	cropland	0.45 [ha/cap.]
rangeland:		0.26 [ha/cap.]
forest:	1.1 [m^3/cap./yr.] corresponds to	0.47 [ha/cap.]
fossil fuel:	210 [GJ/cap./yr.] corresponds to	2.10 [ha/cap.]
degraded land	(settlements and roads):	
	(538,000 [ha] / 15,000,000 [Dutch people])	0.04 [ha/cap.]
Total individual footprint:		**3.32 [ha/cap.]**

Table 2 Assessing the footprint of the Netherlands.

The Netherlands' aggregate Ecological Footprint is:
(15,000,000 [Dutch people] × 3.32 [ha/cap.] × 0.01 [ha/km^2] =) 498,000 km^2.

This is almost 15 times larger than the Dutch territory of 33,920 km^2.

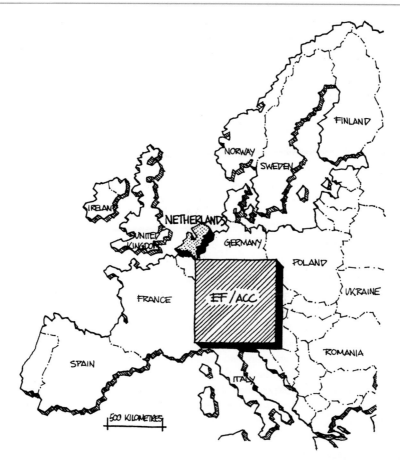

Figure 6 The ecological footprint of the Netherlands. For urbanization, food, forest products and fossil fuel use, the Dutch use the ecological functions of a land area over fifteen times larger than their country.

Source: Illustration by Phil Testemale.

50

"Seeing Change"

from *Looking at Cities* (1985)

Allan Jacobs

Editors' Introduction

One of the most basic tools of planning is an ability to look at the landscape around us – whether urban, suburban, or rural – and see what is going on. This is particularly important for sustainability planning, when it is important to be able to recognize how places have evolved in the past and how they might become more sustainable in the future. Clues always exist about the history of a place, the social and economic dynamics within it, the ways that people are actually using it, and the ways that land use, buildings, and natural environments have changed over time. An experienced urban planner can also quickly determine important pieces of information such as residential density, the dimensions of streets, lots, and buildings, and the presence of culverted streams or remnant bits of wildlife habitat. A sensitivity to the subjective experience of place is also extremely useful – in other words being able to analyze what cities, neighborhoods, or public spaces feel like to different groups of people.

Unfortunately, "looking at cities" is a skill that has often been ignored in planning education and practice. Planners instead have frequently spent their time in front of computers, relying on abstract concepts, numerical data, satellite imagery, and information gathered from secondary sources. These types of information are important, but too often the inability of architects, landscape architects, and urban planners to understand the on-the-ground context of development has helped create sterile places that do not meet the needs of human communities or the natural environment. Luckily, since the early 1990s many planners and citizens have recognized that a new attention to the subjective experience of place is important, as well as detailed, first-hand analysis of urban environments as a basis for planning and design. One proponent of this approach has been Allan Jacobs, the former Planning Director of San Francisco, who taught at the University of California at Berkeley. In this selection from his book *Looking at Cities* (Cambridge, MA: Harvard University Press, 1985), Jacobs describes an approach he has taken with some of his classes of simply walking through cities with the students observing the urban environment firsthand. Jacobs' other work has focused in particular on how to create vibrant, walkable streets. His books include *Great Streets* (Cambridge, MA: MIT Press 1993) and *The Boulevard Book: History, Evolution, Design of Multiway Boulevards* (Cambridge, MA: MIT Press, 2002; with Elizabeth Macdonald and Yodan Rofe).

Other urban designers have also written on this theme. Some are affiliated with the philosophical viewpoint known as phenomenology, which emphasizes direct experience as a basis for knowledge. Examples include David Seamons' volume *Dwelling, Seeing, and Designing: Toward a Phenomenological Ecology* (Albany: State University of New York Press, 1993), Tony Hiss' *The Experience of Place* (New York: Vintage Books, 1990) and *In Motion: The Experience of Travel* (Chicago: American Planning Association, 2010), and *The Role of Place Identity in the Perception, Understanding, and Design of Built Environments* (Dubai UAE: Bentham ebooks, 2011), edited by Hernan Casakin and Fátima Bernardo. MIT urban theorist Kevin Lynch laid much of the groundwork for this point of view in works such as *The Image of the City* (Cambridge, MA: MIT Press, 1960) and *Good City Form* (Cambridge, MA: MIT Press, 1981).

As parents always say to their children, "It's in front of you. Use your eyes." We take messages – or we fail to take them – from urban environments by looking, and we act upon those messages to maintain or change or create places in ways that seem appropriate responses to urban problems and opportunities.

This book calls for getting involved with what we see: learning from what we observe in the urban environment; employing observation more consciously and regularly as an analytic and decision-making tool; and using what we learn to help people live in concert with one another and with the land. If conscious, systematic observation, as opposed to haphazard visual experiencing, does nothing more than help avoid unfortunate decisions and actions that affect people's lives, it will have served well. But it can do much more than that.

I want to summarize briefly some of the more significant findings about observation and to offer additional ideas that can help any interested city dweller go out and do it.

There is nothing quite like walking as a way to observe and get to know a city. Much more than any other mode of transportation, walking allows the observer to control the pace of observation, and there are fewer distractions than there are when driving or riding a bike. It is possible to get to otherwise inaccessible places. Most important, walking allows the observer to be in the environment more fully, and the deliberate pace permits one to integrate what is seen with the knowledge and experiences stored in one's mind. I also think it facilitates recall.

There are problems, to be sure, not the least of which is that the observer feels like an intruder in an unfamiliar environment and therefore is uncomfortable. Because of that feeling, the observer may see things differently, may look too rapidly, may come to conclusions that reflect the discomfort. Women, who can be targets of overt observation, verbal confrontation, and sometimes even physical abuse, are more likely to be uncomfortable as walking observers, a problem that has yet to be overcome. A short, simple explanation of what one is doing can be an adequate response to the "Who are you? What are you doing here?" questions, even when asked with hostility. Once people know what the observer is doing, they are often pleased to talk about or show their neighborhoods.

For some purposes and at certain scales, walking may not be appropriate. A helicopter trip is a good way to find out quickly where major new development is happening or is likely to happen in the future. One flight over the Phoenix area clearly shows that nothing will stop the development of all the presently cultivated land, and maybe more, if someone wants to do so. Donald Appleyard and Kevin Lynch reconnoitered all of the San Diego metropolitan area in one afternoon by helicopter, and the messages they read became an important part of their subsequent work there.

Low-density suburban areas, which were designed for driving, not walking, invite observation by car. The windshield survey has its uses, particularly for getting a general impression of the nature of development and the income status of residents. But I would always suggest getting out of the car at some point and taking a walk, even if only for ten minutes. One begins to experience the area differently.

Buses and other public transportation, bicycle, boat – all can be appropriate for specific purposes. When walking is not possible or appropriate, one should look for clues that are consistent with the speed of looking and the distance from what is being looked at. In a car or helicopter, for example, one should not try to see or interpret detail. Look for the physical qualities of large areas, not the dynamics within them.

Do not try to take photos and observe at the same time. Taking a picture interrupts observing and thinking about and questioning what you see. The photographer is concerned with focus, lens openings, composition, light, and shadow, with how the picture will look. Come back to take pictures later.

Sketching, however, helps looking, makes one more observant. One can think about what one is seeing while sketching, how the elements are arranged and fit together. Sketching facilitates measuring, which, as I have noted, is crucial in making comparisons and understanding the meaning of small and large, good and bad, a lot and a little. Drawing skills are not important, because the sketch will not be shown to anyone.

If possible, walk an area at a time when it is busy. Seeing more people means seeing more clues, and also seeing how people use their city, what is important to them and what is less so. At the same time, the observer should be aware that this is an active time and should try to imagine what it is like at other times. An area may have a very different character in summer and winter, sun and rain, day and night. Understanding these differences is the next best thing to repeated and prolonged observation.

There is no best path, no best place to start or stop a walk. If there is a best way, it is to follow a number of different, overlapping paths, including those that go in back of buildings, along alleys and service lanes. Backs sometimes tell more about maintenance, condition, and space than fronts. People are always surprised at the spaciousness of rear yards in many densely populated eastern cities and in San Francisco.

In this business of looking and interpreting, two people seem to be better than one. Two people can question each other, develop and challenge more hypotheses, bring more knowledge to a situation. Two people may also be an answer to the safety problem women face when alone in some areas. If there is a drawback to having two or more observers, it is that it takes more time to express verbally what one is seeing or thinking, and in that time the observer is less aware of the environment.

When I am observing, I talk with anyone who speaks to me in a friendly way or who, after eye contact and a nod or smile, seems willing to talk. Firemen at leisure, storekeepers, real estate brokers, and librarians know a lot about their areas; so may a person walking along the street. The observer uses everything he can get his hands on to understand and plan for a community; the residents are a very good source of information.

Remember that observation is not a test. No one is forcing the observer to come up with conclusions, except perhaps himself. Don't try to cover too much ground at one time, because one sees less when tired.

A single clue cannot answer questions about an area's historic development and evolution, present state, and problems that exist or may unfold. Taken together, clues are more meaningful, but even then their meanings are more "iffy" than precise. That iffiness is not necessarily a problem; it is a reality that is also true of other research and diagnostic techniques. The lack of certainty may lead to a number of alternative hypotheses about an area, which can be tested if they are important enough. Rather than being a problem, the unsureness of observation makes an area more real, alive, breathing.

Observers see things differently, even to the extent of seeing different clues. But it should come as no surprise that any number of different clues may lead one to similar conclusions about an area. Seeing a rash of bicycles, basketball hoops, caution signs for drivers, all within a tract of ten-year-old three-bedroom homes may lead an observer to conclude that there are many school-aged children, that the families have a particular lifestyle, and even that certain problems will accompany this population group. But an observer who sees none of those clues but sees a neighborhood school with many students might well come to the same conclusions. . . . One need have no fear of not seeing "the right things" – there may be no right things. There is plenty to see, plenty from which to take messages and form hypotheses.

The knowledge one brings to observation can help narrow down the many possible interpretations of what is seen. What knowledge is most helpful? The *social and economic history* of a culture and urban area is crucial; that knowledge is the context for what is observed. When did important social and economic movements take place? What was life like for people here in different periods? What was the timing of reform movements? How have welfare concepts and programs, government roles, technology, and political movements and philosophies changed over time? This knowledge is just as important locally as on a regional or national scale.

Urban planners and others involved in city conservation and development should know *how cities have grown and developed physically*. They should be able to relate that knowledge to the social and economic history of the culture. One ought to know, for example, how streetcars and railroads and highways have structured city development.

Some knowledge of *architectural styles* and their history is important. Experience suggests that to be useful, style periods need not be precise and that they can be longer the further removed they are from the present day: pre-Civil War, late 1800s, turn of the century, the 1920s, the Depression, pre-World War II, 1950s, 1960s, post-1960s. Most people know more than they think they do about when different styles of buildings were constructed. Without studying the subject, however, they are not likely to know enough to consistently understand what these styles tell us about urban areas.

In the same way, knowledge of *artifact history* is profoundly useful. By this I mean the time periods when different types of curbs, street lights, paving materials, signs, curtains, blinds, and building materials were used. This knowledge is more difficult to acquire; there are so many details, and the evolution of any one kind is rarely documented or easily discovered. Reading old photographic journals and technical manuals and looking for dates on the artifacts themselves helps. Perhaps more than for any other category of clues, this information is best learned from experienced professionals. Once a person becomes aware of and starts thinking about the history of a detail, say of the different kinds of curbs that have been used in a city, it becomes an enjoyable pastime to find out more about it.

It is critical to know something of *construction and maintenance*. The condition and maintenance of buildings are important clues to problems and changes that are taking place. Nonspecialist observers often do not understand building construction and what it takes to keep a building in good condition. This can be learned, if not from taking academic courses then from reading books and manuals on construction and renovation, from being a truly attentive sidewalk superintendent at construction sites, or from actually building and maintaining a house.

Almost all of this knowledge can be learned. Effective observation and diagnosis require no special gift, but they are facilitated by all the knowledge one has accumulated and by constant conscious questioning of what is observed.

The next question is, what can you do with what you find out by looking? There are situations where observation may be the only tool available to suggest what to do. A group or an official or a potential client may need to know quickly how to go about planning for a specific area. There may be only enough time for one site visit before making some initial decisions. I was once told by the officials of a large city that they were strongly considering a major development project, which they would be announcing soon. What did I think about their intentions? In the two to three hours I spent in the area I came up with what seemed *obvious* questions about dislocation of people and businesses, traffic circulation, the market for what they were considering, and more. Some of those

questions had not been obvious to the officials; at least they had not thought about them. They decided to find out a great deal more about the area before proceeding.

Usually observation is used, less dramatically, with other research tools in a continuous, back and forth manner. Today a team of observers recognizes that downtown seems to be pushing into a neighboring residential area. This observation generates economic and demographic research, with implications for public policies and programs. Tomorrow some traffic data calls for a field trip to see what the actual conditions are like. Often the local residents' concerns about a particular issue generate coordinated research, including observation. In any case, field observation is used along with other research methods.

Early knowledge of a problem permits early action, if that is appropriate, or early preparation for action. Observation may reduce the number of surprises to be faced. If one knows by looking that a large unused railroad yard near a busy downtown area is a likely site for development, then one can prepare for it. After looking at an area south of San Francisco's downtown, an area of marginal economic uses and boardinghouses for poor, transient men, I was able to advise a potential purchaser of land for a small new office building that the location was presently inappropriate. But having also seen signs that the downtown area was rapidly expanding in that direction, I was also able to advise the buyer that depending on the price and the length of time he could hold an empty lot, the site would soon have potential value for what he had in mind.

Observation also enables planners to take early direct action in response to problems and opportunities. In 1968, toward the end of the Johnson administration in Washington, San Francisco had an unexpected opportunity to receive federal funds for small, neighborhood parks. Within thirty days the city would have to propose specific sites where the money could be appropriately used. Intimate knowledge of the city, gained in large measure through looking, enabled a handful of staff to come up with over a hundred sites almost overnight, from which the final thirty-three locations were then chosen. . . .

I think also that any meaningful plan for a city, including small building projects, should start with an understanding of the nature of the place and should call for respecting and improving the existing physical character of the community. It should respond to important social and economic issues within that framework. . . .

In the end, the whole process of looking, questioning, trying to gain understanding makes a person a more intimate, respectful part of any environment and therefore more likely to be caring of it. That is the basis for good planning and beneficial action.

51

"For Planners, Investment in Social Media Pays Dividends"

from Planetizen, www.planetizen.com (2013)

Jessica Hsu

Editors' Introduction

Many new forms of technology can help promote sustainable urbanism. Among these tools are the following:

- Graphic software tools such as Photoshop, Illustrator, Sketchup, and many others provide ways to creatively illustrate sustainable design proposals in two or three dimensions;
- Online census data and other datasets and portals from public agencies and nongovernmental organizations allow quicker and easier analysis of community characteristics;
- Google Earth and other interactive mapping sites give any internet user the ability to analyze aerial and street-view images of any community;
- Geographic information systems (GIS) software such as ArcGIS helps users create and analyze mapped overlays of environmental and social data in a particular place;
- Modeling and visualization software such as CommunityViz helps communities envision and portray alternative futures for themselves; and
- Social media provide ways for project leaders to gather public input and interact with diverse constituencies.

In the piece below, Jessica Hsu provides additional discussion of how planners are using social media to develop conversations with the public aimed at creating more livable and sustainable communities. She is now Public Information Officer for the city of San Gabriel, California, but wrote this piece while working as a journalist for Planetizen, a leading online source of progressive urban planning news. Interested readers can subscribe to Planetizen's twice-weekly news digest at www.planetizen.com.

As social media strategies recently employed by Los Angeles's Department of City Planning and New York's Metropolitan Transportation Authority demonstrate, social media networks can be useful tools in connecting planners and public agencies to their communities. Their experiences show that an open conversation between decision-makers and the general public is important for creating more livable communities, and offer lessons for those considering utilizing such platforms.

Facebook, Twitter, YouTube and other social media platforms enable planners and organizations to build awareness of causes, provide the most up-to-date news, and engage the community in planning processes. "I think we've seen in the past few years that social media has a really important place in planning," said Jane Choi, Planning Assistant at the City of Los Angeles. "The social media tools we have help us to engage the public in a more dynamic way." In November 2011, she and Claire Bowin, City Planner at the City of Los Angeles, launched an integrated social media strategy utilizing Facebook, Flickr, Twitter, a blog, and an online town hall for LA/2B, the public outreach campaign helping to steer the replacement of the General Plan's Transportation Element with a new "Mobility" Element.

Choi adds, "There is transparency and value in people re-tweeting and re-blogging and helping us get the message out. Having a strong web presence helps people to find us instead of us going out to find them in a city of 3.8 million people." The LA/2B project has collected surveys and feedback from more than a thousand members in the city, and the input has been used to help craft the goals and policies for the Mobility Element and its Complete Streets Network.

"The whole idea is that social media is flexible. It's human. It's direct contact," said Adam Lisberg, Director of External Communications at New York's Metropolitan Transportation Authority (MTA). The subsidiary agencies of the MTA have separate Facebook pages and Twitter feeds that provide basic service information. The @NYCTSubwayScoop Twitter feed is one of their most accessed social media features with over 50,000 followers. "Every weekend, we do a lot of maintenance work because that's when passenger volume is lower. We have to take tracks out of service, and it's an inconvenience for customers who are affected by it. One of the things we're trying to do is to show our customers that there's a real long-term benefit," explains Lisberg. "So we post photos under the heading 'Weekend at Work' where we show crews hammering away at the roadbed, removing old track and installing new track, pouring new concrete, and lifting a switch into place. They're enormously popular to the general public, and it says this is why we have to do what we're doing."

Traditional outreach may be fitting for older generations, but Millennials are more accustomed to laptops and smartphones than to newsletters and workshops. "Twenty years ago, people turned on the television and the radio if they wanted to know what was going on. Now, people look at their Facebook and Twitter feeds to see what their friends are saying," said Lisberg. "This is where our customers are, and this is where we need to be."

On participatory planning for LA/2B, Choi said, "Traditionally, we haven't seen a lot of the 25–40 age group participating in our city-wide planning processes, but through using social media and our online town hall, we've been able to reach that part of the population." The project, which relies on public feedback, also uses social media to reach out to the busy individuals "who may not have time to go to a meeting on a weeknight or Saturday morning" and "allows people to participate on their own time rather than have to show up in-person."

Planners who want to get on social media can start with the most popular networks: Facebook (over 800 million users) and Twitter (over 200 million users). Facebook offers the opportunity for planners and organizations to promote their work to a broad audience. They can build a profile page to explain who they are and what causes they are advocating. Quality posts and status updates can engage people, and "likes" and activity will catch the attention of followers' friends. This allows a community to directly reach the decision-makers, while also interacting with others who may share the same thoughts.

Twitter is a "microblogging service" in which users can send "tweets" up to 140 characters. The limited content makes for fast and dynamic conversations, and hashtags (marked by #) allow users to look up topics of interest. The platform is valuable for planners and organizations interested in providing up-to-date information, learning about trending issues and hearing viewpoints directly from their citizens.

Even though Facebook and Twitter are the most popular networks, status updates and tweets are only the beginning of an effective social media strategy. For example, the LA/2B project also uses a blog, online town hall, and Flickr account; several agencies in the MTA use podcasts, Flickr, Tumblr and YouTube, which can be found on their social media page. The idea of handling so many social media platforms may be daunting to planners, but Facebook and Twitter have the advantage of integration with other social media and blogging services. This means that users do not have to spend the time-consuming process of updating information on each account because they can embed feeds on connected networks. "We decided to start with

Facebook and Twitter, as well as WordPress, because there's easy integration between them," said Choi. "There's also integration with our online town hall so when we publish a blog post, it automatically tells our Twitter and Facebook communities that there's a new blog post up that people can review."

While planners will most likely use social media to collect information from the community or to provide service information, these networking tools were effectively utilized by the MTA to convey emergency information during Hurricane Sandy. "At first, we were putting things out there because we had access that no one else had," said Lisberg. "Our crews are trained and equipped to get into spaces that we couldn't bring thirty television cameras through. This was the most recent information we had, and the most up-to-date images of what was out there. It served our purpose in telling New York what we were facing." The organization used Facebook and Twitter to provide coverage of the high flooding and winds, direct followers to useful information about assistance, and share links to photos on Flickr and videos on YouTube that showed the damage. The timeliness of the updates and the convincing use of visuals were two key parts of the strategy, said Aaron Donovan, a Deputy Director of External Communications for the MTA, and he noted that there was a huge uptake in

followers after the storm to their @MTAInsider feed on Twitter. Two-thirds of their current followers signed up during Sandy. The MTA's Sandy-related social media efforts were praised inlaudatory articles in the New York Times and Buzzfeed.

Social media has proven useful in planning for organizations like LA/2B and MTA, but planners and other relevant organizations do not yet have a strong presence on these networks compared to individual users and businesses. "Planners are getting more well versed in how to use social media. We've come a long way, but we have a long ways to go," said Choi, who added that one of the biggest challenges for LA/2B is a lack of personnel for consistent maintenance of the different accounts. Lisberg also cited the personnel challenge and added that the MTA is grappling with the expectation for one-to-one communication over social media. "I think the next step is to deepen the feeling of a relationship between our customers and the agency," he said. "You can have 10,000 people tweeting a complaint about what happened on their commute in the morning. Is there an expectation that we reply to all of them? I hope not because we certainly don't have the capacity to. But what are the parameters, and what is the nature of that relationship?" It's up to you, planners, to help test the waters.

52

"Multilevel Governance for the Sustainability Transition"

from *Globalism, Localism and Identity: Fresh Perspectives on the Transition to Sustainability* (2001)

Uno Svedin, Tim O'Riordan, and Andrew Jordan

Editors' Introduction

These days global warming and other sustainability threats seem to demand more effective action than most governments are able to take. So the question of how social institutions can better promote sustainable development is a pressing one. In this piece three European planning scholars suggest some important themes:

- Collaborative "governance" that combines action of public and private sectors, rather than traditional forms of state-led "government";
- Development of direct ties between different levels and types of agencies, for example direct connections between cities and the European Union;
- "Self-organizing" initiatives around particular issues, in which multiple public and private organizations come together to work for change;
- Coordination across different scales of governance, in which actions at different levels reinforce one another;
- Informal and flexible lines of authority, in which power shifts and is shared by players;
- "Subsidiarity," or the idea that action should be taken at the lowest level of governance consistent with effective action; and
- "Light green" and "deep green" worldviews.

These authors highlight the growing role of multilevel governance in addressing sustainability needs. It is an open question, however, to what extent often-informal constellations of public, private, and civil-society actors will be able to bring about social change for sustainability, and to what extent old-fashioned regulation by central governmental authorities will be required.

Other resources on governance for urban sustainability include Robyn Eckersley's *The Green State: Rethinking Democracy and Sovereignty* (Cambridge, MA: MIT Press, 2004); *Governing Sustainable Cities* (London: Earthscan, 2005), edited by Bob Evans, Marko Joas, Susan Sundback, and Kate Theobald; *Governing Sustainability* (Cambridge: Cambridge University Press, 2009), edited by W. Neil Adger and Andrew Jordan; Yvonne Rydin's *Governing for Sustainable Urban Development* (London: Routledge, 2010); and John Dryzek's *The Politics of the Earth: Environmental Discourses* (Oxford: Oxford University Press, 2013).

The transition to sustainability will have to take place through a complicated and ever-shifting set of governing structures. These structures underwent a slow but nonetheless radical transformation in the latter part of the last century, as "government" was increasingly replaced by "governance."

According to Stoker (1998, 17), the word "government" refers to activities undertaken primarily or wholly by states' bodies, particularly those "which operate at the level of the nation state to maintain public order and facilitate collective action." Typically these latter functions were performed by the state within its own territory via different parts of the public sector. The term "governance," on the other hand, refers to the emergence of new styles of governing in which the boundaries between public and private sector, national and international, have become blurred. For Stoker, then, "the essence of governance is its focus on governing mechanisms which do not rest on recourse to the authority and sanctions of government." Under a system of governance, more services are supplied by the market, with the state retaining control over core functions such as law and order, regulation and civil defense.

Because of policies pursued by many industrialized states such as privatization, new public management, and cutting the size of the civil service, the operations of the central state have in many countries become gradually reduced, with more and more services provided by government agencies and the private sectors. In consequence, governance involves a search for new means of steering and controlling activities through more indirect mechanisms such as financial control and incentives.

The shift from government to governance is also bound up with the trend towards more internationalized patterns of policy-making, in which important decisions are increasingly being made across a range of different administrative tiers or levels (Rosenau, 1997; Svedin, 1997). These stretch from the supranational down through the sub-national to the local. The term "multilevel governance" is popularly used to describe the increasingly dense set of interconnections between actors who operate at these different levels of governance, sometimes channeled through states, but very often bypassing them. Again, there is no commonly agreed definition of this term and interpretations seem to be numerous and varied (see Hix, 1998).

According to Gary Marks and his colleagues, multilevel governance in Europe has the following essential characteristics (Hooghe and Marks, 1996, 23–24):

■ The state no longer monopolizes policy-making at the European level. Decision-making is shared by actors at different levels, including supranational bodies such as the European Commission (EC) and the European Court.

■ Increasingly, collective decision-making among states involves a significant loss of control for individual states as they are forced to accept decisions adopted by the majority.

■ Levels of governance are interconnected rather than nested: national and sub-national actors (both public and private) act directly at all levels, by-passing the normal channels of interstate negotiation. For example, many local authorities in Europe have their own offices in Brussels so that they can lobby directly at the European level.

Kooiman (1993, 4) summarizes this set of relationships as follows:

These interactions are based on the recognition of interdependencies. No single actor, public or private, has all knowledge and information required to solve complex, dynamic and diversified problems; no actor has sufficient overview to make the application of particular instruments effective; no single actor has sufficient action potential to dominate unilaterally in a particular governing model.

Rhodes (1996, 660) helpfully summarizes four features of governance:

■ interdependence between organizations: governance is broader than government, covering non-state actors with shifting boundaries between public, private and voluntary sectors in a series of networks;

■ continuing interactions between and within these networks, caused by the need to exchange resources and negotiate shared purposes;

■ game-like interactions, rooted in familiarity, trust and shared commitments to legitimacy;

■ a significant degree of autonomy from the state, which seeks to steer the networks to achieve its policy goals, but cannot fully control them.

Another feature is the tendency for more decentralized arrangements to self-organize, that is, to develop their own sense of autonomy and self-interest. This in turn makes the center's traditional job of coordinating policy within and across different sectors and levels of activity – now demanded by the sustainability agenda – ever more difficult. The rise of voluntary, or negotiated, agreements in environmental management is but one manifestation of this. The state sets the aims to be achieved, but important aspects of the implementation process are let to non-state actors. Another is the inevitable tendency for sustainability issues to span a range of different government departments and agencies.

The resulting mismatch between institutional form and the inherently interconnected nature of sustainability problems generates "policy messes" (Rhodes, 1985). These occur in situations when policy problems demand a coordinated response involving several agencies and levels of government activity, but for a variety of political and bureaucratic reasons do not receive it. Another example of the new governance-setting is the increasingly quasi-corporatist pattern of consultative and advisory bodies, commissions of investigation, and lay–professional membership in safety and public-health bodies found in European states. What we are witnessing here is the emergence of a more policy-centered policy process within changed organizational structures.

The new frame also rests on the implicit premise that there is some kind of shared responsibility to make this creative and participatory pattern of decision-making work and to be responsive to new demands. These demands, in turn, are being encouraged by pluralistic pressures arising from the social, economic and environmental conditions of sustainability.

For all its ambiguities and contradictions, sustainability as a norm is creating a sense of common objectives among a varied set of actors by caring for threatened natural resources, for vulnerable and potentially disrupted populations, and for reliable wealth creation that encourages entrepreneurs to flourish while still looking after the good of the planet. Along with all this comes a fresh sense of civic participation. In a world that is both globalizing and fragmenting into more and more governmental units, people

are adapting in order to cope. According to Rosenau (1992, 291):

> Given a world where governance is increasingly operative without government, where lines of authority are increasingly more informal than formal, where legitimacy is increasingly marked by ambiguity, citizens are increasingly capable of holding their own by knowing what, where and how to engage in collective action.

Empowerment comes, in part, through identity and self-esteem. Patterns of multilevel governance can provide a promoting context to such an aim, but the outcome is not necessarily automatic. It is important to heed Rhodes's (1996) warning that without proper systems of democratic control and oversight, governance risks being less, not more, accountable than government if more and more decisions are taken outside the traditional governmental system.

FROM LAYERS WITHIN NATIONS TO LAYERS IN THE EU

Traditionally, there have been three levels of government and public administration in most European countries. The relative importance of these different levels has, however, varied considerably between countries and over time.

The first level is the central national level. All member states of the European Union (EU) are now considered democratic, although a number of them have experienced non-democratic forms of government during the 20th century, some as late as the 1970s. Each member state has a parliament, some have a president, and others are monarchies. The strength of the central government is dependent upon a number of factors among which tradition, the election system, and the country's constitution are the most important.

The second level is the regional level. The importance of this level differs between countries. There are considerable differences between, for example, the power of the German Länder and the power of the new devolved country governments of the UK.

The third level is the local scale. Again, a country's constitution and tradition combine to create varying degrees of autonomy for local government in the different member states of the EU. The Kommun, or municipalities, of Sweden have, for example, the right to collect income tax from their inhabitants and to determine how funds are utilized within a reasonably loose framework laid down by the central government. Municipalities in the UK, on the other hand, lost much of their autonomy during the 1970s and 1980s, and despite the political rhetoric of New Labour, still have to gain much of this back.

With the development of the EU, however, a fourth layer of governance has been added. Individual nation states have given up part of their sovereignty to a new actor – the EC. The rationale for this process is at the core of the paradox of political power. In order to gain political power, political power must be relinquished. The globalization of the political and economic systems has led to the erosion of the power of national governments in favor of a more multilevel pattern of institutional arrangements. In order to regain some of that power, national governments have attempted to pool some of their political resources in the institutions of the European Community, and through these institutions, regain some of the political power that they have lost.

While the creation of the EU has provided national governments with a new political platform, it has also initiated a new layer of governance that has provided sub-central authorities with channels of influence that enable them to by-pass central government. Local authorities in France, Italy and Spain can work together to develop regional interests. Swedish municipalities cooperate with local authorities in Latvia and Estonia and help them to prepare for EU membership. Projects of this kind are often part-financed by the EU, which encourages cooperation between sub-central governmental agencies.

During recent years, sub-central governments in Europe have developed closer relations with the EU and its Commission. These networks of actors have increased in importance to create the processes of "layering" and "delayering." Three types of networks can be identified. These are lobbying and exchange networks, policy networks and inter-governmental relations networks.

Pressure groups, organizations, businesses or sub-central government form lobbying and exchange networks. Their aim is usually to lobby EU institutions or to facilitate the exchange of information among network participants. Four categories of networks can be determined: namely, peak, spatial, thematic and sectoral:

- Peak groups include pan-European umbrella networks, such as the Council for European Regions and Municipalities (CEMR), which aim to coordinate local and regional governmental influence in Europe.
- Spatial networks are networks between regions with specific spatial or geographical characteristics. These networks attempt to advance regional interests and may be initiated by the EU. The INTERREG initiative, aimed at improving relations between richer and poorer parts of Europe, is an example of this kind of network.
- Thematic networks are created to lobby EU political institutions and may be formed in response to a specific EU policy program.
- Sectoral networks are initiated by regions with comparable economic attributes and aim to support economic growth in similar economic sectors, or to alleviate the effects of changing economic conditions through EU aid.

In recognition of these historical transformations, Marks and his colleagues (1996, 343–346) identify two contending theories of governance in the EU. One is that *state-centric* government through nations pools sovereignty in international organizations, as well as devolves power downwards to regional and local authorities. Marks *et al.* contend that the overall direction of policy is that of a national government's political needs. Typically, state-centric government relies upon unanimous voting for key policy issues in the EU. This ensures that no member state needs to be committed beyond what it is willing to tolerate. The core claim of the state-centric model is that "policy-making in the EU is determined primarily by state executives constrained by political interests nested within autonomous state arenas that connect sub-national groups to European affairs" (Marks *et al.*, 1996, 345).

The second theory framework offered by Marks *et al.* (1996, 346–350) is, as we discussed above, *multilevel governance*. Here, the state is not autonomous: decision-making is shared explicitly by actors, and not monopolized by state executives.

Super-national EU institutions such as the EC and the European Court of Justice have power in their collective right. Decision-making typically becomes more multilevel when decisions are taken by qualified majority voting in the EC Council of Ministers. Crucially, this inevitably involves loss of autonomy by national executives. Therefore multilevel governance is interconnected, not nested, and integration takes place through a huge variety of organizational and policy structures. As a result, state level ministers agree to share power and to delegate authority to multilevel structures, thus shielding themselves from the political fallout of unpopular EU-wide policies.

SUBSIDIARITY AND MULTILEVEL GOVERNANCE IN THE EU

If governance in the EU is, as we and others contend, becoming more multilevel, then an obvious question which arises is: what is the best level to address a particular set of issues or policy problems – the supranational, the national or the local?

This may seem an entirely abstract, theoretical matter, but it has hugely important implications in terms of addressing the sustainability agenda, and in terms of the basic democratic accountability of any resulting actions. Clearly, some environmental issues such as climate change are better addressed at the supranational level; however, mechanisms need to be found that make appropriate trade-offs with economic and social concerns at successively lower levels, while ensuring an appropriate level of democratic accountability. The further one moves problem-solving up and away from the local, the more elongated the channels of democratic representation become.

The EU has been trying to resolve complicated governance issues such as these since its inception in the 1950s, but it was not really until the 1990s that the need for a more systematic political response became apparent. The focus for this debate is the concept of subsidiarity. Although it entered popular discourse in the immediate aftermath of the Maastricht Treaty negotiations in 1991, subsidiarity has always been a well-established Eurofederalist principle. It is conventionally understood to mean that decisions in a political system should be taken at the lowest level consistent with effective action. The Oxford English Dictionary

defines it as "the principle that a central authority should have a subsidiary function, performing only those tasks which cannot be performed effectively at a more immediate or local level." Subsidiarity therefore provides a strong presumption in favor of decentralization. However, in the EU, subsidiarity has taken on a meaning that is sufficiently open-ended to satisfy both advocates and opponents of decentralization (Teasdale, 1993). Thus, the British saw subsidiarity as a means of reserving power for national government against the EU.

In German eyes subsidiarity, defined as action at the lowest effective level, coupled with the creation of a Committee of the Regions, promised to safeguard its federal system of government (involving the federal state and the regional Länder) from excessive European intervention. Arguably, Article 3b of the Treaty on European Union (TEU) tries to reconcile these two very different visions of a future Europe: one based around a strong network of independent nation states, the other involving a constitutionally enshrined allocation of powers across multiple levels of government – in a word, federalism.

While governmental networks have become, in many cases, more obscure and remote, the shift from more hierarchically sequenced "vertical" paths of decision-making has also opened up new possibilities for local authorities. New forms of relations have been created that allow patterns of power to leap across traditional, strictly hierarchical, structures. Within this context, a London suburb or a Swedish rural municipality may obtain a grant for developing a community initiative directly from the EU, without too much interference from intervening levels of government. Business opportunities for private companies in small municipalities can be opened up due to direct contacts with world markets and support from the EU's programs for small- and medium-sized industries, for rural development or for old industrial centers. These developments enable some local communities to preserve a satisfactory taxation base and provide service for their inhabitants. We can find new types of drivers not only for economic local prosperity but also for sustainability. For example, the eco-city of Graz or the high-tech city of Linkoping are using EU, national and local funds, along with private-sector participants, to promote clean technology for the new round of economic enterprise.

GREEN POLITICAL THEORY AND MULTILEVEL GOVERNANCE

What can green political theory bring to the debate? Does it provide a means of balancing the democratic and economic elements of subsidiarity? The most well-known typology of green political values ranges from deep green through to very light green (see Table 1).

Typically, light greens have great faith in the ability of humans to manage nature and live in harmony through the application of science and technology. They believe that sustainable development requires only a judicious mix of regulations and market-based instruments such as green taxes to correct market failures and to ensure that the environment is fully considered in decision-making.

Deep greens – or ecocentrists – on the other hand, see science and technology very much as part of the problem rather than the solution. They tend to view humans as one small part of nature rather than a superior resource "manager." Accordingly, humans need to find ways to live with nature, rather than "over" it.

In contrast to legal approaches, green political theory is not afraid to engage with the more normative aspects of subsidiarity. Deep greens in particular openly celebrate local diversity (Dryzek, 1987, 217):

- Small-scale communities are seen to be more reliant on their local environment and therefore more responsive to local disruption.
- Decision paths are therefore shorter, making it easier to respond to new challenges when they arise.
- Local action actively promotes the social responsibility and participation which is deemed necessary to achieve sustainable development.

According to deep-green ecocentrists "decentralization" is a means of redressing the distrust people feel about national and supranational politics. Ecocentrists would argue, moreover, that local participation addresses the widespread feelings of powerlessness and fatalism held by people in the face of an increasingly globalized world (see also Dobson, 1990). These sentiments are captured in the phrases "think global and act local," and "small is beautiful." Lighter greens, on the other hand, are much more agnostic with regard to the question of task assignment: they are happy to leave the matter to experts who will use technical, legal or scientific criteria – just the sort of project upon which the EU is currently engaged.

Does green political theory specify what is the most appropriate level at which to address environmental problems? The key dilemma here is between devolved democracy (which fits the main tenets of green thought but risks dissolving into conflict between localities, NIMBYism,[1] or a deregulatory "race to the bottom") and state control (which is often the only realistic political level at which the competing demands of local, regional and international actors can properly be mediated).

Having interrogated a range of positions and perspectives including socialism, Marxism, individualism and ecocentrism, Robyn Eckersley (1992, 182) concludes that a judicious mixture of international and local-level initiatives offers the best solution. Small, she argues, may not be beautiful if power is devolved to local communities that choose, or are forced, to adopt a development path that degrades the environment.

CONCLUSION

The adding of the EU-level above earlier local, sub-regional and national levels does not necessarily reinforce an old hierarchical layering of power. It may easily, through the added new combinations, erode "the old system."

There are clear examples that this is happening right now. In addition, other eroding factors – for example, due to new technologies, new networking patterns of alliances, new mechanisms in the global economy – help to reinforce such tendencies. The question of who gains and who loses control thus cannot be answered in a simple fashion. It seems as if active actors, even at the most local levels, can move more easily and more dramatically in this new ambiguous setting than passive actors. Thus the globalization processes may deliver the paradoxical effects on local circumstances both to enhance and diminish vulnerabilities.

The way in which local communities – both in a formal institutional sense and in a broader informal sense – operate and how the cultural framing

Table 1 Environmental worldviews and levels of governance.

	Light Green		Deep Greens	
	Cornucopian	Accommodation	Communalist	Deep Ecologist
Green label	Resource exploitative	Resource conservationist	Resource preservationist	Extreme preservationist
Type of Economy	Anti-green: unfettered markets	Green: markets guided by market instruments	Deep green: markets regulated by macro-standards	Very deep green: markets heavily regulated to reduce "resource take"
Management strategy	Maximization of GNP: human-environmental resources infinite substitutable	Modified economic growth preservation of 'critical' environmental resources	Zero-economic growth	Smaller national economy: localized production (bio-regionalism)
Ethical position	Instrumental (man over nature)	Extension of moral considerability: inter- and intra-generational equity	Further extension of moral considerability to non-human entities (bio-ethics)	Ethical equality (man in nature)
Level of decision-making	Unspecified: depends on problem	Unspecified: depends on problem	Regional/local	Local
Sustainability label	Very weak sustainability	Weak sustainability	Strong sustainability	Very strong sustainability

Note: Various ways of comprehending governance according to ideology. The groups on the left tend to prefer decentralized market structures, with increasing degrees of state intervention. The groups on the right look for less hierarchical forms of governance, with more poly-centric modes of division making. In essence, the two wings adopt very different positions in their vision of multilevel governance in a future Europe.

of democracy processes develop is thus of basic importance. The key to a locally based sustainability, which also has to be understood in the same "layering" terms, will thus be less dependent on a general recipe and instead will be seen as the outflow of a pattern of many interplaying factors. Many of these will depend upon the way in which local situations are handled. An interplay between a sufficiently forceful visionary capacity, its consolidation in everyday politics, and its acceptance in the population are all important factors.

These processes also quickly change in their connotations. What one day may be seen as a policy of mismatch versus real needs may soon turn out to be quite reasonable – and the reverse. Thus the time factor and timing aspects have to be more strongly taken into account when considering how to deal with the combinations of constantly changing factors. The aspect of path dependence in the movements towards a more sustainable development therefore grows in importance. The pressure on old rigid political systems to improve their capacities for flexible responses is also part of the reality to create a more sustainable future. The quickly changing landscape of new layering patterns and the erosion of old ones is part of this challenge.

NOTE

1 Ed.: This term refers to "Not In My BackYard" opponents of many local actions.

REFERENCES

Dobson, A (1990) Green Political Thought, Harper Collins, London

Dryzek, J (1987) Rational Ecology, Basil Blackwell, Oxford

Eckersley, R (1992) Environmentalism and Political Theory, UCL Press, London

Hix, S (1998) "The study of the EU II: the 'New Governance' agenda and its rival," Journal of European Public Policy, 5(1), 38–65

Hooghe, L, and Marks, G (1996) "Contending models of governance in the European Union" in A Cafruny, and C Lankowski, (eds) Europe's Ambiguous Unity, Lynne Rienner, Boulder, CO, 161–175

Kooiman, T (1993) "Social political governance: an introduction" in T Kooiman (ed) Modern Governance, Sage Publications, London, 5–20

Rhodes, RAW (1985) The National World of Local Government, Allen and Unwin, London

Rhodes, RAW (1996) "The new governance: governing without government," Political Studies, 44(4), 652–667

Rosenau, JN (1992) "Citizenship in a changing global order" in JN Rosenau and EO Czempiel (eds) Governance Without Government: Ordered Change in World Politics, Cambridge University Press, Cambridge, 291–287

Rosenau, JN (1997) "Global environmental governance: delicate balances, subtle nuances, and multiple challenges" in M Rolen, H Sjoberg and U Svedin (eds) International Governance of Environmental Issues, Kluwer Academic Publishers, Dordrecht, The Netherlands, 19–56

Stoker, G (1998) "Governance as theory," International Social Science Journal, 155, 17–28.

Svedin, U (1997) "International environmental governance – a round up of a discussion" in M Rolen, H Sjoberg and U Svedin (eds) International Governance of Environmental Issues, Kluwer Academic Publishers, Dordrecht, The Netherlands, 173–176

Teasdale, A (1993) "Subsidiarity in post-Maastricht Europe,"Political Quarterly, 32(1), 47–67

"A Progressive Politics of Meaning"

from *The Politics of Meaning: Restoring Hope and Possibility in an Age of Cynicism* (1993)

Michael Lerner

Editors' Introduction

To accomplish the enormous changes that a sustainable society requires, such as ending reliance on fossil fuels within a generation, entirely new ways of thinking about politics and governance may be required. As Albert Einstein is reputed to have said, "Problems cannot be solved within the same mind set that created them." Such a shift requires rethinking the basic values and goals of governance and political institutions, or "shifting the dominant discourse," as Michael Lerner puts it in the classic selection that follows.

In this selection from his book *The Politics of Meaning* (Reading, MA: Addison-Wesley, 1993), Lerner takes a radical view of the process of political change, agreeing that politics and institutions need to be changed, but arguing that the process must at the same time be one of personal and spiritual transformation. He stresses reforming social institutions so as to be more supportive of individuals and community, and to emphasize compassion and caring. In his view this reform can come about both through developing a collective political agenda and movement, and through dedicated work by individuals based on their growing appreciation of the fundamental unity of all beings.

Not surprisingly, Lerner's own roots are in spiritual traditions. A rabbi and psychologist, he is the founder and editor of *Tikkun* Magazine and a leader of the Jewish Renewal movement. Following the publication of *The Politics of Meaning* Lerner was consulted by candidate and then President Bill Clinton and his wife Hillary, and briefly became known as "the guru of the White House." Although the Clintons in many ways failed to follow through on their interest in developing a politics of caring, and their successor George W. Bush relied on a rhetoric of "compassionate conservatism" which was anything but compassionate in practice, it could be argued that the fact that these and other politicians acknowledge the importance of such values represents a step forward.

Other writings along these lines include Frances Moore Lappe's *Getting a Grip 2: Clarity, Creativity and Courage for the World We Really Want* (San Francisco: Small Planet Media, 2010) and *EcoMind: Changing the Way We Think, to Create the World We Want* (New York: Nation Books, 2011); *Active Hope: How to Face the Mess We're in without Going Crazy* (Novato, CA: New World Library, 2012) by Joanna Macy and Chris Johnstone; and Thomas Berry's classic *The Great Work: Our Way into the Future* (New York: Bell Tower, 1999).

A BRIEF DEFINITION

A progressive politics of meaning is a political effort to accomplish the following five goals:

1. *To create a society that encourages and supports love and intimacy, friendship and community, ethical sensitivity and spiritual awareness among people.*
Our economic, political and social arrangements make this kind of sensitivity and awareness more difficult to obtain and sustain. A politics of meaning does not seek to create a particular meaning system, but it does seek to create social and economic arrangements that will be friendly to meaning-oriented communities rather than harmful to their central concerns. In part, it means challenging the instrumental, utilitarian, mechanistic reductionism of thought and the disenchantment of our social experience. In part, it means creating institutions and economic practices that awaken within us our own ethical and spiritual sensitivity and our desire to treat one another with gentleness and compassion.

By spiritual awareness or sensitivity, I mean this: an awareness of the fundamental unity of all being and of our connectedness to one another and to the universe. When our unity and interconnectedness is fully appreciated, the arrogance and egotism that predominates in politics will dramatically decrease. Virtually every religious and spiritual system aims at this awareness. The politics of meaning does not seek to endorse any particular way of achieving it, but seeks to replace political and economic institutions that undermine this kind of spiritual awareness and that encourage human arrogance and ecological insensitivity.

2. *To change the bottom line.*
In most Western societies, productivity or efficiency is measured by the degree to which any individual or institution or legislation or social practice increases wealth or power. To pay attention to the bottom line is thus defined as paying attention to the degree to which the person or the project in question succeeds in maximizing wealth and power. Other goals are ancillary – acceptable only if they help accomplish (or, at least, do not thwart) the material goal.

A progressive politics of meaning posits a new bottom line. An institution or social practice is to be considered efficient or productive to the extent that it fosters ethically, spiritually, ecologically, and psychologically sensitive and caring human beings who can maintain long-term, loving personal and social relationships. While this new definition of productivity does not reject the importance of material well-being, it subsumes that concern within an expanded view of "the good life": one that insists on the primacy of spiritual harmony, loving relationships, mutual recognition, and work that contributes to the common good.

3. *To create the social, spiritual, and psychological conditions that will encourage us to recognize the uniqueness, sanctity, and infinite preciousness of every human being, and to treat them with caring, gentleness, and compassion.*

4. *To create a society that gives us adequate time and encouragement to develop our inner lives.*
We seek a society that no longer counterposes the time needed for inner work to develop our spiritual, aesthetic, and psychological sensibilities with the time needed to make ourselves economically secure and successful.

5. *To create a society that encourages us to relate to the world and to one another in awe and joy.*
Instead of rewarding our ability to dominate, manipulate, or control, we seek to build families, communities, and economic and political institutions that encourage our capacity to experience wonder and radical amazement at the grandeur of the universe, and to experience pleasure and celebration of one another as embodiments of the spiritual energy of the universe (or, in religious terms, as creatures made in the image of God).

Large-scale changes of this kind cannot be accomplished quickly. Nevertheless, there are many things we can do in the short run to move our society toward this goal. I discuss these possibilities in greater detail in the last section of this book, but will outline a few of them here.

For example, a program for a progressive politics of meaning in the next few decades would seek:

■ Public-school curricula that integrate the teaching of empathy and caring for others, and reward schools according to their success in creating empathic human beings (a goal quite different from that of traditional liberal demands, which have focused on student–teacher ratios or teacher salaries, but have rarely addressed the values being taught).

- Restructured health-care systems, so that medical care more adequately reflects an understanding of how the frustration of people's needs – for meaningful and nonalienating work, mutual recognition, love and caring, and a spiritually and ethically grounded society – may underlie receptivity to disease. Certainly, a progressive politics of meaning would endorse a single-payer universal health-care plan, on the grounds that we are never going to build a society that rejects selfishness while denying equal access to health care. But what distinguishes a politics-of-meaning approach is that it also links physical health to our ethical and spiritual well-being. So, for example, it also maintains that meaningful work – affording workers the opportunity to use their intelligence, creativity, and cooperative ability, and providing space for their spiritual lives – will decrease vulnerability to disease.
- A pro-family agenda that gives families the social supports they need. In addition to the traditional liberal elements (economic viability, flextime, child care, and so on), a politics-of-meaning approach includes as equally central the psychological and spiritual needs of families. We want to create a society that is safe for love and intimacy – as opposed to contemporary societies that identify sophistication with cynicism, critical intelligence with moral detachment, and maturity with "healthy suspicion of others." Instead, we seek a society that will encourage people to be more caring, sensitive, and empathic to others. So, for example, we want workplaces that encourage cooperation and give all of us an opportunity to use our intelligence and creativity. We want an economy that encourages us to take into account the needs and interests of others. We see these things not primarily as "rights" of the lone individual, but as requisite for shaping societies that nourish loving relationships.
- Annual ethical-impact reports from government and private sector institutions to assess their effect on the ethical, spiritual, and psychological well-being of our society and on the people who work in and with these institutions.
- Reflection within every profession on activities and attitudes that would be possible if the goals were to serve the common good; to heighten ethical, spiritual, and ecological sensitivity; and

to reward loving and caring behavior. Such reflection, for example, has led some lawyers associated with a politics-of-meaning perspective to envision a second stage of trials, in which the adversary system is suspended and the focus is shifted to healing the problems and pain that the initial trial has uncovered in the community.

Anyone who takes these specific examples and regards them as the sum total of the politics of meaning has missed my point. A progressive politics of meaning leads to a rethinking of every aspect of our public and private lives. However, it resists any attempt to impose one particular lifestyle or one particular approach to spirituality.

Most importantly, a politics of meaning is an invitation to transcend all the internalized messages that tell us that it is unrealistic to base our lives on our highest ideals and to fight for their realization. It is no longer appropriate to fight for instrumental goals that do not really express our vision of the kind of world we actually seek. Doing politics in this limited way turns out to be unrealistic and ineffective, because the political Right advocates its full vision enthusiastically. Progressives, to compete successfully, must present their highest, most compelling vision.

THE DESTRUCTIVE WAYS IN WHICH PEOPLE FIND MEANING

The primary dynamic of politics in the twentieth century has been the alternation between repressive communities of meaning and the alienation and loneliness produced by market societies. In reaction to the disintegration of existing communities, and with nothing to protect them from the alienation and loneliness produced by market societies, many people have become so hungry for human connection that they have even been willing to join repressive communities to fill that need. Some have turned back to traditional religious communities. Others have been attracted to the vision of community being offered by xenophobic nationalist movements.

Yet, to the extent that these various communities embody patriarchal privilege and class oppression, they frequently generate their own internal opposition. Just as, in the past, liberalism was created in opposition to the repressive nature of feudal societies, so today we find groups of people within these repressive communities who have begun to question the necessity of sacrificing for the community, once they realize that the sacrifices being sought primarily benefit the interests of small male elites. Disillusioned with such communities, these people turn back to the market, happy to be rid of what they have discovered: that their community is not truly a community, and that it does not really operate according to its proclaimed values. By contrast, in the market there is no similar hypocrisy: everybody really does try to maximize his or her own self-interest, and says so.

However, pursuing the market option and its logic of individualism also proves problematic for many people, since most of them *don't* "make it," and the ethos of selfishness sanctified by market-dominated societies soon yields deep unhappiness. So the cycle continues, and once again, people begin to grasp for communities of meaning, however repressive.

[. . .]

ARE PEOPLE READY TO BE PART OF A MEANING-ORIENTED MOVEMENT?

There already are millions of people effectively engaged in challenging the de-meaning of the world. But so far, most of these people do not see themselves as part of a larger meaning-oriented movement.

There already are millions of people involved in the pursuit of meaning and in challenging the atomistic, meaning-denying, economistic, or reductionist accounts of reality that have dominated public discourse. These people recognize that there is something fundamentally wrong with the dominant paradigm in the West, and they are building a more holistic view of our relationships to one another and to the natural world.

Some are involved in alternative approaches to health and healing, to nutrition and diet, to exercise and sports. Some are involved in a worldwide ecological movement which understands that, in order to save the planet, we may need to relate to the world with greater amounts of spiritual energy (awe, wonder, and radical amazement). Some are involved in developing spiritual practices, either in connection with religious communities or in connection with secular approaches to meditation and inner spiritual work.

Many are engaged in creating new relationships between men and women, either through the organized women's movement and the pro-feminist men's movement, or through more personal experiments in creating gentle and caring ways of relating. Many are engaged as teachers, nurses, social workers, and counselors; as rabbis, priests, ministers, and imams; and in other pursuits explicitly dedicated to reconnecting people to one another, to the spiritual energies of our inner selves, and to God's presence in the universe.

Currently, however, most people who are working for meaning operate in relative isolation from one another; they do not share a unified analysis. Some of them are involved in New Age philosophies that I find unpersuasive, or even in right-wing politics. Others resolutely assure one another that they are apolitical and uninterested in social issues. But what all these people do share is an ultimate concern with transcending the mechanistic, atomistic, antispiritual, and nonrelational ways of understanding the universe that have so crippled our thinking in the past.

These people already are challenging the dominant discourse, already are discounting what the official spokespeople have been saying. Some of these challengers turn to self-help and inspirational books, creating best-sellers as they grasp (sometimes indiscriminately) at accounts of reality that affirm our unity and innerconnectedness. Some people, to be sure, want nothing more than a personal solution, and will never be interested in any transformative vision that takes them beyond the confines of their own personal situation. But many more people understand that there is something deeply wrong with our world. Their alienation from the meaning-deadening world in which we live turns them toward new paradigms and new ways of conceptualizing reality. These people, who have faced the spiritual impoverishment of our contemporary world, will form the vanguard of social and political change in the twenty-first century.

One subgroup of this coming vanguard includes the millions of people who were ethically engaged with the social change movements of the 1960s and 1970s, but who today live private lives in part because they can see no political organization that plausibly speaks to their sensibilities. The political struggles of this generation of baby boomers awakened millions of Americans to an ethos of caring for others that was manifested in the civil rights movement, the antiwar movement, the women's movement, and the social justice and environmental movements.

Conservatives have correctly pointed out the ways in which these groups sometimes flirted with a countercultural ethos of individual fulfillment and unchecked self-indulgence that often undermined the moral content of the social change movements. There were moments when "liberation" was construed to mean freedom to do whatever one wanted to do, without regard for the consequences to others. To the extent that countercultural and political movements fell into this way of thinking or acting, they were quintessentially mainstream, providing yet another way for the dominant ethos of the market to permeate and shape mass consciousness. I myself and some of my friends sometimes fell into this self-indulgent definition of liberation, and we were deeply mistaken. I have learned from this experience to be more self-critical and also to listen carefully to the criticisms of people who have conservative politics with which I disagree, since they are sometimes correct and can see things that I cannot see. In fact, I believe that my entire generation has learned to approach politics with the humility that we tragically lacked several decades ago.

Indeed, a politics of meaning runs counter to this earlier tendency toward moral relativism and immediate gratification without moral standards. Yet, I must hasten to add that, even in the 1960s and 1970s, there was within these social change movements a countertendency, often explicitly challenging countercultural indulgence, that emphasized social solidarity and caring for others, that rejected moral relativism, and that articulated a powerful moral critique of the alienation and injustice of the contemporary world. Millions of people who went through that experience remain deeply committed to social justice and to building a more humane and loving society. Many of them have despaired that it ever would be possible to achieve those ends, and have become involved in lifestyles that on the surface seem superficially unconcerned about larger issues. Yet, like so many people in the religious world and in the labor movement, they could be mobilized to a new politics of love and caring were they to learn about it and come to believe that it was possible. Having been burnt by past failures, these former activists will not quickly jump into new political movements. Yet, as a meaning-oriented movement gains momentum, many of them will feel a homecoming that reconnects to their deepest hopes.

These are some of the groups from whom the movement for a politics of meaning will draw its initial support. They will become the transformative agents who move these ideas into the mainstream of American society. These people respond out of a real inner need, not from a commitment to an abstract idea, nor out of a sense that someone else ought to be treated differently. These people know that they cannot secure the kind of life they deeply desire, unless much changes in our society – its structure of values, its relationship to spiritual values, and its opportunities for mutual recognition. These are radical needs. Unlike needs for economic well-being or political rights, these cannot be fulfilled inside our society as it currently is constructed. Nor can these be fulfilled by buying off any one group. In that sense, the condition for the fulfillment of our needs for meaning is the condition for the liberation of our entire society from a materialist and individualist ethos.

Today, people involved in the pursuit of meaning do not yet form a coherent movement, and if they did, it seems unlikely that it would be a politically progressive movement. Nevertheless, the problems that have sensitized them to the crisis of meaning will not be solved by any other contemporary political movements. For this reason, I believe that people who recognize the crisis of meaning will provide the basis for a socially transformative movement in the twenty-first century. This movement will be at the center of creating a different kind of politics in America. That such a movement would be progressive is not guaranteed. It depends on whether the crisis of meaning remains the property of the Right, as it has become in the past decades, or whether liberals and Progressives allow themselves to move beyond the limits of their current conceptual schemes and seriously begin to address meaningful issues.

One function of bringing these people together through the framework of a politics of meaning is to help them recognize their potential power as a transformative and healing force. At this moment, their potential social power appears invisible even to themselves, and their voices remain marginalized.

Nevertheless, their marginality applies only to the conventional political arena, whereas the strategy of social transformation articulated here is not narrowly political in that sense. Changes in consciousness and in the ways people lead their lives will, in the long run, be far more important than who wins this or that election.

Every person engaged in acts of aesthetic and spiritual creativity, every person engaged in acts of mutual recognition that reach beyond the conventions of contemporary isolation, every person engaged in prayer that is spiritually alive, every person who refuses to be cowed by the dominant materialism or ethos of selfishness, every person who rejects technocratic accounts of reality, every person who affirms humor and playfulness and awe and wonder, is herself or himself part of the transformative process that will eventually break through the stranglehold of a meaning-deadening society.

THE IMMEDIATE TASK: SHIFTING THE DOMINANT DISCOURSE

Society's dominant discourse shapes not only its politics but the way people think about their personal lives and choices. Just as John F. Kennedy helped legitimize a discourse of idealism that gave impetus to the social movements of the 1960s, so Ronald Reagan managed to legitimize a discourse of selfishness and insensitivity that has had profound social consequences, far beyond his administration's legislative successes.

Shifting society's discourse – from one of selfishness and cynicism to one of idealism and caring – is the first and most important political goal of a politics of meaning in the next several decades. Long before we can reshape American society in any practical way, we must shift the way we think about our social institutions, politics, and economic practices. Therefore, one of the first priorities of a campaign for a politics of meaning will be to challenge the bottom-line assumption that people must always give priority to looking out for number one.

Changing the dominant discourse will change the messages we give to ourselves and to one another. The more we are able to support the part of ourselves that wishes to commit to a higher vision of who we can be, the likelier we are to take the steps in our personal lives that will make that possible. We are likelier to care for our souls and for our own spiritual and moral development when we live in a society where these kinds of concerns are publicly validated rather than ridiculed or marginalized. We are likelier to find ways to repair the psychic damage done by early childhood misrecognition and the forced denial of our desire for meaning when we live in a society which publicly values our ability to care as much as it values our ability to dominate or control.

The ultimate test of a politics-of-meaning movement, however, will always be the degree to which it liberates us to engage in small acts of caring and spiritual sensitivity. The more we engage in such acts, the more we are actually building a politics of meaning rather than just talking about it.

54

"The Story of Change"
from The Story of Stuff Project (2012)

Annie Leonard

Editors' Introduction

The question of how power structures of societies can be changed to better promote sustainability has no easy answers. Lerner and other writers in this book such as Leopold and Meadows imply that inner change is essential – the evolution of people's values and worldviews. Other previous authors in this section suggest more pragmatic tools for social change related to the institutions of governance, sustainability indicators, educational tools such as the ecological footprint, and social media. The following piece by Annie Leonard adds yet another dimension: organizing.

Leonard argues that the successful social change movements of recent generations – Gandhi's Indian independence movement, the US civil rights movement, the 1970s environmental movement, and the campaign to end apartheid in Africa (and we could add others such as the gay rights movement) – have changed the rules of the game by developing a large, powerful idea and organizing specific actions in support of it. Although personal change and small-scale incremental reform were important too, the focus on the big idea was essential.

Is urban sustainability such a "big idea"? If so, how might we organize around it? Are traditional protest marches and rallies still useful, or are there new and creative strategies for bringing people together on behalf of social change? Those are questions for each of us to answer in our own way. It may be useful to look at organizations such as 350.org, MoveOn.org, and Greenpeace as well as recent campaigns around slow food, gay rights, and social equity (such as the early-2010s Occupy movement).

With a graduate degree in Urban and Regional Planning, Leonard found her own creative way to promote change through a widely viewed series of videos beginning with *The Story of Stuff* (2007). *The Story of Change* (2012) and others in this series are available from The Story of Stuff Project at www.storyofstuff.org and on YouTube. Other materials considering strategies for long-term social change include *Great Transition: The Promise and Lure of the Times Ahead* (Boston: Stockholm Environment Institute and Tellus Institute, 2002) by Paul Raskin *et al.*; James Gustave Speth's book *The Bridge at the End of the World: Capitalism, the Environment, and Crossing from Crisis to Sustainability* (New Haven: Yale University Press, 2009); Gar Alperovitz' *What Then Must We Do?: Straight Talk about the Next American Revolution* (White River Junction, VT: Chelsea Green, 2013); and materials from the magazine *Yes!*

Ever since I learned where our stuff really comes from – and how this system is trashing people and the planet[i] – I've been trying to figure out how we can change it.

I've read a lot of these: *700 Ways to Save the Planet Without Leaving Your House, 50 Simple Things You Can do to Save the Earth, The Little Green Book of Shopping.*[ii]

I thought they might have the answers, but their tips all start here – with buying better stuff[iii] – and they all end here – with recycling[iv] all that stuff when I'm done with it.

But when it comes to making change, this story of "going green" – even though we see it everywhere – has some serious shortcomings.[v]

It says that if I become a smarter shopper, and tell my friends to do the same, I've done my part. And if I don't buy all this green stuff, then it's my fault that the planet's being destroyed.

Wait a minute. My fault? I didn't choose to put toxic products[vi] on the shelves or to allow slave labor in factories around the world.[vii] I didn't choose to fill stores with electronics that can't be repaired and have to be thrown away.[viii] I didn't choose a world in which some people can afford to live green, leaving the rest of us to be irresponsible planet wreckers!

Of course when we do shop we should buy the least toxic and most fair products we can,[ix] but it's not bad shoppers who are the source of the problem, it's bad policies[x] and bad business practices.[xi] And that's why the solutions we really need are not for sale at the supermarket.

If we actually want to change the world, we can't talk only about consumers voting with our dollars.[xii] Real change happens when citizens[xiii] come together to demand rules that work.

Look, it is important to try to live green. As Gandhi said, "be the change.[xiv]" Living our values in small ways shows ourselves and others we care. So it is a great place to start.

But it's a terrible place to stop. After all, would we even know who Gandhi was if he just sewed his own clothes and then sat back waiting for the British to leave India?[xv]

So how do we make big change?

To answer that question, I went back and looked at Gandhi,[xvi] the anti-apartheid movement in South Africa,[xvii] the U.S. Civil Rights Movement,[xviii] and the environmental victories in the 1970s.[xix] They didn't just nag people to perfect their day-to-day choices. They changed the rules of the game.

It turns out, there are three things you find whenever people get together and actually change the world.

First, they share a big idea[xx] for how things could be better. Not just a little better for a few people, a whole lot better for everyone. And they don't just tinker around the edges; they go right to the heart of the problem, even when it means changing systems that don't want to be changed. And that can be scary!

Hey, millions of us already share a big idea for how things can be better. Instead of this dinosaur economy that focuses only on corporate profits, we want a new economy[xxi] that puts safe products,[xxii] happy people,[xxiii] and a healthy planet[xiv] first. Duh, isn't that what an economy should be for?

Trying to live eco-perfectly in today's system is like trying to swim upstream, when the current is pushing us all the other way. But by changing what our economy prioritizes,[xxv] we can change the current so that the right thing becomes the easiest thing to do.

Second, the millions of ordinary people who made these extraordinary changes didn't try to do it alone. They didn't just say, "I will be more responsible." They said, "We will work together until the problem is solved."[xxvi]

Today it's easier than ever to work together. Can you imagine how hard it was to get a message across India in 1930? We can do it now in less than a second.

And finally, these movements succeeded in creating change because they took their big idea, and their commitment to work together, and then they took action!

Did you know that when Martin Luther King junior organized his march on Washington, less than a quarter of Americans supported him?[xxvii] But that was enough to make change – because those supporters took action – they did stuff. Today 74% of Americans support tougher laws on toxic chemicals.[xxviii] 83% want clean energy laws.[xxix] 85% think corporations should have less influence in government.[xxx]

We've got the big idea and the commitment. We just haven't turned it all into massive action yet. And this is our only missing piece. So let's do it.[xxxi]

Making real change takes all kinds of citizens – not just protestors. When you realize what you're good at and what you like to do, plugging in doesn't seem so hard. Whatever you have to offer, a better future needs it.

So ask yourself, "What kind of change maker am I?" We need investigators, communicators, builders, resisters, nurturers, and networkers.

Being an engaged citizen starts with voting.[xxxii] That's one of those basic things that everyone's just gotta do. But it gets way more exciting – and fun[xxxiii] – when we put our unique skills and interests to work alongside thousands of others.

I know that changing a whole economic system is a huge challenge. It's not easy to see a clear path from where we are today to where we need to go. And there's no ten simple things we can do without leaving our couches!

But the path didn't start out clear to all these guys either. Doctor King said, "Faith is taking the first step even though you don't see the whole staircase."

So, they worked hard to get organized, practiced the small acts that built their citizen muscles and kept their focus on their big idea – and when the time was right, they were ready.

It's time for us to get ready too – ready to make change and write the next chapter in the story of stuff.

REFERENCE

Very extensive notes are available by downloading the annotated script at http://storyofstuff.org/movies/story-of-change/.

PART FOUR

Sustainable urban development internationally

INTRODUCTION TO PART FOUR

Although the character of cities and towns varies enormously around the world, many urban sustainability issues are similar. Communities almost everywhere these days must decide how to limit or reduce automobile congestion, clean up pollution and contaminated lands, ensure decent, affordable housing for residents, provide infrastructure that enhances rather than degrades natural environments, foster steady local sources of jobs, and promote equity and quality of life. While a few societies remain remote, urban areas throughout Europe, Asia, Africa, and Latin America are experiencing rapid western-style development. Such transformation is fueled by the expansion of global corporations, ubiquitous technologies such as the motor vehicle, advice from development agencies and First World planning consultants, and the spread of materialist culture generally. As growth occurs and industrialization spreads, similar sets of development problems mount.

That being said, many differences do exist in less-developed countries. Third World local governments are frequently overwhelmed with the challenge of meeting demands for basic infrastructure and services. Urban population growth in many developing nations is far more rapid than in the First World. Mexico City, for example, grew from 3.5 million people in 1950 to 17.6 million in 2000, and is projected to expand to 24.6 million by 2025.[1] Lagos, Nigeria, grew from 1 million in 1950 to 12.2 million in 2000, and is projected to reach 18.9 million in 2025. Mumbai grew from 2.8 million in 1950 to 16.9 million in 2000, and is expected to be flooded by 26.6 million people in 2025. The existence of enormous informal settlements (in which residents lack title to land and have usually constructed dwellings themselves out of available materials) also presents unique problems in less-developed countries. At the same time, cities in Europe and elsewhere often benefit from far older urban traditions than in North America, with substantial advantages in terms of compact urban centers, walkable streets, strong public transit systems, and wonderful historic architecture and indigenous design traditions.

The following selections are intended to give a flavor of the issues related to sustainable urban development in different parts of the world. Those wishing more detail on international sustainability planning initiatives are encouraged to visit the web sites of the United Nations Division for Sustainable Development (www.un.org/esa/sustdev/), the United Nations Human Settlements Programme (www.unchs.org), the International Council on Local Environmental Initiatives (ICLEI; www.iclei.org), the European Sustainable Cities and Towns Campaign (www.sustainable-cities.org), the International Institute for Sustainable Development (www.iisd.org), the World Wide Web Library on Sustainable Development (www.ulb.ac.be/ceese/meta/sustvl.html), the World Business Council for Sustainable Development (www.wbcsd.ch), and other organizations.

NOTE

1 Figures from United Nations Department of Economic and Social Affairs. 2011. "World Urbanization Prospects: The 2011 Revision"; National Intelligence Council. 2000. Global Trends 2015. Washington, D.C.

55

"Urban Planning in Curitiba"

from *Scientific American* (1996)

Jonas Rabinovich and Joseph Leitmann

Editors' Introduction

More creative or sweeping sustainability-related initiatives can sometimes take place in developing nations than within industrialized countries, because of urgent crises demanding immediate action, centralization of government authority, more dynamic political leadership, or a lack of established bureaucratic tradition. For such reasons the Brazilian city of Curitiba, a metropolis of 1.6 million in the southern part of the country below Sao Paolo, has emerged as one of the world's leading examples of creative urban development, and received the Sustainable City Globe Award in 2010. Although certainly not without problems, Curitiba has managed to innovate consistently over 40 years to help manage its extremely rapid growth and provide a more livable and sustainable environment for its residents. It did so largely under the leadership of former mayor Jaime Lerner, an architect who helped found the Research and Urban Planning Institute of Curitiba, a strong independent city planning agency.

This selection from *Scientific American* provides an overview of some of Curitiba's achievements. Jonas Rabinovitch is a Brazilian planner who has worked for the city's research and urban planning institute as well as the United Nations. Joseph Leitmann is a senior manager with the World Bank and the author of *Sustaining Cities: Environmental Planning and Management in Urban Design* (New York: McGraw-Hill, 1999). Additional information about Curitiba is available from the website of the city's urban planning institute at www.ippuc.org.br (English version available). See also Steven A. Moore's *Alternative Routes to the Sustainable City: Austin, Curitiba, and Frankfurt* (Lanham, MD: Lexington Books, 2007), Clara Irazabal's *City Making and Urban Governance in the Americas: Curitiba and Portland* (Burlington, VT: Ashgate, 2005), Bill McKibben's *Hope, Human and Wild: True Stories of Living Lightly on the Earth* (Minneapolis: Milkweed Editions, 2007), and the video *Curitiba: City of Dreams* (Journeyman Pictures, 2006), available at www.youtube.com/watch?v=hRD3l3rlMpo.

In contrast to Curitiba's model, which has been led by a strong mayor and planning agency and criticized at times for its top-down nature, the nearby Brazilian city of Porto Alegre has pioneered one of the world's best examples of participatory planning. The latter city has pioneered "participatory budgeting" through which each neighborhood gets to vote on spending priorities for a percentage of the budget. In the years since reforms began in 1989 the city has made enormous gains in literacy, sanitation, housing, and other public services. More information on Porto Alegre and other Latin American examples is available from *Widening Democracy: Citizens and Participatory Schemes in Brazil and Chile* (Leiden: Koninklijke Brill NV, 2009), edited by Patricio Silva and Herwig Cleuren, and from U.N. Habitat's Best Practices database at www.unhabitat.org.

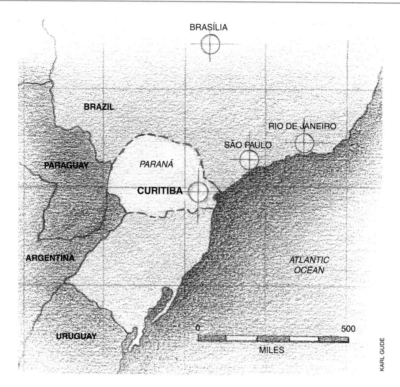

Figure 1 Location map of Curitiba.

Source: Illustration by Karl Gude.

As late as the end of the nineteenth century even a visionary like Jules Verne could not imagine a city with more than a million inhabitants. Yet by the year 2010 over 500 such concentrations will dot the globe, 26 of them with more than 10 million people. Indeed, for the first time in history more people now live in cities than in rural areas.

Most modern cities have developed to meet the demands of the automobile. Private transport has affected the physical layout of cities, the location of housing, commerce and industries, and the patterns of human interaction. Urban planners design around highways, parking structures and rush-hour traffic patterns. And urban engineers attempt to control nature within the confines of the city limits, often at the expense of environmental concerns. Cities traditionally deploy technological solutions to solve a variety of challenges, such as drainage or pollution.

Curitiba, the capital of Paraná state in south-eastern Brazil, has taken a different path. One of the fastest-growing cities in a nation of urban booms, its metropolitan area mushroomed from 300,000 citizens in 1950 to 2.1 million in 1990. Curitiba's economic base has changed radically during this period: once a center for processing agricultural products, it has become an industrial and commercial powerhouse. The consequences of such rapid change are familiar to students of Third World development: unemployment, squatter settlements, congestion, environmental decay. But Curitiba did not end up like many of its sister cities. Instead, although its poverty and income profile is typical of the region, it has significantly less pollution, a slightly lower crime rate and a higher educational level among its citizens.

DESIGNING WITH NATURE

Why did Curitiba succeed where others have faltered? Progressive city administrations turned Curitiba into a living laboratory for a style of urban development based on a preference for public transportation over the private automobile, working with the environment instead of against it, appropriate

Figure 2 Twenty-four-hour street, an arcade of shops and restaurants that never close, helps to keep Curitiba's downtown area vital. The city has also regulated the location of banks, insurance companies, and other nine-to-five businesses to prevent the district from becoming a ghost town after working hours.

Source: Photo by Jonas Rabinovitch.

rather than high-technology solutions, and innovation with citizen participation in place of master planning. This philosophy was gradually institutionalized during the late 1960s and officially adopted in 1971 by a visionary mayor, Jaime Lerner, who

was also an architect and planner. The past 25 years have shown that it was the right choice; Rafael Greca, the current mayor, has continued the policies of past administrations and built on them.

One of Curitiba's first successes was in controlling the persistent flooding that plagued the city center during the 1950s and early 1960s. Construction of houses and other structures along the banks of streams and rivers had exacerbated the problem. Civil engineers had covered many streams, converting them into underground canals that made drainage even more difficult – additional drainage canals had to be excavated at enormous cost. At the same time, developers were building new neighborhoods and industrial districts on the periphery of the city without proper attention to drainage.

Beginning in 1966 the city set aside strips of land for drainage and put certain low-lying areas off-limits for building. In 1975 stringent legislation was enacted to protect the remaining natural drainage system. To make use of these areas, Curitiba turned many riverbanks into parks, building artificial lakes to contain floodwaters. The parks have been extensively planted with trees, and disused factories and other streamside buildings have been recycled into sports and leisure facilities. Buses and bicycle paths integrate the parks with the city's transportation system.

This "design with nature" strategy has solved several problems at the same time. It has made the costly flooding a thing of the past even while it allowed the city to forgo substantial new investments in flood control. Perhaps even more important, the use of otherwise treacherous floodplains for parkland has enabled Curitiba to increase the amount of green space per capita from half a square meter in 1970 to 50 today – during a period of rapid population growth.

PRIORITY TO PUBLIC TRANSPORT

Perhaps the most obvious sign that Curitiba differs from other cities is the absence of a gridlocked center fed by overcrowded highways. Most cities grow in a concentric fashion, annexing new districts around the outside while progressively increasing the density of the commercial and business districts at their core. Congestion is inevitable, especially if most commuters travel from the periphery to the

Figure 3 Lakeside parks serve multiple functions in Curitiba. In addition to providing green space for citizens and forming part of the metropolitan bicycle path network, they help to control the floods that once plagued the city. The artificial lakes, created during the 1970s, are designed to facilitate drainage and to hold excess rainwater and keep it from inundating low-lying areas.

Source: Photo by Jonas Rabonovitch.

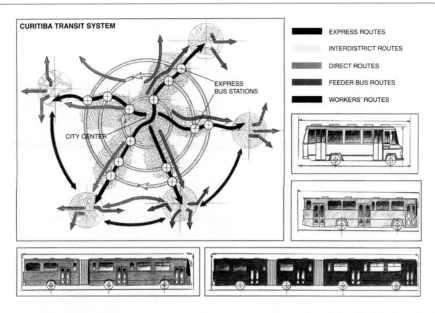

Figure 4 Transit system. Bus routes have grown with the city. Express bus routes define Curitiba's spoke-shaped structural axes; interdistrict and local lines fill in the space between spokes. Each route is served by a bus of appropriate scale, from minibuses that carry forty people.

Source: Illustration by Karl Gude.

center in private automobiles. During the 1970s, Curitiba authorities instead emphasized growth along prescribed structural axes, allowing the city to spread out while developing mass transit that kept shops, workplaces and homes readily accessible to one another. Curitiba's road network and public transport system are probably the most influential elements accounting for the shape of the city.

Each of the five main axes along which the city has grown consists of three parallel roadways. The central road contains two express bus lanes flanked by local roads; one block away to either side run high-capacity one-way streets heading into and out of the central city. Land-use legislation has encouraged high-density occupation, together with services and commerce, in the areas adjacent to each axis.

The city augmented these spatial changes with a bus-based public transportation system designed for convenience and speed. Interdistrict and feeder bus routes complement the express bus lanes along the structural axes. Large bus terminals at the far ends of the five express bus lanes permit transfers from one route to another, as do medium-size terminals located approximately every two kilometers along the express routes. A single fare allows

passengers to transfer from the express routes to interdistrict and local buses.

The details of the system are designed for speed and simplicity just as much as the overall architecture. Special raised-tube bus-stops, where passengers pay their fares in advance (as in a subway station), speed boarding, as do the two extra-wide doors on each bus. This combination has cut total travel time by a third. Curitiba also runs double- and triple-length articulated buses that increase the capacity of the express bus lanes.

Ironically, the reasoning behind the choice of transportation technology was not only efficiency but also simple economics: to build a subway system would have cost roughly $60 million to $70 million per kilometer; the express bus highways came in at $200,000 per kilometer including the boarding tubes. Bus operation and maintenance were also familiar tasks that the private sector could carry out. Private companies, following guidance and parameters established by the city administration, are responsible for all mass transit in Curitiba. Bus companies are paid by the number of kilometers that they operate rather than by the number of passengers they transport, allowing a balanced

BOX 1 Integrated design makes busways work

Curitiba's express bus system is designed as a single entity, rather than as disparate components of buses, stops, and roads. As a result, the busways borrow many features from the subway system that the city might otherwise have built had it a few billion dollars to spare. Most urban bus systems require passengers to pay as they board, thus slowing loading. Curitiba's raised-tube bus-stops eliminate this step: passengers pay as they enter the tube, and so the bus spends more of its time actually moving people from place to place.

Similarly, the city installed wheelchair lifts at bus-stops rather than on board buses. Easing weight restrictions and simplifying maintenance – buses with built-in wheelchair lifts are notoriously trouble-prone, as are those that "kneel" to put their boarding steps within reach of the elderly. The tube-stop lifts also speed boarding by bringing disabled passengers to the proper height before the bus arrives.

Like subways, the buses have a track dedicated entirely to their use. This right-of-way significantly reduces travel time compared with buses that must fight automotive traffic to reach their destinations. By putting concrete and asphalt above the ground instead of excavating to place steel rails underneath it, however, the city managed to achieve most of the goals that subways strive for at less than 5 percent of the initial cost.

Some of the savings have enabled Curitiba to keep its fleet of 2,000 buses – owned by ten private companies under contract to the

Bus tracks. Like subways, the buses have a track dedicated entirely to their use. This right of way significantly reduces travel time compared with buses that must fight automotive traffic to reach their destination. By putting concrete and asphalt above the ground instead of excavating to place steel rails underneath it the city managed to achieve most of the goals that subways strive for at less than 5 percent of the initial cost.

Source: Photo by Jonas Rabinovitch.

city – among the newest in the world. The average bus is only three years old. The city pays bus-owners 1 percent of the value of a bus each month; after ten years it takes possession of retired vehicles and refurbishes them as free park buses or mobile schools.

Companies are paid according to the length of the routes they serve rather than the number of passengers they carry, giving the city a strong incentive to provide a service that increases ridership. Indeed, more than a quarter of Curitiba's automobile owners take the bus to work. In response to increased demand, the city has augmented the capacity of its busways by using extra-long buses – the equivalent of multicar subway trains. The biarticulated bus, in service since 1992, has three sections connected by hinges that allow it to turn corners. At full capacity, these vehicles can carry 270 passengers, more than three times as many as an ordinary bus.

Figure 5 Bus tube. Most urban bus systems require passengers to pay as they board, slowing loading. Curitiba's raised-tube bus stops eliminate this step: passengers pay as they enter the tube, so the bus spends more of its time actually moving people from place to place.

Source: Photo by Jonas Rabinowitch.

distribution of bus routes and eliminating destructive competition.

As a result of this system, average low-income residents of Curitiba spend only about 10 percent of their income on transport, which is relatively low for Brazil. Although the city has more than 500,000 private cars (more cars per capita than any Brazilian city except the capital, Brasilia), three quarters of all commuters – more than 1.3 million passengers a day – take the bus. Per capita fuel consumption is 25 percent lower than in comparable Brazilian cities, and Curitiba has one of the lowest rates of ambient air pollution in the country. Although the buses run on diesel fuel, the number of car trips they eliminate more than makes up for their emissions.

In addition to these benefits, the city has a self-financing public transportation system, instead of being saddled by debt to pay for the construction and operating subsidies that a subway system entails. The savings have been invested in other areas. (Even old buses do not go to waste: they provide transportation to parks or serve as mobile schools.)

The implementation of the public transport system also allowed the development of a low-income housing program that provided some 40,000 new dwellings. Before implementing the public transport system, the city purchased and set aside land for low-income housing near the Curitiba Industrial City, a manufacturing district founded in 1972, located about eight kilometers west of the city center. Because the value of land is largely determined by its proximity to transportation and other facilities, these "land stocks" made it possible for the poor to have homes with ready access to jobs in an area where housing prices would otherwise have been unaffordable. The Curitiba Industrial City now supports 415 companies that directly and indirectly generate one fifth of all jobs in the city; polluting industries are not allowed.

PARTICIPATION THROUGH INCENTIVES

The city managers of Curitiba have learned that good systems and incentives are as important as good plans. The city's master plan helped to forge a vision and strategic principles to guide future developments. The vision was transformed into reality, however, by reliance on the right systems and incentives, not on slavish implementation of a static document.

One such innovative system is the provision of public information about land. City Hall can immediately deliver information to any citizen about the building potential of any plot in the city. Anyone wishing to obtain or renew a business permit must provide information to project impacts on traffic, infrastructure needs, parking requirements and municipal concerns. Ready access to this information helps to avoid land speculation; it has also been essential for budgetary purposes, because property taxes are the city's main source of revenue.

Incentives have been important in reinforcing positive behavior. Owners of land in the city's historic district can transfer the building potential of their plots to another area of the city – a rule that works to preserve historic buildings while fairly compensating their owners. In addition, businesses in specified areas throughout the city can "buy" permission to build up to two extra floors beyond the legal limit. Payment can be made in the form of cash or land that the city then uses to fund low-income housing.

Figure 6 The historic center of Curitiba has received special planning protection, including incentives to build elsewhere, that preserves old buildings. Many of the district's streets have been converted to pedestrian use, reducing pollution and fostering a sense of neighborhood. Ceremonial gates mark sections of the central city that were once enclaves of particular immigrant groups. (The entrance to the former Italian quarter is shown at the right).

Source: Photo by Jonas Rabinovitch.

Incentives and systems for encouraging beneficial behavior also work at the individual level. Curitiba's Free University for the Environment offers practical short courses at no cost for homemakers, building superintendents, shopkeepers and others to teach the environmental implications of the daily routines of even the most commonplace jobs. The courses, taught by people who have completed an appropriate training program, are a prerequisite for licenses to work at some jobs, such as taxi driving, but many other people take them voluntarily.

The city also funds a number of important programs for children putting money behind the often empty pronouncements municipalities make about the importance of the next generation. The Paperboy/Papergirl Program gives part-time jobs to schoolchildren from low-income families; municipal day care centers serve four meals a day for some 12,000 children; and SOS Children provides a special telephone number for urgent communications about children under any kind of threat.

Curitiba has repeatedly rejected conventional wisdom that emphasizes technologically sophisticated solutions to urban woes. Many planners have contended, for example, that cities with over a million people must have a subway system to avoid traffic congestion. Prevailing dogma also claims that cities that generate more than 1,000 tons of solid waste a day need expensive mechanical garbage-separation plants. Yet Curitiba has neither.

The city has attacked the solid-waste issue from both the generation and collection sides. Citizens recycle paper equivalent to nearly 1,200 trees each day. The Garbage That Is Not Garbage initiative has drawn more than 70 percent of households to sort recyclable materials for collection. The Garbage Purchase program, designed specifically for low-income areas, helps to clean up sites that are difficult for the conventional waste-management system to serve. Poor families can exchange filled garbage bags for bus tokens, parcels of surplus food and children's school notebooks. More than 34,000 families in 62 poor neighborhoods have exchanged over 11,000 tons of garbage for nearly a million bus tokens and 1,200 tons of surplus food. During the past three years, students in more than 100 schools have traded nearly 200 tons of garbage for close to 1.9 million notebooks. Another initiative, All Clean, temporarily hires retired and

(a)

(b)

Figure 7 Recycling in Curitiba takes many forms. (a) As in other cities, families sort their garbage to ease recovery of glass, metal, and plastic. Old buses find a second or third career as free transportation to city parks or as mobile offices or classrooms. (b) Even the city's old electricity utility poles find a new life as part of park buildings and public offices, including the Free University for the Environment.

Source: Photos by Jonas Rabinovitch.

Figure 8 The transport network includes bicycle paths integrated with streets and the bus network for most efficient travel. The bicycle paths also connect the city's main parks.

Source: Photo by Jonas Rabinovitch.

unemployed people to clean up specific areas of the city where litter has accumulated.

These innovations, which rely on public participation and labor-intensive approaches rather than on mechanization and massive capital investment, have reduced the cost and increased the effectiveness of the city's solid-waste management system. They have also conserved resources, beautified the city and provided employment.

LESSONS FOR AN URBANIZING WORLD

No other city has precisely the combination of geographic, economic and political conditions that mark Curitiba. Nevertheless, its successes can serve as lessons for urban planners in both the industrial and the developing worlds.

Perhaps the most important lesson is that top priority should be given to public transport rather than to private cars and to pedestrians rather than to motorized vehicles. Bicycle paths and pedestrian areas should be an integrated part of the road network and public transportation system. Whereas intensive road-building programs elsewhere have

Figure 9 The main boulevard of Curitiba, now devoted to pedestrian traffic, is the site of a weekly celebration of children gathering to paint. The ceremony began more pragmatically in 1972: when motorists threatened to ignore the traffic ban and drive on the street as usual, city workers blocked them by unrolling enormous sheets of paper and inviting children to paint water colors.

Source: Photo by Jonas Rabinovitch.

led paradoxically to even more congestion, Curitiba's slighting of the needs of private motorized traffic has generated less use of cars and has reduced pollution.

Curitiba's planners have also learned that solutions to urban problems are not specific and isolated but rather interconnected. Any plan should involve partnerships among private-sector entrepreneurs, nongovernmental organizations, municipal agencies, utilities, neighborhood associations, community groups, and individuals. Creative and labor-intensive ideas – especially where unemployment is already a problem – can often substitute for conventional capital-intensive technologies.

We have found that cities can turn traditional sources of problems into resources. For example, public transport, urban solid waste, and unemployment are traditionally considered problems, but they have the potential to become generators of new resources, as they are in Curitiba.

Other cities are beginning to learn some of these lessons. In Brazil and elsewhere in Latin America, the pedestrian streets that Curitiba pioneered have become popular urban fixtures. Cape Town has recently developed a new vision for its metropolitan area that is explicitly based on Curitiba's system of structural axes. Officials and planners from places as diverse as New York City, Toronto, Montreal, Paris, Lyons, Moscow, Prague, Santiago, Buenos Aires and Lagos have visited the city and praised it.

As these planners carry Curitiba's examples back to their homes, they also come away with a crucial principle: there is no time like the present. Rather than trying to revitalize urban centers that have begun falling into decay, planners in already large cities and those that have just started to grow can begin solving problems without waiting for top-down master plans or near fiscal collapse.

Figure 10 The botanical gardens were once a city dump. In addition to providing space for recreation the gardens serve as a research center for studies of plant compounds.

Source: Photo by Jonas Rabinovitch.

56

"Planning for Sustainability in European Cities: A Review of Practice in Leading Cities" (2003; updated 2013)

Timothy Beatley

Editors' Introduction

With their compact land development patterns, well-developed transit systems, and pedestrian-oriented mix of small shops and housing, many older European cities possess many natural advantages in terms of sustainable urban development. (The same can be said for historic Asian and Latin American cities.) A range of recent innovations by local governments, progressive developers, and civic organizations has also put many of these communities in the forefront of sustainability planning. Yet at the same time European cities and towns face many of the same problems of suburban sprawl, pollution, rising automobile use, persistent inequities, and growing non-local control over their economies that affect communities elsewhere in the world.

   In this selection, Tim Beatley, one of the co-editors of this volume, surveys main themes of European sustainability planning initiatives. Beatley has studied European sustainability efforts intensively; his other writings include *The Ecology of Place: Planning for Environment, Economy, and Community* (Washington, D.C.: Island Press, 1996), *Green Urbanism: Learning from European Cities* (Washington, D.C.: Island Press, 2000), *Native to Nowhere: Sustaining Home and Community in A Global Age* (Washington, D.C.: Island Press, 2004), and *Green Cities of Europe: Global Lessons on Green Urbanism* (Washington, D.C.: Island Press, 2012).

INTRODUCTION: LEARNING FROM EUROPEAN CITIES

In few other parts of the world is there as much interest in sustainability as in Europe, especially northern and northwestern Europe, and as much tangible evidence of applying this concept to cities and urban development. For approximately the last six years this author has been researching innovative urban sustainability practice in European cities. The findings from the first phase of this work are presented in the book *Green Urbanism: Learning from European Cities* (Island Press, 2000). What follows is a summary of some of the key themes and most promising ideas and strategies found in the 30 or so cities, in 11 countries, described in this book, as well as more recent case studies and field work.

   An initial observation from this work is just how important sustainability is at the municipal level in Europe, especially evident in the cities chosen. "Sustainable cities" resonates well and has important political meaning and significance in these cities, and on the European urban scene generally. One measure of this is the success of the Sustainable Cities and Towns campaign, an EU-funded informal network of communities pursuing sustainability begun in 1994. Participating cities have signed the so-called Aalborg Charter (from Aalborg, Denmark, the site of the first campaign conference), and more than 1800 cities and towns have done so. Among the activities of this organization are the publication of a newsletter, networking between cities, and initiation of conferences and workshops. The organization has also created the annual European Sustainable City award (with the first of these awards issued in 1996), and it is clear that they have been coveted and highly valued by politicians and city officials.

Many European cities have also gone through, or are currently going through, some form of local Agenda 21 process (including many of the same cities that have signed the Aalborg charter), and this is another important indicator of the relevance of local sustainability. Indeed, in the countries studied, high percentages of municipal governments are participating (for instance, in Sweden 100 percent of all local governments are at some stage in the local Agenda 21 process). Often these programs represent tremendous local efforts to engage the community in a dialogue about sustainability, and typically involve the creation of a local sustainability forum, sustainability indicators, local state-of-the-environment reports, and the preparation of comprehensive local sustainability action plans. European cities and towns demonstrate serious commitment to environmental and sustainability values and what follows are a few of the more important ways in which these concerns are being addressed.

Compact cities and regions

Urban form and land use patterns are primary determinants of urban sustainability. While European cities have been experiencing considerable decentralization pressures, they are typically much more compact and dense than American cities. Peter Newman and Jeffrey Kenworthy have monitored and tracked average density in a number of cities throughout the world. Western European cities like Amsterdam and Paris have substantially higher densities, as measured in persons per hectare, than typical American cities. Overall or whole-city densities for European cities are typically in the 40–60 persons per hectare range; American cities are much lower, commonly under 20 persons per hectare (Newman and Kenworthy 2000). Even American cities that we tend to think of as particularly dense, for example New York, are comparatively less dense when the entire metropolitan wide pattern is considered. Density and compactness directly translate into much lower energy use, per capita, and lower carbon emissions, air and water pollution, and other resource demands compared with less dense, less compact cities.

Many of these European examples, moreover, show that compactness and density need not translate to skyscrapers and excessive high-rise. Density and compactness in cities like Amsterdam happens through a building pattern of predominately low-rise structures. While many sustainability proponents advocate the need for the green high-rise development (e.g. see Ken Yeang's designs for bio-climatic skyscrapers), these European cities demonstrate convincingly that tremendous compactness and density can be accomplished at a clearly human scale. The European model is appealing to many precisely because of its more traditional form of density and compactness, and many believe its more human scale.

These characteristics of urban form make many other dimensions of local sustainability more feasible, of course (e.g. public transit, walkable places, energy efficiency). There are many factors that explain this urban form, including an historic pattern of compact villages and cities, a limited land base in many countries, and different cultural attitudes about land. Nevertheless in the cities studied there are conscious policies aimed at strengthening a tight urban core. Indeed, the major new growth areas in almost every city studied are situated in locations within or adjacent to existing developed areas, and are designed generally at relatively high densities.

Exemplary and for the most part effective efforts at maintaining the traditional tight urban form can be seen in many cities. Cities like Amsterdam are actively promoting urban redevelopment and industrial reuse (e.g. through its eastern docklands redevelopment). Berlin's plan calls for most future growth to be accommodated within its urbanized area through a variety of infill and re-urbanization strategies. Freiburg, Germany, has been able to effectively steer relatively compact, high-density new growth along the main corridors of its tram system, as well as to protect existing housing supply in the center (there is now a prohibition on the conversion of housing to offices and other uses).

European cities are utilizing a variety of planning strategies to promote compactness and to maintain a tight urban form. These include strict limits on building outside of designated development areas, a strong role for municipal governments in designing and developing new growth areas, extensive public acquisition and ownership of land (especially in Scandinavian cities like Stockholm), and a willingness to make significant transportation and other infrastructure investments that facilitate and support compactness.

GREEN URBANISM: COMPACT AND ECOLOGICAL URBAN FORM

Growth areas and redevelopment districts in these European cities are incorporating a wide range of ecological design and planning concepts, from solar energy to natural drainage to community gardens, and effectively demonstrate that *ecological* and *urban* can go together. Good examples of this compact green growth can be seen in the new development districts planned for or recently completed in Utrecht (Leidsche Rijn), Frieburg (Rieselfeld), Amsterdam (e.g. IJburg), Copenhagen (Orestad), Helsinki (Viikki), and Stockholm (Hammarby Sjöstad).

Leidsche Rijn, for example, is an innovative new growth district in the Dutch city of Utrecht. In addition to incorporating a mixed-use design, and a balance of jobs and housing (30,000 dwelling units and 30,000 new jobs), it will include a number of ecological design features. Much of the area will be heated through district heating supplied from the waste energy of a nearby power plant, a double-water system which will provide recycled water for non-potable uses, and storm water management through a system of natural swales (what the Dutch call "wadies"). Higher-density uses will be clustered around several new train stations and bicycle-only and bicycle/pedestrian-only bridges will provide fast, direct connections to the

city center. Homes and buildings will meet a low-energy standard and only certified sustainably harvested wood will be allowed.

European cities also provide excellent and generally successful examples of redevelopment and adaptive reuse of older, deteriorated areas within the center-city. Good examples include Amsterdam's eastern docklands, where 8000 new homes have been accommodated on recycled land. In *Java-eiland*, one major piece of this project, an overall plan (prepared by urban designer Sjoerd Soeters) lays out broad density, massing, and circulation for the district. Diversity and distinctiveness in actual design of the buildings, however, was encouraged through a restriction on the number of buildings that could be designed by a single architect. The result is a stimulating community where buildings have been created by scores of different designers. This island district successfully balances connection to the past (a series of canals and building scale reminiscent of historic Amsterdam) with unique modern design (each of the pedestrian bridges crossing the canals offers a distinctive look and design). *Java-eiland* demonstrates that city building can occur in ways that create interesting and organically evolved places, and which also acknowledge and respect history and context, overcoming sameness.

European cities on the whole (and especially the cities examined in this study) have been able to maintain and strengthen their center cities and urban cores. In no small part this is a function of historic density and compactness; they also the result of numerous efforts to maintain and enhance the quality and attractiveness of the city-center. In the cities studied, the center has remained a mixed-use zone, with a significant residential population. *Groningen*, for instance, has undertaken a host of actions to improve its center including the creation of new pedestrian-only shopping areas (creating a system of two linked circles of pedestrian areas), and installation of (yellow) brick surfaces and new street furniture in walking areas, among other actions. Committed to a policy of compact urban form, Groningen has also made strong effort to keep all major new public buildings and public attractions close-in. As one example, a new modern art museum has been sited and designed to provide an important pedestrian link between the city's main train station and the town center.

SUSTAINABLE MOBILITY

Achieving a more sustainable mix of mobility options is a major challenge, and in almost all of the cities studied in *Green Urbanism* a very high level of priority is given to building and maintaining a relatively fast, comfortable and reliable systems of public transport.

There are impressive examples of cities that have been working hard to expand and enhance transit, in the face of rising auto use in many areas. Zurich implements an aggressive set of measures to give priority to its transit on streets. Trams and buses travel on protected, dedicated lanes. A traffic control system gives trams and buses green lights at intersections and numerous changes and improvements have been made to reduce the interference of autos with transit movement (e.g. bans on left turns on tram line roads; prohibiting stopping or parking in certain areas; building pedestrian islands; etc.) A single ticket is good for all modes of transit in the city (including buses, trams, and a new underground regional metro system). The frequency of service is high and there are few areas in the canton that are not within a few hundred meters of a station of stop. Cities like Freiburg and Copenhagen have made similar strides.

In these European cities transit modes are integrated to an impressive degree. This means coordination of investment and routes so that transit modes complement each other. In most of the cities studied, for instance, regional and national train systems are fully integrated with local routes. It is easy, as well, to shift from one mode to another. Local transit centers are viewed in these cities as multi-modal, mixed-use centers of activity. Arnhem's new central train station in the Netherlands is a case in point. It integrates in a single location high-speed and conventional train service, local transit, bicycle parking, rental, and repair, as well as shops, offices and housing. These uses are all within a few hundred meters of the city center.

The ease of traveling throughout Europe is aided tremendously by the commitment on this continent to high-speed rail. Cross-national movement by high-speed train is increasingly comfortable and easy, and investments in dedicated tracks and infrastructure reflect impressive forward thinking on this issue. And increasingly it is not just the northern and northwestern European nations leading the way. Major new high-speed rail systems are under construction in Italy and Spain for instance. Overall, plans are on the books to double the length of dedicated high-speed rail track in Europe over the next eight years. And, the newest generation of trains will travel faster – on average 300 kph or higher.

Importantly, investments in transit complement, and are coordinated with, important land use decisions. Virtually all the major new growth areas identified in this study have good public transit service as a basic, underlying design assumption. The cities studied here do not wait until after the housing is built, but rather the lines and investments occur contemporaneously with the projects. The new community growth area *Rieselfeld* in Freiburg, for instance, has a new tram line even before the project has been fully built. In Amsterdam, as a further example, at the new neighborhood of *Nieuw Sloten*, tram service began when the first homes were built. In the new ecological housing district *Kronsberg*, in Hannover, three new tram stops ensure that no resident is further than 600 meters away from a station. There is a recognition in these cities of the importance of providing new residents with options, and establishing mobility patterns early.

Car sharing has become a viable and increasingly popular option in European cities. Here, by joining a car sharing company or organization residents have access to neighborhood-based cars, on an hourly or per-kilometer cost. There are now some 100,000 members served by car sharing companies or organizations in 500 European cities. Some of the newest car sharing companies, such as *GreenWheels* in the Netherlands, are also pursuing creative strategies for enticing new customers. This company has been developing strategic alliances, for example with the national train company, to provide packages of benefits at reduced prices. One of the key issues for the success of car sharing is the availability of convenient spaces, and a number of cities, including Amsterdam and Utrecht, have been setting aside spaces for this purpose. In cities such as Hannover, Germany, the car sharing organization there (a non-profit called Okostadt) has strategically placed cars at the stations of the Stadtbahn, or city tram, furthering enhancing their accessibility.

FOUR

Thinking beyond the automobile

Many of these cities are in the vanguard of new mobility ideas and concepts and are working hard to incorporate them into new development areas. Amsterdam, for example, has taken an important strategy in developing *IJburg*. It is working to develop a comprehensive mobility package that all new residents will be offered and which includes, among other things, a free transit pass (for a certain specified period) and discounted membership in local car sharing companies. Minimizing from the beginning the reliance on automobiles, and giving residents more mobility options, are the goals. Eventually this new area will be served both by an extension of the city's underground metro and fast tram.

An increasing number of carfree housing estates are also being developed in these cities, as a further reflection of the commitment to minimizing auto-dependence. The *GWL-Terrein* project, also in Amsterdam, built on the city's old waterworks site, incorporates only very limited peripheral parking. An on-site car sharing company, in combination with good tram service, are part of what makes this concept work there. The interior of the project incorporates extensive gardens (and 120 community gardens available to residents) and pedestrian environment, with key-lock access for fire and emergency vehicles.

Another carfree experiment is the new ecological district *Vauban*, in Freiburg. Built on the former site of an army barracks, this project is unique because it gives new residents the opportunity to declare their intentions to be carfree, and rewards them financially for doing so. Specifically, if residents choose to have a car, they must pay approximately $13,000 for the cost of a space in the nearby parking garage (a bit less than one-tenth the cost of the housing units). In this way there is a strong financial incentive to choose to be carfree and so far about half the residents have taken the carfree path. Projects like *Vauban* challenge new residents to think and act more sustainably and reward them for doing so.

Bicycles are an impressive mobility option in almost all of the cities studied in *Green Urbanism*, and many of these cities have taken tremendous efforts to expand bicycle facilities and to promote bicycle use. Berlin has 800 km of bike lanes, and Vienna has more than doubled its bicycle network since the late 1980s. Copenhagen now has a policy of installing bike lanes along all major streets, and bicycle use in that city has risen substantially. Few have gone as far, of course, as the Dutch cities, with cities like Groningen, where more than half of the daily trips are made on bicycles. In virtually all new growth areas in the Dutch cities, as well as many Scandinavian and German cities, bicycle mobility is an essential design feature, including providing important connections to existing city bicycle networks.

A number of actions have been taken by these cities to promote bicycle use. These include separated bike lanes with separate signaling and priority at intersections, signage and provision of extensive bicycle parking facilities (e.g. especially at train stations, public buildings), and minimum bicycle storage and parking standards for new development. Many cities are gradually converting spaces for auto parking to spaces for bicycles. Utrecht has discovered that it can fit 6–10 bicycles in the same space it takes to park one automobile. Tilburg, in the Netherlands, has recently built an underground valet bicycle parking facility in the heart of that city's shopping district. Freiburg's mobility center combines two levels of bicycle parking, with car-sharing cars on the ground level, a café, travel agency, and office of the Deutsche Bahn (and the structure has a green roof and a photovoltaic array generating electricity!).

These cities are also innovating in the area of public bikes. The most impressive program is Copenhagen's "City Bikes," which now makes available more than 2000 public bicycles throughout the center of the city. The bikes are brightly painted (companies sponsor and purchase the bikes in exchange for the chance to advertise on their wheels and frames), and can be used by simply inserting a coin as a deposit. The bikes are geared in such a way that the pedaling is difficult enough to discourage their theft. The program has been a success, and the number of bikes has been expanding. These sustainable European cities have discovered that bicycles are an important and legitimate alternative mode of transport to the car and with modest planning and investments substantial ridership can be achieved.

BUILDING PEDESTRIAN CITIES; EXPANDING THE PUBLIC REALM

European cities represent, as well, exemplary efforts at creating walkable, pedestrian urban environments. Relatively compact, dense, and mixed-use urban environments make cities much more walkable, of course. And most European cities and regions benefit from having a compact historic core, designed and evolved around walking and face-to-face commerce. The vitality, beauty, and attraction of European cities is in no small part a function of the impressive public and pedestrian spaces. Cities like Barcelona and Venice remain positive and compelling models of pedestrian urban society. The uses of these spaces are varied and many: they are outdoor stages, the "living rooms" in which citizens socialize, interact, and come together, places where political events occur and democracy plays itself out. These areas are now the social heart of these communities – places where children play, casual conversations and unexpected meetings take place, and people come to watch and be seen.

The overall land use pattern in these cities, and the priority given to maintaining their compact form, certainly make a walking culture more feasible. What is especially impressive, however, is the continued attention given to this issue and the continued expanding of pedestrian areas and the strengthening of the public and pedestrian realm. Cities like Copenhagen have set the stage, beginning in the early 1960s, gradually taking back their urban centers from cars. That city pedestrianized the Stroget, one of its main downtown streets, in 1962. Copenhagen continues this pedestrianizing in a gradual way each year. The city has adopted the policy of converting 2–3 percent of its downtown parking to pedestrian space each year, to dramatic effect over a 20–40 year period. Today the amount of pedestrian space is tremendous. Eighteen pedestrian squares have been created in Copenhagen where there was once auto parking – some 100,000 square meters in all. Had proponents of public space in Copenhagen attempted to convert this amount of space all at once it would have been very politically difficult to do so.

Many other cities have followed suit, especially Dutch and German cities, but examples can be found throughout Europe. Cities like Vienna and Groningen have pedestrianized much of their centers, creating delightful, highly functional public spaces. Groningen's compact city policy ensures that major new public buildings and facilities are kept in the center, and accessible through walking – it is a compact city of "short distances." In cities like Leiden, emphasis has been given to installing new pedestrian bridges over canals connecting major streets, and every new residential area is designed to include a grocery, post office, and other shops within an easy walk. The greater mixing of uses means that residents of these cities typically have many shops, services, cafés within a walkable range.

The experience of these European cities in pedestrianizing much of their urban centers has been a positive one, both economically and in terms of quality of life. The spaces created commonly contain fountains, sculptures and public art, extensive seating and, of course, many reasons for being there – restaurants, cafés, shops. Each city has its own unique history and features that can be used to strengthen the unique character of its pedestrian environment. Freiburg's "backle," or urban streams that run through the streets of its old center, as well as its pebble mosaics are delightful and special and this city has done an excellent job expanding and adding on to these unique qualities of place.

Good public transit appears a major factor strengthening the pedestrian realm in these cities, as well as commitments to bicycles, as in the case of Copenhagen (Hass-Klau *et al.* 1999). Extensive efforts to calm urban traffic, to restrict auto access, and to raise the cost of parking and auto mobility are also important elements. A number of European cities have experimented with or are anticipating some form of road pricing. The City of London is the most recent notable example, now charging a fee of five pounds for cars wishing to enter central London (and already resulting in a significant reduction in car traffic there). These European experiences support that a pedestrian culture and community life is indeed possible, even where the climate may be harsh and that these spaces serve an incredible range of social, cultural, and economic functions.

GREENING THE URBAN ENVIRONMENT

Ensuring that compact cities are also green cities is a major challenge, and there are a number of impressive greening initiatives among the study cities. First, in many of these cities there is an extensive greenbelt and regional open space structure, with a considerable amount of natural land actually owned by the cities. Extensive tracts of forest and open lands are owned by cities such as Vienna, Berlin, and Graz, among others. Cities such as Helsinki and Copenhagen are spatially structured so that large wedges of green nearly penetrate to the center of these cities. Helsinki's large *Keskuspuisto* central park extends in an almost unbroken wedge from the center to an area of old growth forest to the north of the city. It is 1000 hectares is size and 11 km long (City of Helsinki, 1995).

In Hannover an extensive system of protected greenspaces exists, including the *Eilenriede*, a 650 hectare dense forest located in the center of the city. Hannover has also recently completed an 80-kilometre long *green ring* (der grüne Ring) which circles the city, providing a continuous hiking and biking route, and exposing residents to a variety of landscape types, from hilly Borde to the river valleys of the Leineaue river.

There is a trend in the direction of creating and strengthening ecological networks within and between urban centers. This is perhaps most clearly evident in Dutch cities, where extensive attention to ecological networks has occurred at the national and provincial levels. Under the national government's innovative Nature Policy Plan, a national ecological network has been established consisting of core areas, nature development areas, and corridors, which must be more specifically elaborated and delineated at the provincial level. Cities in turn are attempting to tie into this network and build upon it. At a municipal level, such networks can consist of ecological waterways (e.g. canals), tree corridors, and green connections between parks and open space systems. Dutch cities like Groningen, Amsterdam, and Utrecht have full time urban ecology staff, and are working to create and restore these important ecological connections and corridors.

Many examples exist of efforts to mandate or subsidize the greening of existing urban areas. There is a continuing trend, for instance, towards installation of ecological or green rooftops, especially in German, Austrian, and Dutch cities. Linz, Austria, for instance, has one of the most extensive green roof programs in Europe. Under this program, the city frequently requires building plans to compensate for the loss of green space taken by a building. Creation of green roofs has frequently been the response. Also since the late 1980s the city has subsidized the installation of green roofs – specifically, it will pay up to 35 percent of the costs. The program has been quite successful and there are now some 300 green roofs scattered around the city. They have been incorporated into many different types of buildings including a hospital, a kindergarten, a hotel, a school, a concert hall, and even the roof of a gas station. Green roofs have been shown to provide a number of important environmental benefits, and to accommodate a surprising amount of biological diversity. Many other innovative urban greening strategies can be found in these cities from green streets, to green bridges, to urban stream daylighting.

RENEWABLE ENERGY AND CLOSED-LOOP CITIES

A number of the cities have taken action to promote more closed-loop urban metabolism, in which, as in nature, wastes represent inputs or "food," for other urban processes (e.g. Girardet, 2000). The city of Stockholm has made some of the most impressive progress in this area, and has even administratively reorganized its governmental structure so that the departments of waste, water, and energy are grouped within an eco-cycles division. A number of actions in support of eco-cycle balancing have already occurred. These include, for instance: the conversion of sewage sludge to fertilizer and its use in food production, and the generation of biogas from sludge. The biogas is used to fuel public vehicles in the city, and to fuel a combined heat and power plant. In this way, wastes are returned to residents in the form of district heating. Another powerful example of the closed-loop concept can be seen in Rotterdam, in the Roca3 power plant that supplies district heating and carbon dioxide to 120 greenhouses in the area. A waste product becomes a useful input, and in this case prevents some 130,000 metric tonnes of carbon emissions annually.

Energy is very much on the planning agenda, and these exemplary cities are taking a host of serious measures to conserve energy and to promote renewable sources. The heavy use of combined heat and power (CHP) generation, and district heating, especially in northern European cities, is one reason for typically lower per capita levels of CO_2 production here. Helsinki, for instance, has one of the most extensive district heating systems: more than 91 percent of the city's buildings are connected to it. The result is a substantial increase in fuel efficiency, and significant reductions in pollution emissions. District heating and decentralized combined heat and power plants are now commonly integrated into new housing districts in these cities. In *Kronsberg*, in Hannover, for instance, heat is provided by two CHP plants, one of which, serving about 600 housing units and a small school, is actually located in the basement of a building of flats.

Many cities, including Heidelberg and Freiburg, have set ambitious maximum energy consumption standards for new construction projects. Heidelberg has recently sponsored a low-energy social housing project, to demonstrate the feasibility of very low-energy designs (specifically a standard of 47 kwh/m^2 per year). The Dutch are promoting the concept of energy-balanced housing – housing that will over the course of a year produce as much energy as it uses – and the first two of these units have been completed in the *Nieuwland* district in Amersfoort.

Many cities such as Heidelberg have undertaken programs to evaluate and reduce energy consumption in schools and other public buildings. Incentive programs have been established which allow schools to keep a certain percentage of the savings from energy conservation and retrofitting investments. Heidelberg has engaged in an innovative system of performance contracts, in which private retrofitting companies get to keep a certain share of the conservation benefits.

There is an explosion of interest in solar and other renewable energy sources in these cities (and countries). Cities like Freiburg and Berlin have been competing for the label "solar city," with each providing significant subsidies for solar installations. In the Netherlands, major new development areas, such as *Nieuwland* in Amersfoort and *Nieuw Sloten* in Amsterdam, are incorporating solar energy, both passive and active, into their designs. In Nieuwland, described as a "solar suburb," there are more than 900 homes with rooftop photovoltaics, 1100 homes with thermal solar units, and a number of major public buildings producing power from solar (including several schools, a major sports hall, and a childcare

facility). What is particularly exciting is to see the effective integration of solar into the architectural design of homes, schools and other buildings.

The degree of public and governmental support in these European cities, financial and technical, for renewable energy developments is truly impressive. Reflecting a generally overall level of concern for global warming issues and energy self-sufficiency, significant production subsidies and consumer subsidies have both been given. The degree of creativity in incorporating renewable energy ideas and technologies in many of these cities is also quite impressive. Oslo's new international airport, for example, provides heating through a bark/wood bio-energy district heating system. This system provides heat for buildings through 8 km of pipes, as well as the airport's de-icing system. The moist bark fuel is a local product, and costs only one-third as much as fuel oil. In Sundsvall, Sweden, snow is collected, stored, and used as a major cooling source for the city's main hospital. In Copenhagen, twenty 2 MW wind turbines have been installed offshore which will together generate enough energy for about 30,000 homes.

Green cities, green governance

Many of these cities are taking a hard look at ways their own operations and management can become more environmentally responsible. As a first step, many local governments have undertaken some form of internal environmental audit. Variously called green audits or environmental audits, they represent attempts to study comprehensively the environmental implications of a city's policies and governance structure. A number of local governments are now going through the process of becoming certified (the London borough of Sutton being the first) under the EU's Eco-Management and Audit Scheme (EMAS), an environmental management system more commonly applied to private companies. Several German cities are preparing environmental budgets, under a pilot program. The cities of Den Haag and London have calculated their ecological footprints and are using these measures as policy guideposts (e.g. see Best Foot Forward Ltd, 2002). Albertslund, Denmark, has developed an innovative system of "green acounts," used to track and evaluate key environmental trends at city and district levels, and many of the study cities have developed sustainability indicators (e.g. Leicester, London, and Den Haag). Cities like Lahti, Helsinki, and Bologna have gone through extensive in-house education and involvement of city personnel, often as part of the local Agenda 21 process, in examining environmental impacts and in identifying ways that personnel and city departments can reduce waste, energy, and environmental impacts.

Municipal governments have taken a variety of measures to reduce the environmental impacts of their actions. A number of communities have adopted environmental purchasing and procurement policies. Cities like Alberstlund have adopted policies mandating that only organic food can be served in schools and child care facilities, and restricting use of pesticides in public parks and grounds. Other cities are aggressively promoting the development of environmental vehicles. Stockholm's environmental vehicles program is one of the largest (a pilot program under the EU-funded initiative ZEUS), with over 300 vehicles. A number of cities have sought to modify the mobility patterns of employees, for instance by creating financial incentives for the use of transit or bicycles. Cities like Saarbrücken, Germany, have made great strides in reducing energy, waste, and resource consumption in public buildings.

Communities have also engaged in extensive public involvement and outreach on sustainability matters. A variety of creative approaches have been taken. Leicester, for instance, has developed alliances with the local media and has sponsored a series of educational campaigns on particular community issues. As a further example, it has established (with its NGO partner Environ) an environmental center and cyber-cafe called the Ark, as well as a demonstration ecological home. Officials in these exemplary cities often express the belief that it is essential to set a positive example for the community and that before they could ask citizens to change their behaviors and lifestyles, the municipal government must have its environmental house in order.

Understanding European cities: some concluding thoughts

To be sure, many European cities are facing some serious problems and trends working against sustainability, in particular a dramatic rise in automobile ownership and use, and a continuing pattern of deconcentration of people and commerce. And, with their relatively affluent populations consuming substantial amounts of resources, European cities exert a tremendous ecological footprint on the world. Yet, these most exemplary cities provide both tangible examples of sustainable practice, and inspiration that progress can be made in the face of these difficult pressures.

The lessons are several. These cities demonstrate the critical role that municipalities can and must play in addressing serious global environment problems, including reliance on fossil-fuels and global climate change. Innovations in the urban environment offer tremendous potential for dramatically reducing our ecological impacts (European cities produce about half the per capita carbon emissions of American cities), while at the same time enhancing our quality of life (e.g. by expanding personal mobility options with bicycles and transit).

Many, indeed most, of the ideas, initiatives, strategies undertaken in these innovative cities serve, in addition to reducing ecological footprints, to enhance livability and quality of life. Taking back space from the auto and converting it to pedestrian and public space does much to enhance the desirability of these cities. Investments in public transit reduce dramatically energy consumption, CO_2 emissions, and urban air quality problems, but at the same time provide tremendous levels of independence and mobility to the youngest and oldest members of society. Making bicycling riding safer and easier helps the environment, but also provides a badly needed form of physical exercise.

These experiences demonstrate clearly that it is possible to apply virtually every green or ecological strategy or technique, from solar and wind energy to greywater recycling, in very urban, very compact settings. Green Urbanism is not an oxymoron. Moreover, the lesson of these European cities is that municipal governments can do much to help bring these ideas about, from making parking spaces available for car sharing companies to providing density bonuses for green rooftops, to producing or purchasing green power.

There are also process lessons here. Key among them is an understanding of the great power of partnerships and collaboration between different parties with an interest in sustainability. While not always easy, success at achieving sustainability will depend on them. This means getting different departments to talk to each other and to work together (as in Stockholm), and getting different public and private actors to join together in common initiatives that demonstrate that green urban ideas are possible and desirable.

It is important to recognize, to be sure, the differences in governmental structure. The economic and planning frameworks in place in these countries (compared with, say, the United States) often facilitate many of the exemplary urban sustainability projects described here. The role of economic incentives and the economic incentive structures is critical and undeniable. High prices at the gas pump (typically $4–5 per gallon in Western Europe) have been a conscious policy decision in European countries, and in countries like Germany have provided essential funding for public transit. Such high prices, relative to countries like the United States, undoubtedly help to encourage more compact land use and personal choices in favor of more sustainable modes of mobility. Also, carbon taxes in countries like Denmark help to substantially level the economic playing field between conventional fossil-fuel energy and more sustainable, renewable forms of energy. Higher energy prices generally help to promote greater conservation and energy efficiency improvements. The important role of adjusting incentives and economic signals is itself a key lesson from the European scene. Rather than being seen as a pre-existing background condition, raising gasoline and energy taxes can be seen as an example of an important strategic societal and political choice.

There are other political, social, and cultural conditions, to be sure, that favor many of the exemplary ideas discussed here. Parliamentary governmental structures that give relative voice and power to green party and other social and environmental views (with local representation of these views as well) have been important. Historically stronger planning and land use control systems are helpful also, as well as generally stronger and more proactive roles afforded to government. Many of the important (more activist) urban sustainability activities undertaken in these European cities – as market stimulators, promoters of innovation, and financial underwriters for innovative urban sustainability practices and projects – are common and accepted roles for local governments to play.

But there are also certainly many underlying value differences (compared with the U.S.) that further explain good practice. Prevailing European views of land are less imbued with a sense of personal use and freedom, and there is little expectation, for instance, on the part of a rural landowner or farmer that his or her land will eventually be convertible to urban development.

There are also a number of more regionally unique cultural values and differences, each with significant planning and land use shaping implications. A stronger desire to live within a city or town center clearly exists throughout much of Europe, borne undoubtedly from an older, more developed urban culture. Importance given to strolling, spending time in public places, and to the values of the public realm more generally, in countries like Spain and Italy, certainly help explain the success of pedestrian spaces in these countries. Pace of life, cultural organization of the day, and the number of hours in the work week are also clearly important. In Italy, public and pedestrian spaces are used in part because there is time to use them – the culture organizes its day so as to support the early evening stroll, after the shops close but before the evening meal. To many observers of the European scene there are also lessons to emulate – suggestions and ideas for humanizing cities and strengthening their livability and sociability, as well as their sustainability. The lessons are many and profound on many levels.

REFERENCES

Beatley, Timothy. (2000) *Green Urbanism: Learning from European Cities*, Washington, DC: Island Press.

Best Foot Forward, Ltd. (2002) *City Limits*, London: Best Foot Forward Ltd.

Gehl, Jan and Lars Gemzoe. (2000) *New City Spaces*, Copenhagen: The Danish Architectural Press.

Girardet, Herbert. (1999) *Creating Sustainable Cities*, Devon, UK: Green Books.

Hass-Klau, Carmen, Graham Crampton, Clare Dowland, and Inge Nold. (1999) *Streets as Living Space: Helping Public Places Play Their Proper Role*, London: Lander Publishing Ltd.

Newman, Peter and Jeffrey Kenworthy. (1999) *Sustainable Cities: Overcoming Automobile Dependence*, Washington, DC: Island Press.

57

"Collective Action Toward a Sustainable City: Citizens' Movements and Environmental Politics in Taipei"

from *Livable Cities?: Urban Struggles for Livelihood and Sustainability* (2002)

Hsin-Huang Michael Hsiao and Hwa-Jen Liu[1]

Editors' Introduction

While Curitiba provides an inspiring example of visionary planning in a developing world city, the reality is usually much more mixed. As in wealthier industrialized nations, developing-world urban policies often favor wealthy elites and corporations over less affluent residents and natural ecosystems. To counter these repressive forces, urban social movements have often arisen to fight for a wide variety of environmental and social goals. Urban politics has become increasingly complex in many places; the emerging environmental concerns of middle- and upper-class residents for example may run counter to the equity needs of less-well-off communities.

In the following selection, Hsin-Huang Michael Hsiao and Hwa-Jen Liu provide a detailed analysis of such dynamics in Taipei, Taiwan's 3 million-person capital city. They show how the government-led "growth machine" bent on making the country part of the global economy has produced great environmental and social harm. Wealthy developers have built on many fragile hillside lands within the city, power plants and polluting facilities have been constructed next to low-income neighborhoods, and the city at times has bulldozed informal settlements to create parks for the middle and upper classes. Several generations of environmental activism have arisen in opposition to such forces. Their successes have been limited, but the slow building of citizen movements such as these is vitally important for sustainable development in both developed and developing nations.

Hsiao and Liu are both professors of sociology at the National Taiwan University in Taipei. This selection comes from the excellent collection of articles on Third World cities edited by Peter Evans, *Livable Cities: Urban Struggles for Livelihood and Sustainability* (Berkeley: University of California Press, 2001). For additional material on urban sustainability challenges in the developing world, see Mike Davis' *Planet of Slums* (New York: Verso, 2006), the Worldwatch Institute's *State of the World 2007: Our Urban Future: a Worldwatch Institute Report on Progress Toward a Sustainable Society* (New York: Norton, 2007), and *Climate Change, Disaster Risk, and the Urban Poor: Cities Building Resilience for a Changing World* (Washington, D.C.: World Bank, 2012), edited by Judy L. Baker.

We begin with two stories of community mobilization in Taipei. The first is about defending living space. It involves a well-to-do middle-class community named Ching-Cheng, which indefatigably fought against the state-owned power company (Taipower), a large department store backed by overseas capital, and finally the city authorities.

From 1988 to 1989, this community faced the imminent construction of a nearby power substation. Residents found out later that the increase in demand for electricity in this district resulted mainly from a newly opened department store, which also planned to rent its basement to a brothel. Unhappy, the community's residents decided to implement a boycott. They fought the department store directly by calling it "a lousy neighbor" in the media. They protested against Taipower for selecting a site without having first informed or consulted them, but they ultimately were unable to overturn the decision. The self-organized Ching-Cheng residents were not defeated, however. In 1991 they successfully saved their community park from becoming a parking garage, and in 1992 they interrupted the municipal government's plan to change the area from a residential zone into a commercial one (Shen 1994). The case of Ching-Cheng, like many others in Taipei, stands in direct opposition to business interests and the top-down decision-making system of government. The statement that the community made was loud and clear: using economic growth to justify abusing citizen rights and the urban environment is no longer acceptable, even if it once was. The mixed success of the Ching-Cheng community over the years is the most frequently cited case of community mobilization among urban reformers and community advocates, and it has also become an exemplar for other mobilized communities in Taiwan.

The second story concerns the birth of a five-hectare city park, Nos. 14–15 park. It illustrates the conflicting interests that can exist between the urban poor and other communities when improving the physical environment of a city as a whole is at stake. The predesignated park site was a Japanese-only cemetery under colonial rule and then was turned into one of the biggest slums in the city, the Kang-Le neighborhood, in the postwar development of Taipei. Kang-Le slum was an enclave surrounded by five-star hotels and skyscrapers in one of Taipei's most prosperous areas. The graphic portrait of urban poverty and the cohabitation of living humans and the dead served only to surprise visitors and offend nearby business owners and middle-class neighbors.

On March 4, 1997, despite strong opposition from urban planners and slum residents, the Taipei municipal authorities sent out bulldozers, escorted by police, and forcibly evicted thousands of slum dwellers, over one-third of whom were handicapped, aged, or extremely poor, and tore down their illegal but long tolerated shacks (Hsiao *et al*. 1997, 3). Overnight, 961 households and the community networks that supported their subsistence were destroyed. After the eviction, the first-ever freely and directly elected mayor of Taipei, Chen Shui-bian, proudly announced that citizens' urgent need for more green space would soon be satisfied. One year later, the city administration revealed its proposal to name this park "International Plaza" and to use it to host carnivals and celebrations of such holidays as Halloween and St Patrick's Day for foreign residents (Yang 1998). The whole process, as an example of the more vulgar side of "globalization," was praised by bureaucrats and certain media outlets as an effort to create a foreigner-friendly atmosphere and to enhance the visual presentability of the city.

These two vignettes illustrate both the successes and failures of urban environmental activism in Taiwan's capital city, Taipei. On the one hand, in the 1980s and even more in the 1990s the environmental movement epitomized mobilization from below in a society in which top-down politics had predominated since colonial times. On the other hand, community participation in this movement still reflected the inequalities of the society at large. Middle-class communities were often able to protect or even improve their quality of life, even in the face of the ecologically blind developmental programs pushed by the state and private business groups. Poor and slum communities were as likely to be the victims of the environmental movement as they were to be its beneficiaries. . . .

URBANIZATION AND ENVIRONMENTAL CRISES IN TAIPEI

Every city has its own experience of environmental degradation incurred by rapid urban growth.

Usually, a city starting its "developmental career" as a manufacturing center accepts the costs of industrial production: noise from mechanization, soil and water pollution, harmful fumes, and damage to the landscape beyond remedy (Mumford 1989, 458–74). Even if the rise of a service economy and the reduction in manufacturing gradually cause change in the industrial structure in the later stages of urban development, early industrial pollution has already left permanent scars on the urban landscape. Classic industrial cities such as Detroit and Manchester have paid for their unsustainable land use by falling in the world city hierarchy and finding themselves incapable of reviving their economic prominence even with enormous reinvestment (Friedmann 1997). But that has not been the case with Taipei.

In contrast to another major metropolis in southern Taiwan, Kaohsiung, Taipei has suffered relatively little harm from extensive industrial pollution because, over the past hundred years, Taipei never developed into a manufacturing city. Industrial products accounted for 88.7 percent of the total products manufactured in Taipei between 1938 and 1940, but they came from the relatively nonpolluting sectors of food and tea processing and rice milling. In fact, in the 50 years of Japanese colonial rule (1895–1945), aside from its political function as the colony's capital city, Taipei served as only a light industrial center, processing and exporting agricultural products to boost the Japanese economy (Chiang and Hsiao 1985, 194).

When the Kuomintang (KMT) took over Taiwan in 1945, 75 percent of the industrial infrastructure had been destroyed as a result of World War II, and the engine of industrialization had to be restarted (Huang 1983, 489). In contrast to the exporting of agricultural commodities under colonial rule, the pattern of industrial development was shaped by ISI (import-substituting industrialization) in the 1950s and EOI (export-oriented industrialization) between 1960 and 1972, then shifted to the second ISI and second EOI strategies from 1973 onward (Gereffi 1990, 17–18; Haggard 1990, ch. 4). Along the path of development, the picture of the domestic economy dramatically changed. The drastic decline of the agricultural sector mirrored the significant expansion of the industrial sector, and nationwide environmental degradation has in general been triggered by the growth of certain

industries, namely steel and petrochemicals, that were encouraged by state sponsorship during the second ISI phase. While the coastal areas and cities of west and south Taiwan have suffered greatly from these two environmentally devastating industries, however, Taipei has been spared.

If Taipei has benefited from the absence of extensive industrial pollution, then what is damaging Taipei's eco-social system? It is fair to say that the single-minded pursuit of economic prosperity, under pressure from competition to achieve regional economic hegemony during global economic restructuring, and as mediated through local mechanisms of politics, should be held responsible. Since the early 1980s, Taipei has been made into "a real estate profit machine" and is a "world city" candidate in the Asia-Pacific region that, much like its cohort cities (Hong Kong, Singapore, Seoul, Manila, Bangkok, Kuala Lumpur, and Jakarta), strives to attract global capital, expand its hinterland, and invest in vast urban construction projects (Berner and Korff 1995; Douglass 1998a, 1998b; Friedmann 1986; Knight 1993; Boon Thong Lee 1998; Machimura 1992; Yeung 1996).

The prosperity of the real estate market started with Taipei's demographic growth. In 1946, the Taipei metropolitan area (including Taipei city and Taipei county) had a population of less than 800,000 (Tseng 1993, 85), but that figure had jumped to 6.1 million by 1999, amounting to nearly 30 percent of the total population of Taiwan. With such a concentration of population, the problem of housing and the demand for urban infrastructure soon loomed large. Unfortunately, because Taipei is situated in a basin, the geographical nature of the surrounding environment imposes limits on urban growth (Chiang and Hsiao 1985, 203). The sacrifice of environmentally sensitive areas to unbridled urban growth became "inevitable" (Chang 1993, 450). Though more than 70 percent of the hillsides around Taipei had been assessed as "improper" places to pursue construction work, a "mountain-removing, then town-building" movement still took place (Sun 1997, 76).

As major developers, the conglomerates and local factions, mainly consisting of political bosses in town, initiated the first wave of purchasing hillside lands around the Taipei basin in the early 1970s. In the next ten years, by working through their political connections, they played with the loose hillside-development regulations to change the designated land use from observed to developable lands, and they acquired permits to build. Though the national government implemented stricter hillside-development regulations in 1983, the new bylaws did not apply to those permits that had already been issued. Since the late 1980s, in response to the booming housing market around the Taipei metropolitan area, developers have side-stepped the new regulations, insisting their old permits still stand, and actually have built large hillside housing projects.

In the Taipei metropolitan area, the total investment from the private sector on hillside development projects in 1997 alone was estimated at more than US$360 million, an amount almost identical to the total expenditure on environmental protection from national revenue in the same year. It also was estimated that solely in one subdistrict of Taipei county, more than 100,000 housing units would be built through the hillside development projects approved in the previous ten years (Chen and Chen 1997). However, this massive amount of investment and development could not guarantee the quality of the housing units being built. Because of the corruption and incompetence of contractors and government officials, many newly built estates became life threatening. As people moved into the new hillside housing in the mid-1990s, tragedies followed. Incidents involving collapsing buildings were widely reported during the wet season each year. The two most notorious tragedies happened in 1997 and 1998. A mudslide in August 1997 took twenty-eight lives, injured fifty people, and completely or partially destroyed two hundred housing units (Wan 1998). As a result, twelve government officials received 5- to 9-year prison sentences for their abuse of power in illegally facilitating improper hillside development (Chen 1998). In October 1998, more than 10,000 households in Taipei county suffered from mudslides and floods and three lives were lost. The cost of urban expansion – land subsidence, deforestation, landslides, soil erosion, disturbance of the watershed, flooding, air pollution – was paid for by the suffering of ordinary citizens.

Unsustainable urban expansion does not happen naturally but is mediated through specific political mechanisms – namely, the unholy alliance between a pro-development state and the agents of the market (conglomerates and local factions). Tracing the history of the strategic coalition between the KMT government and conglomerates and local factions, formerly clients and now competitive partners of the KMT, one finds a clear pattern of "trading political loyalty for economic privileges." In certain economic fields, such as real estate, the government exclusively opened political channels for conglomerates to seek economic advantage in exchange for their political support (Chu 1989, 151; Hsiao and Liu 1997, 51; Chen 1995). Under such circumstances, public goods are inevitably privatized. Without the politically motivated laissez-faire attitude of an authoritarian regime toward land and housing policies, it is hardly possible that the investing of conglomerates and local factions in the real estate market could be so extensive and unconstrained. Taipei is a case in point.

In the last decade, one-fifth of the top 100 conglomerates in Taiwan have invested in the housing market, and local factions in Taipei county, on the outskirts of Taipei city, have owned a total of 161 construction companies and become involved in at least 230 large-scale development projects (Chen 1995, 161–7, 207). In 1998, 42 percent of Taipei county council members (25 out of 65) were directly engaged in real estate, redevelopment, and constructions. By controlling the urban planning and land use committees, council members affiliated with local factions and conglomerates facilitated unsound development projects, which brought in huge profits for the businesses of politicians and these local factions and conglomerates. The real estate coalition gained huge profit margins by obtaining title at low prices to protected areas not included in urban development plans (hillside or waterfront land, or tillage), then having the land status changed to residential or commercial; bidding for and obtaining newly released public land through political connections; and purchasing land adjacent to designated zones for national construction projects (public transit systems, highways, industrial parks) in advance.

Through manipulating urban planning agencies, conglomerates and local factions have been actively involved in the real estate market since the 1970s (Hsiao and Liu 1997, 46–53). The value of land multiplied by 184.7 times from 1952 to 1975, while the overall value of goods increased by only 4.2 times during the same period (Mi 1988, 112–13). From

1973 to 1981, the index of family income increased by only 3.1 times, but housing prices multiplied by 4 to 5 times, and land value increased by more than 30 times (Hsu 1988, 171). But these figures are not the worst part of the story. The most dramatic instance of speculation in Taiwan's history occurred in the late 1980s. The market price of housing multiplied by 2.5 times in various districts of Taipei between 1987 and 1988 alone. A typical two-bedroom apartment would cost a middle-income white-collar worker more than 20 years' salary (Hsiao and Liu 1997). Constrained by rising real estate prices, urban slums became the solution for low-income families to satisfy their need for affordable housing.

Taipei's land speculation fever in the late 1980s is an excellent example of the maximizing of urban land use at the expense of social and ecological considerations. This value-maximizing ideology is perpetuated in Taiwan's shift from a labor-intensive economy to a capital- and brain-intensive economy. It is also embedded in the concept that, according to the national plan of APROC (the Asia-Pacific Regional Operations Center, a project involving multiple cabinet-level agencies), Taipei is projected to change its role from the leading city in a single country to a regional economic capital whose hinterlands include Taiwan, part of Southeast Asia, and the coastal area of China. It aims to host the headquarters of transnational corporations, provide advanced services such as high-speed information exchange, and be capable of competing with and then superseding Hong Kong and Singapore.

In Taipei, as in other cities that strive to climb the hierarchy of world cities, both national and municipal governments have invested trillions of dollars in transportation infrastructure since the late 1980s: public transit systems, high-speed trains, expressways, connection lines to airports and seaports, and much more. Though environmental impact assessment procedures, required by Taiwanese environmental laws, were conducted in all cases, not one project was turned down. Moreover, the quality of public construction projects was no better than that of private development. Some major projects, such as the public transit system and highways, were filled with rumors of scandals and corruption that disgusted most Taipei residents.

As with profit-oriented hillside development initiated by the private sector, the single-minded

ambition of the Taiwan and Taipei governments to gain a bigger share in global prosperity has caused numerous nightmares. Excessive construction has not only heavily changed the physical appearance of the city but also altered its cultural and social landscape. In recent years, several officially designated cultural heritage sites (archaeological sites, landmarks or architecture with historical significance) have been legally and illegally damaged, if not totally destroyed. For example, the city administration once intended to tear down a first-class national relic, a fortress dating back 100 years, to facilitate the construction of an inner-city highway. After strong protests, the building was finally saved, but a strange, even heartbreaking, juxtaposition of the highway and the relic has resulted – the highway was built directly over the roof of the fortress, and a 50-centimeter distance between the two has been left to safeguard the integrity of the building.

Furthermore, to fulfill the requirements of a world city, regulations on land development and urban zoning have been loosened to create space for high-tech science parks; trade centers; locations for financial, banking, and other service industries; "smart" buildings; and exposition sites for hosting major world events. Not surprisingly, because of the government's redevelopment and rezoning efforts, traditional economic activities and residential areas have been disregarded to make way for developers, transnational corporations, and public agencies. In the process of urban restructuring stimulated by global economic forces, there have been within the last two years numerous public and private development and redevelopment projects, changing large-scale residential zones into commercial zones (Chi 1997).

Despite the increasingly detrimental impact on the environment of traffic congestion, overcrowding, and noise brought about by such changes, the billions of dollars in profits behind them made them attractive to the city government's policy makers. Still, it is because of this very urban restructuring process that the social and economic map of Taipei has been redrawn and that many urban dwellers either have undergone a painful loss of cultural identity or have been totally rubbed off the map. Some development projects commenced with the eviction of a whole village or community that had existed for an extended period of time. In most cases, monetary compensation was provided, but usually without a proper resettlement plan. While the material needs of the city are being met, the social bond, a rare treasure in an overcrowded human setting, and one that residents built over a long period, is in jeopardy and may even be lost forever.

LOCAL RESISTANCE: THE PROFILE OF TAIPEI'S ENVIRONMENTAL MOVEMENT

As a result of urban restructuring, a perception that living in this city has become both difficult and unbearable is widespread among its residents. For those who are affluent enough to own a house, buildings might crumble overnight thanks to wet-season landslides. Among those who have lower incomes, inadequate housing and LULUs (locally unwanted land uses) are prominent issues: the city authorities have threatened to sue the residents of slums for illegally occupying public land, and facilities that have raised health concerns (landfills, incinerators) have been situated near poor communities. Regardless of wealth, air pollution causes respiratory diseases across all generations, and particularly among children. To satisfy the booming demand for office space, recreational area is minimized and skyscrapers are built one after another. Furthermore, within the city proper everything is either overcrowded or under construction. Feeling suffocated by a lack of open space, urbanites desperately look for somewhere with green space and fresh air. Their search usually yields two results: first, such a place is hard to find, since the slopes of the hills surrounding the Taipei basin have been cleared on a massive scale; second, the traffic on the way to assured green spaces is appalling, as reflected in the statistic that the number of vehicles in Taipei increased from 772,297 in 1987 to 1,443,630 in 1996 (Department of Budget, Accounting, and Statistics 1997, 458–9).

Public concern over hillside development was also raised by the frequent mudslides and floods. According to a nationwide poll conducted in 1998, 93 percent of interviewees considered Taiwan's hillsides in general to be overdeveloped; 67 percent believed the government and the developers should be held responsible for the detrimental effects of overdevelopment; and 77 percent supported stricter development regulations. In another poll conducted that same year among Taipei residents only,

68 percent urged the government to completely prohibit further development on the Taipei metropolitan hillsides, and 70 percent believed that governmental officials and developers jointly engaged in land speculation (Fei 1998; Tung 1998). All of these factors – hyperurbanization, horrendous air pollution, traffic congestion, housing shortages, inadequate transportation, insecure housing conditions – have led to a diminished quality of urban life in Taipei and increased physical and mental stress for the city's residents.

Alienated and blasé city dwellers . . . have found ways to collectively vent their pain and anger by taking part in nationwide environmental struggles. It is well noted that during the 1980s, Taiwan experienced a surge of massive civil protests in which citizens attempted to gain political rights (freedom of speech, association, and demonstration) repressed under colonial and authoritarian rule for nearly a century and to express their concerns on pressing social issues such as the environment, gender inequalities, labor, welfare, and human rights. Collective action against environmental deterioration is no doubt one of the "early risers" in the process of democratization and has played a critical role in both demonstrating "the vulnerability of authorities" and diffusing "a propensity for collective action" to other social groups with different concerns (Tarrow 1994, 155). . . .

Local environmental protesters have severely questioned the state's pro-development and growth-centered policies and have fully developed a repertoire of collective action, which includes sit-ins, hunger strikes, blockades, religious parades, and theatrical performances. Environmental protests soon became a model for other movements to follow, playing a role similar to that of the civil rights movement in the United States (McAdam 1988, 1999).

As part of the nationwide environmental movement, hundreds of protests took place in the Taipei metropolitan area between 1980 and 1996, and citizens have made explicit demands for a livable environment and a higher-quality urban life. . . . Of the 274 protests in the Taipei metropolitan area between 1980 and 1996, 46 percent concerned government sponsored development projects; 16.8 percent opposed development and construction projects proposed by the private sector. Protests against all development projects in Taipei amounted to more than 60 percent of total protests. . . . It is clear that the evolution of protest in the Taipei metropolitan area tracks the changing macroeconomic context described in the previous section. Despite the nonindustrial characteristic of Taipei's economy, the chief target of environmental protests in the early 1980s (55.6 percent) was industrial pollution from the private sector, especially the chemical industry. Between 1986 and 1990, the percentage of protests against industrial pollution remained high (60 percent), but the private business sector no longer was the sole target (32.5 percent): state-owned enterprises and public construction projects emerged as other important adversaries (27.5 percent).

In the 1990s, protests began to confront the attempts of real estate developers, politicians, and bureaucrats to make Taipei into a world city. The percentage of protests against specific industrial pollution dropped significantly, but protests against development projects, whether state sponsored or private, increased from around 40 percent in the late 1980s to nearly 70 percent between 1991 and 1996.

Though the issues and the strategies of each protest case may have differed, local and community-based mobilization is nevertheless the common characteristic. This confirms the repeatedly highlighted significance of "locality" in environmental movements worldwide (Castells 1997; Diani 1995). The citizens' "war for survival" mainly responds to excessive development and the obvious deterioration of the urban habitat. In the context of Taipei's struggle, the conflicting interest between citizens and global-domestic capital surfaced, and the tension over redistributing spatial resources among diverse social groups exploded. Citizens openly claimed their right to control over their immediate environment and strove to discover and elaborate on the meaning of community.

By means of these protests, Taipei residents have not only expressed their discontent over a deteriorating urban environment, but also challenged the top-down political structure that has long deprived urban dwellers of their right to participate in the decision-making process of public affairs. In most cases, the residents were "informed" by the government about changes in the designation of land use or about the introduction of locally unwanted facilities after the decision had been made. The citizens were outraged not only by the perceived immediate threat brought about by these development projects, but also by the undemocratic, top-down decision-making machinery that had approved them. Thus, these protests also reflect the residents' demands for institutional change toward participatory democracy in the city's public life. In the process of struggle, local communities have served as agents of environmental action, reflecting not only the immediate interest of their communities, but also the potential for their action to contribute to a more sustainable and just urban environment overall.

COMMUNITY AND GOVERNMENTAL RESPONSIVENESS

In responding to the negative effects caused by urban restructuring, communities rebelled. Ching-Cheng and Kang-Le, both protest cases cited at the beginning of this paper, articulated the idea of "a city for citizens" and aimed to defend their living spaces. Both the middle-class neighborhood and the slum community were pressured by urban redevelopment and the overdue need for public facilities and infrastructure.

Other communities also have been sensitive about hillside development and landscape preservation issues. Wan-Fang, another middle-class neighborhood, campaigned to prevent the Public Housing Office from situating a large-scale housing project on a nearby hillside that experienced landslides during the wet season. Chihshan Yen, with a high proportion of female residents who are members of the Homemakers' Union Environmental Protection Foundation (a female-dominated national environmental NGO), showed strong opposition to the construction of a gas station that, it was feared, might severely damage a historical ruin nearby and lower the quality of life of the community (Yu 1994, 43–56).

The rise of the community-based protests illustrates the efforts of community residents to preserve social bonds and place-specific identity, and to fight against unwanted development projects that usually favor the few over the many. Their demands for stricter environmental sanctions, cultural preservation, livable habitats, and institutional access to decision-making processes have put enormous pressure on government at all levels and finally changed the landscape of local politics. Facing ongoing direct action from these communities and electoral competition among political parties in Taiwan's fledgling democracy, the incumbent parties of both national and municipal governments could not afford to ignore these explicit demands to halt unwelcome urban development and unsound land use on hillsides.

At the national level, island-wide environmental protests have directly contributed to a wave of legislative and institutional reform since the late 1980s. The Pollution Disputes Resolution Law and Environmental Impacts Assessment Act were passed, and the Environmental Protection Basic Law is being reviewed in the legislature (Chi *et al.* 1996). Also, the Environmental Protection Administration (EPA) was upgraded to a cabinet-level agency in 1987. Due to pressure from lobbying by many environmental NGOs, stricter environmental regulations were adopted. In the case of deforestation and landslides, for example, the Ministry of Interior lowered the permitted angle of hillside development twice in one year and thus largely reduced the total area permitted for development.

At the municipal level, party competition in each local election has increased voter sensitivity (note the campaign slogan from various candidates, "citizen first"). This new trend has been more salient since the December 1994 change from appointment by the national government of the Taipei city mayoral post to direct election of this official. Any dubious connection with conglomerates and other local business interests is now viewed as a liability, if not poison, for a candidate in any local election. "Greening" the political platform has become inevitable, and it accurately reflects the rising public demand for a higher quality of urban life. As a result, in the first direct mayoral election, a pro-environment candidate from the Democratic Progressive Party (DPP, the biggest opposition party in Taiwan) was elected. At the same time, in the city council election, quite a number of candidates who had never been involved in plutocracy defeated the longtime representatives of or collaborators with conglomerate interests. The political efficacy of community protests is thus being successfully channeled through the system of representative politics. The question, then, is: What do elected officials do to prove to their voters that they are working toward a livable city? The magic answer is: Build more city parks.

According to a statistic from 1993, the area of park per citizen in Taipei at the time was 2.5 square meters, smaller than in other metropolises such as New York City (13.95), Paris (12.70), Seoul (8.70), and Tokyo (4.70) (Hsiao *et al.* 1997, 12). It is thus not surprising that in almost all campaign packages in recent elections, city parks have been one of the favorite topics of all candidates. With the urgent need for open space, a formula of "parks equal votes" was adopted. The current and previous mayors have strenuously worked toward building city parks during their tenures, no matter what conflicts the process might attract. The construction of the No. 7 city park in the early 1990s resulted in the eviction of thousands of slum dwellers, as did the Nos. 14–15 park. Between 1991 and 1996, the two mayors built sixty-seven parks, and the total area of parkland increased from 805.8 hectares to 979.4 (Department of Budget, Accounting, and Statistics 1997, 452). The city government was also expected to add at least ten more parks in 1997 (Yang 1997).

In addition to city parks, several pro-community programs were set up. Following his election victory, Chen Shui-bian appointed the first environmentalist with a national reputation to take charge of the local EPA. Also, the city government's Department of Urban Development proposed various redevelopment projects that showed concern for the preservation of traditional landscape and historical architecture. For example, two old and historically significant districts of Taipei city (Tihua Street and the Tatung District) are expected to be economically revitalized and to attract cultural tourism. Each year, ten to fifteen communities are officially sponsored to improve their physical environment and to build stronger community networks. Tougher measures also have been enacted to regulate hillside development at the city government level.

Protest, then, has affected the political arena both by changing the character of political discourse around environmental issues and by forcing modification of the formal rules governing urban development. The municipal government has indeed responded to demands of urban environmental protests, stressing community autonomy and the importance of protecting environmentally sensitive areas by initiating various programs to facilitate community involvement. These effects, however, must be considered only a first step toward making Taipei more livable. Changes in rules and discourse do not necessarily mean changes in practice and behavior "on the ground." Equally important, efforts in the direction of livability continue to have very different implications for poor and middle-class communities.

THE POLITICS OF PROTEST

At first glance, it seems as if citizens' concerns over urban land use were addressed through the tougher measures adopted to inhibit hillside development and by several other governmental initiatives. Tempered by democratization, the top-down nature of urban planning has somehow been softened by the creation of new mechanisms for community participation. The mayor and county magistrate, pressured by the consideration of re-election, must take community voice into account and prevent urban-planning bureaucrats from exhibiting behavior unfriendly to communities.

However, after closer examination, one finds that problems still remain. Unequal treatment of communities prevails, the government's capacity to implement its new policies is highly problematic, the "development first" mentality still lingers, and the vision of becoming a world city has yet to be realized. The positive policy responses toward environmental causes have been tempered by an increasing number of huge infrastructure projects and district redevelopment plans that were approved and quickly and aggressively commenced. . . .

SOCIAL CLASS AND THE EFFICACY OF MOBILIZED COMMUNITIES

Since urban environmental protest is location-specific and community-based, the social attributes of each community to some extent affect its bargaining power with the authorities and adversaries in the course of protest. From various case studies of Taipei's community and environmental movements, it is clear that the class composition of a community is closely associated with the readiness of available resources, the span of information

networks, the extent of media coverage, and the level of government responsiveness (Shen 1994; Yu 1994). Middle-class communities are usually in a better strategic position to attract media coverage and to generate public support. They are also more likely than poor communities to receive a fast government response and to seek assistance from outside NGOs.

In contrast, a poor or slum community is indeed more vulnerable when it encounters powerful social institutions. If poor communities get any attention from the media, it is likely to take the form of distorted and negative coverage, and they are much less successful in dealing with the authorities. Let us take as an example the case of the Kang-Le community. In opposing the eviction of thousands of residents, Kang-Le formed a self-help organization in 1992 and desperately looked for a way to balance the citywide demands for public parks and the needs of slum residents. Before the professors and students from the Graduate School of Architecture and Urban–Rural Planning Studies at National Taiwan University got involved to help organize the anti-bulldozer campaign in early 1997, very few media outlets had paid attention to this story. When the press began to look into the dispute, the government authorities occupied center stage in the news coverage and marginalized the voice of the shantytown residents. Some municipal government officials even used the media to distort the Kang-Le community's image. Slum residents were portrayed as a group of "greedy rich people" who had illegally resided on public land for a long time and who were going to be generously paid up to 52 million US dollars in compensation out of the taxpayers' pockets. They were also criticized for having deprived decent citizens of the right to a park.

Similar problems have been faced by urban Aborigines. Throughout the history of Taiwan, the Aborigine rights to land, natural resources, and cultural heritage have been largely ignored. On the one hand, the officially "reserved" homelands of the Aborigines have borne the weight of public construction projects (mostly dams and, in the case of Orchid Island, a nuclear-waste storage facility). Aboriginal opposition to these projects resulted in very few concessions from the state. On the other hand, having undertaken the process of rural–urban migration in the hope of escaping the troubled mountain economy, urban Aborigines still face the nightmare of eviction. Aboriginal migrants who moved to Taipei illegally built their ethnic settlements either on riverbanks or on hills, where the natural environment was similar to their homeland. These small, scattered ethnic settlements suffered from extremely poor living conditions; some were even without electricity and water, let alone more luxurious amenities. The rationale to evict the Aboriginal shantytown dwellers was quite similar to that which evicted the Kang-Le community: public safety, the retrieval of state-owned property, and the need for public works (in this case, high-speed trains). The Aboriginal struggle with the government received even less media coverage and public attention than did Kang-Le's, and the government response to their concerns was very unfavorable.

In sum, although in principle environmental protests have consistently advocated the right of people to have a say regarding their immediate environment, in practice the claim to that right by different communities is weighed unequally by politicians, the media, and the public. The way each community's requests are dealt with depends very much on the status of the community in the pyramid of power, which is structured around class and ethnicity.

THE MINIMAL INVOLVEMENT OF TRANSLOCAL NGOS

It is intriguing to note that translocal NGOs, whose membership is mainly of the middle class, have played a limited role in the urban struggle, in contrast to the aggressiveness of local communities. According to a survey of environmental organizations conducted in 1997, roughly one-third of Taiwan's environmental NGOs had established their headquarters in Taipei, giving this metropolis the highest density of organizational networks in Taiwan (Hsiao 1997). However, three-quarters of Taipei's NGOs identified themselves as "national organizations" that focus on broader issues, and they are therefore inattentive to local affairs and urban protests in Taipei. Only a few NGOs had half an eye on urban environmental issues. The major interests of these NGOs were recycling, vehicle usage, and hillside development, and furthermore, they unanimously adopted soft-line approaches to

change, such as lobbying, instead of actively allying themselves with local communities. Even in the case of the Chihshan Yen community, the involvement of the Homemakers' Union Environmental Protection Foundation derived more from an overlap in membership than from a primary concern on the part of the NGO for the preservation of Taipei's cultural landscape.

It is not unusual that a division of labor exists between translocal NGOs and local communities within the larger picture of environmental movements. Over the years, Taiwan's NGOs, backed by lobbyists and by think tanks consisting of academic recruits, have played a translating role to reframe local concerns into significant issues in the public sphere. In many cases, combining NGO access to the political system with the pressure derived from residents' direct action has yielded effective and favorable results. For example, communities in southern Taiwan and translocal NGOs have allied with one another to fight successfully against an incoming petrochemical and steel-refining industrial park, to secure the livelihoods of local fish farmers and fishermen, and even to protect endangered wetlands and a threatened species of bird, the blackfaced spoonbill. However, there is no link between NGOs and mobilized communities in Taipei's urban struggle.

Eighteen out of 33 of Taipei's NGOs said that the quality of the urban environment was not even one of their major concerns. Why have Taipei's NGOs shown negligence on urban environmental issues and maintained such a distance from urban communities? What has made Taipei a special case in contrast to the alliances between NGOs and anti-pollution communities elsewhere? To answer these questions we need to consider the following issues: the pro-DPP sentiment among environmental NGOs and these NGOs' perception of "what counts as an environmental issue."

In Taiwan, visible nationwide environmental NGOs fall into roughly two categories: research-oriented groups that specialize in public policy and present their perspective through moderate means such as lobbying and environmental education campaigns, and action-oriented groups that engage in community mobilization and are prone to express their environmental concerns through noninstitutional and confrontational strategies. Both types of group have similarly decent annual budgets and a membership disproportionately made up of academics, but the real difference lies in organizational size. The first type, usually with an elite membership of fewer than 50, acts as a body of consultants for policy makers and never directly participates in local protests. The second type, with a much larger membership pool, tends to bring local concerns into higher-level policy debates and actively engages in protests against large, government-sponsored development projects.

During the struggle against the authoritarian KMT state in the early 1990s, DPP and many action-oriented NGOs developed and cultivated an issue-based comradeship and an overlapping membership. It is well documented that many outspoken leaders of these NGOs are in fact members or supporters of the DPP (Weller and Hsiao 1998). In the 1990s, when the DPP and its incumbents occasionally did not maintain their firm stand on environmental causes, those action-oriented NGOs found it difficult to fight against old friends, except in extreme cases. For example, a DPP county magistrate was strongly attacked because the construction of Taiwan's seventh naphtha cracker (an industrial complex where crude-oil products are refined into the precursors of plastics), a project that environmental NGOs had strongly opposed for years, proceeded partially at his discretion. Besides this extreme case, most NGOs have been friendly toward DPP administrations, and the cooperation between the two has been pervasive. In Taipei's case, instead of working with protesting communities, quite a few environmental NGOs came forward to support DPP mayor Chen Shui-bian's campaign to increase green space. They also endorsed a citywide recycling project and community improvement plans proposed by the city government.

To most environmental NGOs, urban restructuring and changes in urban land use are not really environmental issues but rather technical matters related to urban planning. Compared to the concern of NGOs for forests, rivers, the ocean, wetlands, wildlife, and much more, the urban habitat is rather orphaned – a point that has been nicely elucidated by Harvey (1996, chs 13–14). While making a firm stand against hillside development because of resulting deforestation and soil erosion, NGOs still have difficulty placing other urban issues, such as slum clearance, on their list of priorities. It is thus not surprising that they have remained silent on issues relating to the urban poor. Attentive to a narrowly defined "natural environment," NGOs somehow neglect the "unnatural" urban setting and remain largely uninvolved in many crucial urban battles. It is rather ironic to note that a movement and numerous NGOs, which "exhort the harmonious coexistence of people and nature, and worry about the continued survival of nature (particularly loss of habitat problems), somehow forget about the survival of humans (especially those who have lost their 'habitats' and 'food sources')" (Harvey 1996, 386).

COMMUNITY INTEREST, SUSTAINABILITY, AND SOCIAL JUSTICE

The third and probably the most important issue in the politics of protest is how Taipei's locally driven protests relate to the long-term sustainability of the city. Is it hypocritical that middle-class communities oppose highway construction and refuse to sacrifice their community parks for parking garages while their residents intensify their use of private automobiles? Is it narrow-minded to boycott the nearby construction of incinerators while the quantity of garbage that each citizen produces increases every year? Is it selfish for slum dwellers to try to save illegal shacks from being torn down regardless of the need for green space? Is it a public-safety problem that Aborigines build their ethnic settlements on the riverbank? In other words, are local communities waving the flag of environmentalism just to protect their own interests? Or can local interest be congruent with ecological concern across a city? On the road toward a more livable city, how is it possible to balance the subsistence of the urban poor and the sustainability of the city?

To praise all community protests as heroic and progressive is as problematic as to blame them for being insular and self-serving. Either romanticism or cynicism may generate an emotional response, but neither provides us with a vision. The motivation behind local protests might be led by self-interest, but once channeled through proper means, these local interests may be translated into larger issues of public concern. The fact that most local protests are interest-driven does not necessarily prevent them from contributing to urban sustainability.

Let us take highways and parking garages as examples. As Friedmann has put it (1997, 18): "for an Asian city to replicate Los Angeles' love affair with freeways . . . is to commit collective suicide." In a city like Taipei, with an average density of 9,586 persons per square kilometer (Department of Budget, Accounting, and Statistics 1997), and with some districts even reaching 25,000 persons per square kilometer (Chang 1993, 432), to protest the excessive construction of highways and parking garages is to try to rein in the overuse of private vehicles and encourage the use of more environmentally sound methods of mass transportation. The same reasoning may apply to incinerators. When people refuse to accept an incinerator as their neighbor, we should not immediately denounce them as engaging in NIMBY behavior or as selfish, because their recalcitrance could point to better ways of solving the garbage problem. Instead of endlessly building incinerators and landfills, a task that Taiwan's EPA has been undertaking, one can argue that in the face of insufficient garbage processing facilities, it is more sensible to strengthen the system of recycling and to persuade citizens to voluntarily reduce their garbage output. It is undeniable that public discussion of stringent controls on private vehicles and a citywide recycling plan "was aroused in the wake of these selfish" community protests.

Another issue concerns the conflicting interests of the urban poor and society as a whole. In order to shed light on this sensitive issue, we would first like to go back to the conception of "sustainability." Regardless of its wide variety of connotations, one basic point of consensus on the meaning of sustainability is the concept of "intergenerational justice": in pursuit of a livelihood, the happiness of future generations should not be compromised or sacrificed (Friedmann 1997; Piccolomini 1996). If the pursuit of sustainability should take into account the quality of life of those who are not yet born, then the subsistence and quality of life of the urban poor should not be disregarded in satisfying the needs of other citizens. Like two sides of a coin, in theory or in practice, sustainability should not and cannot be separated from a more just social system. To increase the area of green space might help to mitigate a city's air pollution, but to use that as the justification for dismantling a poor community in its entirety is untenable. If it does not

take into account the unequal distribution of economic goods and power among different groups in a society, an ostensible environmental or ecological solution might unintentionally lead to dreadful social consequences (Beck 1995; Harvey 1992, 1996; criticisms on thoughtless slum demolition projects are also detailed in Gans 1962).

By re-examining the case of the Kang-Le struggle, we discover that the effort to balance both social and environmental demands has failed and that, even more serious, a fast-food approach to improving the urban environment has overwhelmed considerations of social justice, a value that must be maintained on the path toward a sustainable city.

Learning from the classic case of Orcasitas, in Madrid, the urban planners who supported the Kang-Le community's cause made a proposal that would have kept the city park project on course but also saved the community. They said that "the occupants of the illegal shantytown settlements who were . . . responsible for utilizing the area and increasing its value should be the first to benefit once the area had been redeveloped" (Castells 1983, 2–53). In this proposal, 6 percent of the designated park area would have been used to build affordable government-sponsored housing to relocate the aged and poor residents living on the same site, and the design would have turned the park into an ecomuseum, evoking memories of the area's past by including features of the Japanese cemetery and slum architecture. However, the proposal was turned down by the city's top officials. Their refusal rested on the argument that the resettlement proposal favoring slum dwellers had no legal basis and would only stir up "unrealistic expectations" among slum dwellers in other areas. Immediately after the demolition, there was a 15 to 20 percent increase in rental and land value surrounding the Nos. 14–15 park (Coalition against the City Government Bulldozers 1975, 11). This further confirms that an ecological vision such as a park system "was easily co-opted and routinized into real-estate development practices for the middle classes" (Harvey 1996, 417).

CONCLUSION

We have tried to set out the central dynamics of the evolution of Taipei's environmental movement, showing both how the movement reflects the changing character of the larger Taiwanese political economy and how the character of the movement itself limits the kinds of changes that it can achieve. The task of enacting change has not been an easy one, since the movement's overall character emerges out of myriad small, heterogeneous actions. Nonetheless, we have tried to show that there is an overall logic to the movement, one whose limitations must be balanced against its undeniably positive potential.

A simplified version of Taipei's struggle can be described as follows: Both the national and municipal governments engaged in a large project of urban space restructuring in order to attract foreign capital, and the KMT government was keen to provide the private sector with incentives to maximize land use on hillsides. Residents later found out that this impressive government-inspired project promised everything but a livable urban environment. Foreign and domestic capital was placed in the "growth machine," leading to assaults on the natural environment, historical buildings, and the living space of various groups of residents. Feeling betrayed and hurt, unhappy residents mobilized neighbors, friends, and relatives to call their local representatives, to vote for other parties, and even to demonstrate in the street. Some translocal environmental NGOs noted this trend and passively participated in a few cases of community mobilization.

Though the governments initially had been unresponsive in meeting resident demands, the voice of the community grew too powerful and the threat of political backlash was serious; thus, several policy revisions were made. Both the national and municipal governments decided to inhibit excessive hillside development, but vast urban infrastructures would proceed as planned. To meet rising social grievances, the municipal government tried hard to build more city parks as an extra bonus for citizens. By doing so, it turned slum dwellers into the victims of urban improvement. . . .

Taipei's experience of urban deterioration is by no means unique. Leading cities in East Asia have been haunted by land speculation and unsound land use. They compete among themselves in the construction of attention-grabbing but wasteful urban infrastructure projects. What makes Taipei special is that civil protests against these macro-economic processes are facilitated by Taiwan's democratization. Institutionally consolidated party competition provides various communities with more means to address their locally driven interests. Through these means, mobilized communities were finally able to put the long-neglected issue of "urban sustainability" on the national and city agendas. The responsiveness of both the national and municipal governments has proven the political efficacy of community action. In turn, community mobilization has been encouraged by a more responsive and democratic regime – that is why the number of protest cases has increased with time.

One benefit of democracy has been that the voice of Taipei's residents has been heard. However, because of the inequality of power between different social groups, the demands of some communities have been met and those of others blocked. If modern cities are usually portrayed as the center of wealth, power, information, and cultural production, we should not forget that the sharpest disparity between the rich and the poor and between the powerful and the powerless also takes place in urban settings. While demands in the form of environmental protests for a higher quality of urban life have brought positive institutional and legal reforms, poor and slum communities have suffered. The extent to which these proenvironment decrees can restrict excessive public and private development is still unknown; in the meantime, shantytown settlements have already been torn down and traded for city parks.

The dynamics of Taipei's urban environmental struggles have important implications for other Third World urban centers that aspire to be "world cities." In order for their environmental movements to pursue a broader vision of livability, they must structure their actions and their ideology to incorporate more horizontal links between mobilized communities and NGOs. The NGOs themselves need to acknowledge the central importance of dealing with complex urban systems. Most important, the livelihood of poor and slum communities must be incorporated into the concept of sustainability; the broader ecological movement must be more sensitive to equality. The idea that the improvement of the urban environment cannot be divorced from the realization of social justice must become a more integral part of movement ideology. In the end, a city cannot become more sustainable without also being more just.

NOTE

1 The authors' names are in alphabetical order, and each author contributed equally to this collaboration. The authors thank Peter Evans, Mike Douglass, Dungsheng Chen, and two anonymous reviewers for their constructive comments. They are also grateful for editorial assistance from Martin Williams and Fan Chang.

REFERENCES

Beck, Ulrich. 1997. *Ecological Politics in an Age of Risk.* Cambridge: Polity Press.

Berner, Erhard and Rudiger Korff. 1995. "Globalization and Local Resistance: The Creation of Localities in Manila and Bangkok." *International Journal of Urban and Regional Research*, 19, pp. 208–222.

Castells, Manuel. 1983. *The City and the Grassroots: A Cross-cultural Theory of Urban Social Movements.* Berkeley, CA: University of California Press.

—— 1997. *The Power of Identity.* Vol. 2 of *The Information Age: Economy, Society, and Culture.* Oxford: Blackwell.

Chang, Shih-chiao. 1993. "The Experience of and Out-look for Taiwan's Urban Environment and Development" (in Chinese). In *Chinese Cities and Regional Development: Prospect for the Twenty-first Century.* Ed. Yueman Yeung, pp. 425–542. Hong Kong: Hong Kong Institute of Asia-Pacific Studies, Chinese University of Hong Kong.

Chen, Cheng-Tzeng and Chun-Kuang Chen. 1997. "The Minister and County Magistrate Calls a Stop to Hillside Development" (in Chinese). *China Times*, August 19.

Chen, Chung-Hsiung. 1998. "Heavy Sentences for County Officials and Developers: The Case of Lincoln Hillside Development" (in Chinese). *China Times*, August 19.

Chen, Dung-Sheng. 1995. *The City of Money and Power: A Sociological Analysis of Local Factions, Conglomerates, and Urban Development in Metropolitan Taipei* (in Chinese). Taipei: Ju-Liu Publishing.

Chi, Chih-ko. 1997. "Urban Planning: We Should Listen to the Voice of the People" (in Chinese). *China Times*, July 21.

Chi, Chun-Chieh, H.H. Michael Hsiao, and Juju Chin-Shou Wang. 1996. "Evolution and Conflict of Environmental Discourse in Taiwan." Paper presented at the Association of Asian Studies annual meeting, Honolulu, Hawaii.

Chiang, Nora Huang and Hsin-Huang Michael Hsiao. 1985. "Taipei: History and Problems of Development." In *Chinese Cities: The Growth of the Metropolis since 1949.* Ed. Victor F.S. Sit, pp. 188–209. New York: Oxford University Press.

Chu, Yun-han. 1989. "The Oligarchic Economy and the Authoritarian Political System" (in Chinese). In *Monopoly and Exploitation: The Political Economic Analysis of Authoritarianism.* Ed. Hsin-Huang Michael Hsiao *et al.*, pp. 139–160. Taipei: Taiwan Research Fund.

Coalition against the City Government Bulldozers. 1975. *My Home is at Kang-Le District: The Documents of the Movement against Bulldozers of the City Government* (in Chinese). Taipei: Coalition against the City Government Bulldozers.

Department of Budget, Accounting, and Statistics. 1997. *The Statistical Abstract of Taipei City, 1997.* Taipei: Department of Budget, Accounting, and Statistics, Taipei City Government, Republic of China.

Diani, Mario. 1995. *Green Networks: A Structural Analysis of the Italian Environmental Movement.* Edinburgh: Edinburgh University Press.

Douglass, Mike. 1998a. "World City Formation on the Asia Pacific Rim: Poverty, 'Everyday' Forums of Civil Society, and Environmental Management." In *Cities for Citizens: Planning and the Rise of Civil Society in a Global Age.* Eds. Mike Douglass and John Friedmann, pp. 107–137. New York: Wiley.

Fei, Kuo-Jen. 1998a. "Seventy percent of Taipei Residents urge a Complete Prohibition of Hillside Development" (in Chinese). *China Times*, October 27.

—— 1998b. "East Asian Urbanization: Patterns, Problems, and Prospects." Discussion paper, Institute for International Studies, Asia/Pacific Research Center, Stanford University.

Friedmann, John. 1986. "The World City Hypothesis." *Development and Change*, 17, pp. 69–83.

—— 1997. *World City Futures: The Role of Urban and Regional Policies in the Asia-Pacific Region.* Occasional Paper no. 56. Hong Kong: Hong Kong Institute of Asia-Pacific Studies, Chinese University of Hong Kong.

Gans, Herbert. 1962. *The Urban Villagers: Group and Class in the Life of Italian-Americans.* New York: Free Press.

Gereffi, Gary. 1990. "Paths of Industrialization: An Overview." In *Manufacturing Miracles: Paths of Industrialization in Latin America and East Asia*. Eds. Gary Gereffi and Donald L. Wyman, pp. 3–31. Princeton, NJ: Princeton University Press.

Haggard, Stephan. 1990. *Pathways from the Periphery: The Politics of Growth in the Newly Industrializing Countries*. Ithaca, NY: Cornell University Press.

Harvey, David. 1992. "Social Justice, Post-modernism, and the City." *International Journal of Urban and Regional Research*, 16 (4), pp. 589–601.

—— 1996. *Justice, Nature, and the Geography of Difference*. Malden, MA: Blackwell.

Hsiao, Hsin-Huang Michael. 1997. *A Symbiotic Relationship with Tension: The Relationship between EPA and Local Environmental Groups* (in Chinese). Taipei: Environmental Protection Agency.

Hsiao, Hsin-Huang Michael and Hwa-Jen Liu. 1997. "Land–Housing Problems and the Limits of the Non-homeowners Movement in Taiwan." *Chinese Sociology and Anthropology*, 29, pp. 42–65.

Hsiao, Hsin-Huang Michael, *et al.* 1997. "Re-creating the New Functions of Urban Development: An Analysis on the Eviction Case of the Nos. 14–15 Park in Taipei" (in Chinese). Unpublished report sponsored by the Department of Social Affairs, Taipei Municipal Government.

Hsu, K'ung-Jung. 1988. "A Sociological Analysis of the Housing Market in Peripheral Taipei" (in Chinese). *Taiwan: A Radical Quarterly in Social Studies*, 1 (2–3), pp. 149–210.

Huang, U.-An. 1983. *The History of Taipei's Development* (in Chinese). Taipei: Taipei Archive Committee.

Knight, Richard V. 1993. "Sustainable Development – Sustainable Cities." *International Social Science Journal*, 45 (1), pp. 35–54.

Lee, Boon Thong. 1998. "Globalization, Telerevolution and the Urban Space." Paper presented at the Workshop on Southeast Asia under Globalization, the Program for Southeast Asian Area Studies (PROSEA), Academia Sinica, Taipei.

Machimura, Takashi. 1992. "The Urban Restructuring Process in Tokyo in the 1980s: Transforming Tokyo into a World City." *International Journal of Urban and Regional Research*, 16 (1), pp. 14–28.

McAdam, Doug. 1988. *Freedom Summer*. New York: Oxford University Press.

—— 1999. *Political Process and the Development of Black Insurgency, 1930–1970*. 2nd ed. Chicago: University of Chicago Press.

Mi, Fu-Kuo. 1988. "Public Housing Policy in Taiwan" (in Chinese). *Taiwan: A Radical Quarterly in Social Studies*, 1 (2–3), pp. 97–147.

Mumford, Lewis. 1989 [1961]. *The City in History: Its Origins, its Transformations, and its Prospects*. New York: MJF Books.

Piccolomini, Michele. 1996. "Sustainable Development, Collective Action, and New Social Movements." *Research in Social Movements, Conflict, and Change*, 19. pp. 183–208.

Shen, Yao-pin. 1994. "Community Mobilization and the Transformation of Urban Meaning: The Case Analysis of Taipei's Ching-Cheng Community" (in Chinese). Master's thesis, Graduate School of Architecture and Urban–Rural Studies, Nation of Taiwan University, Taipei.

Sun, Hsiu-huei. 1997. "They Didn't Know the Buildings Were Dangerous, They Only Saw Them Collapse" (in Chinese). *Global Views Monthly*, 13 (5), pp. 76–77.

Tarrow, Sidney. 1994. *Power in Movement: Social Movements, Collective Action, nd Politics*. New York: Cambridge University Press.

Tseng, Hsu-Cheng. 1993. "The Formation of the Taipei ren [person]" (in Chinese). In *Research on the Immigration into Taipei County*. Eds. Hsin Huang Michael Hsiao *et al.*, pp. 79–121. Taipei: Taipei County Culture Center.

Tung, Mong-Lung. 1998. "Ninety percent of People Interviewed consider that Hillsides have been Overdeveloped" (in Chinese). *China Times*, July 22.

Wan, Zen-Quai. 1998. "No Hope for Monetary Compensation and Resettlement: The Victims of Mudslide Area Trapped" (in Chinese). *China Times*, August 19.

Weller, Robert P. and Hsin-Huang Michael Hsiao. 1998. "Culture, Gender, and Community in Taiwan's Environmental Movement." In *Environmental Movements in Asia*. Eds. Arne Kalland and Gerard Persoon, pp. 83–109. Honolulu: University of Hawaii Press.

Yang, Chin-yen. 1997. "Chungshan No. 1 Park: Completed a Year Ahead of Schedule" (in Chinese). *United Daily*, October 23.

58

"Sustainable Urban Development in China" (2013)

Kang-Li Wu

Editors' Introduction

Nations such as China and India are urbanizing rapidly, and with their large populations and enormous migrations of rural residents into cities raise particular challenges for sustainable urban development. In this selection, Kang-Li Wu highlights both challenges and opportunities for China. He points out many ways that the Chinese government is taking action, but also some of the obstacles to ecological city development. Wu is a professor at the Harbin Institute of Technology in northern China, and Executive Director of the HIT-UC Berkeley Joint Research Center of Sustainable Urban Development.

Other materials related to sustainable urban development in China include *Sustainable Low-Carbon City Development in China* (Washington, D.C.: World Bank, 2012), edited by Axel Baeumier, Ede Ijjasz-Vasquez, and Shomik Mehndiratta; Paul G. Harris' *Environmental Policy and Sustainable Development in China: Hong Kong in Global Context* (Bristol, UK: The Policy Press, 2012); John Friedmann's *China's Urban Transition* (Minneapolis: University of Minnesota Press, 2005); Thomas J. Campanella's *The Concrete Dragon: China's Urban Revolution and What It Means for the World* (Princeton: Princeton Architectural Press, 2011); and Jingzhu Zhao's *Towards Sustainable Cities in China: Analysis and Assessment of Some Chinese Cities in 2008* (New York: Springer, 2011).

With more than a quarter of the population of the world, extremely rapid urbanization, urban sprawl, and rapid economic growth, China faces more serious sustainable urban development challenges than most other countries. However, with the growth of domestic consumer demand, the increase of citizens' environmental consciousness, and the strong motivation of new political leaders, the road to sustainable urban development in China is full of opportunities as well.

During the past 20 years, China has transformed itself from a relatively poor country to a middle-income nation and global economic power. During its change from a planned economy to a largely market-driven system, the country has experienced extremely rapid urbanization, massive migration from rural areas to cities, institutional changes of planning and political organization, and growing gaps between urban and rural quality of life. It is hard for anyone to imagine the speed and scale of urban development in China if they have not witnessed the situation for themselves. Nor can any modern planning theory fully describe and interpret this kind of urban development. Behind the physical construction of modern cities in China there are of course growth pains. In fact, the whole process is full of conflicts caused by various interests as well as the adjustment of planning and social values.

In March 1994 the executive meeting of the State Council of China passed *The China's Agenda 21 – The White Paper on China's Population, Environment and Development in the 21st Century*,[1] and submitted this document to the UN Sustainable Development Commission. Subsequently the government submitted *The People's Republic of China National Report on Sustainable Development*[2] to the Special Session of the U.N. General Assembly in 1997. This report clearly states that the Chinese government wants to use comprehensive development and management of the urban environment and construction of residential districts as key tasks of sustainable development.

At that time China's large cities had begun to experience rapid urbanization fostered by rapid economic growth. Shanghai, Beijing, and Tianjin were experiencing rapid urban sprawl, and Shenzhen had grown like a meteor, changing from a small fishing village to a large city with a population of more than 10 million in less than 20 years. By the end of 2010, a total of 14 cities in China had a population of more than 10 million. In 2013 the population of Chongqing was 28.85 million, Shanghai's was 23.02 million, and Beijing's was more than 20 million – making these cities some of the largest human settlements in history.[3]

CHALLENGES OF URBANIZATION

Many Chinese cities have shown dazzling achievements in physical urban construction, such as large-scaled urban streets, big parks and plazas, and enormous numbers of new buildings in the newly developed urban areas. However, critical problems in environmental management and social equity have also emerged. One of these is the disequilibrium between urban and rural areas. By the end of 2011 China's urban population had reached 691 million, with more than 50 percent of the population urbanized. As a result of this rapid change, urban and rural areas became polarized in terms of public services, incomes, opportunities, and infrastructure investment. Much of the population of rural villages lived in poverty, while the rush of rural dwellers into big cities resulted in a heavy

burden on infrastructure and public services. Traffic jams and congestion became part of daily life in many cities, and additional tie-ups occurred as the new urbanites attempted to return home to their ancestral villages during long holidays.

A second main set of problems has centered on increasing resource consumption. The nation became dependent on external suppliers for more than 50 percent of oil and iron, and China's consumption of cement grew to more than 50 percent of the world's total.[4] Mismatches between supply and demand of land and water resources also became serious.

A third type of problem concerns environmental pollution. Amounts of solid waste, motor vehicle pollution, persistent organic pollutants, and heavy metal pollution are increasing continuously. The problem of poor air quality in cities such as Beijing has become serious. At present, the air quality of 90 percent of Chinese cities hits "unhealthy" levels frequently according to EU and WHO standards, and respiratory diseases caused in part by poor air quality have increased dramatically.[4]

Fourth, social equity problems have emerged. The gaps among different population groups in terms of living quality, incomes, welfare, and development opportunities are key challenges to urban planning and management, and are sources of social contradiction. The boom in housing prices has also presented problems for many Chinese.

Lastly, weaknesses exist within the urban economy and industrial base. The traditional manufacturing industries of China, characterized by pollution and high energy use, are insufficient to support continued national development and economic development of cities. In view of this dilemma, the Chinese government actively supports industrial development in six selected categories: information industries, biological industries, new material industries, energy industries, environmental resource industries, and modern manufacturing industries. However, the technology of these star industries is still insufficient. Therefore, redefining the economic roles and functions that cities can play in the process of regional industrial transformation has become a critical unsolved question.

The above-mentioned problems require us to think about the roles urban planning and management can play in the process of moving toward sustainable development in China. Now is an important time for China to move towards sustainable urbanism with a better balance of environmental integrity, economic efficiency, and social equity. Given that the major cities are still indispensable hubs for capital accumulation, labor markets, information communication, and technological innovation, promoting the sustainability of Chinese cities will no doubt have a significant influence on the sustainable development of China as a whole as well as on global society.

POLICIES AND ACTION PLANS OF SUSTAINABLE URBAN DEVELOPMENT IN CHINA

Sustainable urbanism in China is influenced by governmental policies (especially national policies), the motivation of political leaders, programs and implementation ability of local government, and support from big state-owned enterprises and private companies. Although implementation has been limited, an array of relevant policies and action plans over the past 20 years has demonstrated the ambitions of the Chinese government. In addition to the 1994 *China's Agenda 21* document, the country signed the United Nations-sponsored *Kyoto Protocol*[5] in 1998, giving it some pressure to reduce the heavy use of fossil fuels. During its 11th Five-Year Period (2006–2011), the Chinese government adjusted its national policy direction to include the concept of sustainable development in order to promote balanced growth. In 2006 it adopted an "Energy Saving and Emission Reduction Measures" policy. Also in that year the Ministry of Housing and Urban–Rural Development (MOHURD) and the National Development and Reform Commission published new editions of the *Reporting and Evaluation Measures for Water Saving Cities*[6] and the *Monitoring Criteria of Water Saving Cities*.[7] These documents require that each city implement an incentive and punishment system, financial investment system, and detailed statistical system to bring about urban water saving.

Likewise in 2006 the State Council of China approved *The National Plan of the Sum Total Control of Major Pollutants During the 11th Five-Year Period*.[8] In 2007 MOHURD approved a *City Domestic Wastes Management Measures* policy[9] and attempted to enhance the treatment level of harmless municipal solid waste. Meanwhile that same year the Office of State Council issued the *11th Five-Year Plan of Comprehensive Work Plan on Energy Conservation and Emission Reduction*. Also, the *Urban and Rural Planning Law* issued in October 2007 provided legal and executive supports to conduct planning and zoning control in rural areas and to a certain degree help mitigate the problem of the polarization between urban and rural environment.[10]

During the 12th Five-Year Period (2012–2016), the national leadership continues to pay much attention to environment protection and sustainable development. The government has announced key policy goals of promoting ecological civilization (a policy term which means to promote modern civilization with ecological concerns taken into account), fostering sustainable urbanization, and maintaining balanced development between urban and rural regions. In 2010 MOHURD worked with the Shenzhen Municipal Peoples Government to come up with a cooperation agreement for building a low carbon ecological city in Shenzhen. This program will use a selected region within that city as a national pilot site for exploring the possibilities of new green technologies and innovative planning know-how.

Air quality control and emission reduction have become important policies in the past five years. In view of the deterioration of air quality, the Ministry of Environmental Protection of the People's Republic of China issued *The Ambient Air Quality Standards* in 2012, requiring stronger actions to mitigate smog problems in big cities.[12] This was followed by the nation's government presenting *The National Sustainable Development Report of the People Republic of China* to the UN Conference on Sustainable Development held in Rio de Janeiro, Brazil. In this conference, China reported its efforts toward implementing the strategies of sustainable development since 2001, and admitted that there are still gaps between its planning goals and what has actually been implemented.[13]

As a big country with more than a billion people, energy consumption is a key issue. In 2012 the State Council of China enacted *The Energy Policy of the People Republic of China* that stresses the importance of encouraging clean energy, renewable energy resources, and the development of green technologies.[14] The government also selected several pilot cities for promoting new energy technology and energy-saving lifestyles, including Dezhou and Baoding. In 2013 MOHURD issued *The Plan of Green Buildings and Green Ecological Districts During the Period of the 12th Five-Year Plan*[15] as part of the effort to promote ecocities and green building planning, design, construction, management, and certification in China, and later that year approved *The Sustainable Urban Development Planning of Resource-based Cities (2013–2020)*. The latter planning document attempts to build on considerations of comprehensive national development to solve problems of unreasonable profit allocation resulted from the existing financial and taxation system, and also establishes some ecological compensation funds.

From the above-mentioned policies and measures, one can see that the government has made many efforts during the past decade using the current command-and-control, top-down planning approach. But when evaluating actual implementation results, the performance of each city varies. Some key factors such as the degree of support from the central government, whether financial support and external investment is stable or not, and individual promotion considerations of local high-level decision-makers play important roles in influencing implementation of sustainable development policies and plans. More often than not, there are gaps between national policies and local execution. Motivation, attitude towards execution, and political relationships are all very important. Moreover, too much emphasis on physical construction and policy development but too little on management and sustainable maintenance is another problem that needs to be overcome.

THE PROGRESS OF URBAN SUSTAINABILITY PRACTICES BY SECTOR

Urbanization and demographic trends

Rapid urbanization, rural pauperization, and an aging population are major barriers to promoting sustainable urban development in China. Urban population exceeded rural population for the first time in 2011.[16] The nation has ended its time in which rural society was dominant, and the flow of population from the country into cities has caused lagging development in the rural areas. This author has witnessed an urban father telling his child that she would be sent to a rural village if she did not behave herself, upon which the child suddenly became quiet. It is probably very hard for people in western countries to appreciate the extent of the Chinese rural–urban divide because of the relatively comfortable country life in those other nations.

China has strongly implemented the one child policy for more than two decades, and has achieved remarkable success in controlling the increase of national population (while of course causing pain for many families). The population growth rate decreased from 1.35 percent in 1978 to 0.48 percent in 2010. Although by 2010 China's population had reached 1,339,724,852 (excluding Hong Kong, Macao and Taiwan), this figure represents a decreased share of the world's population.[17] Thus as the world's largest country China has made a significant contribution in mitigating the pressure of global population growth.

Poverty and enlarged gaps between rich and poor are critical problems demanding immediate solutions. According to the 2011 Chinese national standard of poverty (2300 RMB per capita yearly for rural residents), there are still 128 million poor Chinese, and the problem of urban residents "returning to poverty" occurs in many cities. These inequities have become an obstacle for developing a harmonious and well-to-do socialist society. In addition, with the increase of its aging population China has become the only country in the world that has more than 100 million elderly people.

Urban water resource management

Water shortages are becoming a great challenge in city management that needs to be addressed actively. The water resources of China are distributed unequally across regions and per capita water availability is less than 1/3 of the world level.[18] Moreover, problems of water pollution and water waste are also very common. Agriculture is the main consumer of fresh water, accounting for 67.7 percent of use. Industry uses 25.7 percent, while domestic water use only accounts for 6.6 percent.[19]

With the rapid pace of urbanization, the uneven distribution of water resources among different cities and the competition between domestic water use and economic water use is increasingly apparent. Out of China's 600-plus cities more than 400 suffer chronic water shortages. One hundred and ten cities are facing serious shortages.[20] Although this situation has been gradually improved, progress varies by regions. Water shortages have limited the urban development of cities such as Tangshan.[20]

Wastewater treatment and the pollution of rivers are also critical problems. In the past ten years China has made great efforts in promoting the construction of urban wastewater treatment plants and improving the operation efficiency of already established facilities. By the end of 2004, 661 cities in China had wastewater treatment plants, with a capacity 43 percent greater than in 2000. Still, the overall treatment rate of domestic sewage was only 45.7 percent. Fortunately, with the efforts of the responsible government agencies, this situation has been significantly improved. By 2013, almost all Chinese cities had built their own wastewater treatment plants.[21]

River pollution control and the development of comfortable waterfront spaces are also critical issues influencing sustainable urban development in China. In recent years, following the worldwide trend, Chinese cities have been actively promoting the control of urban river pollution and the development of waterfront spaces. Some cities and towns, such as Nanning, Shanghai, Tianjin, Zhangjiagang, and Wuzhen (a famous water town of China) have made good progress in these areas. But many other cities still suffer from obstacles in implementation. For example, some 75 percent of urban river water in Wuxi was under level V in 2005, showing poor water quality. In order to solve this problem, the city's government has put forward a series of urban water management strategies to control urban floods, promote rational use of urban water resources, and restore ecological environments.[22] China is also promoting efforts to balance the consumption of domestic water and industrial water, separate rain water and sewage, treat river pollution, mitigate agriculture pollution, integrate water management affairs, and encourage water conservation and rainwater reuse. Moreover, many planning agencies also encourage the construction of livable urban waterfront spaces to improve the value of surrounding real estate and the quality of urban life.[23]

Air quality control

Deterioration of air quality is a critical issue that needs to be explicitly addressed during the process of urban development. According to the information from MEPC (the Ministry of Environmental Protection), in the first half of 2013 air quality fell below standards 45 percent of days for 74 selected large cities. The level of air pollution varies by regions. The percentage of days meeting air quality standards is 65 percent in Shanghai and 80 percent in the Pearl River Delta region. However, only 31 percent of days in the Beijing-Tianjin-Hebei region met the national standard. The major air pollutants in these cities are sulfur dioxide, nitrogen oxide, ozone, and particulates (PM2.5 and PM10).[24]

The air quality of China is strongly influenced by sandstorms and smog. Sandstorms are caused in large part by inappropriate land development and agricultural practices, and have serious impacts on human health and livelihood. Sandstorms occur mainly in the northern part of China, including south Xinjiang Basin, southwest of Qinghai, west of Tibet, middle and western parts of Inner Mongolia, and the north central part of Gansu. On average, these regions encounter sandstorms more than 10 days in one year, and there are normally more than 20 days of sandstorms in south Xinjiang Basin and the west of Inner Mongolia. The range of influence of sandstorms can be very large, sometimes affecting all of China. Smog is mainly caused by air pollution and fog, and often lasts for several days. The main cities affected are Beijing, Nanjing, Wuhan, Jinan, Shanghai, Guangzhou, Qingdao, and Urumchi. It is difficult to see blue skies in these cities.

To solve these problems, many cities are actively implementing air pollution control measures. For example, the Beijing Municipal Government proposed a series of measures to improve air quality between 1998 and 2008. These included efforts to move high-polluting industries out of cities, enhance industrial technology, increase vehicle emission standards, strengthen public transport, increase renewable energy use, improve heating systems, and convert coal-fired boilers to other fuels. During the Beijing Olympic Games in 2008, the city's government adopted short-term measures to improve air quality, including restricting motor vehicles from entering certain sectors, stopping highly polluting construction, halting production in certain industries, and encouraging the use of buses and taxis.[25] In another China megacity with poor air quality – Tianjin – air quality efforts are focusing on improving the efficiency of coal use, controlling for fugitive dust, and reducing exhaust from motor vehicles. Other air quality-related programs in the country include forest protection projects, programs to convert cropland into forestland, and sandstorm source treatment projects. However, given climate change and rapid urbanization the ultimate way to solve these problems is likely to include promoting sustainable industrial development, encouraging the use of green transport, optimizing urban planning layout and scale, and strengthening urban greening as a way to absorb pollutants and purify the air.

The urban transportation system

Urban transportation infrastructure in China has been developed rapidly in the past decade. Main roads in large cities are often very spacious with nice landscaping. Some regions, such as the Pearl River Delta area, have already built an extensive network of parkways. However, many large urban roads are far from human-scaled, and road designs in China are often big and monumental, quite different from the "road diet" philosophies often practiced now in western countries. With rapid urbanization and economic development, the growth of motor vehicle ownership has been amazing in the past ten years. By 2012 national motor vehicle ownership had reached 219 million,[26] representing an annual growth rate of over 18 percent.[27] Since the mileage of urban roads is growing about 6 percent annually, road infrastructure and parking facilities can barely meet the demand. Many cities have serious problems of traffic congestion during rush hours. The Chinese situation bears out the old wisdom in urban planning that traffic problems can never be solved by building more roads solely, without taking full consideration of the linkage of transportation and land use planning.

Many Chinese cities have embraced mass transit systems to provide safe, fast, convenient, and low-pollution solutions to traffic jams and transport-related problems. The concept of Transit Oriented Development (TOD) has also received much attention in big cities. By the end of 2012, 35 cities had been approved to build urban rail transit systems, with a total length of 5720 km nationally.[28] Nineteen cities in China already have modern bus rapid transit and/or Metro systems. Beijing and Shanghai are far ahead with rail transit construction, while Tianjin, Wuhan, Shenzhen, Nanjing, Dalian, and Guangzhou are also vigorously developing systems. However, compared to the speed of urban development, the mileage of rail transit construction is still insufficient. In many places lagging development of public transit systems, under-investment in public transport, and the lack of suitable planning strategies to promote urban bus systems have resulted in poor quality of public transport service, thus further encouraging urban residents to shift to other modes of travel such as private vehicles.[29]

Urban greening

For more than a decade, urban greening in China has been strongly promoted by several methods, including urban regeneration planning for old inner-city areas, demolition of walls surrounding private yards, addition of parks and green spaces in central city areas and new development areas, and reforestation of urban fringe areas. In 2011 the per capita amount of park and green space within China cities was 11.18 square meters, and green space represented 39 percent of the built-up area of cities. There were a total of 183 cities named "national garden city" and 31 cities designated as "national forest city" for their urban greening efforts.[30] In order to promote urban greening, each city was required to establish its own urban green space planning goals. For instance, Shanghai set up a goal of expanding its forest coverage ratio to over 30 percent by 2020, the citywide greening coverage ratio to 35 percent, and the per capita share of public green space in the urban area to at least 10 square meters.[31]

Protecting ecosystems and developing low-carbon cities

Since the beginning of the 21st century, the Chinese government has paid much more attention to ecological protection and restoration. Several public policies have achieved successes in converting idle farmlands to forests, preserving natural forests, maintaining the ecological environment of the three major river source areas, enhancing the management of ecological preservation areas, identifying natural reserve areas, and enhancing the control of soil erosion.

Given the trend of rapid urbanization, the concept of giving priority to ecological preservation in controlling non-urban land development has received increasing attention. Local authorities and urban planners have adopted urban green lines and ecological control lines as measures to control urban sprawl and protect ecological resources.[32] In 2002 the executive meeting of MOHURD adopted an *Urban Green Line Management Measures* policy that provides clearer implementation support for setting up public green spaces, parks, and greenbelts surrounding the city. At about the same time, a number of regions and cities such as Shenzhen, Wuxi, Dongguan, Guangzhou, and Changsha have developed the concept of ecological control lines, an approach similar to setting up urban growth boundaries and green fingers around cities in western countries. The ecological control lines are defined on the basis of relevant laws, regulations, the importance of local ecological resources, and urban development conditions.

In addition, establishment of urban ecological corridors has been gradually promoted. Quite a number of such plans have been developed, including the ecological greenbelt plan around Beijing and the green corridor on the western part of the city area, the planning of an urban forest corridor in Nanjing, the ecological corridor planning of four mountains and two river areas of Chongqing, and the ecological corridor planning of Hangzhou. Moreover, the Shenzhen International Low Carbon Ecological City is also actively promoting ecological corridor planning and related sustainable land management.[33]

China has actively promoted the construction of low-carbon cities in recent years. In 2009 Dr. Qiu Baoxing, the Deputy Minister of MOHURD, proposed this concept at the International Urban Planning and Development Forum held in Harbin, 2009. About the same time MOHURD began to study an indicator system for Eco-cities, and cooperated with Shenzhen City and Wuxi City to build national pilot Low-Carbon ecological city areas. Five provinces (Guangzhou, Liaoning, Hubei, Shanxi and Yunnan) and 8 cities (Tianjin, Chongqing, Shenzhen, Xiamen, Hangzhou, Nanchang, Guiyang and Baoding) were selected as pilot testing places. Enthusiasm for building low-carbon ecological cities is spreading widely in China. At present, related projects are being promoted in Shenzhen, Guangzhou, Shanghai, and Beijing. Among them, Shenzhen currently leads in low-carbon development. "Shenzhen Speed" is the commonly used term to describe the extremely high execution ability and efficiency of the urban construction in that city.

Economic performance, housing affordability and social equity

There are significant differences in economic performance and income level among various cities of China. Except for some resource-based cities which have very good economic performance, the cities with high per-capita income in 2010 were mainly located in the Yangtze River Delta, the Pearl River Delta, the Beijing-Tianjin-Hebei region, Shandong Peninsula, the south of Liaoning Province, and coastal areas of Fujian Province.[34] A boom of housing prices in many cities has become a common trend. Between 2001 and 2007 prices increased more than 250 percent in many first-tier and second-tier cities. In 2008, influenced by the international financial storm, the Chinese real estate market took a downturn. But after 2009 rapid housing price jumps and speculative house buying began occurring in many cities once again.

The recent housing boom has produced significant impacts on people's lives. The middle class in these cities cannot afford a 100m^2 apartment unit after ten years' of hard work because salary increases can never catch up with increases in house prices. According to a commonly used housing affordability indicator in China – the ratio of house price to income – the most affordable cities in 2013 are Guiyang (4.7), Hohhot (4.7), Yinchuan (5.8), Shijiazhuang (6.1), and Changsha (6.1), while the most unaffordable cities include Beijing (13.3), Shenzhen (12.9), Hangzhou (12.2), Shanghai (11.7), and Fuzhou (11.5).[35]

There are many causes of the housing price boom in China's large cities, including things such as the massive release of land for development by local governments to raise money, speculative activities of investors and real estate firms, and expectations shared by many urban homebuyers that housing prices will never fall and that if they don't buy immediately they will lose the opportunity forever. The rapid increase in housing prices and the ineffective distribution of housing services has significantly impacted the price of other urban commodities and social equity. In order to prevent social problems the State Council of China placed strong controls on the real estate market three times in 2010. The measures limited housing loans for second homes, implemented a housing credit-and-loan policy, placed limits on the number of housing units that households can buy during a specified time period, and halted bank loans for those who want to purchase three housing units or more. In order to solve the housing problems of middle- and low-income groups and of civil servants, China's government makes efforts to provide a certain amount of social housing. These units are sold or rented to the above-mentioned groups for at least 40 percent below the market price. Between 2005 and 2010, the Chinese government built eleven million units of low-cost social housing. However, by the end of 2010, there were still more than 20 million low-income and below-average-income families that lacked affordable housing. The quantity and speed of construction of low-cost social housing in many cities is far below demand, and the equity of social housing distribution is sometimes questioned.

CONCLUSION: CHINA'S OWN PATH TO SUSTAINABLE URBAN DEVELOPMENT

After more than a decade of effort, China is trying hard to overcome obstacles in promoting urban sustainable development. Because of its unique characteristics, the country is confronted with some challenges and barriers that are quite different from those in western developed nations. Obstacles include the following:

1 Current urban planning practices cannot keep pace with rapid urbanization, urban sprawl, and population migration.

2 Concerns for personal promotion have led many local political leaders to focus on massive physical construction projects and short-term economic development, but to ignore social planning, equity planning, and ecological environment management.

3 Significant differences exist between development in various parts of China. Coastal cities and provincial primary cities find it much easier to gather capital, labor, and governmental support, while the cities and towns in some disadvantaged regions, such as the midwest and the northeast, lack these advantages.

4 Lack of environmental consciousness in certain population groups hinders progress in building greener cities.

5 Lack of sufficient understanding of the concepts of community planning and citizen participation likewise hinders progress. Compared with western developed countries, China has just started in this area. With an increasingly open and diverse urban society, the country needs to provide educational programs to help the general public as well as decision-makers learn about the true meaning of community empowerment and public involvement.

6 Uniform townscape and residential development which copies typical urban planning models of big cities, and leads to a loss of local place identity.

The above-mentioned challenges, together with an aging population due in part to the one-child policy, will test the wisdom of the Chinese leadership as well as the general public in dealing with social planning and educational reform. Fortunately, many within the middle-aged-to-senior population groups have the ability to think self-reflectively and make adjustments since all of them have experienced the hard years and rapid changes of the past. These age groups are deeply concerned with the welfare of their children and are willing to sacrifice their time and enjoyment to devote themselves to environmental protection activities and ecological construction. Moreover, with improved quality-of-life as a result of economic development and modernization, many people are beginning to think about what they can do for the society and future generations. As a consequence, civil society and volunteer environment protection actions have received increasing attention in many cities. These trends are likely to promote the accumulation of social capital, which can benefit urban sustainability efforts.

Based on this author's recent experiences in college teaching and planning practice in China, the country's young planning professionals and students likewise have the ability to think reflectively about planning practice. For instance, the younger generation of planners is starting to think seriously about how to efficiently and effectively resettle original residents during urban redevelopment projects taking equity and empowerment issues into account. The traditional approach was just to move people out, clear away old neighborhoods, and construct new buildings and infrastructure. They are also starting to think more seriously about questions of who pays and who gains from planning processes.

Finally, of course, China's political leaders are the ones who play a leading role in fulfilling our common vision. The new generation of leadership (such as President Xi Jinping) is emphasizing honesty, "anti-corruption" pragmatism, and ecological civilization. With this approach from the top and growing bottom-up grassroots citizen participation, China may develop the capacity to "walk its own way" in terms of sustainable urban development.

REFERENCES

[1] The State Council of People's Republic of China (PRC). 1994. *China's Agenda 21 – White Paper on China's Population, Environment, and Development in the 21st Century.* (in Chinese)

[2] National Sustainable Development Task Group. 1997. *The People's Republic of China National Report on Sustainable Development.* (in Chinese)

[3] National Bureau of Statistics of PRC. 2013. *China Statistical Yearbook (2013).* (in Chinese)

[4] Sustainable Development Strategy Study Group of Chinese Academy of Sciences. 2013. *China Sustainable Development Report 2013 – the Road to Ecological Civilization: the Next Decade.* (in Chinese)

[5] The United Nations. 1998. *Kyoto to the United Nations Framework Convention on Climate Change.*

[6] The Ministry of Housing and Urban–Rural Development (MOHURD). 2006. *Reporting and Evaluation Measures for Water Saving Cities.* (in Chinese)

[7] MOHURD. 2006. *The Monitoring Criteria of Water Saving Cities.* (in Chinese)

[8] The State Council of PRC. 2006. *The National Plan of the Sum Total Control of Major Pollutants During the 11th Five-Year Period.* (in Chinese)

[9] MOHURD. 2007. City Domestic Wastes Management Measures. (in Chinese)

[10] The State Council of PRC. 2007. *The Urban and Rural Planning Law of the People's Republic of China.* (in Chinese)

[11] See for example The State Council of PRC. 2011. *Comprehensive Work Program for Energy Conservation and Emission Reduction During the Period of the 12th Five-Year Plan.* (in Chinese)

[12] The Ministry of Environmental Protection of the PRC. 2012. *Ambient Air Quality Standards.* (in Chinese)

[13] National Report of Sustainable Development Task Group. 2012. *The People's Republic of China National Report on Sustainable Development.* (in Chinese)

[14] The State Council Information Office of the PRC. 2012. *China's Energy Policy.* (in Chinese)

[15] MOHURD. 2013. *The Plan of Green Buildings and Green Ecological Districts During the Period of the 12th Five-Year Plan.* (in Chinese)

[16] Chinese Academy of Social Sciences. 2012. *Annual Report on Urban development of China.* (in Chinese)

[17] Sustainable Development Strategy Study Group, op. cit.

[18] The Ministry of Water Resources of the People's Republic of China. 2012. *The 2011 China Water Resources Bulletin.* (in Chinese)

[19] Ibid.

[20] Chen, Hui. 2013. "A review on the reform of urban water management system in China." *On Economic Problems* 5: 15–19. (in Chinese)

[21] MOHURD. 2013. *The Construction and Operation of Urban Sewage Treatment Facilities in the First Quarter of 2013.* (in Chinese)

[22] Feng, Xiaohong. 2005. "Some thoughts about promoting the construction of water environment in Wuxi." *Jiangsu Water Resources* 10: 2–30. (in Chinese)

[23] Qiu, Baoxing. 2013. "Situation of and measures for water security on Chinese cities." *Urban Development Studies* 12: 1–11. (in Chinese)

[24] Ministry of Environmental Protection of the PRC. 2013. *Air Quality of the Key Regions and 74 Selected Cities in the First Half of Year of 2013.* (in Chinese)

[25] Greenpeace organization. 2008. *Environmental Assessment Report of the 2008 Beijing Olympic Games.* (in Chinese)

[26] National Bureau of Statistics of PRC. 2013. *Statistical Communique of the People's Republic of China on the 2012 National Economic and Social Development.* (in Chinese)

[27] National Bureau of Statistics of PRC. *China Statistical Yearbook (2009–2012).* (in Chinese)

[28] The Institute of Comprehensive Transportation of National Development and Reform Commission. 2013. *Annual Report of China Urban Mass Transit (2012–2013).* (in Chinese)

[29] Sun, Ping. 2013. "Analysis of current urban transportation situation and strategies." *Theory Research*: 88–89. (in Chinese)

[30] The National Greening Committee of China. 2012. *Bulletin of the Greening Condition of National Land Use in China for 2011.* (in Chinese)

[31] Shanghai City Government. 2001. *Urban Green Space Planning of Shanghai (2002–2020).* (in Chinese)

[32] Zhou, Zhican. 2011. "Research on the planning arrangement of setting basic ecological lines of China." in *the Proceeding of Annual National Planning Conference.* (in Chinese)

[33] Wu, Kang-Li, et al. 2014. "Developing a planning model for the construction of ecological corridor for low-carbon ecological cities: case study of the Shenzhen International Low-Carbon City." *Advanced Materials Research* 838: 2823–2829.

[34] Chinese Academy of Social Sciences. 2012. *China City Sustainable Report (2012).* (in Chinese)

[35] E-house China R&D Institute. 2013. Housing information report (distributed on the website). (in Chinese)

59

"Sustainable City: Crisis and Opportunity in Mexico"

from *Sustaining Cities: Urban Policies, Practices, and Perceptions* (2013)

Alfonso Iracheta

Editors' Introduction

In many parts of the developing world national, state, and local governments have far less power than in China, and several decades of neoliberal policies have left the public sector weak, the private sector little regulated, poverty rampant, and urban development a disaster. Here, senior Mexican scholar Alfonso Iracheta argues that the idealistic visions of progressive urban planning that he and others held with in the 1970s were lost when "free market" ideologies, imposed in part by the World Bank, International Monetary Fund, and developed nations such as the United States, fundamentally changed his country's politics and government. The necessary alternative, Iracheta believes, is a new form of governance that reasserts public control over market processes, based on greater civic participation and democracy.

Iracheta is head of the Urban and Environmental Studies Program at El Colegio Mexiquense in Zinacantepec, Mexico and president of the National Urban Land Congress. Other resources related to sustainable urban development in Latin America include Keith Pozzoli's *Human Settlements and Planning for Ecological Sustainability: The Case of Mexico City* (Cambridge, MA: MIT Press, 2000); Julie Cupples's *Latin American Development* (London: Routledge, 2013); Shawn William Miller's *An Environmental History of Latin America* (New York: Cambridge University Press, 2007); Eduardo Silva's *Challenging Neoliberalism in Latin America* (New York: Cambridge University Press, 2009); and Lynn R. Horton's *Grassroots Struggles for Sustainability in Central America* (Boulder: University of Colorado Press, 2007).

I begin with two hypotheses: Mexican cities as now constituted are unsustainable; and, the problem of urban sustainability has not been adequately formulated by the Mexican government or society.

Within the dominant development model in Mexico, there are in place public policies to address not only environmental sustainability, but also many other problems related to the development process. However, proposed solutions do not reflect the sustainable city problem per se but rather the dominant interests of capital accumulation. This is why urban and environmental planning and public policy in Mexico have not achieved the positive results possible within this economic-political model in the developed world.

Mexico's sustainable city problem has to be understood as a confrontation between two different perspectives: that of economic development and that of environmental sustainability. Analysts agree that in most Latin American countries, since the early 1980s, economic growth has increased, whereas social and environmental sustainability has decreased or stagnated.[1] In developed countries, there is a growing tendency to view economic growth and environmental policies as mutually compatible and noncontradictory[2] to such an extent that the latter becomes an important incentive for the former.[3] But in less developed countries environmental problems are mainly a consequence of poverty and underdevelopment. Poverty reduces people's capability to use natural resources in a sustainable fashion, thereby intensifying pressure on the environment.[4] Poor people are forced to forgo future needs to meet those of the present.[5] Therefore, what concerns the countries of the South – poverty, inequality, potable water, land desertification, and the like – are problems profoundly different from those besetting the countries of the North.[6]

Even though economic and political interests are at stake, a country like Mexico is debating environmental and urban public policies with almost no success in defining sustainable cities' development priorities. To understand this crisis, we must begin with a discussion of the unfair, unbalanced relationship between society and the natural environment that has been imposed by Mexico's dominant model of development.[7]

CRITICAL ENVIRONMENTAL PROBLEMS IN MEXICAN CITIES

Natural resource destruction and misuse, along with the systematic violence wielded against the environment in many Mexican cities, justify a search for new ways of thought that allow us to resolve the irrationality of the economic-political development model. The more significant problems in cities and metropolises include low productivity and competitiveness, poverty and inequality, chaotic land use, poor mobility, environmental deterioration, lax governance, and misguided economic policies.

National and regional development depends greatly on cities, especially on more populated ones such as the Mexico City Metropolitan Area (MCMA), which accounts for over 30 percent of the GNP, with less than 20 percent of the country's population. As has been recorded since the early 1990s

by the Fideicomiso de Estudios Estrategicos sobre la Ciudad de Mexico (Mexico City's Strategic Study Trust), the MCMA's productivity is declining.[8] The reasons are higher transaction costs and negative effects caused by land-use disorder, lower mobility, and increasing land, water, and air pollution. All these problems directly affect the productivity and global competitiveness of Mexico's national capital – none more so than poverty and socioeconomic inequality.

Depending on the source, between 50 and 70 percent of Mexico's urban population is poor.[9] It is not only that poverty is increasingly concentrated in cities and metropolises, but that the economic crisis is reducing the number of formal jobs, so that about 60 percent of new urban jobs are in the informal sector. Further, as the Mexican economy becomes increasingly global, the distribution of wealth becomes less equitable. The result is larger urban populations of low- and very low-income workers. And these workers place greater demands on housing, infrastructure, services, and the environment.

Mexican cities have been growing since the early 1950s without solving the issues of an adequate urban form and land-use distribution. There are also land speculation and land tenure irregularities. That is why the last five decades have been called the period of wild urbanization. When new housing policies were developed at the turn of the millennium, their main objective was to offer as many houses as possible without regard for social and environmental issues. The consequence is that, after nearly a decade of more than half a million new houses annually, most of the construction has contributed virtually nothing to orderly urban growth. On the contrary, many cities are expanding so widely that they have thousands of empty hectares within the city limits and the immediate periphery, and are at risk of having a permanently chaotic urban pattern. Add the extent of poorly constructed, irregular, and precarious human settlements on the outskirts of most cities, and the problem of urban sustainability and spatial ordering becomes paramount, along with issues of impeded vehicular and pedestrian mobility.

The number of vehicles in Mexico has been growing constantly. In 2003 there were approximately 21.2 million registered vehicles. Each year, there are around 1.16 million new motor vehicles, representing an annual growth rate of around 7.4 percent. In turn, the MCMA concentrates 25 percent of the nation's vehicles, with a total of 4.5 million vehicles in 2001. This number rose by approximately 300,000 annually between 1997 and 2001.[10]

Since the early 1950s Mexican policy concerning urban mobility has been, following the United States' and other countries' postwar strategy, clearly oriented toward promoting the use of private cars. However, on the one hand, most Mexican cities have a colonial road pattern unsuitable for a huge number of private cars, and, on the other hand, there are no clear public transport and other mobility strategies. The consequence is that cities are filled with cars, but they lack the appropriate pedestrian and bicycle infrastructure and functional, adequate public mass transport systems that would increase mobility for all inhabitants. Mexico City's Metropolitan Environmental Commission has estimated that MCMA's traffic congestion costs seven billion dollars (US) per year. As with land-use and urban structure policies, Mexican cities need an integrated mobility strategy. An adequate mass transit infrastructure would also help reduce automobile emissions.

Mexico's cities have misused vital natural resources. From Central Mexico northward, almost all cities face a huge water shortage and, to some degree, water pollution, because water treatment and reuse are relatively recent policies and many cities have failed to meet deadlines proposed by the federal government. The problem of waste management is also behind schedule, and urban waste disposal has been subject to various modes of privatization rather than being managed as a public service under local government responsibility.

As a result of this situation many urban projects have failed during the past few years, leaving local society with environmental problems and local government without the necessary tools for solving such challenges internally. Moreover, due to urban expansion, the existence of agricultural and forestry land in urban peripheries has become an environmental problem. Because the developer-dominated urbanization process typically is seen as a question of housing or other land-use supply mechanisms rather than as something to be approached in a holistic, community-oriented manner, the result is a disorganized and unsustainable urban structure at war with nature.

From a global perspective, the world is facing a revolution that started in 1989 with the fall of the Berlin Wall. The principal consequences have been twofold: a new unipolar view of global politics offers a vision of free trade and market-driven national politics, while at the same time neoliberal economics casts aside many public policy traditions by reducing citizens to consumers and recipients of government social services. Similarly, urban land has become primarily a real estate commodity, implying that those who can't afford land prices are expelled to the city's outskirts. In Mexico such persons constitute more than half the total urban population. The (relative) abandonment of urban and environmental planning by the Mexican government since the early 1980s has shown that, with certain exceptions, no tier of government (federal, state, or municipal) has sufficient political strength, technical skill, and resources to solve urban community needs, particularly those of the urban poor.

Beginning in the early 1980s when economic and political neoliberalism was installed in Mexico, the federal government reduced its own resources, power, capabilities, and legitimacy, making way for the private sector to take over an important share of public decisions. Without any national debate, suddenly the market became the leading agent in economic, spatial, and environmental policies. With no clear rules of translation from the private to the public domain, new forms of planning and of public-policy design and decision-making were put into practice.

As a result, community-based planning principles were rejected, and the market was left to decide, for instance, about the location of most infrastructure, housing, and other urban facilities. The 1992 constitutional amendment that gave birth to a new agrarian law paved the way for the privatization of ejidal land [communal land used for agriculture], which represents more than three quarters of urbanizable land in most Mexican cities. This ejidal land[11] was the natural social escape valve for the urban poor (more than 60 percent of urban informal settlements are located there). Once most ejidatarios could privatize their plots of land, prices began going up, to match average real estate prices, in each city. The consequence was an increase in the supply of peripheral urbanizable land, with the possibility of price reductions in favor of the formal land market and, paradoxically, a decrease in land supply for the urban poor, who have been forced to pay higher land prices or move to settlements farther from the city limits.

The relative withdrawal of the government from spatial and environmental decision making – even though spatial and environmental planning have continued to exist – led to rampant land speculation. This has been one of the most important reasons why the urban poor have been expelled from the formal city. An estimated three million families now live in informal settlements. These people cannot afford a plot of land and other, associated costs (urban services, land taxes, and the like),[12] so they must reside in environmentally unsound and remote areas.

All these problems have exacerbated a breakdown in social arrangements and reduced local governance capabilities. Increasingly, employment, housing, transit, and public services tend to be informal in the poorest areas of cities. The social actors who pleaded in the late 1970s for more market, and less state, intervention and planning are now asking for strong government intervention in critical urban matters.

NEW PRINCIPLES FOR URBAN PLANNING AND GOVERNANCE IN MEXICO

Any conceptual contribution toward understanding sustainability and transforming Mexico's urban development should address some essential ideas: energy conservation, socio-spatial equality, integrated land-use strategies, and participatory governance.

Public policy strategies for addressing urban problems should consider how to direct resources to poorer and less educated inhabitants within cities and metropolises. From this standpoint, it is important to realize that urban informal settlements and informal markets, although constituting a huge social and environmental problem, have also served as a sort of solution both because they have evidenced an ability to offer housing suited to the income constraints of the urban poor and because they have diminished urban social unrest. What informal housing solutions lack is land tenure and housing certainty. Thus, a transformation of Mexico's urban land regularization strategy is needed.

In addition, academic and social working groups of the National Housing Council have amply demonstrated that the promotion of, and incentives for, mass supply of serviced land suitable for the urban poor (in price, location, basic services, and terms of payment) must be features of the national strategies if the government and society are to be truly committed to reducing urban inequality and poverty. There should also be public support for self-building and socially concerned housing production – until now, housing policy has been solely directed to the less poor of the poor and has not improved urban-metropolitan sustainability. Finally, Mexico needs a national urban-metropolitan land policy that requires certification of property rights and compliance with land obligations.

Mexican cities also need a national, sustainable urban mobility policy that would be oriented to the long term, integrated with urban development, and widely participatory. The aims of this policy should be to promote sustainable urban mobility as a civil right, and to integrate mobility strategies with those of urban planning, environmental protection, and public health. Other objectives should be to reduce the number of cars within cities by offering high-quality public transport (mainly bus, rapid transit, light rail, and nonmotorized means of transport), as well as by mandating fuel-efficient cars and buses.

The search for sustainable cities in Mexico calls for new principles based on a different form of government, through which citizens and social actors can collectively solve their problems and attend to their social needs, using government as the main instrument to attain their goals. This new government is called governance (*gobernanza*).

In some Latin American countries, and particularly in Mexico, governance is a new sociopolitical model that places greater value on the local social resources and capabilities of a territory and encourages politics to go beyond public administration and political parties to include social actors and their organizations as well. Representative and participatory democracy is a fundamental principle of governance, and the source of citizenship building. This requires greater co-responsibility in urban policy decisions between the government and social actors, all of which necessarily involves a reevaluation of politics and the creation of a new relationship between government and society. For governance to work it must distinguish between public and private spheres, rebuild social networks, view urbanization as a socio-spatial construct, encourage participant diversity, and be sensitive to local concerns.

A CLOSING NOTE

I remember from my days as a very young scholar in the early 1970s how the urban-metropolitan sustainability crisis was envisioned through many research projects and the methods we used to alert the government about the magnitude of the risks that Mexican society was to face. Many intelligent, deeply committed civil servants from federal and state governments shared this vision and supported the proposals that emerged from it. Regrettably, all of this was lost when urban space and the environment became a mere commodity to be dealt with by the market with practically no government intervention.

Now, the limitation to seriously confronting the unsustainable city problem in Mexico is not merely loss of momentum. Rather, due to nearly thirty years of market dominance in this sphere, and little government effort to order and control urbanization processes, the issue has become a very low priority. There is a real risk that it will no longer be possible to solve many urban sustainability problems and that many other social issues may follow similar paths downward. For all these reasons, Mexicans should resolve to politicize urban space and the environment in the sense of putting both at the highest level of the national political debate. The time has come to make cities and sustainability mutually compatible and of very high priority.

ACKNOWLEDGEMENTS

The author wishes to thank Susan Beth Kapilian for her careful editorial revision of the text.

NOTES

1 Victor Ramiro Fernandez and Sergio Boisier, *El vuelo de una cometa: Una metáfora para una teoría del desarrollo territorial* (Santiago de Chile: Instituto Latinoamericano de Planificacion Economica y Social, 1997), 558.

2 David Pearce *et al.*, *Blueprint for a Green Economy* (London: Earthscan, 1989), *passim*. See also Michael Jacobs, *The Green Economy: Environment, Sustainable Development, and the Politics of the Future* (London: Pluto Press, 1991), *passim*.

3 Andrew Blowers, "Environmental Policy: The Quest for Sustainable Development," *Urban Studies* 30, no. 4/5 (1993): 775–796.

4 Brundtland Commission, *The Brundtland Report, Our Common Future*, World Commission on Environment and Development (Oxford: Oxford University Press, 1987), 49.

5 Johan Holmberg *et al.*, *Defending the Future: A Guide to Sustainable Development* (London: IIED/Earthscan, 1991), 32.

6 Blowers, "Environmental Policy," 779.

7 Eduardo Galeano, "Naturaleza muerta," *La Jornada*, April 10, 1995, 20.

8 Fideicomiso de Estudios Estratégicos sobre Ia Ciudad de Mexico, "La Ciudad de Mexico hoy, bases para un diagn6stico" (Mexico: Gobierno del Distrito Federal, 2000).

9 CONEVAL, "Reporta CONEVAL cifras de pobreza por ingreso 2008," press release no. 006/09, July 18, 2009. See also Julio Boltvinik and Araceli Damián, coordinators, *La pobreza en México y el mundo: Realidades y desafíos* (Mexico: Siglo XXI Editores, 2004), 541.

10 Alfonso Iracheta, *La necesidad de una politica pública para el desarrollo de sistemas imegrados de transporte en grandes ciudades mexicanas*, INE, Centro Mario Molina, CTS (Zinacantepec, Mexico: El Colegio Mexiquense, 2006), 14–16, 48.

11 An *ejido* is a form of "social" land tenure that has existed in Mexico since the pre-Hispanic era. It was modernized after the 1910–1920 agrarian revolution giving birth to an agrarian reform, through which around one million hectares were distributed among peasants, who became *ejidatarios*, and their community land, *ejidos*. The new Agrarian Law of 1992 represented the end of agrarian reform and the beginning of *ejido* privatization.

12 Alfonso Iracheta and Martim Smolka, "Access to Serviced Land for the Urban Poor: The Regularization Paradox in Mexico," *Economía, Sociedad y Territorio*, 2, no. 8 (Zinacantepec, Mexico: El Colegio Mexiquense, 2000), 757–789.

60

"Climate Change in the Context of Urban Development in Africa"

from *Climate Change and Sustainable Urban Development in Africa and Asia* (2011)

Kempe Ronald Hope, Sr.

Editors' Introduction

In the following selection Kempe Ronald Hope, Sr., a professor of Development Studies at the University of Botswana, addresses the impacts of climate change on urban sustainability in Africa. The changing climate there will affect urban food systems, water availability, flooding, public health, and rural–urban migration. As usual in developing countries, the poor will suffer the most, which is all the more unfair in Hope's view since Africa has contributed relatively little to creating global warming. He argues that climate adaptation planning is urgently needed and must include improved municipal governance to ensure that actions are effective.

Hope is also author of *The Political Economy of Development in Kenya* (New York: Continuum, 2012) and editor of *AIDS and Development in Africa: A Social Science Perspective* (New York: The Hayworth Press, 1999). Other resources on urban sustainability in Africa include *Disposable Cities: Garbage, Governance, and Sustainable Development in Urban Africa* (Burlington, VT: Ashgate, 2005) by Garth Andrew Myers; *Green Infrastructure for Sustainable Urban Development in Africa* (London: Earthscan, 2012) by John Abbott; *From Understanding to Action: Sustainable Urban Development in Medium-Sized Cities in Africa and Latin America* (New York: Springer, 2005), edited by Marco Keiner, Christopher Zegras, Willy A. Schmid, and Diego Salmeron; and *Environmental Problems in an Urbanizing World: Finding Solutions in Cities in Africa, Asia, and Latin America* (London: Routledge, 2001), edited by Jorge E. Hardoy, Diana Mitlin, and David Satterthwaite.

Although Africa contributes the least to global climatic change, it will be the region most affected by climate change. The continent is a victim of circumstances that are beyond her influence and is, therefore, one of the most climate change vulnerable regions in the world. Climatic change has consequently emerged as a major threat to sustainable growth and development in Africa.

Given the rapid urbanization on the continent, climatic change will impact greatly on urban areas. Urban centers contain a large proportion of people who are highly vulnerable to the effects of climate change. Climate change impacts have the potential to undermine, and even undo, the progress made in improving the socio-economic wellbeing that Africa has been experiencing in the past several years.

The negative impacts associated with climate change are also compounded by factors such as poverty, weak capacity, diseases, and high population density. In addition, while Africans can take some steps to mitigate climate change, the most important aspects of relief for Africa are the implications of the mitigation strategies adopted by the developed countries. Consequently, the rich developed countries have the ability to subvert the viability of the livelihoods of millions of Africans if they do not make serious efforts to tackle climate change.

AGRICULTURE AND FOOD SECURITY

Agriculture and fisheries are very sensitive to climate change (FAO 2008). By the 2080s, climate change is estimated to place an additional 80–120 million people at risk from hunger, and 70–80% of these will be in Africa (Parry et al. 2004). This will, in turn, worsen the state of food insecurity and malnutrition while demonstrating agriculture's position as the most vulnerable sector to climate change (Nyong 2005, 2009).

The majority of the African population derive their livelihoods from agriculture, which represents the single largest economic activity on the continent. In sub-Saharan Africa for instance, it contributes at least 40% of exports, 34% of GDP (more than 50% in some countries), up to 30% of foreign exchange earnings, and 64–80% of employment (Hope 2008).

Overall in Africa, both arid and semi-arid areas are expected to expand by between 5% and 8% by 2080. This corresponds to a reduction of approximately 60–90 million hectares of agriculturally productive land (Boko et al. 2007). Climate change is expected to considerably reduce cereal production in countries such as Nigeria, Ethiopia, Zimbabwe, Sudan, and Chad. In East Africa, the declining rainfall during 1996–2003 has resulted in declining production in crops such as maize and sorghum (Case 2006). In South Africa, crop revenues are estimated to fall by as much as 90% by 2100, and wheat production is likely to disappear entirely from the continent by 2080 (Boko et al. 2007).

Climate change may also impact Africa's fisheries. The region's fish supply is already in a crisis as it is the only continent where fish supply per capita is on the decline. By 2020, the continent will need 61% more fish per year just to maintain current consumption levels (APF 2008). Since significant numbers of Africans depend on fish as a source of protein, employment, and revenue, climate change will have an impact not only on nutrition, but also on livelihoods. In Namibia and Senegal, for example, fisheries contribute more than 6% of GDP (Boko et al. 2007). Nutritional reliance on fish as a source of protein is very high in sub-Saharan Africa, with countries such as Sierra Leone, Ghana, and The Gambia deriving 59–67% of their animal protein from fish (Allison et al. 2009).

Climate change can result in food access issues that are likely to become more and more significant in urban areas over time. Not only will the availability of certain basic local agricultural foods be affected in the urban areas, but crop failures can further increase the rural–urban migrant flows as more and more rural residents seek alternative opportunities in the cities.

WATER, SEA LEVEL RISE AND COASTAL ZONES

Among the future potentially devastating impacts of climate change in Africa will be changes in water availability. Changes in precipitation and evaporation translate directly into deficits in water supply. Some analyses suggest that the population at risk of increased water stress in Africa is 75–250 million by 2020, and 350–600 million by 2050 (Boko et al. 2007). Water is critical to the realization of the development potential of Africa. Apart from being an essential input for agriculture and other productive activities, including the provision of hydroelectric power, safe water is also critical to health and well-being (APF 2008).

In addition, some African countries are also prone to vagaries of climate change such as droughts and floods. About one-third of people in Africa live in drought-prone areas and are vulnerable to the consequences of droughts, which have contributed to, among other things, migration to urban areas (Bates et al. 2008). Droughts have occurred in the Sahel, Southern Africa, and the Horn of Africa since the end of the 1960s. These droughts have visited untold destruction on productive assets and have had severe impacts on the size of livestock populations. For instance, between 1975 and 1997, cattle owners in Southern Ethiopia lost 46% of their cattle and 41% of their sheep and goats (FAO 2008). This state of affairs leads to meat scarcity and higher meat prices in urban areas.

In Mali's Bamako city, the availability of water is declining as groundwater levels fall. In Nouakchott, Mauritania, large areas are buried in sand due to prolonged droughts. In Diourbel, Senegal, the River Sine has dried up completely since the 1970s, while groundwater salinity has been increasing with the over-exploitation of underground water (Dodman 2008).

Flooding also tends to have a direct impact on urban populations and urban areas. Indeed, many of the urban poor in Africa are at risk of severe flooding due to increased storm frequency and intensity. Rapid urbanization and urban population increases have forced large numbers of people, especially the poor, to settle in flood-prone areas such as floodplains usually found in and around urban areas (Douglas et al. 2008). This pattern of settlement, which has resulted in the emergence and concentration of slum communities, has rendered the region more vulnerable to flooding.

Climate change induced sea-level rise along coastal zones where there are high human populations is likely to disrupt economic activities such as tourism, fisheries, and mining. Rising sea levels and the resulting coastal erosion and destruction of coral reefs will also threaten human settlements and mangrove forests. More than a quarter of Africa's population lives within 100 km of the coast, and 12% of the urban population live within the low elevation coastal zones. It is projected that the number of people facing the risk of coastal flooding will increase from one million in 1990 to 70 million in 2080 (APF 2008; McGranahan et al. 2007).

HUMAN HEALTH

The effects of climate change on human health in Africa and elsewhere continue to be a matter of scientific debate, primarily due to the mixed results of research on climate suitability for malaria transmission. Nonetheless, there is a sufficient body of research that suggests that climate change will have significant negative effects on human health in Africa.

The continent is already vulnerable to several climate-sensitive diseases and altered temperatures and rainfall patterns, factors which are expected to increase incidences of vector-borne diseases such as malaria. A number of studies have demonstrated a correlation between climatic change and malaria incidence, as well the incidence of dengue fever. Africa's urban poor generally live in slum-like conditions, which tend to be opportunistic breeding grounds for disease carriers such as mosquitoes. These slums are also characterized by high population densities, supplying a large pool of susceptible individuals (Campbell-Lendrum and Corvalán 2007). Currently, it is estimated that the majority of the almost one million people who die of malaria each year are poor African children (Boko et al. 2007; WHO 2008).

RURAL–URBAN MIGRATION

The influence of future climatic changes on rural–urban migration in Africa can, disturbingly, have potential compound effects. For instance, a decrease in the viability of farming in rural areas may encourage people to migrate to urban areas. Yet the economies of African cities may also be affected by climate change impacts and thereby diminish the prospect of rural migrants finding gainful employment in urban areas.

Droughts and floods in rural areas have also forced many rural people to migrate to towns and cities, adding large new populations to existing slums. These new migrants, in turn, add to the urban activities that increase the flow of rainwater to rivers, and consequently exacerbate the intensity of urban flooding (ActionAid International 2006).

ADAPTATION POLICY IMPLICATIONS FOR AFRICA

Adaptation strategies must aim at the eventual ability of African cities to cope with the impacts of climate change as an outcome of harmonious urban development. Policies for urban development therefore need to address the multiple challenges of supporting sustainable, climate-resilient growth, along with good local governance, better jobs, better urban infrastructure, and better basic urban public services.

In addition, national governments need to put much greater emphasis on rural development so as to allow for the emergence of multiple economic opportunities in these rural areas. This will reduce the attraction of cities for rural residents (Hope 1999). Adapting to climate change affords governments the ability to implement and/or scale-up appropriate plans for urban development by drawing from available international assistance.

Among other things, appropriate adaptation strategies must involve a relationship between (1) the accurate assessments of current vulnerabilities to climate change impacts; (2) use of appropriate technologies; and (3) information on traditional coping practices, diversified livelihoods, and current government and local interventions. It is generally agreed among stakeholders that adaptation strategies are best implemented through, and within, National Adaptation Programs of Action (NAPAs). A NAPA provides an important platform for a country to prioritize urgent adaptation needs. It draws on existing information and data as well as community-level input to both identify and determine the priority adaptation approaches and projects that are required to enable a country to cope with the impacts of climate change.

As of May 2009, some 41 countries had submitted a NAPA to the United Nations Framework Convention on Climate Change (UNFCCC) Secretariat. This group included 29 African nations classified as least developed countries (UNFCCC Secretariat n.d.). A NAPA is one of the requirements for accessing funding from the various funding sources available under the UNFCCC and the Kyoto Protocol.

While most of the NAPAs are national in scope, as they ought to be, there is still need to disaggregate their strategies in terms of the urban–rural divide, given what we know now about the differential impacts of climate change on these two distinct geographic areas in each country. In fact, what is being advocated here is the need to think nationally but to act locally in a more community-oriented manner. In some countries such as South Africa, for example, the approach to adaptation to climate change obliges municipalities to respond to and implement the objectives of the National Climate Change Response Strategy. South African municipalities have taken up the challenge, representing a good model for other African countries to follow.

Adaptation strategies for moderating the impacts of climate change in urban areas must include a number of elements and projects that deal specifically with urban public services (including health, water, and sanitation), settlements, coasts, and good local governance. The latter is often overlooked. As convincingly argued by Moser and Satterthwaite (2008), the quality of local government influences the levels of risk from climate change, especially for those with limited incomes and assets. In addition, good local governance is absolutely necessary to drive and sustain effective pro-urban-poor actions and projects to reduce their vulnerability to climate change (Satterthwaite *et al.* 2007).

To date, adaptation planning in most African countries has been focused primarily at the national level, and has not adequately addressed urban adaptation. However, it is at the urban level that many people are directly affected by climate-induced impacts, and it is at that level that appropriate institutional approaches that target large numbers of people should be introduced.

The South African approach for urban adaptation to climate change provides a good framework for other African countries to emulate. In Cape Town, for example, a municipal adaptation plan (MAP) for climate change has been developed. It covers a number of adaptation initiatives that include water supplies (restrictions, tariffs, reducing leaks, pressure management, regulations, awareness campaigns, etc.); storm water; bushfires; and coastal zones. The MAP is to be integrated into on-going strategic plans for all municipal departments (Mukheibir and Ziervogel 2007).

Similarly, in Durban (eThekwini Municipality), another coastal South African city, a locally rooted climate change adaptation strategy has been developed. Durban's MAP is being rooted in initiatives that include, for example, human health, water and sanitation, coastal zone, food security and agriculture, infrastructure, and cross-sectoral activities (Roberts 2008; eThekwini Municipality 2007).

African central governments need to move with speed to empower and encourage their municipal governments to develop urban adaptation plans to moderate the impacts of climate change in urban areas. These plans must also of necessity include elements for building local adaptive capacity (Phalatse 2008), and be facilitated with the cooperation and support of the central government so as to seek out all potential opportunities available for funding of adaptation projects under the UNFCCC and the Kyoto Protocol, as well as from other international and bilateral donors. South Africa has led the way in this regard and, in this author's view, represents an excellent model for other African countries to pursue.

CONCLUSION

Undoubtedly, climate change poses a serious threat to urban development in Africa. It will add to the burdens of those urban residents who are already poor and vulnerable, as well as burden the capacity of governments to cope with and adapt to the severe impacts that this work has outlined. Many of the adverse consequences of climate change are already apparent. However, given that Africa's contribution to climatic variation is marginal at best, the concern for the continent is not the reduction of carbon emissions (*mitigation*) but *adaptation*. Strategies for facilitating adaptation need to be continuously developed, and African governments (both national and municipal) need to be vigilant in their policy approaches for controlling the impacts of climate in their cities as well as nationally.

REFERENCES

ActionAid International (2006) *Climate change, urban flooding and the rights of the urban poor in Africa: key findings from six African cities.* Action Aid International, London.

Allison EH *et al.* (2009) Vulnerability of national economies to the impacts of climate change on fisheries. *Fish Fisheries* 10(2): 173–96.

APF (Africa Partnership Forum) (2008). Climate challenges to Africa: a call for action. http://www.africapartnershipforum.org. Accessed 15 March 2009.

Bates BC *et al.* (2008) Climate change and water: IPCC technical paper VI. IPCC Secretariat, Geneva.

Boko M *et al.* (2007) Africa. In: Parry M *et al.* (eds) *Climate change 2007: impacts, adaptation and vulnerability: Working Group II contribution to the Fourth Assessment Report of the Intergovernmental Panel on Climate Change.* Cambridge University Press, Cambridge, UK, 433–67.

Campbell-Lendrum D, Corvalán C (2007) Climate change and developing-country cities: implications for environmental health and equity. *J Urban Health* 84(1): i109–i117.

Case M (2006) Climate change impacts on East Africa: a review of the scientific literature. World Wide Fund for Nature, Gland, Switzerland.

Dodman D (2008) Against the tide: climate change and high-risk cities. IIED Briefing. International Institute for Environment and Development, London.

Douglas I *et al.* (2008) Unjust waters: climate change, flooding and the urban poor in Africa. *Environ Urban* 20(1): 187–205.

eThekwini Municipality (2007) Climate change: what does it mean for eThekwini Municipality? eThekwini Municipality, Durban.

FAO (2008) Climate change and food security: a framework document. FAO, Rome.

Hope KR (1999) Managing rapid urbanization in Africa: some aspects of policy. *J Third World Stud* 16(2):47–59.

Hope KR (2008) *Poverty, livelihoods, and governance in Africa: fulfilling the development promise.* Palgrave Macmillan, New York.

McGranahan G, Balk D, Anderson B (2007) The rising tide: assessing the risks of climate change and human settlements in low elevation coastal zones. *Environ Urban* 19(1):17–37.

Moser C, Satterthwaite D (2008) Towards pro-poor adaptation to climate change in the urban centres of low- and middle-income countries. International Institute for Environment and Development, London.

Mukheibir P, Ziervogel G (2007) Developing a municipal adaptation plan (MAP) for climate change: the City of Cape Town. *Environ Urban* 19(1): 143–58.

Nyong A (2005) The economic, developmental and livelihood implications of climate induced depletion of ecosystems and biodiversity in Africa. Presented at the Scientific Symposium on Stabilization of Greenhouse Gases, Exeter, UK, 1–3 Feb 2005.

Nyong A (2009) Climate change impacts in the developing world: implications for sustainable development. In: Brainard L, Jones A, Purvis N (eds)

Climate change and global poverty: a billion lives in the balance? Brookings Institution Press, Washington, DC, pp 43–64.

Parry ML *et al.* (2004) Effects of climate change on global food production under SRES emissions and socio-economic scenarios. *Global Environ Change* 14(1): 53–67.

Phalatse L (2008) Capacity building needs and opportunities from the perspective of municipal government. Prepared for the Workshop to Assess Needs and Opportunities, African Climate Change Fellowship Programme, Dar es Salaam, Tanzania, 11–13 March 2008.

Roberts D (2008) Thinking globally, acting locally – institutionalizing climate change at the local government level in Durban, South Africa. *Environ Urban* 20(2): 521–37.

Satterthwaite D *et al.* (2007) Adapting to climate change in urban areas: the possibilities and constraints in low- and middle-income nations. International Institute for Environment and Development, London.

UNFCCC Secretariat (n.d.). Submitted NAPAs http://www.unfccc.int. Accessed 15 March 2009.

WHO (2008) Protecting health from climate change: World Health Day 2008. WHO, Geneva.

61

"Protecting Eden: Setting Green Standards for the Tourism Industry"

from *Environment* (2003)

Martha Honey

Editors' Introduction

Tourism can be both a powerful economic development mechanism for impoverished nations and communities, and a main cause of environmental damage, social disruption, and loss of cultural heritage. One the one hand it can provide jobs for local residents, customers for shops, and much-needed cash for local and national governments. On the other, it can result in out-of-control development, escalating local land and housing prices, air and water pollution, privatization of attractive locations and resources, invasion of local communities without their consent by large numbers of tourists, and replacement of businesses serving local residents with mass-market tourist stores and hotels.

Ecotourism is widely seen as a way to gain the advantages of tourism in a more sustainable manner. First appearing in the late 1980s, ecotourism spread rapidly in the 1990s and early 2000s, and proved such an attractive concept that the United Nations designated 2002 as the "International Year of Ecotourism." Ecotourist projects generally emphasize protecting ecosystems and biodiversity, generating financial benefits for local populations, building environmental awareness and cultural respect, and creating mechanisms to use tourist revenues to bring about local environmental restoration and economic development.

One main challenge though is in determining what qualifies as ecotourism, given that many existing tourist businesses would like to market themselves in this way if they could. There have been reports of mismanagement of ecotourist sites, and of deceptive or fraudulent marketing. In the following article, Martha Honey addresses the challenge of developing ecotourism standards that can help ensure the protection and sustainable development of local places. Honey is Executive Director of the Center on Ecotourism and Sustainable Development, worked for 20 years as a journalist in Latin America and Africa, and is author of *Ecotourism and Sustainable Development: Who Owns Paradise?* (Washington, D.C.: Island Press, 2008) and *Ecotourism and Certification: Setting Standards in Practice* (Washington, D.C.: Island Press, 2002). Other resources include David A. Fennell's Ecotourism (London: Routledge, 2008), Stephen Wearing and John Neil's *Ecotourism: Impacts, Potentials, and Possibilities* (Oxford: Elsevier, 2009), and *Ecotourism and Sustainable Tourism: New Perspectives and Studies* (Oakville, ON: Apple Academic Press, 2012), edited by Jaime A. Seba. Further information is available on the web through The Center on Ecotourism and Sustainable Development, www.responsibletravel.org, and The International Ecotourism Society, www.ecotourism.org.

Behind the front desk at Lapas Rios – a 1,000 acre private rainforest reserve in Costa Rica's Osa Peninsula where guests take bird walks, night hikes, and boat trips to a botanical garden – there hangs a plaque that reads "Sustainable Tourism" in both English and Spanish. Below this the plaque bears four white leaves, indicating that the lodge is one of the top-rated hotels under the Costa Rican government's Certification for Sustainable Tourism (CST) program. As General Manager Andrea Bonilla explains in her welcome speech, Lapa Rios's owners Karen and John Lewis built the lodge with the aim of supporting conservation and the local community through, for instance, putting the land they bought under protection, using low impact designs and technologies, hiring and training local people, and helping to build and support a community primary school.[1]

On the other side of the globe, a high-speed catamaran operated by Quicksilver in Port Douglas, Australia, carries up to 400 tourists out to a section of the Outer Barrier Reef, where it moors at the company's large permanent diving platform. On board are some dozen certified dive instructors and marine biologists who give lessons on the ecology and biology of the reef and recite the do's and don'ts of diving near coral. Some passengers sign up for scenic helicopter rides over the reef, while others choose to stay on board and view the reef from an enclosed semi-submersible tank connected to the catamaran's hull. In the glossy brochure that tourists are handed when they queue to get on board, there is a small logo that reads "ECO Tourism Advanced Accreditation." Quicksilver has been named Australia's best tour operator and, as the ECO logo indicates, has received the highest rating under Australia's internationally respected tourism certification scheme, NEAP (Nature and Ecotourism Accreditation Program). General Manager Max Shepherd tells guests that his company is committed to conservation of the reef – a World Heritage Site – as exemplified by their efforts to educate visitors, generate revenue (each passenger pays an A$4 environmental management charge), and provide marine biologists and scientific equipment to help monitor the reef.[2]

Lapa Rios is a picture postcard of ecotourism – small scale, low impact, low key – while Quicksilver's catamaran, with its 72 ft hull, triple decks, and spacious air-conditioned salon, bar, and restaurant, borders on mass tourism. Lapa Rios accommodates a maximum of thirty-two guests at a time and averages 7,600 visitors per year. Quicksilver's boats carry a million visitors a year to the Great Barrier Reef. But despite their differences, these two businesses share much in common. Their similarities are summed up in the eco-logos they each have earned.

Costa Rica's CST and Australia's NEAP are two of the best-known "green" certification programs, designed to measure sustainability within the tourism industry. Certification is defined as a procedure that assesses, audits, and gives written assurance that a facility, product, process, or service meets specific standards.[3] It awards a marketable logo to those that meet or exceed baseline standards. Increasingly, companies like Lapa Rios and Quicksilver are seeing voluntary certification programs as a way to help ensure they are following the best practices in the industry, give them market advantage with consumers, and raise standards within the industry.

COMPLEXITIES OF THE TOURISM INDUSTRY

Travel and tourism is widely estimated to be the world's largest industry,[4] employing directly and indirectly almost 200 million people.[5] This amounts to 11 percent or one in twelve jobs globally, 10.2 of the world's Gross Domestic Product, and 11.2 percent of global exports.[6] If tourism were a country, it would have the world's second-largest economy, surpassed only by the United States.[7]

Unlike other green and socially responsible certification programs for a single product – wood, bananas, coffee, cut flowers, aquarium fish – where the chain of custody can be fairly easily established from the point of origin to wholesalers, retailers, and the consumer, tourism is found in virtually every country, is multifaceted and nonlinear, and involves a wide variety of both services and products. According to the United Nations Environment Programme (UNEP), tourism-related businesses include:

- Travel: travel agents, tour operators, airlines, car rental companies, buses, railways, and taxies;
- Accommodation, catering, and retail establishments: hotels, guesthouses, hostels, camping

sites, cafes and restaurants, shops (clothing, souvenir, and handicraft, for example);

- Leisure and entertainment: theaters, museums, theme parks, cinemas, and spectator sports; and
- Sports and recreation: athletic centers, diving clubs, chartered transport, safaris and other guided visits.[8]

Before the development of green eco-labels, there were other tourism certification programs in the United States, Canada, and Europe designed to measure professional and business quality, service, and safety standards. The oldest, the Certified Travel Counselor (CTC), was introduced in 1965 by the Institute of Certified Travel Agents as a voluntary program to rate and recognize the competence of individual travel agents.

Much older than professional certification programs are those that are linked to the growth of automobile travel and family vacations, which rate quality, price, and service of accommodations along major roadways. Beginning in 1900, the French tire company Michelin published its first guidebook rating hotels and restaurants. Shortly afterward, the American Automobile Association (AAA), made up of US automobile clubs, also began producing motorist handbooks that used a series of stars to rate the quality and cost of accommodations and restaurants located along highways. Gradually the five-star quality and safety-rating system for accommodations spread around the world, although then, as now, the criteria varied from country to country. Today, in Europe, Costa Rica, Australia, and elsewhere, these five-star certification programs often coexist with newer green certification programs.

As engineer and consultant Robert Toth states, tourist certification programs can be described as a three-legged stool.[9] One leg measures and rates health, hygiene, and safety; a second, quality, service, and price; and a third, sustainability. According to Toth, government generally regulates health and safety standards and most tourists take them for granted (or as with the SARS virus, government warnings dramatically affect travel and the travel industry). Certification programs either include compliance with government health and safety standards or, in countries where standards are weak or poorly enforced, include standards that go beyond government regulations. The second

leg – price and quality standards – have typically been most important to travelers and are those most often set and measured by industry associations such as AAA or Michelin. While the focus of the mass or conventional tourism industry has historically been on rating these first two legs, the newer green certification programs also measure environmental and socioeconomic impacts and consider the satisfaction of the host community as well as of the traveler.

During the past fifteen years, there has been a flowering of green certification programs. According to a World Tourism Organization study, by 2001, there were fifty-nine "very comprehensive state-of-the-art" tourism certification schemes.[10] Of these, the majority are for accommodations (68 percent), but there are a growing number of certification programs covering other sectors of the tourism industry, including sports facilities, destinations, transport, tour operators, and naturalist guides. The largest concentration of programs, 78 percent, is in Europe, while Latin America has the largest number of new programs in development.

The United Nations (UN) declaration of 2002 as the International Year of Tourism, which included a series of regional workshops that culminated in the World Ecotourism Summit in May 2002, gave further impetus to the expansion of certification efforts. During the year several new green certification programs were launched, including the Eco-Rating System in Kenya, the first program of its kind in Africa, and the Swedish Ecotourism Society's Nature's Best. At the World Ecotourism Summit itself, held in Quebec City, at least nine new programs, including ones in Fiji, Ecuador, and Japan, were announced. In the United States, the Boulder-based organization, Sustainable Tourism International, began to develop what is to be the first green certification program in the country. Currently, the most ambitious and best financed green certification program is in Brazil, where the InterAmerican Development Bank (IDB) has put up $1.6 million, and local nongovernmental organizations (NGOs) and businesses are raising matching funds to create and launch an environmentally and socially responsible eco-label for hotels.[11]

The origins of this third leg of certification, sustainability, can be traced directly to the rise of the ecotourism movement.[12]

ECOTOURISM AND CERTIFICATION

The term "ecotourism" first appeared in the 1970s, a decade that saw the rise of a global environmental movement and a convergence of demand for sustainable and socially responsible forms of tourism. It initially grew in scattered experiments and without a name, in response to deepening concerns about the negative effects of conventional tourism. Countries in Latin America, Africa, and Asia, which viewed tourism as a development tool and foreign exchange earner, were becoming increasingly disillusioned with the economic leakage of tourist dollars and the negative social and environmental impacts of mass tourism. Simultaneously, scientists, parks officials, and environmental organizations in various parts of the world were becoming increasingly alarmed by the loss of rainforest and other habitats and of rhino, elephant, tiger, and other endangered wildlife. They began to argue that protected areas would only survive if the people in and around these fragile ecosystems saw some tangible benefits from tourism.

Mounting criticism of the collateral damage caused by tourism led the World Bank and IDB, which had invested heavily in large tourism projects, to conclude that tourism was not a sound development strategy. In the late 1970s, both institutions closed down their tourism departments and ceased lending for tourism. (They only moved back into providing loans for tourism projects in the 1990s, this time under the rubric of ecotourism.) Parallel with these trends, a portion of the traveling public was becoming increasingly turned off by packaged cruises, overcrowded campsites, and high-rise beach hotels and began seeking less crowded and more unspoiled natural areas. Spurred by relatively affordable and plentiful airline routes, increasing numbers of nature lovers began seeking serenity and pristine beauty overseas.

Gradually these different interests began to coalesce into a new concept that was labeled "ecotourism." Ecotourism, as most popularly defined by The International Ecotourism Society (TIES), is "responsible travel to natural areas that conserves the environment and improves the welfare of local people."[13] While nature tourism and adventure tourism focus on what the tourist is seeking or doing, ecotourism focuses on the impact of this travel on the traveler, the environment, and the people in the host country – and posits that this impact must be positive. As such, ecotourism is closely linked to the concept of sustainable development. Rather than being simply a niche market within tourism or a subset of nature tourism, properly understood, ecotourism is a set of principles and practices for how the travel industry should operate.

During the 1990s, propelled in part by the UN's 1992 Earth Summit in Rio and a rapidly growing tourism industry, ecotourism exploded. By the mid-1990s, ecotourism (together with nature tourism) was being hailed as the fastest-growing sector of the travel and tourism industry. The International Ecotourism Society estimated that in the year 2000, ecotourism was growing by 20 percent annually, compared with 7 percent for tourism overall.[14] In 1999, Hector Ceballos-Lascurain, the well-known Mexican architect and conservationist, declared, "Ecotourism is no longer a mere concept or subject of wishful thinking. On the contrary, ecotourism has become a global realty . . . .

There seem to be very few countries in the world in which some type of ecotourism development or discussion is not presently taking place."[15] More than anything else, this "global reality" was signified by the UN's declaration of 2002 as the International Year of Ecotourism.

But parallel to ecotourism's global reach and recognition have been concerns, most articulately and persistently voiced by those in the global South, that the radical tenets of ecotourism would not continue to take root and grow in this new century. There is ample evidence that, in many places, ecotourism's principles and core practices are being corrupted and watered down, hijacked and perverted. Indeed, what is currently being served up as ecotourism includes a mixed grill with three rather distinct varieties: ecotourism "lite" businesses, which adopted a few environmental practices (such as not washing sheets and towels each day or using water-saving shower heads); "green washing" scams, which use green rhetoric in their marketing but follow none of the principles and practices; and genuine ecotourism, or those businesses that are striving to implement environmentally and socially responsible practices.[16]

During the last decade it has become increasingly clear that if ecotourism is to fulfill its revolutionary potential, it must move from imprecision to a set of clear tools, standards, and criteria. Ecotourism needs to not just be conceptualized, but codified, and it is here that green certification programs are viewed as having a central role to play. While ecotourism seeks to provide tangible benefits for both conservation and local communities, certification that includes socioeconomic and environmental criteria seeks to set standards and measure the benefits to host countries, local communities, and the environment.

Today tourism, via the concept of ecotourism, is viewed perhaps more than any other global industry as a tool for both conservation and local community development. "[E]cotourism embraces the principles of sustainable tourism, concerning the economic, social, and environmental impacts of tourism," states the Quebec Declaration, the document approved in May 2002 by delegates to the UN's World Ecotourism Summit. It goes on to affirm that "different forms of tourism, especially ecotourism, if managed in a sustainable manner, can represent a valuable economic opportunity for local and indigenous populations and their cultures and for the conservation and sustainable use of nature for future generations." The Quebec Declaration endorses the use of certification as a tool for measuring sound ecotourism and sustainable tourism, while recognizing that "certification systems should reflect regional and local criteria."[17]

COMMON COMPONENTS OF CERTIFICATION PROGRAMS

In analyzing the current array of green certification programs within the tourism industry, it can be seen that they are all united by some common components. However, these programs are divided by their methodology – process versus performance – and by the sector of the industry they cover – conventional or mass tourism, sustainable tourism, and ecotourism. Examining the common components and broad distinctions helps to illuminate the strengths and weaknesses of these programs and to lay out the basic framework and principles that need to be part of any environmentally and socially responsible programs.

While certification programs within the travel and tourism industry vary widely, they do all have several common features.

Voluntary enrollment

At present, all "green" certification programs in the travel and tourism industry are voluntary, even though governments are involved in financing and in some cases running many of these programs. In contrast, some governments do require hotels to be certified under the five-star rating system to have a license to operate.

Standards and criteria

Certification requires that businesses be assessed by measuring their level of compliance with prescribed criteria and standards. This can be done by using one of the two broad methods: a process that sets up an environmental management system tailored to the business, or a performance that measures every building against a common set of environmental and socioeconomic criteria or benchmarks. Understanding the process–performance distinction is crucial in evaluating the effectiveness of socially and environmentally responsible certification programs. Increasingly, programs are using a combination of these two methodologies.

Assessment and auditing

Certification involves first-party assessment, by the company itself, which typically involves applicants completing a written questionnaire; a second-party assessment, usually by an industry association like AAA; or a third-party audit, by an authorized, independent auditor who is not connected with either the company seeking certification or the body that grants certification. On-site, third-party auditing is considered the most rigorous and credible because it avoids any conflict of interest.

Logo

All programs award a selective logo, seal, or brand designed to be recognizable to consumers. Most permit the logo to be used only for a specified period of time before another audit is required. Many certification programs give logos for different levels of achievement – one to five suns, stars, or leaves, for instance. This is considered superior to a single logo because it encourages business to improve and helps customers to distinguish among certified products.

Membership and fees

While many programs are initially financed by governments, aid agencies, or NGOs, the long-run aim is to make them self-supporting through, at least in part, charging an enrollment fee to businesses seeking certification. Many times a sliding scale is used, with larger and more profitable businesses paying more. These fees vary widely and tend to be highest for certification based on environmental management systems that typically require outside consultants. One of the major challenges for green certification programs is how to make themselves self-supporting over the long haul.

METHODOLOGIES: PROCESS VERSUS PERFORMANCE

While certification programs all share these common components, they are distinguished by whether they use a process or performance methodology and by the sector of the tourism industry they cover.

PROCESS-BASED CERTIFICATION PROGRAMS

Process-based certification programs are all variations of environmental management systems (EMS). The EMS method is widely used, particularly for large hotels or hotel chains, to help management conduct baseline studies, train staff, and set up systems for ongoing monitoring and attainment of set environmental targets such as pollution, water, and electricity reduction. The best known EMS standard for "green" hotel certification is the ISO 14001 (or one of its variants[18]), developed in the wake of the 1992 Rio Earth Summit as one of industry's responses to increasing public interest in sustainable development. It contains the specification and framework for creating an EMS for any business, regardless of its size, product, service, or sector. ISO 14001 can be applied corporate-wide,

at an individual site, or to one particular part of a firm's operations, with its exact scope left to the discretion of the company.[19] Certification to ISO standards is based on having an acceptable process for developing and revising an EMS; it is not based on implementation of an EMS or achieving any declared benchmarks.[20]

ISO and other forms of process-based certification fit well with how large hotels and chains are organized. The advantages of ISO 14001 are that it is internationally recognized and has standards tailored to the needs of the individual business. However, the drawbacks are considerable. It is costly (setting up an EMS usually requires hiring commercial consultants and can cost $20,000 to $40,000 for a medium-sized company and as much as $400,000 for large hotels). It is also complicated and heavily engineering-oriented; focused on internal operating systems, not a company's social and economic impact on the surrounding area or on how a business compares with others in the field; and concerned only with how a company operates, not what it does. The ISO certification does not guarantee certain standards have been met and does not allow comparisons among resorts. Therefore ISO and other types of process-based management systems are insufficient, by themselves, to guarantee sustainable tourism practices.[21] There is a growing awareness about the shortcomings of this methodology and growing agreement that, to be credible, certification programs must include performance-based standards. For instance, in 2002, Green Globe, which had been a largely process-based certification program, teamed up with NEAP to create an international ecotourism standard based mainly on performance criteria.

PERFORMANCE-BASED CERTIFICATION PROGRAMS

Today, an increasing number of certification programs are performance-based, meaning they include a set of benchmarks, often in the form of yes/no questions, against which a business is measured. These programs focus on what a business does in a variety of environmental, sociocultural, and economic areas. While process-based programs set up a system for monitoring and improving performance, performance-based methodology states the goals or targets that businesses must achieve to receive certification and use of a logo. Programs that are largely performance-based, such as CST and NEAP, tend to be less costly and permit comparisons among businesses.

Performance-based certification programs are typically easier to implement because they do not require setting up complex and costly environmental management systems. They are therefore more attractive to small and medium-size enterprises that comprise as much as 90 percent of tourism enterprises worldwide.[22] In addition, although EMS programs are typically devised by management and outside consultants, the most effective performance-based programs are created and implemented by a large range of stakeholders (including representatives from industry, government, NGOs, host communities, and, often, academics) and can solicit and integrate opinions from tourists.

Performance-based programs do, however, present some challenges. The yes/no format can be harsh because many questions are better answered with nuances. There is no agreed-upon standard for what should or should not be included. NEAP has come under criticism, for instance, for not censoring Quicksilver for its use of helicopters on the Great Barrier Reef.[23] Even more prevalent, many standards and criteria are qualitative, subjective, imprecise, and undefined and are therefore difficult to measure. For instance, CST does not specify how large a protected area a hotel must have, permitting a hotel that has a small garden to receive the same points as one that has an extensive private reserve. The question "The hotel's protected area is appropriately managed" can be open to wide interpretation. Despite these difficulties, there is a growing consensus among tourism certification experts that performance standards better measure sustainability, that is, the environmental and socio-economic impacts of a business.

As a study commissioned by the Worldwide Fund for Nature (WWF) in Britain concludes, "Only where universal performance levels and targets that tackle sustainability (environmental, social, and economic) are specified within and by a standard, and where criteria making their attainment a prerequisite are present, can something akin to sustainability be promised by certification."[24]

The WWF study also concludes that a combined approach is useful because it "encourages businesses to establish comprehensive environmental management systems that deliver systematic and continuous improvements, include performance targets and also encourage businesses to invest in technologies that deliver the greatest economic, and environmental benefits within a specific region."[25] Blue Flag for beaches, one of the oldest and most successful programs, sets concrete performance criteria in three areas – water quality, safety, and environmental education and information – but also requires that beach authorities create environmental management systems.[26]

Increasingly, many of the newer or revamped programs such as Green Globe, NEAP in Australia, Green Deal in the Peten region of Guatemala, the Nordic Swan in Scandinavia, and Green Keys in France represent a hybrid of process-based environmental management systems and performance-based standards or benchmarks.[27] This type of hybrid appears certain to become the norm in the future.

NOTES

1 Author's visit to Lapa Rios and interviews with Andrea Bonilla and John Lewis, August 2002; interviews with Karen Lewis, Washington, DC, April 2003; K. Lewis, *The Lapa Rios Story, With Costa Rican Recipes of Intrigue* (San Jose: Costa Rica, 2002).

2 Author's visit to Great Barrier Reef and interview with Max Shepherd, December 2002.

3 "Glossary of Terms," in M. Honey, ed., *Ecotourism and Certification: Setting Standards in Practice* (Washington, DC: Island Press, 2002), p. 380.

4 Honey, ibid., p. 9.

5 World Tourism Organization, Sustainable Development of Tourism Section, *Voluntary Initiatives for Sustainable Tourism* (Madrid: WTO, 2002).

6 World Travel and Tourism Council measurements cited in L. Mastay, *Traveling Light: New Paths for International Tourism*, Worldwatch Paper 159 (Washington, DC: Worldwatch Institute, December 2001), p. 16.

7 Honey, note 3 above, p. 9.

8 "Ecotourism: Facts and Figures," *UNEP Industry and Environment*, Vol. 24, No. 3–4 (Paris, France: United Nations Environment Programme

Division of Technology, Industry and Economics, July–December 2001), p. 5.

9 World Tourism Organization, note 5 above.

10 Author's interview with various people, including Brian Mullis, Sustainable Tourism International; Juan Luna, InterAmerican Development Bank; Ron Sanabria, Rainforest Alliance; and Oliver Hillel, UNEP.

11 R. Toth, "Exploring the Concepts Underlying Certification," in Honey, note 3 above, p. 74.

12 For more on the origins of ecotourism see M. Honey, "Treading Lightly? Ecotourism's Impact on the Environment," *Environment*, June 1999, pp. 4–9, 28–33; and M. Honey, *Ecotourism and Sustainable Development: Who Owns Paradise?* (Washington, DC: Island Press, 1999).

13 Author's interviews, cited in *Honey, Ecotourism and Sustainable Development: Who Owns Paradise?*, ibid., p. 21.

14 Cited in Mastay, note 6 above, p. 37.

15 H. Ceballos-Lascurain, "A National Ecotourism Strategy for Yemen" (paper presented at Report for World Tourism Organization/UNDP and Government of Yemen, Madrid, Spain, 1998).

16 Honey, *Ecotourism and Sustainable Development: Who Owns Paradise?*, note 12 above, pp. 47–55 and elsewhere.

17 World Ecotourism Summit, "Quebec Declaration on Ecotourism," Quebec City, Canada, 22 May 2002, at http://www.uneptie.org/pc/tourism/documents/ecotourism/WESoutcomes/Quebec-Declar-eng.pdf.

18 ISO 14001 variants include ISO 14001 Plus, EcoManagement and Audit System (EMAS), life cycle assessment, and the Nature Step.

19 ISO stands for the International Organization for Standardization, a world federation of standards bodies that develops voluntary standards designed to facilitate international manufacturing, trade, and communications. Another ISO standard, also widely used by the tourism industry, especially hotels, is the 9000 family that sets up management systems for quality and service. Still another cluster – ISO 28, 65, 66, and 67 – contains guidance for establishing and managing certification systems, while ISO 61 contains the requirements for assessment and accreditation of certification bodies. (These are important for creating uniformity in how certification programs function and are accredited.) See Toth, note 11 above; M. Honey and A. Rome, *Protecting Paradise: Certification Programs for Sustainable Tourism and Ecotourism* (Washington, DC: Institute for Policy Studies, 2001), pp. 25–30, and R. Krut and H. Gleckman, *ISO 14001: A Missed Opportunity for Sustainable Global Industrial Development* (London: Earthscan, 1998).

20 Krut and Gleckman, ibid.; and Synergy, "Tourism Certification: An Analysis of Green Globe 21 and other Tourism Certification Programmes," prepared for WWF-UK (London: Synergy, August 2000).

21 For a full discussion see Honey and Rome, note 19 above, pp. 23–32.

22 Mastay, note 6 above, p. 16.

23 Author's interview with Dr. Alice Crabtree, one of the founders of NEAP, Port Douglas, Australia, December 2002.

24 Synergy, note 20 above, p. 10.

25 Synergy, note 20 above, pp. 19–20.

26 X. Font and T. Mihalie, "Beyond Hotels: Nature-based Certification in Europe," in Honey, note 3 above, p. 216.

27 Honey, note 3 above, table 6.1, "Tourism Certification and Ecolabeling Programs in Europe," pp. 190–194.

PART FIVE

Visions of sustainable community

INTRODUCTION TO PART FIVE

Vision is a key ingredient for long-term change. It can inspire, motivate, spread ideas, and provide guidance for more pragmatic activities. Vision can be developed in many ways, including through community planning documents, blueprints for reform, manifestos, speeches, visual images, films, and a variety of forms of utopian literature. Not for nothing was crusading urban critic Lewis Mumford's first book entitled *The Story of Utopias* (New York: Boni and Liveright, 1922). The very idea of "sustainable communities" represents a long-term vision, albeit one that needs much definition and detail. Documents such as the Brundtland Commission report and *Agenda 21* also fall into the category of vision documents, as do the Charter of the New Urbanism, McDonough's Hannover Principles, and many other writings of authors represented in this book.

In this part we present three fictional visions of sustainability. Ernest Callenbach's *Ecotopia* achieved a wide distribution in the 1970s, and has staked out a position as the most fully developed ecological utopia of recent decades. Ursula K. LeGuin's science fiction novel *The Dispossessed* probes the political and social dynamics behind the establishment of a utopian civilization on another planet several centuries from now, drawing on anarchist political philosophy as well as recent environmental sensibilities. And an original piece by one of us (Wheeler) asks what life would be like after three centuries of global warming, and how humanity might get through the transition to sustainability.

Visions of better places or societies tend to emerge at historical times in which a creative ferment exists and substantial numbers of people are exploring new ideas. The turn of the last century was one such time, with the dawn of the Progressive Era, a growing movement for women's suffrage, and the rapid spread of new technologies (the electric light, the telephone, the streetcar, etc.). This period of rapid change stimulated utopian fiction such as Edward Bellamy's *Looking Backwards* (Boston: Ticknor, 1888), Howard's *Garden Cities of To-morrow* (1898), and Charlotte Perkins Gilman's *Herland* (New York: Pantheon Books, 1915 [1979]). It also gave rise to the organized city planning profession, which began to take shape in the 1900s and 1910s.

The late 1960s and 1970s was another such time when many people's conceptions of possible futures for humanity changed rapidly, with the growth of environmentalism, feminism, civil rights movements, and "small planet" consciousness. Notable utopian writings from this second period include Callenbach's *Ecotopia*, Marge Piercy's *Woman on the Edge of Time* (New York: Knopf, 1976), Dorothy Bryant's *The Kin of Ata Are Waiting for You* (San Francisco: Moon Books/Random House, 1976), and a number of LeGuin's works, especially *Always Coming Home* (New York: Harper & Row, 1984). At the conclusion to *The Granite Garden* (New York: Basic Books, 1984, pp. 268ff), Spirn presents a vision of "the celestial city" rather similar to Callenbach's, in which nature "is everywhere evident and cultivated in the city." Paul and Percival Goodman's nonfiction *Communitas* (New York: Vintage Books, 1947) presaged these later writings by presenting a very thoughtful consideration of different urban development alternatives. Kevin Lynch's "A Place Utopia" at the end of *Good City Form* (Cambridge, MA: MIT Press, 1981) likewise proposed an ecologically oriented urban utopia, in the form of a decentralized, web-like network of human communities. Such writings coincided with further growth in the agenda of urban planning, especially the rapid development of environmental planning, the emergence of the environmental design field, and growing public participation in planning.

Since that time there have been relatively few utopian visions of future cities, Richard Register's *EcoCities: Rebuilding Cities in Balance with Nature* (Gabriola Island, BC: New Society Publishers, 2006) being one main exception. Yet visionary or utopian writings help expand the framework of permissible ideas for a generation or more, and let us hope more are forthcoming soon, along with pragmatic efforts to bring such visions into reality.

F
I
V
E

62

"The Streets of Ecotopia's Capital," and "Car-Less Living in Ecotopia's New Towns"

from *Ecotopia* (1975)

Ernest Callenbach

Editors' Introduction

In retrospect, given the enormous growth of environmentalism in the 1960s and 1970s it seems inevitable that someone would write an ecological utopia and call it "Ecotopia." But it was a University of California Press editor named Ernest Callenbach who made the idea reality. Published in 1975, *Ecotopia* (New York: Bantam Books, 1975) sold half a million copies and was translated into eight languages. It told the story of a journalist named William Weston who visited the new nation of Ecotopia – northern California, Oregon, and Washington – some years after it had seceded from the United States and sealed its borders. The changes were, of course, astonishing.

Callenbach followed up *Ecotopia* with a prequel, *Ecotopia Emerging* (Berkeley, CA: Banyon Tree Books, 1981), which told the story of the ecotopian revolution. This event began, appropriately enough, in the anarchic coastal hamlet of Bolinas on a peninsula in Marin County north of San Francisco, when local residents rolled stones across the only access road and declared themselves independent from the United States. The revolution spread, the rest of the country being distracted by economic crisis, toxic contamination, nuclear meltdowns, and dysfunctional politics. The resulting grassroots democracy in Ecotopia featured a strong female President who, unlike other politicians, took ideas seriously and spoke directly and honestly to the public. And with such leadership, people from all walks of life joined into the challenge of creating an ecological society.

THE STREETS OF ECOTOPIA'S CAPITAL

San Francisco, May 5. As I emerged from the train terminal into streets, I had little idea what to expect from this city – which had once proudly boasted of rising from its own ashes after a terrible earthquake and fire. San Francisco was once known as "America's favorite city" and had an immense appeal to tourists. Its dramatic hills and bridges, its picturesque cable cars, and its sophisticated yet relaxed people had drawn visitors who returned again and again. Would I find that it still deserves its reputation as an elegant and civilized place?

I checked my bag and set out to explore a bit. The first shock hit me at the moment I stepped on to the street. There was a strange hush over everything. I expected to encounter something at least a little like the exciting bustle of our cities – cars honking, taxis swooping, clots of people pushing about in the hurry of urban life. What I found, when I had gotten over my surprise at the quiet, was that Market Street, once a mighty boulevard striking through the city down to the waterfront, has become a mall planted with thousands of trees. The "street" itself, on which electric taxis, minibuses, and delivery carts purr along, has shrunk to a two-lane affair. The remaining space, which is huge, is occupied by bicycle lanes, fountains, sculptures, kiosks, and absurd little gardens surrounded by benches. Over it all hangs the almost sinister quiet, punctuated by the whirr of bicycles and cries of children. There is even the occasional song of a bird, unbelievable as that may seem on a capital city's crowded main street.

Scattered here and there are large conicalroofed pavilions, with a kiosk in the center selling papers, comic books, magazines, fruit juices, and snacks. (Also cigarettes – the Ecotopians have not managed to stamp out smoking!) The pavilions turn out to be stops on the minibus system, and people wait there out of the rain. These buses are comical battery-driven contraptions, resembling the antique cable cars that San Franciscans were once so fond of. They are driverless, and are steered and stopped by an electronic gadget that follows wires buried in the street. (A safety bumper stops them in case someone fails to get out of the way.) To enable people to get on and off quickly, during the fifteen seconds the bus stops, the floor is only a few inches above ground level; the wheels are at the extreme ends of the vehicle. Rows of seats face outward, so on a short trip you simply sit down momentarily, or stand and hang on to one of the hand grips. In bad weather fringed fabric roofs can be extended outward to provide more shelter.

These buses creep along at about ten miles an hour, but they come every five minutes or so. They charge no fare. When I took an experimental ride on one, I asked a fellow passenger about this, and he said the minibuses are paid for in the same way as streets – out of general tax funds. Smiling, he added that to have a driver on board to collect fares would cost more than the fares could produce. Like many Ecotopians, he tended to babble, and spelled out the entire economic rationale for the minibus system, almost as if he was trying to sell it to me. I thanked him, and after a few blocks jumped off.

The bucolic atmosphere of the new San Francisco can perhaps best be seen in the fact that, down Market Street and some other streets, creeks now run. These had earlier, at great expense, been put into huge culverts underground, as is usual in cities. The Ecotopians spent even more to bring them up to ground level again. So now on this major boulevard you may see a charming series of little falls, with water gurgling and splashing, and channels lined with rocks, trees, bamboos, ferns. There even seem to be minnows in the water – though how they are kept safe from marauding children and cats, I cannot guess.

Despite the quiet, the streets are full of people, though not in Manhattan densities. (Some foot traffic has been displaced to lacy bridges which connect one skyscraper to another, sometimes fifteen or twenty stories up.) Since practically the whole street area is "sidewalk," nobody worries about obstructions – or about the potholes which, as they develop in the pavement, are planted with flowers. I came across a group of street musicians playing Bach, with a harpsichord and a half dozen other instruments. There are vendors of food pushing gaily colored carts that offer hot snacks, chestnuts, ice cream. Once I even saw a juggler and magician team, working a crowd of children – it reminded me of some medieval movie. And there are many strollers, gawkers, and loiterers – people without visible business who simply take the street for granted as an extension of their living rooms. Yet, despite so many unoccupied people, the Ecotopian streets seem ridiculously lacking in security gates, doormen, guards, or other precautions against crime. And no one seems to feel our need for automobiles to provide protection in moving from place to place. . . .

Ecotopians setting out to go more than a block or two usually pick up one of the sturdy whitepainted bicycles that lie about the streets by the hundreds and are available free to all. Dispersed by the movements of citizens during the day and evening, they are returned by night crews to the places where they will be needed the next day. When I remarked to a friendly pedestrian that this system must be a joy to thieves and vandals, he denied it heatedly. He then put a case, which may not be totally farfetched, that it is cheaper to lose a few bicycles than to provide more taxis or minibuses.

Ecotopians, I am discovering, spout statistics on such questions with reckless abandon. They have a way of introducing "social costs" into their calculations which inevitably involves a certain amount of optimistic guesswork. It would be interesting to confront such informants with one of the hard-headed experts from our auto or highway industries – who would, of course, be horrified by the Ecotopians' abolition of cars.

As I walked about, I noticed that the downtown area was strangely overpopulated with children and their parents, besides people who apparently worked in the offices and shops. My questions to passersby (which were answered with surprising patience) revealed what is perhaps the most astonishing fact I have yet encountered in Ecotopia: the great downtown skyscrapers, once the headquarters of far-flung corporations, have been turned into apartments! Further inquiries will be needed to get a clear picture of this development, but the story I heard repeatedly on the streets today is that the former outlying residential areas have largely been abandoned. Many three-story buildings had in any case been heavily damaged by the earthquake nine years ago. Thousands of cheaply built houses in newer districts (scornfully labeled "ticky-tacky boxes" by my informants) have been sacked of their wiring, glass, and hardware, and bulldozed away. The residents now live downtown, in buildings that contain not only apartments but also nurseries, grocery stores, and restaurants, as well as the shops and offices on the ground floor.

Although the streets still have an American look, it is annoyingly difficult to identify things in Ecotopia. Only very small signs are permitted on the fronts of buildings; street signs are few and hard to spot, mainly attached to buildings on corners. Finally, however, I navigated my way back to the station, retrieved my suitcase, and located a nearby hotel which had been recommended as suitable for an American, but still likely to give me "some taste of how Ecotopians live." This worthy establishment lived up to its reputation by being almost impossible to find. But it is comfortable enough, and will serve as my survival base here.

Like everything in Ecotopia, my room is full of contradictions. It is comfortable, if a little old-fashioned by our standards. The bed is atrocious – it lacks springs, being just foam rubber over board – yet it is covered with a luxurious down comforter. There is a large work table equipped with a hot plate and teapot. Its surface is plain, unvarnished wood with many mysterious stains – but on it stands a small, sleek picturephone.

(Despite their aversion to many modern devices, the Ecotopians have some that are even better than ours. Their picturephones, for instance, though they have to be connected to a television screen, are far easier to use than ours, and have much better picture quality.) My toilet has a tank high overhead, of a type that went out in the United States around 1945, operated by a pull chain with a quaint carved handle; the toilet paper must be some ecological abomination – it is coarse and plain. But the bathtub is unusually large and deep. Like the tubs still used in deluxe Japanese inns, it is made of slightly aromatic wood.

I used the picturephone to confirm advance arrangements for a visit tomorrow with the Minister of Food, with whom I shall begin to investigate the Ecotopian claims of "stable-state" ecological systems, about which so much controversy has raged.

[. . .]

CAR-LESS LIVING IN ECOTOPIA'S NEW TOWNS

San Francisco, May 7. Under the new regime, the established cities of Ecotopia have to some extent been broken up into neighborhoods or communities, but they are still considered to be somewhat outside the ideal long-term line of development of Ecotopian living patterns. I have just had the opportunity to visit one of the strange new minicities that are arising to carry out the more extreme urban vision of this decentralized society. Once a sleepy village, it is called Alviso, and is located on the southern shores of San Francisco Bay.

You get there on the interurban train, which drops you off in the basement of a large complex of buildings. The main structure, it turns out, is not the city hall or courthouse, but a factory. It produces the electric traction units – they hardly qualify as cars or trucks in our terms – that are used for transporting people and goods in Ecotopian cities and for general transportation in the countryside. (Individually owned vehicles were prohibited in "carfree" zones soon after Independence. These zones at first covered only downtown areas where pollution and congestion were most severe. As the minibus service was extended, these zones expanded, and now cover all densely settled city areas.)

Around the factory, where we would have a huge parking lot, Alviso has a cluttered collection of buildings, with trees everywhere. There are restaurants, a library, bakeries, a "core store" selling groceries and clothes, small shops, even factories and workshops – all jumbled amid apartment buildings. These are generally of three or four stories, arranged around a central courtyard of the type that used to be common in Paris. They are built almost entirely of wood, which has become the predominant building material in Ecotopia, due to the reforestation program. Though these structures are old-fashioned looking, they have pleasant small balconies, roof gardens, and verandas – often covered with plants, or even small trees. The apartments themselves are very large by our standards – with ten or fifteen rooms, to accommodate their communal living groups.

Alviso streets are named, not numbered, and they are almost as narrow and winding as those of medieval cities – not easy for a stranger to get around in. They are hardly wide enough for two cars to pass; but then of course there are no cars, so that is no problem. Pedestrians and bicyclists meander along. Once in a while you see a delivery truck hauling a piece of furniture or some other large object, but the Ecotopians bring their groceries home in string bags or large bicycle baskets. Supplies for the shops, like most goods in Ecotopia, are moved in containers. These are much smaller than our cargo containers, and proportioned to fit into Ecotopian freight cars and on to their electric trucks. Farm produce, for instance, is loaded into such containers either at the farms or at the container terminal located on the edge of each minicity. From the terminal an underground conveyor belt system connects to all the shops and factories in the minicity, each of which has a kind of siding where the containers are shunted off. This idea was probably lifted from our automated warehouses, but turned backwards. It seems to work very well, though there must be a terrible mess if there is some kind of jam-up underground.

My guides on this expedition were two young students who have just finished an apprenticeship year in the factory. They're full of information and observations. It seems that the entire population of Alviso, about 9,000 people, lives within a radius of a half mile from the transit station. But even this density allows for many small park-like places: sometimes merely widenings of the streets, sometimes planted gardens. Trees are everywhere – there are no large paved areas exposed to the sun. Around the edges of town are the schools and various recreation grounds. At the northeast corner of town you meet the marshes and sloughs and saltflats of the Bay. A harbor has been dredged for small craft; this opens on to the ship channel through which a freighter can move right up to the factory dock. My informants admitted rather uncomfortably that there is a modest export trade in electric vehicles – the Ecotopians allow themselves to import just enough metal to replace what is used in the exported electric motors and other metal parts.

Kids fish off the factory dock; the water is clear. Ecotopians love the water, and the boats in the harbor are a beautiful collection of both traditional and highly unorthodox designs. From this harbor, my enthusiastic guides tell me, they often sail up the Bay and into the Delta, and even out to sea through the Golden Gate, then down the coast to Monterey. Their boat is a lovely though heavy-looking craft, and they proudly offered to take me out on it if I have time.

We toured the factory, which is a confusing place. Like other Ecotopian workplaces, I am told, it is not organized on the assembly line principles generally thought essential to really efficient mass production. Certain aspects are automated: the production of the electric motors, suspension frames, and other major elements. However, the assembly of these items is done by groups of workers who actually fasten the parts together one by one, taking them from supply bins kept full by the automated machines. The plant is quiet and pleasant compared to the crashing racket of a Detroit plant, and the workers do not seem to be under Detroit's high output pressures. Of course the extreme simplification of Ecotopian vehicles must make the manufacturing process much easier to plan and manage – indeed there seems little reason why it could not be automated entirely.

Also, I discovered, much of the factory's output does not consist of finished vehicles at all. Following the mania for "doing it yourself" which is such a basic part of Ecotopian life, this plant chiefly turns out "front ends," "rear ends," and battery units. Individuals and organizations then connect these to bodies of their own design. Many of them are weird enough to make San Francisco minibuses look quite ordinary. I have seen, for instance, a truck built of driftwood, almost every square foot of it decorated with abalone shells – it belonged to a fishery commune along the coast.

The "front end" consists of two wheels, each driven by an electric motor and supplied with a brake. A frame attaches them to a steering and suspension unit, together with a simple steering wheel, accelerator, brake, instrument panel, and a pair of headlights. The motor drives are capable of no more than thirty miles per hour (on the level!) so their engineering requirements must be modest – though my guides told me the suspension is innovative, using a clever hydraulic load-leveling device which in addition needs very little metal. The "rear end" is even simpler, since it doesn't have to steer. The battery units, which seem to be smaller and lighter than even our best Japanese imports, are designed for use in vehicles of various configurations. Each comes with a long reel-in extension cord to plug into recharging outlets.

The factory does produce several types of standard bodies, to which the propulsion units can be attached with only four bolts at each end. (They are always removed for repair.) The smallest and commonest body is a shrunken version of our pick-up truck. It has a tiny cab that seats only two people, and a low, square, open box in back. The rear of the cab can be swung upward to make a roof, and sometimes canvas sides are rigged to close in the box entirely.

A taxi-type body is still manufactured in small numbers. Many of these were used in the cities after Independence as a stop-gap measure while minibus and transit systems were developing. These bodies are molded from heavy plastic in one huge mold.

These primitive and underpowered vehicles obviously cannot satisfy the urge for speed and freedom which has been so well met by the American auto industry and our aggressive highway program. My guides and I got into a hot debate on this question, in which I must admit they proved uncomfortably knowledgeable about the conditions that sometimes prevail on our urban throughways – where movement at any speed can become impossible. When I asked, however, why Ecotopia did not build speedy cars for its thousands of miles of rural highways – which are now totally

uncongested even if their rights of way have partly been taken over for trains – they were left speechless. I attempted to sow a few seeds of doubt in their minds: no one can be utterly insensitive to the pleasures of the open road, I told them, and I related how it feels to roll along in one of our powerful, comfortable cars, a girl's hair blowing in the wind. . . .

We had lunch in one of the restaurants near the factory, amid a cheery, noisy crowd of citizens and workers. I noticed that they drank a fair amount of the excellent local wine with their soups and sandwiches. Afterward we visited the town hall, a modest wood structure indistinguishable from the apartment buildings. There I was shown a map on which adjacent new towns are drawn, each centered on its own rapid-transit stop. It appears that a ring of such new towns is being built to surround the Bay, each one a self-contained community, but linked to its neighbors by train so that the entire necklace of towns will constitute one city. It is promised that you can, for instance, walk five minutes to your transit station, take a train within five minutes to a town ten stops away, and then walk another five minutes to your destination. My informants are convinced that this represents a halving of the time we would spend on a similar trip, not to mention problems of parking, traffic, and of course the pollution.

What will be the fate of the existing cities as these new minicities come into existence? They will gradually be razed, although a few districts will be preserved as living museum displays (of "our barbarian past," as the boys jokingly phrased it). The land will be returned to grassland, forest, orchards, or gardens – often, it appears, groups from the city own plots of land outside in the country, where they probably have a small shack and perhaps grow vegetables, or just go for a change of scene.

After leaving Alviso we took the train to Redwood City, where the reversion process can be seen in action. Three new towns have sprung up there along the Bay, separated by a half mile or so of open country, and two more are under construction as part of another string several miles back from the Bay, in the foothills. In between, part of the former suburban residential area has already been turned into alternating woods and grassland. The scene reminded me a little of my boyhood country summers in Pennsylvania. Wooded strips follow the winding lines of creeks. Hawks circle lazily. Boys out hunting with bows and arrows wave to the train as it zips by. The signs of a once busy civilization – streets, cars, service stations, supermarkets – have been entirely obliterated, as if they never existed. The scene was sobering, and made me wonder what a Carthaginian might have felt after ancient Carthage was destroyed and plowed under by the conquering Romans.

63

"Description of Abbenay"
from *The Dispossessed* (1974)

Ursula K. LeGuin

Editors' Introduction

A number of leftist thinkers have historically been attracted to a decentralized, anarchist model of society in which very little centralized power exists and there is a strong emphasis on equality and individual responsibility toward others. In theory such a social structure avoids the oppressive concentration of wealth and power found in other societies. Thus it would emphasize the equity goals of sustainable development, as well as potentially the environmental and economic goals. Although commentators frequently dismiss anarchism by equating it with acts of terrorism against the state, it has a rich heritage as a political philosophy, having attracted thinkers such as Peter Kropotkin, Paul Proudhon, Bertrand Russell, Emma Goldman, and Paul Goodman.

In her science fiction novel *The Dispossessed* (New York: Avon Books, 1974), Ursula K. LeGuin describes a utopian, quasi-anarchist society that has left behind a corrupt and oppressive world of warring nation-states not unlike our own to colonize the inhabitable moon of the parent planet. On the home world, Urras, a wealthy class lives in luxury while its servants inhabit vast slums. Women are denied rights and professional roles, social conformity is extreme, and free thought is discouraged. On the renegade world, Annares, a new social structure has been established in which men and women are completely equal, no one is rich, decisions are made by committee, and everyone partakes in the labor of providing the essentials of life. This new society follows the principles of a female philosopher, Odo, who had led the rebellion on the mother planet. Its culture is somewhat reminiscent of the kibbutz model followed in the early Israeli state.

LeGuin goes beyond simply describing this utopia to explore the problems it faces in living sustainably on a dry, barren terrain and the inherent tensions between anarchic decentralization and the need for some sort of organizational structure. The society has difficulty being completely self-sufficient, and by cutting itself off from other worlds it has lost much of its own heritage and the benefits of trade and communication. In the course of the book a pioneering physicist, Shevek, makes the first voyage back to Urras and seeks to rebuild ties between the two worlds in ways that can have benefits for both.

This science fiction vision portrays a society struggling for many qualities that we might equate with sustainability. Though not perfect by any means, the residents of Annares have sought to live harmoniously on their planet, to minimize depletion of natural resources, to promote equality between individuals, and to foster a sense of mutual responsibility. Their philosophy is also based on principles of organic unity that reflect those of many current earthly environmentalists. Although set in a far distant time and place, LeGuin's beautifully written novel describes many dynamics that our own human communities will need to contend with if they are to become more sustainable in the long run.

Decentralization had been an essential element in Odo's plans for the society she did not live to see founded. She had no intention of trying to de-urbanize civilization. Though she suggested that the natural limit to the size of a community lay in its dependence on its own immediate region for essential food and power, she intended that all communities be connected by communication and transportation networks, so that goods and ideas could get where they were wanted, and the administration of things might work with speed and ease, and no community should be cut off from change and interchange. But the network was not to be run from the top down. There was to be no controlling center, no establishment for the self-perpetuating machinery of bureaucracy and the dominance drive of individuals seeking to become captains, bosses, chiefs of state.

Her plans, however, had been based on the generous ground of Urras. On arid Anarres, the communities had to scatter widely in search of resources, and few of them could be self-supporting, no matter how they cut back their notions of what is needed for support. They cut back very hard indeed, but to a minimum beneath which they would not go; they would not regress to preurban, pretechnological tribalism. They knew that their anarchism was the product of a very high civilization, of a complex diversified culture, of a stable economy and a highly industrialized technology that could maintain high production and rapid transportation of goods. However vast the distances separating settlements, they held to the ideal of complex organicism. They built the roads first, the houses second. The special resources and products of each region were interchanged continually with those of others, in an intricate process of balance: that balance of diversity which is the characteristic of life, of natural and social ecology.

But, as they said in the analogic mode, you can't have a nervous system without at least a ganglion, and preferably a brain. There had to be a center. The computers that coordinated the administration of things, the division of labor, and the distribution of goods, and the central federatives of most of the work syndicates, were in Abbenay, right from the start. And from the start the Settlers were aware that that unavoidable centralization was a lasting threat, to be countered by lasting vigilance.

O child Anarchia, infinite promise
infinite carefulness
I listen, listen in the night
by the cradle deep as the night
is it well with the child

Pio Atean, who took the Pravic name Tober, wrote that in the fourteenth year of the Settlement. The Odonians' first efforts to make their new language, their new world, into poetry, were stiff, ungainly, moving.

Abbenay, the mind and center of Anarres, was there, now, ahead of the dirigible, on the great green plain.

That brilliant, deep green of the fields was unmistakable: a color not native to Anarres. Only here and on the warm shores of the Keran Sea did the Old World grains flourish. Elsewhere the staple grain crops were ground-holum and pale mene-grass.

When Shevek was nine his afternoon schoolwork for several months had been caring for the ornamental plants in Wide Plains community – delicate exotics, that had to be fed and sunned like babies. He had assisted an old man in the peaceful and exacting task, had liked him and liked the plants, and the dirt, and the work. When he saw the color of the Plain of Abbenay he remembered the old man, and the smell of fish-oil manure, and the color of the first leafbuds on small bare branches, that clear vigorous green.

He saw in the distance among the vivid fields a long smudge of white, which broke into cubes, like spilt salt, as the dirigible came over.

A cluster of dazzling flashes at the east edge of the city made him wink and see dark spots for a moment: the big parabolic mirrors that provided solar heat for Abbenay's refineries.

The dirigible came down at a cargo depot at the south end of town, and Shevek set off into the streets of the biggest city in the world.

They were wide, clean streets. They were shadowless, for Abbenay lay less than thirty degrees north of the equator, and all the buildings were low, except the strong, spare towers of the wind turbines. The sun shone white in a hard, dark, blue-violet sky. The air was clear and clean, without smoke or moisture. There was a vividness to things, a hardness of edge and corner, a clarity. Everything stood out separate, itself.

The elements that made up Abbenay were the same as in any other Odonian community, repeated many times: workshops, factories, domiciles, dormitories, learning-centers, meeting halls, distributaries, depots, refectories. The bigger buildings were most often grouped around open squares, giving the city a basic cellular texture: it was one subcommunity or neighborhood after another. Heavy industry and food-processing plants tended to cluster on the city's outskirts, and the cellular pattern was repeated in that related industries often stood side by side on a certain square or street. The first such that Shevek walked through was a series of squares, the textile district, full of holum-fiber processing plants, spinning and weaving mills, dye factories, and cloth and clothing distributories; the center of each square was planted with a little forest of poles strung from top to bottom with banners and pennants of all the colors of the dyer's art, proudly proclaiming the local industry. Most of the city's buildings were pretty much alike, plain, soundly built of stone or cast foamstone. Some of them looked very large to Shevek's eyes, but they were almost all of one storey only, because of the frequency of earthquake. For the same reason windows were small, and of a tough silicon plastic that did not shatter. They were small, but there were a lot of them, for there was no artificial lighting provided from an hour before sunrise to an hour after sunset. No heat was furnished when the outside temperature went above 55 degrees Fahrenheit. It was not that Abbenay was short of power, not with her wind turbines and the earth temperature-differential generators used for heating; but the principle of organic economy was too essential to the functioning of the society not to affect ethics and aesthetics profoundly. "Excess is excrement," Odo wrote in the *Analogy*. "Excrement retained in the body is a poison."

Abbenay was poisonless: a bare city, bright, the colors light and hard, the air pure. It was quiet. You could see it all, laid out as plain as spilt salt.

Nothing was hidden.

The squares, the austere streets, the low buildings, the unwalled workyards, were charged with vitality and activity. As Shevek walked he was constantly aware of other people walking, working, talking, faces passing, voices calling, gossiping, singing, people alive, people doing things, people afoot. Workshops and factories fronted on squares or on their open yards, and their doors were open. He passed a glassworks, the workman dipping up a great molten blob as casually as a cook serves soup. Next to it was a busy yard where foamstone was cast for construction. The gang foreman, a big woman in a smock white with dust, was supervising the pouring of a cast with a loud and splendid flow of language. After that came a small wire factory, a district laundry, a luthier's where musical instruments were made and repaired, the district small-goods distributory, a theater, a tile works. The activity going on in each place was fascinating, and mostly out in full view. Children were around, some involved in the work with the adults, some underfoot making mud pies, some busy with games in the street, one sitting perched up on the roof of the learning center with her nose deep in a book. The wiremaker had decorated the shopfront with patterns of vines worked in painted wire, cheerful and ornate. The blast of steam and conversation from the wide-open doors of the laundry was overwhelming. No doors were locked, few shut. There were no disguises and no advertisements. It was all there, all the work, all the life of the city, open to the eye and to the hand. And every now and then down Depot Street a thing came careering by clanging a bell, a vehicle crammed full of people and with people festooned on stanchions all over the outside, old women cursing heartily as it failed to slow down at their stop so they could scramble off, a little boy on a homemade tricycle pursuing it madly, electric sparks showering blue from the overhead wires at crossings: as if that quiet intense vitality of the streets built up every now and then to discharge point, and leapt the gap with a crash and a blue crackle and the smell of ozone. These were the Abbenay omnibuses, and as they passed one felt like cheering.

Depot Street ended in a large airy place where five other streets rayed in to a triangular park of grass and trees. Most parks on Anarres were playgrounds of dirt or sand, with a stand of shrub and tree holums. This one was different. Shevek crossed the trafficless pavement and entered the park, drawn to it because he had seen it often in pictures, and because he wanted to see alien trees, Urrasti trees, from close up, to experience the greenness of those multitudinous leaves. The sun was setting, the sky was wide and clear, darkening to purple at the zenith, the dark of space showing through the thin atmosphere. He entered under the trees, alert, wary. Were they not wasteful, those crowding leaves? The tree holum got along very efficiently with spines and needles, and no excess of those. Wasn't all this extravagant foliage mere excess, excrement? Such trees couldn't thrive without a rich soil, constant watering, much care. He disapproved of their lavishness, their thriftlessness. He walked under them, among them. The alien grass was soft underfoot. It was like walking on living flesh. He shied back on to the path. The dark limbs of the trees reached out over his head, holding their many wide green hands above him. Awe came into him. He knew himself blessed though he had not asked for blessing.

Some way before him, down the darkening path, a person sat reading on a stone bench.

Shevek went forward slowly. He came to the bench and stood looking at the figure who sat with head bowed over the book in the green-gold dusk under the trees. It was a woman of fifty or sixty, strangely dressed, her hair pulled back in a knot. Her left hand on her chin nearly hid the stern mouth, her right held the papers on her knee. They were heavy, those papers; the cold hand on them was heavy. The light was dying fast but she never looked up. She went on reading the proof sheets of *The Social Organism*.

Shevek looked at Odo for a while, and then he sat down on the bench beside her.

He had no concept of status at all, and there was plenty of room on the bench. He was moved by a pure impulse of companionship.

He looked at the strong, sad profile, and at the hands, an old woman's hands. He looked up into the shadowy branches. For the first time in his life he comprehended that Odo, whose face he had known since his infancy, whose ideas were central and abiding in his mind and the mind of everyone he knew, that Odo had never set foot on Anarres: that she had lived, and died, and was buried, in the shadow of green-leaved trees, in unimaginable cities, among people speaking unknown languages, on another world. Odo was an alien: an exile.

The young man sat beside the statue in the twilight, one almost as quiet as the other.

At last, realizing it was getting dark, he got up and made off into the streets again, asking directions to the Central Institute of the Sciences.

64

"The View from the Twenty-Third Century" (2008)

Stephen M. Wheeler

Editors' Introduction

Perhaps we live in an age that is suspicious of utopias, or perhaps the editors have just not found the right recent examples (if you know of good ones, please let us know). But there seem to have been few recent works that address the future in an optimistic way, especially concerning the issue of climate change.

One view of a post-climate-change future is Ursula LeGuin's 1984 anthropology of the future, *Always Coming Home*. In this work she portrays a tribal society in the far distant future living a life similar to that of Native Americans before the arrival of Europeans upon the continent. These future residents of northern California marvel at the buildings of San Francisco far beneath the water, and benefit from an Internet-like communication system. However, in general they live close to the land with an extremely low level of technology. It is a view of the future that assumes the collapse of industrial civilization, not something most of us would wish.

Recent films have taken an even more apocalyptic view of global warming. *The Day after Tomorrow* (2004) portrays the catastrophic onset of a new ice age brought about as a perverse effect of global warming. In the film the warming melts polar ice, which floods the oceans with fresh water and brings about the collapse of the conveyor belt of oceanic currents, including the Gulf Stream, known as the thermohaline circulation. Without warm water moving towards the North Pole, a global winter quickly descends. Most of the film consists of characters battling the sudden weather change, with few glimpses of a brighter future. *Waterworld* (1995) is an even more extreme example of a climate change dystopia. The planet's icecaps have melted (though from a tilting of the earth's axis which warms the poles, not human-caused greenhouse gas emissions), and the earth is covered with water. The survivors fight for existence on the surface of the oceans, and "Dryland" is seen as a mythical belief. At the end a few survivors find this solid ground (the peak of Everest), but, as following the biblical flood, they will be starting civilization over again from scratch.

We would like to think that a much brighter future is possible than in these examples, one in which humanity does in fact head off the worst of global warming and learn to live sustainably in the process. So we've added this final piece to our visions section. Here, rather than bringing about the collapse of civilization, global warming motivates humanity to take a quick leap forward in its social evolution. We certainly hope that such a future is what will in fact happen.

It was April 22, 2270, the three hundredth anniversary of that first Earth Day long ago, which had been celebrated with rallies and teach-ins across the United States. Chitra Shivani, Secretary General of the United Nations, was sitting in her upper-floor office gazing out at the steamy landscape of Costa Rica, reworking the conclusion of her State of the Earth speech. In a few hours it would be time to deliver that address at the podium of the General Assembly. The full text and, if desired, her image reading it would then be available on the personal communication devices of most of the world's three billion people.

Because of this momentous anniversary the responsibility felt unusually heavy to her this year. After all that humanity had been through in its process of learning to live sustainably on the planet, what more could she possibly say? She shivered, even though the warm, humid breeze of the Central American morning flowed through the open windows.

By now, of course, Earth Day was the most sacred of international holidays. It was a time for reflection and rededication to the task of healing human societies and the natural world, somewhat similar on a collective level to the ancient Jewish holiday of Yom Kippur, the Day of Atonement. It was the catastrophic heating of the planet that had lent its present importance to what had been a fringe observance in the late twentieth century. Earth Day was now symbolic of so much – failures in the past, different ways of doing things in the present, continuing challenges in the future. Though the story was well known, it still bore constant reflection, she thought. The story not of the day itself, but of the events behind it. Of how, despite increasingly frantic warnings from scientists and environmental activists, industrial nations had continued along the trajectory known as BAU – Business As Usual – well into the twenty-first century. Of how they had consumed vast quantities of fossil fuels, emitted huge volumes of greenhouse gases, and spread their gospel of consumption to every corner of the Earth. Of how the world's people had finally awoken to their predicament, though by then it had been too late to stop much of the damage.

Originally thought to be slow and gradual, the process of climate change had instead turned out to be sudden and abrupt. As the planet warmed

positive feedback mechanisms had accelerated the change. The melting of the North Polar icecap had meant that instead of ice reflecting solar energy back into space, open water had absorbed the sun's rays and warmed the sea and atmosphere in turn. The thawing of Siberian tundra had released methane, one of the most potent greenhouse gases. The Amazon rain forest had dried out due to fires and logging, releasing billions of tons of carbon. By 2050 the Earth had warmed by 4°F; by 2100 the temperature had soared 10°F. Large portions of the Greenland Ice Sheet had collapsed into the Atlantic, starting in the 2020s, followed by the disintegration of most of the West Antarctic Ice Shelf in the 2030s. By 2050 sea level had risen 3.5 m; by 2100 it was well on its way to its present level 25 m above twentieth-century coastlines.

The original United Nations headquarters in New York had had to be abandoned, of course, and was now partially submerged, along with most of the rest of Manhattan. Bangladesh had vanished beneath the waves, causing enormous loss of life. Much of the Netherlands was gone, despite heroic Dutch efforts at raising dikes and levees. In a great tragedy for historic art and architecture, the lovely city of Venice had been covered by the waters of its lagoon. Water now lapped close to the ceiling of the San Marco Baptistry, dating from the year 1096, covering thousand-year-old mosaics and frescos.

Too late humanity had realized that BAU was no longer possible. Faith in markets and voluntary corporate action had finally evaporated even in the United States, the last holdout. With the reality of a changed planet confronting them, people simply had no choice but to adopt different values. Making a virtue out of necessity, a low-carbon lifestyle had become a point of honor, the mark of a good citizen. What one day had seemed like wild idealism – that human and ecological well-being rather than profit should be the goal of life – was the next day's conventional wisdom.

In recognition of its role in manipulating public opinion and forestalling action on climate change, Exxon/Mobil was the first large corporation to be dismantled. There was no longer much need for it anyway: oil had run out, and fossil fuels were banned by international treaty after 2070. Hundreds of other large corporations were broken up in the years that followed. The capitalism that emerged in the twenty-second century had been small-scale, locally and regionally based, tightly monitored by governments, and balanced by the power of the nonprofit cooperatives that now dominated economic activity.

In line with the new spirit of contrition, nations of the developed world had gone along with the move of the UN headquarters to a developing nation as a symbol of the world's commitment to places long neglected. Latin America was known by this point for its tradition of democratic socialism, begun around the year 2000. Asia had become known for innovation and technical knowhow, with China becoming the world's largest supplier of renewable energy technologies. Africa had become self-sufficient in food, a notable accomplishment, considering that the Sahara desert had doubled in size. Stable governments, many of them led by women, had replaced corrupt dictatorships on that continent.

In an amazing feat of collective will, societies had converted all energy use to renewable sources by 2050. What had once been known as "green building" became universal; there was no longer any need for special standards. By the mid twenty-first century it was the norm for structures to be carbon-neutral, requiring so little artificial heating and cooling that these needs could be met with power generated on site from photovoltaic panels. People now marveled that twentieth-century architects had not bothered to orient buildings toward the south, or to use night breezes and the chill of the Earth 2 m down for cooling.

To the extent that private motor vehicles were still used, they were vastly different, powered by electricity generated from rooftops and the walls of buildings. Ancient sports utility vehicles proved a prime attraction in museums and had become a universal symbol of waste. Life had become simpler and more local. The idea that one would commute dozens of miles every day to a job now seemed outrageous, as much for the loss of time and diminished quality of life as for the waste of fuel and emission of carbon. Builders designed communities so that everything their residents needed was within a fifteen-minute walk. Long-distance travel could be accomplished by rail or hydrogen-powered bus if the need arose.

Initially people had wondered where the money would come from for the massive changes necessary to create a carbon-neutral society. But it had turned out that ample resources had always existed; they had simply been poorly distributed. The wealth controlled by the top one percent of the world's population, once redistributed back to society, had been more than sufficient for the new initiatives. The extremes of wealth present at the turn of the millennium were soon seen as a form of antisocial behavior. As one of its first actions, the reorganized and strengthened United Nations had established the Global Convention on Wealth, under which maximum wages were allowed to be no higher than three times the local minimum wage. Upon the death of individuals excessively large estates were now returned to the public coffers. Mandatory reductions in national military expenditures had freed up enormous sums as well.

Perhaps most challenging had been the decision to quickly stabilize and then reduce the world's population. Nations had recognized that even at lower levels of consumption the planet could not support more than two to three billion people in the long run. Luckily, the adoption of the landmark United Nations Treaty on Population in 2150 had coincided with the reign of Pope John XXIV, the "Green Pope." On grounds that the need for stewardship of God's creation took precedence over biblical commands to "go forth and multiply," he had ended the Catholic Church's long opposition to contraception. The sharp outcry from traditionalists was soon dampened by an outpouring of support from millions of believers tired of hiding their personal behavior from their religious advisors.

Pope John had also led the joint effort by leaders of the world's faiths – Buddhists, Jews, Christians, Hindus, and Muslims among them – to develop the Earth Creed. This document was now universally accepted as humanity's ethical foundation, setting forth the understanding that had been brewing within industrial society at least since John Muir and Aldo Leopold that humans are only part of a much larger web of life and have a responsibility to care for and nurture it. In addition to asserting values of love, charity, and compassion as the foundation of human existence, this document had established new forms of environmentally based spiritual observance, including periods of reflection, prayer, and gratitude three times a day. Only by reinforcing the new mind-set on such a daily basis, these leaders realized, could people truly cultivate and maintain a changed awareness. These precious moments were now a part of daily life worldwide.

Chitra reflected on this long history of crisis and redemption as the sun climbed higher outside her window and then disappeared behind the protective overhang. Global problems were not over, she knew. Oceans were still rising, since even with the cessation of human greenhouse gas emissions enough of those substances were in the atmosphere to melt the remaining Antarctic ice sheets over a millennium or so. Species were still vanishing into extinction despite human efforts to recreate their habitats in other places. Hunger still persisted in some parts of the world, in part because of enormous local climate changes, including drought and flood. The institutions of the global government were still cumbersome, with too much misunderstanding and distrust. But still, so much had been accomplished.

The draft in front of her contained a review of the usual environmental and social progress indicators, a summary of the next year's budget, and an outline of the year's legislative program. She would also remind listeners in the strongest possible terms of the history behind the holiday. But what, she wondered, could she finish with that would best mark this great anniversary?

The main problem, she decided, was that the qualities of the spirit that had helped humanity surmount its ecological crisis had to be continually relearned now that a new plateau had been reached. Eventually memory of the crisis would fade. The sense of urgency that had brought about such rapid change would no longer be present. Temptations of greed, materialism, nationalism, and violence would reemerge. She had seen it happen in small ways within the meetings of the United Nations itself, when representatives had started to take national interest too seriously again and had forgotten their resolutions to cooperate.

"Remember Earth!" would be her exhortation. "Remember what we've been through! Remember how we were able to learn and grow! Remember our past! Remember our future! Let us be worthy of them both!" She typed furiously for a few minutes. Then she read her words through twice with a growing sense of determination, and made her way toward the outside rooms where her staff had begun to gather.

PART SIX

Case studies of urban sustainability

URBAN SUSTAINABILITY AT THE BUILDING AND SITE SCALE

1 Commerzbank Headquarters, Frankfurt, Germany

Designed by Norman Foster architects, this commercial office building is the tallest building in Germany and the world's first ecological skyscraper. This 53-story, nearly 300-meter-tall building, completed in 1997, is designed to incorporate creatively a number of environmental and sustainability features. The building takes a triangular form, with a continuous atrium in its center, extending the entire height of the structure.

A major design element is a series of sky gardens. The building incorporates these gardens every four floors, with a larger garden open to the elements on the 43rd floor. Nine gardens in all are provided, each with trees and vegetation, serving both as places for employees to visit and relax, and as climate control regulators for the building. Convection in summertime draws air through the sky gardens and the atrium providing natural cooling and serving as a "natural chimney."

Other ecological elements of the structure include reliance on natural ventilation during 60 percent of the year (operable windows in all offices and on all floors), extensive daylighting, sensors that automatically adjust artificial lighting inside, and low-emissity windows (windows coated to allow in short

Figure 1 Commerzbank, Frankfurt, Germany (Norman Foster & Partners).

Source: Photograph by Timothy Beatley.

Figure 2 Sky garden in the Commerzbank.

Source: Norman Foster & Partners.

wavelength solar energy but reduce radiant loss of interior heat). The building is enclosed in a high tech double skin with the space between acting as a thermal buffer. The city's district heating system provides the building with heat. The building was designed to use 30 percent less energy than a conventional office building but is actually using 50 percent less. Beginning in 2008 it has used only electricity from renewable sources.

For more information, see: www.fosterandpartners.com and www.commerzbank.com/en/hauptnavigation/konzern/commerzbank_im__berblick/hochhaus_1/zahlen_fakten.html.

2 Menara Mesiniaga Bio-Climatic Skyscraper, Kuala Lumpur, Malaysia

Designed by architect Ken Yeang, the IBM headquarters building in Kuala Lumpur, Malaysia represents an important example of what Yeang calls "bio-climatic" skyscrapers. Such buildings are designed from

Figure 3 Menara Mesiniaga, a bio-climatic skyscraper designed by architect Ken Yeang.

Source: Photograph by K.L. Ng from ArchNet.

Figure 4 Menara Mesiniaga, garden terraces.

Source: Photograph by K.L. Ng from ArchNet.

the start to take full advantage of local climate, to incorporate plants and vegetation, and to substantially reduce their energy and resource consumption levels compared with typical high-rise structures.

The exterior of the Menara Mesiniaga tower serves as an "environmental filter" rather than a hard façade; a permeable membrane that allows movement of air and natural ventilation and breaks up the visual monotony of the exterior. Extensive exterior louvers provide shading on the east and west sides of the building and also add to the building's visual distinctiveness. Perhaps most impressive is the "vertical landscaping," as Yeang calls it, that spirals up and around the structure and connects with a series of recessed "sky courts." These sky courts facilitate ventilation and act as thermal buffers. A sun-shaded roof is designed as important habitable space, and includes a gym and pool. A partially louvered sunroof also acts as a wind scoop, directing air back into the interior of the structure. Other elements of the design include placement of the elevators and core services on the hottest side of the structure. The building rises out of a green terraced base, and is intended to connect the 15-story building to earth and land.

Completed in 1992, this is perhaps Ken Yeang's best example of a completed Bio-Climatic skyscraper design. It was awarded the Aga Khan Award for Architecture in 1995. Other important examples include the UMNO Tower in Penang, the planned EDITT Tower in Singapore, and a design for a 40-story eco-tower as part of a comprehensive redevelopment scheme for the Elephant and Castle area of London.

For more information, see: Ken Yeang, *Bioclimatic Skyscrapers*, Revised Edition, (London: Ellipsis London Press Ltd, 2000); *The Green Skyscraper: The Basis for Designing Sustainable Intensive Buildings* (New York: Prestel, 2000) and *Reinventing the Skyscraper: A Vertical Theory of Urban Design* (London: John Wiley & Sons Ltd, 2002). See also http://www.akdn.org/architecture/project.asp?id=1356.

3 Adelaide Eco-Village (Christie Walk), Australia

The Southern Australian city of Adelaide features an ecologically designed 27-unit cohousing project known as Christie Walk, built in three phases over seven years. (Cohousing is a form of cooperative living in which residents have their own dwelling units but share a common house and other facilities, and usually eat many meals together.)

Green city ideas have a long history in Adelaide, under the advocacy of Paul Downton and the organization Urban Ecology Australia. This urban infill project relied on extensive community participation in its design and development. Community facilities include kitchen, dining, reading, and laundry rooms. Stormwater is retained onsite and used for toilet flushing and gardening purposes. Initial plans called for both graywater and sewage to be recycled onsite, but the tight site and lack of support from the utility prevented this initially.

The homes at Christie Walk are designed to be very energy-efficient, and include both active and passive solar. Units use 60 percent less energy than the average one-person all-electric household and 50 percent less energy than the average two-person household. The building designs take advantage of high thermal mass, extensive insulation, and a natural ventilation system. All buildings use solar hot water, and the community generates some of its own electricity with two types of photovoltaic cells. Stairwells act as ventilation flues. Vegetation and landscaping using native plants cool the air.

Extensive use has been made of recycled materials (e.g. flyash in concrete, recycled timber in windows, reuse of brick and stone from demolished buildings), as well as straw bales in cottage walls and non-toxic paints and finishes. The outer shell of buildings has been designed to last longer than 100 years, with interior doors and walls made from renewable resources. A rooftop garden is included, as well as a community garden where food is produced for the neighborhood.

Siting this development on a T-shaped parcel in the heart of Adelaide reflects its sustainability values as well. Its urban location permits living with little or no dependence on cars. Public transit and shopping are nearby. In recognition of the project's location, some relief from the city's parking requirement was given – only 11 parking spaces were required for the 27 units.

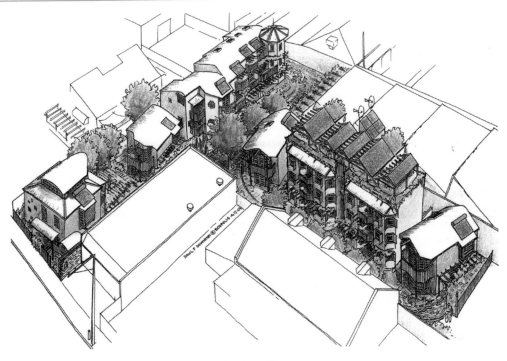

Figure 5 Artist's rendering of Christie Walk, Adelaide, Australia.

Source: Courtesy of Paul Downton, Urban Ecology Australia.

Figure 6 Christie Walk, Adelaide.

Source: Photograph by Paul Downton.

For more information, seen http://www.urbanecology.org.au/eco-cities/christie-walk/, http://www.aila.org.au/SustainableCanberra/009-christie/default.htm, http://www.yourhome.gov.au/technical/fs92.html, and http://www.yourhome.gov.au/technical/fs101.html. Suburban Adelaide is also home to the Aldinga Arts Eco Village, with more than 100 units, a small farm, and onsite sewage treatment. More information about this project is at http://www.aaev.net/about/index.html.

4 Condé Nast Building (4 Times Square), New York

Nicknamed the "green giant," the Condé Naste building, also known as 4 Times Square, is the first major office structure in New York City designed and built around sustainability principles. Completed in 1999, the building is 48 stories in height and includes 1.6 million square feet of space. Designed by the architectural firm Fox and Fowle, the structure incorporates many impressive sustainability features. These include a very energy efficient building design, utilizing large low-emissity windows that capture sunlight

Figure 7 Condé Nast building, 4 Times Square, New York.

Source: Photograph by Timothy Beatley.

Figure 8 Condé Nast building, neon frontage on Times Square, with sign powered by photovoltaic production.

Source: Photograph by Timothy Beatley.

and retain heat and provide extensive daylight to the building, natural gas absorption chiller/heaters, added insulation, thin film photovoltaic panels (on the south and east façades of the top nine floors of the building, producing about 15 kW at peak), and two 200 kW fuel cells that produce enough energy to operate the building in the evening.

The building uses 40 percent less energy than the New York State code minimum. The building's unique ventilation system delivers much more fresh air to building occupants than a typical building – five times the amount required by code.

Careful planning of construction deliveries (reducing engine idling) and management and recycling of construction waste were also important elements. Other green elements of the building include non-CFC air-conditioning, use of energy-efficient, variable-speed motors and pumps, use of low-water-use fixtures, and extensive use of recycled materials in its construction. All floors are equipped with waste recycling chutes. A set of tenant guidelines have been prepared to suggest ways in which tenants can reduce their environmental impacts (e.g., by selecting environmentally-friendly furniture).

An important urban sustainability dimension of the building is its location in the center of Manhattan. Built on the foundations of a former building, this is an urban infill project, embedded in a very pedestrian urban environment, with great access to transit. In fact, the building provides no parking.

For further information, see the US Department of Energy Buildings Database entry at http://eere.buildinggreen.com/overview.cfm?projectid=32.

5 Pearl River Tower, Guangzhou, China

Proclaimed by its designer, Skidmore, Owings and Merrill (SOM) to be "the world's most energy-efficient supertall office tower," the Pearl River Tower has raised the bar for sustainable skyscraper designs following its completion in 2011. The 71-story, 2.3 million square foot tower is located in the Chinese City of Guangzhou and is the corporate headquarters of the Guandong Tobacco Company.

Figure 9 Pearl River Tower.

Source: SOM.

The building incorporates more than 30 different energy strategies and technologies and produces a large portion of the energy it needs. The most dramatic energy feature of the building are its four vertical-axis wind turbines, integrated into the interior of the structure. Wind is directed into four visually dramatic openings in the building, accelerating significantly over the curved funnels guiding the air. The building is oriented to take advantage of the prevailing winds from the south most of the year (and from north for two months of the year). As it flows over the building's surfaces, wind is estimated to accelerate some 2.5 to 3 times the ambient speed, in turn dramatically increasing the power that will be produced from the turbines (as power generation is the cube of wind velocity). It has been estimated that each of the internal turbines will produce some 15 times the power of a conventional stand-alone wind turbine.

The unique curving form of the building helps to guide wind into the turbines from the street below as well as from the upper portions of the structure above. The curvature of these wind funnels leading to the building's turbines also creates a visually dramatic result, making the energy-production dimension very apparent and visible.

But while the vertical-axis turbines are the most unusual and dramatic energy feature of the building, there are many other features. Many of the design features that will result in a low-energy building have to do mostly with saving energy, including use of chilled beams and radiant cooling, extensive daylighting, a double-layer curtain wall system, and use of a geothermal heat sink (routing the building's hot water to be cooled underground).

Energy needs are also reduced through natural day-lighting, solar hot water heating, integration of photovoltaic cells into the building façades, and careful ventilation of excess heat. Together these techniques reduce the building's energy demand by some 65 percent and do much to help it reach its low-energy goal. Production of energy then occurs through the innovative vertical-axis wind turbines and also through the façade-integrated photovoltaics, producing electricity directly from the sun. The extra costs associated with the building's energy features are estimated to pay for themselves in five years through substantially reduced operating costs.

For more information, see https://www.som.com/project/pearl-river-tower or http://www.josre.org/wp-content/uploads/2012/09/Pearl-River-Case-Study-China.pdf.

6 Via Verde, New York

Is it possible to build densely in cities but also ensure access to nature? A terrific new development in the South Bronx, in New York City, is showing the way.

There was a time not that long along when the Bronx was literally burning. It was a place where, in the 1970s and 1980s, high foreclosure rates and tax delinquencies left the city owning much of the land. Much has changed since then, and increasingly the Bronx is a testing ground for ideas that merge poverty reduction and affordability with what is green and sustainable.

Via Verde (Spanish simply for the "Green Way") is one such inspiring example, a very unique affordable housing project that opened its doors to new residents in 2012. It all began with a city-sponsored design competition, and with the winning design co-developed by Phipps Houses and Jonathan Rose Companies. A key aspect of Via Verde is that it doesn't look at all like an affordable housing project. There is use of a varied set of materials, including prefabricated panels of cement board, metal and wood laminate. It is a visually interesting exterior. The large windows and distinctive sun shades are also contrary to the usual look of housing for low- and moderate-income families.

Situated on a relatively skinny lot, running from north to south, the design response is creative indeed – 222 units in total, stepping up from three-story townhouses on the south end to a 20-story residential tower on the north, and maximizing sunlight as a result.

When it is fully occupied, more than 400 residents will live in Via Verde. And they will have an unusual green living environment. Perhaps most distinctive about this project is multi-layered green

Figure 10 Via Verde.

Source: Rose Development.

rooftops. Beginning in a grassy ground-level courtyard, residents can ascend first to an evergreen forest on the third floor, then an orchard of dwarf apple and pear trees on the fourth floor, then onto extensive raised-bed vegetable gardens on the fifth floor. Higher floors have more traditional sedum-covered extensive green roofs. What results is an impressive set of connected rooftop community spaces, for gardening but also just for walking and strolling, providing attractive areas for residents to be and spend time.

A common question is whether the structural loads required by the trees and green elements posed a major problem. The answer is a surprising "no," as the building's block and plank construction had only to be modified marginally: replacing 10-inch planks with 12-inch planks to accommodate the extra loads.

Via Verde has enlisted the nonprofit GrowNYC to initially plant and care for the gardens and trees for the first two years. It will be engaging residents and holding gardening workshops with the goal of turning over the gardens and fruit trees to the loving care of residents at the end of this period.

For developer Jonathon Rose, Via Verde represents the new ways in which we need to design and work in the city, and especially the importance of integrating nature and density in cities. People in cities need that nature, Rose believes: "I think it's because of the biophilic nature of people. We're just seeing a hunger for it."

There are many features in Via Verde aimed at enhancing the health of residents. The relatively narrow building allows for fresh air and cross ventilation, and with ceiling fans there is no need for summer air conditioning. The stairwells were intentionally designed to be on the outside, and brightly painted, to encourage their use (and discourage use of the elevators). And there is some not-so-subtle messaging to residents such as the placard in the lobby imploring residents to "take the stairs – burn calories, not electricity." There are stores and shopping nearby, as well as an onsite medical clinic and space for a community-based pharmacy.

Figure 11 Via Verde community gardens.

Source: Photograph by Timothy Beatley.

From the beginning the project was conceived as a partnership with the city, and has been shepherded along by them (an important lesson). Via Verde was "deeply supported by all the city agencies who collaborated together with us," Rose notes. This has been a helpful arrangement when special waivers and approvals have been required. There is no parking provided by the project, for example, something that required a special mayoral override (and makes a lot of sense given the nearby access to very good transit).

There are other sustainability features as well. Much of the south façade is covered in angled PV panels, producing enough power for all of the common lighting. A large cistern collects stormwater fall in the courtyard and on the roofs and is used for watering the gardens and trees. And the apartments have low-flow water fixtures and energy-star appliances and bamboo countertops.

It may be years before the health and other benefits of Via Verde can be demonstrated. However, there is a research project underway that will compare how healthy the lives and lifestyles of residents of Via Verde are compared with others who were unable to secure a unit there. And if the attractiveness of the project is to be judged by interest among those who want to live there, it is already a huge success.

More information on Via Verde is available at http://viaverdenyc.com/press and http://www.rosecompanies.com/all-projects/via-verde-the-green-way.

7 Barclay Ecological Park, Tainan, Taiwan

By Kang-Li Wu

The development of Barclay Memorial Park is an example of grassroots, bottom-up community planning by self-organized residents. Covering 33,038 square meters or 8 acres, the park is located in the east of Tainan city in Taiwan. In spite being designated as an urban park in the city's land use plan, the site remained vacant for many years in the 1990s due to land ownership conflict and the city's financial problems. Developers and residents used it to dump construction debris and garbage, and the park was a dirty site which depressed housing prices in the neighboring block. Finally, former neighborhood head Li Ren-Ci decided to clean up the rubbish with the help of two of his mountain-climbing friends in 2001. They removed more than 20 truckloads of garbage and construction waste, and their voluntary action motivated other community residents to join them, finally leading to action by the city government as well.

Figure 12 Restored wetlands at Barclay Ecological Park, Taiwan.

Source: Photograph by Kang-li Wu.

After the grassroots-initiated environmental clean-up, a series of action plans were developed for the park, focusing on cleaning and maintenance, restoration of ecological systems, and the creation of a public ecological park accessible to all. The park has become one of the best-known tourist attractions in the Tainan region. A local NGO named the Tainan Enterprise Culture and Arts Foundation provides funding and cooperates with many local volunteers to maintain the park on a daily basis.

The Barclay Park focuses on restoration of the natural environment instead of man-made facilities. A variety of ecological design methods were employed, such as restoring ecological corridors and waterfront space, preserving wetlands, stabilizing slopes with eco-engineering methods, and preserving old and native trees. Park designers and managers also created sites for ecological education and worked with local elementary schools to design an easy-to-understand vegetation interpretation system. Artificial lighting was controlled to avoid affecting the growth of plants and the activities of animals. The beautiful Lotus Pond in the southern part of the park, which used to be the water source of the Zhu Stream (a major stream running through Tainan city), is preserved and has become a tourist attraction.

The effects of the ecological development of the park spilled out to neighboring communities. The outdoor space of the adjoining Cultural Center was redeveloped using ecological design ideas to transform its artificial waterfront space into a natural waterfront space with a comfortable, human-scaled urban plaza. The long-vacant neighboring Durban Department Store began operations again after the park gained popularity, followed by the opening of the environmentally friendly Evergreen Plaza Hotel next door. Car repair shops and some light industries moved out, while popular restaurants, famous bookstores, and various retail services moved in. The area has become a more mixed-use, multi-functional neighborhood in which a variety of daily services can be obtained within walking distance.

With the above-mentioned achievements, the Barclay Memorial Park has won several awards within Taiwan, was a finalist for the FIABCI *Prix d'Excellence* (on the finalist list), and was awarded second place in the Barcelona International Building and Public Works Contest.

URBAN SUSTAINABILITY AT THE NEIGHBORHOOD OR DISTRICT SCALE

8 Hammarby Sjöstad, Stockholm, Sweden

Few places have done as much to put the idea of a sustainable circular metabolism into practice as Swedish cities, and Stockholm is the leading case. This city's different municipal departments and agencies have sought to coordinate their work, and to take a comprehensive material and resource flows approach. Stockholm's new urban ecological district Hammarby Sjöstad is the best demonstration to date of these efforts, the successful outcome now being commonly described as the "Hammarby model."

Hammarby Sjöstad provides an extremely powerful example of how this metabolic flows view can manifest in a new approach to urban design and building in a new dense urban neighborhood. From the beginning of the planning of this new district, an effort was made to think holistically and to understand the inputs, outputs and resources that would be required.

Once it was understood as a system of resource and material flows, the designers of Hammarby found ways to connect these flows, with substantial energy and conservation benefits. For instance, about 1,000 flats in Hammarby Sjöstad are equipped with biogas stoves that utilize biogas extracted from wastewater generated in the community. Biogas also provides fuel for buses that serve the area. Collection of solid waste in the neighborhood happens through an innovative vacuum-based underground collection system that allows efficient separation of organic, recyclable and other wastes. Combustible waste is burned and returned to the neighborhood in the form of electricity and hot water, the latter delivered through a district heating grid. Stormwater from streets is directed into a special purification

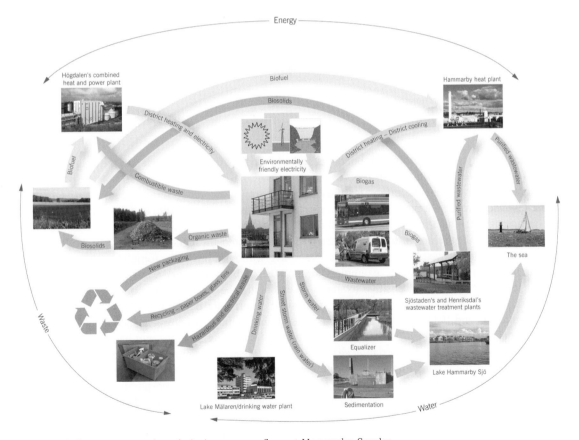

Figure 13 Systems approach to designing resource flows at Hammarby, Sweden.

Source: Courtesy of Erik Freudenthal.

Figure 14 Tram stop in Hammarby, Sweden.

Source: Photo by Timothy Beatley.

and filtration system, and runoff from buildings and other living areas is channeled to the neighborhood's green features (wetlands and green roofs).

There are many other important features at Hammarby that reduce energy demand and carbon emissions. Most important perhaps is the close proximity to central Stockholm and the installation (from the beginning) of a high-frequency light rail system (the Tvärbanan) and an extensive pedestrian and bicycle network. There are also 30 car-sharing cars distributed throughout the neighborhood. These transportation alternatives make it truly possible to live without a private automobile.

Hammarby Sjöstad represents exemplary urban sustainability in other ways as well. Located on former industrial land, it is a positive example of extending the City of Stockholm through relatively dense brownfield development. Eventually the district when fully built-out will accommodate some 25,000 residents and 10,000 jobs, in a mixed-use design that emphasizes physical and visual connections to the water (Hammarby Lake). Substantial restoration of shoreline biotopes and habitat has occurred, and a land-bridge provides pedestrian access to a nearby forest. Environmental education has also been emphasized and there is an impressive onsite environmental information center, the GlashusEtt. Here classes are taught and ecology displays can be found, and the structure itself incorporates and demonstrates a number of green technologies (including a fuel cell, green rooftop and photovoltaic panels). All in all, Hammarby represents an especially valuable new example of sustainable community development, requiring a degree of interdisciplinary and inter-sectoral collaboration that is unusual in most cities.

For more information, see http://www.hammarbysjostad.se/ or http://www.sweden.se/eng/home/society/sustainability/reading/hammarby-sjostad---living-green-in-central-stockholm/.

9 Kronsberg Ecological District, Hannover, Germany

This model ecological housing district is Hannover's newest growth area. Designed and built as a model development for the 2000 World Expo, it incorporates almost every urban sustainability or ecological design element imaginable. The sustainability dimensions begin with its basic form: relatively high density, multi-family housing, sited along a new line of the city's tram system (with three very accessible new tram stops), and with a car-minimal grid street pattern. The entire district is a traffic-calmed (30 km restricted) zone, with extensive bike lanes and onsite car-sharing providing additional alternatives to the automobile. A parking space ratio of 0.8 per apartment was set to reduce parking space needs. The district's new town hall takes sustainability as a key theme. This building is constructed from sustainable materials, with PVs on its rooftop, and houses social service offices, meeting space, and a library specializing in the environment.

The district captures and contains all stormwater onsite through an innovative system of treatment bioswales that feed into two long stormwater retention boulevards, serving also as important green features and delightful pathways. The Germans refer to this as the Mulden-Rigolen-System, or gulley and trench system. Through this stormwater collection system and a number of other water design elements, water is present and made visible in this community. Other sustainability features include extensive use of green rooftops, green courtyards and water features, and community gardens.

The district was designed and built with attention to green space and landscape. During construction, more than 60 hectares were reforested and a soil management plan was created to guide the reuse of excavated soil. The Kronsberg district has 5 to 10 percent more open green space than conventional urban areas. There are five transverse green corridors and a hilltop woodland surrounding the development as well as a system of interconnected public, semipublic and private green space areas within the development.

Figure 15 Natural stormwater collection system at the Kronsberg Ecological District in Hannover, Germany.

Source: Photo by Timothy Beatley.

Figure 16 Centralized solar hot water heating system, Kronsberg, Hannover, Germany.

Source: Photo by Timothy Beatley.

This model district demonstrates a number of important energy features. Homes are designed to meet an impressive low-energy standard, and a goal of a 60 percent reduction in CO_2 emissions compared to standard construction practices was set for Kronsberg. Two very efficient combined heat and power plants provide heat for the roughly 3,000 units in the district's first phase of development (one of these power plants is actually in the basement of a building of flats!). Three wind turbines have also been built, including one large 1.8 MW turbine, and all are but a few hundred meters away from housing. A number of solar energy technologies and design ideas are being tried, including a centralized solar hot water heating system which serves one portion of the district (and stores hot water in a partially underground 2,800 square meter tank, which doubles as a children's play area). Other sustainability elements include a demonstration ecological farm, a sustainable landscape management plan, and a green elementary school.

For more information, see: KUKA, *Living on Kronsberg* (Hannover: Kronsberg-Umwelt-Kommunikations-Agentur GmbH (KUKA), 2000); and http://connectedcities.eu/downloads/showcases/kronsberg_hannover_handbook.pdf.

10 Beddington Zero Energy Development (BedZED), London

Bedding Zero-Energy Development (or BedZED for short) is an ecological housing project in the Hackbridge neighborhood in South London. Designed by Bill Dunster architects, it was intended as the first carbon-neutral development in the UK. The project, completed in 2002, includes a mix of housing and workspace (82 homes, including 14 home/work units), on a reclaimed sewage works. These three-story buildings are oriented with living space to the south to capture the sun, with most

Figure 17 Beddington Zero Energy Development, Hackbridge, London.

Source: Photo by Timothy Beatley.

Figure 18 Dramatic wind cowls that scoop up air ventilating BedZED buildings and giving the neighborhood a distinctive and colorful look.

Source: Photo by Timothy Beatley.

units provided with rooftop sky-gardens. Well-insulated, and with floors and walls providing extensive thermal mass, a unique system of natural ventilation is provided through a visually distinctive set of windcowls that rotate into the wind, capturing fresh air but also extracting heat from outgoing air. Most construction materials were obtained within a 35-mile radius of the site, and wood both for construction and to burn as fuel in the new combined heat and power plant comes from sustainable local forests. Other elements include photovoltaic panels to provide enough electricity to recharge a fleet of electric car-sharing cars, and a combined heat and power plant fueled by wood chips, although problems with this facility have led to a gas boiler being employed instead, undermining the project's carbon-neutral billing.

The BedZED project required a legally binding Green Transportation Plan as a requirement for planning permission, and offers car-sharing onsite. The project was planned with public transportation in mind, including railway, bus, and tram. As a result, BedZED has reduced residents' car mileage by 65 percent compared to conventional development.

Much of the credit for this bold project goes to the Peabody Trust, a London housing association known for its support of innovative and sustainable design. A new larger extension of the BedZED ideas is now in the works also in London: Ladbroke Green is a mixed-use project located on a former gas works, where 308 units of ecological housing and 16,000 square meters of commercial space will be built on about 10 acres. With photovoltaic panels covering the rooftops of this project, this mixed-use development would be the largest application of solar panels in the UK. However, in May 2007 the Kensington and Chelsea council announced it would not recommend approval of the project because its proximity to gasholders is a health and safety concern.

For more information see http://www.oneplanetcommunities.org/communities/bedzed/ or http://www.zedfactory.com/projects_mixeduse_bedzed.html.

11 Greenwich Millennium Village, London

This new ecological neighborhood in the heart of the city of London is being built on a former industrial site on the south bank of the Thames. The original plan consisted of 1,377 homes and 5,000 m² of office space on about 29 hectares. Also included are office and commercial space, a community center, cafes, and a large village green. The project is projected to be completed in 2015. Residents of varying incomes live in close proximity, and are served by good transit (the Jubilee line of the London Tube) and extensive bicycle and pedestrian routes.

Buildings in the village, including the Ralph Erskine-designed phase one housing and Proctor Matthews-designed phase two housing, incorporate many green features, including sustainably harvested wood and climate-sensitive designs to maximize use of the sun and minimize the effects of prevailing winds. The building process itself has been sustainable, with 30 to 40 percent of the wood and aluminum waste being recycled. The goal of an 80 percent reduction in energy in comparison with conventional development has been set, as well as a 30 percent reduction in water consumption. A CO_2 emission standard has been set for the village; emissions are not to exceed 20kg/square meter. The insulation standard for the village is 10 percent higher than the national standard. A graywater system is used for flushing toilets, which has helped to reduce water consumption by 20 percent, based on consumption performance. The Village incorporates a combined heat and power plant (the first in Great Britain as part of a private housing development). The power plant, along with the high insulation, standard has reduced primary energy by 65 percent.

An interesting feature of the houses is the design of flexible living spaces. The units utilize sliding interior walls which allow adjustment and reorganization of the spaces as family needs change. Structural systems are designed, as well, to accommodate upgrades and modifications later (e.g. the addition of balconies).

For more information see http://gmv.gb.com/ or http://greenwichmillenniumvillage.tumblr.com/.

Figure 19 Greenwich Millennium Village bike path.

Source: Photo by Timothy Beatley.

Figure 20 Greenwich Millenium Village ecological housing.

Source: Photo by Timothy Beatley.

S
I
X

12 Nieuwland (Solar Suburb), Amersfoort, Netherlands

One of the largest demonstrations of a community organized around solar energy, this new district of 4,500 houses in the Dutch city of Amersfoort has been nicknamed the "solar suburb." The sustainability of this unique town begins with orientation – some 85 percent of the buildings are oriented to the south in this otherwise circular urban form with a town center (with grocery, post office, shopping) at the core. Everything is within walking distance, and the community is served by a good bus service. Both photovoltaic and solar hot water panels are used throughout. PVs on building rooftops and façades generate 1.35 megawatts. Major institutional buildings, including two elementary schools, a sports complex, a kindergarten, and multi-family social housing, all have photovoltaics on their rooftops. Two pilot energy-balanced homes have also been constructed here. These are grid-connected homes that generate as much energy as they need over the course of a year.

The unique energy features of this community have benefited from substantial technical and financial support from the regional energy company REMU (Regionale Energiemaatschappij Utrecht), as well as from the Dutch Energy Agency (Novem) and the European Union. The positive environmental impacts of this project are tremendous, with an estimated reduction in carbon dioxide of almost 89,000 kilograms/year (approximately 98 tons).

For more information, see: REMU, *Building Solar Suburbs: Renewable Energy in a Sustainable City* (Utrecht: REMU, 1999) or http://www.secureproject.org/download/18.360a0d56117c51a2d30800078414/Nieuwland_Amersfoort_NL.pdf.

Figure 21 Nieuwland solar community, Amersfoort, Netherlands.

Source: Photo by Timothy Beatley.

Figure 22 Solar homes at Nieuwland.

Source: Photo by Timothy Beatley.

13 Village Homes, Davis, California

The brainchild and creation of husband and wife team Judy and Michael Corbett, this sustainable neighborhood in Davis, California has been an inspiring model since the late 1970s. This mostly single-family community (240 units on 60 acres) is organized around a series of interior greenways, which collect the development's stormwater (there is no conventional stormwater collection system), and provide a beautiful network of community greenspaces and walkways. Relatively small homes on small lots are grouped in clusters of eight, with small green spaces connecting to larger spaces. Homes are oriented south and employ passive solar design; units of different sizes and types are mixed together. Narrower-than-typical streets provide access to the rear of the homes (homes front on the connected greenspaces), and the east–west orientation allows for solar access which most of the homes take advantage of. Fruit trees and edible landscaping pervade the neighborhood. Pocket orchards produce large quantities of apricots, cherries, peaches, pears, persimmons, plums, and citrus for residents, and two extensive community garden areas provide ample space for urban agriculture.

While the Corbetts faced significant obstacles in building Village Homes (such as objections to the natural drainage system by the city's public works department and the fire department's opposition to the narrow streets), they were able to creatively overcome them. For example, a compromise on street width was reached, providing for a 3-foot clear zone on each side to allow emergency vehicle movement. The project has been a success on virtually every measure: the homes are highly desired, residents tend to know each other, and crime in the neighborhood is very low compared with nearby conventional

Figure 23 Natural drainage swale at Village Homes, Davis, California.

Source: Photo by Timothy Beatley.

Figure 24 Green storm water swale at Village Homes.

Source: Photo by Timothy Beatley.

suburban-style development. The natural stormwater system has worked well, and incidentally resulted in a $600 per home savings, providing much of the funding for the neighborhood's impressive landscaping and green features.

For more information, see Michael Corbett and Judy Corbett, *Toward Sustainable Communities: Learning from Village Homes* (Washington, D.C.: Island Press, 1996), or go to http://www.villagehomesdavis.org/.

14 U.C. Davis West Village, Davis, California

Only one mile from Village Homes in Davis, California is a second pioneering eco-community which may become the first truly zero-net-energy (ZNE) neighborhood in the U.S. and perhaps the world. U.C. Davis West Village aims to house more than 3,000 people in 662 apartments and 343 single-family homes on 130 acres of land. The new neighborhood also includes 42,500 square feet of office space, a community college building, a recreation center, a village square, and courtyards and playing fields. The project's first phase opened in 2010, and the apartment housing was completed in 2013.

Central to the zero-net-energy strategy was constructing buildings that are more than 50 percent more energy efficient than the State of California's already-strict Title 24 energy code. Designers achieved this energy performance through added insulation, radiant barrier roof sheathing, solar reflective roofing, high-efficiency lighting fixtures, high-efficiency appliances, and high-efficiency heating, ventilation, and air conditioning (HVAC) units. Extra-thick exterior walls use 2″ × 6″ framing to accommodate the additional insulation. Buildings also use passive solar design features extensively, including southern orientation of most structures, large south-side windows with summer overhangs or moveable shades, and vertical corrugated metal on south and west sides of buildings to ventilate the facades and act as a thermal shield. In an effort to promote energy-conserving behavior, residents receive real-time information on their unit's energy consumption.

Four megawatts of photovoltaic panels on building roofs and parking lot shades then provide electricity to offset building use of electricity and gas. An anaerobic waste digester may eventually add to this renewable energy by composting food waste, landscape clippings, and manure to produce methane and

Figure 25 U.C. Davis West Village showing village square and rooftops angled south to maximize sun for photovoltaic panels.

Source: Photo by Robert Segar.

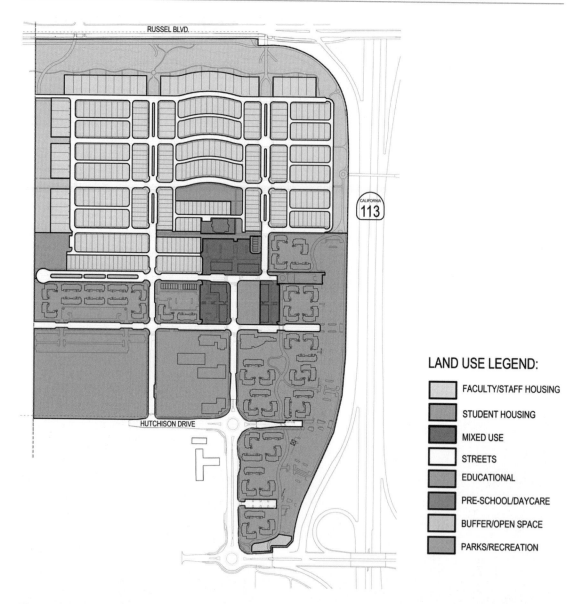

RUSSEL BLVD.

CALIFORNIA 113

HUTCHISON DRIVE

LAND USE LEGEND:

- FACULTY/STAFF HOUSING
- STUDENT HOUSING
- MIXED USE
- STREETS
- EDUCATIONAL
- PRE-SCHOOL/DAYCARE
- BUFFER/OPEN SPACE
- PARKS/RECREATION

Figure 26 U.C. Davis site plan showing highly connected street pattern that maximizes bicycle and pedestrian circulation as well as southern exposure for buildings.

Source: U.C. Davis.

hydrogen, which can in turn be burned to generate electricity. However, during initial operation the PV panels appear to be producing enough energy by themselves to give the project zero-net-energy status.

West Village is intended to meet urban sustainability goals in a number of other ways as well. Most stormwater runoff will be accommodated onsite, and the landscape design features native and/or drought-tolerant plant species. Car use is minimized by providing good bike, pedestrian, and transit connections to the main University of California, Davis (UCD) campus, and by not allowing residents to purchase campus parking permits. Many building components are recycled or otherwise environmentally sustainable. And the neighborhood's faculty and staff housing will be offered on a limited-equity, below-market basis to try to meet affordability goals.

Since the University did not have the capital to build West Village itself, a third-party developer is building the neighborhood under a 65-year ground lease, within a strong planning and design framework established by UCD. A primary goal of financial modeling was to reach ZNE status without placing any additional financial burden on residents. Consequently, residents pay an "energy bill" similar to what they would have paid for energy within a conventional development, but which instead goes toward paying off the additional energy efficiency and solar investments.

For more information, see http://westvillage.ucdavis.edu/ or Stephen M. Wheeler and Robert B. Segar, "Zero Net Energy at a Community Scale: U.C. Davis West Village." 2013. In Fereidoon Sioshansi, ed. *Energy Efficiency: The End of Demand Growth*. New York: Elsevier.

15 Cheonggyecheon Restoration Project, Seoul, South Korea

The Cheonggyecheon Restoration Project in downtown Seoul, South Korea, took down a 5.6 kilometer elevated freeway in the city's downtown to restore an urban stream and create a linear greenway. Rather than producing gridlock, removing a four-lane traffic artery that carried 168,000 vehicles daily and re-placing it with a bus rapid transit line has improved travel times. Urban designer Kee Yeon Hwang and colleagues dreamed up the idea of restoring the stream corridor in 2001, and shopped it around to mayoral candidates in the city of 10 million people. One candidate, Lee Myung-bak, took on the idea and made it a centerpiece of his winning campaign. With a large public engagement program, the city government assuaged the fears of merchants and vendors and designed and built the project in two-and-a-half years.

Figure 27 The Cheonggyeong Restoration Project. Seoul, South Korea removed a 5.6-kilometer elevated freeway to restore green spaces and walkways along this river.

Source: Wikimedia Commons.

Completed in 2005, the $281 million project has helped catalyze civic pride and central business district revitalization. Property values nearby have increased 300 percent, and another elevated freeway in the city has also been taken down in favor of a surface street. About 90,000 people visit the parkway daily. The number of fish species in the river has increased from 4 to 25, and the number of bird species from 6 to 36. Stream restoration has become widespread elsewhere in the country. The success of the project had political benefits as well for its sponsor: Lee Myung-bak was elected president of South Korean in 2007.

For further information, see *Deconstruction/Construction: The Cheonggyecheon Restoration Project in Seoul* (Cambridge: Harvard Graduate School of Design, 2011), edited by Joan Busquets, and websites such as http://www.preservenet.com/freeways/FreewaysCheonggye.html or http://grist.org/infrastructure/2011-04-04-seoul-korea-tears-down-an-urban-highway-life-goes-on/.

16 Weilai City Ecological Community, Dezhou, China

By Kang-Li Wu

Weilai City is a pilot ecological community in the Hedong district of the city of Dezhou, in Henan province, China. The 13.3-hectare residential neighborhood integrates a variety of solar energy technologies into buildings arranged in a garden-like community environment with abundant greening and a beautiful artificial lake located in the center of the community. A prominent Chinese solar energy company has built the community since 2006 to demonstrate ways that high-tech photovoltaic (PV) products can be integrated into building and community design.

Figure 28 The Welai district in the Chinese city of Dezhou integrates solar energy technologies into buildings around a greenway system.

Source: Photo by Kang-li Wu.

Weilai City employs the concept of Building Integrated Photovoltaics (BIPV) in roofs and building façades to generate electricity and to show how PV can be integrated with building design. Wavy-shaped solar PV plates on rooftops help create a unique skyline for the community. Photovoltaics are also used to power community facilities, outdoor lighting, and fountains, and are integrated into the design of a variety of public artworks. A community solar hot water system supplies 300 liters of hot water per household 24 hours per day. Ground source heat pumps help to meet the need for heating and conditioning by conveying the earth's relative warmth into buildings. More than 60 percent of the site is green space, and designers have employed multiple-layer planting methods and native plants. Rainwater is collected and recirculated within the neighborhood's water collection system and the central lake. Buildings use a variety of high-tech energy-saving products, such as better-insulation Energy Efficiency Glass and in-window ventilators to purify air.

The advanced solar energy technology and natural ecological environment attract visitors and home-buyers, and make housing prices of the Weilai community the highest in the city. The Chinese Ministry of Construction and the Chinese Ministry of Finance selected the community as one of the renewable energy demonstration projects in 2006. It also won the Swedish international "Correct Lifestyle" award in 2011. In addition, Dezhou City itself is a pilot solar city selected by the Chinese government. The 2010 World Solar Energy Congress was held in the city, and a variety of photovoltaic products have been used in infrastructure and buildings throughout the city, including solar water heaters, solar building with PV in façades, solar street lighting, street furniture, and public sculptures assembled with PV equipment.

17 Shenzhen Overseas Chinese Town, China

By Kang-li Wu

The Shenzhen Overseas Chinese Town (OCT) is a relatively sustainable urban village that employs a "recreational oriented development" planning model. It is located in the extremely fast-growing city of Shenzhen, near Hong Kong, whose population increased 500 times and GDP 2,000 times between the 1980s and the 2010s. In such a rapidly growing "instant city," the OCT has evolved into a livable, mixed-use, slow-paced, human-scaled eco-town within a big metropolitan city.

The 5-square-kilometer OCT site was a farm in the 1970s, while Shenzhen was just a small fishing village. In the early 1980s, Shenzhen was selected by former Chinese leader Deng Xiaoping as the first special economic zone to help implement an open door policy to the outside. To promote the economic development of Shenzhen, the OCT was originally positioned as a special manufacturing district designed to attract investment from overseas Chinese. The OCT's early development emphasized jobs–housing balance while clustering housing and amenities along with industrial development.

The early planning team headed by Singapore planner Meng Ta Cheang employed ecological planning and design principles such as "building a city in a garden," "planting trees before development," and "no cutting on mountains, no filling in rivers." Thanks to such concepts, much of the original terrain and many unique landscape elements such as hills, lakes, wetlands, and native trees were preserved, and became important public recreational resources later. Moreover, the concept of eco-streets was used in the early planning stage. For example, two-way, single-lane, curving streets discouraged through traffic and thereby maintain a slow and peaceful atmosphere here, while comfortable pedestrian space with good greening was provided along streets connecting to major service facilities. Now the trees planted 20 years ago have grown up, making the entire district full of green.

When the original industrial activities became less competitive, the Shenzhen Overseas Chinese Town Co., Ltd. decided to develop cultural tourism theme parks, including the Splendid China Miniature Scenic Spot in 1989, the Chinese Folk Culture Village in 1990, and the later Window of the World. It also built an elevated tourism light rail transit system linking to these recreational sites. Tourism-oriented land

Figure 29 The Shenzhen Overseas Chinese district integrates housing, transit, and extensive green spaces.

Source: Photo by Kang-li Wu.

development has increased housing prices, and supports hotels, shopping malls, and a variety of retail services. In addition, some of the old industrial factories have been transformed using green building methods into culturally innovative workshops, special stores and restaurants. This district of Shenzhen now provides many attractive public places and buildings to support creative businesses.

Shenzhen OCT illustrates many successful eco-community planning practices. These include:

1 Preserving natural landscape elements and controlling building density and development intensity to allow for better urban greening;
2 Employing recreational land development strategies and introducing diverse commercial and retail activities for both tourists and residents;
3 Promoting jobs–housing balance by providing suitable housing for people working in the region;
4 Encouraging mixed land uses and the development of a pedestrian-friendly environment that allows for all kinds of daily services to be obtained within walking distance;
5 Providing comfortable public spaces for people to gather in; and
6 Developing green streets that fit into the local terrain, limiting the width of the streets to reduce car traffic, and using multiple-layer planting on both sides of the streets to construct comfortable pedestrian space.

Figure 30 Green streets in the Shenzhen Overseas Chinese district.

Source: Photo by Kang-li Wu.

After 20 years' implementation of ecological planning and design concepts, OCT has established certain types of ecological infrastructure. The integration of traditional industries, cultural tourism theme parks, retail stores, hotel industries, and cultural creative businesses also enables OCT to maintain a diverse industrial base, mixed land uses, and a high degree of urban livability. Currently, the residents of OCT share a strong sense of identity which makes it a symbol of success to live here, and housing price increases in this district show it to be among the most sought-after locations in Shenzhen.

URBAN SUSTAINABILITY AT
THE CITY AND REGIONAL SCALE

18 Vancouver, British Columbia

Vancouver, BC, with a population of 600,000 in the City of Vancouver and 2.5 million in the region, has long been a model of relatively compact, sustainable urban development. It has been able to guide much of its growth into compact, dense, walkable, urban neighborhoods, and has pioneered a model of high-rise development in which tall, thin residential towers are set back from the street allowing light and air to reach the streetscape, while a lower level of buildings provides a sense of urbanity along the sidewalk. The city now boasts the lowest per capita carbon footprint in North America.

Growth in Vancouver is governed by a comprehensive set of urban design guidelines that, among other things, stipulate a minimum number of affordable units, and a minimum number of family-friendly units (e.g. units where day care and schools are within a certain walking distance). Buildings are oriented towards the street, and great importance is given to promoting a vibrant street life and amenity-rich urban environment. High-rise buildings must be flanked by four-to-six story structures that serve to soften the visual effects of the skyscrapers, and advance a human street scale.

In 2009 the city adopted a Greenest City 2020 Action Plan, which sets both medium-term (2020) and long-term (2050) goals in a number of areas. The 2020 targets include doubling the number of green jobs and companies, reducing greenhouse gases by 33 percent, reducing energy use in existing buildings by 20 percent, and reducing the average distance that residents drive by 20 percent.

A larger-scale regional livability strategy has been in place since 1996, developed by the Greater Vancouver Regional District (GVRD), now rebranded as Metro Vancouver. The plan calls for steering development in the region into designated town centers (one of these being the City of Vancouver)

Figure 31 New high-rise development along the waterfront of False Creek, downtown Vancouver, British Columbia.

Source: Photo by Timothy Beatley.

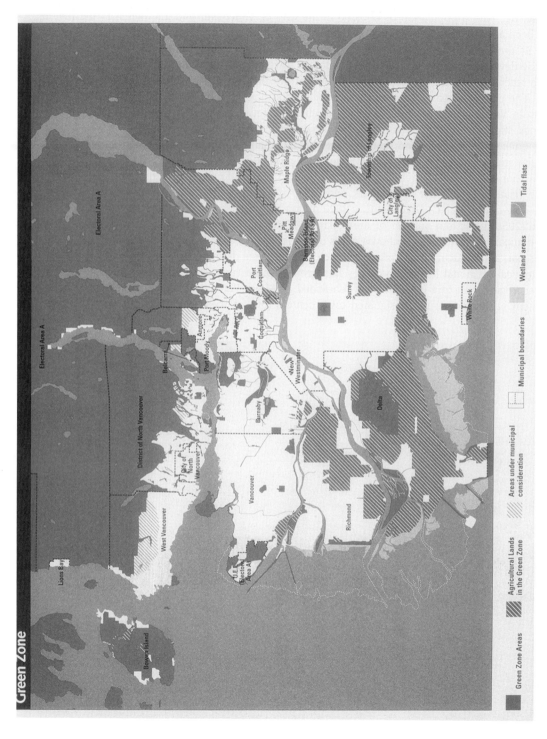

Figure 32 Map of the Vancouver, BC region showing protected Green Zone.

Source: Courtesy Greater Vancouver Regional District.

lying along the route of the Skytrain, an elevated train system. Much rural land in the region is contained in an Agricultural Land Reserve (ALR) and is thus off-limits to suburban development. In 2002 Metro Vancouver adopted a *Roadmap to Sustainability* plan which established sustainability principles for the region, including collaboration, protecting ecosystems, building community capacity, and using full-cost accounting to include environmental and social externalities within economic decision making. To implement the *Roadmap* 10 specific plans focus on energy, water, food systems, parks and greenways, housing, air quality, liquid and solid waste, ecological health, and regional growth. In 2011 Metro Vancouver approved the latest version of its Regional Growth Strategy, called *Metro Vancouver 2040 – Shaping Our Future*. This plan established an Urban Containment Boundary, continues the focus on urban centers and transit-oriented development, and promotes diverse and affordable housing types.

There are many explanations for Vancouver's success, including its relatively strong planning framework and support from the province of British Columbia. The fact that Vancouver has no freeways is also helpful to growth management efforts, since this lessens fast road access to suburban areas (Canada has never had an interstate highways program like in the U.S.). For more information, see https://vancouver.ca/green-vancouver.aspx and www.metrovancouver.org.

19 Bogotá, Colombia

Bogotá, a city of 7.5 million in Colombia, has emerged as a city of innovation and inspiration, forging ahead with an impressive agenda to make the city more livable and sustainable. Largely through the leadership of former mayors Antanas Mockus and Enrique Peñalosa, the city has undertaken a number of creative transportation strategies. Like Curitiba, it is now using a bus-only rapid transit system, with

Figure 33 Bogotá TransMilenio transit system.

Source: TransMilenio SA.

Figure 34 Bogotá Ciclovia.

Source: Instituto Distrital de Recreación y Deporte de Bogotá.

articulated buses operating like a subway. Called TransMilenio, the transit system opened in late 2000. Travel times fell by one third in the TransMilenio corridors and accidents declined by 84 percent. The system expanded steadily to 11 lines in 2012. Because of the transit system, Bogotá has made sufficient air quality improvements to qualify to sell greenhouse gas emission credits on world carbon markets.

Other accomplishments include 300 km of bicycle paths, hundreds of new neighborhood parks, and a new greenway. One of the boldest actions was to convert a major street in the downtown to pedestrians only. This 17-kilometer long street, which includes pedestrian amenities, has become cherished public space. Bogotá has also started something it calls its CICLOVIA – the closing every Sunday of the city's main road arteries, some 120 km in all, in order to make them available to bicycles, walking, and socializing. Former mayor Peñalosa, a champion of this idea, calls it a "marvelous community building celebration," and it attracts an incredible 1.5 million residents each week. Even more impressive is the number of people who participate when the city closes those same streets on a Christmas evening – some 3 million residents participate.

Bogotá's accomplishments are all the more amazing because until the 1990s it was one of the more dystopian developing-world cities, with high levels of violence, corruption, and social inequality (Birney, 2011). Increased national stability, strong leadership, and creation of successful public spaces and transportation infrastructure helped rebuild civil society and leverage dramatic change in the city.

Reference: Birney, Rachel (2011). "Pedagogical Urbanism: Creating Citizen Space in Bogatá, Colombia." *Planning Theory* 10 (1): 16–43.

For more information, see the United Nations Habitat case study at http://www.unhabitat.org/downloads/docs/GRHS.2013.Case.Study.Bogota.Colombia.pdf, and Néstor Sáenz Saavedra, "The Evolution of Transportation Planning in Bogota (2010), available at http://www.docentes.unal.edu.co/nsaenzs/docs/TheEvolutionOfTransportionPlanning_in%20Bogota_Ver_ENG_Stuttgart.pdf. Bogatá has been the subject of several films, including *The People's City: How Bogatá Succeeded in Reducing Traffic Congestion and Smog* (2009), from Cambridge Educational, *Bogatá Change*, a 2010 Danish film from Upfront Films, and *Bogatá: Building a Sustainable City* (2012), available at http://www.youtube.com/watch?v=IjhMQM8eaVY.

20 Gaviotas, Colombia

This small community, a 16-hour jeep ride from Bogotá, is a very different model ecological development in Colombia. About 200 people live in this village founded by Paolo Galori, who thinks of it not as a utopia, but a real place illustrating creative ways to live sustainably. The emphasis is on low-tech ideas and tools, and on designing, building, and creating things uniquely suited to the wet savanna environment. Groundwater for the village is extracted through a specially designed hand pump, so easy to operate that it has been designed into a children's seesaw. There is extensive use of solar energy – for providing hot water, for generating electricity, and for purifying drinking water. Specially designed small windmills capture the modest wind energy. Travel is by way of a specially designed bicycle. Food is produced through a hydroponics system. Gaviotas is, essentially, a solar-powered, self-sufficient ecological village.

It is also an ecologically restorative village. One important economy-building step was the planting of some 20,000 acres of new forest on barren lands using Caribbean pines, which were discovered to be unusually suited to the region's acidic soil. Turpentine is distilled from the bark resin of these trees

Figure 35 Paolo Lugari sitting on one of Gaviotas's famous water pumps, Gaviotas, Colombia.

Source: ZERI Foundation, Gunter Pauli.

Figure 36 Self-sufficient hospital, Gaviotas.

Source: ZERI Foundation, Gunter Pauli.

and has become a major economic product for the village. As well, a rich undergrowth has reestablished in this new forest, with a flourishing of plants and wildlife. Eventually, the pines will be replaced by this regenerated rainforest.

In 2004, Gaviotas became completely independent from fossil fuels through the development of a biodiesel plant, the first in Colombia. The plant was built with support and technical assistance from Boulder Biodiesel, the University of Colorado, and the Friends of Gaviotas. The plant is the first in the world to utilize crude palm oil on an industrial scale. It can produce 400,000 gallons of fuel a year.

The village's hospital, the only one within half a day's car ride, is also designed to function ecologically. Its unique ventilation system brings cool air from a nearby hillside, its electricity is generated from photovoltaic panels, and hot water and distilled drinking water are provided from solar hot water panels. Methane is extracted from livestock dung and sent to the hospital.

Gaviotas has become a model of sustainable village building, in Colombia as well as other parts of Latin America. The water pump for instance is now in use in some 700 other Colombian villages, and the windmills, solar water heating systems, and other technical innovations, have been replicated elsewhere also. Gaviotas shows the power of creative innovation and place-based sustainable living.

For more information see: Alan Weisman, *Gaviotas: A Village to Reinvent the World* (Post Mills, VT: Chelsea Green, 2008), Richard E. White's "Las Gaviotas: Sustainability in the Tropics," published by *World Watch Magazine* in 2007 and available at http://www.worldwatch.org/node/5020; and the website http://www.friendsofgaviotas.org/.

21 Paris, France

Few cities have risen as quickly as Paris to become an exemplar of urban sustainability and sustainable innovation. Beloved for its urban qualities and charm, the city has long illustrated many existing elements of a sustainable city, including dense, mixed-use neighborhoods, great parks, and an extensive underground Metro system. But new initiatives have made it a leader particularly in the area of transportation.

The election of socialist mayor Bertrand Delanoë in 2001 stimulated much of this progress; he campaigned and was elected on a platform of controlling cars and traffic in the city. He was re-elected in 2008. Especially notable under his leadership have been transport reforms including the conversion of roadway space in the city to transit and bicycles (bus-only lanes along major boulevards, and major expansion of bike lanes in the city), and the building of a new tram line around the periphery of the city. Sidewalks have been widened, streets narrowed, trees planted, and thousands of parking spaces eliminated in the city.

Most dramatically, Delanoë closed the Georges Pompidou expressway to car traffic for a month and then eventually installed a beach in its place, complete with sand, parasols and palm trees, along the Seine River. Now called the *Paris-Plage* (Paris Beach) this has become an annual event, every July and August. Complete with concerts, volleyball games, sunbathers, and strolling Parisians, the popular idea has now been embraced by other European cities.

And there have been other innovations designed to help build a more vibrant city life and public realm. For example, every year in October Paris celebrates a *Nuit Blanche* (literally "white night"), an all-night city party, where from 7pm to 7am the city is alive. Museums, restaurants, the Metro are all open and operating the entire evening.

In 2007 Paris made the news for its extremely ambitious public bikes initiative. Operated by the advertising company JCDecaux, the system consists of an extensive network of bike stations around the city – more than 1,500 when the system is fully running, offering the use of more than 20,000 bicycles for short periods. It is called the "Velib" and residents or visitors use a credit card to access the bikes; the first half hour is free, after that period a charge is imposed. The system has already been extremely

Figure 37 Paris Velib bike share system.

Source: Photo by Stephen Wheeler.

popular, with an estimated 2 million journeys in the first 40 days the system operated. A similar car-sharing program known as "Autolib" was launched in late 2011.

In 2007 the city adopted the Paris Climate Plan, which aims to reduce GHG emissions by 25 percent between 2004 and 2025. This plan led to further efforts to reduce road traffic and lower building energy use. Beginning in 2009 the city used aerial thermography to identify particular building rooftops and façades that were in need of insulation. Paris also has developed 290 megawatts of electricity produced from geothermal plans, and operates Europe's second-largest district heathing system, efficiently supplying steam and hot water to 500,000 housing units. As a way to speed climate adaptation, a "Plan for Heat Waves" helps building owners increase insulation, shutters, sun shading, and ventilation. This effort also assists residents in preparing for flooding from the River Seine.

The city of Paris and the Paris metropolitan area certainly still have their share of problems and challenges, of course, including social and economic disparities, inadequate housing, bad air quality, and traffic. However, these recent innovations show especially the possibilities of joining urban sustainability with improvements in the quality and joys of urban life.

For more information, see Lucie Laurien's chapter "Paris, France: A 21st-Century Ecocity," in Beatley, Timothy, ed. (2012). *Green Cities of Europe: Global Lessons on Green Urbanism*. Washington, D.C.: Island Press.

22 Auroville, India

This ecological, utopian new town is located in southeast India, about 160 kilometers south of Madras. Planned to eventually have a population of 50,000, there are now about 1,600 residents, representing about 30 different countries. Auroville has been conceived as a "universal township," inspired by the thinking of philosopher and yogi Sri Aurobindo, as a place where "men and women of all countries are able to live in peace and progressive harmony, above all creeds, all politics and all nationalities."

Human unity is a central ideal behind the town. The town's physical form is in the shape of a galaxy, with a matrimandir, or large golden temple, in the very center. This spiritual center of the community, or Peace Area, also contains a Banyan tree and an amphitheater for community events. The community's more elaborated master plan sets out a series of zones, essentially concentric circles, that extend out from the town's center. These include residential, international, industrial and cultural zones, as well as a large Green Belt. The nearly 1,500-hectare Green Belt is home to agriculture, forests, and biodiversity.

Auroville's many other ecological elements include extensive efforts at environmental regeneration of the landscape (planting of more than 2 million trees, extensive soil and water conservation, training programs in sustainable farming), the use of earth construction techniques in building (use of onsite, locally produced compressed earth blocks), and extensive use of renewable energy (e.g. 30 windmills, a solar community kitchen producing 900 meals a day with 50 percent of the energy coming from a "giant mirrored solar dish"), among others.

The region around Auroville contains 13 villages and 50,000 people. Much of the work of Auroville has focused on improving the quality of life and sustainable practices in these surrounding villages. This has happened through training and workshops, for instance, and through cooperative projects.

For more information see: www.auroville.org.

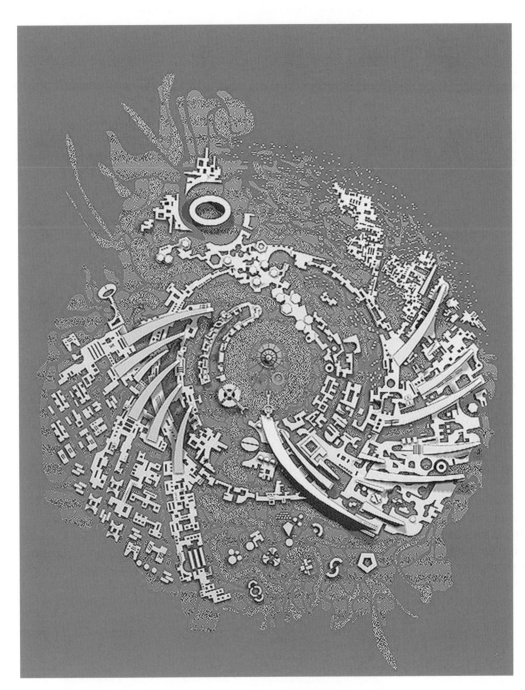

Figure 38 Auroville Galaxy Plan, Auroville, India.

Source: Photograph by Pino Marchese.

Figure 39 *Matrimandir* (temple) and Peace Area, Auroville.

Source: Photograph by Pino Marchese.

23 IBA Emscher Park, Germany

The Ruhr Valley in northwestern Germany is a landscape heavily scarred by industrialization and littered with the remnants of coal mining and steel production. In 1989 the state of North-Rhine Westphalia along with the federal government embarked on an ambitious project to promote long-term rehabilitation and reuse of this immense industrial area covering 800 square kilometers (307 square miles) along a 70 kilometer (43 mile) urban corridor. Called the International Building Exhibition (IBA) Emscher Park, more than 100 demonstration projects were funded over a 10-year period, with a remarkable impact on the cities and landscape, and a new appreciation and pride for an industrial past that most in the region sought to hide or forget. After 1999 a successor "Project Ruhr" continued to clean up the Emscher River, an effort which is expected to be completed in 2014.

An impressive array of projects and creative examples of land and landscape recycling have resulted. Projects have included the establishment of 11 technology centers, new ecological housing on reclaimed

Figure 40 Landscape Park, Duisburg-North, former steel mill, design by Peter Latz.

Source: Photo by Timothy Beatley.

Figure 41 Major foundation pillars used for climbing and rapelling, Landscape Park, Duisburg-North.

Source: Photo by Timothy Beatley.

brownfield sites, and the conversion of industrial buildings to cultural uses (e.g. conversion of the gasometer in Oberhausen to an exhibition hall). The Park also includes a network of hiking paths and bicycle trails, recreational areas and nature reserves. The core of the environmental design is the recovery of the existing watercourses. One of the most creative projects is the Landscape Park in Duisberg-North. Here, a former steel mill has been converted into a regional park and industrial monument, where formal gardens have been created and trees planted amidst the remnant buildings, and massive foundation pillars have been converted into areas for climbing and rapelling.

Monte-Cenis Academy, in Herne-Sodingen, Germany, is one of the most impressive buildings sponsored through IBA. This government training center, designed by Jourda and Perrandin, takes the form of a large glass building, a "vast timber framed hanger" that contains offices, seminar spaces, a library, town hall, cafes, restaurant, and hotel. Nine wooden buildings lie protected under this "glass weather shield." A series of computer-controlled flaps and louvers open and close, depending on outside weather patterns, to ensure comfortable temperature and ventilation inside the glass shield. The structure permits use much of the year with minimal heating, and allows natural cross ventilation during the summer. Most dramatic of all are the photovoltaic (PV) cells, 10,000 square meters in all, on the building's rooftop. More power is produced than is needed by the structure. Careful thought was given (and simulation done) on how to configure or array the PV panels. In the end they were grouped in such a way that they "form cloud-like patterns that magically diffuse light into the great greenhouse" (Kugel, 1999). It is the largest solar roof in the world.

Sited on former coalfields, the development builds on this history in several ways. The architecture evokes its mining and industrial past; according to one observer, "its monumental scale and repetitive structure evoke the big sheds, furnaces and factories of the Ruhr's old industrial landscape" (Kugel, 1999). A combined heat and power plant, moreover, takes advantage of this port in even more tangible ways, as it utilizes escaping methane to produce heat and electricity.

For more information, see Kugel, Claudia (1999). "Green Academy." *Architectural Review*. October, http://www.fedenatur.org/docs/docs/238.pdf, and http://www.dac.dk/en/dac-cities/sustainable-cities/all-cases/green-city/emscher-park-from-dereliction-to-scenic-landscapes/?bbredirect=true.

24 London, England

With the reestablishment of city-wide governance in London at the end of the 1990s (the new Greater London Authority), an impressive new emphasis has been given to sustainability in this metropolitan area of 7.5 million people. The stated vision is to "develop London as an exemplary, sustainable world city." Already, a number of specific sustainability plans and actions have been developed in the city. Especially impressive are city's Biodiversity Strategy and Energy Plan (including greenhouse gas emission targets). London also made its commitment to sustainability a central part of its hosting of the 2012 Olympics, and incorporated green features into facilities for that event.

Historically, London has been a sustainability leader through its establishment of an urban green belt in 1938, which sharply restricts development in a large area around the central city. This buffer has successfully preserved both countryside and existing small villages near the city. To house its growing population, the city also established a variety of new towns in the early and mid-twentieth century, many of which still serve as excellent examples of transit-oriented development today.

Under the leadership of mayors Ken Livingstone and Boris Johnson in the 2000s, new institutional structures were formed to promote and consider sustainability, including the London Sustainable Development Commission. The key charge of the Commission is to establish the London Sustainable Development Framework. Elements of this framework include the development of a comprehensive set of sustainable indicators, and review of the proposed spatial plan for London. The city is making dramatic efforts to reduce automobile traffic in its center, through a bold and controversial road pricing scheme. The Congestion Charge Zone was begun in 2003 by charging motorists five pounds per car entering Central London during the hours of 7am to 6pm Monday through Friday. In 2005 the charge was raised to eight pounds, and in 2007 it was extended to West London as well. Net revenues from the system have allocated hundreds of millions of pounds to public transit.

A city-commissioned study entitled *City Limits* presents a detailed analysis of the material flows required by the city and calculates its ecological footprint. The first such detailed study for a city of this size, the report concludes that London's footprint is extremely large, nearly 300 times its own land area. The study documents a tremendous material flow of inputs and outputs. Londoners consume, for instance, 154,407 gigawatts of energy each year, 2 million tons of wood products, and 730,000 tons of vegetables. At the same time, they emit 50 million tons of carbon dioxide and produce 8 million tons of sewage sludge.

A 2011 London Plan adds new twenty-first-century emphases on climate planning and strengthening of diverse neighborhoods. A Blue Ribbon Network aims to link rivers, ponds, reservoirs, and canals into a more integrated system of waterways with improved habitat value. All new development is expected

Figure 42 New London City Hall, a green building designed by Norman Foster.

Source: Photo by Timothy Beatley.

to be zero-carbon by 2016 (for residential buildings) or by 2019 (for nonresidential buildings). The city is focusing in particular on the development of decentralized energy systems providing district heating, cooling, and electric power, with a goal of supplying 25 percent of these needs in this way by 2025.

For more information see: Camilla Ween's chapter "London, England: A Global and Sustainable Capital City," in Beatley, Timothy, ed. (2012). *Green Cities of Europe: Global Lessons on Green Urbanism.* Washington, D.C.: Island Press. Also see Greater London Regional Authority, "Regeneration," at https://www.london.gov.uk/priorities/regeneration, and the London Sustainable Development Commission website at http://www.londonsdc.org/.

Figure 43 The Swiss Reinsurance Building in London, popularly known as "The Gherkin," uses half the energy that similar towers consume.

Source: Photo by Timothy Beatley.

25 Masdar, United Arab Emirates

One of the most ambitious sustainable city projects in the world is the new city of Masdar, in the United Arab Emirates. Designed by the British firm Norman Foster and Partners, the project is intended as a low-rise, mixed-use, high-density district for 50,000 people about a dozen miles from the city of Abu Dhabi, near an international airport. Nearly one-mile square, Masdar is becoming home to alternative energy research institutes, green businesses, the International Renewable Energy Agency, and a new Masdar Institute of Science and Technology, which opened in 2010 and is collaborating with MIT in the US.

Perhaps the most creative aspect of Masdar is its integration of modern technology with ancient architectural practices from the desert environments of the Middle East. Buildings shade one another as well as the narrow streets. Perimeter walls keep out hot desert winds. Hollow "wind towers" funnel air to street level. Oasis-like interior gardens provide users with respite from the glaring sun and strong

Figure 44 Artist's conception of Masdar, United Arab Emirates.

Source: Foster and Partners.

Figure 45 Courtyard at Masdar showing façade screening, ventilation tower, and photovoltaics.

Source: Wikimedia Commons.

winds. Building façades feature mesh screens that shade surfaces, minimize direct sunlight to interiors, and also reflect traditional Islamic designs.

Initial plans called for Masdar to be zero-carbon, car-free, and zero-waste, but by the early 2010s developers were aiming for low-carbon status with transportation by foot, very energy-efficient private vehicles, and a light rail system connecting the new district to Abu Dhabi. Living and working are intended to take place in close proximity, reducing transportation needs. Ninety percent of the community's electricity is to come from photovoltaics, with the remainder produced by waste incinerator plants. A 10-megawatt solar plant provides most of the power for the first phase of development. Water use will be much lower than in other desert communities, with much of this resource provided by desalination plants. About 80 percent of water will be recycled.

Built by a government-owned development company, the $20 billion project has encountered skepticism and criticism since plans were first announced in 2007. Many have noted the basic irony of a low-carbon city being built atop the world's seventh-largest oil and gas reserves and financed with petro-dollars. The socially stratified, nondemocratic nature of the Emirates is also often seen as incompatible with sustainable city ideals. After Masdar's first phase opened in 2010, *New York Times* architecture critic Nicolai Ouroussoff commented that the design "reflects the gated-community mentality that has been spreading like a cancer around the globe for decades. Its utopian purity, and its isolation from the life of the real city next door, are grounded in the belief . . . that the only way to create a truly harmonious community, green or otherwise, is to cut if off from the world at large." Still, despite such drawbacks Masdar seems likely to pioneer many green development practices for desert environments.

For more information, see http://www.masdar.ae/en/#city/all or http://www.fosterandpartners.com/projects/masdar-development/.

26 Songdo, South Korea

Somewhat similar to Masdar in terms of being a new, high-tech green city developed outside existing urban areas, the district of Songdo is rising on 1,500 acres of former marshland and bay fill 40 miles from Seoul, South Korea. Near Incheon International Airport, the new town is intended as a model of sustainable, city-scale development, albeit in large-scale, corporate, "instant city" fashion. Its first phase opened in 2009 and includes about 100 buildings. Many are high-rises; the 65-story Northeast Asia Trade Tower is South Korea's tallest building.

Designated a "Free Economic Zone," Songdo is seeking to become home to green-tech companies. The $35 billion project is being designed and developed by New York City-based Gale International and Korea's Posco E&C. Cisco Systems is working with the developers to give the entire town a high-tech communications infrastructure. By 2016 the city is expected to be home to 65,000 residents and boast 45 million square feet of office space. Structures are LEED-certified and the whole community is seeking LEED-ND (Neighborhood Development) certification. There will be no garbage trucks; an automated waste disposal system sucks trash from households through pipes to processing centers. The town will also include an 18-hole golf course and a 100-acre Central Park. Two American universities, George Mason University and the University of Utah, are expected to open campuses in Songdo.

The Songdo development was initially criticized by South Korean environmentalists, due to loss of wetland habitat for threatened bird species. Like Masdar, it is also open to criticism for being an over-scaled, corporate-dominated enclave far removed from existing cities. It remains to be seen whether residents and businesses will relocate to the new community; much of the commercial office space is currently empty.

For more information, see http://www.songdo.com/songdo-international-business-district/why-songdo/sustainable-city.aspx; Sheridan, Mike (2010). "Songdo: Sustainable City of the Future." *Urban Land*, October 28; Williamson, Lucy (2013). "Tomorrow's Cities: Just How Smart is Songdo?" *BBC News*, 1 September, available at http://www.bbc.co.uk/news/technology-23757738.

Figure 46 Greenway leading to Songdo International Business District.

Source: Gale International.

27 Austin, Texas

Although it has long been a leader in promoting green building, Austin, Texas began its Community Sustainability Initiative (CSI) in 1997 by establishing the City's Office of Sustainability. This new branch of city government spearheaded the city's climate action program beginning in 2007, and expanded sustainability efforts in 2012 with an intensive, two-year "Rethink Austin" campaign. That program includes more than 40 initiatives in different city departments. In 2011 Austin became the largest local government in the US to have all civic facilities powered by 100 percent renewable energy, and in 2013 it received a national Climate Leadership Award from the US Environmental Protection Agency.

Austin's Green Building program is one of the oldest such efforts in the United States. Under this program, residential and commercial buildings are rated according to the extent and number of green features (receiving from one to five stars). The city provides technical assistance, convenes workshops and training for builders and designers, helps market green homes, and provides a variety of rebates and subsidies for green construction and rehabilitation (e.g. energy-efficiency loans). In the early 2010s Austin added incentives for green roofs, and created green roofs for Austin City Hall and other showcase buildings.

At the turn of the millennium Austin set a goal of making all new single-family construction zero-energy capable by 2015. Homes built by that time are to be 65 percent more efficient and solar-power capable, so that adding photovoltaics could produce the remaining 35 percent of energy needs. As of 2012 the city was on track toward this target, with the city council adopting a series of increasingly tough energy conservation codes to meet the efficiency goal. To encourage energy efficiency improvements for existing homes, a 2008 ordinance requires that homeowners have energy audits performed on their properties prior to sale.

Especially notable in the past have been Austin's Smart Growth Initiative and Green Builder Program. Under its Smart Growth Initiative, Austin prepared a Smart Growth Map indicating "desired development zones" where future growth is encouraged through a system of incentives. The goal is to discourage sprawl and to promote more compact, higher density within the existing urbanized areas of the city. Development proposals are evaluated through the city's Smart Growth Matrix and the resulting score (based primarily on the proposed location) determines the extent of financial incentives available in the form of infrastructure fee reductions or waivers. Fee waivers and expedited permit review are also available under the city's S.M.A.R.T. housing incentives for the provision of affordable housing.

In 2012 the city adopted an Imagine Austin Comprehensive Plan, which while not overtly a "sustainability plan" continues many of those same policies. Following City Council mandate, sustainability is "the central policy direction" of the document. In addition to growth management it tackles issues of social segregation, livability, public health, and climate change. The plan outlines urban infill and revitalization strategies more specifically than in the past, establishes a network of centers and corridors within the city, and proposes an expanded urban trail system with recreational routes along many creeks.

For more information, see http://austintexas.gov/department/sustainability and http://www.austintexas.gov/department/our-plan-future.

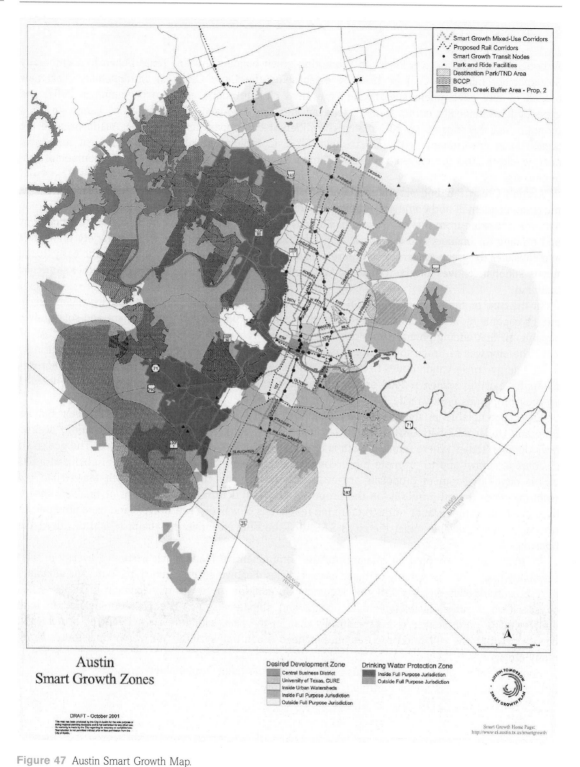

Figure 47 Austin Smart Growth Map.

Source: Courtesy Department of Growth Management, City of Austin.

28 New York City

Few cities, certainly few American cities, have been as successful at creating the basic conditions of sustainable urban living as New York City has: relatively dense, compact, mixed-use and walkable neighborhoods, good public transit and high rates of transit use (the highest percentage of workers traveling to their jobs by transit), and access to impressive parks and green space, among other qualities. New York City continues to exert important leadership in urban sustainability and has been expanding upon this history of planning and environmental innovation. Under the leadership of Mayor Michael R. Bloomberg the city adopted innovative green building standards, known as High Performance Building Standards, along with High Performance Infrastructure Guidelines. The City's Parks and Recreation Department developed a similar set of high performance guidelines for landscape and park management.

Under Bloomberg sustainability was given greater visibility and importance throughout city government. In 2006, the Mayor created an Office of Long-Term Planning and Sustainability, as well as a Sustainability Advisory Board and a partnership between the City and the Earth Institute at Columbia University, through which the Institute serves as scientific advisor on sustainability matters. The mayor's sustainability office helped develop *PlaNYC: A Greener, Greater New York* in 2007, and updated it in 2011. An impressive vision of a sustainable world city, this document contains a dizzying array of urban sustainability measures and proposals. In unveiling the plan, Bloomberg declared the city's intention of becoming "the first ecologically sustainable city of the 21st Century."

Figure 48 The City of New York has experimented with pedestrianizing Times Square and Herald Square, leading to increased tourism, business satisfaction, and commercial rents in those locations. Photo by Mario Roberto Durán Ortiz.

Source: Wikimedia Commons.

PlaNYC is quite comprehensive, though implementation of many of its parts remains an uncertainty. It contains elements that address land use and housing, open space, water, transportation, energy, air, and climate change. Its housing element actively embraces accommodation of nearly a million new residents in the city by 2030, and emphasizes the need to find creative places to house this additional population. New housing is envisioned through the rezoning of land and redevelopment of brownfield sites within the city. Perhaps most creatively, the plan also envisions decking over rail yards, rail lines and highways to create new housing sites.

A number of measures are proposed for enhancing and expanding green spaces within the city, and the plan declares the goal of having a park or playground within a one-quarter-mile walk of every New Yorker. The plan proposes opening schoolyards as parks and expanding the City's Green Streets program. A major new initiative to plant trees is intended to add 1 million new trees by 2030 (planting 23,000 additional new trees each year). The plan calls for the greening of parking lots, the use of vegetated swales and other green stormwater management techniques, and new incentives to encourage the installation of green rooftops. The city also began offering a property tax abatement to offset up to 35 percent of the cost of installing a green rooftop in the city.

Some of the plan's most controversial proposals are found in its sections on energy and climate change. The plan sets the goal of reducing the city's greenhouse gas emissions by 30 percent by 2030 through a variety of means including strengthening its energy code, retrofitting city buildings, expanding district energy (i.e. combined heat and power), encouraging renewable energy (e.g. a property tax rebate for solar panels), and the creation of a new city energy planning board. The most controversial element was a proposal to impose a congestion pricing system for cars traveling into Manhattan, similar to the system employed in London. This idea was scuttled by the state legislature in Albany.

Following Hurricane Sandy in 2012, which seriously damaged many coastal areas within the region, New York developed a strong climate resilience program. Announced by Mayor Bloomberg in 2013, the rebuilding and resiliency effort includes more than 250 actions to reduce risk from future storms and climate change. Among these initiatives are storm surge barriers on creeks and waterways, levees or dune restoration on beachfronts, mandates for large buildings to undertake flood retrofits, and revised building codes throughout the city. The complete plan is expected to cost $19.5 billion.

For more information, see http://www.nyc.gov/html/sirr/html/report/report.shtml or Freedman, Andrew (2013). "New York Launches $19.5 billion Climate Resiliency Plan." Climate Central, http://www.climatecentral.org/news/new-york-launches-20-billion-climate-resiliency-plan-16106. For info on PlaNYC see http://www.nyc.gov/html/planyc2030/html/home/home.shtml.

29 Portland, Oregon

Few American cities have accomplished as much as Portland, Oregon in the area of urban sustainability. This region of about 2.3 million people has developed a deserved reputation as an American testing ground for innovative urban planning ideas.

Promoting a more compact urban form is one of this region's key accomplishments. In part this was a response to the passage of Senate Bill 100 in Oregon in 1973. This law mandated, among other things, that all cities adopt urban growth boundaries (UGBs) and that land use plans and regulations protect productive forest and farmland. UGBs are intended to delineate between developable or urbanizable lands and areas that are to be conserved and protected from development. Around Portland this UGB has been delineated on a regional basis, encompassing 23 smaller cities and portions of three counties. This regional-scale growth management was aided substantially through the establishment of a strong regional government, the Portland Metropolitan Services District, or "Metro" for short. The only popularly elected regional government in the United States, it has real powers especially in the areas of land use, transportation, parks and green spaces, and solid waste management.

Figure 49 Portland MAX light rail system.

Source: Photo by Timothy Beatley.

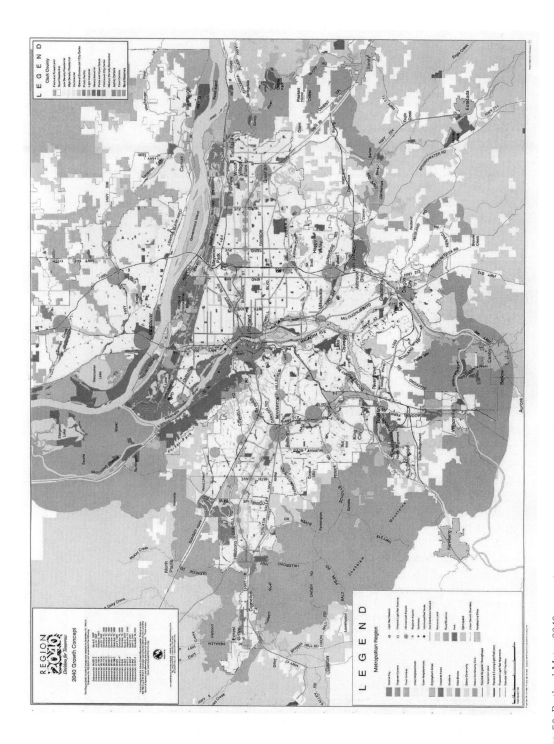

Figure 50 Portland Metro 2040 concept map.

Source: Courtesy of Portland Metropolitan Services District.

The City of Portland has itself taken many actions to reduce the reliance on private automobiles, to promote a more walkable city, and to protect its environment. A series of exemplary downtown plans have sought to increase the amount of housing in the center, to re-connect the city with its riverfront, and to create a highly attractive urban environment. A percent-for-art program has resulted in delightful street sculptures and public art. The city, moreover, has a long history of recapturing space from the automobile and giving it back to pedestrians. Dramatic examples include tearing out a riverfront highway and putting a park in its place – Tom McCall Waterfront Park – and removing a downtown parking garage to create Pioneer Courthouse Square, scene of many concerts, festivals, and other public events. A highly popular regional light rail system, the Metropolitan Area Express, or "MAX," connects much of the region, in addition to an exemplary bus system, and much regional planning has occurred to guide growth along transit corridors. The city and region have also increased significantly the number of bikeways, now extending for more than 240 kilometers (149 miles).

Together these planning tools have led to a somewhat more compact regional development pattern than in most other American cities, with a much more vibrant urban core and relatively effective public transit. While the UGB has been extended several times, the overall pattern of development is gradually becoming denser and more pedestrian friendly. A long-range regional plan, Region 2040, developed through an extensive public process, envisions maintaining the region's tight urban form and steering much new growth into a series of centers along the spine of the MAX system. The goal is that 85 percent of new residents will be within a 5-minute walk of a transit station.

While Portland has secured its reputation on these successful land-use and transportation achievements, in recent years it has embraced a broader, more expansive urban sustainability agenda. It now boasts a city Office of Sustainable Development (formerly Portland Energy Office), housed in the Jean Vollum Natural Capital Center. The city has adopted a set of sustainable city principles and has been promoting a variety of other green city ideas. These have included green building, sustainable technologies and practices, energy conservation and efficiency, and solid waste and recycling. The City's Green Investment Fund has provided important funding to support green building, and its green building policy mandates minimum standards for city-funded facilities. Portland's 2009 Climate Action Plan, updated in 2013, calls for a 40 percent reduction in carbon emissions by 2030 and an 80 percent reduction by 2050, compared to 1990 levels. Through its land-use and transportation planning, in addition to improved vehicle fuel efficiency and electricity generation, Portland and Multnomah County reduced greenhouse gas emissions per capita by 26 percent between 1990 and 2010. Total carbon emissions fell 6 percent, taking population growth into account.

Portland has not been without its critics. Sharp rises in the price of housing in recent years have been blamed on its UGB and growth controls, for instance (though it is far from clear that this has been the cause). Others have been critical that while greater density and compactness have been achieved, the overall form this development takes is not dramatically different than other places; much of it is still very suburban and car dependent.

For more information see the Portland Office of Sustainable Development's website at http://www.portlandonline.com/osd/ or the Metro Council's at http://www.oregonmetro.gov/.

30 Burlington, Vermont

This progressive city of about 40,000 has made a name for itself for embracing and implementing a host of community sustainability measures. The city has adopted the Earth Charter, an international declaration of environmental principles, and has made significant commitments to sustainability. It boasts one of the few successful pedestrian malls in the nation, and its town center is vibrant, walkable, and mixed use.

Burlington is the center of many creative urban sustainability ideas. A nonprofit called The Good News Garage, for instance, repairs donated automobiles, provides them to low-income families at cost, and also provides job training. The Burlington Community Land Trust, formed in 1984 with seed funding

Figure 51 McNeil generating plant, Burlington, Vermont.

Source: Burlington Electric Department.

Figure 52 Young and old together at the McClure Intergenerational Center, Burlington, Vermont.

Source: Photograph by Nicole Craft.

from the City, is an innovative model for both conserving land resources and providing affordable housing. The first of its kind in the nation, the Trust buys land, builds housing, and then rents or sells the homes to low- and moderate-income families. The Trust has also been instrumental in securing new parks, health care, and other community facilities, and in restoring and recycling older buildings in the city, for instance recently converting old bus and trolley barns into apartments.

Burlington residents give strong support to local organic agriculture. The Intervale Farm Area has become an important national model. Consisting of about 800 acres of floodplain land along the Winooski River, the Intervale is now home to two community-supported agriculture (CSA) operations and several other commercial organic farms. Managed now by a nonprofit foundation, the Intervale is also home to the Burlington Compost program, a living machine (a biological system for purifying wastewater) producing a Tilapia fish farm, a gardeners' supply store, and a small farms incubator program. A Food Enterprise Center is currently under development. The Burlington Food Council has created an action plan aimed at connecting school students to their food sources and serving local produce in the meal program. The School District has embraced this program, increasing its purchasing of local produce by 50 percent.

Energy conservation and renewable energy are also important topics in Burlington. The McNeil Generating Station, run by the city's Electric Department, generates 50 megawatts of electricity from burning wood from local private woodlands and mill wood waste. This plant generates nearly enough electricity for the city. In 2005, 46 percent of the energy for the city's energy portfolio came from renewable sources.

Perhaps most impressive is the Burlington Legacy Project, an extensive community-based process created in 1999 that developed a vision and plan for the city for the year 2030. The Burlington Legacy document lays out a common future vision, identifies and discusses key themes (economy, neighborhoods, governance, youth and life skills, and environment), sets goals and priority actions, and provides profiles of projects already underway. The bottom-up planning process involved unusually extensive opportunities for community participation and some unique elements, including community surveys, focus groups convened at the neighborhood level, active involvement of the city's youth (e.g. poster and essay contests, school focus groups), public hearings, and a "Summit on the City's Future." Some recent projects undertaken by Burlington Legacy include a no-idling campaign (a public awareness campaign to reduce unnecessary vehicle idling), a sustainable schools project which is establishing sustainability as part of the curriculum for elementary schools, and the Youth on City Boards and Commissions Project, which is a pilot program which allows high school students to serve on city and nonprofit boards and commissions. A 2010 Report Card on the Legacy Project showed that the city had made substantial progress towards goals related to economy, youth/life skills, and environment.

In terms of climate planning, Burlington developed one of the nation's earliest climate plans in 2000, which was updated in 2008 and replaced by a new Climate Action Plan in 2013. The latest plan shows that the city had made progress in reducing per capita emissions from electricity and natural gas, but not transportation. New policies call for improved public transportation and bicycle and pedestrian infrastructure, as well as higher prices for parking.

For more information, see http://www.burlingtonvt.gov/City/Sustainability/ or http://www.burlingtonvt.gov/Legacy/.

31 Oslo, Norway

Oslo's motto is "the Blue and the Green and the City in Between." With the Oslo Fjord to the south and large forested areas to the north (two-thirds of the city is in protected forest), there is immense nature near to where Oslo residents live. It is estimated that 94 percent of the city's residents live within 300 meters of a park or green space, a very good sign. There is especially a culture of visiting and recreating in the city's forests, what Norwegians affectionately refer to as the "marka."

The City strives to, at one and the same time, promote a denser, more compact urban form, while also strengthening connections to nature. The City's green plan lays out the bold vision of restoring and

Figure 53 Oslo river and greenway.

Source: Photograph by Timothy Beatley.

bringing back to the surface all eight of the main rivers that run through the city. The Akerselva is already mostly restored, and serves as an amazing green corridor, where hikers and bicyclists, and families, picnicking can hear and see nature in the very heart of this city.

Oslo is also making a concerted effort to get its residents out of their cars. The city has created a 365-kilometer trail system, and is investing heavily in bicycle infrastructure, for example adding 1,000 new bike parking spaces in the last decade. Meanwhile the city is gradually reducing parking spaces in the center, and was one of the first cities to impose a congestion charge for driving into the downtown. Forty-five percent of the revenues from that charge go towards improving public transportation.

The proximity to water in Oslo also promotes a biophilic connection between the residents and natural landscapes. The waters of the inner fjord have become much cleaner in recent years, and are now swimmable and fishable. Many Oslo residents own their own boats. In terms of climate planning, Oslo is the best-performing European city on carbon emissions, due to its expanding district heating system and public transit improvements as well as the large percentage of hydroelectric power in the country.

For further information, see http://matadornetwork.com/change/green-guide-to-oslo/ or Mark Luccarelli and Per Gunnar Røe, eds., *Green Oslo: Visions, Planning and Discourse* (Aldershot: Ashgate, 2012).

32 Singapore

Singapore has impressively set the goal and vision of becoming "A City in a Garden." As a highly populated, dense island city – 5 million residents on 700 square kilometers, most living in high-rise towers – incorporating nature is a challenge. But the city is setting a world model for combining density and nature and has already accomplished much. It has extensive park and green areas, tied together by 200-kilometers of Park Connectors, much of it in the form of elevated walkways and canopy walks. It has given priority to planting trees, and provides support for community gardens and the installation of green walls and rooftops. Landsat Images show that while the city grew in population by some 2 million between 1986 and 2007, the percentage of the island in green area actually increased as well, from 36 percent to 47 percent. Few dense cities can truly boast being "in a garden" in the way that Singapore can.

Singapore also leads in a number of other sustainable urban development innovations as well. For example since 2009 it has featured a single IBM-designed payment card for all forms of transportation, including buses, taxis, the metro, and road tolls. Travel data from card users allows transportation planners to adjust the system for the most efficient performance. The city-nation also has a sophisticated congestion pricing scheme that charges motorists for road use, and is developing a new urban district called Punggol Eco-Town that may become zero-net-energy.

Coordinating sustainability actions across different departments of government is often a problem for cities. In 2008 Singapore established an Inter-Ministerial Committee on Sustainable Development to

Figure 54 The Kaliang River now flows freely through Bishan Park.

Source: Photograph by Timothy Beatley.

Figure 55 Seven-story green wall at 158 Cecil Street, Singapore.

Source: Photograph by Timothy Beatley.

do this while establishing a national sustainable development strategy. The resulting Sustainability Blueprint sets 2030 goals in many areas, such as opening 100 kilometers of waterways for recreational activities, improving energy efficiency by 35 percent, reducing CO_2 emissions by 30 percent, and attaining a recycling rate of 70 percent.

Public involvement is one area in which outside observers have historically faulted Singapore. The country is ruled autocratically, and in the past there has been little public input into decision making. Still, in developing the country's recent Blueprint the government solicited ideas from thousands of people, and held focus groups, public forums, and dialogue sessions. More than 700 citizens participated in these events. "Encourage Community Ownership and Participation" is one of four key priorities in the resulting sustainability plan. How significant this progress is remains to be seen, but the government does at least seem to be aware of the issue.

For further information, see "Sustainable Singapore" at http://app.mewr.gov.sg/web/contents/ContentsSSS.aspx?ContId=1034; "Designing Our City: Planning for a Sustainable Singapore." 2011. Singapore Urban Redevelopment Authority; and Flemmich Webb (2012). "Sustainable Cities: Innovative Urban Planning in Singapore." *The Guardian*, October 11.

PART SEVEN

Sustainability pedagogy and class exercises

INTRODUCTION TO PART SEVEN

How best to teach the skills and knowledge that future sustainability professionals will need is still far from clear. There is consensus on some elements of such an education: it will help students become better at solving complex problems; it will address a wide range of social, environmental, and economic issues; and it will be highly interdisciplinary (some would argue transdisciplinary, i.e. transcending particular specialties in favor of a problem-based approach). However, detailed teaching strategies are still in a process of development.

In an introductory reading for this section, two British academics, Debbie Cotton and Jennie Winter, argue that "many of the core principles of integrating sustainability into higher education require substantial shifts in thinking and practice." Particularly important in their view are participatory and inclusive educational processes, transdisciplinary cooperation, experiential learning, and using environment and community as a context for learning. These strategies often face an uphill battle within traditional institutions of higher education. Too often the word "sustainability" is just appended to existing programs instead.

"Active, experiential learning" is one of the strategies recommended by Cotton and Winter. Consequently, we present a number of active learning exercises following their commentary that are appropriate either for classroom use or for individuals or groups to complete on their own. They have been developed by one of us (Wheeler) in conjunction with courses taught at the University of California at Berkeley, the University of New Mexico, and the University of California at Davis. An entire undergraduate or graduate class can even be structured around a series of such exercises, which examine sustainability planning at different scales and in different contexts. Bringing in guest speakers undertaking sustainability projects in the real world can help supplement class material as well.

Interested readers are encouraged to modify or expand these teaching exercises to fit their own needs. Generally these tasks are done in groups, on the theory that we can all learn enormously from each other and from the process of testing our ideas against those of others and combining ideas to reach new syntheses. Many of these exercises require some graphic representation – simple diagrams, plans, or maps to convey ideas. Although many of us are not used to presenting ideas visually, and may not have drawn anything since kindergarten, graphic work has been very much a part of the urban planning profession from the start. It is of course integral to architecture and landscape architecture as well. We also live in an increasingly visual age, when much of the population learns through images on electronic media. What matter in this case are not fine works of art but relatively simple diagrams, maps, or plans that can convey concepts effectively, especially for presentation to an audience. Clear labels that can be read at a distance are important for these poster-size graphics. Even students or community members with no background in graphic representation should be encouraged to experiment with different ways of illustrating their sustainable urban planning ideas.

"Sustainability Pedagogies"

from *Sustainability Education: Perspectives and Practice across Higher Education* (2010)

Debbie Cotton and Jennie Winter

Sustainability pedagogies have emerged from an initial focus on environmental education compartmentalized as *about*, *in* and *for* the environment. Education *about* the environment focuses on declarative knowledge and provides the learner with information about environmental systems and issues using approaches designed to investigate and discover. Education *in* the environment capitalizes on the environment as a real-world resource for enquiry and discovery that can enhance the learning process and challenge traditional understandings of metacognition. Education *for* the environment conceptualizes the transformative (and contentious) component of environmental education. It requires the development of a "personal environmental ethic"; the values and attitudes that motivate behavioral change in favor of the environment (Palmer and Neal, 1994, 19). From these beginnings, alternative pedagogic approaches have emerged, promoting inclusive forms of communicating knowledge based primarily on dialogue and experience (Scott and Gough, 2003) that have remained a central feature through the movement away from the traditional interests of environmental education towards the wider concept of ESD.

There are a number of general principles regarding sustainability pedagogies including participatory and inclusive education processes, transdisciplinary cooperation, experiential learning, and the use of environment and community as learning resources; all of which involve student-centered and interactive enquiry-based approaches to teaching and learning (Fien, 2006; Rasmussen, 2008; Sterling, 2004). It is clear that many of the core principles of integrating sustainability into higher education require substantial shifts in thinking and practice that may be out of reach of the individual lecturer and more challenging for some disciplines than others (see Table 1).

The difficulty of negotiating transformative changes to curricula – which are themselves within the boundaries of a wider (and largely traditional and conservative) educational system – has been raised by Sterling (2001). However, recent research (Cotton *et al.*, 2007 and 2009), carried out with the University of Plymouth's Centre for Sustainable Futures (CSF), shows that academics in a wide range of disciplines are making changes to their teaching to incorporate sustainability into the content of their curricula, despite highly variable institutional support for such activities. What is perhaps more difficult is promoting the changes in pedagogies that sustainability seems to require. Tilbury (2007, 119) notes that, "more and more we are seeing the word sustainability being added to the titles of programs, projects, activities, departments or units – however, few have actually been redesigned to address new social learning approaches." This is perhaps simply because the pressures on higher education militate against such changes – the increasing marketization of higher education on a mass scale makes participatory, collaborative approaches problematic. However, there may be wider impediments at work: previous research has identified lack of curriculum time, perceived irrelevance of sustainability to some disciplines, and lack of a shared understanding of the terminology as barriers to the growth of sustainability pedagogies (see Dawe *et al.*, 2005, for a discussion of some of the issues).

From	To
Transmissive learning	Learning through discovery
Teacher-centred approach	Learner-centred approach
Individual learning	Collaborative learning
Learning dominated by theory	Praxis-oriented learning linking theory and experience
Focus on accumulating knowledge and a content orientation	Focus on self-regulative learning and a real issues orientation
Emphasis on cognitive objectives only	Cognitive, affective, and skills-related objectives
Institutional, staff-based teaching/learning	Learning with staff but also with and from outsiders
Low-level cognitive learning	Higher-level cognitive learning

Table 1 Integration of sustainability with higher education implies shifts.

Source: Sterling, 2004:58; adapted from Van den Bor *et al.*, 2000.

WHY DO WE NEED DIFFERENT PEDAGOGIES FOR SUSTAINABILITY?

The need for different approaches for teaching about sustainability (and previously environmental education) has been under discussion for some time and is often linked with the potentially controversial nature of environmental or sustainability issues. However, research on how to teach controversial issues reveals a far from straightforward situation. It has long been assumed that a neutral or balanced perspective is required to avoid indoctrination of vulnerable students, and this belief is still held by many teachers in secondary and tertiary education. For example, research in England (Oulton *et al.*, 2004, 415) suggests that teachers identify three underpinning beliefs about teaching controversial issues. These are:

- a focus on rationality, reasoning and sticking to the facts;
- presenting a balanced view;
- teacher neutrality.

All three of these underpinning beliefs, however, are problematic if investigated further. For example, the facts relating to sustainability may be less than clear and depend significantly on the values of the individual describing them. An understanding of the transitory nature of facts, of knowledge and of what can be known undermines this position particularly strongly in the research-led environment of higher education. Moreover, the notions of maintaining neutrality and balance may be constrained by a number of practical limitations including: premature consensus, entrenched positions or apathy on the part of students involved in discussion; inadvertent projection of the teacher's views while attempting to convey a neutral position; and reduction of complex arguments to dichotomies and polarized positions in an effort to provide balance.

Oulton *et al.* (2004) offer a range of possible suggestions for the teacher, including:

- helping students to distinguish between sound and unsound reasoning, developing a respect for evidence and open-mindedness;
- being open about the fact that true balance is an unachievable goal, but
- helping students develop "a critical awareness of bias and make this one of the central learning objectives";
- declaring their own position explicitly so that students can be aware of potential bias in the teaching.

These approaches potentially offer an attractive proposition for higher education across the disciplines, drawing as they do on notions of logic, reasoning and criticality. Alongside an attempt to incorporate sustainability content into the curriculum, it is perhaps helpful to think in terms of developing the knowledge and skills of sustainability literacy – these would include open-mindedness and critical awareness of bias. A set of key questions might help to scaffold students' critical thinking skills in terms of issues such as:

- Where did this information or view come from?
- Who provided it?
- How are they funded?
- Whose interests do they represent?
- What values are they expressing (explicitly or covertly)?
- What evidence do they present?
- Do they evaluate it?

These questions could be utilized in a range of different contexts to explore the basis of decision-making, including decisions on sustainability issues, as appropriate. In this way, ESD is viewed as a different lens through which to view the discipline that focuses on the implications for economy, environment and society, rather than an imposed set of constructs and beliefs.

WHAT KINDS OF TEACHING METHODS HAVE BEEN ADVOCATED FOR SUSTAINABILITY?

The literature includes a wide range of suggestions for appropriate approaches to teaching about sustainability and also for specific teaching methods. Underlying many of these approaches is support for active, experiential learning, interdisciplinarity and use of the local (and regional) environment for educational purposes. Potential learning approaches are participative inquiry/action research, where students investigate an issue which is of importance to them personally (Tilbury, 2007); transformative sustainability learning (TSL), where tutors attempt to use the three domains of learning – cognitive, psychomotor and affective, or head, hands and heart as they have been described – to engage students in a transformative educational experience (Sipos *et al.*, 2008); and action competence, where students are encouraged to envisage alternatives and solutions to unsustainable practices (Breiting and Mogensen, 1999).

Specific teaching strategies advocated for environmental or sustainability education include those listed below. It is likely that utilizing a range of these strategies would be most appropriate.

Role-plays and simulations

Role-plays have long been recommended for teaching about environmental issues and sustainability, although there is a surprising lack of evidence in terms of effective outcomes (Oulton *et al.*, 2004). Potential advantages of role-plays are that they provide an opportunity for students to gain an in-depth understanding of another person's perspective and to empathize with others; disadvantages are the amount of time and organization required to enable effective role-playing and the difficulties of managing the role-play, particularly with large groups. Role-plays are used rarely in university education, possibly because of the practical difficulties or because the pedagogy is poorly aligned with the learning culture of higher education.

Group discussions

Group discussions were frequently mentioned by both school teachers and lecturers when asked to describe an appropriate pedagogy for sustainability (Cotton, 2006; Cotton *et al.*, 2007). The use of a discussion may be an attempt to counteract the risk of the teacher taking a transmissive or authoritarian approach, thereby enabling students to discuss their own and others' views. Discussions potentially enable a range of perspectives to be aired, but they may be confrontational and prove difficult to control, especially if the topic is a controversial one. The teacher needs to be able to encourage listening and self-reflection rather than argument and should be clear about their own role in the discussion. Structured questions to scaffold students' learning may be helpful, as may explicit meta-cognitive instructions as to the purpose of the discussion and the rules of engagement. Without such guidance, many students – accustomed to the transmissive nature of much of their educational experience – may be uncertain how to respond.

Stimulus activities

A stimulus activity might involve watching a video or looking at photos, poems or newspaper extracts to initiate reflection or discussion (Oulton et al., 2004). Students may even be involved in producing their own work such as photos taken around the campus to stimulate a discussion on campus greening. Use of videos or externally produced documents potentially enables the teacher to bring in a wide range of viewpoints for critical analysis, and this approach is feasible even with very large groups.

Debates

Debates in which two groups of students put forward opposing arguments on an issue are often recommended as a method of teaching about sustainability since they encourage students to gather information about the topic and develop an argument. However, they can become confrontational and students may be discouraged from engaging or empathizing with others' views. Authors such as Oulton et al. specifically warn against asking students to vote on an issue as this may lead to them making up their minds too soon, hardening their attitudes and leaving them feeling committed to the stance that they have taken (Oulton et al., 2004).

Critical incidents

The use of critical incidents to teach about sustainability is described in a paper by Nott and Wellington. Students are given an example and asked what they would do, what they could do, and what they should do (Nott and Wellington, 1995). This allows them to consider their personal perspectives and actions in the light of a moral or ethical stance. The approach can also be used with groups to promote awareness about multiple perspectives on sustainability.

Case studies

Another popular choice of pedagogy for teaching about sustainability described by lecturers in our research was the case study approach. Teachers described using case studies to bring sustainable development into areas of the curriculum that had not traditionally involved a clear focus on sustainability (Cotton et al., 2009), and to provide students with an holistic view of an issue. Case studies enable students to investigate issues that affect their local area, to work with private enterprises and community groups and to work together in finding solutions to local issues. They may take a variety of forms, but one possible approach is to place strong emphasis on "reflection, research, participation and action" (Junyent and de Ciurana, 2008, 769).

Reflexive accounts

Considering their own position in relation to new knowledge about sustainability can help students understand how individual actions contribute to sustainability. Although contentious in higher education (Knight, 2005), behavior change is a cross-cutting priority of the UK sustainable development strategy (DEFRA, 2005) and education is identified as a core vehicle for achieving this. Therefore, pedagogies that provide opportunities for students to reflect on personal roles, attitudes and responsibilities in relation to a range of sustainability issues are potentially advantageous.

Personal development planning (PDP)

PDP has been embedded in UK higher education since 2000 (Quality Assurance Agency for Higher Education [QAA], 2000). "PDP is a structured and supported process undertaken by a learner to reflect upon their own learning, performance and/or achievement and to plan for their personal, educational and career development. It is an inclusive process, open to all learners, in all higher education provision settings, and at all levels" (QAA, 2009, 3). PDP can provide an opportunity for students to learn about and reflect on sustainability (John Forster Associates, 2006). Sustainability literacy may be a set of skills, development of which is encouraged throughout the student experience of higher education and recorded through the PDP process. Students may also be able to integrate relevant informal learning activities and volunteering into the PDP record.

Critical reading and writing

Reading and writing are often downplayed in favor of more interactive pedagogies. However, these are important social practices and the key to progressing sustainability and literacy. Stibbe suggests students can gain from deconstructing destructive, alternative, or counter-discourses to identify the possible motivation of the author. They may also be able to envisage alternative futures, and write a contrasting account based on a differing set of values (Stibbe, 2008).

Problem-based learning

Problem-based learning is an iterative learning process that can be used to teach a whole range of subject matter. A sustainability-related issue may be identified and students asked to research this to generate a body of knowledge. They can then develop a vision of alternative actions and potential solutions to the problem, which they use to devise a plan of action. The action may then be carried out, followed by a period of reflection and evaluation. This process can be extremely useful because it promotes both the conceptual and practical aspects of sustainability literacy. Brunetti *et al.* (2003) describe a specific example of the use of problem-based learning to teach about social, economic and environmental sustainability issues.

Fieldwork

Fieldwork is an example of experiential pedagogy that can influence students' emotions (Sivek, 2002) and help develop the critical thinking skills so essential to understanding the complexity of sustainability (Jones, 2003; Scott and Gough, 2003). Fieldwork for sustainability can be based on issues in the local community and environs, linking theory to real-world examples (Hope, 2009), which can help students to understand multiple stakeholder perspectives in situ. There is also evidence that outdoor experience is an important precursor to understanding sustainability (Palmer and Suggate, 1996) and that fieldwork promotes broader benefits for learning by encouraging active and reflective learning among students (Hope, 2009).

Modelling good practice

Despite the focus here on teaching strategies, the importance of learning taking place implicitly through the hidden curriculum and outside the classroom should not be underestimated. In our research, many lecturers talked of reducing paper (by provision of online resources, for example) and turning lights out at the end of a teaching session, and it is clear that students will react cynically to any indication that the lecturers' expressed views conflict with their own behavior. Moreover, the role of higher education as a transformative experience could, and perhaps should, go well beyond turning off lights and providing recycling facilities:

> Social education for me is a big part of the university. It is not just coming out with a degree, it is how you change and what your values are when you finish. I think there is so much to sustainable development for me in the University of Plymouth that is not just about bits of paper and light bulbs (lecturer in health).

POTENTIAL FOR FUTURE DEVELOPMENT OF SUSTAINABILITY PEDAGOGIES IN HIGHER EDUCATION

Looking at the pedagogies advocated above, it is immediately obvious that many of these approaches require a significant amount of prior preparation, as well as small groups and a reasonable time allocation. The barriers to engaging in such teaching approaches in the constraints of the current higher education system should not be underestimated.

However, it is also clear that if students are to be given the opportunity to reflect seriously on sustainability issues, time needs to be made available within the curriculum for serious exploration and discussion. To provide a successful learning environment that encompasses sustainability, both students and lecturers should feel free to express their views in a supportive environment, but one where self-reflection and change of viewpoint is encouraged. As a general rule, students should be encouraged to evaluate critically any information provided, to identify potential sources of bias and to reflect on their own views and prejudices to help them make decisions about complex issues both within and beyond the sustainability debate.

If the use of sustainability pedagogies is to become more widespread, then those lecturers who are already committed to sustainability principles will need support and resources to enable them to develop their teaching along these lines. For those who are yet to be convinced, an understanding of the potential of sustainability pedagogies to form part of a wider move towards student-centered, active learning approaches may act as an incentive.

Pedagogies that help students develop critical thinking skills also provide an important contribution to the wider skills sector. The link between skills for sustainable development and the needs of the economy to provide a sustainable future has only recently been made (DfES, 2003); however, skills for sustainable development are fast becoming viewed as a 'vital national asset' (DfES, 2003, 8) with the potential to enhance social mobility and tackle exclusion as well as achieve economic objectives. Although this may be more easily accommodated in vocational degrees, the profile of employability skills continues to rise throughout academia and may provide a useful framework for including sustainability literacy in subjects where its relevance is not clear-cut. In all cases, there is an undoubted need for professional development to embed, value and reward good (sustainability) pedagogic practice and to enable lecturers to engage with new ways of thinking about teaching and learning in higher education.

A further requirement is what has been described as "space for pedagogical transformation" – the creation of "spaces on campus where transformative and transdisciplinary learning is supported and encouraged" (Moore, 2005, 337). This should include physical space to support the development of more student-centered, collaborative (and interdisciplinary) approaches, a space where reflection can take place, and where project or action planning can occur. It should be innovative, exciting, technology-rich and, crucially, flexible in terms of lighting, seating and presentation areas to encourage lecturers to consider variation in modes of working and interacting with students – in contrast to the formal teacher-focused pedagogy which is encouraged by the traditional stepped lecture theatre.

E-learning contexts can also provide a space for sustainability-related learning to occur. Online forums can provide a suitable environment for constructing discussions and debates and enable participants to cross disciplinary boundaries by offering a neutral space for students and staff from different subject areas. Online modules about sustainability can be used as an aid by lecturers with little experience of sustainable development.

The perception of students as consumers, together with the marketization of higher education, has further consequences for sustainability. Students may be reticent to sign up to courses that feature sustainability, not viewing it as applicable to their personal or career trajectories. However, Sterling and Scott (2007) identify ways in which student demand can put pressure on academics to include sustainability in their courses, and the HEFCE's [Higher Education Funding Council for England] 2008 review notes that "There is a clear niche in the academic marketplace for institutions that wish to champion sustainability." (35). In this way, pressure from both staff and students may combine to move this agenda forward.

REFERENCES

Breiting, S. and Mogensen, F. (1999) "Action competence and environmental education", *Cambridge Journal of Education*, 29 (3), 349–353.

Brunetti, A.J., Perrell, R.J. and Sawada, B. (2003) "SEEDing sustainability: Team project-based learning enhances awareness of sustainability at the University of British Columbia, Canada", *International Journal of Sustainability in Higher Education*, 4 (3), 210–217.

Cotton, D.R.E. (2006) "Teaching controversial environmental issues: Neutrality and balance in the reality of the classroom", *Educational Research*, 48 (2), 223–241.

Cotton, D.R.E., Warren, M.F., Maiboroda, O. and Bailey, I. (2007) "Sustainable development, higher education and pedagogy: A study of lecturers' beliefs and attitudes", *Environmental Education Research*, 13 (5) 579–597.

Cotton, D., Bailey, I., Warren, M. and Bissell, S. (2009) "Revolutions and second-best solutions: Education for sustainable development in higher education", *Studies in Higher Education*, 34 (7).

Dawe, G., Jucker, R. and Martin, S. (2005) *Sustainable Development in Higher Education: Current Practice and Future Developments*, HEA, York www.heacademy.ac.uk/assets/York/documents/ourwork/tla/sustainability/sustdevinHEfinalreport.pdf. Accessed 6 March 2009.

DEFRA (2005) *Securing the Future: The UK Government Sustainable Development Strategy*, The Stationery Office, London, available at www.defra.gov.uk/sustainable/government/publications/uk-strategy/documents/SecFut_complete.pdf, accessed 6 March 2009.

DfES (2003) *White Paper Twenty-First Century Skills: Realising Our Potential. Individuals, Employers, Nation*, HM Treasury: Department of Trade and Industry, London.

Fien, J. (2006) Education for Sustainable Development: A Perspective for Schools, Raja Roy Singh Lecture, 10th Asian Programme of Educational Innovation for Development International Conference, Bangkok, Thailand, 6–8 December 2006, available at www.unescobkk.org/education/apeid/apeid-international-conference/10-th-apeid-international-conference/speakers-and-speeches/john-fien/raja-roy-singh-lecture/, accessed 23 June 2009.

HEFCE (2008) *Strategic Review of Sustainable Development in Higher Education in England*, HEFCE, Bristol, available at www.hefce.ac.uk/pubs/rdreports/2008/rd03_08, accessed 6 March 2009.

Hope, M. (2009) "The importance of direct experience: A philosophical defense of fieldwork in human geography", *Journal of Geography in Higher Education*, 33 (2), 169–182.

John Forster Associates (2006) Embedding Sustainability into the Curriculum of Scotland's Universities and Colleges, available at www.sfc.ac.uk/reports_publications/reports_publications.aspx?Search=john%20forster&Type=Reports%20and%20publications&Sector=-1, accessed 22 February 2010.

Jones, V. (2003) Young People and the Circulation of Academic Knowledges, Department of Geography, University of Aberystwyth.

Junyent, M. and de Ciurana, A.M.G (2008) "Education for sustainability in university studies: A model for reorienting the curriculum", *British Educational Research Journal*, 34 (6), 763–782.

Knight, P. (2005) "Unsustainable developments", *The Guardian*, 8 February, available at www.guardian.co.uk/education/2005/feb/08/highereducation.administration, accessed 6 March 2009.

Moore, J. (2005) "Seven recommendations for creating sustainability education at the university level: A guide for change agents", *International Journal of Sustainability in Higher Education*, l6 (4), 326–339.

Nott, M. and Wellington, J. (1995) "Critical incidents in the science classroom and the nature of science", *School Science Review*, 76 (276), 41–46.

Oulton, C., Dillon, J. and Grace, M. (2004) "Reconceptualising the teaching of controversial issues", *International Journal of Science Education*, 26 (4), 411–423.

Palmer, J. and Neal, P (1994) *The Handbook of Environmental Education*, Routledge, London.

Palmer, J. and Suggate, J. (1996) "Influences and experiences affecting pro-environmental behaviour of educators", *Environmental Education Research*, 2 (1), 109–121.

QAA (2000) Policy Statement on a Progress File for Higher Education, available at www.qaa.ac.ukjacademic infrastructure/progressFiles/archive/policystatement/default.asp, accessed 23 June 2009.

Rasmussen, H. (2008) The Earth Charter and the Educative Principles of Sustainability, Earth Charter International, available at www.cartadellaterra.org/media/File/PROGETTI_E_PERCORSI/terra%20che%20cura/intervento%20henriette.pdf, accessed 23 June 2009.

Scott, W. and Gough, S. (2003) *Sustainable Development and Learning: Framing the Issues*, Routledge, London.

Sipos, Y., Battisti, B. and Grimm, K. (2008) "Achieving transformative sustainability learning: Engaging head, hands and heart", *International Journal of Sustainability in Higher Education*, 19 (1), 68–86.

Sivek, D.J. (2002) "Environmental sensitivity among Wisconsin high school students", *Environmental Education Research*, 8 (2), 155–170.

Sterling, S. (2001) *Sustainable Education: Re-visioning Learning and Change*, Green Books, Darrington.

Sterling, S. (2004) "An analysis of the development of sustainability education internationally: Evolution, interpretation and transformative potential", in Blewitt, J. and Cullingford, C. (eds) *The Sustainability Curriculum: The Challenge of Higher Education*, Earthscan, London, pp 43–62.

Sterling, S. and Scott, W. (2007) England contribution, First International Meeting on Implementation of Education for Sustainable Development for Higher Education Institutes, 8–9 February 2007, Amsterdam, available at www.eauc.org.uk/first_international_meeting_on_implementation_of_e, accessed 10 March 2008.

Stibbe, A. (2008) "Words and worlds: New directions for sustainability literacy", *Language and Ecology*, 2 (3), 1–10.

Tilbury, D. (2007) "Monitoring and evaluation during the UN decade of education for sustainable development", *The Journal of Education for Sustainable Development*, 1 (2), 239–254.

1 Cognitive Mapping Exercise

The purpose of this exercise is to gain understanding into the nature of the environments we grew up in, how they affected us, and how they might relate to our views of sustainable urban development. This exercise draws upon cognitive mapping techniques developed by Kevin Lynch at MIT and others in the field of environmental design research. Such researchers have systematically asked particular groups to draw maps of their home environments to see what can be learned about how people perceive neighborhoods, cities, or the region as a whole.

INSTRUCTIONS TO PARTICIPANTS

Close your eyes and imagine yourself as a child at an age that was particularly important to your development. Put yourself in a location that you most identified with at that age and picture the landscape around you, including any features that were important to you such as your family's house and yard, locations where your friends lived, places where you played or worked, and the surrounding city, town, or countryside. What elements in the human or natural environment were important to your daily life? Did any parts of the landscape feel pleasant or unpleasant, safe or threatening, nurturing or draining? Who did you interact with on a daily basis? Where could you go? How did you get around?

Using colored markers on a large sheet of paper, map this environment as it appeared to you at that time (not as it really is). Put yourself in the center (stick figures are fine) and work outwards, adding those features that were most influential in your daily life. These elements can be represented symbolically in whatever way works for you and do not have to be to scale (i.e. in realistic spatial relationship to one another). Dramatize them as much as possible. Express how things felt to you then, rather than trying to reconstruct a realistic map. Fill the paper. Use bright color and thick lines so that someone else can understand this map from a distance. Label your drawing clearly so that others can understand it. Drawing skill does not matter.

After 15 to 20 minutes get together with several others and take turns explaining what you've drawn. The others can ask clarifying questions but should not debate anything you've shown. Focus on how this environment affected you and your family personally, and what can be learned from it regarding how this place functioned, how different people or cultures related to one another, and how they related to the natural world. If you are doing this exercise individually at home, you might try writing up a couple of pages analyzing this environment and its social, environmental, and economic implications.

2 Future Visions Exercise

This second exercise can be combined with the first to analyze how formative urban environments from our past influence our views on how cities should be developed in the future. In this case, participants can draw both maps one after the other, and then discuss them in groups. For some, ideal environments may be rather similar to childhood environments. For others, they will be quite different, perhaps in order to respond to some of the problems experienced while growing up.

INSTRUCTIONS TO PARTICIPANTS

Close your eyes again and imagine your ideal community – a place you would really like to live in if you could, and that is sustainable in however you want to define that. How is this place laid out in terms of streets, buildings, public spaces, parks, homes, workplaces, and shopping areas? How do people get around (what transportation networks exist)? What types of people live there? How do they interact or relate? How does the built environment relate to the natural landscape?

On a large sheet of paper draw a plan of this place. Do it at whatever scale seems appropriate to you. Include symbols and labels to identify key elements. Again, fill the paper and use lots of color, so that someone can read this map from a distance. Drawing quality does not matter. At one side list four to five key characteristics of this place that will help convey its unique qualities.

As before, discuss your vision in small groups. Each person should explain their image in a few minutes, with others asking only clarifying questions, not debating issues or components. Consider whether or how this ideal relates to the formative environment(s) you experienced when you were young. If desired, write up a couple pages on this relationship.

3 Definitions of Sustainable Development

This is a brief exercise to stimulate discussion of different perspectives toward sustainability. In three to five minutes have participants write down their own personal definition of sustainable development. They should keep this definition as simple as possible, something that they could use to explain the concept to a parent or grandparent. Then ask participants to repeat their formulations for the others. Write down key themes from each on the board. Then go through the different definitional approaches in a group discussion, highlighting the pros and cons of each approach, and discussing the perspective or worldview that might most endorse it.

For comparison, here are some definitions from various international sources:

Meeting the Needs of Future Generations
"Sustainable development is development that meets the needs of the present without compromising the ability of future generations to meet their own needs." (*Brundtland Commission, 1987*)

Carrying Capacity of Ecosystems
Sustainable development means "improving the quality of human life while living within the carrying capacity of supporting ecosystems." (*World Conservation Union, 1991*)

Maintain Natural Capital
"Sustainability requires at least a constant stock of natural capital, construed as the set of all environmental assets." (*British environmental economist David Pearce, 1988*)

Maintenance and Improvement of Systems
"Sustainability . . . implies that the overall level of diversity and overall productivity of components and relations in systems are maintained or enhanced." (*American ecological economist Richard Norgaard, 1988*)

Positive Change
Sustainable development is "any form of positive change which does not erode the ecological, social, or political systems upon which society is dependent." (*Canadian planner William Rees, 1988*)

Sustaining Human Livelihood	Sustainability is "the ability of a system to sustain the livelihood of the people who depend on that system for an indefinite period." (*Indonesian economist Otto Soemarwoto, 1991*)
Protecting and Restoring the Environment	"Sustainability equals conservation plus stewardship plus restoration." (*Ecological architect Sim Van der Ryn, 1994*)
Oppose Exponential Growth	"Sustainability is the fundamental root metaphor that can oppose the notion of continued exponential material growth." (Ecotopia *author Ernest Callenbach, 1992*)
Composite Approach	"Sustainable development seeks . . . to respond to five broad requirements: (1) integration of conservation and development, (2) satisfaction of basic human needs, (3) achievement of equity and social justice, (4) provision of social self-determination and cultural diversity, and (5) maintenance of ecological integrity." (*International Union for the Conservation of Nature, 1986*)

4 Role Plays to Analyze Points of View and Sustainability Decision Making

Role plays can be an excellent way for students to learn about different perspectives on sustainability debates, and to then think more critically about those issues. Each student is asked to play a particular character, putting themselves in that person's shoes. The material they are given includes specific information about what that character thinks and why. The group of students then interacts, playing out the roles within a prescribed scenario, usually for 10–30 minutes. Afterwards, the full class debriefs the exercise with the instructor, with emphasis on sustainability themes such as how environmental, economic, and equity viewpoints are represented.

Two main types of role plays may be particularly useful:

1 Small group role plays within a large class (30 students and up). Break the class up into groups of 4–7, and hand out a role to each participant within each group (usually a name, title, and paragraph of description on a single sheet of paper). Usually the role play is a stakeholder meeting of some sort concerning a planning or design project. One person is the facilitator (often a government official). Their job is to try to develop consensus between the other stakeholders, making sure everyone gets heard, and suggesting compromises or alternatives where appropriate. The others play their roles, embroidering on their characters if possible.

A small group role play one of us has often run concerns the Cape Wind project, a large proposed wind farm off the eastern shore of the US between Cape Cod and the islands of Nantucket and Martha's Vineyard. The facilitator is a representative of the US Department of the Interior trying to broker consensus, and stakeholders include the wind farm developer, the president of the local fishermen's association concerned about livelihoods, a wealthy property owner who doesn't want her views spoiled, a representative of a local Indian tribe concerned that these are sacred waters, and several representatives of environmental groups with different points of view (for example, arguing that decentralized energy systems are better than huge facilities even if renewable). The initial handout for this exercise includes a map of the area and a few paragraphs of background. Parameters of the debate are set so that there is no easy way out for participants (in this case, the proposed site has the best wind and is shallow enough to make construction feasible).

2   A more complex role play format within a smaller class or sections of a large class that assigns many roles to members of a single large group. Again, character assignments are handed out along with background information. The group then reenacts a public process aimed at reaching an official decision.

  For example, the class can re-enact a recent urban planning decision by the governing body in your community, such as approval of a large new building or subdivision. The students play roles of developers, landowners, city council members, environmentalists, neighbors, business owners, and other constituencies. Provide the participants with a few pages of background on the selected project (from press clippings, local government materials, or web sources) as well as a map or site plan. Make up as many roles as needed (this exercise can be done with groups as large as 30). Prepare a few sentences describing the background and views of each participant, and give a sheet with this description to the individual assigned this role. Often it works best to have those participants who consider themselves the strongest environmentalists be given the role of developers, and vice versa.

Roles might include the following:

- City Council or County Board of Supervisors (usually 5 to 9 members; one is Mayor or Chair)
- County Staff (Planning or Community Development Director, Transportation Planner, Housing Planner)
- Development Team (Landowner, Principal of Development Firm, Architect, Landscape Architect or Environmental Consultant)
- Public Commenters (Representatives of environmental groups such as the Sierra Club, Audubon Society, and local "Friends of" groups; representatives of a local Citizens Transportation Alliance; local affordable housing advocates; Chamber of Commerce or local Business Development Association representatives; owners of nearby stores; the president of the regional Homebuilders Association; local union representatives; landowners in or near the project area; other nearby residents or landowners concerned about noise, views, traffic, loss of open space, effects on property values, etc.)

INSTRUCTIONS TO PARTICIPANTS

This exercise will recreate a development approval decision by a local city council or board of supervisors.[1] In our simplified version of this appeal hearing, the Council Members or Supervisors will hear from the project applicants, municipal staff, and the public in turn, and will then vote on whether to approve the project and if so what conditions to attach. You will each be assigned a role. Feel free to make up additional details that you think would fit with your character, and play the role as much as possible. A prize will be given for best actor/actress, as voted by your classmates.

  Exercise Format (Adapt this format to reflect local procedures):

- 10 minutes – Participants review roles and huddle with allies.
- 10 minutes – The Mayor or Chair of the Board of Supervisors opens the hearing. Project backers present their proposal and describe its benefits. Council members or Supervisors may ask brief clarifying questions.
- 5 minutes – The city or county's planning staff presents its recommendations, usually to approve the project with mitigations to reduce its environmental impacts.

- 15 minutes – Public hearing. Any member of the public wishing to make a statement must fill out a speaker card (distribute index cards) and hand it to the chair before the hearing begins. The chair calls these cards in random order. Speakers identify themselves and their organizations, if any, and have 2–3 minutes to make a statement.
- 15 minutes – Council members or Supervisors state their own positions when recognized by the chair. Then any member may make a motion, in this case to approve the project, approve it with conditions, deny it, or postpone action pending some further process. If the motion is seconded by another supervisor, discussion on it takes place. Other supervisors can offer amendments to the motion, which must be accepted by the primary sponsor in order to be incorporated. After discussion on the motion has taken place the chair calls for a vote. Motions must be approved by majority vote.

Afterwards, prepare a short write-up analyzing how the Three E's were represented within this planning debate. Who speaks for each perspective? Which perspectives are underrepresented, if any? How might alternative decisions come about that better meet goals such as the Three E's in the long run? Which individuals or groups might develop such alternatives, and how?

NOTE

1 In real life, this hearing would probably be on an appeal of an earlier decision by a planning commission or zoning board.

5 Sustainability Indicators Exercise

This exercise explores how progress toward sustainable development might be measured in your particular community or region. First, review the factors making for good sustainability indicators, as discussed in the Maclaren reading or other sources. Have small groups brainstorm six key indicators of urban sustainability for your area. These indicators should integrate a range of themes such as environment, economy, and equity, reflect meaningful trends, targets, or goals, reflect specific regional conditions, use readily available and updatable information, and be easily understandable to the general public. You may need to advise the groups on what types of information are available from which agencies or other sources.

The groups should take 30–40 minutes to come up with their lists, writing them on large poster paper. Then each team should present their recommendations to the rest of the participants in 5–10 minutes. Follow-up discussion can focus on what types of public policies might affect these indicators, what agencies or levels of government have authority over each area, and how significant change might come about.

6 Personal Ecological Footprints/Household Sustainability Audit/Carbon Calculators

Sustainability begins at home, so this exercise asks participants to evaluate their own lifestyle and household, and to come up with recommendations for changes. The exercise is probably best completed individually by participants or housemates and then presented in the form of a short paper or class presentation.

First, have participants complete the personal ecological footprint analysis online at www.rprogress.org. This will give them a figure for the number of acres that would be required to offset their personal resource consumption, and the number of earths that would be required if everyone in the world lived at their consumption level. Share and discuss these findings. Are there ways that these individual footprints might be reduced? Online carbon calculators such as the Cool Climate calculator at http://coolclimate.berkeley.edu/carboncalculator or water calculators such as at http://www.wecalc.org/ are also useful.

Then participants should prepare an audit (a careful and systematic survey) of their own households to determine what improvements could be made to reduce resource usage and otherwise improve sustainability. Background readings on home energy and resource conservation measures may be helpful. Your local utility company's website may have extensive information on these, as well as websites of nonprofit organizations such the American Council for an Energy Efficient Economy (www.aceee.org).

INSTRUCTIONS TO PARTICIPANTS

Systematically examine your home (both the building and lot) to determine how your use of energy and resources might be reduced, and sustainability otherwise improved. Review data such as utility bills (for electricity, gas, and water consumption) and if possible compare these with past bills or averages for your community to see how your consumption varies and what factors might be affecting it. Tabulate all uses of energy and water in your home, and list what conservation measures have been applied or might be applied to each. Examine also the landscaping of your lot, calculate the amount of paved surfaces, and investigate possibilities for changes in either. You may want to consider the following questions in your analysis:

- Where and how extensively have energy and water conservation measures been used?
- What other conservation measures are possible?
- Are there ways to improve the recycling of various materials?
- Is solar hot water or electricity a possibility? Where might such devices go?
- Is use of graywater a possibility? If so, how might that be done?
- How might landscaping be improved in terms of water consumption, use of native or drought-tolerant species, and/or creation of habitat?
- Is urban agriculture a possibility on your lot? How might space for that be created?
- How much of the site is covered by impermeable surfaces? Where does the runoff go? How might runoff be reduced, made cleaner, or made less severe after storms?
- Does this site contribute to an urban heat island effect? How might that be reduced?
- Might the planting of trees or changes to the building reduce air conditioning needs in the summer?
- Does the existing development on this parcel best contribute to the city and region's overall needs? Might additional housing units be created on this site? If so, how? Is the site zoned in the best way

to meet urban sustainability goals? Should the zoning be changed? (Zoning codes are often online these days at your city's website.)

■ Should measures be adopted to ensure the long-term affordability of this housing?
■ Is there anything else about your lifestyle that you would change to promote sustainability?

Write a short analysis of key steps to improve the sustainability of your home. Be comprehensive in your analysis, but focus on what you feel are the most promising strategies. Include any graphics or photos you feel are necessary.

7 First-Hand Analysis of Urban Environments

In line with Allan Jacobs' piece "Looking at Cities," a walking tour is an indispensable laboratory for teaching sustainability planning. It is an opportunity for students to practice first-hand analysis of urban environments, seeing how particular places have evolved in the past and how they might change in more sustainable directions in the future. A tour probably works best if it is in a relatively dense urban area where pedestrian travel is relatively easy and a number of interesting sites can be viewed within a 3–4 hour stretch. Bike tours to cover larger distances are also possible. It is a good idea to choose an area with some historical sites or context, some diversity of building types and land uses, some parks or remnants of natural ecosystems, and some new building or streetscape projects that the students can evaluate.

Prepare a map of the tour route, and if desired structure in some mini-exercises such as calculating residential densities in several sites (by pacing off lot sizes, determining numbers of units, and calculating the net density figure), brainstorming traffic calming strategies at dangerous streets and intersections, and making use observations of public squares or parks. Read local histories beforehand to gain historical information that can be passed along to students. Much of the value of a walking tour depends on how good the tour leader is at pointing out detail and history, and at raising questions for the participants.

INSTRUCTIONS TO PARTICIPANTS

On this walking tour we will practice observing urban environments first hand, to see how they have evolved in the past and how they might become more sustainable in the future. Walk slowly and see how much information about the urban environment you can gather through careful observation. See if you can identify opportunities for infill development, ecological restoration, and improved street design. Also note how people use particular places, and what groups will feel welcome or not in different contexts. Try to imagine how particular blocks and streetscapes might look in the future, and think about what urban planning strategies might help these areas change.

Questions for discussion during the walking tour:

■ What was this place like 100 years ago?
■ What might it be like 10, 20, or 100 years from now?
■ How does this place feel to you? What exact features make it feel that way? How would you change those if you could?

- Whom do we observe using this place? What are they doing? Are there any groups who would not feel comfortable here?
- When was this place built? What clues tell us the age of the place?
- What kinds of people live/work/play here? What clues do we see about them?
- How big are the lots? The houses? How are the buildings positioned on the lots? How big are the setbacks? (Dimensions of the neighborhood can be paced off; one pace = two steps = about six feet, a little more for tall people, a little less for not-so-tall ones.)
- How wide are the streets? The sidewalks? How are the streets designed? Are there particular features that make them more or less pedestrian friendly? Are there ways that the street could be retrofitted to slow traffic and be more pedestrian and bike friendly? To add landscaping?
- Where is the nearest park or natural area? Are there opportunities for creation of additional green spaces?
- What elements of the original ecosystem of this place remain? Are there any ways that habitat could be restored or recreated?
- Where is the nearest waterway? How would water drain in this place? Are there ways that the hydrology of this place could be restored?
- Are there particular elements that give this place historic or cultural significance? What exactly are they? Should they be protected or preserved in some way?

8 Regional Vision Exercise

The purpose of this exercise is to have participants think about physical planning on a regional scale – not a level most citizens are used to considering. In particular, they should focus on how land use, transportation, environmental, and equity planning at this scale can help improve urban sustainability. Participants will consider how they would structure development in your region if they could start over from scratch. This exercise is not as far fetched as it may sound; many metropolitan regions these days are developing Smart Growth plans which seek to shape regional form in such ways, and new towns or large subdivisions not infrequently undertake physical planning on very large scales.

This exercise requires obtaining or developing a large-format base map of the region, preferably with just natural landscape features on it (rivers, streams, shorelines, hills, mountains, wetlands, etc.). A simple but workable base map can be created by tracing these features onto vellum or mylar from any good existing regional map, then printing copies onto bond paper for the students to use (one per group). Alternatively, student teams can create their plans by placing tracing paper on top of any decent large-format regional map, ignoring existing roads and urbanization.

INSTRUCTIONS TO PARTICIPANTS

In this exercise your team will determine a sustainable development scenario for your region. Assume that you can design the region from scratch (i.e. there is no existing development). What would be the form of communities? What mixture of land uses and urban densities would you suggest? What types of transportation? What kind of street network? What sort of park and open space system? How would the physical development reflect economic and equity goals?

Be aware of the natural landscape – which parts of it you might recommend for development, and which you wouldn't. You will be given a large base map to use for this exercise. Illustrate your concept on this map. Don't try to be extremely detailed; you probably won't be able to show every street, for example (you may want to show a sample neighborhood at a larger scale off to the side to give viewers an idea of what details you would incorporate). The overall concepts are what is important.

First, decide as a group on an overall theme and a short list of physical design strategies that will carry out this theme. Write these in large print off to the side of your map. Next, map basic land uses and transportation systems with standard city planning land-use colors. Either colored pencils or markers are fine, although the former may look better. Show the following:

- Developed areas. (Use several shades of yellow and brown to show different residential densities, with brown being the densest. Use red for mixed-use development or centers of activity, blue or purple for heavy industry, and gray for large institutional land uses like airports and universities.)
- Parkland and agricultural land. (Use two shades of green.)
- Transportation systems. (Show major transit routes and highways in black).

Label as much as you can on the map, including types of economic activity you want to promote in particular places. Include a key showing which color corresponds to which type of land use.

You should plan your vision to accommodate the current population of the region. This means that you'll have to make some tradeoffs with density. If you plan a low-density environment with single-family homes and big yards, you'll probably have to cover most of the map with development. If you're planning medium-density communities with a mixture of housing types (including townhouses and small apartment buildings), you'll have to cover only about 1/3 as much land area. If you're planning a higher-density environment (with some high-rise apartment buildings, many townhouses and 3–5 story apartment buildings, and a few single-family homes), you'll only need to cover about 1/6 of the area. But then be able to say how you're going to make this higher-density environment livable for people who don't currently like density.

Make sure everyone on your team participates in each part of the project. Be prepared to describe your vision to the class in a few minutes.

9 Economic Development Exercise

After covering economic development topics in lectures and readings, prepare brief case studies of four different cities or towns in your area, including mention of historical, cultural, demographic, and economic circumstances. These towns should illustrate different scenarios such as:

- a wealthy, fast-growing new suburb with both jobs and housing
- a declining central city area with falling tax base
- a stable, built-out inner-ring suburb with concerns about future stagnation
- a far-flung, working-class bedroom community which has allowed rapid housing development but little employment base.

Then divide the class into four teams and give each the following instructions, along with a map of the region.

INSTRUCTIONS TO PARTICIPANTS

Congratulations! Your consulting firm has been hired to produce an economic development strategy for a local city that has become interested in sustainable development. The background and challenges facing your particular city will be handed out separately. Based on this information and what you know of the region and sustainability planning, your firm must come up with a carefully thought-out strategy for the mayor and city council.

This strategy should include:

- An overall theme that will be catchy and particularly appropriate for the unique context of this city;
- A recommendation on particular types of business or industry the city should focus on (the history, culture, and location of the city should come into play here, as well as your judgment about what might be the most dependable and ecologically appropriate sources of jobs in the future);
- Particular incentives, programs, or regulatory changes by which the city might support these;
- Land use, infrastructure, and urban design changes that can complement these forms of economic development.

You will have 30 minutes to develop your recommendations. Then your team will make a brief presentation to the city council and planning staff (the rest of the class). Short, punchy concepts will be easy for the city council to understand and will prevent them from going to sleep. Be as visionary as you like, but back up your recommendations with pragmatic suggestions for how these can be implemented. You will also need to keep your city's fiscal situation in mind. A particularly important question is: How can this community improve its fiscal, social, and ecological health in the long run without seeking continual growth in its land area?

Some potential economic development strategies

- *Business recruitment.* City tax breaks and subsidized infrastructure or land to lure businesses from other cities (a traditional strategy).
- *Business retention.* Support for existing businesses in the form of loans, technical assistance, and efforts to help them find larger spaces within the city.
- *Microenterprise development.* A special focus on small start-up companies to help them get going, often through loans, technical assistance, and subsidized office space and services within a business incubator (a building with flexible office space and shared facilities).
- *Eco-business development.* A focus on environmentally protective and restorative types of businesses (recycled products; pollution control and cleanup; alternative energy; alternative building materials).
- *Eco-industrial parks.* A strategy to link industrial and manufacturing businesses so that they can use one another's waste products as inputs.
- *Sustainable agriculture.* If your city or town still has farmland, a strategy to preserve this while encouraging ecologically appropriate production for local markets.
- *Local self-sufficiency.* A strategy emphasizing locally owned businesses using local labor and providing products and services for local markets. This might imply policies to keep out chain stores and big-box retailers.
- *Cooperative businesses.* Support for worker- or consumer-owned businesses.
- *Conservation-based development.* Efforts in rural areas to support sustainable resource production and agriculture.
- *Tourism.* A leading economic development strategy in many places, taking advantage of environmental amenities and cultural heritage as economic assets (implies support for historic preservation).

■ *Place-oriented strategies.* Focus on creating particular centers or hubs of economic activity that will then generate further economic growth, for example through efforts to create 24-hour downtowns that attract entertainment-related businesses.

10 Creek Mapping Exercise

This exercise is appropriate for participants with some background in environmental issues, or for classes that include background material on creek restoration, biodiversity, landscape ecology, or related topics.

Divide the group into small teams. Assign each team one creek or other watershed feature within your community. Provide each team with the best possible map of this feature, preferably one showing the original route of the waterway in cases where it is now culverted, and/or locations of buried culverts. Such maps are often available from local environmental organizations or city public works departments.

INSTRUCTIONS TO PARTICIPANTS

1 Study your creek's route on the map provided, and if desired enlarge portions of the map to assist in field observation and note taking. You may also find additional information about this creek or local creek restoration activities on various local websites.
2 Follow the current or original route of the creek as best you can (preferably by foot or bicycle) and make systematic notes of present conditions and restoration opportunities. Where the creek is underground, follow the approximate route of the culvert or original channel, and identify locations where the creek might potentially be unearthed. Where the creek is above ground, follow the channel as best you can, observing the condition of the stream (flow, erosion, pollution, trash, form of the channel, vegetation, wildlife, and human use). Map these conditions as accurately as you can, and take photos if possible. Don't try to get through brushy parts of the creek corridor or trespass on private property without permission. Often creek channels can be observed where they cross streets.
3 Write up a concise set of recommendations for restoring the creek, including near-term activities (trash clean-up, removal of non-native vegetation, revegetation) and longer-term possibilities (channel restoration, unearthing from culverts). Add photos of existing conditions if possible. Present this to the larger group, and share copies with local environmental groups and city planners.

Potential opportunities to look for:

■ Clean-up of trash or debris.
■ Removal of invasive or non-native species and replanting with native riparian species.
■ Restocking with native fish and creation of good habitat for these.
■ Replacement of concrete channel stabilization devices with more naturalistic measures.
■ Unearthing of culverted creeks.
■ Creation of parks, greenways, ecological educational facilities, or other community amenities in conjunction with creek restoration.

11 Neighborhood Planning Exercise

In this exercise, teams of students meet in a charette format to develop sustainable development proposals for a particular neighborhood. Ideally this workshop can replicate a public meeting held during a recent local Specific Plan, Area Plan, or Neighborhood Plan process, using the same background materials acquired from the city. Or it can focus on another neighborhood likely to be well known to participants. To start with, the instructor should prepare a packet of background information and maps, potentially including a printout of the relevant zoning for the area, traffic statistics, and demographic data from the census bureau. A walking tour of the neighborhood also provides a good starting point.

Divide students into groups of five or six and prepare this number of roles, with a short description of each character. Roles might include a city planner (facilitating the workshop), several residents of differing backgrounds, the owner of a local store or automobile dealership, a representative of an environmental group, and a local developer or community development corporation staff person. Unlike the role plays described earlier, participants are not assigned a viewpoint, but must make up their own point of view based on the character.

INSTRUCTIONS TO PARTICIPANTS

During the next 30 minutes, city staff will lead a public workshop in which members of the public will sit down to develop a potential plan of action for this neighborhood. Informational material and maps of the area are provided. Each group should come up with a list of proposed strategies for sustainable development of this area, and show these graphically on the maps or flip-chart paper provided. You will be assigned a role for this exercise. Feel free to improvise based on what you think this character might be concerned about.

Issues the city suggests groups focus on include the following:

- Ways to promote reuse of vacant or underutilized sites.
- The lack of a pedestrian-friendly street environment.
- Heavy traffic and dangerous street crossings.
- Attracting new locally owned and local-serving businesses to the area.
- Restoration of deteriorated housing in nearby blocks.
- The city's general need for more housing, particularly affordable housing.
- Creating more pocket parks and attractive public spaces, including some sort of focal point or community center.
- Ecological restoration of parks, creeks, median strips, and other public green space.

Possible strategies that you might consider include the following:

- Changing zoning, for example to require taller or higher-density development, to reduce setbacks from the street, or to require mixed-use buildings (housing on top of storefronts).
- Trying to create a new neighborhood center by zoning so as to create a dense cluster of buildings with public amenities and services, and/or by having the city acquire land and carry out improvements.
- Providing grants or loans to affordable housing providers if they will build on sites in the area.
- Purchasing or acquiring through eminent domain certain sites that would be used for public purposes such as parks, civic facilities, or affordable housing.

- Requiring owners to fix up properties with building or housing code violations, and/or providing loan funds for them to do this.
- Prohibiting certain types of businesses from locating in the area through zoning changes (for example, drive-through fast-food restaurants, liquor stores, or automobile dealerships).
- Daylighting parts of creeks which run through this area in culverts, or restoring native habitat on city parks.
- Adding street trees and widening sidewalks to improve the pedestrian environment.
- Adding stop signs or lights to allow pedestrians to cross (you will need to balance this with motorists' desire not to stop too often!).
- Adding or removing other types of traffic/calming anywhere in the neighborhood (the fire and police departments will still need reasonably fast access everywhere).
- Adding attractive lighting, planters, banners, or public art to improve the streetscape.
- Establishing urban design guidelines or design review standards for new buildings.
- Designating the area a redevelopment district to be able to issue bonds (to be repaid by the projected increase in tax revenues) to raise money for improvements.
- Creating ongoing public-private partnerships or organizations to improve the area.
- Subsidizing new initiatives such as a car-sharing co-op.

The City does not have unlimited resources, however, and your proposals should take this into account. If possible, propose funding mechanisms through which your recommendations can be implemented.

12 An Ecological Site Plan

This exercise allows participants to investigate sustainable development possibilities for a particular site. Choose a vacant lot, parking lot, or other redevelopable site relatively close by, in a location where participants can easily visit it for first-hand observation. Obtain a base map for this parcel from the city or other sources and make copies for each team. The exercise is probably best done in small groups of two or three.

INSTRUCTIONS TO PARTICIPANTS

Your consulting firm has been hired by the property owner to come up with sustainable development ideas for this site. This owner is interested in any development scenarios that would generate an economic return and enhance local sustainability. Key planning challenges include:

- Appropriate intensity and mix of uses for this site. (Check the city's zoning for the parcel, usually available online through the municipal website. You can, however, recommend that the developer seek a variance from local zoning requirements if appropriate.)
- Ways that development can help create more pedestrian-friendly streets.
- Orientation and design of buildings so as to reduce energy and resource consumption.
- Appropriate landscaping as well as preservation or restoration of any natural landscape features on the site.
- Location of walkways, open spaces, plazas, or mini-parks.
- Ways that people will enter the site, pass through it, park, and otherwise circulate.
- Meeting overall city needs, especially for housing and reduced automobile use.

Procedure

1 Decide on a "program": a set of uses that you think is most appropriate for this site.
2 Decide on a set of sustainable design principles to help implement your program, related to the general size and mass of buildings, their relation to the natural landscape and climate, their relation to streets and the surrounding context, and their use of materials and construction techniques.
3 Using tracing paper or copies of the base map, sketch out several different possible building layouts using these strategies.
4 Decide on one and depict it on the base map provided. Don't worry about architectural detail. Simply show building outlines (indicate the height and uses of each structure), open spaces, walkways, landscaping, parking (if any), and street design features. Use several different colors of pencil or marker. Add arrows or other symbols to show how sunlight hits the site, how people will pass through the spaces, key entrances or gathering places, flows of water and drainage, and other important characteristics. Label your site plan clearly.
5 Be prepared to present your proposal in 5–10 minutes.

13 International Development Exercise

Although perhaps the most difficult to do, an international case study can help students explore how sustainability principles might apply at a large scale to a developing country. While in the real world sustainability policies should be grounded in a detailed understanding of a given culture and place, it is possible for students to explore many issues in two to three class periods with background reading beforehand.

Choose a country that you have first-hand knowledge of, or a place with an interesting and timely range of development problems, such as Afghanistan, Iraq, Colombia, or Thailand. Prepare a background reading packet for the students, and if possible show a video or slides to help them gain a feeling for the country (Bullfrog Films, at www.bullfrogfilms.com, has many great videos on developing-world issues). An introductory lecture can then briefly outline the nation's history, geography, and development issues, as well as discussing some historical strategies of development taken by the World Bank, other international institutions, and national governments.

For the exercise itself, split the students up into teams and have each develop a sustainable development strategy for the nation. This can either be done as a role play, with students assigned roles representing a cross-section of major national constituencies, or with students simply acting as foreign advisors or consultants. The teams can meet outside of class, or they can be given part of several class periods to caucus and develop their recommendations. During the final class period they will then make presentations, outlining key recommendations and displaying their scenario on large sheets of paper. These recommendations might include an overall theme or set of long-term goals for the country, a policy paradigm that can best achieve those goals (see below), a handful of specific strategies or technologies that should be promoted, and a half-dozen key indicators to assess progress toward sustainability.

If desired, teams can be assigned one of four scenarios:

1 A relatively stable political situation, substantial international assistance.
2 A relatively stable political situation, little international assistance.
3 Unrest and instability, substantial international assistance.
4 Unrest and instability, little international assistance.

INSTRUCTIONS TO PARTICIPANTS (ONE VERSION)

Your team has been appointed as a special advisor to the national government on sustainable development. Your job is to recommend an appropriate development strategy for the country, given the nation's context and the resources and political situation determined in your scenario. You should focus primarily on government policies, programs, and investment strategies that can meet long-term economic, environmental, and equity needs. Key questions might include the following:

■ How should the government assist people in obtaining shelter long term?
■ How can the nation combat environmental problems such as soil degradation, overgrazing, deforestation, and desertification?
■ How should the country generate electric power?
■ What industries or types of agriculture should the country concentrate on?
■ How can the nation best provide jobs for those currently unemployed?
■ What strategies can ensure that the country's population can feed itself?
■ Should the country aggressively develop its gas, oil, coal, or mineral reserves?
■ How might the country promote equity, in particular the rights of minorities and women?
■ How can the nation effectively make use of returning refugees and émigrés from abroad?
■ Should the country aim for a strategy of self-sufficiency in basic products, an export-oriented economy tied in closely with the global economic and financial system, or some other system?
■ Should the country take extensive loans from international institutions such as the World Bank in order to develop rapidly, even though it will then have to repay the loans with interest and may be subject to pressure to develop natural resources, allow unrestricted operations by multinational corporations, and reduce government spending and social programs?

Each group should appoint a facilitator to coordinate the discussion, a note-taker to write down strategies on large paper, and several spokespersons to make its final presentation (or you can all do this). Your group should decide on the following:

■ An overall theme or set of long-term goals for the country;
■ A policy paradigm that can best achieve those goals (see below);
■ A handful of specific strategies or technologies that are most appropriate given the context;
■ Priorities for government investment;
■ A handful of key indicators to assess progress toward sustainability.

If the country in question is basically going through nation-building from scratch, you should decide on a broad policy paradigm that makes most sense for the government. Would you recommend a free-market approach with minimal government intervention in the economy and privatized public services? A social democratic approach with a strong government role in directly building infrastructure, providing services, and regulating the economy? A decentralized model in which the government seeks to empower local or regional communities and the nonprofit sector to do the work? Some other paradigm?

Outline your recommendations on a large sheet of poster paper and be prepared to present them to the class in ten minutes or so.

14 Mapping Your Own Block

Individually or in teams of two, students create a large-format map of the block they live on, drawing it more-or-less to scale at a scale of 1 inch = 20 feet (or 1 inch = 30 or 40 feet if the block is very large). At this scale they can show building footprints, approximate lot lines, sidewalks, planting strips, vegetation (especially canopies of large trees), and other details. They should include the adjoining streets that frame the block. This drawing can be made on a sheet of paper or tracing paper 24 in. by 36 in. or larger, using several pen weights to differentiate different features, and markers to add color.

The most challenging part of this exercise is to analyze the conditions and opportunities of this place. Students should add labels or symbols in large print to convey information about character and use (e.g. user-created paths, places people gather or kids play, distinctive architecture or landscaping, problem houses, locations of barking dogs, tagging, trash, fast traffic, safety problems, etc.), and should also indicate particular sustainability problems and opportunities. Are there ways that the environment of this place can be restored or enhanced? That sociability can be improved? That alternatives to motor vehicles can be promoted? That more appropriate land uses and densities can be encouraged? That needed businesses or services can be brought into the area? That this place can become more supportive of women, children, the elderly, the disabled, and people from diverse racial, ethnic, and income groups?

15 Using YouTube Videos on Sustainability

A great resource for classroom discussion is YouTube, and other similar websites such as Vimeo. YouTube in particular has a wealth of material that can be used in several ways for sustainability-related discussions.

1 *Short case studies of sustainability-related actions in different parts of the world.* These are often clips from films, documentaries, or news programs. Subjects we have shown in classes include urban planning in Curitiba, Brazil; microenterprise lending in India; local workforce training programs in Ho Chi Minh City, Vietnam; safe drinking water systems development in Honduras; ecodistrict development in Sweden; and environmental justice initiatives in Camden, New Jersey.

2 *Different points of view on controversial issues.* YouTube has many short clips of political leaders, pundits, comedians, and ordinary citizens talking about contemporary issues. Frequently one can show several such clips in turn, taking a few minutes after each to have the class deconstruct the rhetoric and content of arguments. For example, in a class on climate change, one can pair climate-change deniers such as Lord Monckton, Glenn Beck, and Michelle Bachmann with mainstream politicians, climate scientists, and activists such as Bill McKibben, Al Gore, and James Hansen. One can even find newsreel footage from the 1950s warning people about the potential for global warming.

3 *Class discussion on communications strategies and environmental messaging.* It often helps students' critical thinking skills to look at how different organizations or constituencies are trying to influence public opinion. In addition to deconstructing the rhetoric of "talking heads," we can look at political commercials, including those by environmental organizations trying to raise awareness about global warming or the need to save energy. We can also look at photos or videos of political art, including performance art. Sometimes these are humorous and lead to a good laugh, as with a video of a

man trying to ski downhill after global warming has eliminated snow from his favorite mountain. Sometimes they are disconcerting, as in the video of polar bears falling from the sky (the message: every airplane trip generates per capita GHG emissions equal to the weight of a polar bear, thus warming the planet and eventually killing these creatures). But almost always we can have good discussion around how sustainability-related topics can be most effectively communicated to the public.

16 Class Debates on Urban Sustainability Themes

In order to help students think critically about sustainability issues, it is often useful to have debates in the classroom or within sections (usually debates work best with a relatively small number of participants, so that each can speak more often). There are various ways of running these contests. But usually the instructor puts forth some proposition, such as that highly dense cities are more sustainable or that capitalism is not compatible with a sustainable society. Then the class is divided into pro and con groups. If the class is small, two groups of 4–6 may suffice; if it is a bit bigger, two pro-groups and two con-groups can take turns making arguments and rebuttals. A set of judges can also be appointed to decide on the winners of each round. An odd number of judges ensures that a vote will always produce a winner.

The groups have 5–10 minutes to decide among themselves the strongest arguments they think support their position. Then the instructor (or the team of judges) begins the debate, allowing one team to make a brief argument (a 30-second time limit is often good), then another to rebut. The judges then state their opinions of which team won the round and why, and the team roles are then reversed. A new team member takes each turn until everyone has participated. The instructor can create a matrix on the board to keep score, and can also call out key terms or themes on the board as the debate moves along.

17 Studio or Service Learning Classes

Traditional studio classes in urban planning or the design professions can be an invaluable way to learn sustainability-related skills, because they typically take an in-depth, iterative, contextual approach to developing solutions for particular places. However, studios are not always taught in a way that emphasizes sustainability themes or skills. Such classes have often been conducted in ways that do not consider all dimensions of the context, that do not actively consult or anticipate the needs of all stakeholders, or that do not think far enough into the future. Often social equity issues do not receive sufficient consideration, or some environmental issues are considered but not others.

To be most effective within sustainability education, studio instructors can adopt the following strategies.

1   Emphasize a very holistic background analysis for the project. It's important for the students to think beyond the particular site or neighborhood in question, to understand how it fits into the broader town, city, region, state, nation, and world. It's also important that the needs of all stakeholders be anticipated and researched. Students should have a sense of how the place in question has evolved in the past – starting with what the native landscape was like before human settlement – and how it might evolve in the future, in line with the surrounding community. Class members should also be able to articulate environmental, economic, and equity opportunities related to the site. (Seeing sustainability opportunities is a crucial skill to develop.)

2   Engage members of the public if possible. The students will learn greatly from communicating with community members. Hopefully they will develop their listening and public engagement skills, will learn about different stakeholders' values and assumptions, and will produce better plans and designs. Many if not most studio classes have a client, and are configured as service learning classes in which the client receives useful products. But the amount of community contact is often limited. In the best scenario an ongoing relationship is established between the university and a particular community in which trust is built up, classes are able to build on the work of previous cohorts, and opportunities emerge for hands-on involvement. Such an arrangement is not easy for many reasons, including that the transitory nature of classes makes continuity difficult. But as much community engagement as possible is desirable, and students should be encouraged to interact with many individuals and groups on and off campus.

3   Push participants to decide sustainability priorities for the place in question. Not all sustainability strategies are equally important in any given time and place. In fact, some will be much more important than others. It is crucial that students gain the ability to set priorities, and to justify them to their peers, members of the public, and decision makers. In the early twenty-first century climate change mitigation and adaptation is a leading priority virtually everywhere, but many more local priorities can be identified as well.

4   Use sustainability tools within studios. Often classes can estimate credits for their design proposals under LEED or other rating systems. The instructor can provide an Excel worksheet for that purpose, and can talk through ways to estimate some of the more challenging credits. Calculators for estimating energy usage and generation, or an economic "pencil-out" of development projects can also be useful.

5   Help students push the envelope in terms of sustainability planning. "Make no little plans. They have no magic to stir men's blood and probably will not themselves be realized" Daniel Burnham said more than 100 years ago, and this is doubly true for sustainable development. A sustainability studio should be concerned with dramatic goals such as zero-net-energy, zero-net-water, ecological site restoration, major improvements to social equity within communities, and the like. But students also need to figure out pragmatic strategies for reaching these far-reaching goals. Often, encouraging students to develop phasing plans for their proposals can help, in that they begin with more doable proposals and show clients how projects can then step up towards larger ultimate goals. Talking with students about the politics of situations can also help, in that they can improve their understanding of how to leverage resources, interest, and commitment.

FURTHER READING

Since the 1970s a huge literature has emerged on sustainable urban development. The following are some of the most useful books on the subject from the past three decades.

Adams, W.M. (1990) *Green Development: Environment and Sustainability in the Third World*. New York: Routledge.

Agyeman, Julian. (2013) *Introducing Just Sustainabilities: Policy, Planning, and Practice*. London: Zed Books.

Agyeman, Julian, Robert Bullard and Bob Evans, eds. (2003) *Just Sustainabilities: Development in an Unequal World*. London: Earthscan.

Alkon, Alison Hope and Julian Agyeman, eds. (2005) *Cultivating Food Justice: Race, Class, and Sustainability*. Cambridge, MA: MIT Press.

Audirac, Ivonne. (1997) *Rural Sustainable Development in America*, New York: John Wiley & Sons.

Baker, Susan. (2006) *Sustainable Development*. New York: Routledge.

Bartlett, Peggy F. and Geoffrey W. Chase, eds. (2004) *Sustainability on Campus: Stories and Strategies for Change*. Cambridge, MA: MIT Press.

Beatley, Timothy. (1994) *Habitat Conservation Planning*. Austin: University of Texas Press.

Beatley, Timothy. (2000) *Green Urbanism: Learning from European Cities*, Washington, D.C.: Island Press.

Beatley, Timothy. (2004) *Native to Nowhere: Sustaining Home and Community in a Global Age*. Washington, D.C.: Island Press.

Beatley, Timothy. (2011) *Biophilic Cities: Integrating Nature into Urban Design and Planning*. Washington, D.C.: Island Press.

Beatley, Timothy, ed. (2012) *Green Cities of Europe*. Washington, D.C.: Island Press.

Beatley, Timothy and Kristy Manning. (1997) *The Ecology of Place: Planning for Environment, Economy, and Community*, Washington, D.C.: Island Press.

Beatley, Timothy and Peter Newman. (2008) *Green Urbanism Down Under: Learning from Sustainable Communities in Australia*. Washington, D.C.: Island Press.

Benfield, F. Kaid, Matthew D. Raimi, and Donald D.T. Chen. (1999) *Once There Were Greenfields*, Washington, D.C.: Natural Resources Defense Council.

Beynus, Janine M. (1997) *Biomimicry: Innovation Inspired by Nature*. New York: HarperCollins.

Blowers, Andrew, ed. (1993) *Planning for a Sustainable Environment: A Report by the Town and County Planning Association*. London: Earthscan.

Braidotti, Rosi *et al.* (1994) *Women, the Environment and Sustainable Development: Towards a Theoretical Synthesis*. London: Zed Books.

Brown, Lester R. (2009) *Plan B 4.0: Mobilizing to Save Civilization*. New York: Norton.

Brown, Lester R. (2011) *World on the Edge: How to Prevent Environmental and Economic Collapse*. New York: Norton.

Brown, Lester R. (2012) *Full Planet, Empty Plates: The New Geopolitics of Food Scarcity*. New York: Norton.

Bullard, Robert D. (2000) *Dumping in Dixie: Race, Class, and Environmental Quality*, Boulder: Westview Press.

Bullard, Robert D., ed. (2007) *Growing Smarter: Achieving Livable Communities, Environmental Justice, and Regional Equity.* Cambridge, MA: MIT Press.

Calkins, Meg. (2012) *The Sustainable Sites Handbook: A Complete Guide to the Principles, Strategies, and Best Practices for Sustainable Landscapes.* Hoboken, NJ: Wiley.

Callenbach, Ernest. (1975) *Ecotopia.* New York: Bantam Books.

Calthorpe, Peter. (1993) *The Next American Metropolis: Ecology, Community and the American Dream.* New York: Princeton Architectural Press.

Calthorpe, Peter. (2010) *Urbanism in the Age of Climate Change.* Washington, D.C.: Island Press.

Calthorpe, Peter and William Fulton. (2001) *The Regional City: Planning for the End of Sprawl,* Washington, D.C.: Island Press.

Campagna, Michele, ed. (2006) *GIS for Sustainable Development.* Boca Raton, FL: CRC Press.

Cervero, Robert. (1998) *The Transit Metropolis: A Global Inquiry,* Washington, D.C.: Island Press.

Condon, Patrick. (2007) *Design Charrettes for Sustainable Communities.* Washington, D.C.: Island Press.

Condon, Patrick. (2010) *Seven Rules for Sustainable Communities: Design Strategies for the Post Carbon World.* Washington, D.C.: Island Press.

Cooper, Phillip J. and Claudia Maria Vargas. (2008) *Sustainable Development in Crisis Conditions: Challenges of War, Terrorism, and Civil Disorder.* Lanham, MD: Rowman & Littlefield.

Corbett, Judy and Michael Corbett. (2000) *Designing Sustainable Communities: Learning from Village Homes.* Washington, D.C.: Island Press.

Costanza, Robert, ed. (1991) *Ecological Economics: The Science and Management of Sustainability.* New York: Columbia University Press.

Coyle, Stephen J. (2011) *Sustainable and Resilient Communities: A Comprehensive Action Plan for Towns, Cities, and Regions.* Hoboken, NJ: Wiley.

Daly, Herman. (1996) *Beyond Growth: The Economics of Sustainable Development.* Boston: Beacon Press.

Daly, Herman and John B. Cobb, Jr. (1989) *For the Common Good: Redirecting the Economy toward Community, the Environment, and a Sustainable Future.* Boston: Beacon Press.

Dietz, Rob and Dan O'Neill. (2013) *Enough is Enough: Building a Sustainable Economy in a World of Finite Resources.* San Francisco: Berrett-Koehler.

Duany, Andres *et al.* (2000) *Suburban Nation: The Rise of Sprawl and the Decline of the American Dream.* New York: North Point.

Dunham-Jones, Ellen and June Williamson. (2011) *Retrofitting Suburbia: Urban Design Solutions for Redesigning Suburbs.* Hoboken, NJ: Wiley.

Elkin, Tim *et al.* (1991) *Reviving the City: Towards Sustainable Urban Development.* London: Friends of the Earth.

Elliot, Jennifer. (2006) *An Introduction to Sustainable Development.* New York: Routledge.

Engwicht, David. (1993) *Reclaiming Our Cities and Towns: Better Living with Less Traffic.* Philadelphia: New Society Publishers.

Evans, Peter, ed. (2002) *Livable Cities? Urban Struggles for Livelihood and Sustainability.* Berkeley: University of California Press.

Farr, Douglas. (2007) *Sustainable Urbanism: Urban Design with Nature.* Thousand Oaks: Wiley.

Frank, Lawrence D., Peter O. Engelke, and Thomas L. Schmid. (2003) *Health and Community Design: The Impact of the Built Environment on Physical Activity.* Washington, D.C.: Island Press.

Frumkin, Howard, Lawrence Frank, and Richard Jackson. (2004) *Urban Sprawl and Public Health: Designing, Planning, and Building for Healthy Communities.* Washington, D.C.: Island Press.

Girardet, Herbert. (1999) *Creating Sustainable Cities,* Devon, UK: Green Books.

Girardet, Herbert. (2008) *Cities People Planet: Urban Development and Climate Change.* Hoboken, NJ: Wiley.

Global Cities Project. (1991) *Building Sustainable Communities: An Environmental Guide for Local Government.* San Francisco: The Center for the Study of Law and Politics.

Goldsmith, Edward. (1993) *The Way: An Ecological World View.* Boston: Shambhala.

Haas, Tigran, ed. (2012) *Sustainable Urbanism and Beyond: Rethinking Cities for the Future*. New York: Rizzoli.

Hamm, Bernd and Pandurang K. Muttagi, eds. (1998) *Sustainable Development and the Future of Cities*. London: Centre for European Studies.

Hardoy, Jorge E. *et al.* (1992) *Environmental Problems in Third World Cities*. London: Earthscan.

Hawken, Paul. (2007) *Blessed Unrest: How the Largest Social Movement in History is Restoring Grace, Justice, and Beauty to the World*. New York: Penguin.

Hawken, Paul. (2010) *The Ecology of Commerce: A Declaration of Sustainability*. New York: HarperCollins.

Hester, Randolph T. (2006) *Design for Ecological Democracy*. Cambridge, MA: MIT Press.

Hesterman, Oran B. (2011) *Fair Food: Growing a Healthy, Sustainable Food System for All*. New York: Public Affairs.

Holmberg, Johan, ed. (1992) *Making Development Sustainable: Redefining Institutions, Policy, and Economics*. Washington D.C.: Island Press.

Honey, Martha. (1998) *Ecotourism and Sustainable Development: Who Owns Paradise?* Washington, D.C.: Island Press.

Jacobs, Allan. (1985) *Looking at Cities*. Cambridge, MA: Harvard University Press.

Jacobs, Allan. (1993) *Great Streets*. Cambridge, MA: The MIT Press.

Jacobs, Jane. (1961) *The Death and Life of Great American Cities*. New York: Random House.

Johnston, Sadhu Aufochs, Steven S. Nicholas, and Julia Parzen. (2013) *The Guide to Greening Cities*. Washington, D.C.: Island Press.

Kelbaugh, Douglas. (1997) *Common Place: Toward Neighborhood and Regional Design*. Seattle: University of Washington Press.

Kunstler, James Howard. (1993) *The Geography of Nowhere: The Rise and Decline of America's Man-Made Landscape*. New York: Simon & Schuster.

Layard, Antionia, Simin Davoidi, and Susan Batty, eds. (2001) *Planning for a Sustainable Future*. New York: Spon.

Lerch, Daniel. (2008) *Post Carbon Cities: Planning for Energy and Climate Uncertainty*. Santa Rosa, CA: Post Carbon Institute.

Lyle, John Tillman. (1994) *Regenerative Design for Sustainable Development*. New York: John Wiley & Sons.

Lyle, John Tillman. (1999) *Design for Human Ecosystems: Landscape, Land Use, and Natural Resources*. Washington, D.C.: Island Press.

Mazmanian, Daniel A. and Michael E. Kraft, eds. (2009) *Toward Sustainable Communities: Transition and Transformations in Environmental Policy*. Cambridge, MA: MIT Press.

McDonough, William and Michael Braungart. (2002) *Cradle to Cradle: Remaking the Way We Make Things*. New York: North Point Press.

McDonough, William and Michael Braungart. (2013) *The Upcycle: Beyond Sustainability – Designing for Abundance*. New York: North Point Press.

McHarg, Ian L. (1969) *Design with Nature*. Garden City, NY: The Natural History Press.

McHarg, Ian L. and Frederick Steiner. (1998) *To Heal the Earth: Selected Writings of Ian L. McHarg*. Washington, D.C.: Island Press.

Meadows, Donella, Dennis L. Meadows, and Jörgen Randers. (1992) *Beyond the Limits: Confronting Global Collapse, Envisioning a Sustainable Future*. Post Mills, VT: Chelsea Green.

Meadows, Donella, Dennis L. Meadows, Jörgen Randers, and William W. Behrens III. (1972) *The Limits to Growth*. New York: Universe Books.

Merchant, Carolyn. (1992) *Radical Ecology*. New York: Routledge.

Mitlin, Diana. (1992) Sustainable Development: A Guide to the Literature. *Environment and Urbanization* 4(1): 111–124.

Monbiot, George. (2007) *Heat: How to Stop the Planet from Burning*. Cambridge, MA: South End Press.

Newman, Peter and Isabella Jennings. (2008) *Cities as Sustainable Ecosystems: Principles and Practices*. Washington, D.C.: Island Press.

Newman, Peter and Jeffrey Kenworthy. (1999) *Sustainability and Cities: Overcoming Automobile Dependence.* Washington, D.C.: Island Press.

Norgaard, Richard. (1994) *Development Betrayed: The End of Progress and a Coevolutionary Revisioning of the Future.* New York: Routledge.

O'Connor, Martin, ed. (1994) *Is Capitalism Sustainable? Political Economy and the Politics of Ecology.* New York: The Guilford Press.

Pearce, David *et al.* (1990) *Sustainable Development: Economics and Environment in the Third World.* London: Edward Elgar.

Pearce, David and Edward B. Barbier. (2000) *Blueprint for a Sustainable Economy.* London: Earthscan.

Pearce, David, Edward Barbier and Anil Markandya. (1989) *Blueprint for a Green Economy.* London: Earthscan.

Pearce, David and Jeremy J. Warford. (1993) *World without End: Economics, Environment and Sustainable Development.* New York: Oxford University Press.

Peck, Sheila. (1998) *Planning for Biodiversity: Issues and Examples.* Washington, D.C.: Island Press.

Pickett, S.T.A., Mary L. Cadenasso, and Brian McGrath, eds. (2013) *Resilience in Ecology and Urban Design.* New York: Springer.

Porter, Douglas R. (2007) *Managing Growth in America's Communities*, Second Edition. Wahsington, D.C.: Island Press.

Portney, Kent. (2013) *Taking Sustainable Cities Seriously: Economic Development, Quality of Life, and Environment in American Cities.* Cambridge, MA: MIT Press.

Pratt, Rutherford H., Rowan A. Rountree, and Pamela C. Muick. (1994) *The Ecological City: Preserving and Restoring Urban Biodiversity.* Amherst: University of Massachusetts Press.

Prugh, Thomas, Robert Costanza, Herman E. Daly. (1998) *The Local Politics of Global Sustainability.* Washington, D.C.: Island Press.

Pugh, Cedric. (2001) *Sustainable Cities in Developing Countries.* London: Routledge.

Randolph, John and Gilbert Masters. (2008) *Energy for Sustainability: Technology, Planning, Policy.* Washington, D.C.: Island Press.

Redclift, Michael. (1987) *Sustainable Development: Exploring the Contradictions.* London: Methuen.

Register, Richard. (1987) *Ecocity Berkeley: Building Cities for a Healthy Future.* Berkeley: North Atlantic Books.

Register, Richard. (2006) *Ecocities: Building Cities in Balance with Nature.* Berkeley: Berkeley Hills Books.

Riley, Ann L. (1998) *Restoring Streams in Cities.* Washington, D.C.: Island Press.

Roseland, Mark. (2012) *Toward Sustainable Communities: Resources for Citizens and their Governments.* Gabriola Island, BC: New Society Publishers.

Satterthwaite, David, ed. (1999) *The Earthscan Reader in Sustainable Cities.* London: Earthscan.

Shoup, Donald. (2011) *The High Cost of Free Parking.* Chicago: APA Planners Press.

Sperling, Daniel and Deborah Gordon. (2009) *Two Billion Cars: Driving Towards Sustainability.* New York: Oxford University Press.

Spirn, Anne Whiston. (1984) *The Granite Garden: Urban Nature and Human Design.* New York: Basic Books.

Steiner, Frederick. (2008) *The Living Landscape*, Second Edition. Washington, D.C.: Island Press.

Stivers, Robert L. (1976) *The Sustainable Society: Ethics and Economic Growth.* Philadelphia: Westminster Press.

Stren, Richard *et al.*, eds. (1992) *Sustainable Cities: Urbanization and the Environment in International Perspective.* Boulder: Westview Press.

Suzuki, Hiroaki, Robert Cervero, and Kanako Iuchi. (2012) *Transforming Cities with Transit: Transit and Land-Use Integration for Sustainable Urban Development.* Washington, D.C.: The World Bank.

Thompson, J. William and Kim Sorvig. (2007) *Sustainable Landscape Construction: A Guide to Green Building Outdoors.* Washington, D.C.: Island Press.

Todd, Nancy Jack and John Todd. (1993) *From Eco-Cities to Living Machines.* Berkeley: North Atlantic Books.

Todd, Nancy Jack and John Todd. (2006) *A Safe and Sustainable World: The Promise of Ecological Design.* Washington, D.C.: Island Press.

Tumlin, Jeffrey. (2012) *Sustainable Transportation Planning: Tools for Creating Vibrant, Healthy, and Resilient Communities*. Hoboken, NJ: Wiley.

Urban Ecology, Inc. (1996) *Blueprint for a Sustainable Bay Area*. Oakland.

Van der Ryn, Sim. (2013) *Design for an Empathic World: Reconnecting People, Nature, and Self*. Washington, D.C.: Island Press.

Van der Ryn, Sim and Peter Calthorpe, eds. (1984) *Sustainable Communities*. San Francisco: Sierra Club Books.

Van der Ryn, Sim and Stuart Cowan. (2007) *Ecological Design*. Washington, D.C.: Island Press.

Wheeler, Stephen M. (2012) *Climate Change and Social Ecology: A New Perspective on the Climate Challenge*. London: Routledge.

Wheeler, Stephen M. (2013) *Planning for Sustainability: Creating Livable, Equitable, and Ecological Communities*. London: Routledge.

World Commission on Environment and Development. (1987) *Our Common Future*. New York: Oxford University Press.

Zetter, Rober and Georgia Butina Watson, eds. (2006) *Designing Sustainable Cities in the Developing World*. Aldershot, Hampshire: Ashgate.

ILLUSTRATION CREDITS

Every effort has been made to contact copyright holders for their permission to reprint plates and figures in this book. The publishers would be grateful to hear from any copyright holder who is not here acknowledged and will undertake to rectify any errors or omissions in future editions of this book. The following is copyright information for the figures and plates that appear in this book.

PART ONE ORIGINS OF THE SUSTAINABILITY CONCEPT

7.1 Human perspectives. *Limits to Growth* (New York: Universe Books, 1972) by Donella H. Meadows, Dennis L. Meadows, Jörgen Randers, and William W. Behrens III, © 1972 by Dennis L. Meadows, reproduced by permission of Dennis L. Meadows.

PART TWO DIMENSIONS OF SUSTAINABLE URBAN DEVELOPMENT

13.1 Wedges. "Stabilization Wedges: Solving the Climate Problem for the Next 50 Years with Current Technologies" © 2004 by *Science* magazine. Reproduced by permission of *Science* magazine.

14.1 "Global Greenhouse Gas Abatement Cost Curve." From "Impact of the Financial Crisis on Carbon Economics: Version 2.1 of the Global Greenhouse Gas Abatement Cost Curve. © 2010 McKenzie & Company.

16.1 Conventional Suburban Development (Sprawl) vs. Traditional Neighborhood Development. *The Next American Metropolis: Ecology, Community, and the American Dream* by Peter Calthorpe. © 1993 by Princeton Architectural Press, Inc. Reproduced by permission of Princeton Architectural Press, Inc.

16.2 Transit-Oriented Development. *The Next American Metropolis: Ecology, Community, and the American Dream* by Peter Calthorpe. © 1993 by Princeton Architectural Press, Inc. Reproduced by permission of Princeton Architectural Press, Inc.

16.3 Housing Types. *The Next American Metropolis: Ecology, Community, and the American Dream* by Peter Calthorpe. © 1993 by Princeton Architectural Press, Inc. Reproduced by permission of Princeton Architectural Press, Inc.

16.4 Secondary Areas. *The Next American Metropolis: Ecology, Community, and the American Dream* by Peter Calthorpe. © 1993 by Princeton Architectural Press, Inc. Reproduced by permission of Princeton Architectural Press, Inc.

16.5 Relationship to Transit. *The Next American Metropolis: Ecology, Community, and the American Dream* by Peter Calthorpe. © 1993 by Princeton Architectural Press, Inc. Reproduced by permission of Princeton Architectural Press, Inc.

16.6 Residential Density Mix. *The Next American Metropolis: Ecology, Community, and the American Dream* by Peter Calthorpe. © 1993 by Princeton Architectural Press, Inc. Reproduced by permission of Princeton Architectural Press, Inc.

16.7   Street and Circulation System. *The Next American Metropolis: Ecology, Community, and the American Dream* by Peter Calthorpe. © 1993 by Princeton Architectural Press, Inc. Reproduced by permission of Princeton Architectural Press, Inc.

16.8   Regional Form. *The Next American Metropolis: Ecology, Community, and the American Dream* by Peter Calthorpe. © 1993 by Princeton Architectural Press, Inc. Reproduced by permission of Princeton Architectural Press, Inc.

16.9   Quality of Outdoor Spaces vs. Activities. *The Next American Metropolis: Ecology, Community, and the American Dream* by Peter Calthorpe. © 1993 by Princeton Architectural Press, Inc. Reproduced by permission of Princeton Architectural Press, Inc.

16.10   Open Space. *The Next American Metropolis: Ecology, Community, and the American Dream* by Peter Calthorpe. © 1993 by Princeton Architectural Press, Inc. Reproduced by permission of Princeton Architectural Press, Inc.

16.11   Urban Growth Boundaries. *The Next American Metropolis: Ecology, Community, and the American Dream* by Peter Calthorpe. © 1993 by Princeton Architectural Press, Inc. Reproduced by permission of Princeton Architectural Press, Inc.

16.12   Waste Water. *The Next American Metropolis: Ecology, Community, and the American Dream* by Peter Calthorpe. © 1993 by Princeton Architectural Press, Inc. Reproduced by permission of Princeton Architectural Press, Inc.

16.13   Drainage. *The Next American Metropolis: Ecology, Community, and the American Dream* by Peter Calthorpe. © 1993 by Princeton Architectural Press, Inc. Reproduced by permission of Princeton Architectural Press, Inc.

16.14   Landscaping. *The Next American Metropolis: Ecology, Community, and the American Dream* by Peter Calthorpe. © 1993 by Princeton Architectural Press, Inc. Reproduced by permission of Princeton Architectural Press, Inc.

16.15   Energy Conservation. *The Next American Metropolis: Ecology, Community, and the American Dream* by Peter Calthorpe. © 1993 by Princeton Architectural Press, Inc. Reproduced by permission of Princeton Architectural Press, Inc.

16.16   Ancillary Units. *The Next American Metropolis: Ecology, Community, and the American Dream* by Peter Calthorpe. © 1993 by Princeton Architectural Press, Inc. Reproduced by permission of Princeton Architectural Press, Inc.

16.17   Residential Densities. *The Next American Metropolis: Ecology, Community, and the American Dream* by Peter Calthorpe. © 1993 by Princeton Architectural Press, Inc. Reproduced by permission of Princeton Architectural Press, Inc.

16.18   Building Setbacks. *The Next American Metropolis: Ecology, Community, and the American Dream* by Peter Calthorpe. © 1993 by Princeton Architectural Press, Inc. Reproduced by permission of Princeton Architectural Press, Inc.

16.19   Garages. *The Next American Metropolis: Ecology, Community, and the American Dream* by Peter Calthorpe. © 1993 by Princeton Architectural Press, Inc. Reproduced by permission of Princeton Architectural Press, Inc.

16.20   Streets. *The Next American Metropolis: Ecology, Community, and the American Dream* by Peter Calthorpe. © 1993 by Princeton Architectural Press, Inc. Reproduced by permission of Princeton Architectural Press, Inc.

16.21   Intersections. *The Next American Metropolis: Ecology, Community, and the American Dream* by Peter Calthorpe. © 1993 by Princeton Architectural Press, Inc. Reproduced by permission of Princeton Architectural Press, Inc.

18.1   Infill multifamily housing. Reproduced by permission of Stephen M. Wheeler.

18.2   Stapleton main street, part of an infill neighborhood. Reproduced by permission of Stephen M. Wheeler.

18.3   Stapleton residential street, part of an infill neighborhood. Reproduced by permission of Stephen M. Wheeler.

19.1   Graphic representation of the relationship between the quality of outdoor spaces and the rate of occurrence of outdoor activities. *Life between Buildings* by Jan Gehl, English translation © 1987 by Jan Gehl. Reproduced by permission of Jan Gehl.

19.2   A Sociable Street. *Life Between Buildings* by Jan Gehl. English translation © 1987 by Jan Gehl. Reproduced by permission of Jan Gehl.

21.1   Energy use per capita in private passenger travel versus urban density in global cities, 1990. Peter Newman and Jeffrey Kenworthy, Sustainability and Cities: Overcoming Automobile Dependence (Washington, DC: Island Press, 1999). Reproduced with the permission of Island Press.

22.1   Bicycling rates in selected countries. Reproduced with the permission of John Pucher and Rralph Buehler.

29.1   Arcata sewage treatment marsh. *Regenerative Design for Sustainable Development* by John Tillman Lyle. Copyright © 1994 by John Wiley & Sons, Inc. Reproduced by permission of Wiley-Liss, Inc., a subsidiary of John Wiley & Sons, Inc.

31.1   Running the Gauntlet. *Redesigning the American Dream* (New York: Norton, 1984). Reproduced with the permission of Dolores Hayden.

31.2   Inhospitable Environments. *Redesigning the American Dream* (New York: Norton, 1984). Reproduced with the permission of Dolores Hayden.

38.1   Green and decent jobs. Reproduced with the permission of Michael Renner *et al.*

41.1   Code Obstacles to Green Design: Ms. Truly Green's home. In "Sustainability and Building Codes," *Environmental Building News*, 10(9), September 2001. Reproduced with the permission of Bruce Coldham.

46.1   Obesity trends. Reproduced with the permission of Howard Frumkin *et al.*

PART THREE SUSTAINABILITY PLANNING TOOLS AND POLITICS

48.1   Sustainable Seattle Indicators. Reproduced by permission of Sustainable Seattle.

49.1   Ecological Footprint. *Our Ecological Footprint* by Mathis Wackernagel and William Rees. Copyright © 1996 by Mathis Wackernagel and William Rees. Reproduced by permission of New Society Publishers.

49.2   Living in a Terrarium. *Our Ecological Footprint* by Mathis Wackernagel and William Rees. Copyright © 1996 by Mathis Wackernagel and William Rees. Reproduced by permission of New Society Publishers.

49.3   What Is an Ecological Footprint? *Our Ecological Footprint* by Mathis Wackernagel and William Rees. Copyright © 1996 by Mathis Wackernagel and William Rees. Reproduced by permission of New Society Publishers.

49.4   Our Ecological Footprints Keep Growing While Our per capita "Earthshares" Continue to Shrink. *Our Ecological Footprint* by Mathis Wackernagel and William Rees. Copyright © 1996 by Mathis Wackernagel and William Rees. Reproduced by permission of New Society Publishers.

49.5   Wanted: Two (Phantom) Planets. *Our Ecological Footprint* by Mathis Wackernagel and William Rees. Copyright © 1996 by Mathis Wackernagel and William Rees. Reproduced by permission of New Society Publishers.

49.6   The Ecological Footprint of the Netherlands. *Our Ecological Footprint* by Mathis Wackernagel and William Rees. Copyright © 1996 by Mathis Wackernagel and William Rees. Reproduced by permission of New Society Publishers.

PART FOUR SUSTAINABLE URBAN DEVELOPMENT INTERNATIONALLY

55.1 Location Map of Curitiba. Scientific American. March 1996. Reproduced by permission of Karl Gude.

55.2 24-Hour Street. *Scientific American*, March 1996, pp. 46–53. Reproduced with permission. Copyright © 1996 by Scientific American, Inc. All rights reserved.

55.3 Lakeside Parks. *Scientific American*, March 1996, pp. 46–53. Reproduced with permission. Copyright © 1996 by Scientific American, Inc. All rights reserved.

55.4 Curitiba Transit System. Scientific American. March 1996. Reproduced by permission of Karl Gude.

55.5 Bus Tubes. *Scientific American*, March 1996, pp. 46–53. Reproduced with permission. Copyright © 1996 by Scientific American, Inc. All rights reserved.

55.6 Historic Center. *Scientific American*, March 1996, pp. 46–53. Reproduced with permission. Copyright © 1996 by Scientific American, Inc. All rights reserved.

55.7a and b Recycling. *Scientific American*, March 1996, pp. 46–53. Reproduced with permission. Copyright © 1996 by Scientific American, Inc. All rights reserved.

55.8 Transport Network. *Scientific American*, March 1996, pp. 46–53. Reproduced with permission. Copyright © 1996 by Scientific American, Inc. All rights reserved.

55.9 Main Boulevard. *Scientific American*, March 1996, pp. 46–53. Reproduced with permission. Copyright © 1996 by Scientific American, Inc. All rights reserved

55.10 Botanical Gardens. *Scientific American*, March 1996, pp. 46–53. Reproduced with permission. Copyright © 1996 by Scientific American, Inc. All rights reserved.

Box 55.1 Bus Tracks. *Scientific American*, March 1996, pp. 46–53. Reproduced with permission. Copyright © 1996 by Scientific American, Inc. All rights reserved.

PART SIX CASE STUDIES OF URBAN SUSTANABILITY

All photos by Timothy Beatley unless otherwise credited.

2 Sky garden in Commerzbank, Frankfurt, Germany. Photo by Nigel Young. Permission from Norman Foster and Partners, London.

3 Menara Mesiniaga, a bio-climatic skyscraper designed by architect Ken Yeang. Photo by K.L. Ng. General permission given by ArchNet, online slide library.

4 Menara Mesiniaga, garden terraces. Photo by K.L. Ng. General permission given by ArchNet, online slide library.

5 Artist's rendering of Christie Walk, Adelaide, Australia. Permission to reproduce given by Paul Downton, Urban Ecology Australia.

6 Photo of Christie Walk, Adelaide, Australia. Photo by Paul Downton. Permission for use granted by Paul Downton, Urban Ecology Australia.

12 Photo of Barclay Ecological Park. Photo by Kang-li Wu. Permission for use granted by Kang-li Wu.

13 Systems approach flow-chart. Permission for use granted by Erik Freudenthal.

27 The Cheonggyeong Restoration Project. Wikimedia Commons.

28 Photo of Weilai Ecological Community. Photo by Kang-li Wu. Permission for use granted by Kang-li Wu.

29 Photo of Shenzhen Overseas Chinese District. Photo by Kang-li Wu. Permission for use granted by Kang-li Wu.

30 Photo of green streets in the Shenzhen Overseas Chinese District. Photo by Kang-li Wu. Permission for use granted by Kang-li Wu.

32 Map of Vancouver, BC, region, showing protected Green Zone. Permission to reproduce given by the Greater Vancouver Regional District (GVRD).

33   Bogotá TransMilenio, transit system. Photo by TransMilenio S.A. Permission for use given by Enrique Penalosa.

34   Bogotá Ciclovia. Photo by Institute Distrital de Recreacion y Deporte de Bogota. Permission for use granted by Enrique Penalosa.

35   Paolo Lugari sitting on one of Gaviotas's famous water pumps, Gaviotas, Colombia. Photo by ZERI Foundation. Permission for use given by Gunter Pauli, ZERI Foundation.

36   Self-sufficient hospital, Gaviotas, Colombia. Photo from ZERI Foundation. Permission for use given by Gunter Pauli, ZERI Foundation.

38   Image of Auroville Galaxy Plan, Auroville, India. Photo by Pino Marchese. Permission for use given by Auroville Planning Office.

39   Photo of Matrimandir (temple) and Peace Area, Auroville, India. Photo by Pino Marcheses. Permission for use granted by Auroville Planning Office.

47   Austin Smart Growth Map. Permission to reproduce given by City of Austin, Department of Growth Management.

50   Map of Portland Metro 2040. Permission to reproduce given by the Portland Metropolitan Services District.

51   Photo of the McNeil Generating Plant, Burlington, Vermont. Permission for use of photo given by the Burlington Electric Department.

52   Photo of young and old together at the McClure Intergenerational Center, Burlington, Vermont. Photo by Nicole Craft. Permission given by the McClure Intergenerational Center.

COPYRIGHT INFORMATION

PART FIVE VISIONS OF SUSTAINABLE COMMUNITY

62   "The Streets of Ecotopia's Capital" and "Car-Less Living in Ecotopia's New Towns" from *Ecotopia: The Notebooks and Reports of William Weston* by Ernest Callenbach. © 1975 by Ernest Callenbach. Reprinted by permission of the author.

63   "Description of Abbenay" from *The Dispossessed* by Ursula K. LeGuin. © 1974 by Ursula K. LeGuin. Reprinted by permission of HarperCollins Publishers Inc.

64   "The View from the Twenty-Third Century," by Stephen M. Wheeler. © 2008 by the author.

PART SIX CASE STUDIES OF URBAN SUSTAINABILITY

"Barclay Ecological Park, Tainan, Taiwan" by Kang-li Wu. © 2014 by the author.
"Weilai Ecological Community, Dezhou, China" by Kang-li Wu. © 2014 by the author.
"Shenzhen Overseas Chinese Town, China" by Kang-li Wu. © 2014 by the author.

PART SEVEN SUSTAINABILITY PEDAGOGY AND CLASS EXERCISES

"Sustainability Pedagogies" by Debby Cotton and Jennie Winter, from *Sustainability Education: Perspectives and Practice across Higher Education*, edited by Paula Jones, David Selby, and Stephen Sterling. © 2010 by the authors. Reprinted by permission of the authors.

INDEX